Reconstructing Quaternary Environments

Reconstructing Quaternary Environments

Second Edition

J.J. LOWE
Professor of Geography and Quaternary Science
Centre for Quaternary Research
Royal Holloway
University of London

M.J.C. WALKER
Professor of Physical Geography
University of Wales
Lampeter

Addison Wesley Longman Limited
Edinburgh Gate, Harlow
Essex CM20 2JE
England
and Associated Companies throughout the World

First published 1984
Second edition 1997

ISBN 0 582 10166 2

British Library Cataloguing-in-Publication Data
A catalogue record for this book is
available from the British Library.

Library of Congress Cataloguing-in-Publication Data
A catalogue entry for this title is
available from the Library of Congress

Set by 32 in 9/11 Times
Produced by Longman Asia Limited, Hong Kong

For Professor Russell Coope

John Lowe is Professor of Geography & Quaternary Science and Director of the Centre for Quaternary Research, Royal Holloway, University of London, UK. He is editor of *Quaternary Proceedings*, Vice-President of the UK Quaternary Research Association and a member of several committees of the Natural Environment Research Council with remits covering Earth Science and palaeoclimatology. He has edited or co-edited five volumes of research papers, including *Radiocarbon Dating: Recent Applications and Future Potential* (Wiley, 1991) and *Climate Changes in Areas Adjacent to the North Atlantic during the Last Glacial–Interglacial Transition* (Wiley, 1994).

Mike Walker is Professor of Physical Geography and Head of the Department of Geography, University of Wales, Lampeter, UK. He is editor of *Journal of Quaternary Science*, Chairman of the Natural Environment Research Council Radiocarbon Dating Laboratory Steering Committee, co-author of *Late Quaternary Environmental Change: Physical and Human Perspectives* (Longman, 1992) and co-editor of *Records of the Last Deglaciation around the North Atlantic* (Pergamon, 1993).

John Lowe and Mike Walker are co-ordinators of INTIMATE (Integration of Ice Core, Marine and Terrestrial Records), an international collaborative research programme within the INQUA Palaeoclimate Commission. They have 25 years of research experience and between them have published more than 100 papers in the field of Quaternary science.

Contents

Preface to the first edition

The Quaternary is the most recent geological period, spanning the last two million years or so of the earth's history, and extending up to the present day. Despite the fact that it is one of the shortest formal episodes of geological time, the Quaternary has attracted the interest of a great many scientists from a wide range of disciplines, and there is now a large and rapidly-expanding literature on topics as diverse as glacial geology, climatic history, oceanic circulation and sedimentation, floral and faunal changes, and human evolution. Surprisingly, however, in spite of the wealth of material that is available, relatively few texts have so far been concerned specifically with landscapes of the Quaternary, and with the way in which different forms of evidence can be integrated to provide an insight into both spatial and temporal changes in Quaternary environments. That is the aim of this book, for it contains a description and assessment of the principal methods and approaches that can be employed in the reconstruction of Quaternary environments. The book is designed for undergraduate and first-year postgraduate students who may have been introduced to certain aspects of Quaternary studies, but whose training has not focused specifically on palaeoenvironmental reconstruction. The text is very wide-ranging and reflects the multi-faceted nature of Quaternary research. Although written by geographers, therefore, it is anticipated that students of archaeology, anthropology, botany, geology and zoology will find within the following pages material that is of use to them.

A large number of people have contributed either directly or indirectly towards the writing of this book. We owe a considerable debt to Dr J. B. Sissons who was our research supervisor in the Department of Geography, University of Edinburgh, for not only did he stimulate our interest in the Quaternary, he also taught us to seek out the flaws in arguments and to think clearly and argue logically. Moreover, he established a research school with whose members we have continued to enjoy rewarding academic contact. We are particularly grateful to Dr R. Cornish, Dr R. A. Cullingford, Dr A. G. Dawson, Dr J. M. Gray, Dr D. E. Smith and Dr D. G. Sutherland for discussing material with us, and for their assistance and companionship in the field over a number of years. We should also like to thank those who have read and commented upon drafts of the chapters of this book. Dr R. A. Cullingford (University of Exeter), Dr B. D'Olier (City of London Polytechnic), Dr P. Gibbard (University of Cambridge), Professor F. Oldfield (University of Liverpool), Dr J. D. Peacock (Institute of Geological Sciences), Dr R. C. Preece (University of Cambridge), Dr J. E. Robinson (University of London), Mr J. Rose (University of London), Dr D. E. Sugden (University of Aberdeen) and Dr D. G. Sutherland (University of Edinburgh) all provided useful suggestions, and large sections of the text have been much improved as a result of their constructive critical appraisal. Professor G. I. Meirion-

Jones made the cartographic facilities of the Geography Section, City of London Polytechnic, available to us, and we are grateful to Don Shewan, Susannah Hall, Connie Flewitt, Ed Oliver, Mavis Teed and Clare Terry for their efficient production of the line drawings.

We also wish to express our gratitude to colleagues who have kindly supplied photographs: Dr N. F. Alley (Figs 2.15, 2.16, 2.21, 2.22, 2.23, 3.8, 5.14, 5.15 and 5.24); Dr J. Crowther (Fig 3.22); Dr. J. M. Gray (Fig 2.7a); Mr S. Lowe (Fig 3.1); Dr J. A. Matthews (Figs 2.1 and 5.19); and Mr R. Tipping (Fig 3.15).

Overall, however, our greatest debt is to our wives. Not only have they helped with the usual routine chores of typing, correcting, editing and proof-reading, but they have provided encouragement when we began to despair of the book ever seeing the light of day, and they have usually managed to keep Kathy and Stephen away from their harassed fathers when the mountains of typescript threatened to engulf us. They, probably more than we, will be glad to see the manuscript safely in the hands of the publishers. This book, therefore, is for them.

John Lowe
Mike Walker
October 1983

Preface to the second edition

The gestation period of the second edition of *Reconstructing Quaternary Environments* has bordered on the elephantine. We began work over four years ago, and the book should have appeared in 1994. That it did not do so is partly due to the intrusion on our time of other research activities and academic commitments. However, the principal cause of the delay in delivering a final manuscript to the publishers has been the enormous volume of Quaternary literature that has appeared since the first edition was completed in the early 1980s, and which we have felt obliged to work our way through and to attempt to digest. At the outset, we believed that minor revisions would suffice, but it rapidly became clear that large sections of the original text required a complete overhaul. Furthermore, in order to broaden the scope of the book, we have included a completely new Chapter 7, the collection of material for which has led us into areas which are exciting but, at the same time, relatively unfamiliar. The bibliography might seem excessive, but it could easily have been three times the length, and agonising over which sources were essential to provide adequate background further prolonged our task. We readily accept that the end product will not please all sections of the Quaternary community, for some will undoubtedly feel that important aspects have been either inadequately dealt with, or even completely overlooked. We have, however, made every effort to identify those ideas and lines of evidence that we believe to be fundamental to Quaternary palaeoenvironmental reconstruction and, in particular, to present the material in such a way that it will both inform and stimulate the next generation of Quaternary science students.

We could not have completed this book without the assistance of a large number of friends and colleagues. We are deeply indebted to Colin Ballantyne, Keith Barber, Rick Battarbee, Svante Björck, Russell Coope, Ed Derbyshire, Phil Gibbard, Rob Kemp, Dick Kroon, Jon Pilcher, Richard Preece, Eddie Rhodes, Jim Rose and Ian Shennan, all of whom have read and commented on particular sections and, in some cases, on whole chapters of the book. Their input has been considerable and the manuscript has been significantly improved by their perceptive and constructive critical appraisal. Any mistakes that remain are entirely our responsibility, not theirs! The line drawings have been efficiently and skilfully produced by Don Shewan and his colleagues, Gareth Owen and Drew Ellis, at London Guildhall University, and we would like to express our sincere gratitude to each of them. We also acknowledge the help and encouragement that we have received from Sally Wilkinson, Tina Cadle and their team at Addison Wesley Longman Ltd. We are sure that they will both be pleased (and relieved!) to see the manuscript arrive on their desks. Once again, we should like to express our warmest thanks to our wives and families. They have, as on previous occasions, shown a remarkable forbearance while yet another book has intruded on domestic life, and we are especially grateful for the support and encouragement that they have continued to provide.

Finally, we would like to pay tribute to one colleague in particular. In the preface to the first edition of this book, we acknowledged our debt to Brian Sissons, who had a profound influence on our early careers, and who retired (prematurely in our view) from academic life in 1984. It was difficult to see anyone fulfilling a similar role, but someone has. For several years now we have enjoyed the intellectual stimulation, academic collaboration and personal friendship of Russell Coope. In many ways he has become our new mentor and, in recognition of this fact, it is our privilege and pleasure to dedicate this book to him.

John Lowe
Mike Walker
November 1995

Acknowledgements

We are grateful to the following for permission to reproduce copyright material:

Fig. 1.4 reprinted from *Palaeogeography, Palaeoecology, Palaeoclimatology* **64**, 223, 227, Williams *et al.*, Copyright © 1988, with kind permission from Elsevier Science Ltd, The Boulevard, Langford Lane, Kidlington 0X5 1GB, UK; Figs 1.5 and 6.15 reproduced by permission of the Royal Society of Edinburgh and N. J. Shackleton from *Transactions of Royal Society of Edinburgh: Earth Sciences* **81**(4) (1990), pp. 255, 265, 270; Fig. 1.8 reprinted by permission of Kluwer Academic Publishers from *Milankovitch and Climate* (Imbrie *et al.*, 1984); Fig. 2.1 from Quaternary Research Association, *The Quaternary of the Isle of Skye* (after Ballantyne & Benn, 1991), © Quaternary Research Association, reproduced with permission; Figs 2.3 and 2.4 reprinted from A mountain icefield of Loch Lomond Studial age, Western Grampians, Scotland by P. W. Thorpe from *Boreas*, 1986, **15**, 88, 94, 95, by permission of Scandinavian University Press; Fig. 2.6 from Pergamon Press, Elsevier Science Ltd, *Quaternary Science Reviews* (Bowen *et al.*, 1986); Fig. 2.7 reprinted from 'The glaciation of north-eastern Kansas' by Aber from *Boreas*, 1991, **20**, 298, by permission of Scandinavian University Press; Fig. 2.7 (inset) from The University of Minnesota, *Late-Quaternary Environments of the United States, Volume 1: The Late Pleistocene* (Mickelsen *et al.*, 1983); Fig. 2.8 reprinted from 'Deglaciation pattern indicated by the ice-marginal formations in Northern Karelia, eastern Finland' by M. Eronen and H. Vesajoki from *Boreas*, 1988, **17**, 319, by permission of Scandinavian University Press; Fig. 2.11 reprinted from 'Genesis of the Woodstock drumlin field, southern Ontario, Canada' by J. A. Piotrowski from *Boreas*, 1987, **16**, 250, by permission of Scandinavian University Press; Fig. 2.12 reprinted from 'Influence of Southern Upland ice on glacioisostatic rebound in Scotland: the Main Rock Platform in the Firth of Clyde' by J. M. Gray from *Boreas*, 1995, **24**, 32, by permission of Scandinavian University Press; Fig. 2.13 from Geological Society Publishing House and G. S. Boulton, *Journal of the Geological Society* **142**, 447–474 (Boulton *et al.*, 1985); Fig. 2.14 from Geological Society of America, *The Geology of North America, Volume k3: North America and Adjacent Oceans during the Last Deglaciation* (Andrews, 1987); Fig. 2.15 from Blackwell Publishers, *The Changing Global Environment*, edited by N. Roberts (Sugden & Hulton, 1994); Fig. 2.17 from *Journal of Quaternary Science* **4** p 99 (Ballantyne, 1989), reprinted by permission of John Wiley & Sons, Ltd; Fig. 2.19 from Pergamon Press, Elsevier Science Ltd, *Quaternary Science Reviews* (Sutherland, 1984); Fig. 2.21 reprinted from 'The Holocene glacial history of Lyngshalvoya, northern Norway: chronology and climatic implications' by C. K. Ballantyne from *Boreas*,

1990, **19**, 112, by permission of Scandinavian University Press; Fig. 2.26 from *Earth Surface Processes and Landforms* **11** p 669 (Ballantyne & Kirkbride, 1986), reprinted by permission of John Wiley & Sons, Ltd; Fig. 2.29 reprinted from *Palaeogeography, Palaeoecology, Palaeoclimatology* **44**, 60, Harmon *et al.*, Copyright © 1983, with kind permission from Elsevier Science Ltd, The Boulevard, Langford Lane, Kidlington 0X5 1GB, UK; Fig. 2.30 reprinted with permission from *Nature* (Bard *et al.*, 1990), Copyright © 1990 Macmillan Magazines Limited; Fig. 2.31 from *Journal of Quaternary Science* **9** p 281 (Shennan *et al.*, 1994), reprinted by permission of John Wiley & Sons, Ltd; Fig. 2.32 and 2.34 from Chapman & Hall, *Sea Surface studies* (Devoy Ed., 1987); Fig. 2.38(a) from 'Chronology, palaeogeography and palaeoclimatic significance of the late and post-glacial events in eastern Canada' (Hillaire-Marcel & Occhietti, 1980) *Zeitschrift fur Geomorphologie* **24**, 373–392; Fig. 2.38(b) from John Wiley & Sons Ltd, (Morner, 1980) 'Eustacy and geoid changes as a function of core/mantle changes' in *Earth Rheology, Isostacy and Eustacy*, edited by N. A. Morner, pp. 535–553; Fig. 2.38(c) from Quaternary Research Association, *Quaternary Proceedings* **3**, 6 (Firth *et al.*, 1993), © Quaternary Research Association, reproduced with permission; Fig. 2.40 from 'Glacial rebound and sea level change in the Britich Isles' by Lambeck, *Terra Nova* **3** (1991) 379–389; Fig. 2.42 reprinted by permission of Kluwer Academic Publishers, *Geologie e Mijnbouw* **72** (Ruegg, 1994); Fig. 2.43 reprinted from Lang, G. & SchlÅchter, Ch. (eds), *Lake, Mire and River Environments during the Last 15 000 Years,* Proceedings of the INQUA/IGCP 158 meeting on the palaeohydrological changes during the last 15 000 years, Bern, June 1985. 1988, 248 pp., A. A. Balkema, PO Box 1675, Rotterdam, Netherlands; Fig. 2.47 from The University of Minnesota, *Late-Quaternary Environments of the United States, Volume 1: The Late Pleistocene* (Spaulding *et al.*, 1983); Fig. 2.49 from Geographical Society of New South Wales, *Australian Geographer* **19**, 96 (Wasson *et al.*, 1988); Figs 2.50 and 6.11 from Edward Arnold (Publishers) Ltd, *Quaternary Environments* (Williams *et al.*, 1993); Table 2.1 from Cambridge University Press and P. L. Gibbard, *The Pleistocene History of the Middle Thames Valley* (Gibbard, 1995); Table 2.2 from The Geologists' Association, *Proceedings of the Geologists' Association* (Bridgland, 1988); Table 2.3 from GebrÅder Borntraeger, *Zeitschrift fÅr Geomorphologie* **34** (Vîlkel and Grunert, 1990); Fig. 3.45 from Pergamon Press, Elsevier Science Ltd, *Quaternary Science Reviews* (Shackleton, 1987); Fig. 5.4 reprinted from 'Late-Glacial radio-carbon and palynostratigraphy on the Swiss Plateau' by B. Ammann and A. F. Lotter from *Boreas*, 1989, **18**, 119, by permission of Scandinavian University Press; Fig. 5.7 from The University of Arizona, *Radiocarbon* **35**, 192–193 (Bard *et al.*, 1993); Fig. 5.8 and Tables 5.4 and 5.5 from Quaternary Research Association, *Quaternary Dating Methods* (Smart, 1991), © Quaternary Research Association, reproduced with permission; Fig. 5.9 from American Association for the Advancement of Science, *Science* **248**, 1531 (Phillips *et al.*, 1990), Copyright © 1990 by the AAAS; Fig. 5.12 from Routledge (Baillie, 1992) *Tree-ring Dating and Archaeology* **117**, 310–313; Fig. 5.13 reprinted by permission of Kluwer Academic Publishers from *Climatic Changes on a Yearly to Millennial Basis* (Bartholin, 1984); Fig. 5.14 reprinted from 'Radiodensitometric-dendroclimatological conifer chronologies from Lapland (Scandanavia) and the Alps (Switzerland)' by F. H. Schweingruber *et al.* from *Boreas*, 1988, **17**, 563, by permission of Scandinavian University Press; Fig. 5.15 from Dr N. F. Alley; Figs 5.16 and 5.22 from *Quaternary Research* **35**, 3-5 (Stuiver *et al.*, 1991); Fig. 5.17 from The University of Arizona, *Radiocarbon* **35**, 49, 55, 130 (Kromer & Becker, 1993; Stuiver & Becker, 1993); Fig. 5.19 from *Geographia Polonica* **55** (Bjîrck *et al.*, 1988); Fig. 5.20 from Pergamon Press, Elsevier Science Ltd, *Quaternary Science Reviews* (Lundqvist, 1986); Fig. 5.21 from Dr R. Thomas,

University of Edinburgh; Fig. 5.23 reprinted from 'Neoglaciation in South Norway using lichenometric methods' by Erikstad & S llid, from *Norsk Geografisk Tidsjkrift* **40** (1986), by permission of Scandinavian University Press; Fig. 5.24 from Routledge, *Quaternary Environments* (Andrews, 1985); Fig. 5.25 from *Handbook of Holocene Palaeoecology and Palaeohydrology*, edited by Berglund (Thompson, 1986), reprinted by permission of John Wiley & Sons, Ltd; Fig. 5.27 from *Quaternary Research* **36**, 32–33 (An *et al.*, 1991); Fig. 5.30 from *Quaternary Research* **21**, 305 (Beget, 1984); Fig. 5.31 from Geological Society of America, *Geological Society of America Bulletin* **96** (Bogaard & Schminke, 1985); Fig. 5.32 and Table 5.7 from *Quaternary Research* **27**, 19–22 (Martinson *et al.*, 1987); Fig. 5.34 reprinted with permission from *Nature* (Davies, 1983), Copyright © 1983 Macmillan Magazines Limited; Fig. 5.35 reprinted with permission from *Nature* (Bowen *et al.*, 1989), Copyright © 1989 Macmillan Magazines Limited; Fig. 5.36 from The Royal Society, *Philosophical Transactions of the Royal Society, London*, **B318**, 624 (Bowen & Sykes, 1988); Fig. 5.38 from Pergamon Press, Elsevier Science Ltd, *Quaternary Science Reviews* (Gellatly *et al.*, 1988); Fig. 5.39 from *Quaternary Research* **25**, 34 (Colman & Pierce, 1986); Table 5.3 from *Journal of Quaternary Science* **5** p 139 (Walker & Harkness, 1990), reprinted by permission of John Wiley & Sons, Ltd; Table 5.8 reprinted from Ehlers, JÅrgen, Philip L. Gibbard & Jim Rose (eds), *Glacial Deposits in Great Britain and Ireland*, 1991, 589 pp., A. A. Balkema, PO Box 1675, Rotterdam, Netherlands; Fig. 6.9 from The Royal Society, *Philosophical Transactions of the Royal Society, London*, **B318**, 196–205 (De Jong, 1988); Fig. 6.10 from Pergamon Press, Elsevier Science Ltd, *Quaternary Science Reviews* (Pillans, 1991); Fig. 6.13 reprinted with permission from *Nature* (Guiot *et al.*, 1989), Copyright © 1989 Macmillan Magazines Limited; Fig. 6.14 from *Quaternary Research* **30**, 173 (Lao & Benson, 1988); Fig. 6.16 from The Royal Society, *Philosophical Transactions of the Royal Society, London*, **B301**, 155 (Gascoyne *et al.*, 1983); Figs 7.2 and 7.10 from Pergamon Press, Elsevier Science Ltd, *Quaternary Science Reviews* (Behre, 1989); Fig. 7.3(b) from American Association for the Advancement of Science, *Science* **219**, 169 (Adam & West, 1983), Copyright © 1983 by the AAAS; Figs 7.4 and 7.26 reprinted from *Quaternary International* **10-12**, 164, Sejrup & Larsen, Copyright © 1991, with kind permission from Elsevier Science Ltd, The Boulevard, Langford Lane, Kidlington 0X5 1GB, UK; Fig. 7.5 reprinted with permission from *Nature* (GRIP Members, 1993), Copyright © 1993 Macmillan Magazines Limited; Fig. 7.5 reprinted with permission from *Nature* (Thouveny *et al.*, 1994), Copyright © 1994 Macmillan Magazines Limited; Fig. 7.7 reprinted with permission from *Nature* (Dansgaard *et al.*, 1993), Copyright © 1993 Macmillan Magazines Limited; Fig. 7.8 from Geological Society of America, *The Geology of North America, Volume k3: North America and Adjacent Oceans during the Last Deglaciation* (Ruddiman, 1987); Fig. 7.9 reprinted with permission from *Nature* (Bond *et al.*, 1993), Copyright © 1993 Macmillan Magazines Limited; Fig. 7.11 from American Association for the Advancement of Science, *Science* **261**, 199 (Grimm *et al.*, 1993), Copyright © 1993 by the AAAS; Fig. 7.12 from Pergamon Press, Elsevier Science Ltd, *Quaternary Science Reviews* (Clark, 1993); Fig. 7.13 reprinted with permission from *Nature* (Bond *et al.*, 1992), Copyright © 1992 Macmillan Magazines Limited; Figs 7.15 and 7.17 from *Journal of Quaternary Science* **9** p 191–195 (Lowe *et al.*, 1994), reprinted by permission of John Wiley & Sons, Ltd; Fig. 7.18 from Pergamon Press, Elsevier Science Ltd, *Quaternary Science Reviews* (Levesque *et al.*, 1993); Fig. 7.20 from Koá Karpuz & Jansen, *Paleoceanography* **11**, 510, 1992, copyright by the American Geophysical Union; Fig. 7.21 reprinted with permission from *Nature* (Kapsner *et al.*, 1995), Copyright © 1995 Macmillan Magazines Limited; Fig. 7.22 from *Journal of Quaternary Science* **10** p 179 (Lowe *et al.*, 1995), reprinted by permission of John Wiley & Sons, Ltd; Fig. 7.23 reprinted from *Quaternary International* (in press), Walker, Copyright © 1996, with kind permission from Elsevier Science Ltd, The Boulevard, Langford Lane, Kidlington 0X5 1GB, UK; Fig. 7.24 reprinted with permission from *Nature* **317**, 130–134, Copyright © 1985 Macmillan Magazines Limited; Fig. 7.25 from American Association for the Advancement of Science, *Science* **241**, 1043–1052, Copyright © 1988 by the AAAS; Fig. 7.27 from Pergamon Press, Elsevier Science Ltd, *Quaternary Science Reviews* (Broecker and Denton, 1990); Fig. 7.28 from American Association for the Advancement of Science, *Science* **259**, 928 (Raynaud *et al.*, 1993), Copyright © 1993 by the AAAS; Fig. 7.29 from American Association for the Advancement of Science, *Science* **207**, 18 (Stuiver & Quay, 1980), Copyright © 1980 by the AAAS; Fig. 7.30 from Springer-Verlag GmbH & Co. KG, *Start of a Glacier*, edited by Kukla & West, NATO ASI Series (Kukla & Gavin, 1992), Copyright 1992 by Springer-Verlag GmbH & Co. KG.

Whilst every effort has been made to trace the owners of copyright material, in a few cases this has proved impossible and we take this opportunity to offer our apologies to any copyright holders whose rights we may have unwittingly infringed.

CHAPTER 1 The Quaternary record

1.1 Introduction

The **Quaternary** is the most recent major subdivision (**period**) of the geological record, and it extends up to, and includes, the present day (Figure 1.1). Together with the **Tertiary** it forms the **Cenozoic**, the fourth of the great geological eras. In the geological timescale, periods are conventionally divided into **epochs**, and the Quaternary includes two formally designated intervals of epoch status (Hedberg, 1976): the **Pleistocene** (originally meaning 'most recent'), which ended around 10 ka BP,[1] and the **Holocene** ('wholly recent'), which is the present warm interval within which we live.[2] However, since there is now a considerable body of evidence to suggest that the current temperate period is simply the latest of a number of warm episodes forming part of a long-term climatic cycle (see below), the last 10 ka can be seen as part of the Pleistocene epoch (West, 1977), and the Pleistocene can therefore be regarded as extending up to the present day. This interpretation is adopted here, and throughout this book the terms 'Quaternary' and 'Pleistocene' are used interchangeably, since they refer to the same interval of geological time.[3]

1.2 The character of the Quaternary

The Quaternary has long been considered to be synonymous with the 'Ice Age', a view that can be traced back to the writing of Sir Edward Forbes who, in 1846, equated the Pleistocene with the 'Glacial Epoch'. One of the most distinctive features of the Quaternary has certainly been periodic glacier activity during cold periods, with the build-up of major continental ice sheets and the expansion of mountain glaciers in many parts of the world. However, these cold or **glacial** stages were interspersed with warm episodes (**interglacials**) during which temperatures in the mid- and high latitude regions were occasionally higher than those of the present day. During the last interglacial in Britain around 120 ka BP, for example, contemporary tropical creatures such as hippopotamus swam in the River Thames, while lions and elephants roamed the present site of Trafalgar Square in central London! What makes the Quaternary distinctive, however, is not simply the occurrence of repeated warm or cold episodes, for fluctuations in global climate are apparent throughout the Cenozoic (Raymo & Ruddiman, 1992). Rather it is a combination of the high amplitude and frequency of climatic oscillations, coupled with the intensity of the colder periods in particular, that gives the Quaternary its distinctive character. In some parts of the world, temperatures may have fluctuated through more than 15°C between warm and cold episodes, temperature change was frequently rapid, and the last 800 ka alone have witnessed as many as ten full glacial/interglacial cycles. The precise number of Quaternary climatic cycles remains to be established, but evidence from the deep-ocean sediment record (see below) suggests that over the course of the full range of Quaternary time, the world may have experienced as many as fifty cold or glacial stages and a corresponding number of temperate or interglacial periods (Shackleton *et al.*, 1990).

The effects of these climatic changes were dramatic. In the mid- and high latitudes, ice sheets and valley glaciers advanced and retreated, and the areas affected by periglacial (cold climate) processes expanded and contracted. In low latitude regions, the desert and savannah margins shifted through several degrees of latitude as phases of aridity alternated with episodes of higher precipitation. Throughout the world, weathering rates and pedogenic processes varied with changes in temperature and precipitation, river régimes fluctuated markedly, sea levels rose and fell over a vertical range of *c.* 150 m, and plant and animal populations were forced to migrate and adapt in response to these environmental changes.

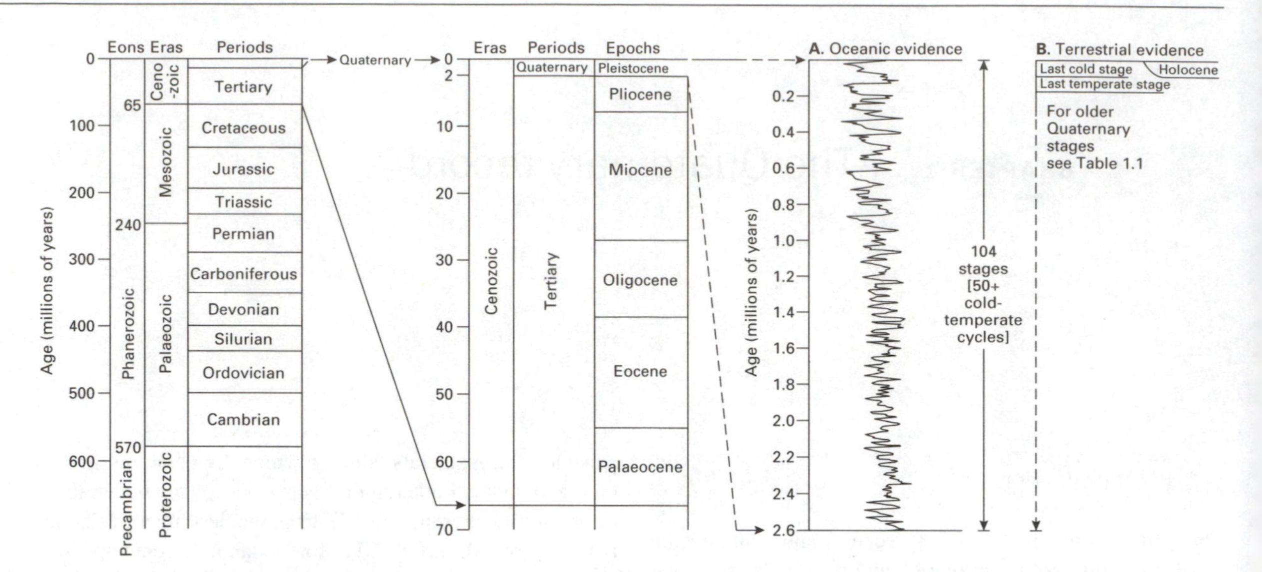

Figure 1.1 The Quaternary relative to the geological timescale. The oxygen isotope trace (section 3.10) from deep-ocean sediments (A) is after Shackleton *et al.* (1990).

The repeated climatic changes that have occurred throughout this latest chapter of earth history have given rise to a rich, but highly complex, record of landforms, sediments, biological (including human) remains and assemblages of human artefacts. From this legacy, it is possible to reconstruct, often with great clarity and in considerable detail, the environmental conditions and associated palaeogeography of particular intervals of Quaternary time. There are a number of separate stages in this process of palaeoenvironmental reconstruction: first, the establishment of the stratigraphy at each site in order to develop a geological framework for the investigation; second, the analysis of **proxy records**[4] from those stratigraphic sequences to produce the basic palaeoenvironmental information; third, the establishment of a chronology of events which involves the development of a **dating framework**; fourth the linking of individual sequences from different locations by means of **correlation**; and finally the integration of different lines of evidence to produce an overall palaeoenvironmental synthesis. Each one of these stages contains its own set of problems. The terrestrial stratigraphic record is often highly fragmented; evidence is absent from many areas, while detailed sequences are only locally preserved. Moreover, the cyclical nature of climatic change has produced similar environmental conditions at different times and, because many records cannot be dated precisely, the process of correlation is frequently beset with difficulties. In a single exposure of Quaternary sediments, therefore, there may be much to perplex the geologist, the geomorphologist, the botanist, the zoologist or the archaeologist, and an explanation of observed geological changes will often require the combined expertise of all of these disciplines. The purpose of this book is to illustrate the very wide range of methods that are currently employed in Quaternary research, and to demonstrate that both a **multidisciplinary** and an **interdisciplinary** approach are required if a proper understanding of the complexities of the Quaternary environment is to be achieved.

1.3 The duration of the Quaternary

The beginning of the Quaternary is very difficult to establish. A view that long held sway was that the Quaternary lasted for approximately one million years (a figure derived from crude extrapolations based on weathering profiles), and that it could be differentiated from the preceding Tertiary on the basis of evidence for widespread glaciation. It is now apparent, however, that many areas of the high latitude regions supported glaciers long before the onset of the Quaternary. There is evidence, for example, of repeated glaciations during the Late Tertiary in Alaska (Hamilton 1986) and Greenland (Larsen *et al.*, 1994), while in Antarctica the Cenozoic glacial record can be traced back to at least 38 Ma BP (Webb & Harwood, 1991). These data reflect the fact that although global temperatures had oscillated, there was a gradual worldwide cooling throughout the Tertiary period (Williams *et al.*, 1993a). In the geological column, therefore, the Pliocene–Pleistocene boundary cannot be drawn simply on the basis of direct terrestrial glacial evidence. Instead, the boundary has usually been located at that point in the stratigraphic record where there are the first indications of climatic cooling, reflected either in the fossil evidence or in some other climatic proxy. In Britain, for example, the Pliocene–Pleistocene boundary in the early East Anglian sequence has traditionally been placed at the stratigraphic discontinuity between the Coralline and Red Crag deposits, a horizon which coincides with an increase in abundance of marine molluscs characteristic of northern seas, and with the first appearance of elephant and horse in the vertebrate fauna (Jones & Keen, 1993). In both the British Isles and elsewhere, however, different indicators of climate change (including various types of fossil) have been employed in different areas, and because precise dating is seldom possible, general agreement on both the position and age of the stratigraphic boundary between the Pliocene and the Pleistocene has been difficult to achieve.

The type or reference section for the Pliocene–Pleistocene boundary is located at Vrica in southern Italy. There the boundary has been placed at the first appearance of the cold-water marine ostracod *Cytheropteron testudo*, and dated on the basis of palaeomagnetic evidence to *c.* 1.64 Ma BP (Aguirre & Pasini, 1985; Harland *et al.*, 1990). This date, which is close to the end of the Olduvai geomagnetic event (Figure 5.26), has been accepted as the geochronometric boundary for the beginning of the Pleistocene in North America (Richmond & Fullerton, 1986a), and in deep-ocean cores from the North Atlantic (Ruddiman *et al.*, 1986). A major environmental change around the end of the Olduvai event has also been recorded in lake records from the western USA, South America and Israel, and in loess sequences from China (Adam *et al.*, 1989; Kukla & An, 1989). Recent revision of the palaeomagnetic timescale based on ocean core evidence, however, suggests an older age for the Olduvai event (section 5.5.1.2), and hence a date for the Pliocene–Pleistocene boundary of 1.81 Ma BP (Hilgen, 1991). This age has been proposed for the base of the Pleistocene in southeast England (Funnell, 1995), while in a revised composite record from different ocean locations the Pliocene–Pleistocene boundary has been placed at *c.* 1.88 Ma (Williams *et al.*, 1988).

Other interpretations suggest a much earlier date for the onset of the Quaternary. In the Netherlands, the climatic deterioration reflected in the biostratigraphic record that is considered to mark the Pliocene–Pleistocene boundary lies close to the Gauss–Matuyama boundary (Figure 5.26) on the geomagnetic timescale (de Jong, 1988). The base of the Pleistocene in the southern North Sea basin is considered to lie close to the same geomagnetic event (Long *et al.*, 1988). This boundary is traditionally dated at *c.* 2.3–2.4 Ma, although a revised age estimate based on ocean core evidence places it at 2.6 Ma (Shackleton *et al.*, 1990). A major global cooling of broadly comparable age is also reflected in other proxy data. For example, microfaunal evidence from cores from the North Atlantic and North Pacific Oceans suggest significantly colder conditions after *c.* 2.4–2.5 Ma BP. These are accompanied by changes in the oxygen isotope content of marine microfossils (see below and Chapter 3), and also by increased quantities of ice-rafted detritus, both lines of evidence indicative of the first major build-up of continental ice masses (Shackleton *et al.*, 1984; Morley & Dworetzky, 1991). In China and in Central Europe, the onset of loess deposition is associated with a major environmental shift around 2.3–2.5 Ma BP (Kukla, 1987a; Kukla *et al.*, 1990), while in lake sequences in Israel, Japan and Colombia, an environmental change of significantly greater magnitude than that at *c.* 1.6 Ma BP appears to have occurred around 2.4 Ma BP (Kukla, 1989).

Opinion on the duration of the Pleistocene is therefore divided between advocates of a 'shorter' chronology in which the Pleistocene lasted for around 1.6–1.8 Ma, and a 'longer' timescale of *c.* 2.5–2.6 Ma. Although there is perhaps a slight majority in favour of the longer chronology, it seems that there is unlikely to be a consensus view within the geological community, and that for some time yet the Pliocene–Pleistocene boundary will have to be chosen in an essentially arbitrary manner (e.g. Kukla, 1989; Williams *et al.*, 1993a). What is apparent, however, is that the Quaternary spans considerably more time than the one million years that is still frequently quoted, although even then it covers no more than about 0.04 per cent of the total age of the earth, or approximately 0.3 per cent of the Phanerozoic, the period of geological time during which fossils have been found in rocks.

1.4 The development of Quaternary studies

1.4.1 Historical developments

The term 'Quaternary' can be traced back to the work of the French geologist Desnoyers who, writing in 1829, differentiated between the strata of 'Tertiary' and 'Quaternary' age in the rocks of the Paris basin. The Quaternary was redefined by Reboul in 1833 to include all strata characterised by the remains of flora and fauna whose counterparts could still be observed in the living world. The term 'Pleistocene' (most recent) was first used by Lyell some six years later to refer to all rocks and sediments in which over 70 per cent of the fossil molluscs could be recognised as living species. Only after the writings of Forbes in the 1840s did the term 'Pleistocene' become synonymous with the glacial period.

Quaternary studies represent one of the youngest branches of the geological sciences, with a history that goes back less than 200 years (Chorley *et al.*, 1964; Davies, 1968). Prior to that it was generally believed that the earth had been created in 4004 BC, a figure based on genealogical calculations from biblical sources by Archbishop Ussher of Armagh and first published in 1658. Hence, early views on geological and environmental changes were constrained by the Ussher timescale of around 6000 years. As a consequence, a **Catastrophist** philosophy held sway in which the form and character of the earth's surface were explained largely through the operation of great floods and other cataclysmic events. Around the turn of the eighteenth century, however, the work of the famous Edinburgh geologists James Hutton and John Playfair began to indicate that the features of the earth's surface could more reasonably be explained by the operation, over a protracted timescale, of processes similar to those of the present day. This significant departure in geological thinking gave rise to the principle of **Uniformitarianism**, first expounded by Hutton, but subsequently popularised by Charles Lyell in his famous dictum 'the present is the key to the past'. Uniformitarian reasoning, in which present-day analogues are used as a basis for the interpretation of observed features within the stratigraphic record, is still fundamental to many aspects of palaeoenvironmental reconstruction (Bell & Walker, 1992).

The nineteenth century saw a number of significant advances in Quaternary studies, many of which stemmed directly from the introduction and gradual acceptance of the **Glacial Theory**. Although for many years there had been speculation that certain Swiss and Norwegian glaciers had formerly been more extensive, it was not until the 1820s that credence was given to the notion of a glacial epoch. The work of Esmark in Norway (Andersen, 1992), of Bernhardi in Germany, and particularly the investigations of the two engineers de Venetz and Charpentier in Switzerland, produced evidence for former glacier activity far beyond the limits of present-day glaciers. However, it fell to the Swiss zoologist Louis Agassiz to expound, in 1837, the first coherent theory of 'the great ice period' involving worldwide climatic changes. Subsequently, Agassiz visited both Britain and North America and in both areas demonstrated that surficial deposits that had previously been interpreted as the products of marine inundation during the flood (**diluvium**) could more reasonably be regarded as the results of extensive glaciation in the relatively recent past.

Although the Glacial Theory did not immediately gain widespread acceptance, its adherents rapidly refined and developed the concept. By the 1850s evidence was beginning to emerge for two glaciations in parts of Britain and Europe and, as early as 1877, James Geikie was describing evidence for four separate glaciations in East Anglia. The strata between the glacial deposits (**drift**) were referred to as '**interglacial**', and hence the idea of oscillating warm (**interglacial**) and cold (**glacial**) episodes emerged. By the end of the nineteenth century, drift sheets of four separate glaciations, the Nebraskan, Kansan, Illinoian and Wisconsinan, along with deposits of three intervening interglacials (in descending order of age, the Aftonian, Yarmouthian and Sangamon) had been identified in North America, while evidence began to emerge for multiple glaciations in different parts of Europe. Probably the most influential work in this respect, however, was that of Penck and Brückner (1909) who resolved the river terrace sequences in the valleys of the northern Alps into four separate series, each relating to a glacial episode. The phases of glaciation were named (from oldest to youngest) Günz, Mindel, Riss and Würm, after major rivers of southern Germany. In both Europe and North America, the maximum limits of Quaternary glaciations were first mapped around the turn of the twentieth century and have subsequently been modified only in detail (Figures 1.2 and 1.3), although views on the terminology adopted and on the number of glacial/interglacial stages experienced during the Quaternary have changed dramatically (see below).

Other effects of glacier expansion and contraction were also recognised at a relatively early stage. The relationship between glaciers and sea level was first considered in a systematic manner by MacLaren who, in 1841, reasoned that at times of glacier build-up sea levels would fall as water was extracted from the ocean basins and locked up in the expanding ice sheets whereas, following ice melting, sea levels would rise as water was returned to the oceans. This was the first statement of the **Glacio-Eustatic Theory** of sea-level change (section 2.5.2). MacLaren suggested that sea levels would fall by 350–400 ft (*c.* 110–130 m) during a glacial phase, a figure that is in remarkably close agreement with more recent estimates. In addition to its effects on global sea levels, the results of the build-up of ice on the earth's surface were also noted. A number of authorities, including Playfair and Lyell, had described the raised shoreline sequences in Scandinavia and around the coasts of Scotland, and had inferred that in both regions crustal uplift had occurred. The mechanism involved in crustal warping, however, remained unclear. In 1865, the Scottish geologist Jamieson finally made the link between the raised shoreline evidence and the Glacial Theory when he deduced that crustal depression would result from the weight of the ice sheets and that uplift would follow deglaciation as the crust was free to rebound to its pre-glacial state. This was the first

Figure 1.2 The maximum extent of Quaternary glaciation in Europe (modified after West, 1977).

clear statement of what are now referred to as **glacio-isostatic effects** (section 2.5.3).

During the later years of the nineteenth century, evidence began to emerge for major environmental changes in areas beyond those directly affected by glacier ice. In the semi-arid southwest of the United States, for example, work by Russell and Gilbert in particular showed that extensive lakes had existed at some time in the past (Figure 1.3) and that phases of higher rainfall (**pluvials**) had alternated with more arid (**interpluvial**) episodes. Moreover, a relationship was postulated (although not clearly articulated) between these climatic oscillations and the glacials and interglacials at higher latitudes. Similar relict drainage features in desert and savannah regions in other parts of the world were described by Victorian explorers and provided further indications of climatic changes in the low latitudes. In the mid-latitude zones, on the other hand, it was gradually recognised that phases of glacier expansion would, in turn, be accompanied by an extension of the tundra belt in which cold-climate (albeit non-glacial) processes would predominate. The term **periglacial** was first used to describe such regions by the Polish geomorphologist, von Lozinski, in 1909.

Biological evidence for Quaternary environmental change also began to emerge soon after the introduction of the Glacial Theory. The writings of Forbes (1846), in which various geographical components of the British flora and fauna were related to successive migrations into the British Isles under different climatic conditions, and of Heer (1865), wherein ecological changes in Switzerland were discussed in the context of Quaternary climatic changes, were particularly important milestones. In the later years of the nineteenth century, the work of the Scandinavian botanists Blytt and Sernander demonstrated the wealth of information

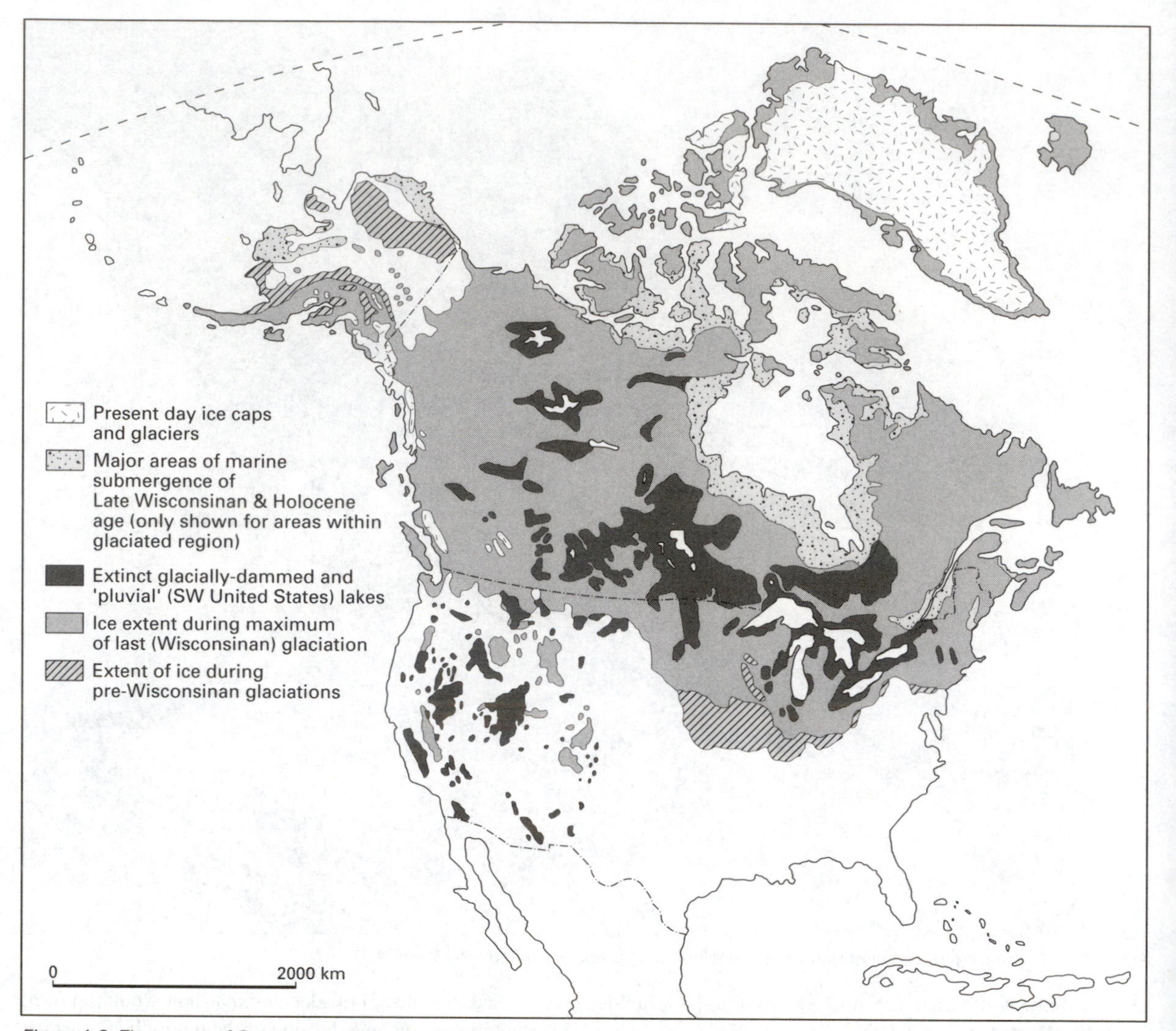

Figure 1.3 The extent of Quaternary glaciers and ice sheets, major ice-dammed lakes, and the principal 'pluvial' lakes in North America (after USGS *National Atlas of the USA*, 1970).

on climatic and vegetational change that could be derived from the stratigraphy and macrofossil content of peat bogs. The scheme of postglacial climatic changes constructed by Blytt and Sernander from Scandinavian peat bog records (Table 3.9) was subsequently refined by the results of pollen analysis, a technique developed in Sweden by von Post (1916) which is still one of the most widely used and successful methods in palaeoecology (Godwin, 1975; West 1977; Birks & Birks, 1980). Systematic investigations of other forms of biological evidence also began during the last century. Important contributions in vertebrate palaeontology included that of Owen (1846), who produced the first comprehensive volume on British fossil mammals and birds, and William Buckland, who not only carried out some of the earliest detailed investigations and analyses of vertebrate assemblages in cave sites (Buckland, 1822), but was also one of the first British converts to the Glacial Theory (Buckland, 1840–41). As early as 1838, James Smith ('Smith of Jordanhill') was using fossil shells to demonstrate that the seas around the coast of western Scotland had been much colder in the past, thereby laying the foundation for subsequent utilisation of marine Mollusca as indicators of former marine temperatures (section 4.8). The seminal works of A.S. Kennard, often in association with B.B.Woodward, in the later part of the nineteenth and early years of the twentieth century provided a similar groundwork for the analysis of land and freshwater Mollusca (e.g. Kennard, 1897; Kennard & Woodward, 1901).

1.4.2 Recent developments

The last fifty years have seen many important developments in Quaternary studies, but five aspects in particular merit attention. The first is the methodological advances that have been made in, and the widespread application of, a range of field and laboratory techniques. Increasingly sophisticated methods of sedimentological analysis have offered new insights into the nature of Quaternary depositional environments, while the interpretation of Quaternary stratigraphy has been greatly assisted by the development of equipment for coring terrestrial, offshore and deep-ocean sequences. Analysis of both terrestrial and marine evidence has been significantly improved by the use of a range of remote sensing techniques, including airborne sensors (e.g. conventional cameras, satellite-mounted imaging systems and radar); ground-based or ship-towed sonar, radar and seismic systems; and tracer methods for the analysis of lacustrine and marine processes. Particularly rapid progress has been made in the mapping, often at very high resolution, of sea-bottom topography and marine sediment architecture through the use of sophisticated sonar and seismic devices. Palaeoecological investigations have also benefited from a range of technological advances, notably in the extraction, recording and analysis of fossil assemblages, and in the fields of both light- and electron-microscopy. These various techniques are considered in more detail in Chapters 2–4.

The second major development has been in the dating of Quaternary events. In the nineteenth century, notions of time were founded largely on estimates of rates of operation of geological and geomorphological processes. Hence, estimated rates of delta construction, cliff retreat, stream dissection, weathering rates and degree of soil development were all used to assess the duration of Quaternary episodes (Flint, 1971). The first, and for many years the only, quantitative method for estimating the passage of time was the varve chronology developed around the turn of the century by the Swedish geologist Gerard de Geer (1912). A major breakthrough came in the years following the Second World War with the discovery, by Willard Libby, of the technique of radiocarbon dating (Libby, 1955). Other radiometric methods, notably potassium/argon and uranium-series dating, were developed in the 1950s and 1960s, along with the techniques of dendrochronology (tree-ring dating) and palaeomagnetism. The 1970s and 1980s have seen the refinement of these various methods and a general increase in levels of chronological precision, particularly as a consequence of the introduction of mass spectrometry into radiometric dating (Linick *et al.*, 1989). In addition, new techniques have been developed, including amino-stratigraphy, electron spin resonance measurement, luminescence measurement, and the use of long-lived cosmogenic radioisotopes such as ^{36}Cl (Aitken, 1990). The principles and applications of the wide range of dating methods now available to the Quaternary scientist are discussed in Chapter 5.

The third important development in Quaternary studies during the course of the twentieth century has been the stratigraphic investigation of sedimentary sequences on the deep-ocean floors. Indeed, it would not be overstating the case to suggest that the results of research into ocean sediments have revolutionised our view of the Quaternary (Imbrie & Imbrie, 1979). In one sense, trying to reconstruct environmental changes from terrestrial evidence is like trying to assemble a jigsaw puzzle and then make sense of the picture when more than 90 per cent of the pieces are missing. This is because much of the evidence has been removed by sub-aerial weathering and erosional processes and, in mid- and high latitudes, by glacial erosion. In parts of the deep oceans of the world, however, sediments have been accumulating in a relatively undisturbed manner for thousands, or even millions, of years, and therefore frequently span the entire range of Quaternary time.

Although the investigation of deep-sea sediments actually began in the nineteenth century with the voyage of the British government research vessel HMS *Challenger* in 1872 (Deacon, 1973), detailed work on the fossil content of core samples from the ocean floors was first undertaken by the German palaeontologist Schott in the 1930s. Prior to the Second World War, only short sediment cores (less than 1 m in length) could be raised from the sea bed. The development of a piston corer by the Swedish oceanographer Kullenberg (1955) heralded the modern phase of deep-sea research, for with the Kullenberg corer and specially equipped research ships, it became possible to take undisturbed sediment cores more than 10 m in length. The changing fossil content of these cores has provided a remarkable record of changes in ocean water temperatures and, by implication, in global atmospheric temperatures during the course of the Quaternary (section 4.11). Many fossils, however, contain other indices of environmental change, most notably variations in oxygen isotope content. Pioneered by Emiliani (1955), oxygen isotope analysis is now regarded as one of the most powerful tools in Quaternary stratigraphy and palaeoenvironmental reconstruction (section 3.10), and continuous isotopic records are now available extending back into the early Quaternary and beyond (e.g. Ruddiman *et al.*, 1989; Shackleton *et al.*, 1990).

A fourth major development over recent decades has been the coring of polar ice sheets and glaciers. Continuous ice-core drilling began on the Greenland ice sheet in the late 1950s, and was followed in the 1960s by the drilling of the first deep polar ice core to reach bedrock at Camp Century, Greenland (Dansgaard *et al.*, 1969). Subsequently, long continuous cores have been recovered from other sites in Greenland, from Antarctica, and from other polar ice caps and mountain glaciers (Oeschger & Langway, 1989). The ice layers revealed in the cores represent annual increments of frozen precipitation, and contain a range of proxy indicators (oxygen isotopes, trace gases, chemical compounds, particulate matter) of past atmospheric and climatic conditions. Ice-core data not only provide a temporal framework for Late Quaternary climatic change (section 3.11), but the upper levels of the ice cores also record the effects of recent industrial activity (e.g. Stauffer *et al.*, 1988). The most recent phase of this research has involved the drilling to bedrock of two cores near the thickest part (>3 km) of the Greenland ice sheet by the European Greenland Ice-core Project (GRIP) and the American Greenland Ice Sheet Project (GISP2). The data from the GRIP and GISP2 programmes provide startling evidence not only of the magnitude of climatic change, but also of the rapidity and frequency with which global climates appear to have oscillated (GRIP Members, 1993).

The fifth significant advance in Quaternary science, particularly during the second half of the twentieth century, has been in the development of increasingly sophisticated computer-based models which simulate a range of aspects of Quaternary environments. This type of work began in the late 1960s with the development of **General Circulation Models (GCMs)**, numerical models that were initially designed to reconstruct patterns of atmospheric circulation during the last cold stage, and possible linkages between terrestrial and atmospheric environments (section 7.7). A range of increasingly sophisticated models has since been developed to explore, in addition to atmospheric circulation, such diverse phenomena as ice-sheet behaviour (e.g. Boulton *et al.*, 1985), glacio-isostatic effects (e.g. Lambeck, 1991a), oceanographical changes (e.g. Broecker & Denton, 1990a), and past vegetation dynamics (e.g. Prentice & Solomon, 1991). Some of the most impressive results have been achieved, however, where scientists from a range of disciplines have collaborated to integrate data on Quaternary environmental change from a variety of different sources, and to use those data as a basis for both descriptive and predictive modelling of Quaternary environments and environmental change. Such an approach is typified by the CLIMAP (Climate/Long Range Investigation Mapping and Prediction) group (CLIMAP Project Members, 1976, 1981), and by the COHMAP (Co-operative Holocene Mapping Project) programme (COHMAP Members, 1988; Wright *et al.*, 1993). These interdisciplinary and multidisciplinary projects are considered in more detail in sections 4.11 and 7.7.

1.5 The framework of the Quaternary

The conventional subdivision of the Quaternary is into **glacial** and **interglacial** stages, with further subdivision into **stadial** and **interstadial** episodes. Glacial stages have traditionally been regarded as protracted cold phases when the major expansions of ice sheets and glaciers took place, whereas stadials have been viewed as shorter cold episodes during which local ice advances occurred. Interglacials are usually recognised as warm intervals when temperatures at the thermal maximum were as high or even higher than those experienced during the Holocene, and which were characterised in the mid-latitudes by the development of mixed woodland. Interstadials, by contrast, are traditionally regarded as relatively short-lived periods of thermal improvement during a glacial phase, when temperatures did not reach those of the present day and, in lowland mid-latitude regions, the climax vegetation was boreal woodland.

These terms are still widely used in Quaternary science, although they clearly lack precision and, as a consequence, are often difficult to apply. Take, for example, the problem of recognising an interglacial as opposed to an interstadial episode on the basis of degree of vegetation development. In northwest Europe, both the interglacials and interstadials of the Late Quaternary were characterised by a range of vegetation types (mixed woodland, boreal woodland, open grassland) depending on latitude, altitude, duration of the warm stage, etc. In more northerly regions, the 'vegetational signature' of an interglacial might be boreal woodland; further south, this type of forest development would be more indicative of an interstadial. Hence, the palaeobotanical distinction between 'interglacial' and 'interstadial' becomes blurred by geographical province. Moreover, this terminology may even be misleading. In the British Isles, for instance, there is no direct evidence for glacier activity during the early Quaternary cold stages (Bowen *et al.*, 1986) and, indeed, this is also the case for many other parts of the world (Dawson, 1992). It is also apparent that during the last cold stage, Southern Hemisphere ice contributed less than 3 per cent to the overall increase in global ice volume, prompting the observation that '...the growth of ice in the Quaternary was essentially a Northern Hemisphere phenomenon' (Williams *et al.*, 1993a, p. 31). The term 'glacial' therefore may have a different connotation in the two hemispheres.

Because of these difficulties, the terms **'temperate stage'** and **'cold stage'** might be considered more appropriate to describe the major climatic episodes of the Quaternary. However, these terms contain their own sets of problems (defining acceptable thresholds between 'warm' and 'cold' episodes; quantifying climatic change; conflicting proxy records for former climate, etc.) and, as a consequence, are equally arbitrary. Moreover, for historical reasons, it is not always possible to avoid the traditional terminology when referring to certain named Quaternary stages. For convenience, therefore, we have opted for the lesser of the evils and have retained the terms 'glacial' and 'interglacial' but, where appropriate, have used these interchangeably with 'cold' and 'temperate' stages. This type of categorisation, based on inferred climatic characteristics, is known as **climatostratigraphy**, and is considered further in Chapter 6.

Attempts to subdivide the stratigraphic record from the land areas of the Northern Hemisphere into a coherent scheme of glacial and interglacial stages that has regional or inter-regional application have hitherto proved to be extremely difficult, principally because of the fragmented nature of most terrestrial sedimentary sequences. Over the past two decades, therefore, reference has increasingly been made to the relatively undisturbed sedimentary sequence in the deep ocean, and particularly to the **oxygen isotope record** in the marine microfossils contained within those sediments. As will be shown in Chapter 3, the oxygen isotope trace (or **'signal'**) obtained from these microfossils reflects the changing isotopic composition of ocean waters over time. Insofar as the marine oxygen isotope balance is largely controlled by fluctuations in volume of land ice (Shackleton & Opdyke, 1973), variations in the isotopic signal in fossils from deep-ocean sediment profiles can be read as a record of glacial/interglacial fluctuations. Working from the top of the sequence, each **isotopic stage** has been assigned a number, even numbers denoting 'glacial' (cold) episodes while the 'interglacial' (warmer) phases are denoted by odd numbers. One of the most impressive features of the deep-sea oxygen isotope record is that the isotopic signal is geographically consistent, and can be replicated in cores taken from different parts of the world's oceans (Figure 1.4). Hence, the marine oxygen isotope sequence provides a climatic signal of global significance.

During the last 800 ka there have been something in the order of ten interglacial and ten glacial stages (Imbrie *et al.*, 1984), and the total number of isotopic stages formally identified in the deep-ocean record of the past 2.5 Ma now exceeds 100 (Figure 1.5). This means that over the course of the Quaternary between thirty and fifty cold/temperate cycles may have occurred (Ruddiman *et al.*, 1989; Patience & Kroon, 1991), depending on the age assigned to the Pliocene–Pleistocene boundary (see above). Even assuming a 'shorter' timescale for the Quaternary, this is many more temperate and cold stages than have been formally recognised and named on the basis of the terrestrial evidence. Hence, the deep-sea sequence provides an independent and unique climatostratigraphic scheme against which individual terrestrial sequences can be compared, and increasingly attempts are being made to establish correlations between these two types of record (Table 1.1). This is discussed more fully in Chapter 6.

The most comprehensive system of designated glacial and interglacial episodes is that for Northern Europe and the British Isles, with less detailed formal schemes for North America and the European Alps. There is general agreement that the **Flandrian** of the British sequence can be equated with the **Holocene** of the European and North American sequences, and that the last cold stage identified in Britain (**Devensian**), northern Europe (**Weichselian**), the Alpine region of Europe (**Würmian**), and North America (**Wisconsinan**), can be considered as broad correlatives. The **Ipswichian, Eemian, Riss–Würmian** and **Sangamon** 'temperate' records from each of these regions are also believed to be essentially coeval and are assigned to the last interglacial, despite the fact that identification is often based on quite different types of proxy evidence. In Europe, a complex sequence of *stadials* and *interstadials* during the last glacial stage (Figure 7.10) is evident in stratigraphic records from areas that lay beyond the limits of the Late Weichselian ice sheet (Behre, 1989).

Throughout northern Europe there is a broad measure of agreement over the brief climatic oscillation that occurred towards the close of the last cold stage (termed the **Devensian Lateglacial** in Britain and the **Weichselian Lateglacial** in northern Europe), for this period can be more precisely dated than older parts of the sequence. However, opinions differ over the extent to which the 'Lateglacial' can be subdivided; in Britain, most scientists accept a twofold division into a **Lateglacial** (or **Windermere**) **Interstadial** and a **Loch Lomond Stadial** (Lowe & Gray, 1980), whereas a more complex sequence with two interstadial episodes, **Bølling** and **Allerød**, separated by a brief cold episode (**Older Dryas**) and followed by the **Younger Dryas Stadial**, has been recognised in records from the European mainland (Mangerud *et al.*, 1974; Figure 1.6). In recent years, a climatic oscillation that appears to be the correlative of the Younger Dryas cold episode has been identified in eastern North America (Wright, 1989).

Prior to the last interglacial, however, the Quaternary records are much more difficult to resolve and, although inter-regional correlations have been attempted (Table 1.1), these are frequently speculative, and become increasingly so as the age of the deposits increases. In the European Alpine

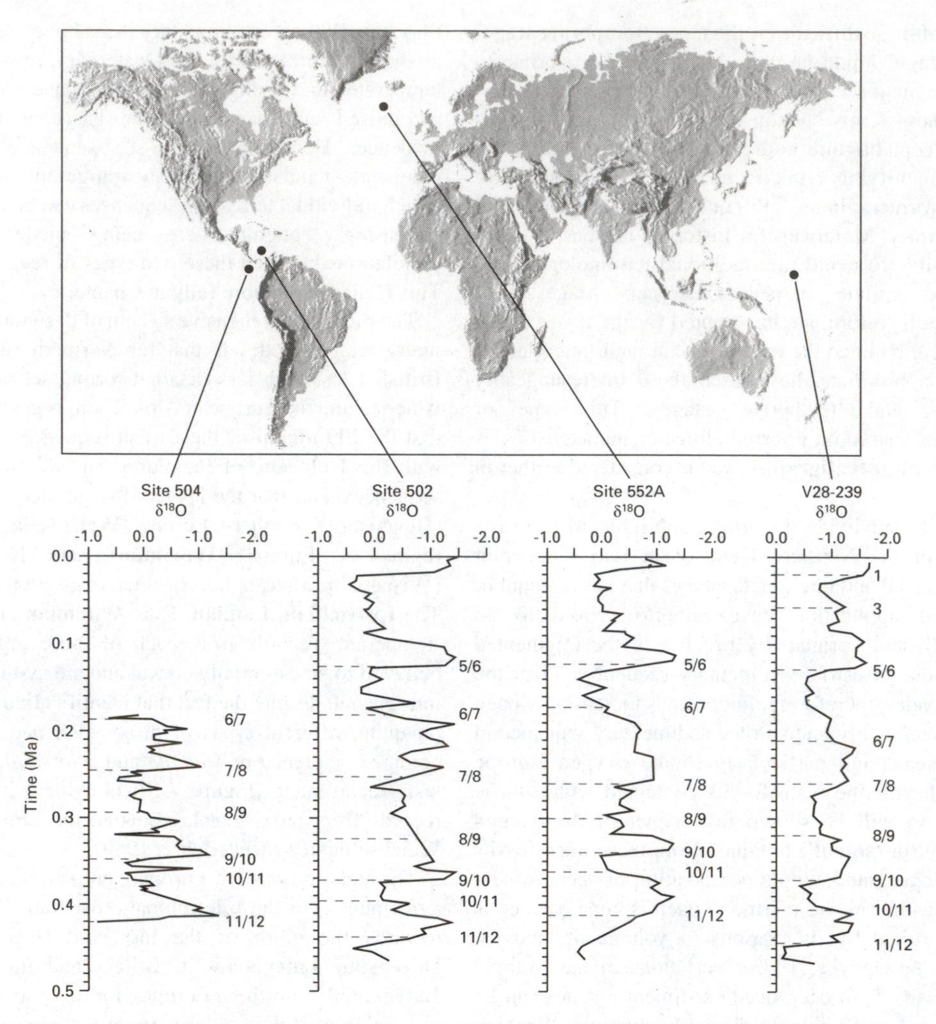

Figure 1.4 Oxygen isotope records for the past 470 ka from the Pacific and Atlantic Oceans (after Williams *et al.*, 1988). The horizontal lines mark the boundaries between isotopic stages (see section 6.2.3.5).

area, the sequence of glacial sediments and 'interglacial' soils now appears to be far more complicated than the classical Alpine model, and hence the standard nomenclature of Günz, Mindel and Riss glacials can be used only in a very general sense (Kohl, 1986; Billard, 1987). Similarly in North America, the classical terms 'Kansan' and 'Nebraskan' glacial periods and 'Yarmouthian' and 'Aftonian' interglacials have been abandoned in favour of a series of stages prior to the Illinoian Glacial that are designated simply by letter (Hallberg, 1986; Richmond & Fullerton, 1986a). In Britain and northern Europe, it is now generally accepted that several of the established named stages must contain a number of separate episodes of cold or temperate character (Table 1.1). Hence, the Cromerian Stage in Britain and in the Netherlands is believed to encompass four or more warm (interglacial?) episodes (de Jong; 1988, Jones & Keen, 1993), while two or maybe three cold or glacial events may have occurred during the classical Saalian Glacial (Šibrava, 1986). A further problem concerns gaps in the terrestrial stratigraphic records. Comparison between the Dutch and British Early and Middle Pleistocene sequences, for example, suggests that as much as a million years of sedimentary history may be missing from the stratigraphic record in southern England between the Beestonian and Cromerian Stages (Gibbard *et al.*, 1991), with other major hiatuses at other times (Table 1.1). Overall, therefore, the individual stages and suggested correlations between those stages shown in Table 1.1 must be regarded

Table 1.1 Quaternary stratigraphy of the Northern Hemisphere. The timescale (left-hand column) is based partly on 'orbital tuning' of the oxygen isotope record (section 5.5.3), and partly on palaeomagnetic stratigraphy (section 5.5.1). Marine oxygen isotope stages (section 6.2.3.5) for the past *c.* 800 ka are shown in the second column from the left; cold (C) and temperate (T) episodes are listed in the right-hand column. Note that beyond OI Stage 21, correlations between the marine oxygen isotope sequence and the terrestrial record are uncertain, and that many of the oxygen isotope stages have yet to be identified in the terrestrial sequence. Based principally on Šibrava (1986), with additional information from Zagwijn (1985), de Jong (1988), Shackleton *et al.* (1990) and Gibbard *et al.* (1991).

Timescale Ma. BP	Marine oxygen isotope stages	NORTHERN EUROPE	THE NETHERLANDS	BRITISH ISLES	EUROPEAN RUSSIA	NORTHERN ALPS	NORTH AMERICA	Cold/Temperate
	1	Holocene	Holocene	Flandrian	Holocene	Holocene	Holocene	T
0.01	2-4d	Weichselian	Weichselian	Devensian	Devensian	Würm	Wisconsinan	C
0.08	5e	Eemian	Eemian	Ipswichian	Mikulino	Riss-Würm	Sangamon	T
0.13	6	Warthe		"Wolstonian"	Moscow (Dneipr Glaciation)	Penultimate Glacial Late Riss ?	Illinoian: Late	C
0.19	7	Saale/Drenthe		"Wolstonian"	Odintsovo (Dneipr Glaciation)		Illinoian	T
0.25	8	Drenthe		"Wolstonian"	Dneipr (Dneipr Glaciation)	Antepenultimate glac. Early Riss / Mindel ?	Illinoian: Early	C
0.30	9	Domnitz [Wacken] (Holsteinian Interglacial)		Hoxnian	Romny			T
0.34	10	Fuhne [Mehleck] (Holsteinian Interglacial)		Hoxnian	Pronya	Pre-Riss ?	Pre-Illinoian A	C
0.35	11	Holsteinian [Muldsberg] (Holsteinian Interglacial)		Hoxnian	Lichvin			T
0.43	12	Elster 1	Elster	Anglian	Oka	Late Mindel ? / Donau	B	C
0.48	13	Elster 1/2	Elster	Anglian	Oka			T
0.51	14	Elster 1	Elster	Cromerian	Oka	Early Mindel ? / Donau	C	C
0.56	15	Cromerian IV	Cromerian IV [Noordbergum]	Cromerian ?				T
0.63	16	Glacial C	Glacial C				D	C
0.69	17	Interglacial III	Interglacial III [Rosmalen]					T
0.72	18	Glacial B	Glacial B				E	C
0.78	19	Interglacial II	Interglacial II [Westerhoven]					T
0.79	20	Helme [Glacial A]	Glacial A			Early Günz ?	F	C
	21	Astern Interglacial I	Interglacial I [Waardenburg]					T
	22 ↓		Bavelian: Dorst				G	C
0.90	↓		Bavelian: Leerdam					T
	↓		Bavelian: Linge					C
0.97	↓		Bavelian: Bavel					T
	↓		Menapian					T/C
	↓		Waalian					T
	↓		Eburonian					C
	↓		Eburonian				H	T/C
	↓		Eburonian	?				T
	↓		Eburonian	Beestonian			I	C
1.65	↓		Tiglian: C5-6	Pastonian				T
	↓		Tiglian: C-4c	Pre-Pastonian/ Baventian				C
	↓		Tiglian: Cl-4b	Bramertonian/ Antian				T
	↓		Tiglian: B	Thurnian			J	C
	103		Tiglian: A	Ludhamian				T
	104		Praetiglian	Pre-Ludhamian				C
2.60			Pliocene	Pliocene				

as no more than a provisional approximation of the Quaternary climatostratigraphic sequence in Europe and North America.

1.6 The causes of climatic change

It is now apparent that the climatic fluctuations of the past 2 million years or so have followed a series of distinctive patterns, and hence explanations of long-term climatic change have, in recent years, tended to focus on the factors that have given rise to both the regularity and frequency of climatic fluctuations (see reviews in, e.g., Imbrie & Imbrie, 1979; Bradley, 1985; Bell & Walker, 1992; Dawson, 1992). The hypothesis that has attracted the greatest attention is undoubtedly the '**Astronomical Theory**', developed by Croll a little over 100 years ago and subsequently elaborated by the Serbian geophysicist Milankovitch. The theory is based on the assumption that surface temperatures of the earth would vary in response to regular and predictable changes in the earth's orbit and axis. Due to planetary gravitional influences, the shape of the earth's orbit is known to change over a period of approximately 100 ka from almost circular to elliptical and back again (Figure 1.7A), a process referred to as the **eccentricity of the orbit**. In addition, the tilt of the earth's axis varies from 21°39' to 24°36' and back over the space of *c.* 41 ka (Figure 1.7B). Because the angle of tilt is measured relative to an imaginary line representing the plane of the ecliptic (the plane described by the earth's elliptical path around the sun), this phenomenon is known as the **obliquity of the ecliptic.** The third variable arises because the gravitional pull exerted by the sun and the moon causes the earth to wobble on its axis like a top (Figure 1.7C). The consequence of this is that the seasons (or the equinoxes) seem to move around the sun in a regular fashion, hence the term **precession of the equinoxes** or **precession of the solstices.**[5] In effect this means that the season during which the earth is nearest to the sun (**perihelion**) varies. At present, the Northern Hemisphere winter occurs in perihelion (Figure 1.7Ci) while the summer occurs at the furthest point on the orbit (**aphelion**). In about 10.5 ka time, the position will be reversed (Figure 1.7Ciii), while *c.* 21 ka hence the cycle will be complete. In fact, it now appears that there are two separate interlocked cycles, a major one averaging around 23 ka and a minor one at *c.* 19 ka.

These variables, in combination, exert a profound effect on global temperatures. The total amount of radiation received is determined largely by the eccentricity of the earth's orbit,[6] while the other astronomical variables affect the way in which that heat energy is distributed at different latitudes. In general it seems that solar radiation receipt in the low and middle latitude regions is governed mainly by precession and eccentricity variations, while in higher latitudes the effects of eccentricity are modulated or amplified by changes in obliquity. Patterns of change through time can be calculated from astronomical data (Figure 1.8A) and Milankovitch was therefore able to obtain

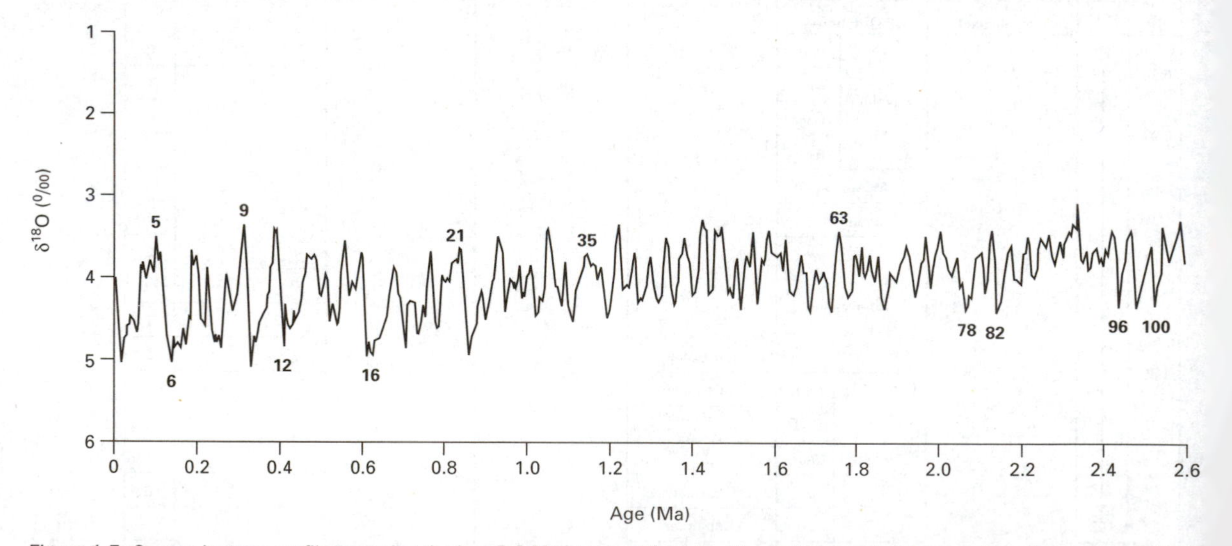

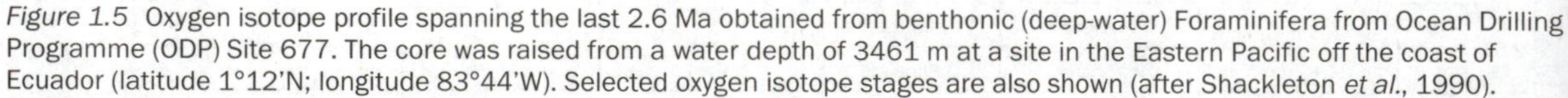

Figure 1.5 Oxygen isotope profile spanning the last 2.6 Ma obtained from benthonic (deep-water) Foraminifera from Ocean Drilling Programme (ODP) Site 677. The core was raised from a water depth of 3461 m at a site in the Eastern Pacific off the coast of Ecuador (latitude 1°12'N; longitude 83°44'W). Selected oxygen isotope stages are also shown (after Shackleton *et al.*, 1990).

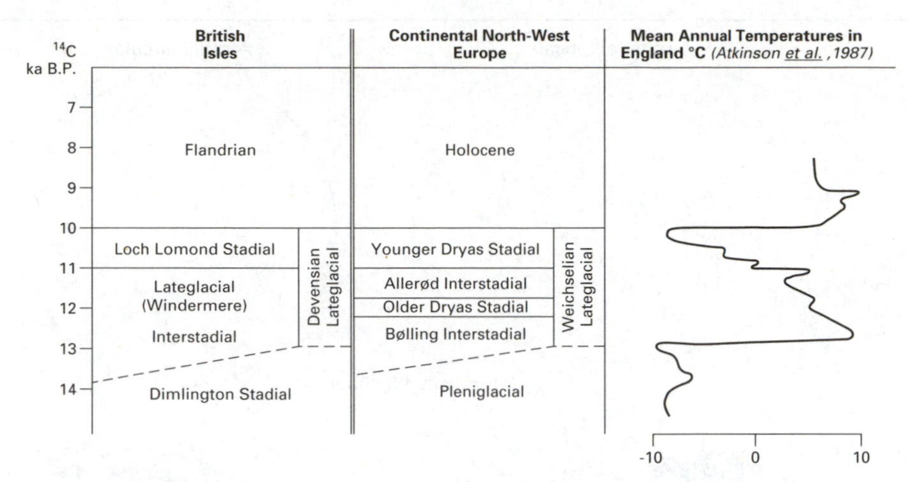

Figure 1.6 The Lateglacial period (*c.* 14–10 k ^{14}C years BP) in northwest Europe.

estimates for radiation inputs at different latitudes, and hence to infer temperature changes through time.

The theory was first published in 1924 and initially found favour with many European geologists, for the sequence of warm and cold stages predicted by the radiation curves appeared to match the record of glacials and interglacials in the classical Alpine region of Penck and Brückner. Increasingly, however, it became apparent that the timing and frequency of glacial episodes during the Late Quaternary did not seem to accord with the pattern of climatic changes predicted by the astronomical variables. This was thrown into sharp relief in the 1940s and 1950s with the development of radiocarbon dating which provided, for the first time, an independent chronology for the Late Quaternary glacial sequence. By the mid 1950s, the Milankovitch hypothesis as an explanation of climatic change had been almost universally rejected. In the late 1960s and early 1970s, however, work initially on sea-level changes and subsequently on deep-ocean sediments reawakened interest in the Milankovitch hypothesis (Imbrie & Imbrie, 1979). Of particular significance was the discovery of oxygen isotope variations in marine microfossils which provided a long-term proxy record of environmental and climatic change (section 3.10; Figure 1.8B). Spectral analysis of ocean core sequences revealed evidence of cycles of 100 ka, 43 ka, 24 ka and 19 ka in the isotopic signal, with the longest cycle driving the glacial/interglacial oscillations of the past 700 ka or so while the others, in combination, modulate or amplify the effects of longer-term changes (Hays *et al.*, 1976). These data provided the first unequivocal evidence of the 100 ka eccentricity cycle, the 41 ka obliquity cycle and the 23 ka and 19 ka precessional cycles in the geological record, and were an impressive demonstration of the role of the astronomical variables in determining patterns of long-term climatic change – hence the title of the seminal paper of Hays *et al.* (1976), 'Variations in the earth's orbit: pacemaker of the Ice Ages'. Subsequently, evidence of the influence of the astronomical variables has been detected in a wide range of proxy records including coral reef sequences (e.g. Aharon 1984), pollen records (e.g. Hooghiemstra *et al.*, 1993), loess sequences (e.g. Kukla, 1987b), ice cores (e.g. Lorius *et al.*, 1990) and tropical lake records (e.g. Kutzbach & Street- Perrott, 1985). Collectively these data would seem to confirm the hypothesis that changes in the earth's orbit and axis, that have become known as **orbital forcing,** are the primary driving mechanism in Quaternary climatic change (Imbrie *et al.*, 1992, 1993).

Although the Astronomical Theory offers a coherent explanation for the sequence of major Quaternary climatic oscillations, it is now apparent that other factors have also influenced the course of global climatic change. For example, although the earliest evidence for the build-up of moderate-sized continental ice sheets dates from around 2.5 Ma BP (Shackleton *et al.*, 1984), proxy data from deep-ocean cores suggest that the climate had cooled, albeit in an oscillatory manner, from around 3.15 Ma onwards (Ruddiman & Raymo, 1988). In addition, the climatic cycles of the Quaternary have not been constant, but have shifted from a periodicity of around 41 ka prior to *c.* 800 Ma BP to a prevailing rhythm of *c.* 100 ka over the course of the last 700–800 ka (Ruddiman *et al.*, 1986). This, in turn, was accompanied by an apparent intensification of glaciation, with the growth of Northern Hemisphere ice sheets to volumes very much larger than those attained over the course of the previous 1.6–1.7 Ma (Ruddiman & Raymo,

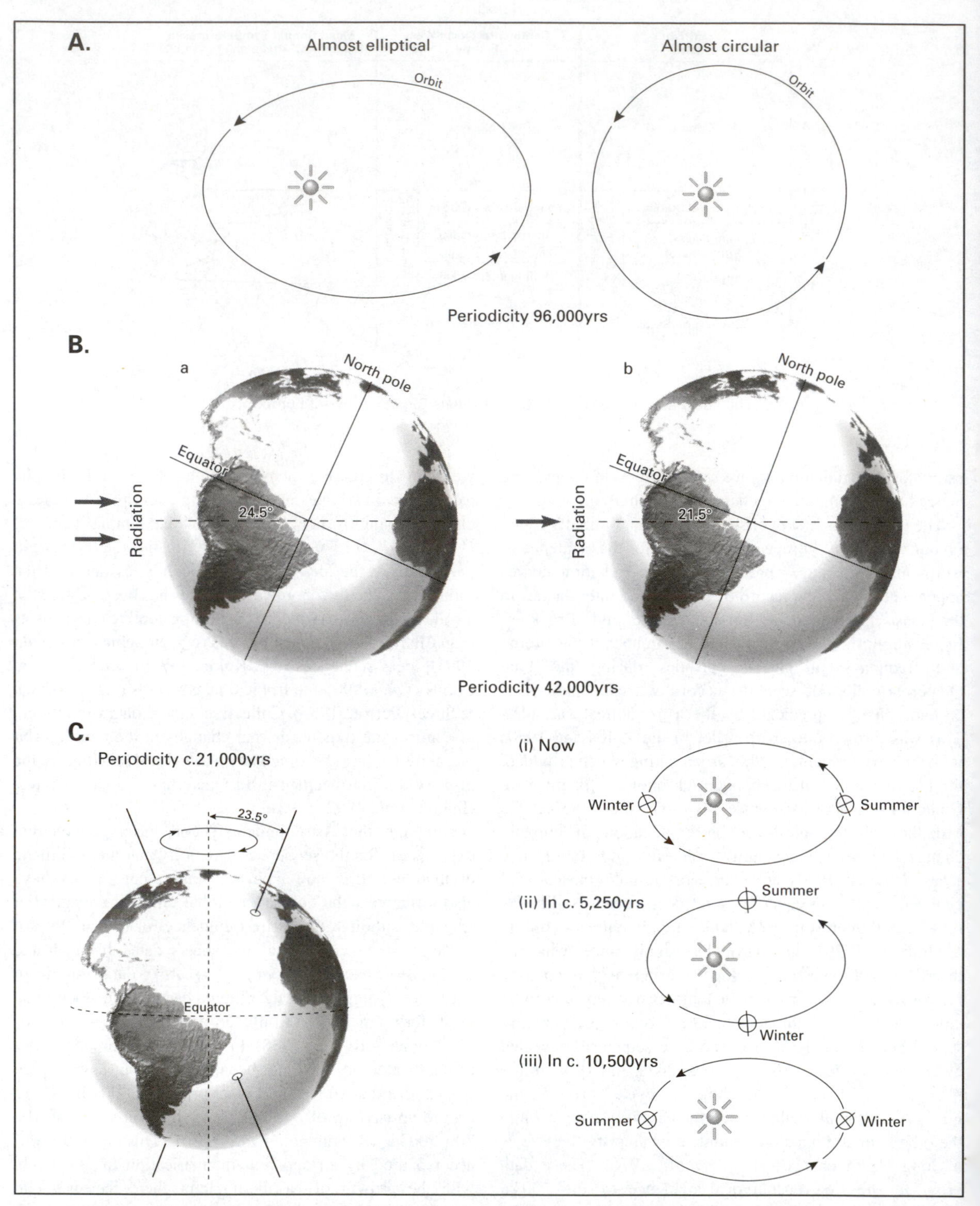

Figure 1.7 The components of the Astronomical Theory of climate change: A, eccentricity of the orbit; B, obliquity of the ecliptic; C, precession of the equinoxes.

1988). This sequence of changes cannot be accounted for solely by the Milankovitch model of orbital forcing.

The major additional elements in the climatic equation which serve to modulate or amplify the effects of the astronomical variables appear to be changes in the disposition of the continental landmasses, tectonic activity, feedback mechanisms caused by oceanic circulation and changes in the extent of ice cover (Broecker & Denton, 1990a, 1990b) and, possibly, variations in the constituents of the atmosphere including, for example, CO_2, methane (CH_4) and dust particles. The long-term global cooling trend referred to above has been widely attributed to the gradual migration towards the polar regions of the major continental land masses (Williams *et al.*, 1993a). In addition, the closing of the Isthmus of Panama some 3.0–3.5 Ma ago (Keigwin, 1978) would have prompted major changes in the oceanic heat and moisture flux in the North Atlantic region, while uplift along the major mountain belts such as the Himalayas, which have risen by some 3000 m over the course of the past 2 Ma alone (Liu *et al.*, 1986), would have led to significant cooling, especially in winter (Birchfield & Weertman, 1983). Land uplift may also have affected the wave structure of the airstreams of the upper atmosphere, the effects of which would have been to cool Northern Hemisphere landmasses, thereby making them especially susceptible to orbitally driven insolation changes. This particular hypothesis, which involves the attainment of key elevation thresholds, has been advanced to explain both the initiation of widespread glaciation in the Northern Hemisphere around 2.5 Ma BP, as well as the intensification of glacial activity during the mid-Quaternary (Raymo & Ruddiman, 1992).

In addition to these long-term variations in climate during the Quaternary, high-resolution proxy records provide evidence of rapid climatic variations, frequently of large amplitude, which are superimposed on the orbitally driven cycles. These short-lived '**sub-Milankovitch**' events occur over timescales varying from centuries to millennia and have been found, *inter alia*, in ice-core records from Greenland (GRIP Members, 1993), terrestial records of the last glacial stage from northwest Europe, and lake records from northern and eastern Africa (Street-Perrott & Roberts, 1983; Gasse *et al.*, 1990). Energy transfer in the world's oceans, driven by salt-density variations ('**thermohaline circulation**'), along with chemical changes resulting from biological activity, are being increasingly considered as major causal factors underlying these events. For example, the abrupt climatic changes that have occurred during the Late Quaternary in many low-latitude regions may be related to subtle changes in the nature and rate of North Atlantic ocean circulation, with fluctuations in sea-surface temperatures (**SSTs**) influencing the pattern and timing of tropical monsoons, in turn leading to marked spatial and temporal variations in precipitation over tropical Africa (Street-Perrott & Perrott, 1990).

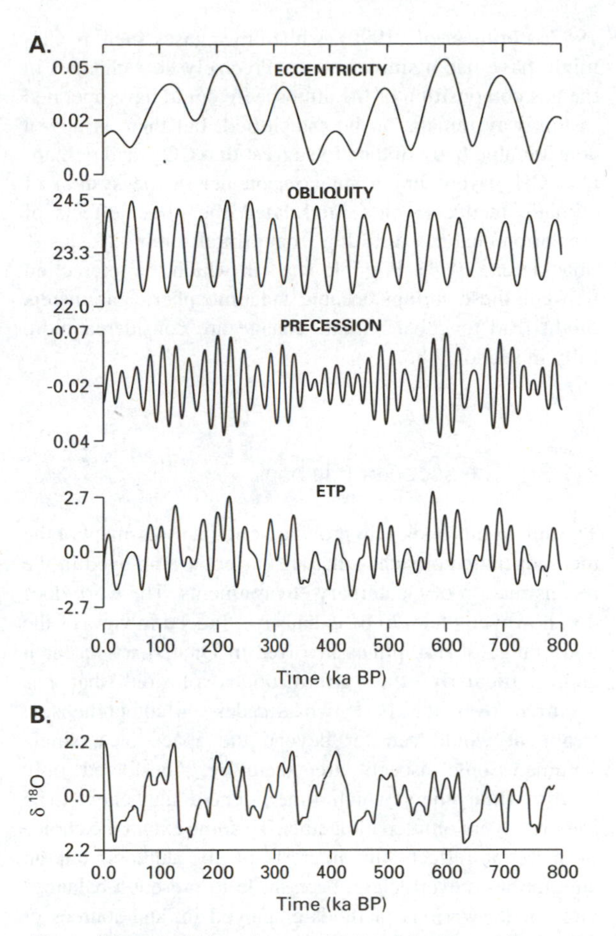

Figure 1.8 A. Variations in eccentricity, obliquity and the precessional index over the past 800 ka. The three time series have been normalised and added to form the composite curve labelled ETP. The scale for obliquity is in degrees and for ETP in standard deviation units. B. Normalised and smoothed variations in the oxygen isotope signal ($\delta^{18}O$) in five deep-sea cores. Note the similarity between this record and the ETP curve above (after Imbrie *et al.*, 1984).

Data from both ocean cores and the polar ice sheets indicate that atmospheric CO_2 levels were significantly lower during the last cold stage than during either the present or last interglacials (Shackleton *et al.*, 1983; Barnola *et al.*, 1987), and a similar pattern has been detected in other gases, notably CH_4 (Chappellaz *et al.*, 1990; Nisbet, 1990, 1992). The close parallels between the CO_2 and temperature profiles from Antarctic ice cores (Chapter 3) has led to the suggestion that CO_2 may also be a major forcing factor in long-term climatic change (Genthon *et al.*,

1987; Lorius *et al*., 1990), while other gases such as CH_4 might have had a similar effect. Precisely how changes in the gas composition of the atmosphere could have operated in this way remains to be established, but there is now a considerable body of data to suggest that CO_2 (and perhaps also CH_4) were important components in the system of climatic feedbacks that **modulated** the direct effects of insolation changes resulting from orbital forcing (Pisias & Shackleton, 1984). Possible ways in which the interaction between these various oceanic and atmospheric parameters might lead to gobal climate change are considered more fully in section 7.8.

1.7 The scope of this book

The aim of this book is to provide a critical assessment of the methods and approaches that are currently employed in the reconstruction of Quaternary environments. The work does not, however, claim to be exhaustive. Indeed in view of the wide range of disciplines involved in Quaternary research and, particularly, the 'information explosion' that has occurred over the past two decades, a comprehensive treatment would run far beyond the space of a single volume. Some aspects are, therefore, considered only briefly, while others (which some will no doubt believe to be important) are omitted altogether. To some extent the choice of material reflects the interests of the authors, but an attempt has, nevertheless, been made to present a balanced view of the various methods employed in, and sources of evidence that form the basis for, Quaternary environmental reconstructions. Some temporal bias is inevitable, as far more is known about the later part of the Quaternary than about the earlier parts of the period, and therefore the majority of examples are drawn from the last interglacial and last glacial stages. The methods, approaches and principles are, however, equally applicable to the analysis of Early and Middle Quaternary environments. In addition, although there is an emphasis on evidence from the Northern Hemisphere mid-latitude regions, particularly from Europe and North America, it is hoped that readers in other parts of the world will find material here that is of interest to them.

The book falls naturally into three parts. In Chapters 2, 3 and 4, the geomorphological, lithological and biological evidence that forms the basis for environmental reconstruction is outlined. Although these are useful general categories within which to describe particular techniques and approaches, they are, to some extent, artificial and there are considerable overlaps between them. Hence, in Chapter 2, where the emphasis is on geomorphology (i.e. surface architecture), certain aspects of the stratigraphy of river terraces and raised shoreline sequences need to be considered also, while in Chapter 3, where sedimentological evidence is being discussed, reference is frequently made to landform evidence as, for example, in the analysis of sand dunes formed in loess and coversand deposits. In all three chapters, field and laboratory techniques are introduced in order to give an indication of the procedures that are involved in generating the basic data.

Chapters 5 and 6 make up the second part of the book. The various dating methods that are currently employed in Quaternary science are described and evaluated in Chapter 5, while the principles of stratigraphy and correlation which enable the researcher to construct meaningful spatial and temporal sequences from often fragmentary evidence are outlined in Chapter 6. The final part (Chapter 7) consists of a discussion of the sequence of events in the North Atlantic region during the last climatic cycle. It illustrates how often diverse evidence can be synthesised into a coherent picture of environmental change and, in particular, the significance of the linkages between the terrestrial, atmospheric and oceanographic components of the natural environment. Insofar as it highlights the gaps in our present state of knowledge, it also serves as a starting point for the next generation of investigations of Quaternary environments.

Notes

1. Throughout this book, the shorthand form is used for **years before present (BP)**: ka – thousand years; Ma – million years. Radiometric dates are quoted in uncalibrated form, and the present is taken as 1950 calendar years AD.

2. Because the end of the Pleistocene is designated as the end of the last cold ('glacial') stage, the period of time since then (the last 10 ka) is often referred to informally as the **'Postglacial'**. In Britain and in some other parts of northwest Europe, however, the term **'Flandrian'** is used as a formal name for the present temperate ('interglacial') stage, in place of the informal and somewhat more equivocal term Postglacial (or the adjectival 'post-glacial'). The name Flandrian derives from the marine transgression which culminated during the present warm stage on the Flemish coastal plain, and follows European practice in naming temperate stages of the Quaternary (e.g. Eemian, Holsteinian) after their characteristic marine transgressions (Hyvärinen, 1978; West, 1979). 'Flandrian' has not been widely adopted outside the British Isles, however, and therefore in this book we use the internationally accepted formal chronostratigraphic term **'Holocene'** for the time interval of the last 10 ka (see Chapter 6).

3. Nevertheless, the International Commission on Stratigraphic Nomenclature (Hedberg, 1976) recommended that, in the

geological record, the Quaternary System should be formally divided into a **Pleistocene Series** and a **Holocene Series,** and this has been endorsed in subsequent international stratigraphic codes (e.g. Whittaker *et al.*, 1991). As a consequence, many Quaternary scientists continue to regard the Pleistocene and Holocene as separate Quaternary intervals each of epoch status.

4. The term **proxy** or **proxy record** is used to refer to any line of evidence that provides an *indirect* measure of former climates or environments. It can include materials as diverse as pollen grains, isotopic records, glacial sediments, tree rings or animal bones (Bell & Walker, 1992).

5. The term '**precession**' describes the slow movement of the axis of rotation of a spinning body (e.g. a gyroscope) about a line that makes an angle with it, so as to describe a cone (Figure 1.7C). It is caused by a torque acting on the rotation axis to change its direction, and is a motion continuously at right angles to the plane of the torque and the angular momentum vector of the spinning body.

6. An alternative explanation of the 100 ka 'glacial' cycle is that it is caused not by eccentricity, but by a previously ignored parameter: **orbital inclination**, or the **tilt of the earth's orbital plane** (Muller & MacDonald, 1995). However, although there appears to be a close correspondence between orbital inclination and the marine $\delta^{18}O$ record, a cause and effect relationship between orbital inclination and long-term climatic change remains to be established.

CHAPTER 2 Geomorphological evidence

2.1 Introduction

The pronounced oscillations in global climate that occurred during the Quaternary led to major changes in the types and rates of operation of geomorphological processes. Undoubtedly the most spectacular manifestations of climatic change were the great ice sheets whose passage resulted in widespread modification of the land surface of mid- and high-latitude regions. The growth and decay of the ice sheets were accompanied by the expansion and contraction of areas affected by periglacial activity, there were fundamental changes in the régimes of many of the major rivers, and the nature and effectiveness of geomorphological processes were strongly influenced by the changes that were continually taking place in the distribution and type of vegetation cover. In low latitudes, phases of aridity were interspersed with periods of wetter climatic conditions so that desert regions were often more extensive during the Quaternary, lake water levels rose and fell, and alluvial processes varied both spatially and temporally. On the global scale, sea level during the glacial phases was more than 100 m lower than that of the present day, but rose to a position above present levels during some of the warm stages.

Landforms that developed under a previous climatic régime have often survived, albeit sometimes in a much modified form, as '**relict**' or '**fossil**' features. Careful analysis of these landforms, and particularly of landform assemblages, can often provide information on the nature of the climatic régime under which they evolved, and also on other environmental parameters such as glacial and fluvial activity, slope stability and groundwater movement. The use of geomorphological evidence in this way, however, requires a proper understanding of the relationships between geomorphological processes and landforms. Moreover, it must be emphasised again that there is frequently a close relationship between geomorphological evidence and lithological evidence (Chapter 3) and that, wherever possible, the two should be used in conjunction in the reconstruction of Quaternary environments.

2.2 Methods

2.2.1 Field methods

2.2.1.1 Mapping

The production of a map illustrating the distribution of the principal landforms is often the first stage in the investigation of the Quaternary history of an area. In some types of analysis, for example the interpretation of drainage characteristics or variation in pedological development, it may be necessary to construct a map showing facets of **morphological mapping** where the aim is to identify and record individual slope elements in the landscape and the nature of the junctions between them (Richards, 1990). A simple hand-held instrument such as an Abney Level or a clinometer can be used for slope measurement. Typical morphological maps are produced at scales of 1:10 000 or larger, for even the most subtle changes in the shape of the land are often recorded. This approach has been widely employed in land survey, but because it is not specifically concerned with landscape evolution, it has found less favour with Quaternary scientists. **Geomorphological mapping**, on the other hand, is one of the most important techniques in Quaternary research, for the maps produced contain not only information on morphology but also on the genesis and, in some cases, on the age of the landforms. This type of mapping can be carried out at a variety of scales ranging from very detailed maps of small areas (typically 1:10 000) to maps at the national scale (e.g. IGS Quaternary Map of the British Isles 1977, scale 1:625 000). Geomorphological mapping is essentially interpretative and therefore requires both an appreciation of the complexity of landform assemblages and a detailed knowledge of their genesis. It also needs an eye for detail, a grounding in field mapping and survey techniques, and a knowledge of the properties of aerial photographs, as the mapping of landforms on aerial photographs often precedes work in the field. Geomorphological mapping, particularly at large scales (i.e. 1:10 000 or greater), has been most effectively employed in the analysis of glacial landscapes, including those resulting from the passage of the last ice sheets and particularly from more recent phases of glacier activity (Figure 2.1A and B).

Further discussion on mapping techniques in geomorphology can be found in Gardiner & Dackombe (1983), Cooke & Doornkamp (1990) and Goudie (1990).

2.2.1.2 Instrumental levelling

In reconstructing the Quaternary history of an area, it is often essential to determine the precise altitude of, and differences in altitude between, particular landforms and landform assemblages. The same applies equally to lithological units (Chapter 3). Altitudinal data can aid in the interpretation of landform assemblages, and may also enable

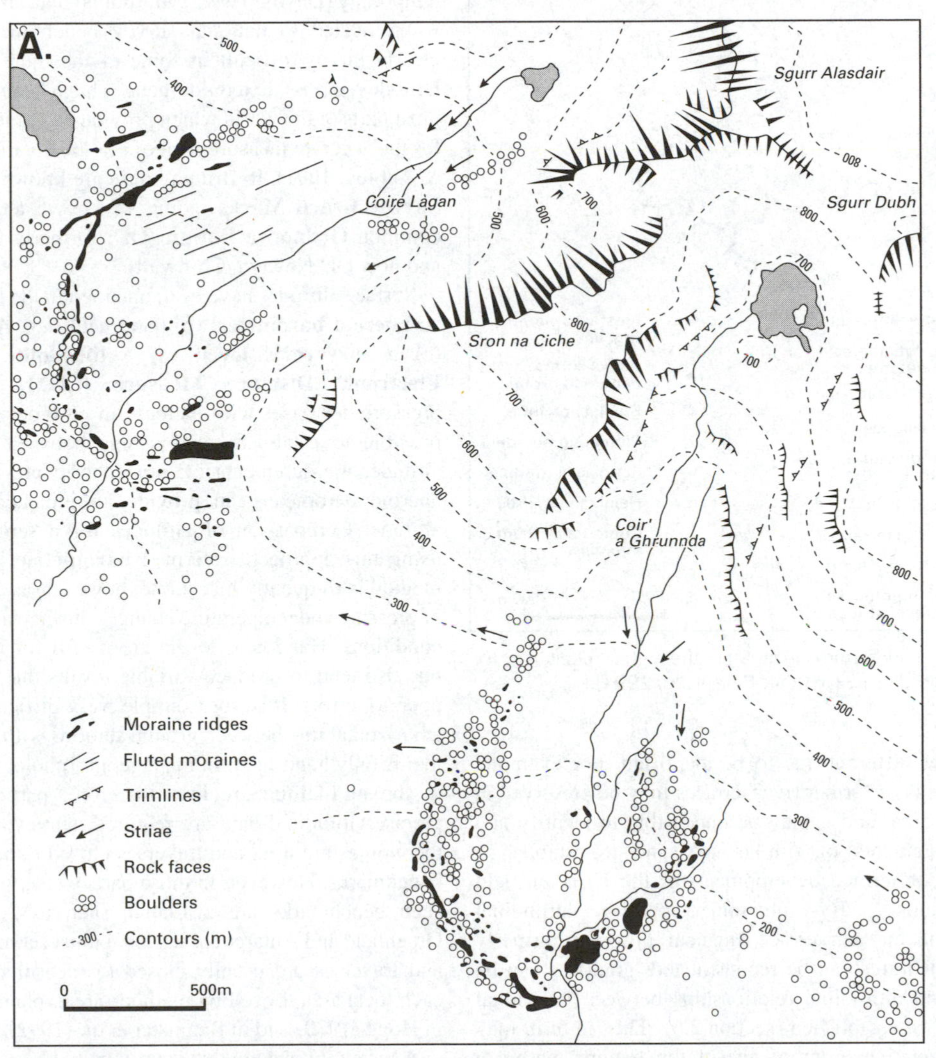

Figure 2.1 Geomorphological maps from the Isle of Skye, western Scotland. A. Landforms relating to the Loch Lomond (Younger Dryas) Stadial glaciers in Coire Lagan and Coir' a Ghrunnda, The Cuillins, south-central Skye (after Ballantyne & Benn, 1991).

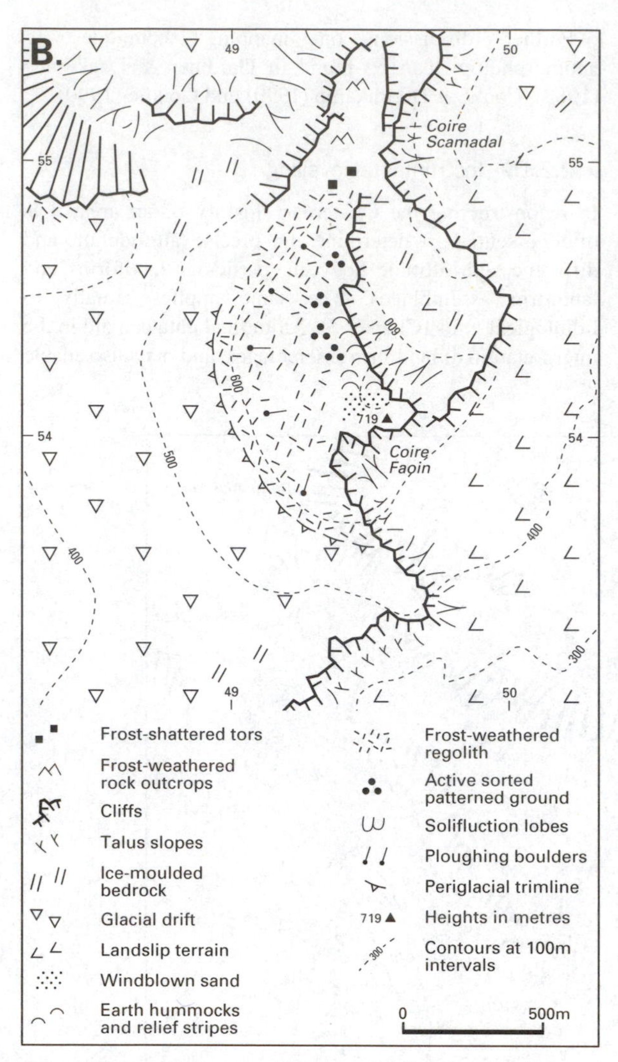

Figure 2.1 B. Landforms on The Storr, the highest peak in Trotternish, northern Skye (after Ballantyne, 1991).

landforms of different age to be identified. For example, only fragments of former river terraces may be preserved in a particular area, and it may be impossible to identify and correlate fragments of similar age, and to establish a chronology of terrace development on the basis of field mapping alone. By obtaining precise altitudinal measurements on each terrace fragment, however, formerly continuous features can be reconstructed, gradients can be measured and altitudinal relationships between individual terraces can be established (section 2.6). This, in turn, may enable the relative order of age of the features within a terrace sequence to be deduced. Similar principles can be applied in the investigation of abandoned shoreline features (section 2.5). Where only a general impression of altitude is required, and where the mapping is being carried out at a relatively small scale, it may be sufficient to obtain the altitudinal data from spot heights and contours on the relevant base maps. Where a more detailed investigation is being conducted, however, the altitudes and surface gradients of landforms must be obtained by instrumental measurements in the field.

The comparison of altitudes of landforms, especially from widely separated localities, requires a common **datum**, a plane of known altitude to which all subsequent measurements can be referred. A frequently employed datum has been sea level, but as this varies both spatially and temporally (Devoy, 1987a), altitudinal data are more reliable when related to national survey bench marks (although clearly this is difficult in some of the more remote high-latitude regions). Fixed tide-gauges have been established in most parts of the world which provide valuable datum points for the accurate measurement of sea-level variations (Emery & Aubrey, 1991). In Britain, these are known as **Ordnance Survey Bench Marks**, points of known altitude above a common **Ordnance Datum** (OD, formerly OD Liverpool and now OD Newlyn, Cornwall).

Surface altitudes have been obtained in the field using: (a) an **aneroid barometer**; (b) hand-held (e.g. **Abney**) **levels**; (c) a **surveyor's level**; (d) a **theodolite**; and (e) an **Electronic Distance Measure (EDM).** Atmospheric pressure decreases with increase in altitude and, since it is possible to establish relationships between pressure and altitude, measurements of atmospheric pressure using an aneroid barometer can provide an indication of ground altitude. Comparison of altitudes for a series of stations using this approach is termed **barometric levelling**. The method is frequently inaccurate, however, as a consequence of pressure and temperature changes during variable weather conditions. Hand-held levels are useful for rapid surveys, but also tend to produce variable results due principally to operator errors. It is, for example, very difficult to maintain a horizontal line between ground stations with an instrument that is only hand-held. In most geomorphological fieldwork in the mid-latitude regions therefore, particularly where precise altitudinal data are required, surveyor's levels and theodolites are used and traverses closed to national survey benchmarks. However, in those parts of the world where no such benchmarks are available, such as Arctic Canada, Greenland and Antarctica, sea level must serve as the datum, and traverses are usually closed to measured sea level in each local area. Levelling methods are explained more fully in Hogg (1980) and in Bannister *et al.* (1992).

A recent innovation that is proving to be an invaluable aid in geomorphological fieldwork, especially in remote terrain distant from survey datum points, is the **Global Positioning**

System (GPS). This is a method of triangulation based on the computation of distance between a point on the earth's surface and a number of earth-orbiting satellites. Using relatively inexpensive hand-held receivers, 'satellite ranging' enables the position of any point on the surface of the earth to be calculated to within a few metres. Indeed, new developments in **differential positioning** can lead to accuracies within a centimetre (Trimble Navigation, 1989). Once the location of a point is known, the altitude above sea level can also be calculated using the same system.

2.2.2 Remote sensing

'Remote sensing' refers to the acquisition of images of earth surface features (and, to a limited extent, of subsurface features) by a variety of devices that receive radiation or sonar information reflecting variations in the albedo, roughness, lithology or other attributes of the earth's surface. It includes conventional **photographic images** obtained from aircraft, using the visible and non-visible (e.g. infra-red) light spectra, **multispectral scanning systems** mounted on satellites, **radar sensors**, **sonar techniques** (echo-sounders) and, more recently (although not yet widely used), **lasers**. Some of the more common remote sensing techniques that are employed in Quaternary research are considered here.

2.2.2.1 Aerial photography

Since the First World War, aerial photographic reconnaissance has increased both in frequency of use and in degree of sophistication. Good quality aerial photographs are now available even for the most inaccessible parts of the world, allowing at least preliminary maps to be made of clearly defined landforms and landform assemblages (e.g. Lagerbäck, 1988; Clapperton, 1993a). A system of grid corrections can be used for transferring details from photographs to maps where scales differ, or where the photographs contain serious distortions (Curran, 1985; Lillesand & Kiefer, 1987). Aerial photographs are especially useful in the mapping of landforms in that:

(a) they direct attention to areas where landforms are most evident or abundant, thereby avoiding much wasted ground reconnaissance;
(b) they reveal larger-scale landform patterns that may go undetected in ground mapping, such as the shorelines of lakes in semi-arid regions;
(c) they may record morphological features subsequently obscured by afforestation programmes or urbanisation;
(d) repeated surveys allow the monitoring of changing landscapes, changing landform assemblages or changes in the positions of such features as glacier termini through time (e.g. Warren, 1991).

Disadvantages of aerial photographs include distortions due to camera tilt or variations in camera altitude, loss of detail due to cloud cover or shadow effects, poor tonal contrasts so that, for example, drift sometimes cannot be distinguished from bedrock surfaces, and difficulty in the detection of small-scale, yet often geomorphologically significant, landforms. Field mapping, therefore, remains essential, and even where mapping is based on large-scale, good quality aerial photographs, the results must be viewed as no more than a *provisional* map of the Quaternary geomorphology of a region until the interpretations can be checked in the field. For areas where good topographic maps are unavailable, aerial photographs provide the only realistic basis for analysing landform assemblages, and enlargements can be made specifically for this purpose.

2.2.2.2 Satellite imagery

In the late 1960s the National Aeronautics and Space Administration (NASA), with the cooperation of the US Department of the Interior, initiated a programme to place in orbit a series of *Earth Resource Technology Satellites* (ERTS), the first of which, ERTS-1, was launched in 1972. Just before the launch of the second ERTS satellite in January 1975, it was renamed **Landsat Thematic Mapper**, ERTS-1 was retrospectively renamed Landsat 1, and all subsequent satellites in the series have carried the Landsat designation (Freden & Gordon, 1983). The first three Landsats were launched into circular orbits at an altitude of 900 km and each made one complete orbit of the earth fourteen times a day, transmitting images continuously to receiving stations in different parts of the USA. The image sensors in Landsats 1–3 can cover the entire globe, with the exception of the 82–90° polar latitudes, every eighteen days. Landsats 4 and 5, launched in the 1980s, also have circular orbits, but are at a lower altitude (705 km), complete just over 14.5 orbits each day, and cover the globe every sixteen days (Lillesand & Kiefer, 1987). The latest satellite in the series, Landsat 6, was launched in the autumn of 1993. Following the Landsat programme, a number of satellites specifically designed for terrain monitoring have been placed in orbit. These include the French SPOT (*Système Pour l'Observation de la Terre*) satellite, the first of which was launched in 1986 and the most recent (SPOT 3) in 1993, and the European Remote Sensing Satellites (ERS-1: 1991; ERS-2: 1995) of the European Space Agency (ESA). Other satellites are also in orbit monitoring aspects of the oceans (e.g. **Seasat,** designed and run by NASA) and the atmosphere. The latter include the **TIROS** and **NOAA** satellites of the American National Oceanic and Atmospheric Administration (NOAA), the **Nimbus** series operated by NASA, and the **Meteosat** satellites of the European Space Agency (Curran, 1985).

The great advantage of satellite imagery over aerial photographs is that distortions are minimised,[1] the process is much more rapid, and repetitive images of large parts of the earth's surface can be obtained. Moreover, a typical Landsat image covers *c.* 185 km by 185 km, a much larger area of ground than normally depicted on aerial photographs. In terms of imaging, conventional photographs result from the simultaneous recording on film of all visible features seen through the lens of the camera. Satellites, on the other hand, always carry a range of sensors and filters which receive and process images in various light-wave bands. Landsats 1, 2 and 3, for example, carried two types of sensor: a set of **Return Beam Vidicon (RBV) Cameras** and a **Multispectral Scanning System (MSS)**. RBVs do not contain film, but images are received and stored on a photosensitive surface in each camera. This surface is then scanned by an internal electron beam to produce a video signal, very similar to that in a conventional television camera (Lillesand & Kiefer, 1987). In MSS systems, a scanner produces a set of corresponding digital images of terrain in different parts of the electromagnetic spectrum. Digital images comprise a grid of cells (PICture ELements or **Pixels**), each of which is assigned a value which corresponds to the intensity of electromagnetic radiation measured by the sensor from a portion of the terrain in the sensor's *instantaneous field of view* (IFOV). All the Landsats carried both the Multispectral Scanning System which measures radiance from an IFOV of roughly 80 m by 80 m, and also a more sophisticated **Thematic Mapper (TM)**, with an IFOV of 30 m by 30 m and which is therefore able to resolve much finer detail. The French SPOT satellites include a high-resolution scanner which can produce images using an IFOV of 20 m by 10 m. This system can also generate stereopair images from which accurate topographic mapping is possible.

The potential of satellite imagery in Quaternary research is considerable, and applications include, *inter alia,* the study of river fan evolution in India (Agarwal & Bhoj, 1992), the analysis of drift lineations as a basis for reconstructing patterns of movement of the Laurentide ice sheet (Boulton & Clark, 1990a), and the examination of recent environmental changes (soil erosion, deforestation) resulting from human activity (Haines-Young, 1994). Further information on both aerial photography and satellite imagery can be found in Curran (1985), Harris (1987), Lillesand & Kiefer (1987), Cracknell & Hayes (1991) and Kramer (1994).

2.2.2.3 Radar

Radio detection and ranging (**radar**) is based on the emission of pulsed signals from a transmitter, usually in the microwave and higher radio frequencies, and the recording of the 'echoes' of these signals as they are bounced back from the ground surface. The returning signals are affected by ground surface roughness, by the orientation of upstanding features, and by the density and electrical properties of ground materials. In dry sediments or cold ice, boundaries between stratigraphic units or ice layers can often be detected. Airborne radar equipment ('**echo sounders**') have been developed that automatically transform received signals into images (**imaging radar**), usually referred to as **side-looking airborne radar (SLAR)** because the radar antenna fixed below the aircraft is pointed to the side. As in satellite scanners, the pulsed signals scan the terrain and the received signals are subsequently converted into electrical impulses that are recorded directly onto magnetic tape or transformed into a photographic image. In practice, however, SLAR systems are limited by resolution problems arising from the length of the antennae to relatively short-range, low-altitude operations (Lillesand & Kiefer, 1987), and in recent years they have been largely replaced by more sophisticated systems known as **synthetic aperture radar (SAR)**. These effectively generate a notional antenna longer than the physical one by a complex process which effectively enhances signal resolution. SAR systems may be mounted on aircraft or satellites, and have been carried, for example, on Seasat and the ERS satellites (see above).

As with some other remote sensing systems, the great advantage of radar is that data can be generated even in cloudy or adverse weather conditions. Radar has proved to be a useful technique in geomorphological mapping and terrain analysis (Blom & Elachi, 1981), and in the investigation of ice thickness (Shabtiae & Bentley, 1988), ice bottom form (Wadhams, 1988) and subglacial topography in currently glacierised regions (Jezek *et al.*, 1993).

2.2.2.4 Sonar and seismic sensing

A number of techniques have been developed that are based on the gravitational, magnetic or electrical properties of the earth. Movements of **acoustic** or **sonic waves** are affected by the density and other characteristics of different materials, and these have formed the basis for **seismic surveys** that have been particularly widely used in geophysical exploration. Quaternary sediments and landforms are now being routinely investigated using seismic equipment (Roberts *et al.*, 1992). For example, the

Sparker sound wave emitter has been employed in studies of submarine geomorphology and Quaternary stratigraphy around the British coasts and in the North Sea region (Davies *et al.*, 1984; Ehlers & Wingfield, 1991). Similarly, **acoustic-reflection profiling** has been used to investigate moraines beneath Lake Superior (Landmesser *et al.*, 1982), submarine terraces in the western Baltic (Healy, 1981), and the Late Quaternary marine stratigraphy of the southern Kattegat (Nordberg, 1989) and North Sea (Salmonsen, 1994). **High-resolution seismic reflection profiling** has also proved valuable in the analysis of lake sediment sequences (Cronin *et al.*, 1993a).

2.3 Glacial landforms

Ever since the general acceptance of the 'Glacial Theory' by the geological community in the middle years of the nineteenth century, it has been recognised that landforms of glacial erosion and deposition are important palaeoenvironmental indicators. When mapped carefully, they reveal a great deal about the extent, thickness and behaviour of former ice masses, the direction of ice movement at both local and regional scales, and the nature and pattern of glacier retreat. In certain circumstances, the evidence may be used to reconstruct the configuration of the former ice sheet or glacier surfaces, and to enable former ice volumes to be estimated. Comparisons with present-day glaciers, particularly those where a close relationship has been established between glacier behaviour and climatic parameters, enable inferences to be made about former climatic régimes. In addition, glacial geomorphological evidence is a key element in the development of computer models of the geometry of former ice sheets and glaciers. The first stage in this process, however, is the production of an accurate map of the extent of Quaternary ice masses.

2.3.1 Extent of ice cover

Establishing the maximal extent of former ice sheets and glaciers has long been regarded as one of the most challenging objectives of Quaternary research. In North America, the systematic field mapping of the outer limit of Pleistocene glaciation began soon after 1860. The maximal extent of ice cover was based largely on the evidence of conspicuous 'end moraines' or on the limits of glacigenic deposits ('Drift Border'), and by 1878 a map had been produced of the southern margin of the glaciated area between Cape Cod and North Dakota (Flint, 1965). Similar investigations of glacial drift cover and end moraines were underway in Europe, Asia and parts of the Southern Hemisphere, so that by 1894 J. Geikie was able to compile maps of the worldwide distribution of glaciers during what he referred to as the 'Great Ice Age'.

The principal types of geomorphological evidence used in the reconstruction of former ice-marginal positions are lateral, terminal, end and retreat moraines, outwash spreads and sandar,[2] ice contact features such as kame terraces, and valley-side or downvalley limits of stagnation moraine, boulder spreads (boulder limits) or drift limits. Lateral and terminal moraines (and, in some cases, marginal meltwater channels) mark the position of the glacier margin at its maximal extent, whereas within those limits, linear moraine ridges will reflect subsequent **recessional stages** as the ice becomes temporarily stabilised during deglaciation (Figure 2.2), while widespread moundy topography (**dump** or **hummocky moraine**) will result from glacier stagnation *in situ* (Eyles, 1983). Kame terraces, which reflect glaciofluvial deposition along a decaying ice margin, may also preserve a record of ice-marginal positions during ice wastage (Gray, 1991). The types of deposits and landform assemblages produced during ice wastage will be determined by a range of often interconnected factors, including manner (e.g. rate) of glacier retreat, debris content of the ice, position of entrainment of debris within the ice, and topographical influences (Shaw, 1994). In general, however, the overall distribution of a variety of glacial landforms will broadly define the extent of the formerly glaciated area, with both lateral (ice marginal) and vertical (valley-side) limits (Sissons, 1967, 1976).

The land lying beyond or above the area directly affected by glacier ice will have been subjected to periglacial activity, with the shattering of exposed rocks by freeze–thaw processes, the development of gelifluction features (lobes, terraces, etc.), and the formation of structures associated with the action of ground ice (wedges, patterned ground, etc.). The distribution of these periglacial features can, in certain cases, provide further evidence of the former extent of glacier ice. In upland areas, for example, the boundary (or, more commonly, the zone of transition) between glacially scoured and frost-shattered bedrock is referred to as the **trimline**, and indicates the approximate positions of the former ice margins (Figure 2.3A and B). Careful measurement of the trimline altitude on different mountain peaks enables the upper limit of glaciers and ice sheets to be established at the regional scale (Nesje *et al.*, 1994; McCarroll *et al.*, 1995; Figure 2.4). The lateral extent of the area affected by frost action can be used in a similar manner to delimit formerly glacierised areas.

A number of difficulties arise in mapping the former extent of Pleistocene glaciers on the basis of geomorphological evidence. First, at the height of the last

Figure 2.2 The Esmark Moraine (named after J. Esmark, one of the founders of the glaciation theory, who first described the ridge in 1824) in southwest Norway. It is an end moraine of Younger Dryas age, the crest of which rises to 30 m above the valley floor (photo: Björn Andersen).

glaciation in both Europe and North America, ice masses submerged many of the upland areas. Hence, although geomorphological evidence can be found in many places to mark the lateral extent of the ice sheets and glaciers, the vertical extent of the ice mass is much more difficult to establish from field evidence alone. In order to obtain estimates of ice thickness, therefore, recourse usually has to be made to models of former ice sheets (see below). Secondly, successive ice sheets covered broadly the same areas, except towards the outer margins (Figures 1.2 and 1.3), so that the geomorphological evidence from earlier glacial episodes has usually been destroyed by later ice advances. Consequently, in many areas of the mid-latitudes that were affected by Pleistocene glaciers, the great majority of the landforms that have been preserved date only from the later stages of the last glaciation. Thirdly, many glacial landforms have been considerably modified by slope processes and periglacial activity, both during and after regional deglaciation (Ballantyne & Benn, 1994a), and this often poses problems in field mapping and interpretation. Fourth, some glacial landforms may resemble ice-marginal features, but may not, in fact, be so. Glaciofluvial landforms (kames, eskers, etc.), for example, often exhibit linear trends and have, on occasion, been used erroneously as evidence for former ice limits (e.g. Charlesworth, 1929; see Sissons, 1967, 1976). In the majority of cases, however, the lineations displayed by such features reflect local patterns of ice disintegration rather than the retreating margins of active ice masses. Hence, a proper understanding of the nature and origin of glacial landforms is necessary if this type of evidence is to be used to determine the former extent of Pleistocene glacier ice. Fifth, the outer margins of large areas of the last great ice sheets lay beyond the present coastline on the continental shelves, and hence establishment of the ice limits requires the use of sophisticated (and expensive) remote sensing techniques (see above). Finally, it should be noted that, in many areas, the outer margins of drift sheets have no distinctive geomorphological expression, and end moraines in particular are often absent. In some cases, the evidence has been destroyed, either by meltwater activity during deglaciation, or by postglacial erosion or subaerial

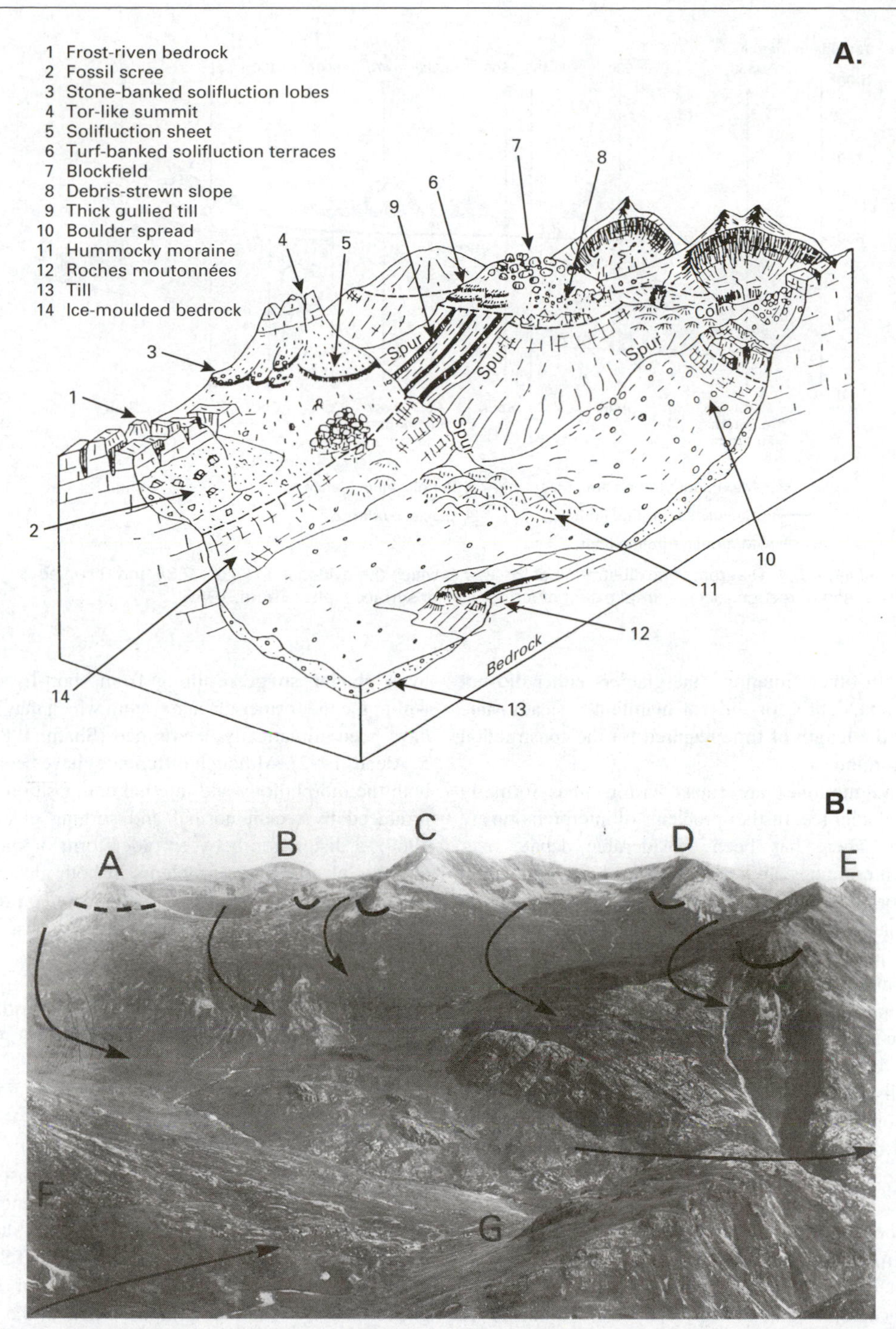

Figure 2.3 A. Idealised features used for identifying trimlines and other types of glacial limits in mountainous terrain. B. The upper Glen Nevis area of western Scotland showing the contrast in appearance of the ground surface above and below a trimline. Note the contrast between the frost-riven nunatak summits at A, B, C, D and E and the ice-moulded and drift-covered bedrock on the lower slopes. Ice movement was from left to right towards the west. A boulder-strewn medial moraine complex extends from F to G across the mouth of the tributary valley in the bottom left of the photograph (after Thorp, 1986).

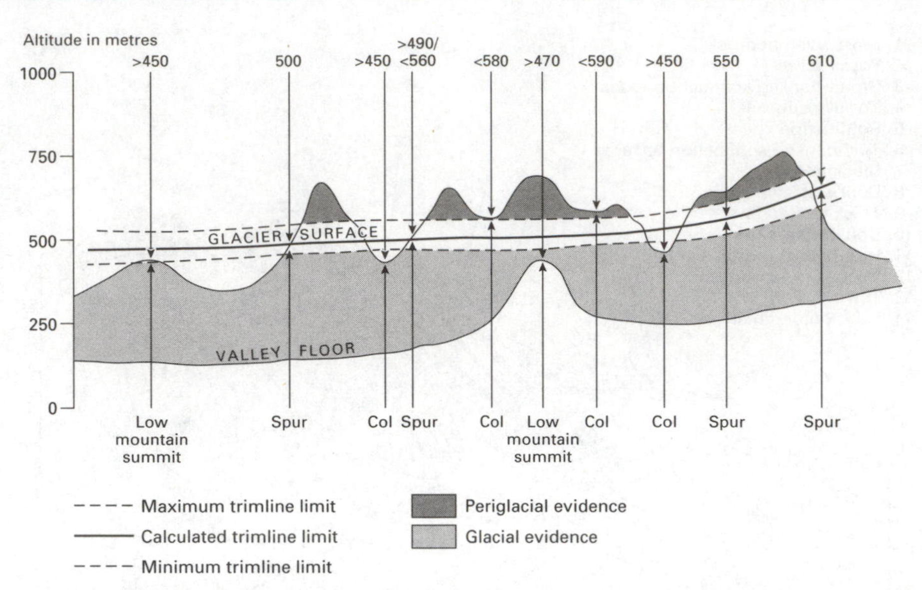

Figure 2.4 Diagrammatic illustration of the way in which the evidence in Figure 2.3A and B can be used to reconstruct the areal extent of former glacier surfaces (after Thorp, 1986).

weathering. In other situations, the glaciers either did not carry sufficient debris, or did not maintain a steady-state position for the length of time required for the construction of an end moraine.

Where end moraines are found within areas formerly covered by glacier ice, further problems of interpretation are encountered. There has been considerable debate over whether such moraines are, strictly speaking, recessional, in that they formed during the stillstand of the ice margin during a phase of overall glacier retreat, or whether they have been produced by a renewed episode of glacier expansion, and therefore reflect a **glacier readvance**. Where prominent end moraines occur, the latter interpretation has usually been adopted, as it has been considered unlikely that large constructional forms would have been produced during a stillstand of the ice margin. The geomorphological evidence, however, is frequently equivocal, and it must be emphasised that conclusive proof of a glacier withdrawal and a subsequent readvance can only be obtained from stratigraphic evidence (e.g. where organic sediments are found interbedded with two glacigenic units) or, in some instances, from other geomorphological features, such as independent indicators of changing directions of ice flow between successive glacial episodes (e.g. Robinson & Ballantyne, 1979). An additional complication in the interpretation of end-moraine evidence is that it is frequently very difficult to distinguish between those landforms that reflect a glacier readvance induced by a deterioration in climate, and constructional forms that have been produced by a **glacier surge** resulting from short-lived instability within the the former glacier system which may, or may not, have been climatically determined (Sharp, 1988; Hambrey & Alean, 1992). Although differences have been detected in both the morphology and internal composition of moraines produced by recent normal and surging glaciers (Rutter, 1969), a distinction between older forms is more difficult, and this clearly poses problems in both glacier modelling and in palaeoclimatic interpretations based on reconstructed Pleistocene glaciers and ice sheets.

2.3.2 Geomorphological evidence and the extent of ice sheets and glaciers during the last cold stage

2.3.2.1 Northern Europe

In many parts of northern Europe, conspicuous end moraines mark the outer limits of the last Fennoscandian ice sheet, and also important recessional stages during glacier decay, while in the mountains of Norway and Sweden there is abundant geomorphological evidence for more recent glacier activity. The southernmost extent of the Weichselian ice sheet is marked by the **Brandenburg Moraine** which can be traced intermittently for several hundred kilometres across the North German Plain (Figure 2.5). Farther east, the maximal position reached by the last ice sheet is indicated by the **Leszno Moraines** in Poland and by the **Bologoye**

Figure 2.5 The extent of the Late Weichselian Fennoscandian ice sheet and inferred recessional stages, based principally on geomorphological evidence (from various sources).

(Bologoe) Moraines in Russia. A series of recessional moraines to the north of the Brandenburg Moraine represent readvances of the ice front during overall deglaciation. These include the **Frankfurt Moraine** of North Germany, the **Poznan Moraine** of Poland and the **Edrovo Moraine** of Russia (the **Outer Baltic Moraines**), while farther north still are the **Inner Baltic Moraines**, with the **Pomeranian Moraine** of the North German plain and the **Pomorze** and **Vepsopo Moraines** of Poland and Russia respectively (Nilsson, 1983; Grube *et al.*, 1986). To the west, in Denmark, the outer limit of Weichselian ice is marked by the **Main Stationary Line**, which has been correlated on geomorphological grounds with the Frankfurt Moraine, while a second ice frontal position, the **East Jutland Line** has been correlated with the Pomeranian Moraine (Lundqvist, 1986a). Large numbers of recessional moraines have been found in southern Sweden, the most prominent of which are the **Halland Coastal Moraines** (*c*. 12.9–13.5 ka BP[3]), the **Göteborg Moraine** (*c*. 12.6–12.8 ka BP) and the **Berghem Moraine** (*c*. 12.4 ka BP) (Berglund, 1979; Lagerlund *et al*., 1983), the last-named of which may be the correlative of the **Hvaler Moraine** in the Oslofjord of southern Norway (Sørensen, 1979) and the **Lille Fiskebank Moraine** of Denmark (Nesje & Sejrup, 1988). On the continental shelf to the west of Norway, the outer limit of the last ice sheet is marked by the **Egga Moraines** (Lundqvist, 1986a), while further north the **Bleik Moraine** marks the Weichselian ice limit in Andøya (Møller *et al.*, 1992) and sequences of submerged moraines mark recessional stages during retreat from the ice sheet maximum (Vorren *et al*., 1983, 1988).

The best-known and most prominent moraines of the Fennoscandian ice sheet formed as a result of a readvance during the Younger Dryas Stadial (*c*. 11–10 k ^{14}C years BP). These have been mapped as almost continuous belts across Southern Norway (**Ra Moraines**), Sweden (**Middle Swedish Moraines**) and Finland (**Salpausselkä Moraines**), while to the west and north the readvance limit is marked by the **Herdla Moraines** of the Bergen area and the **Tromsø–Lyngen Moraines** of northwest Norway (Mangerud *et al*., 1979; Vorren *et al*., 1988). Morphologically and genetically, however, the moraines are very different. In many parts of Norway, for example, constructional forms are small (1–5 m) or are poorly developed (Aarseth & Mangerud, 1974), whereas the Salpausselkä moraines are prominent features over 60 m in height in places, with the two outermost moraines extending as an almost continuous structure for almost 600 km from the coast of southwestern Finland to North Karelia in eastern Finland (Rainio, 1991). All were once considered to be true end moraines, but the Salpausselkäs consist largely of deltas and smaller till ridges which appear to have formed in the Baltic Ice Lake, the extensive water body that was impounded to the south of the Fennoscandian ice sheet during deglaciation (Eronen, 1983). The composition of the Salpausselkä Moraines in eastern Finland was largely determined by the nature of the ice-marginal environment; where the ice terminated in water, massive glaciofluvial sediment sequences accumulated, whereas discontinuous and narrow moraine ridges formed where the ice-front stabilised on dry land (Eronen & Vesajoki, 1988). The form of the Salpausselkä 'Moraines' indicates a lobate ice margin, with the outermost of the Salpausselkä ridges possibly marking the most advanced position of the ice margin during the Younger Dryas (Rainio, 1991). It seems, however, that the Salpausselkäs are not synchronous throughout southern Finland, but represent readvances by different ice lobes at slightly different times (Ignatius *et al*., 1980). The same situation occurs in Norway, for radiocarbon dates associated with the Ra Moraine indicate that the feature is not of the same age throughout its length but that the ice margin was advancing in some areas while retreat was occurring elsewhere (Mangerud, 1980). Hence, although correlations between the Younger Dryas moraines in southern Scandinavia are clear in outline, they become very complicated in detail (Lundqvist, 1986a). These contrasts reflect differing glaciological responses to topographic and climatic factors at different points around the ice sheet and suggest that, on the continental scale at least, former ice limits as shown by geomorphological evidence are more likely to have been metachronous than synchronous. This has obvious implications for ice sheet models derived from such data (see below).

Glacier recession during the early Holocene is marked in many parts of Scandinavia by sequences of recessional moraines, and by the middle of the eighth millennium BP the Fennoscandian ice sheet had virtually disappeared (Lundqvist, 1986a). In many high mountain areas of Norway and Sweden, however, there is widespread geomorphological evidence for renewed glacier activity after that time. This consists principally of 'fresh' and relatively unvegetated terminal and lateral moraines, many of which occur in bands downvalley from present-day active glaciers (e.g. Ballantyne, 1990; Nesje *et al*., 1991). Lichenometric and dendrochronological dating (section 5.4) of the moraines, in association with radiocarbon dates obtained from soils buried beneath glacial materials, indicates that glacier advances occurred on a number of occasions during the Mid- and Late Holocene (Karlén, 1988), but particularly during the **Neoglacial** period (*c*. 3.5–2 ka BP) and during the **Little Ice Age**, the widespread climatic cooling which lasted approximately from the fourteenth to the eighteenth centuries AD (Grove, 1988).

2.3.2.2 British Isles

In Britain, the maximum expansion of glacier ice occurred towards the end of the last (**Devensian**) cold stage, during a period that has been termed the **Dimlington Stadial** (Rose, 1985). The limits of the last ice sheet (Figure 2.6) have been determined using lithostratigraphy, and a range of geomorphological evidence including the extent of 'constructional' glacial topography which divides deposits of the Late Devensian ice sheet ('Newer Drift') from those of earlier glacial episodes (Bowen *et al*., 1986; Ehlers *et al*., 1991). Contrasts in the degree of dissection of surficial deposits and the relative preservation of glacial landforms have been used to establish the position of the ice margin in several regions, while till limits and 'constructional' drift topography, complemented by the distribution of ice-marginal meltwater channels and tors, have provided the principal criteria for establishing the extent of the Late Devensian ice sheet on land (Bowen *et al*., 1986). In many areas, however, the geomorphological evidence is equivocal, and it is clear from stratigraphic data that geomorphological features which have frequently been regarded as marking the outer limits of the 'Newer Drift' do not, in fact, do so. For example, the Escrick and York Moraines in the Vale of York have often been taken to mark the maximum extent of the last ice sheet in this part of

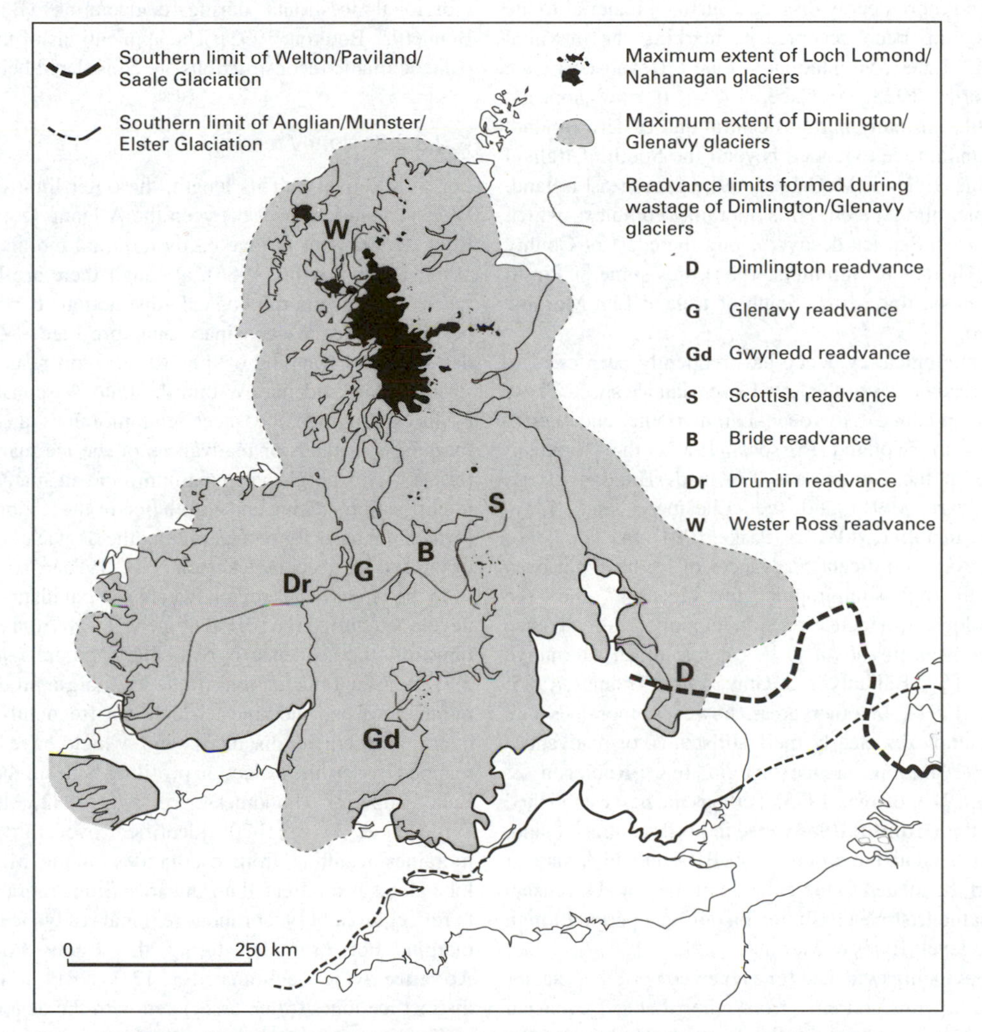

Figure 2.6 Limits of glaciation in Great Britain and Ireland during the last (Devensian) and maximum (Anglian) glacial stages (after Bowen *et al*., 1986).

eastern England (e.g. Embleton & King, 1975). However, more recent evidence suggests that these landforms reflect ice-frontal positions during deglaciation following a surge which carried the ice margin some distance to the south (Catt, 1991; Eyles *et al.*, 1994). Similarly, it is apparent that the last ice sheet occupied a considerable area of southwest Wales beyond the 'South Wales End Moraine' of Charlesworth (1929), many parts of which are now known to be glaciofluvial features which formed within, rather than at the margins of, Late Devensian ice (Bowen, 1981; Harris & Donnelly, 1991). In the Irish lowlands the '**South of Ireland End Moraine**', which can be traced intermittently for over 200 km and which frequently forms a geomorphological divide between relatively fresh and less weathered drift to the north and more deeply dissected surficial material to the south, has long been accepted as marking the maximal extent of Late Midlandian (Late Devensian) ice (Charlesworth, 1928; McCabe, 1985). It now appears, however, that in many parts of central and eastern Ireland, Late Midlandian ice extended beyond the South of Ireland End Moraine (Eyles & McCabe, 1989). In southeast Ireland, for example, the Screen Hills moraine complex, which formed as Irish Sea ice decayed along the coast of County Wexford (Thomas & Summers, 1984), lies some 80 km to the south of the line of the South of Ireland End Moraine (Figure 2.6).

Geomorphological evidence has frequently been cited in support of readvances of the Late Devensian ice sheet. These include the so-called 'Aberdeen–Lammermuir' and 'Perth' Readvances in Scotland (Sissons, 1967), the 'Scottish' Readvance in the Cumberland Lowland (Huddart *et al.*, 1977; Huddart, 1991) and the 'Ellesmere' and 'Llay' Readvances in northeast Wales (Peake, 1961). In a number of cases, however, significant readvances of ice have not been substantiated by stratigraphic investigation, and the geomorphological evidence cited in support of a readvance has frequently been shown to be capable of an alternative explanation (e.g. Ballantyne & Gray, 1984; Thomas, 1985; McCarroll, 1991). In other areas, however, moraines and morainic complexes clearly mark stillstands, or readvances of the Late Devensian ice margin (e.g. Robinson & Ballantyne, 1979; Brown, 1993). Perhaps the best established of these is the **Drumlin Readvance** in north-central Ireland, which has been dated to around 17 ka BP, and which may, at least in part, be related to the collapse of the Late Devensian ice sheet in the Irish Sea basin region during a period of high relative sea level (Eyles & McCabe, 1989).

More convincing evidence for a renewed period of glacier activity following retreat from the Late Devensian maximum can be found in the Scottish Highlands, the Southern Uplands, and the hills of the Lake District, Wales and Southern Ireland (Gray & Coxon, 1991). Terminal and lateral moraines, spreads of hummocky recessional moraine with clear downvalley terminations, well-developed trimlines on mountain sides, intricate meltwater channel systems and the distribution of periglacial features define limits of the **Loch Lomond Readvance**, which is broadly equivalent to the Younger Dryas of Scandinavia (Sissons, 1979). The freshness and relatively unweathered nature of the landforms have enabled the extent of the 'readvance' to be mapped in great detail, and this has not only enabled the ice limits to be established accurately, but has also allowed the individual cirque and valley glaciers to be reconstructed (e.g. Sissons & Sutherland, 1976; Ballantyne, 1989). In many parts of the Scottish Highlands, linear morainic ridges within the Loch Lomond Readvance limits appear to mark individual ice-fronts during deglaciation (Benn, 1992; Bennett & Boulton, 1993). The applications of this work in palaeoclimatic reconstructions are considered below.

2.3.2.3 North America

For a little over half its length, the outer limit of the Late Wisconsinan ice sheet between the Atlantic Ocean and the Rocky Mountains is marked by terminal moraines (Porter, 1983; Šibrava *et al.*, 1986), although there are large areas where no geomorphological distinction can be made between Late Wisconsinan and pre-Late Wisconsinan deposits, and mapping of the last ice limit rests largely on stratigraphic evidence. Within the Late Wisconsinan limits of the Laurentide ice sheet, end moraines are common, marking stillstands or readvances of the ice margin during retreat from the glacial maximum, and in many areas the identification, tracing and correlation of these moraines have formed the basis for reconstructing the deglacial chronology of the last ice sheet (Mickelson *et al.*, 1983).

In the Great Lakes region, where several distinct ice lobes developed during the period of ice wastage from *c.* 15 ka BP onwards, the sequence is particularly complicated (Figure 2.7). Repeated fluctuations of the ice margin produced large numbers of end moraines, which can frequently be traced over considerable distances, and which have also been mapped by seismic reflection profiling beneath the waters of Lake Superior (Landmesser *et al.*, 1982). In Illinois, Willman & Frye (1970) identified over thirty separate moraines resulting from oscillations of the Michigan ice lobe alone as it retreated northwards. Some moraines appear to reflect broadly synchronous regional readvances of the ice margin. For example, during the **Early Port Huron Advance** which culminated *c.* 12.9 ka BP, a number of distinct ice lobes (often associated with the development of end moraines) formed in the Erie–Ontario region, along the southern margins of Lake Huron and in Lake Michigan (Dawson, 1992). A period of ice wastage was followed by a

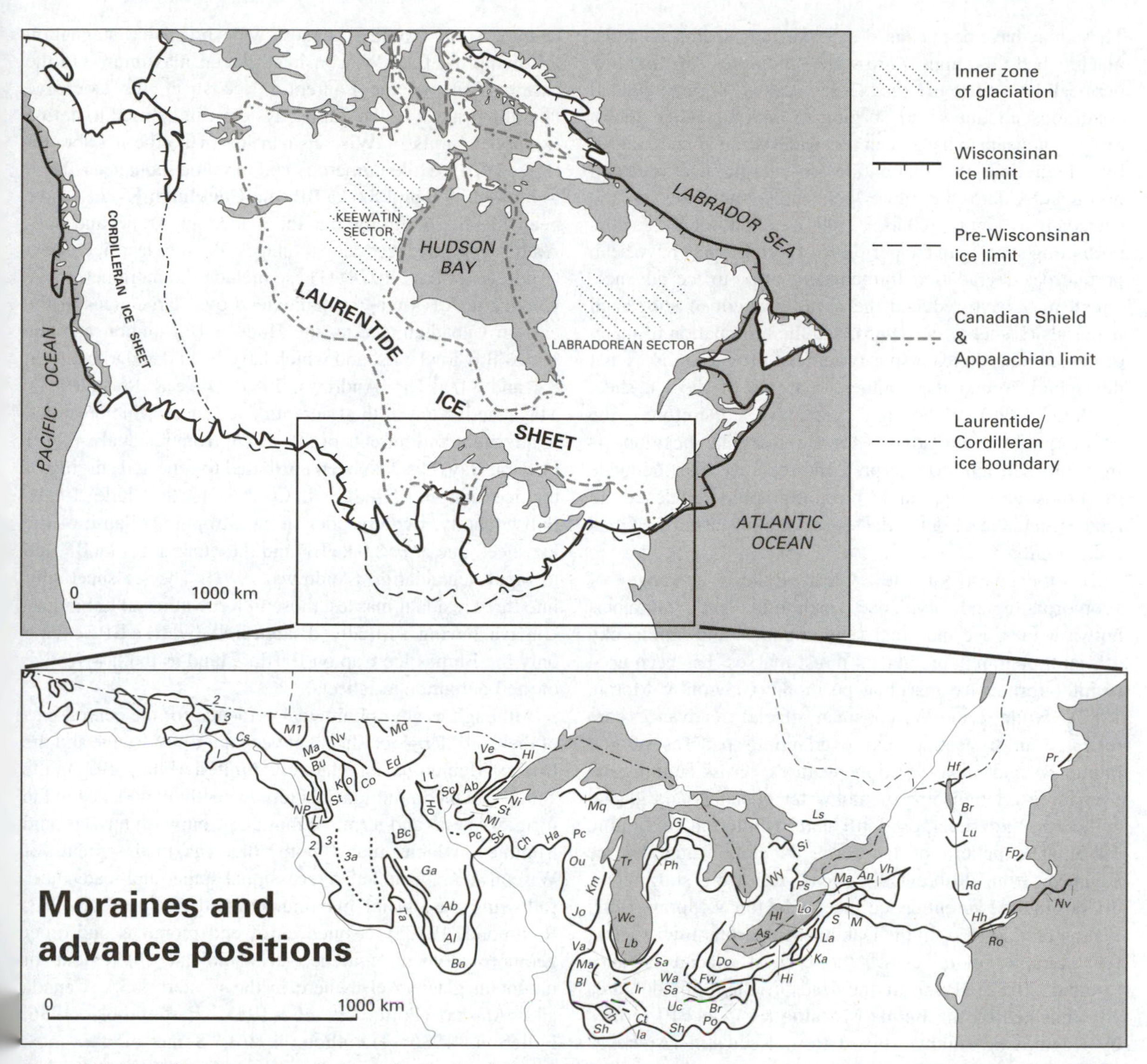

Figure 2.7 Map of North America showing principal ice limits and ice centres (after Aber, 1991). Inset: Wisconsinan moraines and ice limits in the Great Lakes region (after Mickelson *et al.*, 1983).

further readvance around 11.7 ka BP (the **Greatlakean Phase**) in the Michigan, Huron and Green Bay areas. Subsequently, the ice margin readvanced again around 10 ka BP (the **Marquette Advance**) into Lake Superior and northern Michigan (Eschman & Mickelson, 1986). Indeed, this last-named readvance of the southern margin of the Laurentide ice sheet can be traced along a broad front from the prairies (**Cree Lake Moraine**) to Ontario (**Dog Lake and Hartman Moraines**), and along the northern shore of the St Lawrence (Andrews, 1987). There, the prominent **Narcisse and St Edouard Moraines** have been dated to 11–10.4 ka BP and equated with the Younger Dryas climatic cooling of northwest Europe (LaSalle & Shilts, 1993).

In many parts of the Great Lakes region, however, the moraine pattern is complex (Figure 2.7 inset) and younger moraines are frequently found overlapping or cross-cutting older landforms, reflecting a series of non-synchronous ice-marginal fluctuations. These rapid and irregular changes in position of the ice margin do not appear to have been climatically driven, but instead have been interpreted as reflecting inherent instability in the ice sheet which led to repeated glacier surges (Wright, 1973; Clayton *et al.*, 1985).

These may have been caused by deforming sediments at the glacier bed, as high pore-water pressures in the low permeability tills of the Great Lakes region created conditions conducive to surging of individual ice lobes. Surging appears to have been less widespread after 10 ka BP by which time the Laurentide ice margin had retreated northwards onto the more permeable substrates of the Canadian Shield (Clark, 1994). Additional factors promoting surging in the period 15–10 ka BP may have been permafrost degradation immediately prior to ice advance, which may have reduced the shear strength of subglacial materials (Fisher *et al.,* 1985), and the termination of many of the ice lobes in extensive systems of proglacial lakes that developed around the southern margins of the ice sheet (Teller, 1995). Although extensive, therefore, the geomorphological evidence for ice marginal positions is frequently difficult to interpret, and recourse must be made to lithostratigraphic and biostratigraphic evidence to reconstruct in more detail the sequence of events in the Great Lakes region.

To the west of the Great Lakes, a range of geomorphological evidence including end moraines, outwash fans, ice marginal river channels and hummocky glacial topography or 'glacial thrust masses' has been used to infer former ice marginal positions (Clayton & Moran, 1982), while Late Wisconsinan glacial readvances are reflected in truncating and overlapping patterns of end moraines and disintegration features, cross-cutting and overriding relationships of meltwater channels and glacial spillways, and superposed till units (Fullerton & Colton, 1986). The pattern of landforms is complicated by ice advances from different directions, reflecting shifting ice divides in the Laurentide ice sheet (Clayton & Moran, 1982; Fenton *et al*., 1983). In the Dakota–Minnesota–Iowa region, for example, moraines of the earliest glacial advance (around 20 ka BP) from the Hudson Bay/Labrador area are truncated by the **Bemis Moraine** (*c.* 14 ka BP) formed by ice moving southwestward from Keewatin (Andrews, 1987). Subsequent fluctuations of the ice margin, possibly reflecting repeated surging (Clark, 1994), produced a sequence of overlapping and cross-cut moraines (e.g. the **Altamont Moraines**; the **St Croix Moraines**) extending throughout northern Minnesota and northwest Wisconsin (Matsch & Schneider, 1986; Hallberg & Kemmis, 1986).

Farther west on the Canadian Prairies, end moraines and the distribution of hummocky drift and meltwater channels have been used to delimit the extent of Late Wisconsinan ice, although in many parts of the region conspicuous geomorphological features are absent and the limits of the Late Wisconsinan glaciation can only be established on stratigraphic grounds (Fulton *et al*., 1986; Dredge & Cowan, 1989). However, the large expanses of hummocky disintegration moraine suggest widespread ice stagnation following the Late Wisconsinan glacial maximum. On the eastern flank of the Laurentide ice sheet, the extensive lateral moraines of the Saglek system can be used to define the upper limits of Wisconsinan ice in northern Labrador (Ives, 1976), while numerous end moraine sequences, dated to between 9.0 and 8.0 ka BP (the **Cockburn Event**), have been identified throughout the Canadian Arctic and sub-Arctic marking recessional stages of the last ice sheet (Andrews & Ives, 1978). These include the moraines of the **Cochrane Advances**, which extend over large areas of the eastern Canadian Arctic, the Hudson Bay region and the James Bay lowlands, and which have been dated to between 8.4 and 8.0 ka BP (Andrews, 1987; Dyke & Prest, 1987). This reactivation of the Laurentide ice sheet, which resulted in advances and retreats of 50–75 km from residual ice over Hudson Bay, has also been attributed to repeated surging of the ice margin (Dredge & Cowan, 1989; Clark, 1994). Subsequently, two episodes of catastrophic collapse of the ice sheet, one at *c.* 8.0 ka BP and the other at 6.7 ka BP, led to rapid deglaciation (Andrews, 1987). The ice sheet split into three residual masses, those in Keewatin and Labrador–Ungava having virtually disappeared by 6 ka BP leaving only the Barnes Ice Cap on Baffin Island as the last vestige of the Laurentide ice sheet.

Although geomorphological evidence for the central part of the Cordilleran ice sheet between the Coast Range and the Rocky Mountains is relatively limited (Flint, 1971), the valley and piedmont glaciers left a wealth of evidence in the form of lateral and terminal moraines, outwash terraces and trimlines which mark both the maximal extent of Wisconsinan ice and also recessional stages and readvances following the glacial maximum (Waitt & Thorson, 1983; Richmond, 1986a). Sequences of end moraines and other geomorphological evidence also indicate the extent of mountain glaciers elsewhere in the western USA, Canada and Alaska (Porter *et al.*, 1983; Easterbrook, 1986; Fullerton, 1986; Hamilton, 1986). Some of the most prominent moraine sequences date from the mid- and late Holocene, the so-called **Neoglacial** period. The earliest radiocarbon-dated Neoglacial advances occurred around 5 ka BP in the Cascade Mountains of Washington (Davis, 1988), and between 5 and 6 ka BP in the Coast Mountains of British Columbia (Osborn & Luckman, 1988). Subsequent readvances, albeit less securely dated, are reflected in morainic complexes in the mountains of Washington, Wyoming and Colorado (Davis, 1988), while readvances culminating around 3–4 ka BP and 2.5–1.8 ka BP (the **Tiedeman Advance**) occurred in the mountains of Alberta and British Columbia (Osborn & Luckman, 1988; Luckman *et al*., 1993). Farther north, in Alaska, the earliest Holocene glacier readvances date from 7.6–5.8 ka BP, with subsequen readvances around 4.4 ka BP and 3 ka BP (Calkin, 1988). Throughout the western cordillera, the most striking evidence for renewed glacier activity in the form of fresh, unweathered sharp-crested moraines usually adjacent to present ice margins or near headwalls in empty cirques, dates to the Little Ice Age of the past few centuries (Davis, 1988). In the mountains of western Canada, for example, the **Cavell Advance**, which began shortly after 900 BP, culminated in the eighteenth or nineteenth centuries (Osborn & Luckman, 1988), while in Alaska, the Little Ice Age glacial phase, which began around 700 BP, culminated in two major readvances between 50 and 400 BP (Calkin, 1988).

Glacial landforms therefore continue to act as a cornerstone in establishing the extent of former glacier ice. They are of greatest value in the investigation of recent (i.e. Lateglacial and Neoglacial) patterns of glacier activity, especially in highland regions where the features are often well preserved and relatively easily mapped, and from which the vertical and lateral extent of ice can often be reconstructed. For earlier glacial episodes, morainic landforms in particular can still be used to delimit the glacierised area, and as a basis for mapping of readvances, although it is clear that a proper appreciation of glacial and deglacial sequences rests as much (if not more) on stratigraphic verification as on geomorphological evidence. This is discussed more fully in Chapters 3 and 6. Meanwhile, one further line of evidence is required before ice sheets and glaciers can be reconstructed, namely the former directions of ice movement, and it is to this aspect of glacial landforms that we now turn our attention.

2.3.3 Direction of ice movement

In many formerly glaciated areas, a 'grain' or streamlined sculpture is evident in the landscape (Figure 2.3B) reflecting the former direction of ice movement. At a small scale, bedrock protuberances are scratched (striated), fractured, polished and grooved, and at larger scales, whalebacks, *roches moutonnées* and glaciated valleys (troughs) are fashioned by overriding ice. This preferred alignment of erosional forms in a glaciated landscape is often best seen where ice has emphasised local bedrock contrasts, particularly when exploiting the trend of geological weaknesses such as relatively incompetent strata and joint and fault lines (Sugden & John, 1976; Sharp, 1988). Certain glacial depositional landforms, such as drumlins and fluted moraine, may also be aligned in the direction of glacier flow. Careful mapping of landforms that are ice-directed, therefore, allows the dominant patterns of ice movement to be reconstructed.

2.3.3.1 Striations

Striations (or **striae**) form where stones entrained within the basal layers of the ice are dragged across bedrock surfaces, the size of the indentation being determined by the load and relative hardness of the stone and the substrate across which it is dragged (Drewry, 1986). The plotting of striation trends is a relatively straightforward field exercise and, given the availability of exposed striated bedrock in accessible areas, regional directions of ice movement can, at least in theory, be determined fairly rapidly. Where the evidence is abundant and has been mapped over a sufficiently large area (e.g. Hirvas *et al.*, 1988; Peuraniemi, 1989), the dominant directions of ice flow become evident, and the results may often reveal local deflections of ice flow caused by topographic obstructions or interference between competing ice streams (Figure 2.8). In practice, however, the interpretation of striation data is often far from straightforward. Not all 'scratch marks' on bedrock surfaces in formerly glaciated regions have resulted from the passage of ice. Many may simply reflect lines of weaknesses in the rock, accentuated perhaps by sub-aerial weathering, while others may have resulted from fluvial, glaciofluvial, snowcreep or avalanche activity. Where striations are of glacial origin, they may reflect only basal ice movements determined by local bedrock irregularities. They can often be seen, for example, to follow the curvature of the face of bedrock protuberances, and on the lee sides (with respect to ice movement) of rock obstacles are often oblique to the dominant or regional direction of glacier flow. In some areas, diverging sets of striae can be found, reflecting perhaps more than one direction of ice movement, while on certain rock outcrops striations with significantly different trends may be found crossing or superimposed upon one another. **Crossing striations** can arise where glaciological conditions have changed over time, or when ice has readvanced into a region. If the later ice advance has a different direction of flow and the striations resulting from the initial ice advance have not been completely erased, a second set of striations will become superimposed upon the ones that remain from the initial phase of glacier activity (Gray & Lowe, 1982). In some instances, it may be possible to distinguish between different sets of cross-striations; for example, in Snowdonia in North Wales, it has proved possible to distinguish between striae produced by the last ice sheet and those superimposed upon them which resulted from the passage of glacier ice during the later Loch Lomond (Younger Dryas) Stadial (Gemmell *et al.*, 1986; Sharp *et al.*, 1989). In most areas, however, such distinctions are difficult to make. Overall, it would seem that striations are best used in conjunction with other lines of evidence in the reconstruction of regional patterns of ice movement.

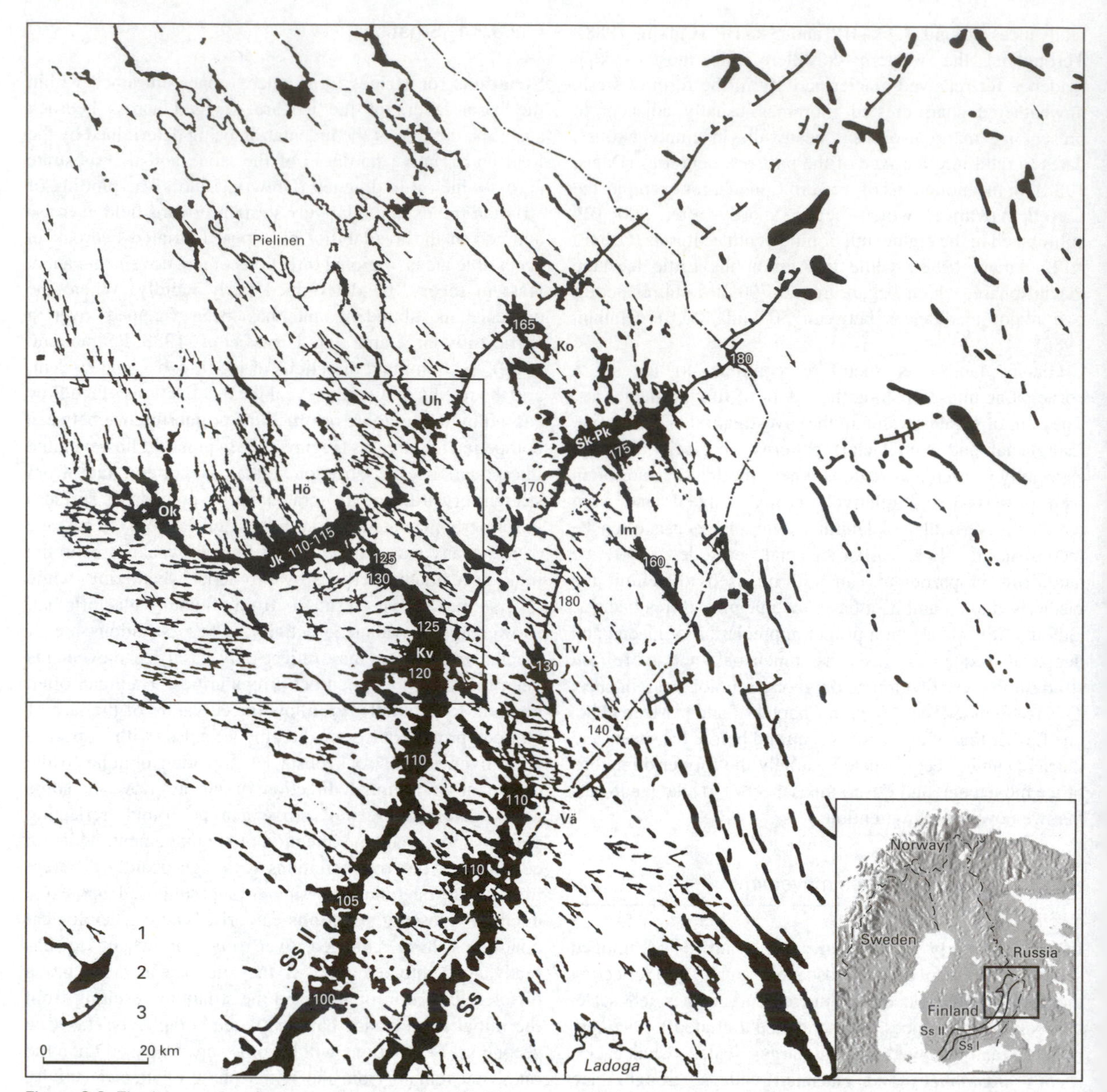

Figure 2.8 The ice-marginal features of northern Karelia. The arrows (1) show directions of glacial striae; shaded areas (2) are glaciofluvial formations of the Salpausselkä Moraines (section 2.3.2.1); continuous lines (3) mark the positions of end moraine ridges. Note the two dominant directions of ice flow (N→S; E→W) and examples of crossing striations in the boxed area on the left (after Eronen & Vesajoki, 1988).

2.3.3.2 Friction cracks

A range of fractures or '**friction cracks**' results from stones in basal ice being forced against underlying bedrock (Boulton, 1974). Perhaps the best known and most widely reported are '**crescentic gouges**' and '**crescentic fractures**' (Figure 2.9). Crescentic gouges are believed to form concave down-ice and crescentic fractures concave up-ice, and the direction of concavity has been used to interpret former direction of ice movement (e.g. Stieglitz *et al*., 1978). Consistent patterns do not always emerge, however, and the use of friction cracks in isolation as ice-directional indicators

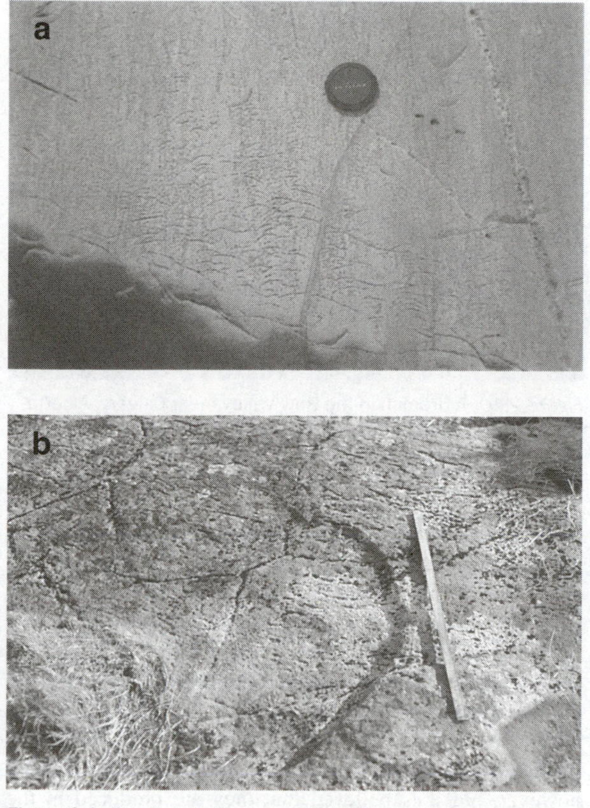

Figure 2.9 Small-scale features of glacial erosion exposed on bedrock surfaces in North Wales: (a) friction fractures (photo: Murray Gray); (b) crescentic scars (photo: John Lowe). These features form either concave up-ice (e.g. fractures) or down-ice, and may be useful ice-directional indicators in areas where other erosional features, such as striae, are not preserved.

now seems a doubtful procedure (Gray & Lowe, 1982). Nevertheless, where employed in conjunction with other evidence (such as striations) they can often lend useful support in the reconstruction of regional ice-flow patterns (Thorp, 1986), and past glacier dynamics (Sharp *et al.*, 1989).

2.3.3.3 Stoss-and-lee forms, *roches moutonnées* and crag-and-tail features

An irregular bedrock surface presents numerous obstacles to the passage of ice. This leads to compression of the ice and increased erosion of the upstream (stoss) sides of the obstruction, whereas plucked and shattered craggy surfaces tend to characterise the downstream (lee) sides. A series of ridges running at right angles to the direction of basal ice flow will, after prolonged glaciation, be smoothed only on the up-ice side to form **stoss-and-lee** landforms, and if a consistent pattern is evident in the landscape, this can be used to infer ice-flow directions. Where a bedrock ridge runs parallel to the direction of ice flow, the bedrock will become smoothed on the up-ice side and also along the flanks to produce the landforms known as **roches moutonnées**. These can vary from knobs with dimensions of a few metres to major prominences 50 to 200 metres high and two to three times that size in length and width (Sharp, 1988). Frequently the whole floor of a valley, or a wide expanse of lowland formerly subjected to glacial erosion, will show a consistent pattern of smoothed ice-scoured surfaces (commonly with striae) on one side, and irregular craggy faces on the other (Gordon, 1981; Figure 2.3B). Where localised outcrops of particularly resistant bedrock occur, such as volcanic plugs, these serve to protect the bedrock on the lee side and glacial erosion results in the development of a **crag-and-tail** feature, the orientation of the often drift-veneered 'tail' indicating the former direction of ice movement (Gordon, 1993). Such ice-moulded landforms, in conjunction with striae and friction cracks, provide good evidence of former directions of ice flow, although the fact that some large-scale ice-moulded features can survive more than one phase of glaciation can lead to problems of interpretation where the direction of ice movement has changed between successive glacial episodes.

2.3.3.4 Glaciated valleys

During the build-up of an ice sheet, glacier flow is concentrated initially in pre-existing valleys and these are progressively modified to form **glacial troughs**. As ice builds up, outlets become impeded and ice from the most constricted valleys overtops the lowest cols on the interfluves to escape into neighbouring valley systems. The original drainage network is therefore modified by **watershed breaching**. In some areas, ice will exploit lines of structural weakness, in which case the trend of glacial troughs may be significantly different from the original drainage pattern. Eventually, individual valley glaciers may coalesce and the mountain summits become submerged beneath an ice sheet, although the major troughs and breaches will still constitute the principal avenues of flow within the ice sheet. By plotting the trends of glacial troughs and interconnecting breaches, therefore, the principal routes taken not only by valley glaciers but also by ice streams beneath the former ice sheets can be established, and the centres of ice accumulation and dispersal can be identified. The major glacial troughs in west–central Scotland, for example, are arranged in a radial fashion around Rannoch Moor, a pattern first noted by Linton (1957) who also drew attention to similar 'radiative dispersal systems' around the uplands of the English Lake District, southern Norway and South Island, New Zealand. In all of these regions, it is assumed that successive ice sheets developed from the same centres of ice accumulation, so that the present morphology of the major glacial troughs is a product of several phases of ice sheet and valley glacier erosion. Problems can arise, however, in such gross interpretations of former glacier movements as a result, for example, of migrating ice sheds and ice centres during the growth of ice sheets over successive glacial phases (Hambrey, 1994), from glacio-isostatic effects (section 2.5.3) which can lead to changes in ice-flow direction, and from the fact that in some areas ice flowed in the opposite direction to the gradients of the pre-glacial river valleys. Examples of such 'intrusive glacial troughs' can be found in Glen Devon in southeast Scotland where ice from the Scottish Highlands pushed southwards into the Ochil Hills (Linton, 1963), and in the Finger Lakes region of New York State where ice flowed south to intrude into the northern escarpment of the Allegheny Plateau (Rice, 1977).

2.3.3.5 Streamlined glacial deposits

Subglacial debris deposited beneath moving ice is frequently streamlined in the direction of ice movement. Streamlined glacial deposits are often found in glaciated landscapes and range in size from small-scale **fluted moraine** with heights of a metre or less (Gordon *et al.*, 1992) to larger flutes and **drumlins**[4] sometimes comprising drift accumulations some tens of metres in thickness (Menzies & Rose, 1987). These depositional landforms invariably record the basal movements of the last ice mass to have affected an area, for any subsequent glacier with a different direction of movement would have erased or at least substantially modified such features. Almost all of the detailed work that has been published on drumlins and fluted moraines relates to forms that have developed in drift of the last glaciation.

Drumlins are undoubtedly among the most intensively studied of all glacial landforms (Menzies, 1978; Menzies & Rose, 1987) and have been particularly widely used as ice-directional indicators (e.g. McCabe, 1991). They frequently occur in 'fields' or 'swarms' in lowland areas where there was little obstruction to the passage of ice, or in piedmont zones where flow was radiative or dispersive. They are also occasionally found on the floors of glacial troughs. Many are ellipsoidal in form, some are almost circular, while at the other extreme, linear drumlins several kilometres in length have been observed. Most possess a prominent stoss end with a trailing distal slope (Figure 2.10). It is generally agreed that the direction of the drumlin long axis reflects local direction of ice movement with the stoss end usually pointing up-glacier. Detailed field mapping of drumlins reveals the dominant local ice-flow paths that prevailed over a particular region and, when the evidence is viewed collectively, macro-scale patterns of ice movement frequently become apparent (Figure 2.11). More than any other line of evidence, perhaps, maps based on drumlins and related landforms have provided the most vivid images of the major arterial flow systems of the last great ice sheets. Glacial geomorphologists remain uncertain as to the precise mode of formation of drumlins (Menzies, 1987, 1989), however, but it is believed that they are produced in the subglacial environment as a response to fluctuating stress and strain conditions within a deforming sediment layer trapped between a rigid bed and overlying mobile ice (Boulton, 1987), although it has been suggested that, in certain instances, drumlin development may be related to catastrophic subglacial flooding (Shaw & Sharpe, 1987; Shoemaker, 1995).

Figure 2.10 A drumlin in the Bow Valley, near Calgary, Alberta, Canada. Ice flow was from left to right across the photograph (photo: Mike Walker).

In addition to the regional ice flow trends displayed by their long axes, the overall shape and distribution of drumlins can provide information on former glacier dynamics such as indications of basal ice pressure, rate and type of ice flow, and basal shear stress variations (Mooers, 1989). Numerous shape indices have been developed using axial and outline ratios (e.g. Reed *et al.*, 1962; Jauhiainen, 1975) which enable comparisons that are independent of scale to be made between drumlins in different areas. Although most measurements of drumlin shapes have been made from maps and aerial photographs, detailed field mapping is essential, for subtleties in form are not always expressed on topographic maps, or may not always be clearly identifiable on aerial photographs (Rose & Letzer, 1975).

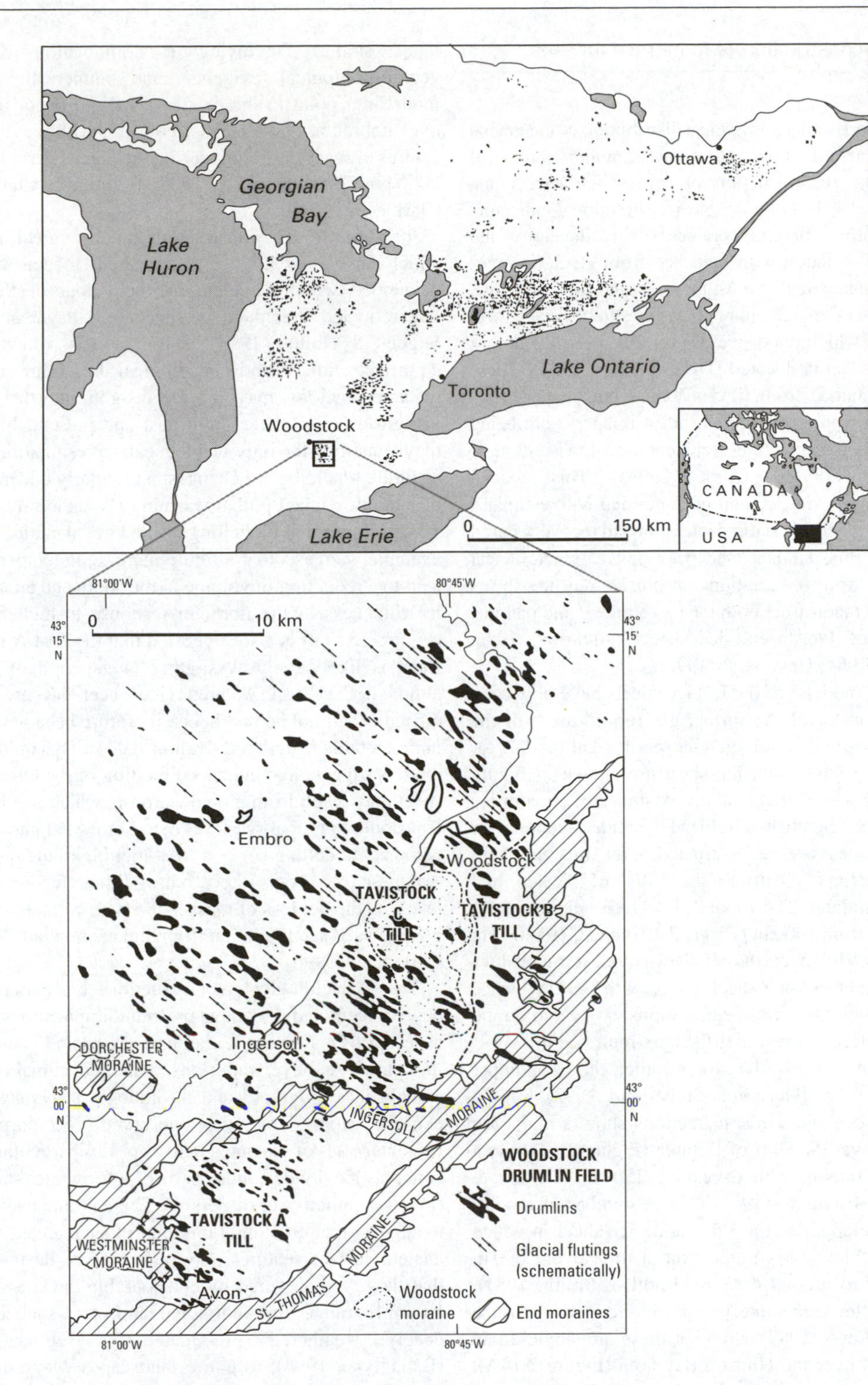

Figure 2.11 The Woodstock drumlin field in southern Ontario, Canada, showing a predominant NW→SE direction of ice movement (after Piotrowski, 1987).

2.3.4 Reconstruction of former ice masses

The general trends and geographical distribution of the glacial landforms discussed above constitute a major source of evidence in the reconstruction of former ice sheets and glaciers (Shaw, 1994). For example, ice-directional indicators such as drumlins, fluted moraines, drift lineations and striations, in association with evidence from glacial erratics and till provenance studies (Andrews *et al*., 1985; Dredge, 1988; see also section 3.2) enable flow lines to be established, on the basis of which major ice-dispersal centres, ice domes and ice divides can be located (Dyke & Prest, 1987). These data can be obtained from field mapping but, increasingly, remote sensing techniques (e.g. Landsat imagery) are being employed, particularly at the continental scale (Boulton & Clark, 1990a, 1990b; Clark, 1993). This glacial geomorphological evidence can be combined with estimates of the extent of glacio-isostatic depression and recovery based on raised shoreline studies (see below), and glaciological theory derived from observations on present-day ice sheets and glaciers, to reconstruct both the morphology and patterns of behaviour of Pleistocene ice sheets (Andrews, 1982; Boulton *et al*., 1984; Hindmarsh, 1993).

Steady-state models[5] of the last ice sheets have now been developed for many of the former glacierised areas of the Northern Hemisphere. Four such morphological models for the British Late Devensian ice sheet are shown in Figure 2.12. All of the models indicate the existence of a major ice dome over the Scottish Highlands, while separate or continuous domes are reconstructed over the Southern Uplands, the Lake District, the hills of Wales and north–central Ireland. The modelled surface heights of the ice sheets vary from 1000 m (Figure 2.12B and C) to 2000 m (Figure 2.12D). However, the relatively modest dimensions of the Late Devensian ice sheet, along with field evidence for former ice heights (trimline elevations, striae and erratic distributions, etc.), suggest that the maximum ice thickness may have been towards the lower rather than the upper reaches of this range (Gordon & Sutherland, 1993). For the Fennoscandian ice sheet, reconstructions show a major ice dome centred over the Gulf of Bothnia (Figure 2.13) with a maximum ice thickness in excess of 2500 m (Denton & Hughes, 1981; Boulton *et al*., 1985). A number of models have been developed for the Laurentide ice sheet in which maximum ice thicknesses range from a little over 2000 m (Peltier, 1981) to around 4000 m (Budd & Smith, 1981), although the latter seems likely to be an overestimate of ice thickness (Andrews, 1987). Some of these are single-dome models centred over the Hudson Bay area (Figure 2.14A), while others have separate ice domes over Keewatin, Labrador/Ungava and Baffin Island (Figure 2.14B). The latter studies, involving a combination of glacial geomorphological evidence and numerical simulation modelling, point to the existence of two major ice centres over Labrador/Quebec and Keewatin, with ice divides and centres of mass migrating during the glacial period by 1000–2000 km (Boulton *et al*., 1985; Boulton & Clark, 1990b; Clark *et al*., 1993).

In addition to reconstructing the last great ice sheets which have now all but disappeared, models have been developed which simulate the behaviour of the present Antarctic and Greenland ice sheets (e.g. Payne *et al*., 1989; Sugden & Hulton, 1994). These models, which integrate empirical climatic and glaciological data from present-day glaciated regions, may provide insights into the long-term evolution of ice sheets both past and present. In addition, they constitute the only viable means of estimating the scale of future glaciological changes, particularly under scenarios of human-induced global warming (Houghton *et al*., 1990, 1992). Numerical modelling of the Greenland ice sheet, for example, suggests that a future increase in temperature will lead to a reduction in volume in the south and an increase in ice thickness in the north in response to higher snowfall (Figure 2.15). It is also suggested that the East Antarctic ice sheet is likely to grow slightly in a period of warming, although the West Antarctic ice sheet has proved more difficult to simulate and hence its future behaviour is more unpredictable (Sugden & Hulton, 1994). Dynamic ice sheet models suggest that enhanced melting of the Greenland ice sheet may result from global warming which would, in turn, contribute to global sea-level rise. Somewhat paradoxically, perhaps, modelling suggests that melting of the much larger Antarctic ice sheet is likely to have little effect on global sea levels as the net loss of mass is likely to be more than offset by increased accumulation from higher snowfall (Warrick & Oerlemans, 1990).

These glaciological reconstructions are a major recent development and they have profound implications for many aspects of Quaternary research. They provide new insights into global sea-level variations, plant and animal migrations and glacial stratigraphy and chronology. They may also help explain many of the observed geomorphological characteristics of glaciated landscapes. A particular example concerns ice divides, those parts of a former ice mass where flow was radiative or dispersive. There is abundant evidence to suggest that erosion is minimal in such areas, and hence ancient surface features may be preserved despite the fact that they were covered by considerable thicknesses of ice. Examples range from individual landforms, such as tors and deeply weathered pre-Quaternary bedrock features (Ballantyne, 1994), to entire landscapes where cold-based ice was frozen to the underlying substrate and hence erosion of the pre-glacial surface was neglible (Kleman, 1994).

Figure 2.12 Morphological models of the British Late Devensian ice sheet at its maximum extent. A. Boulton *et al.* (1977); B. Boulton *et al.* (1985); C. Boulton *et al.* (1991); D. Lambeck (1993b). Surface contours are in metres. Note that the field evidence indicates a maximum ice thickness of less than 1300 m (after Gray, 1995).

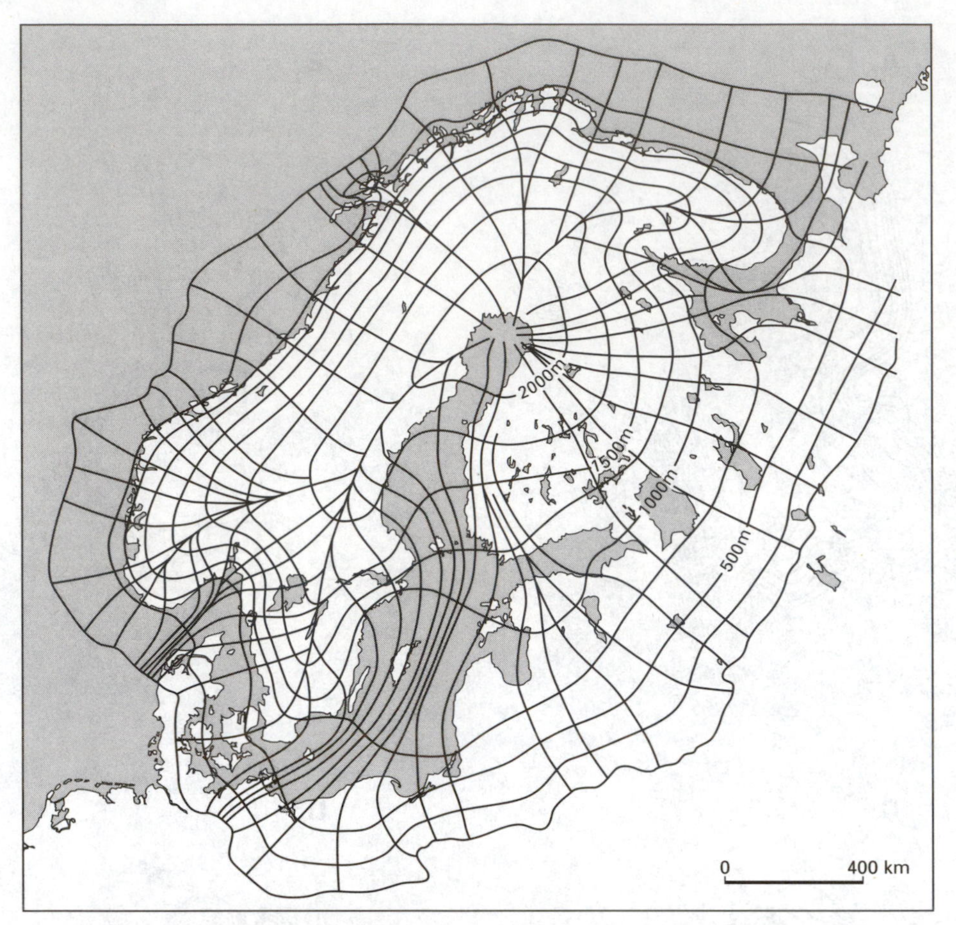

Figure 2.13 A computer model of the Fennoscandian ice sheet at its maximum extent (after Boulton *et al.*, 1985).

However, uneroded pre-glacial features appear to be preserved only in those areas where ice divides remained fixed throughout a glacial cycle. Where ice divide migration occurred, as the ice sheet models predict for parts of the Laurentide ice sheet, for example, no areas seem to have been permanently geomorphologically inactive. Hence, the presence of marked lineations in the area of the final ice divide in Keewatin provides clear evidence of the mobility of the former ice divide in this region (Boulton & Clark, 1990a, 1990b).

Producing models of Quaternary ice sheets, however, is a difficult and complex field, and despite the considerable advances that have been made in recent years, and the increasingly sophisticated nature of the reconstructions, they remain models in the strict sense of the word. They are simplifications of reality and must be regarded as such. The geomorphological evidence, which is fundamental to the reconstructions, is still not always sufficiently well founded to determine the limits of the last ice sheets. Around the coasts of Britain, for example, despite the considerable amount of new data that has emerged from offshore seismographic surveys, the maximal extent of the Late Devensian ice sheet remains to be firmly established off much of western Scotland, in the southern part of the Irish Sea basin, and over large areas of the North Sea basin. The same applies to the Late Weichselian ice sheet off the west coast of Norway. Moreover, throughout the life of an ice sheet, ice thickness, ice flow directions and ice marginal positions will vary, and this clearly poses problems both for dynamic and steady-state modelling.

Less speculative reconstructions are possible for smaller ice caps and for glaciers that were restricted to cirque and valley situations. These include many of the the Loch Lomond Readvance glaciers of Britain (Sissons, 1976; Gray & Coxon, 1991), the Younger Dryas glaciers of Scandinavia (Mangerud *et al.*, 1984) and the Neoglacial glaciers that

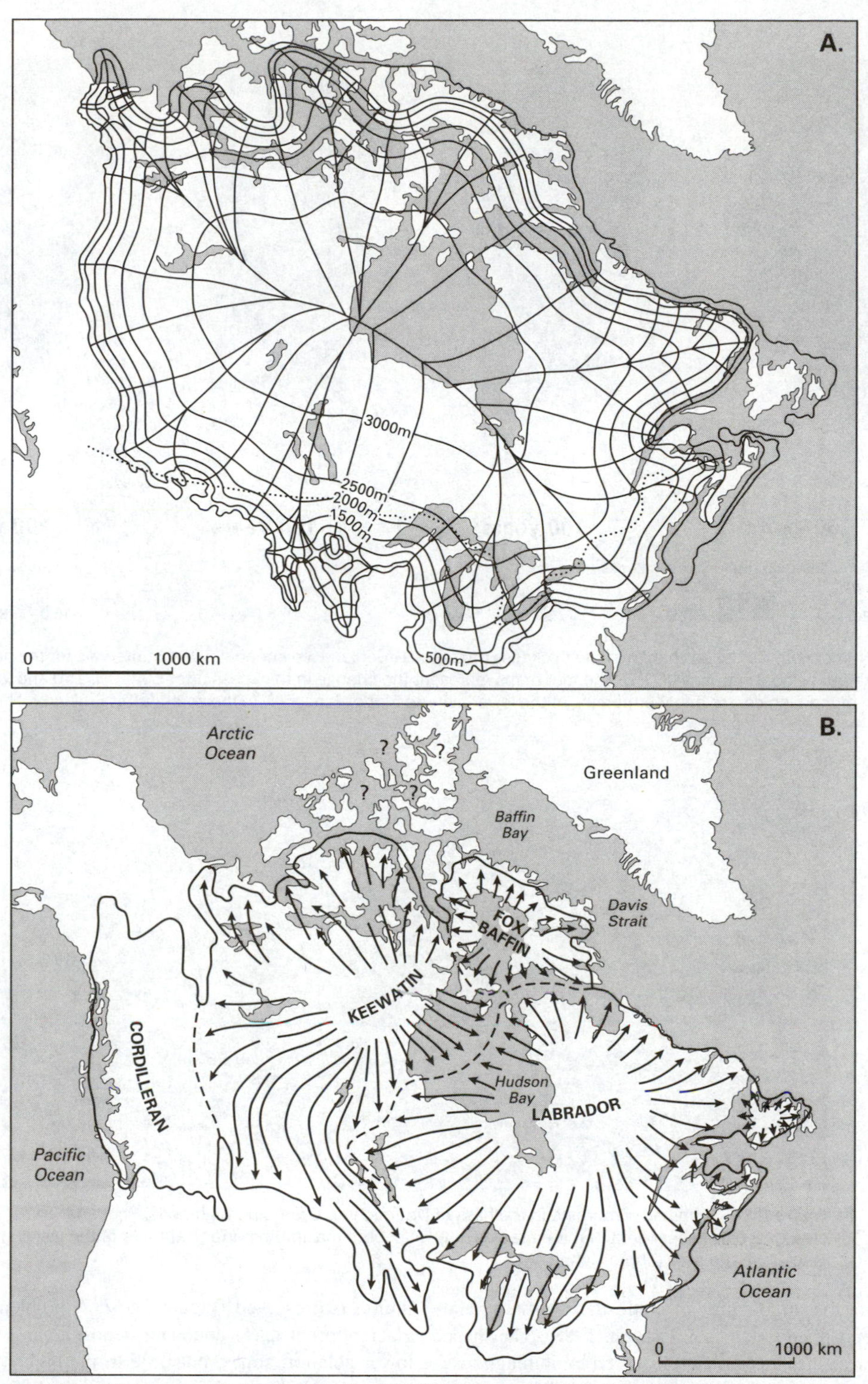

Figure 2.14 Two models of the Laurentide ice sheet at the Late Wisconsinan maximum. A. A computer model of the geometry of the ice sheet (after Boulton *et al.*, 1985). B. Major ice centres and principal flow-lines (from Andrews, 1987).

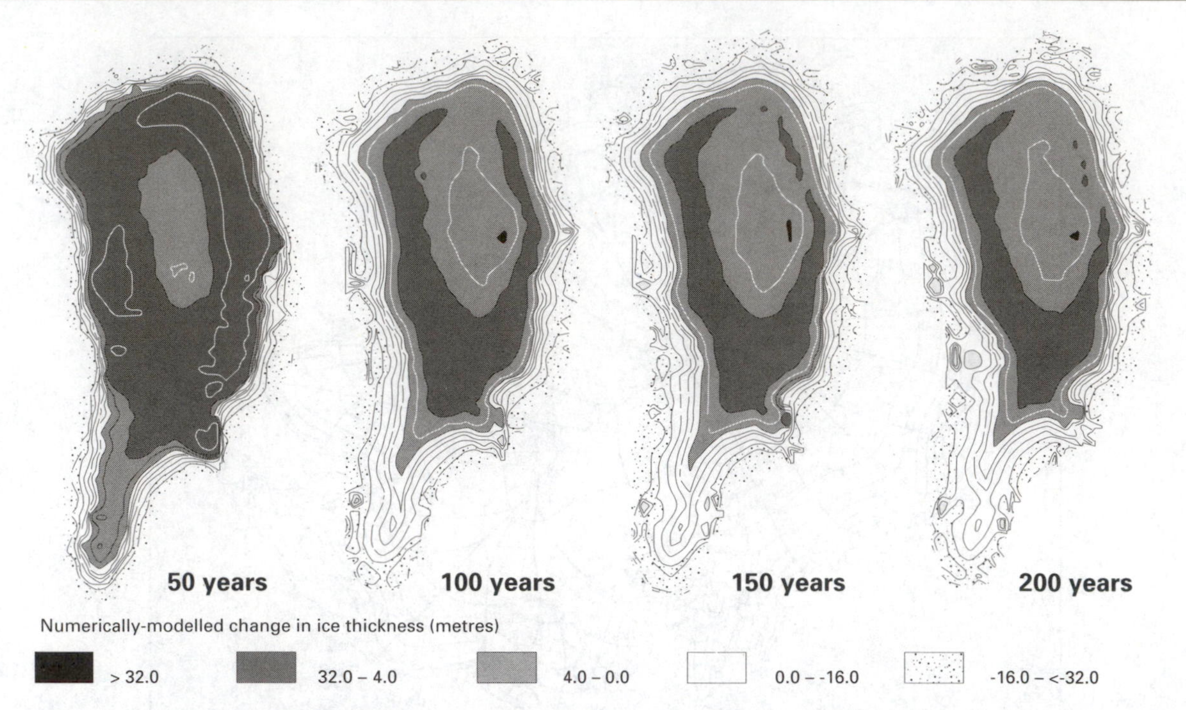

Figure 2.15 Numerically modelled change in the thickness (m) of the Greenland ice sheet at 50-year intervals for the next 200 years assuming a stepped global warming of 6 °C. The four maps represent the change in thickness after 50, 100, 150 and 200 years. Note the thickening of the high centre and the thinning of the lower altitude fringe, especially in the south (after Sugden & Hulton, 1994).

Figure 2.16 Hummocky moraine in the Pass of Drumochter, Grampian Highlands, Scotland. The clearly defined upper limit of the moraines may reflect the maximum surface altitude of the last glaciers in the area (photo: John Lowe).

developed in the mountains of both the Northern and Southern Hemispheres (Davis & Osborn, 1988). The shape of the former glacier margin can be traced by joining those points or areas where clear ice-marginal evidence (e.g. terminal or lateral moraines, trimlines or valley side drift-limits) is preserved (Figure 2.16). A problem here is that glaciers often leave abundant depositional evidence in the lower ablation zones, but little in the higher accumulation zones. If trimline evidence is unavailable for these upper areas, then extrapolation between scattered ice-marginal

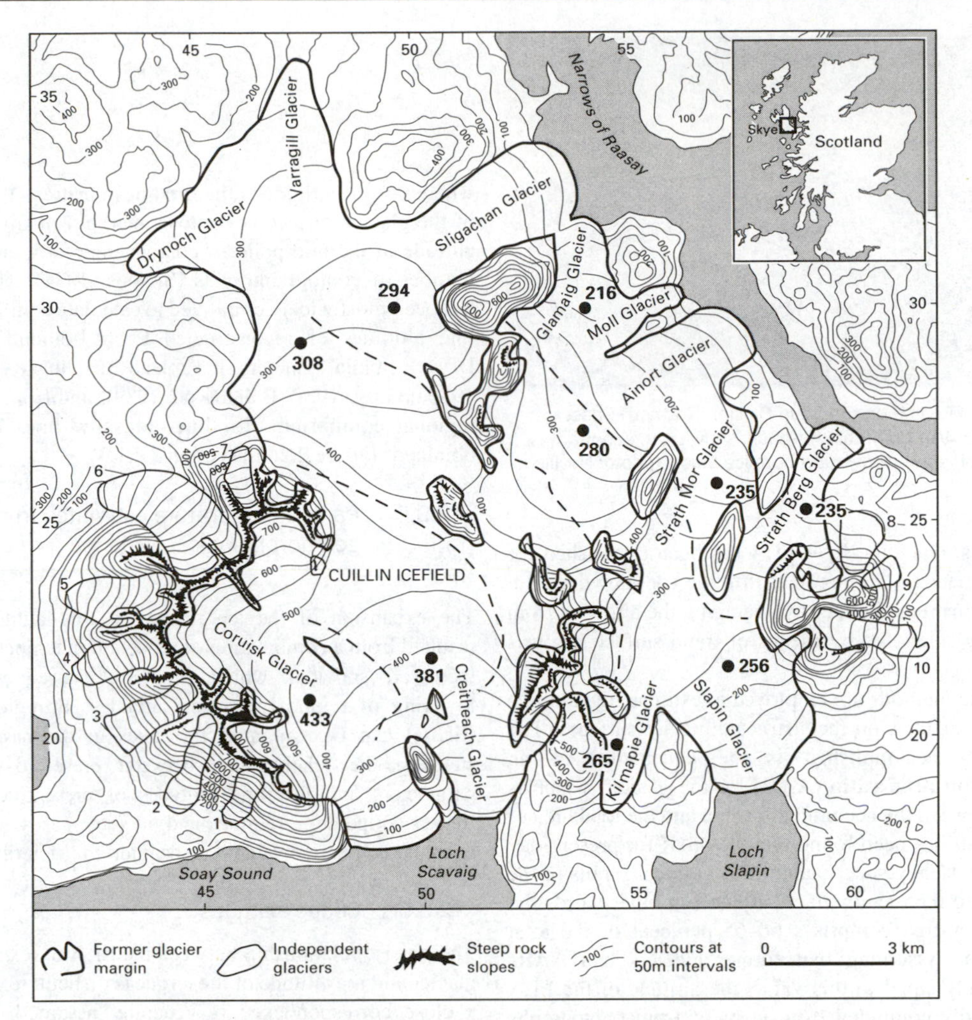

Figure 2.17 Reconstruction of the Cuillin Icefield and other former glaciers that developed during the Loch Lomond (Younger Dryas) Stadial in the area of the Cuillin Hills, Isle of Skye, Scotland (after Ballantyne, 1989). The principal ice-sheds (dotted lines) and ELAs (in metres OD) for the icefield are shown on the map. ELAs for the smaller glaciers (1–10) are as follows:

1. 423 m
2. 437 m
3. 433 m
4. 461 m
5. 459 m
6. 392 m
7. 643 m
8. 274 m
9. 352 m
10. 286 m

indicators becomes necessary in order to estimate the outline of the former glacier. When the glacier outline has been reconstructed, ice-surface contours can be inferred by analogy with typical contour patterns on present-day glaciers (Figure 2.17). Ice-surface contours are commonly normal to valley walls near the median altitude of a valley glacier, and they become progressively more convex towards the glacier terminus and more concave towards the upper reaches (glacier source). In practice, contour drawing is constrained by features indicating direction of ice movement, such as striae and fluted moraines, for contours are drawn normal to ice movement direction.

Once the ice-surface contours have been drawn, the altitude of the **equilibrium line** can be estimated. The equilibrium line is the line on a glacier separating the **accumulation area** (the area where the glacier gains in mass) from the **ablation area** where a net loss of mass occurs. A term that is often used synonymously with equilibrium line is **firn line**, which is the altitude on a glacier surface to which consolidated granular snow (**firn**) recedes on surviving a full summer season's melt (Figure 2.18). In fact, the two are not quite the same,[6] but for the purposes of the present discussion, the **equilibrium line altitude (ELA)** and **firn line altitude (FLA)** can be taken to be

Figure 2.18 Firn line on the Scud Glacier, Northern Coast Mountains, British Columbia, Canada, showing the contrast between fresh snow above and bare ice below (photo: Neville Alley).

synonymous. Once the ELA/FLA has been established for individual glaciers, the regional firn line (or snowline) can then be reconstructed either by averaging the altitudes from individual glaciers, or by means of trend-surface analysis (see below).

Two basic methods are employed to estimate the altitude of the ELA or FLA on reconstructed glacier surfaces. The most easily accomplished is the calculation of an **accumulation area ratio (AAR)** for the glacier, which is the ratio between the accumulation area and the total area of the glacier. It has been found that the AAR for present-day glaciers in a steady state (stable ELA) typically falls in the range 0.60 to 0.65 (Sutherland, 1984a); in other words the accumulation area comprises 60–65 per cent of the total glacier area. Assuming that former glaciers had AARs approximately equal to this value, the altitude of the ELA can be rapidly computed from maps or from photographs where altitudinal distribution of the former glacier surface can be measured. This approach has been used to establish the ELA of Late Pleistocene and Holocene glaciers in many parts of the world, including southern Norway (Dahl & Nesje, 1992), Iceland (Caseldine & Stötter, 1993), the western United States (Leonard, 1989), the Himalayas (Williams, 1983) and New Zealand (Porter, 1975),

The second method is more complicated and involves estimates of the variation in altitudinal distribution of the former ice masses. It assumes that **ablation** and **accumulation gradients** are linearly related to ice-surface altitude, and also that the reconstructed glaciers were at their maximum extent and in equilibrium, so that the firn line at the end of the ablation season marks the line where the total accumulation and ablation were exactly balanced. Where these conditions are satisfied, the equilibrium firn line altitude can then be calculated from the following equation, which reflects the area-weighted mean altitude of the former glacier surface:

$$x = \frac{\sum_{i=0}^{n} A_i h_i}{\sum_{i=0}^{n} A_i}$$

where x = the altitude of the firn line in metres; A_i = the area of the glacier surface at contour interval i in km^2; h_i = the altitude of the mid-point of contour interval i; and n = the number of contour intervals (Sissons, 1974). The method has been most widely employed to calculate equilibrium firn line altitudes of reconstructed Loch Lomond (Younger Dryas) Stadial glaciers in upland Britain (e.g. Sissons, 1980a; Gray, 1982; Ballantyne, 1989), and from these data regional equilibrium firn lines or snow lines have been obtained (Figure 2.26 in section 2.4.1.3).

2.3.5 Palaeotemperature estimates from glacial geomorphological evidence

The expansion of the ice masses in mid-latitude regions resulted from a combination of reduction in temperatures and increased snowfall which, in turn, caused widespread lowering of regional firn-line altitudes. If regional FLAs (ELAs) can be estimated for times in the past, then by studying the relationships between present-day climatic parameters and firn line altitudes of present ice masses, former temperature régimes and, in some cases, seasonal or annual precipitation values can be inferred (Meierding, 1982).

2.3.5.1 Cirque altitudes

The precise relationship between the ELA of a small cirque glacier and the altitude of the cirque is difficult to define, but a close correspondence is generally assumed. Since the regional snowline approximates the ELA of cirque glaciers (the exact relationship can be measured for any particular area) the latter can be estimated for times in the past from measurements of the altitude of cirque floors. The calculation of the **toe-to-headwall ratio (THAR)** is an alternative index which is based on the premisse that ELAs are located a certain fraction (normally 0.4) of the vertical distance between the headwall and the palaeoglacier 'toe' or inferred terminus (e.g. Clapperton, 1986; Rodbell, 1992a). Where the average annual temperature is known for the altitude of the present snowline, the temperature reduction required to lower the snowline to the altitude of cirque floors presently devoid of glaciers can be calculated from regional average temperature lapse rates. A further index used in the same way is the **glaciation limit**, the altitude above which glaciers exist today in mountain ranges. Estimates can be derived for former glaciation limits by comparing mountain summits with cirques with mountain summits showing no geomorphological evidence of cirque glaciation (Sugden &

John, 1976). During the Late Wisconsinan, regional ELAs of mountain glaciers in western North America were 850–900 m lower than at present in the Cascade Range, and about 1000 m lower in the Rocky Mountains. For the Cascades, this is thought to reflect a mean annual temperature lowering of about 4°C to 5°C, while in the Rocky Mountains mean annual temperatures may have been as much as 10–15°C colder than at present (Porter *et al.*, 1983). In the Andes of South America, it has been estimated that the ELA reduction at the Last Glacial Maximum was in excess of 1000 m, implying a temperature reduction of at least 5–6°C (Rodbell, 1992a).

There are, however, a number of problems in the use of cirque floor altitudes as a basis for palaeoclimatic reconstruction. First, the snowline associated with cirque and valley glaciers is normally somewhat lower in altitude than the average altitude on exposed summits and slopes, since wind-drifting and low insolation protect accumulated snow and ice within cirque basins. Second, the method can only provide a means of estimating the snowline elevation, and hence palaeotemperatures, for the time when the glaciers were confined to cirques or individual valleys and it is not suitable for calculating temperatures during periods of more extensive ice cover when the altitudinal relationships between accumulation and ablation areas are much more complex. Third, not all cirques in the landscape were occupied by ice at the same time, and since some variation in the altitude of cirque glaciers can be expected, contemporaneity of cirque glacier development may be very difficult to establish. Finally, it is an implicit assumption in the lapse-rate calculations that the same annual accumulation at the ELA of a modern glacier in a mountain region occurred at the ELA of Late Pleistocene glaciers, which may have been several hundred metres lower. However, empirical data show that precipitation in mountain regions typically decreases with decreasing altitude, and hence the precipitation on the lower altitude Pleistocene glaciers would also have been lower. In the case of the Rocky Mountains, for example, application of the traditional lapse rate (6°C per 1000 m) produces a temperature depression of *c.* 6°C for the Late Pleistocene, whereas estimates corrected for precipitation reduction suggest a Late Wisconsinan temperature depression perhaps twice that value (Porter *et al.*, 1983). Further complications can also arise in mountain regions where marked local variations may be experienced in temperature and precipitation (Zielinski & McCoy, 1987).

2.3.5.2 ELA/FLA method

An alternative approach to deriving temperature estimates from reconstructed glaciers rests on the use of present or recent glacier/climate relationships as an analogue for past glacier/climate conditions. The ELA on a glacier is determined by a combination of the seasonal precipitation and temperature régimes within the glacier catchment. The mean summer ablation season temperature (t) on modern glaciers is closely related to accumulation at the equilibrium line (Figure 2.19A) which, in turn, approximates average accumulation (A) over the whole glacier (Sutherland, 1984a). For ten Norwegian glaciers the form of this relationship has been shown by Ballantyne (1989) to correspond to the regression equation

$$A = 0.915e^{0.339t} \qquad (r^2 = 0.989,\ P < 0.0001)$$

ELA is also linearly related to temperature. For the same data, Sutherland (1984a) calculated an 'ELA temperature gradient' of 0.58°C per 100 m. Using these two relationships, a rise or fall in ELA can be directly related to a change in temperature, or precipitation, or some combination of the two. Thus if either former precipitation values or former temperatures are known or can be estimated within reasonable limits, the appropriate value for the other variable can be calculated. Similar relationships are apparent in graphs that have been produced for glacier groups elsewhere (e.g. Kotlyakov & Krenke, 1979; Ohmura *et al.*, 1992) which demonstrates that glaciers worldwide are restricted to a relatively narrow range of combined mean accumulation and mean summer temperature conditions (Sutherland, 1984a). For reconstructed Quaternary glaciers, therefore, reference to such graphs allows summer temperature values to be estimated where the altitude of the former equilibrium line and associated accumulation data are known.

Two potential difficulties are encountered in the application of this method, however. First, precipitation/temperature graphs have to be employed that are applicable to the area under investigation. In the case of the British Isles, for example, where no glaciers exist at the present day, it has been assumed that the curves calculated for Norwegian glaciers (e.g. Liestøl, 1967) provide the closest analogues for the last glaciers in northern Britain (Sissons, 1974, 1980b). Secondly, in order to derive a temperature value from the graphs, precipitation levels at the firn line have to be estimated. For reconstructed Loch Lomond (Younger Dryas) Stadial glaciers in the central Grampian Highlands, for example, Sissons (1974) assumed that precipitation at the equilibrium firn line was approximately 80 per cent of present-day levels. On that basis, summer temperatures at the firn line (780–790 m) were calculated to have been about 1.5°C. This is equivalent to a sea-level temperature of 7.6°C for July (assuming a lapse rate of 0.6°C per 100 m), about 7°C lower than at the present day. In the English Lake District, similar reasoning produced a July mean temperature at sea level of *c.* 8.0°C

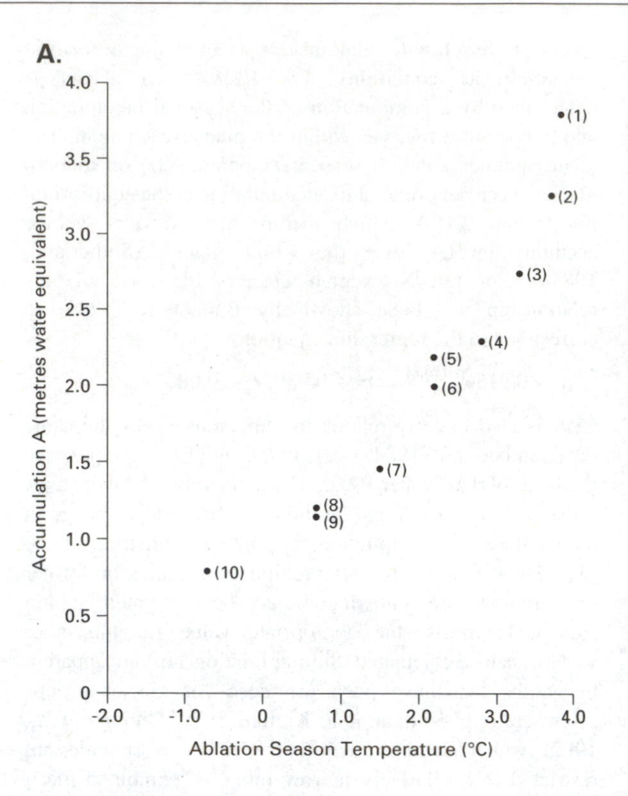

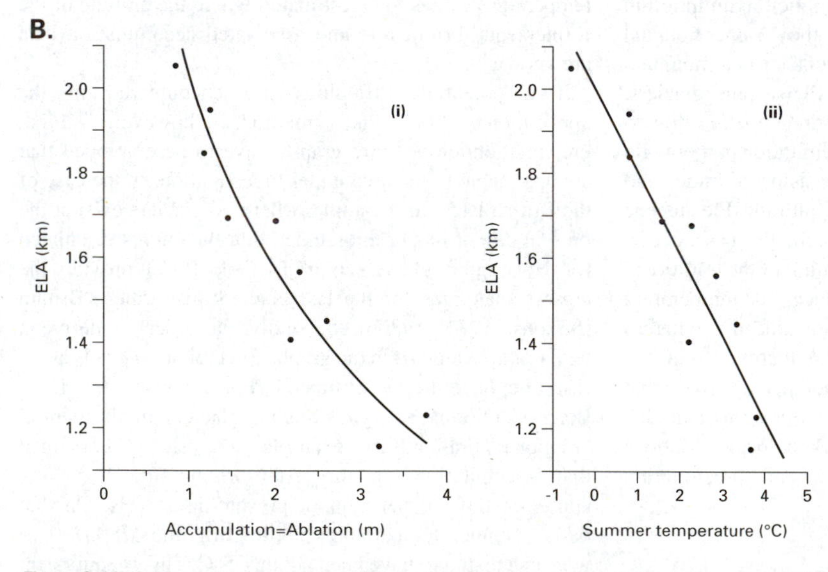

Figure 2.19 A. Mean summer temperatures plotted against accumulation (in m water equivalent) at the equilibrium line for ten Norwegian glaciers: 1. Alfotbreen; 2. Engabreen; 3. Folgefonni; 4. Nigardsbreen; 5. Tunsbergdalsbreen; 6. Hardangerjokulen; 7. Storbreen; 8. Austre Memurubreen; 9. Hellstugubreen; 10. Grasubreen. B. Equilibrium-line altitude plotted against (i) accumulation (in m water equivalent) at the ELA and (ii) mean summer temperature for the same Norwegian glaciers as in A (after Sutherland, 1984a).

(Sissons, 1980a), while a value of *c.* 6.0°C was obtained for the more northerly location of the Isle of Skye (Ballantyne, 1989: Figure 2.17). These palaeotemperature estimates for the glacial maximum of the Loch Lomond Stadial are in good agreement with those based on independent biological evidence (e.g. Coope, 1977a).

A refinement of this method, in which both temperature and precipitation values could be estimated for former glaciers, was described by Sissons & Sutherland (1976). Equilibrium firn-line altitudes were calculated for 27 Loch Lomond Readvance glaciers in the southeast Grampian Highlands of Scotland, and for each glacier the average mass balance was derived from an equation incorporating glacier altitude, regional ablation gradient, direct radiation, the influence of avalanching and blowing snow, and final glacier volume. The influence of direct radiation on the glaciers was calculated using estimates of the effects of the transmissivity of the atmosphere, glacier aspect and surface gradient, and the albedos of ice and snow. From the mass balance equation for 25 of the glaciers, the distribution of annual precipitation was established (Figure 2.20). This suggests that, during the Loch Lomond Stadial, precipitation gradients in eastern Scotland were significantly steeper than at the present day. Average July sea-level temperatures of *c.* 6°C were calculated on the basis of present-day Norwegian glacier–climate relationships. The slightly lower summer temperatures in the eastern Grampians during the Loch Lomond Stadial by comparison to the mountain region

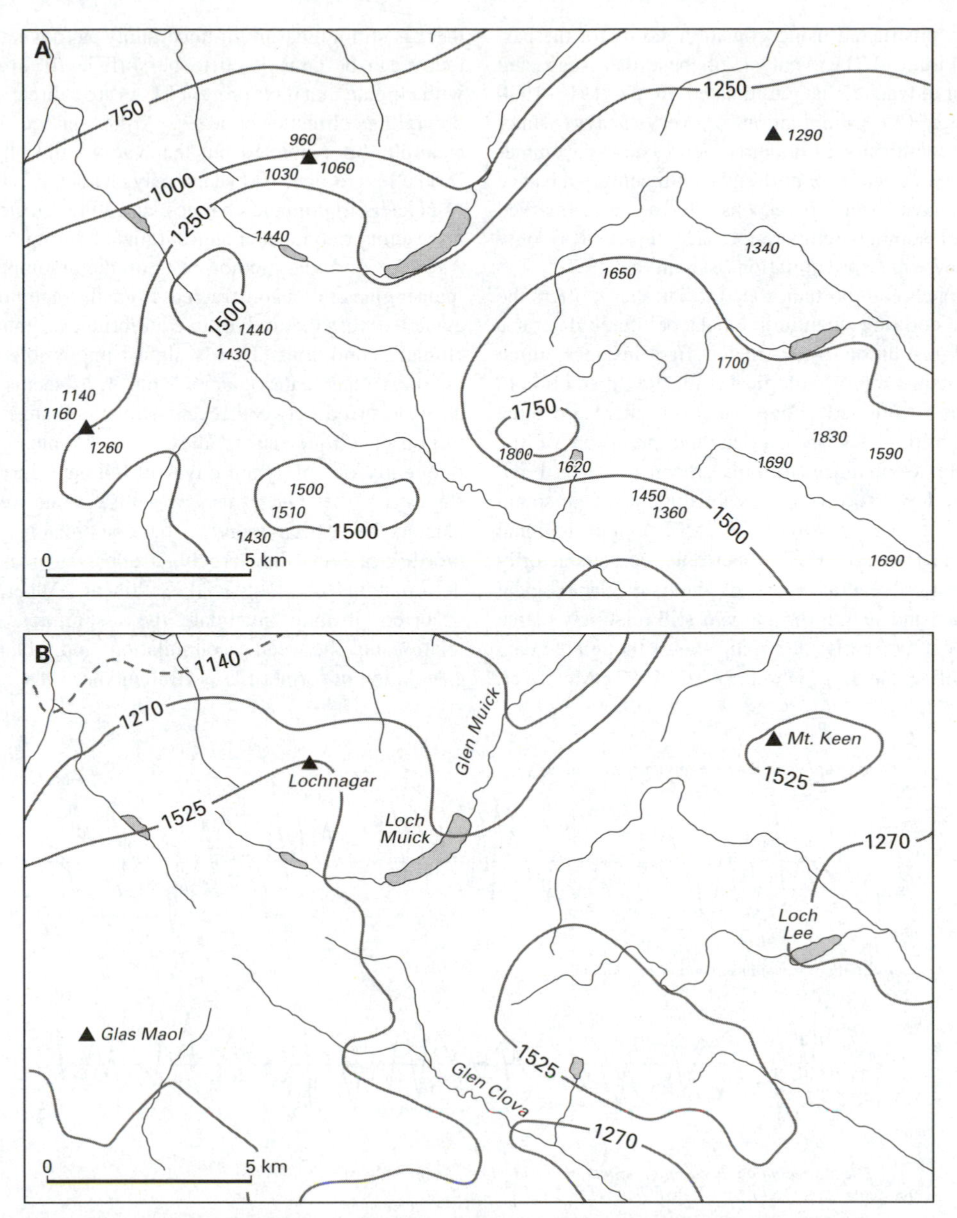

Figure 2.20 A. Inferred precipitation during the Loch Lomond Stadial in part of the Grampian Highlands, Scotland. Calculated precipitation (mm) for 25 glaciers is indicated. B. Present precipitation for the same area. Isohyets for both maps are at 250 mm intervals (after Sissons & Sutherland, 1976).

to the west (see above) was attributed to the heavier summer cloud cover over the former area reflecting the movement of snow-bearing winds from the southeast.

This approach can also be used to investigate the relationship between climatic parameters and glacier response, in particular to examine ways in which the effects of summer temperature and winter precipitation have influenced ELAs during the later part of the Holocene (e.g. Dahl & Nesje, 1992; Caseldine & Stötter, 1993). In the Lyngshalvöya area of northern Norway, for example, geomorphological mapping combined with lichenometric, dendrochronological and historical data enabled a chronology of glacier activity to be established for the Little Ice Age and more recent periods. Potential ELA variations were then reconstructed using climatic records for the past 110 years (Figure 2.21). Analysis of these data suggested that a major advance that culminated in the 1910–1930 period reflected a combination of very heavy winter precipitation with only a moderate depression of summer temperatures. In contrast, a mid-eighteenth century advance appears to have been a response to a more marked depression of summer temperatures accompanied by only modest or low winter precipitation (Ballantyne, 1990).

This approach is important, therefore, in that it offers the potential for deriving quantitative palaeoclimatic data at a high spatial resolution from, in the first instance, often abundant glacial geomorphological evidence. In addition to the problems mentioned above, however, there are three further difficulties that may affect the precision of the climatic estimates obtained from ELA reconstructions. First, the method depends on the establishment of a sound statistical relationship between ELA/FLA and regional climatic data. This relationship can only be satisfactorily determined after several years of detailed glaciological investigations, and such information is still relatively scarce for a number of currently glaciated areas, although the data base is steadily expanding (Ohmura *et al.*, 1992). Moreover, there is still a measure of uncertainty as to whether glaciers today can be considered to be sufficiently in equilibrium with climate that their present ELAs are a direct reflection of prevailing climatic controls. Miller *et al.* (1975), for example, have pointed out that various time lags, ranging from a few to over 100 years, may characterise the response of glaciers to climatic changes, and some glaciers may still be reacting to budget changes initiated during the 'Little Ice Age'. Second, the method rests on the assumption that the 'palaeoglaciers' reconstructed from the geomorphological evidence were themselves in equilibrium with the prevailing climatic conditions. This is almost impossible to establish for individual 'palaeoglaciers' but if, as seems likely, lags have occurred between recent climatic change and glacier response, cirque and valley glaciers must have been frequently out of synchrony with climate during the cold stages of the Quaternary. Finally, some reconstructed glaciers have been found to have anomalously low long-profile gradients due largely, it appears, to the effects of deformation of subglacial sediment. Where this has occurred, it may invalidate the assumption of a linear relationship between accumulation and ablation in the calculation of former ELAs (Ballantyne, 1989).

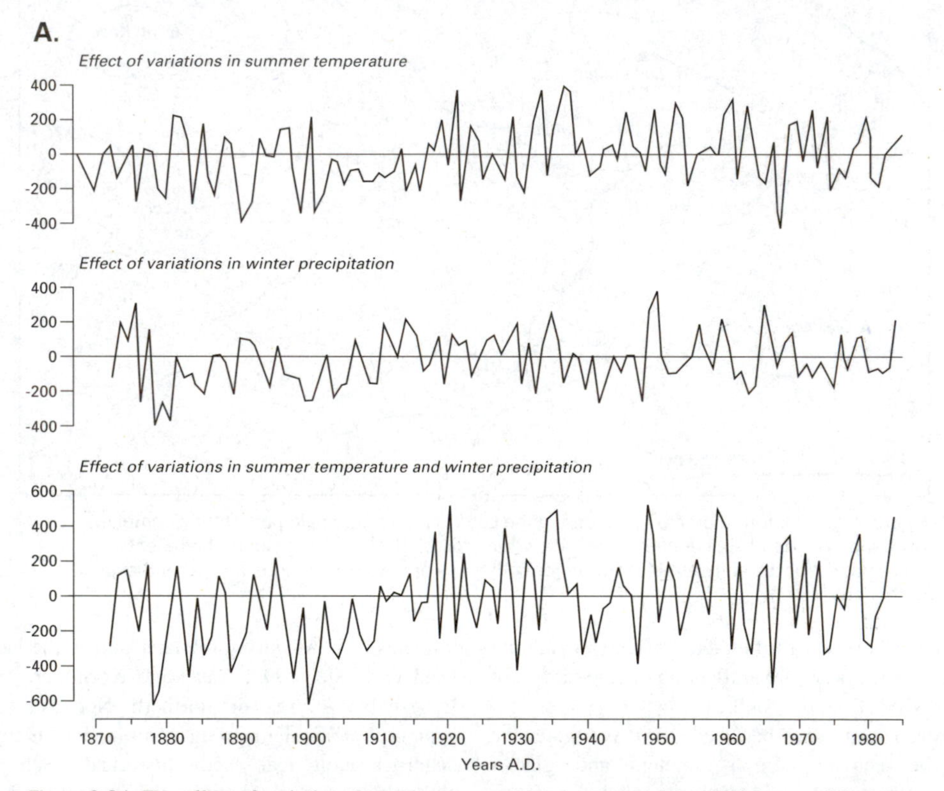

Figure 2.21 The effect of variations in mean summer temperature and mean winter precipitation on potential ELAs of glaciers in Lyngshalvöya, northern Norway, 1870-1985. A: annual values.

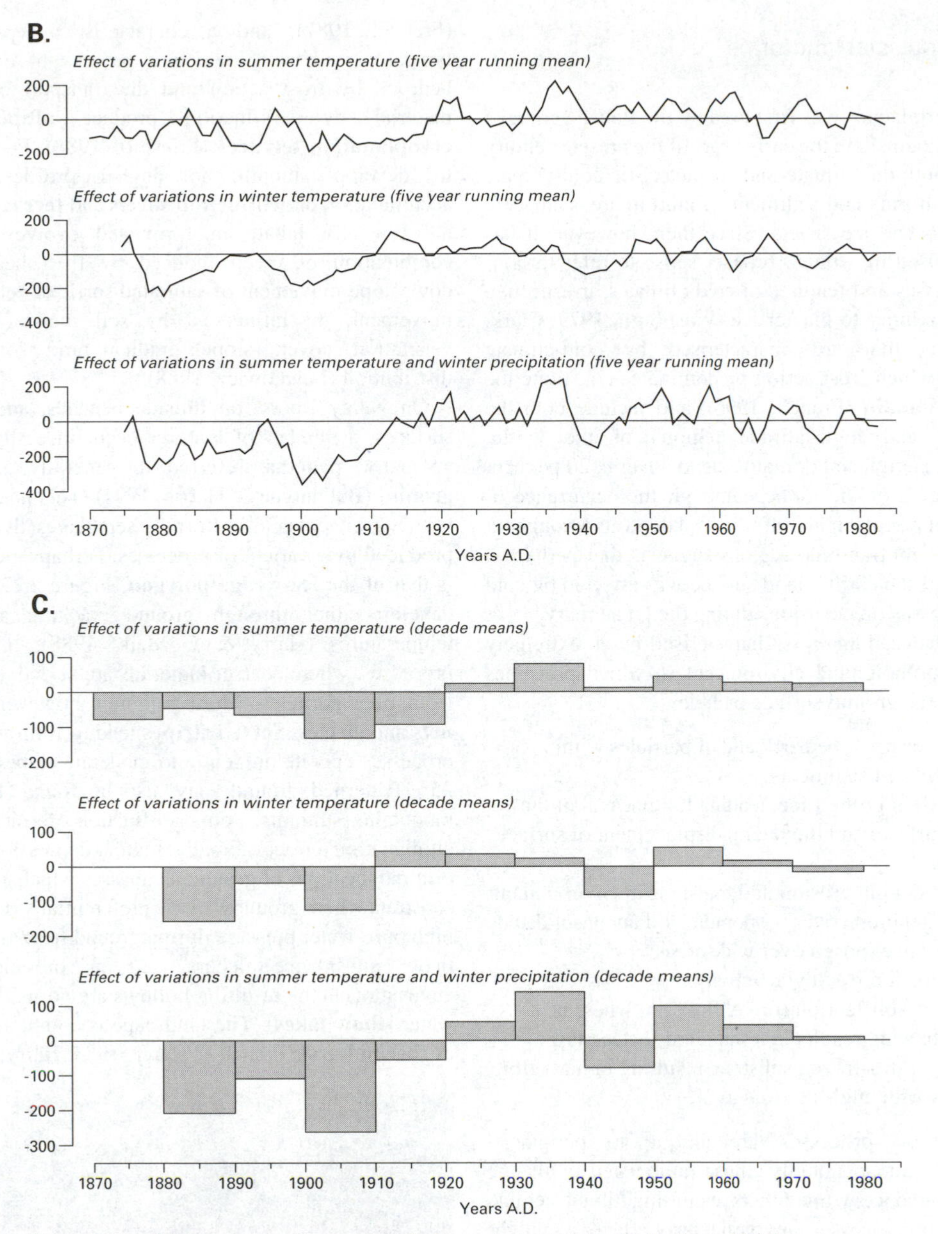

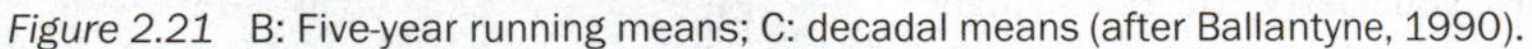

Figure 2.21 B: Five-year running means; C: decadal means (after Ballantyne, 1990).

Despite these problems, however, a broad measure of agreement has emerged between palaeotemperature estimates based on ELA reconstructions, particularly for the Younger Dryas Stadial in northwest Europe. Moreover, many of the palaeoclimatic inferences are very similar to those derived independently using other proxy data (sections 4.5.2 and 7.6.4). This suggests that the ELA method is reasonably reliable and constitutes a valuable and original approach to Quaternary palaeoclimatic reconstruction.

2.4 Periglacial landforms

The term 'periglacial' was first used by the Polish geologist Walery von Lozinski in the early years of the present century to describe both the climate and characteristic cold-climate features (landforms and sediments) found in areas adjacent to the Pleistocene ice sheets. Since then, however, it has come to be used in a much broader sense to refer to non-glacial processes and features of cold climates, irrespective of age or proximity to glacier ice (Washburn, 1979; Clark, 1988). Areas that are characterised by cold-climate processes in which frost action predominates constitute the **periglacial domain** (French, 1996), and include both the high-altitude and high-latitude regions of the world. Currently the periglacial domain extends over *c.* 20 per cent of the land area of the globe, although the occurrence of fossil or relict periglacial landforms and deposits throughout the temperate mid-latitude regions suggests that perhaps a further fifth of the earth's land surface was affected by cold climate processes on occasions during the Quaternary.

The periglacial domain is characterised by an extremely active geomorphological environment in which processes operating on the ground surface include:

(a) frost-shattering of bedrock and of particles within unconsolidated sediments;
(b) the growth of ground-ice, leading to upheaval of the ground surface, and the lateral displacement of surface materials;
(c) accelerated wind erosion and transport in environments where vegetation cover is sporadic and unconsolidated materials are exposed over wide areas;
(d) thermal erosion by fluvial activity;
(e) accelerated solifluction (or gelifluction) where near-surface thawing results in a saturated surface layer overlying a still-frozen substrate resulting in mass flow on slopes with angles as low as 2°.

Some of these processes are unique to periglacial environments, most notably those associated with the growth of ground ice, while others, including fluvial, aeolian and gelifluction processes, are particularly effective in high-latitude and high-altitude regions of the world. Collectively they give rise to a suite of landforms that is highly distinctive and that is characteristic of a periglacial landscape (Harris, 1986; Dixon & Abrahams, 1992; Ballantyne & Harris, 1994).

Within the periglacial domain, three broad morphogenetic landscape units can be recognised. On upper slopes, exposed bedrock is highly fractured, angular and craggy in appearance as a result of frost action. **Tors** (upstanding masses of bedrock) are common on summits and on hillsides (French, 1987), and a characteristic step-like profile typically evolves in which a process of excavation of bedrock by frost action and the transport of frost-riven material by solifluction produce **altiplanation** or **cryoplanation terraces** (Priesnitz, 1988). Footslopes tend to develop smooth, low-angled profiles, with the accumulation of **gelifluction sheets** and **terraces**. The latter are typically lobate in form and evolve through the combination of creep induced by frost heave and the downslope movement of saturated surficial debris. Rates of movement are influenced by soil moisture variations, vegetation cover, slope gradient and soil grain-size distribution (Lewkowicz, 1988).

On valley floors, on hillside benches, and on plateau surfaces, a number of features occur in easily recognised geometric patterns, referred to generally as **patterned ground** (Ballantyne & Harris, 1994). This is an 'umbrella' term which covers landform assemblages that have been produced by a variety of processes. Perhaps the best known is that of the **ice-wedge polygon** (Figure 2.22), which can develop either through ground cracking at very low temperatures (Harry & Gozdzik, 1988), or by sorting processes, where coarser materials are selectively separated from finer particles. More common, however, are **sorted nets** and **circles**. **Sorted stripes** tend to form where sorting processes operate on gentle to moderate slopes up to about 25°. Patterned ground may also be found, however, on mountain summits, on gelifluction terraces and on altiplanation terraces. Finally, regular depressions reflect the original locations of ground ice masses which accumulate in substrata where groundwater is preferentially diverted under high pore-water pressure during ground freezing. Following thaw, subsidence occurs and in present-day arctic environments the resulting hollows are normally filled with water (**thaw lakes**). The landscape is commonly described as **thermokarst** (French & Harry, 1988; Harry, 1988), from

Figure 2.22 Active polygonal patterned ground, Mt Edziza, British Columbia, Canada (photo: Neville Alley).

a visual analogy with karst regions, which are often marked by numerous sink holes. Themokarst is normally associated with the degradation of continuously frozen ground (**permafrost**[7] – Figure 2.23).

Many of these distinctive features of the periglacial landscape have been recognised in relict form in Eurasia and North America, either through field mapping or through the use of remote sensing techniques, and reflect the vast areas of the Northern Hemisphere that were underlain by permafrost during the cold stages of the Quaternary. Patterned ground forms, for example, can frequently be identified on aerial photographs, for the geometrical patterns tend to be emphasised by differences in crop growth resulting from drainage variations. They are particularly clear in areas such as the chalklands of southern and eastern England (Catt, 1987), the flat alluvial or raised marine deposits of the coastal fringes of northwest Europe (Svensson, 1988b), and the Late Wisconsinan drift plains of North America (Johnson, 1990). Patterned ground features provide good evidence for the former existence of periglacial conditions within an area, although in most cases they only allow the most generalised of climatic inferences to be made (Williams & Smith, 1989). Some landforms, however, may form the basis for more detailed palaeoclimatic reconstructions, and these are considered in the next section.

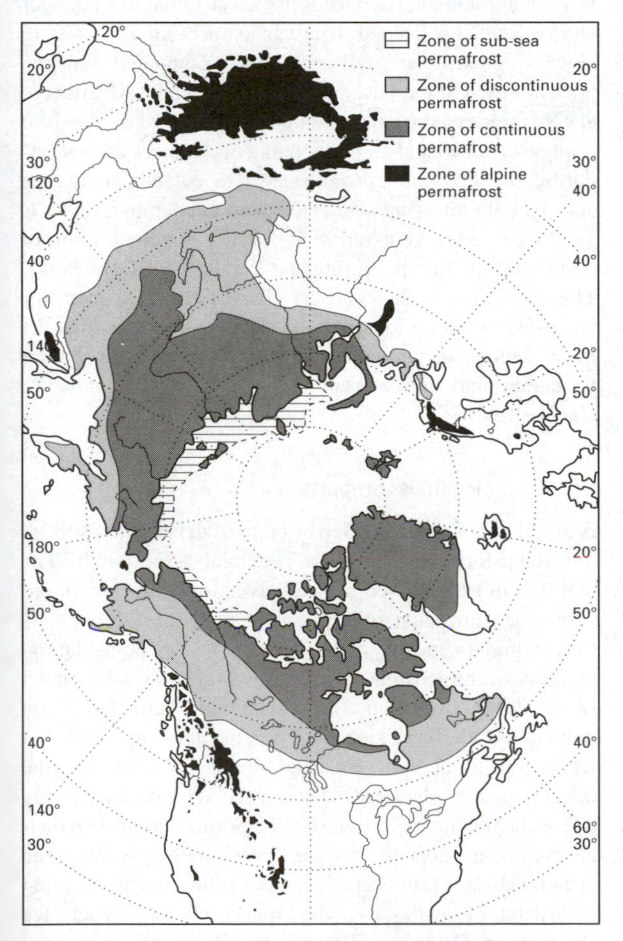

Figure 2.23 Present-day distribution of land and sub-sea permafrost in the Northern Hemisphere (after Péwé, 1983).

2.4.1 Palaeoclimatic inferences based on periglacial landforms

Certain periglacial landforms are unique to present-day arctic and alpine environments, and aspects of the prevailing climate under which they have evolved can sometimes be quantified. Where comparable features can be identified in the fossil form, therefore, they can be used as a basis for estimating climatic parameters for earlier times during the Quaternary. Three examples of periglacial landforms that have been used in this way are rock glaciers, pingos and protalus ramparts. The use of periglacial deposits in palaeoclimatic reconstructions is described in section 3.4.3.

2.4.1.1 Rock glaciers

These are active tongue-shaped or lobate accumulations of rock debris (Figure 2.24) that move slowly downslope as a result of deformation of interstitial ice or frozen sediment. Rates of movement range from 1 cm yr^{-1} to greater than 130 cm yr^{-1} (Giardino & Vitek, 1988). Two types of rock glacier have been recognised. **Morainic rock glaciers** are those that have formed through the burial and subsequent incorporation of a core of glacier ice under a thick cover of morainic debris. Such features are often found spreading downvalley from a cirque or true glacier (Giardino *et al.*, 1987; Barsch, 1988). The second type are referred to as **protalus rock glaciers** (also known as valley wall, lobate or talus-foot rock glaciers) and have developed through the deformation of talus in ice-rich permafrost regions. Typically, they form step-like or lobate extensions of the lower parts of talus slopes (Ballantyne & Harris, 1994). Protalus rock glaciers are characteristic features of the discontinuous permafrost zone in mountain regions, and their formation appears to be governed by a threshold mean annual temperature (MAT) of −2°C (Giardino *et al.*, 1987). Hence relict protalus rock glaciers can be used in palaeoclimatic reconstructions. For example, the large fossil rock glaciers in the western Tyrol, Austria, which lie 500–600 m below their active counterparts, imply a depression of mean annual temperature relative to the present of 3–4°C (Kerschner, 1978). The fossil features were last active

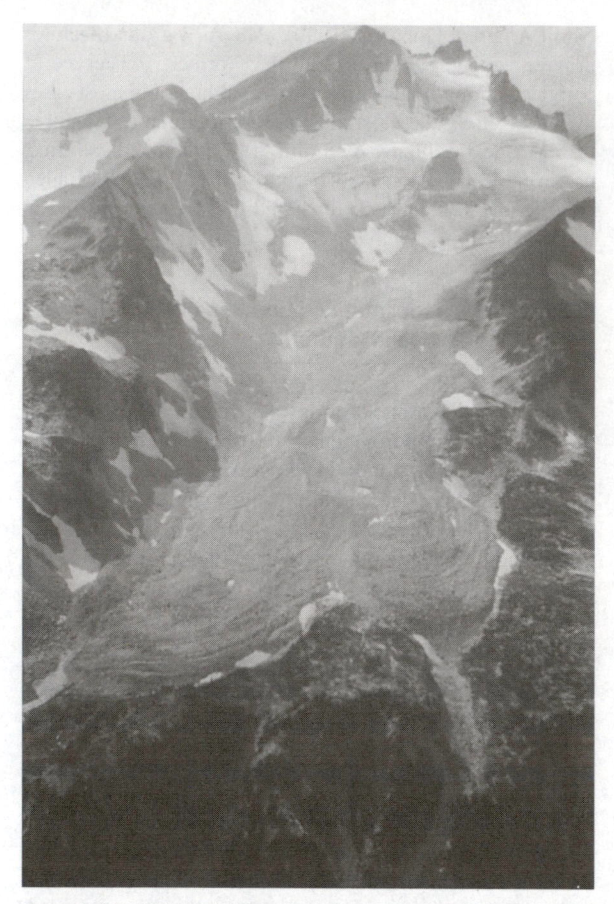

Figure 2.24 Active lobate rock glaciers near Lytton, British Columbia, Canada (photo Neville Alley).

during the Younger Dryas Stadial (*c.* 11–10 ka BP) and from the above evidence it has been inferred that a slightly more continental climatic régime persisted during that period in parts of the eastern Alps, with cooler summers (*c.* 2.5–3.0°C below present), much colder winters (temperatures 5–6°C lower) and, in the more sheltered valleys of the Tyrolean Alps, precipitation about 30 per cent less than present-day values (Kerschner, 1985). In Scotland, the presence in the Cairngorm Mountains of fossil protalus rock glaciers of Loch Lomond (Younger Dryas) Stadial age suggests mean annual air temperatures in the high mountains (*c.* 900 m) of −11.6 to −16.4°C, with precipitation in the range 375–550 mm yr^{-1}, only 19–27 per cent of present values (Ballantyne & Harris, 1994).

2.4.1.2 Pingos

Pingos are dome-shaped hills that occur in permafrost regions as a result of the uplift of frozen ground by the growth of a large mass of ground ice in the substratum. Melting of the ice body leads to ground collapse, forming a central depression or crater, with a characteristic rampart of displaced substrate around the depression (Pissart, 1988). Two broad types of pingos have been recognised: the **'closed-system'** or **Mackenzie Delta** type that form by meltwater expulsion during permafrost aggradation; and the **'open-system'** or **East Greenland** type in which the ice core is fed by subsurface groundwater percolation in areas of discontinuous permafrost (Washburn, 1979). Ramparted ground-ice depressions interpreted as representing the remains of open-system pingos have been recognised in many mid-latitude regions where periglacial conditions once obtained (e.g. Bryant & Carpenter, 1987; Coxon & O'Callaghan, 1987; De Gans, 1988). Using the prevailing climatic conditions in the present permafrost environment of western Alaska where pingos are currently developing as an analogue, Watson (1977) suggested that active pingo growth in England and Wales during the Loch Lomond (Younger Dryas) Stadial reflects a mean air temperature of −4°C to −5°C, representing a fall of 13–14°C compared with the present day. On the basis of similar reasoning, Maarleveld (1976) interpreted fossil pingo remains in the Netherlands to imply a mean annual air temperature of not more than −2°C during the last full glacial stage. He calculated that an approximate lowering of mean annual temperature of 15°C would be required to account for the southern limit of permafrost in Europe during the Weichselian Cold Stage. This may prove to be a conservative estimate, however, for more recent studies of contemporary pingos suggest that the mean annual air temperature (MAAT) threshold for their formation may range from −2 to −8°C (Ballantyne & Harris, 1994).

2.4.1.3 Protalus ramparts

A protalus rampart is a ridge or ramp of debris that forms at the downslope margins of a perennial snow patch. The debris is mostly derived by frost-riving of bedrock exposed above the snowpatch, over which the material slides or rolls to accumulate mostly at the foot, but also at the lateral margins, of the snowpatch (Shakesby *et al.*, 1987). Perennial snowpatches develop in sheltered situations on mountainsides where snow survives the ablation season but where accumulation is insufficient to lead to the development of glacier ice. Their survival is governed partly by local temperature régime, but the primary control on their development appears to be precipitation, since the accumulation of too much snow would result in rapid snowpatch growth and the transition to glacier ice (Ballantyne & Benn, 1994b). Fossil protalus ramparts (Figure 2.25), therefore, mark the positions of former

Figure 2.25 Fossil protalus rampart (middle distance) and stratified scree (foreground: see section 3.4.1) near Cader Idris, Wales (photo: Mike Walker).

snowpatches that accumulated under colder conditions than today, and hence may provide valuable palaeoclimatic information. In the uplands of the British Isles, for example, such features appear to be mostly of Loch Lomond/Younger Dryas Stadial age, and in Scotland their altitudinal distribution is very similar to that of regional glacier ELAs derived from geomorphological evidence (section 2.3.5), with protalus ramparts increasing in altitude from west to east (Ballantyne & Kirkbride, 1986). This appears to reflect heavier snowfall in western Scotland by comparison with the mountains in the east, a pattern that is also apparent in the ELA data (Figure 2.26).

The distribution of fossil periglacial landforms, therefore, not only provides evidence for the former extent of the periglacial domain, but in certain cases their occurrence can be used to derive palaeotemperature and palaeoprecipitation estimates for times during the Quaternary. Care must be exercised in the use of these data, however, for the occurrence of the fossil periglacial landforms indicates only that certain critical temperature thresholds were transgressed. It is possible, and in many cases likely, that temperatures were lower than the threshold values, and hence the derived palaeotemperatures must be regarded as maximal values only. Other problems associated with the use of fossil periglacial evidence in palaeoenvironmental reconstruction are considered in the section on periglacial sediments (section 3.4).

2.5 Sea-level change

There is abundant geomorphological evidence in many parts of the world for variations in sea level during the Quaternary. This includes former coastal landforms now standing above present sea level, such as rock platforms sometimes with well-preserved backing cliffs (Figure 2.27), 'raised beaches'[8] consisting of estuarine or littoral sand and gravel deposits, deltas, spits, shingle ridges, stacks, caves and coral reefs. Evidence for sea-level change can also be found offshore in the form of submerged landforms, including caves, platforms, beaches, reefs and river valleys. Mapping and altitudinal measurement of such features enables the positions of former coastlines to be established and the vertical range of sea-level variations to be estimated. In many situations, however, a complete picture of sea-level change can only be obtained by combining geomorphological data with evidence from the stratigraphic record. Not only do marine deposits form a more continuous record of sea-level change than the often fragmentary geomorphological evidence, but many littoral sedimentary sequences contain fossils which provide an additional data source for the reconstruction of sea-level histories. In addition, sedimentological and biological evidence offers a basis for the dating and correlation of sea-level variations, whereas geomorphological evidence (platforms, beaches, etc.) is more difficult to date precisely. By combining landform data with sedimentary and fossil records from boreholes or exposures, therefore, a sequence of sea-level changes in a particular locality can be established, often in considerable detail (Smith & Dawson, 1983; Devoy, 1987b; Tooley & Shennan, 1987). In this section, therefore, the geomorphological evidence and the lithostratigraphic evidence for changes in sea level are considered together. The biological evidence for sea-level change is discussed in Chapter 4.

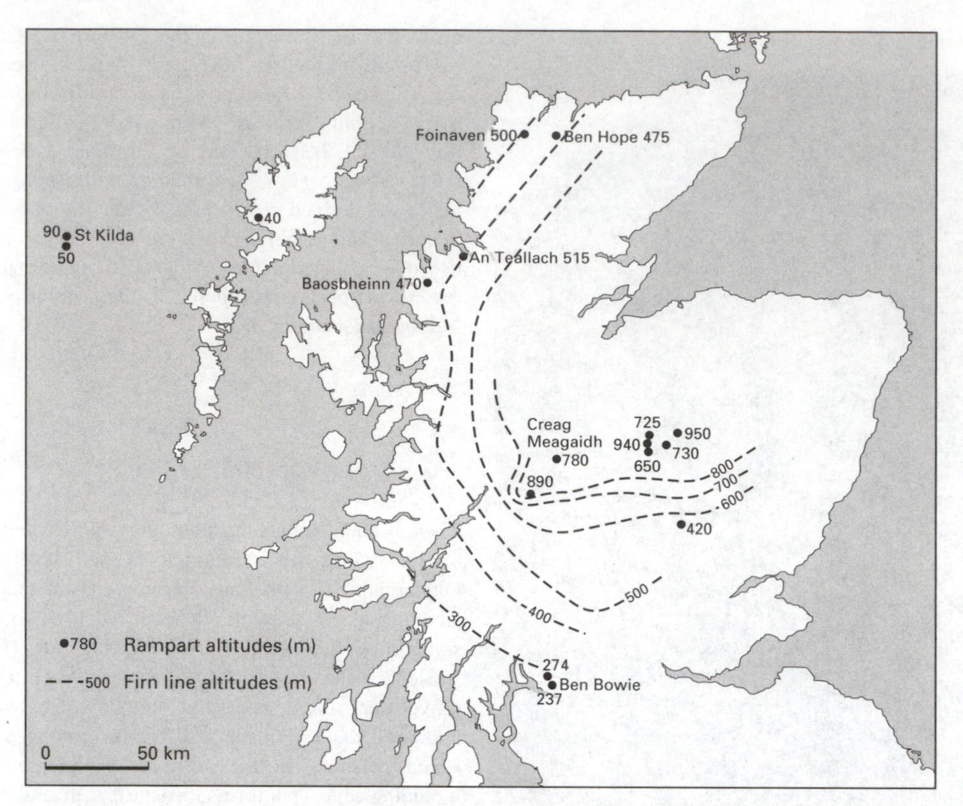

Figure 2.26 Altitudes in metres of the frontal crests of Loch Lomond (Younger Dryas) Stadial protalus ramparts in the Scottish Highlands, and reconstructed ELAs of Stadial glaciers at their maximal extent (see also Figure 2.17) (after Ballantyne & Kirkbride, 1986).

2.5.1 Relative and 'absolute' sea-level changes

The level of the sea relative to the land can vary through the vertical change of *either* the sea *or* the land surface or *both* of these. Where subsidence of the land takes place at a time of stable ocean levels, there will be a local rise in sea level; conversely, land uplift will lead to the elevation of littoral features, and an apparent fall in sea level along a particular stretch of coastline. Where changes in sea level take place either through land or through sea-level movements, they are referred to as **relative sea-level changes**, that is, a change in the position of the sea relative to the land. Such changes are essentially local in effect. Worldwide sea-level changes, on the other hand, which result from fluctuations in the volume of water in the ocean basins, are termed **eustatic**. At one time it was assumed that the extent of any eustatic change would be uniform worldwide, but it is now acknowledged that gravitational influences will result in varying degrees of sea-level rise or fall in different coastal regions (Clark *et al.*, 1978; Shennan, 1987). In effect, this means that while eustatic changes may be broadly similar locally or even regionally, this cannot be assumed when comparing records at the continental or global scale (section 2.5.2). Moreover, even where the amplitude of eustatic change has been the same in two coastal regions, this need not result in the same

Figure 2.27 A raised rock platform formed during the Loch Lomond/Younger Dryas Stadial (*c.* 11–10 ^{14}C ka BP) at Grass Point, Isle of Mull, Scotland (photo: John Lowe).

geomorphological response. For example, a eustatic rise of 50 m over a 1000 year period will result in a relative rise in sea level of 50 m in areas where the land surface is stable, but a 50 m fall in relative sea level along coastlines where the land is being uplifted at a rate of 10 cm per year.

Some land movements are long term and result from tectonic activity associated with the migration of the great lithospheric plates across the surface of the globe. Others may be of shorter duration and are generally more localised in their effects; these are termed **isostatic movements**. The term **isostasy** refers to the state of balance that exists within the earth's crust so that a depression of the crust by the addition of a load (sediment, ice, water, etc.) in one locality will be compensated for by a rise in the crust elsewhere (Fairbridge, 1983). The state of isostatic equilibrium is maintained by viscous flow, perhaps at depth within the mantle, although the nature of the processes involved is still a matter of debate (see e.g. Lambeck, 1993a). In order to understand Quaternary sea level variations, therefore, it is first necessary to establish how the separate effects of isostatic and eustatic changes have affected a region. In more tectonically stable areas (such as the islands of Bermuda and the Bahamas, for example), where the eustatic effect has been the major factor influencing sea levels, it may be possible to reconstruct a sequence of what have been termed **'absolute'** as opposed to relative sea-level changes. **Absolute sea levels** cannot easily be discerned in areas where crustal movements have occurred, however, as it is difficult to distinguish between the isostatic and eustatic effect in shoreline sequences. One way in which an approximation of the extent of isostatic movement might be obtained is to use sea-level curves from relatively 'stable' coastlines, although there are serious methodological limitations to this approach (see section 2.5.2). An alternative strategy is to use an independent measure of sea-level change, such as the oxygen isotope record from deep ocean cores. This line of evidence is discussed in more detail in section 3.10.

2.5.2 Eustatic changes in sea level

Major changes in global sea level have occurred during the Phanerozoic, with most reconstructions suggesting falling eustatic sea levels during the Palaeozoic, a rise during the Mesozoic to a maximum during the Upper Cretaceous when geophysical modelling suggests a relative sea-level rise c. 350 m above present levels, and a subsequent fall throughout the Tertiary (Devoy, 1987c; Haq *et al.*, 1987). The causes of these long-term sea-level trends are complex and reflect a combination of processes including tectonic activity, changes in mass distribution and shape of the earth, changes in the volume and mass of the hydrosphere through the addition of juvenile water, and the effects of variations in the rate of rotation or in the axis of tilting of the earth (Chappell, 1987; Mörner, 1987a). Changes in ocean water volume arising from temperature-induced density changes (thermal expansion of the oceans) will also produce **'steric'** changes in sea level (Warrick & Oerlemans, 1990), while variations in the volume and mass of sea water lead to changes in the hydro-isostatic load on the sea floor beneath (Dawson, 1992). A major influence, however, seems to have been the changes that have occurred in the configurations of the ocean basins, due partly to sediment infill and the consequent displacement of ocean water, but principally to the movement of lithospheric plates and the related phenomenon of sea-floor spreading (Pitman, 1978). A rapid increase in sea-floor spreading would lead to an increase in volume of the mid-ocean ridges resulting in a long-term sea-level rise; conversely, reduction in the rate of sea-floor spreading would lead to a decrease in ridge volume and falling sea levels. This model appears to be supported by geophysical evidence for rates of sea-floor spreading which show a rapid pulse of spreading up to *c.* 85 Ma BP (the Cretaceous sea-level maximum), and a subsequent decrease in rates of spreading resulting in a long-term fall in sea level (Devoy, 1987c).

Superimposed on these long-term trends are major sea-level oscillations that result directly from expansion and contraction of the Quaternary ice sheets. During periods of glacier build-up, water from the oceans was abstracted and stored in the form of ice. It has been estimated that during the growth of the last ice sheet, sufficient water was removed from the oceans to produce a worldwide lowering of eustatic sea level of the order of 130 m (Mörner, 1987a). Complete melting of the present Greenland and Antarctic ice sheets would release meltwater equivalent to a global sea-level rise of more than 70 m (Warrick & Oerlemans, 1990). Sea-level changes which are controlled by the growth and contraction of the ice sheets are termed **glacio-eustatic**.

Glacio-eustatic changes dominate the Quaternary sea-level history of the tectonically stable areas of the world, and in such regions Late Quaternary sequences of sea-level changes can sometimes be reconstructed. On the island of Bermuda, for example, which is a stable mid-oceanic carbonate platform (Hearty & Vacher, 1994), $^{230}Th/^{234}U$ dating and amino-acid racemisation geochronology (sections 5.3.4 and 5.6.1) of corals and speleothems from fossil coral reefs and beach deposits lying both below and above present sea level have revealed a chronology of eustatic sea-level fluctuations spanning almost 250 ka (Figure 2.28). On only two occasions during that period have sea levels been higher than those of the present day; around 200 ka BP when the sea stood at *c* +2 m, and during the last interglacial (125 ± 4 ka BP) when maximum sea levels were 5–6 m above those of the present. Sea levels may

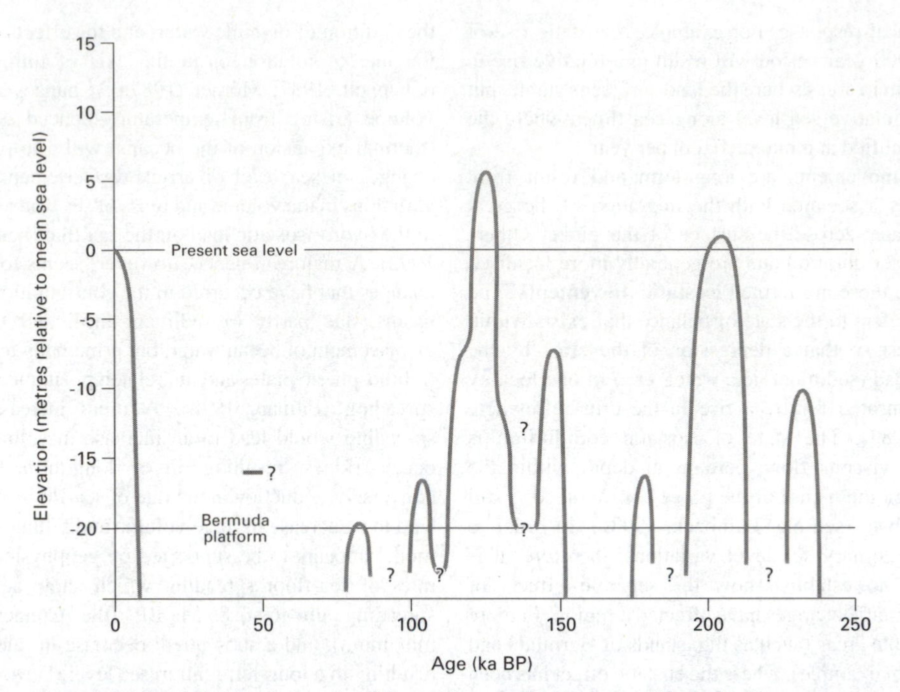

Figure 2.28 Late Pleistocene sea-level fluctuations on Bermuda (modified after Harmon *et al.*, 1983).

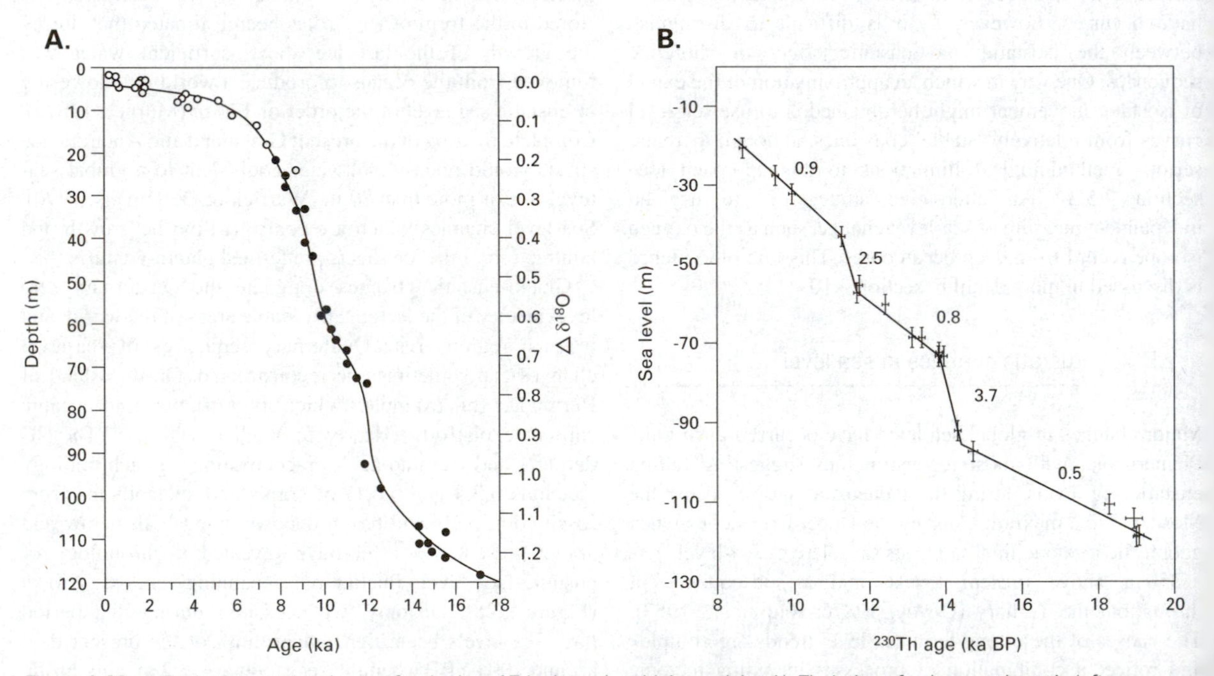

Figure 2.29 A. Barbados sea-level curve for the last 17 ka based on high-precision U–Th dating of submerged corals (after Fairbanks, 1989). B. Sea level during the last deglaciation from Barbados data. The dated samples are of the fossil coral *Acropora palmata*, a species that lives in water depths of less than 5 m. The U–Th ages are quoted at the 2σ level (after Bard *et al.*, 1990a).

have been at comparable elevations to those of the present, however, around 80 ka BP (Vacher & Hearty, 1989). These data are in good agreement with those from coral reef sequences from other parts of the world which show, in particular, a clearly demarcated high stand of sea level at the last interglacial maximum at *c.* 123.5 ka BP (Smart & Richards, 1992). The palaeosea-level curve for Bermuda (Figure 2.28) suggests that these eustatic sea-level changes can occur very rapidly, with transgressions and regressions of the order of 3.5 to 6 mm per year (Harmon *et al.*, 1983).

Glacio-eustatic sea-level changes are known in greatest detail for the last 15 ka, during which time the Laurentide, Fennoscandian and British ice sheets disappeared completely,[9] and the Antarctic and Greenland ice sheets also decreased in mass. Evidence for the rise in sea level that accompanied the wastage of these ice sheets is available from many parts of the world, but the longest and most detailed records are those from coastal regions unaffected by glacier activity. On Barbados, for example, high-precision U–Th dating of fossil coral shows that from a position at around -121 ± 5 m at the last glacial maximum, global sea level rose at an accelerating rate to *c.* -60 m at 10 ka BP and <-20 m at 7 ka BP (Fairbanks, 1989). The steady deceleration in sea-level rise after that time (Figure 2.29A) reflects the fact that by 7 ka BP glacier ice had long disappeared from Britain, only small amounts of ice remained in Scandinavia, while the Laurentide Ice Sheet was restricted to residual remnants on the islands of the Canadian Arctic (Dawson, 1992). Figure 2.29B shows the trend in glacio-eustatic sea level during the period of maximum ice melting. Deglaciation began before 19–18 U–Th ka BP, and the consequent sea-level rise was interrupted by two major meltwater pulses, the first corresponding to a dramatic rate of sea-level change of *c.* 3.7 m per century at *c.* 14 ka BP, which was followed by a sharp reduction in the rate of sea-level rise to <1 m per century. The second abrupt change of *c.* 2.5 m per century occurred at around 11 ka BP (Bard *et al.*, 1990a).

In northwest Europe, the general rise in sea level over the past 10 ka has been calculated at numerous localities around the coastline, largely on the basis of evidence from estuarine sediments and coastal lakes. Sedimentation in estuaries is usually rapid, due to the influx from rivers, while the deposits tend to be protected from erosional processes. Relatively shallow waters are often found in estuaries, so that even minor reductions in sea level can lead to the exposure of marine and brackish water sediments, while a slight rise in sea level can result in the submergence of terrestrial or freshwater deposits. A series of sea-level fluctuations will therefore often result in a stacked sequence of marine, estuarine, freshwater and terrestrial sediments, from which a detailed history of sea-level change can be reconstructed. Figure 2.30, for example, shows a terrestrial peat interbedded between two layers of marine sediment. The boundary between the lower marine unit and the base of the peat marks the emergence of mudflats above local high-tide level, the surface of which was subsequently colonised by land plants whose remains make up the peat. If dated, this horizon would provide a **sea-level index point**, from which not only the age and altitude of relative sea level can be established, but also the **tendency** of relative sea-level change can be deduced (Shennan, 1987). The 'tendency' of an index point refers to the **trend** of relative sea level, in other words whether there is an increase or decrease in marine influence at that locality. This can be inferred not only from changes in the nature of the sediments (from terrestrial to marine or *vice versa*), but also from pollen, diatom or other microfossil evidence (sections 4.3.5 and 4.9.3). In this particular case, the transition from marine to terrestrial deposits indicates a

Figure 2.30 A river bank section in Lower Strathearn, eastern Scotland, showing marine sands at the base, overlain by peat (darker layer) and about 6 m of marine silts and clays. All the deposits lie above present sea level. Radiocarbon dating of the top and base of the peats indicates a marine 'transgression' at c.a 9.6 ka BP, and a marine 'regression' at *c.* 7.5 ka BP (Cullingford *et al.*, 1980) (photo: John Lowe).

decrease in marine influence and hence a *negative sea-level tendency*. In Figure 2.30 the boundary between the top of the peat and the base of the overlying marine deposits reflects the resubmergence of the estuary, and provides a sea-level index point for an increase in marine influence, in other words a *positive sea-level tendency*. In the older literature, such evidence was often interpreted in terms of **marine regressions** and **transgressions** (falls and rises in sea level), although these terms are now less widely used as they imply absolute changes in sea level, whereas only relative sea-level tendencies can really be inferred from such evidence (van de Plassche, 1986; Shennan *et al.*, 1992, 1994).

Lake basins that lie near the coast may also provide important sea-level index points where, for example, such basins have been raised above sea and subsequently begin to accumulate freshwater sediment. These are referred to as **isolation basins** (Svendsen & Mangerud, 1987; Shennan *et al.*, 1994). In some instances, a rise in relative sea level may

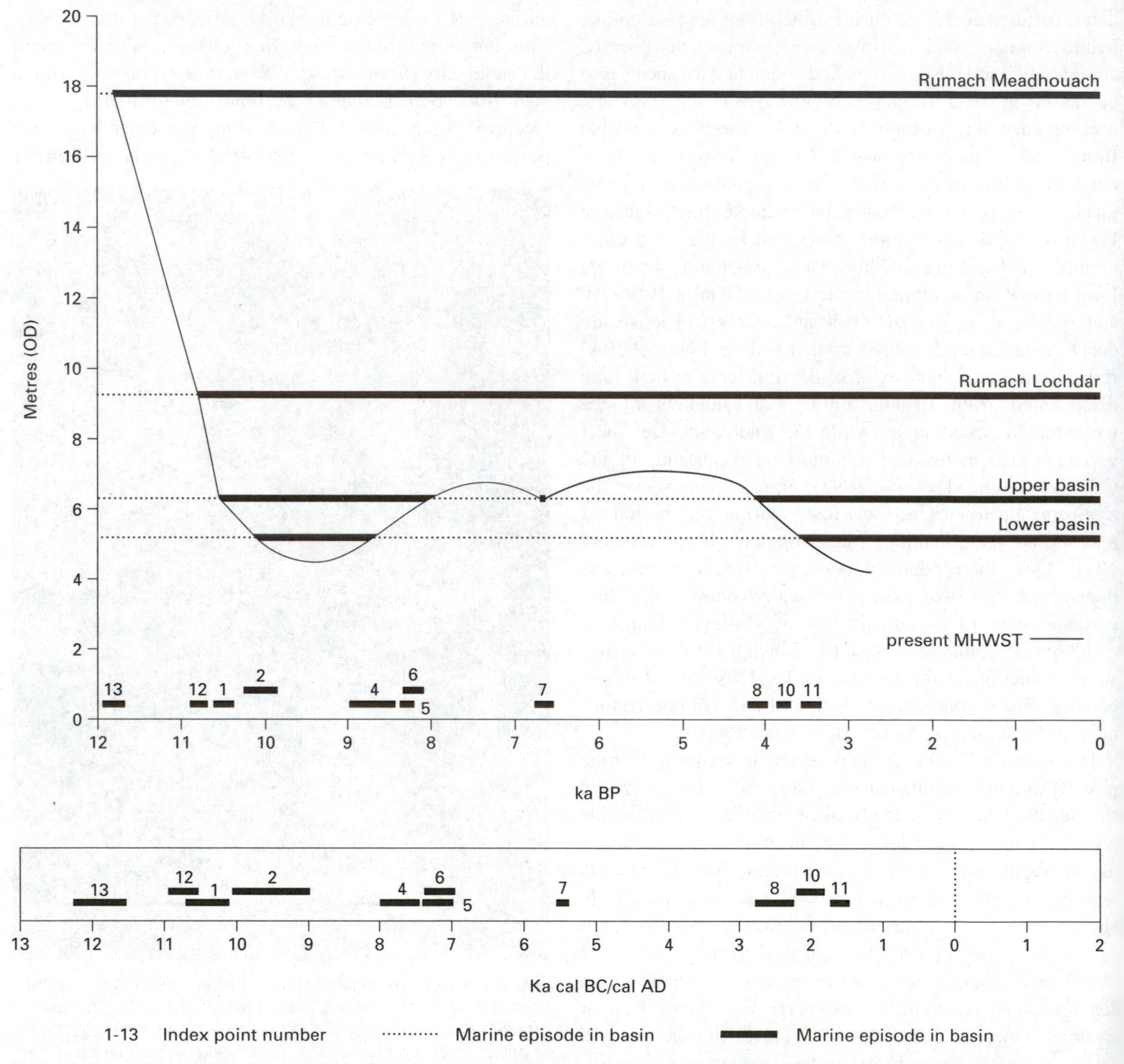

Figure 2.31 A. Chronology of isolations from and connections to the sea for four littoral basins in northwest Scotland: Loch nan Eala upper basin; Loch nan Eala lower basin; Rumach Lochdar; and Rumach Meadhonach. The curve shows the sequence of relative sea-level change.

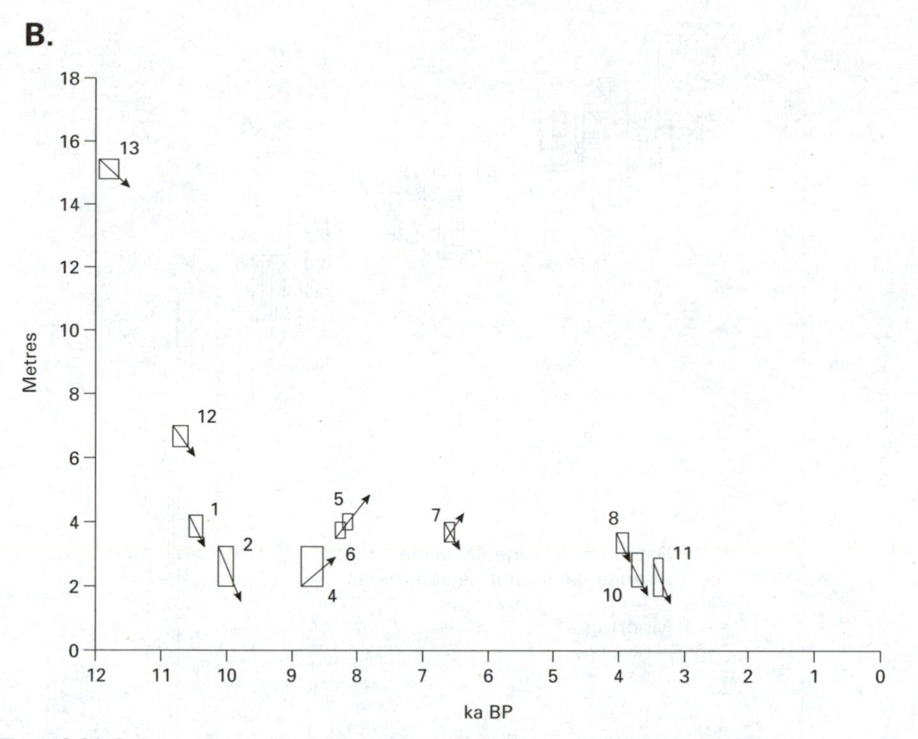

Figure 2.31 B. Age–altitude plot of the sea-level index points in the same area. Altitudes are relative to present-day mean tide level. Error boxes are defined by $\pm 2\sigma$ of the radiocarbon dates. Arrows represent positive (upward) and negative (downward) tendencies of sea-level movement (after Shennan *et al.*, 1994).

lead to the resubmergence of these basins, in which case brackish or marine deposits will accumulate over freshwater sediments beneath. In northwest Scotland, for example, analysis of sediments from a number of littoral basins shows that these were formerly occupied by the sea (Figure 2.31A). In each basin, sea-level index points and tendencies have been identified (Figure 2.31B), a downward arrow indicating a fall in relative sea level, and an upward arrow a rise. The data suggest that, relative to the land, sea level fell by about 14 m between 12 and 10 ka BP as a result of isostatic recovery following regional deglaciation. Subsequently, sea level rose by 2–3 m relative to the land so that between 9 and 4 ka BP sea waters reoccupied some of the basins (Figure 2.31A). In southeast England, by contrast, sea level appears to have risen relative to the land by *c.* 35 m between 10 and 4 ka BP (Figure 2.32). The difference in sea-level tendencies between the English and Scottish data reflects the isostatic effects of the last ice sheet on the landscape of northern Britain. It must be emphasised, however, that unlike the curves in Figure 2.29, those in Figure 2.32 are *relative* sea-level curves, as they are derived from localities where crustal warping has occurred. They therefore provide an indication of local trends in sea level only.

Attempts have been made to separate the isostatic from the eustatic components in relative sea-level curves by using eustatic sea-level data from 'stable' coastlines. However, few areas of the world are completely tectonically stable, and it is now recognised that many coastlines formerly considered to be stable are, in fact, not so. For example, the Mediterranean, with its well-developed horizontal raised shorelines (Figure 2.33), was long regarded as a type area for the study of Quaternary eustatic sea-level change, but it is now known that many of the prominent high-level shorelines have been displaced by earth movements (e.g. Hearty, 1987; Dumas *et al.*, 1993). Similarly, the coast plain of the eastern United States has often been cited as tectonically stable, but evidence is now available of long-term sequences of uplift and subsidence which have varied both spatially and temporally (Dowsett & Cronin, 1990). A further complication arises out of the phenomenon of **geoidal eustasy**. The earth is not spherical, but is flattened at the poles and bulging at the equator, and it has generally been assumed that the free ocean surface (the **geoid**) parallels that of the earth. That this is not so is indicated by maps of the geoid which show startling variations including a 180 m difference in altitude between a high near New Guinea and a trough off the south of India (Mörner, 1976, 1980, 1987b). These irregularities are thought to be caused

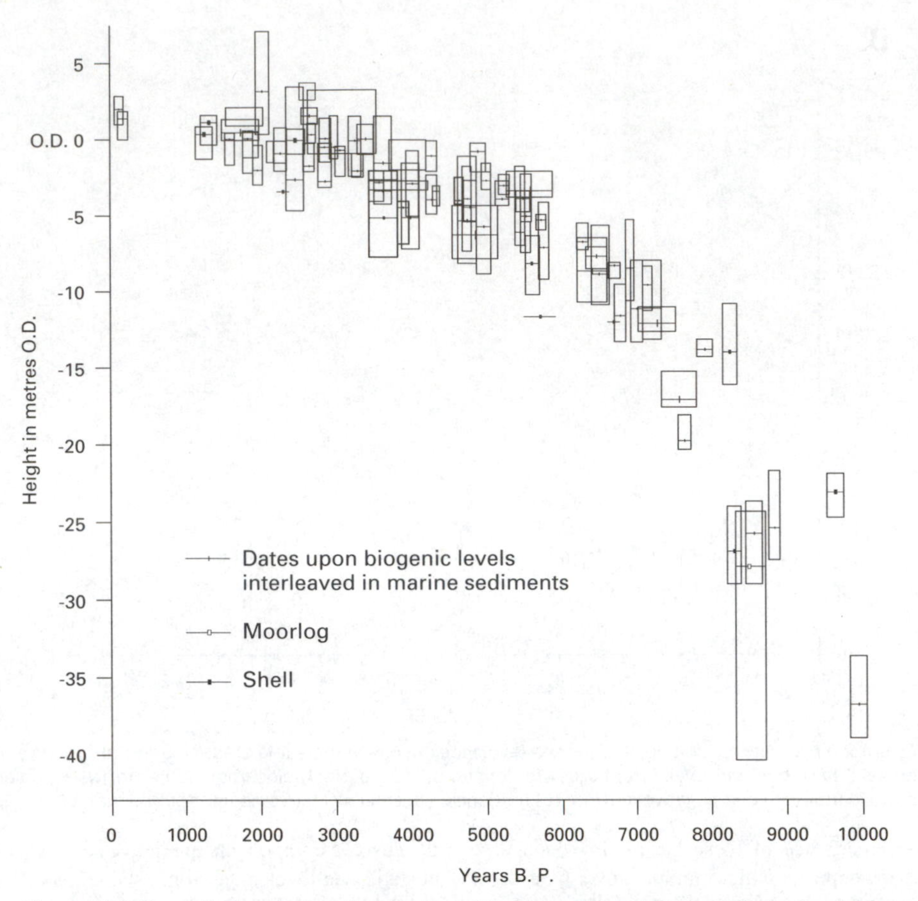

Figure 2.32 Time–depth plot of sea-level index points from southeast England.The horizontal line through each point gives the height of each dated sample. The box around each point represents the uncertainties associated with each age measurement; the horizontal dimension is defined by $\pm 1\sigma$ of the radiocarbon dates while the vertical interval represents the error margins in relating the measured points to mean sea level. The plot illustrates the trend of sea-level rise in the southern North Sea zone (after Devoy, 1987c).

by gravitational variations, which are determined by the earth's pattern of rotation and by its structure and density. Of particular significance is the fact that the change in the distribution of ice during glacial and interglacial phases resulted in gravitational changes which, in turn, would have produced variations in the geoidal surface. Hence, the geoidal ocean surface can intersect different land masses simultaneously at different absolute altitudes (Figure 2.34). Although this does not affect the construction of sea-level curves for individual localities, it does mean that an independent eustatic sea-level curve should be constructed for each region and that those curves remain area-specific (Mörner, 1987c). The corollary, of course, is that a single, universally valid eustatic sea-level curve, so long seen as the 'Holy Grail' of sea-surface studies, is effectively unattainable (Devoy, 1987a).

Figure 2.33 Raised rock platforms, eastern Mallorca, Spain (photo: Mike Walker).

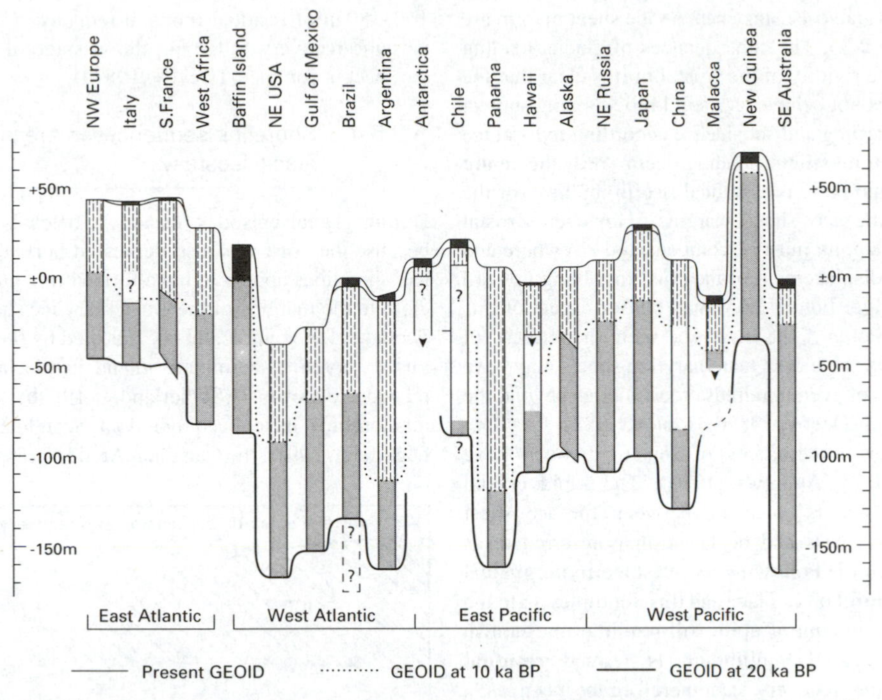

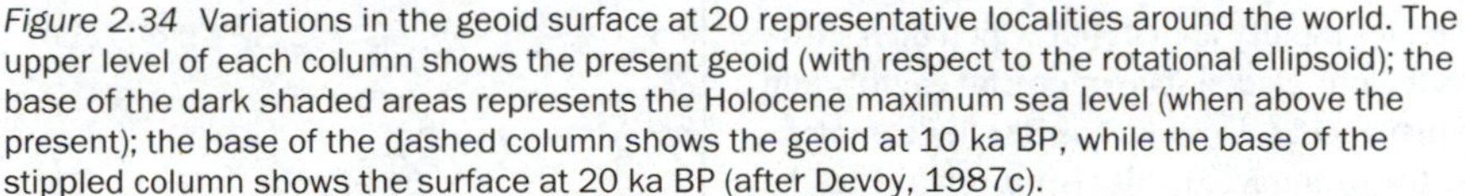

Figure 2.34 Variations in the geoid surface at 20 representative localities around the world. The upper level of each column shows the present geoid (with respect to the rotational ellipsoid); the base of the dark shaded areas represents the Holocene maximum sea level (when above the present); the base of the dashed column shows the geoid at 10 ka BP, while the base of the stippled column shows the surface at 20 ka BP (after Devoy, 1987c).

2.5.3 Tectonic influences

Evidence for shoreline displacement resulting from long-term earth movements is available from many parts of the world. Tectonically displaced shorelines in the Mediterranean region have already been noted, but others include the high-level Holocene shorelines around the coasts of the northwest and western Pacific, some of which have been tectonically uplifted to heights in excess of 30 m (Ota, 1987); the spectacular flights of raised shorelines on the emergent coastline of the Huon Peninsula in New Guinea whose ages span the last 300 ka (Chappell, 1983); and raised coral reefs and marine terraces in Fiji (Miyata *et al*., 1990), the Mariana Islands (Kayanne *et al*., 1993), the Cook Islands (Woodroffe *et al*., 1991) and New Zealand (Ota *et al*., 1991). All of these localities are adjacent to the boundaries of major lithospheric plates, and the various raised marine features are clearly a reflection of long-term plate tectonic activity (Berryman, 1987). Away from the plate margins, shorelines in a number of continental areas have been affected by **neotectonic deformation**[10] (Lambeck, 1988). In Australia, for example, raised shorelines of last interglacial age range in altitude from 6 to 32 m, reflecting differential uplift due to mantle hotspot and associated neotectonic processes (Murray-Wallace & Belperio, 1991).

The effects of isostatic earth movements are also widespread and the most important for the study of sea-level changes are those that resulted from the expansion and contraction of the Quaternary ice sheets (Lambeck, 1993a, 1995). Melting of the ice sheets and the subsequent release of large quantities of meltwater into the ocean basins would, it is believed, have led to crustal warping through the process of **hydro-isostasy** (Hopley, 1983). It has been estimated that following deglaciation, the ocean basins may have been depressed by about 8 m on average, and that hydrostatic warping could have introduced differences of up to 30 per cent in estimates of marine transgression rates between ocean islands and continental crusts (Chappell, 1974). It is worth noting, however, that while hydro-isostatic loading during eustatic sea-level rise may cause depression of the crust, the reverse (i.e. hydro-isostatic unloading following sea-level fall) may not necessarily obtain (Mörner, 1987b). Whatever the long-term effects of hydro-isostasy, however, they in no way compare to the crustal deformation that resulted from the build-up of the great ice sheets, a phenomenon known as **glacio-isostasy**.

The effects of glacio-isostasy near an ice sheet margin are shown in Figure 2.35. The consequences of glacial loading will vary with the rigidity of the crust, but it is clear that the earth's crust does not behave as a solid block, being subject to differential warping and subsidence according to local ice loads. In general, maximum loading occurs near the centre of an ice sheet and there is a gradual rise in the level of the crust towards the ice sheet margins. However, crustal depression at one point must be compensated elsewhere and hence marginal displacement of the crust involving upward bulging (**forebulge:** Figure 2.35) may be one aspect of this compensation (Peltier, 1987). If so, then the effects of glacial loading by the large Quaternary ice sheets may have extended for tens or even hundreds of kilometres beyond the former ice margin (Devoy, 1987d; Hopley, 1987).

Different stages in the process of isostatic recovery have been considered by Andrews (1970). The rapid crustal adjustment that occurs at a site between the ice sheet beginning to lose mass and deglaciation is referred to as **restrained rebound**. Following ice wastage more gradual **postglacial rebound** takes place and this continues up to the present day. The amount of uplift still required to establish the pre-glacial crustal equilibrium is termed **residual rebound**. Isostatic recovery can therefore be seen as a process that accelerates rapidly at first, but which then slows down gradually as the pre-glacial state of crustal equilibrium is approached. In many areas, isostatic uplift is not complete. Highland Britain, for example, is still rising relative to the more stable southern parts of the country (which are experiencing slight subsidence) at rates of almost 0.2 cm per year (Shennan, 1989), the Gulf of Bothnia area of the northern Baltic is rising at 0.8 to 0.9 cm per year (Broadbent, 1979), while in the eastern Hudson Bay region of Canada, current uplift of the order of 1.1 cm per year has been inferred (Hillaire-Marcel, 1980). It has been estimated that 160–180 m of residual rebound remains in this area before isostatic recovery following the wastage of the Laurentide ice sheet is complete (Devoy, 1987d).

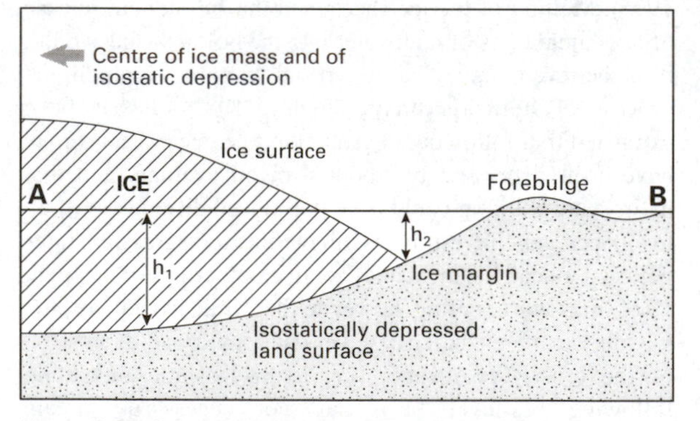

Figure 2.35 Schematic diagram of the effect of an ice mass on a land surface. In general terms, the amount of isostatic depression of the land surface A–B increases with ice load, so that greater depression occurs towards the centre of an ice mass than at the margin (cf. h_1 and h_2)

2.5.4 Shoreline sequences in areas affected by glacio-isostasy

During glacial episodes, eustatic sea levels were low, but because the crust was also depressed beneath the weight of ice, shorelines appear to have formed in a number of places close to the margins of the Quaternary ice sheets. In parts of Scotland, for example, this is indicated by the fact that many raised shorelines terminate inland in glacial outwash and related deposits (Sutherland, 1984b), and similar relationships have been noted in Scandinavia (Björck & Digerfeldt, 1991), the Canadian Arctic (Evans, 1990) and the

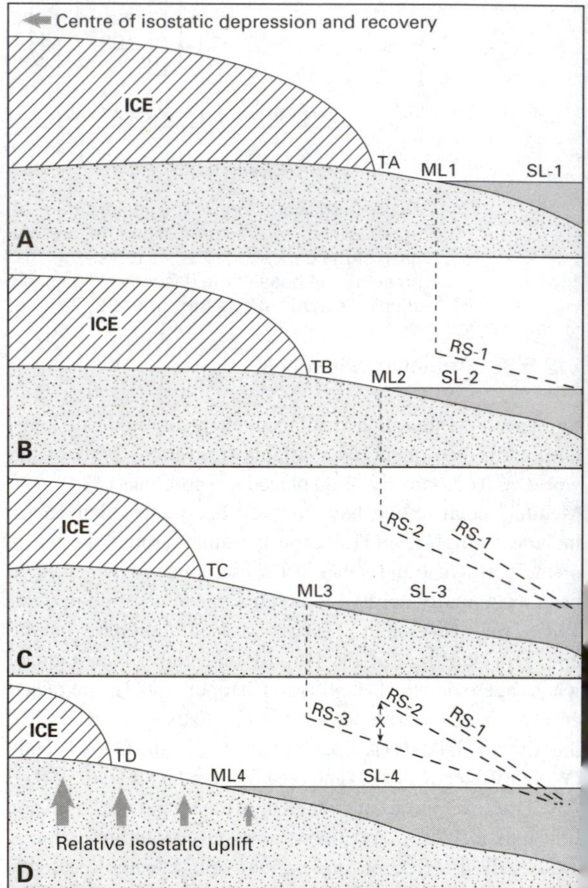

Figure 2.36 The development of a raised shoreline sequence reflecting isostatic recovery and eustatic rise following deglaciation. ML – marine limit; SL – sea level; RS – raised shoreline. The effects of the forebulge (Figure 2.35) have been omitted from the diagram. For further explanation see text.

eastern USA (Koteff *et al.*, 1993). Following deglaciation, both land uplift and sea-level rise occurred and hence sequences of raised shorelines have developed that reflect the often complex interplay of isostatic and eustatic factors.

Figure 2.36 shows the type of shoreline sequence that might be found in an area undergoing isostatic recovery. As the ice front recedes from terminus TA to TB, uplift occurs so that the shoreline formed when the sea level stood at SL-1 is raised above the new sea level, SL-2. Because isostatic recovery decreases with distance from the centre of an ice sheet, the shoreline RS-1 will be tilted away from the ice-sheet centre. During glacier retreat from TB to TC the shoreline that developed while the sea stood at SL-2 is raised and tilted (RS-2), but it will be less steeply inclined than RS-1 which has now been even further deformed. However, because the rate of isostatic recovery at that time was accelerating due to rapid ice wastage, the **marine limit**[11] (ML2) of RS-2 has been raised to a higher altitude than the marine limit ML1 of shoreline RS-1. Subsequently, a third raised shoreline (RS-3) develops, but by this time isostatic recovery has slowed down, so that the marine limit ML3 is found at a lower altitude than either ML1 or ML2. Moreover, a combination of decreased uplift and an increase in the rate of eustatic sea-level rise means that the *relative sea level* in the area is rising and therefore later shorelines will progressively truncate the older and more steeply inclined features (Fig. 2.36D). In general therefore, the oldest and most steeply tilted shorelines will form at the greatest distance from the ice centre, younger shorelines will be less steeply inclined and more extensively developed, and older features will have been destroyed or partly destroyed during the formation of younger shorelines or, in certain cases, will be buried beneath later sediments or will be found below the present sea level. The vertical interval between individual shorelines at particular localities shows the amount of isostatic uplift that has occurred between the times of shoreline formation (X – Figure 2.36).

Raised shorelines can be both depositional and erosional forms, and the extent to which a clear geomorphological feature develops depends on a range of factors including the length of time that sea level remained constant relative to the land, and on the operation of local glacial, fluvial and marine processes. Continuous shoreline features will have evolved in some areas, while in others the geomorphological expression of relative sea-level change may be more sporadic. However, postglacial subaerial and marine activity may have destroyed or extensively modified much of the evidence, even in those localities where coastal landforms were originally well developed. Consequently, only shoreline fragments remain in most areas, and careful mapping and instrumental levelling of each shoreline remnant is necessary before individual shorelines can be reconstructed and inferences made about former sea levels (Sissons, 1976; Rose, 1990). In some sheltered localities, however, because of the complex interplay between land uplift and eustatic sea-level rise, some shorelines may be buried by younger marine sediment. Further isostatic recovery may raise these above present sea level to produce **raised buried shorelines** (Sissons, 1967, 1976). In order to reconstruct the complete history of isostatic and eustatic changes in such areas, therefore, recourse must be made to coring in addition to detailed mapping of visible shoreline fragments.

Raised shoreline data are usually presented in the form of a **height–distance diagram** (Figure 2.37), which is a plot of all the individual data points in a vertical plane running

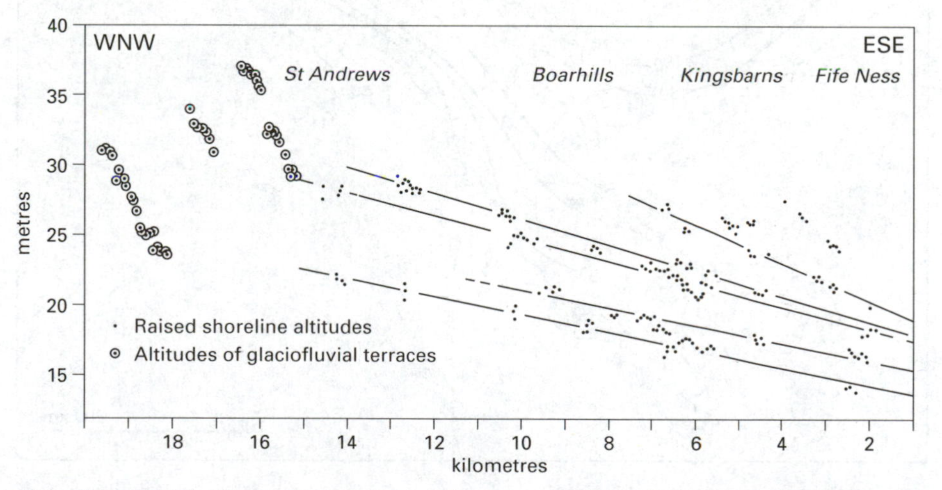

Figure 2.37 Height–distance diagram for shoreline fragments in part of eastern Scotland (after Cullingford & Smith, 1966).

parallel to a line towards the assumed ice centre. Shoreline fragments are resolved into a series of inferred shorelines, and the gradients of the features can then be calculated, usually by means of regression analysis. Where prominent shorelines of the same age are found in different areas, **isobases** can be constructed for these shorelines. Isobases join points of equal altitude (or uplift) on shorelines of the same age. The pattern of isobases, reconstructed either manually or by means of trend surface analysis, gives a three-dimensional image of the deformation of the land surface by the weight of glacier ice (Figure 2.38). It is important to appreciate, however, that while the isobase patterns may indicate in a very general way those areas that experienced maximum glacio-isostatic depression, the values of the isobases themselves bear no relation to the amount of depression (or subsequent rebound) that has actually occurred. Hence isobase maps such as Figure 2.38 show only uplift *relative to present sea level*, and not *absolute* uplift following deglaciation.

A recent development in the study of glacio-isostasy has been in the modelling of glacio-isostatic rebound and associated sea-level fluctuations. Such models can provide information, *inter alia*, on the earth's response to surface loading, on ice sheet dimensions and behaviour (see section 2.3.4), on the ages of raised shorelines where dating by direct means (e.g. radiometric dating) is not possible, and on the extent to which sea-level measurements from different sites can be combined to form representative sea-level curves for a given region (Lambeck, 1991b). In essence, such models combine glaciological evidence, empirical data relating to sea-level change, and geophysical parameters relating to rheological[12] properties of the upper mantle (Lambeck, 1993a, 1993b). A numerical model of uplift for the Baffin Island and the northern Hudson Bay region, for example, showed very good agreement between isobases calculated on the basis of the model and those derived from relative sea-level observations (Figure 2.39), both empirical and model data indicating a single rebound centre over

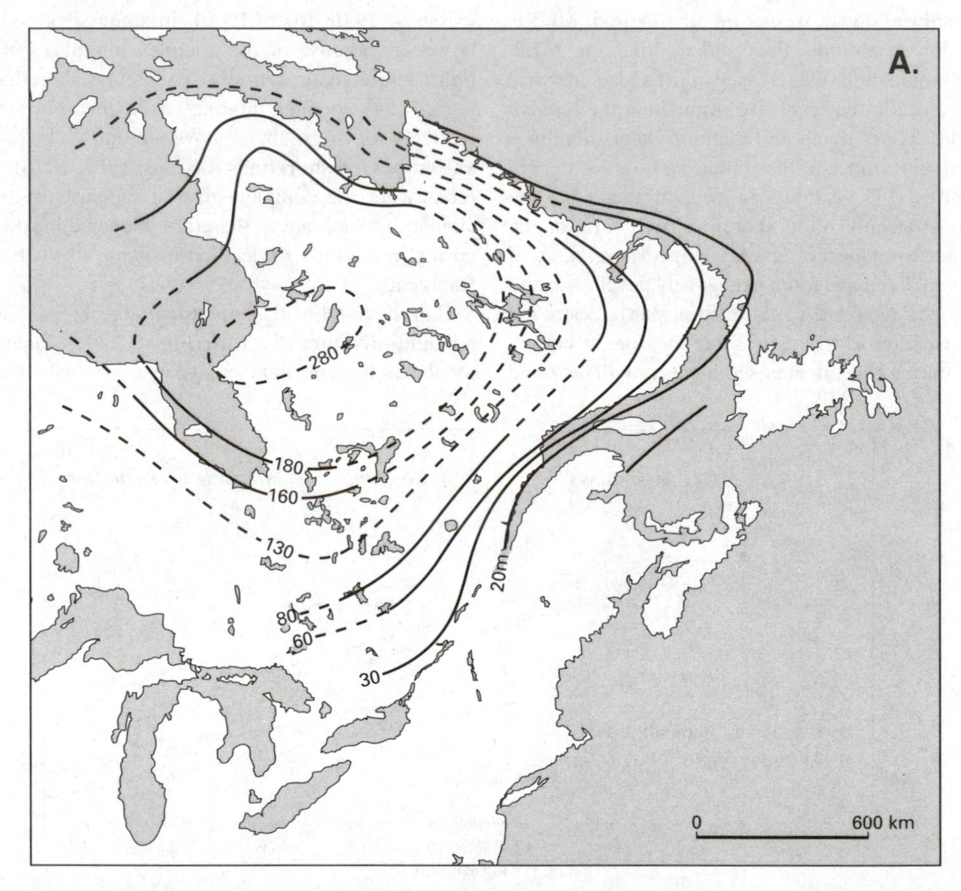

Figure 2.38 A. Isobase map showing uplift (in metres) in eastern Canada since 7.5 ka BP (Hillaire-Marcel & Occhietti, 1980).

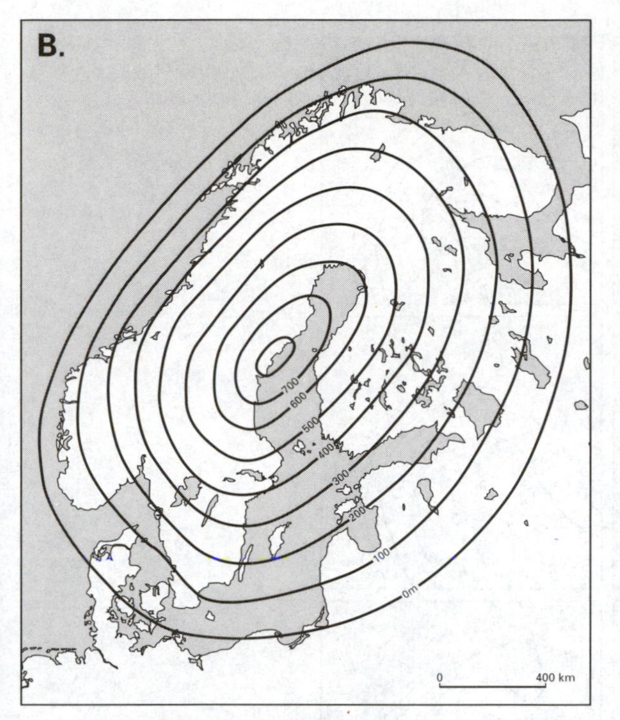

Figure 2.38 B. Isobases (in metres) showing absolute uplift of Scandinavia during the Holocene (Mörner, 1980).

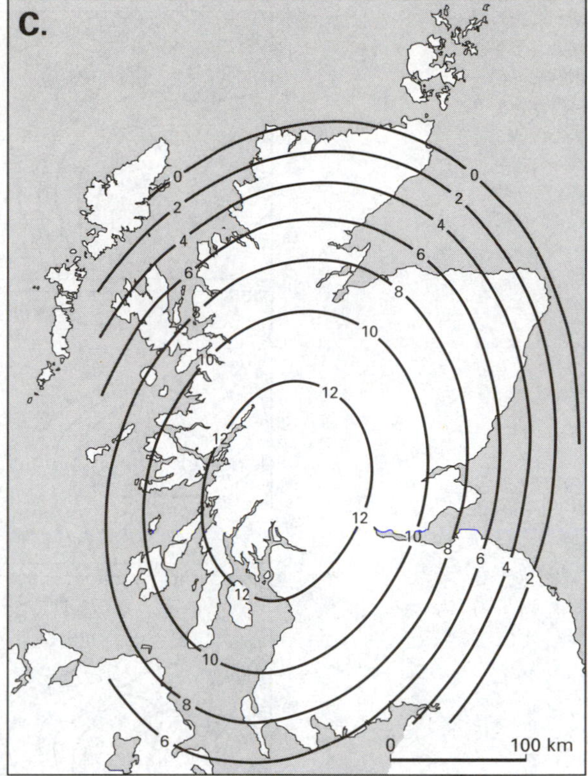

Figure 2.38 C. Quadratic trend surface showing isobases (in metres) on the Main Postglacial Shoreline (c. 6.5 ka BP) in Scotland (Firth et al., 1993).

western Baffin Island and the Foxe Basin (Quinlan, 1985). For the British Isles, Lambeck (1991a, 1991b, 1993a, 1993b) has developed a model for glacial rebound and relative sea-level change over the past 16 ka (Figure 2.40). Again, the model is consistent with both the spatial and temporal patterns of sea-level change derived from empirical observations, and demonstrates that these are explicable entirely in terms of (i) glacio-isostatic rebound in response to unloading of the ice sheet over Britain and, to a lesser extent, over Fennoscandia; (ii) the rise in sea level from the melting of Late Pleistocene ice sheets, including the response of the crust to water loading (hydro-isostasy); and (iii) the fact that the agreement between the model and observations is such that there is no need to invoke vertical crustal movements for Great Britain and Ireland other than those of glacio-isostatic origin. The model also suggests a maximum ice thickness over the British Isles during the last cold stage of less than 1500 m, a figure that is consistent with field evidence and with a number of other glaciological models for the last British ice sheet (section 2.3.4).

Isostatic rebound models therefore provide valuable data for assessing the rheological behaviour of the earth's crust as well as offering a means of testing glaciological reconstructions. However, discrepancies between measured shoreline patterns in the British Isles and model predictions of glacio-isostatic rebound suggest that a number of refinements still need to be made to these models, either by increasing the resolution of the field data by which the models are constrained, or by adjusting the geophysical parameters that are built into the models themselves (Lambeck, 1995).

2.5.5 Palaeoenvironmental significance of sea-level changes

Changing levels of land and sea impinge on many aspects of Quaternary research, and a knowledge of the causes and effects of sea-level change is of fundamental importance in the analysis of Quaternary environments. Information on the sequence of Late Quaternary climatic change can be derived from those areas where a chronology of sea-level fluctuations can be established (e.g. Bard *et al.*, 1990a),

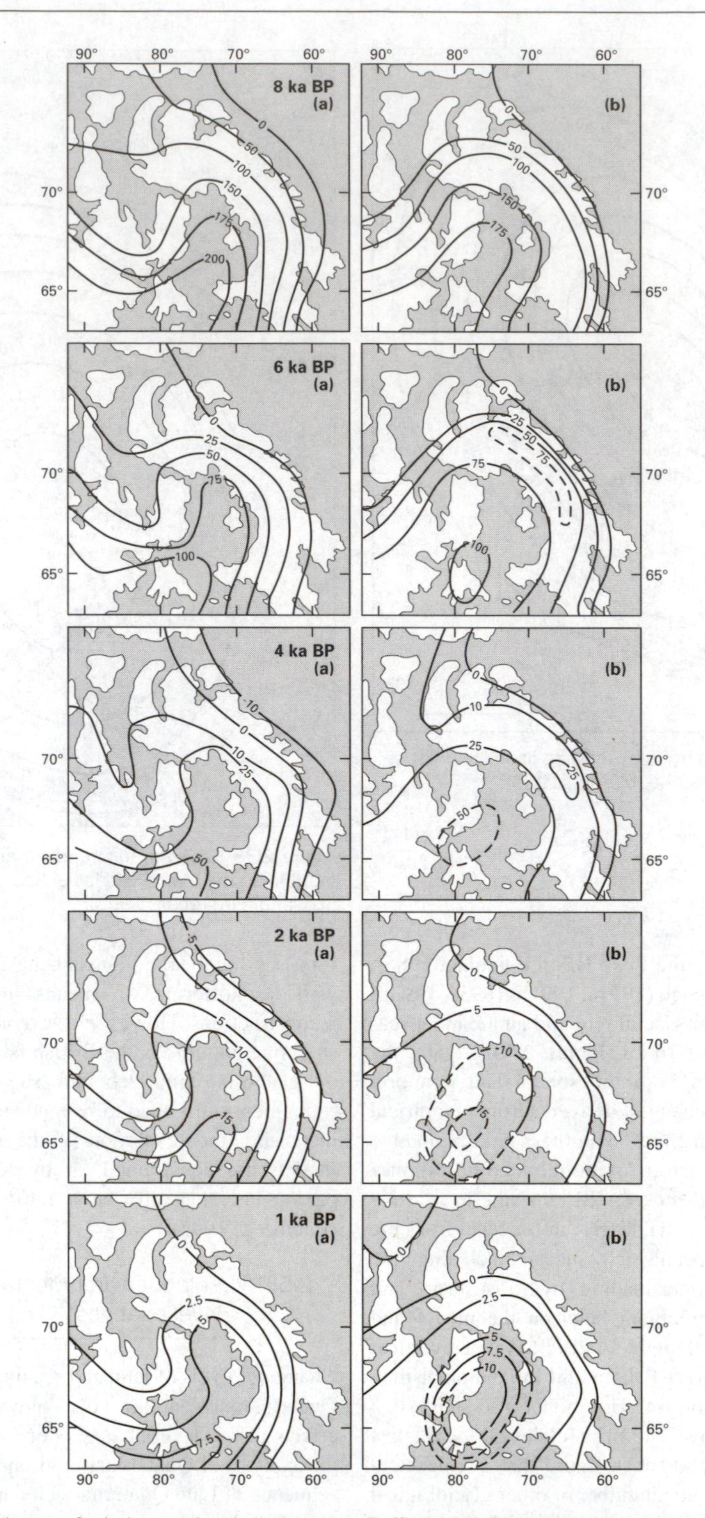

Figure 2.39 Calculated isobases of relative sea-level change near Baffin Island, Canada (a) compared with isobases reconstructed from empirical observations (b). The contours are in metres relative to present sea level (after Quinlan, 1985).

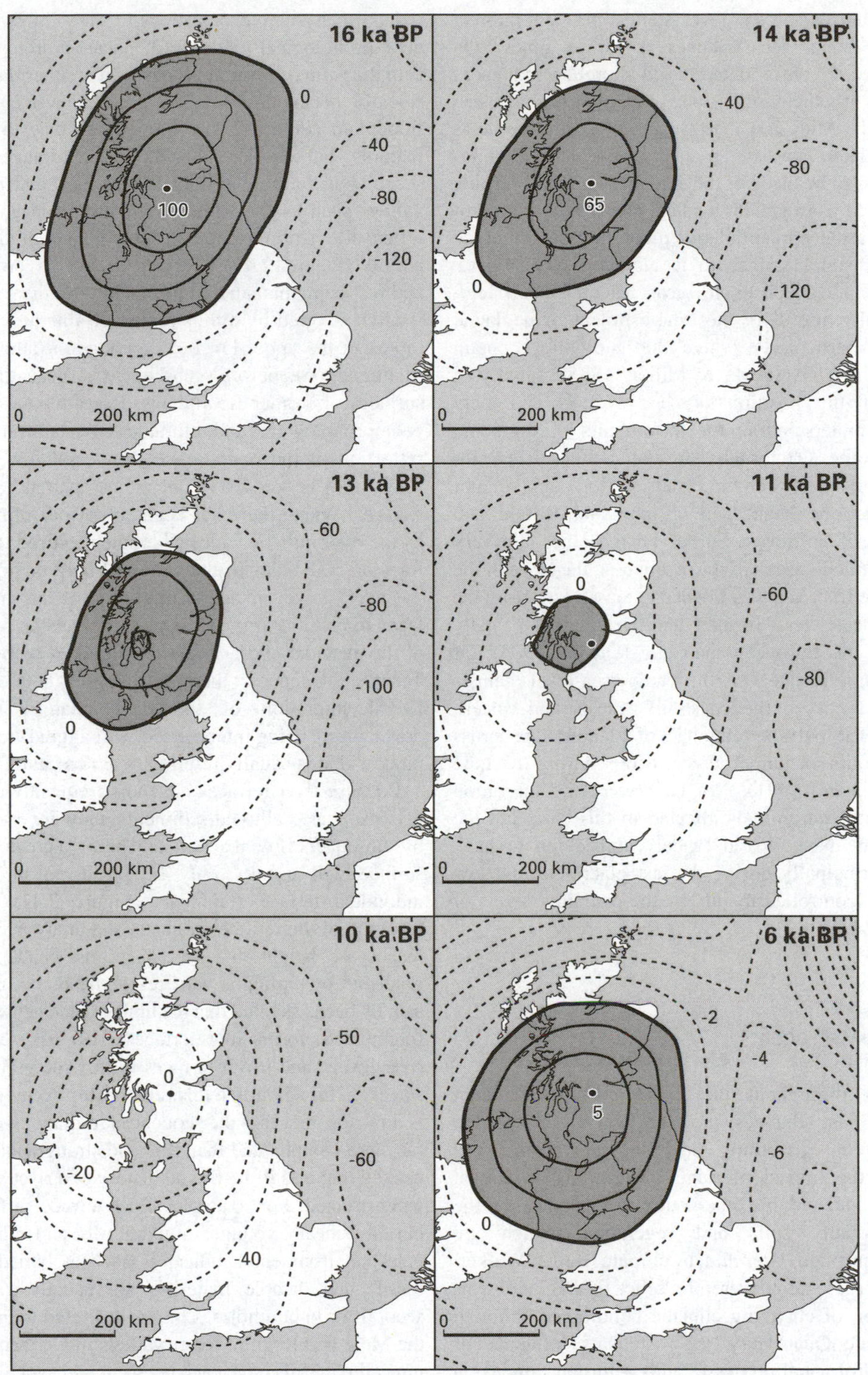

Figure 2.40 The spatial pattern of sea-level change in the British Isles predicted by the isostatic rebound model at selected epochs. The contours represent mean sea level at past epochs relative to present mean sea level. Shaded areas are positive (after Lambeck, 1991b).

while high interglacial sea-level stands provide indirect evidence of global ice volumes at those times. On tectonically active coasts, dated raised shoreline sequences show the rates at which uplift along plate margins has taken place during the Mid- and Late Quaternary. In those areas affected by glacio-isostasy, gradients of tilted shorelines allow estimates to be made of the amount of crustal warping that has resulted from glacial loading, and these data are of particular importance in simulation modelling of former ice sheets and crustal response to ice loading (see above). Raised shoreline sequences in areas affected by glacio-isostasy are also useful in the establishment of deglacial chronologies, particularly where the shorelines contain dating media (e.g. Andrews & Miller, 1985). Finally, a knowledge of the position of sea level is extremely important in understanding the movements of flora and fauna during the Quaternary. In the British Isles, for example, contrasts between the Flandrian flora and fauna of Ireland on the one hand, and of mainland Britain and adjacent parts of northwest Europe on the other, are very largely explicable in terms of the fact that, at the close of the last glacial, the Irish Sea was flooded at least 3 ka before the Straits of Dover were formed and Britain was finally isolated from the European mainland (Devoy, 1985). Of even greater significance were the relative sea-level changes that occurred between the coasts of Alaska and eastern Siberia which led to the development of a land bridge across the Bering Straits on numerous occasions during the Late Quaternary (Fagan, 1991). Not only were the migration routes of plants and animals affected in this case, but the peopling of the New World (widely considered to have taken place principally during the last glacio-eustatic low sea level) was controlled mainly by the changing levels of land and sea.

2.6 River terraces

River valleys throughout the world contain abundant evidence for past changes in river régime. While the dominant factor governing hydrological changes is precipitation, the impact of rainfall variations on fluvial systems is modulated by other other components of the landscape, notably soils and vegetation, which are themselves controlled, *inter alia*, by climate (Bell & Walker, 1992). To a large extent, therefore, relict fluvial landforms are a reflection of changing climatic conditions, although during the Late Quaternary, the increasing influence of people on hydrological processes adds a further dimension to the interpretation of river histories (e.g. Starkel, 1987; Petts *et al.*, 1989; Thornes & Gregory, 1991).

In many river valleys, the most impressive geomorphological evidence of palaeohydrological changes is in the form of **river terraces** which either flank the valley sides or occur alongside the present river channel on the floodplain (Figure 2.41). Sometimes they exist as single features, but on occasions they are arranged in vertical successions forming a flight or 'staircase' and, in major river valleys such as the Rhine or Meuse (Ruegg, 1994), these reflect alluviation over a considerable part of the Quaternary period (Figure 2.42). Terraces may be erosional, with bedrock being planated to form a low-gradient **strath** which is often covered by a thin veneer of alluvium, or they may represent the upper level of aggradation (alluvial sediment accumulation) before subsequent downcutting, i.e. the surfaces of former floodplains. River terraces occur in all geomorphological and climatic environments and are a reflection of the operation of universal fluvial processes. They may be preserved as either 'paired' or 'unpaired' terraces. Where there has been an episode of rapid incision by the river into the valley floor, then '**paired terraces**' may form on both sides of the valley. Where the river begins to meander, however, lateral migration of the stream channel leads to erosion of the floodplain gravels on the outer edge of the meander and a single or **unpaired terrace** develops. Terrace sequences, therefore, reflect both incision and lateral migration of the river channel, episodes of downcutting being interspersed with aggradational phases, a process that is usually referred to as '**cut and fill**'.

Because river terraces are most frequently developed on unconsolidated alluvial sediments, they are easily destroyed by subsequent fluvial action, and hence a previous floodplain surface will usually only be preserved in the form of individual terrace fragments (Figure 2.42). Instrumental levelling of the terrace remnants and analysis of the data by means of height–distance diagrams enable downvalley gradients of particular terraces to be reconstructed. It has usually been assumed that the highest (and generally the most fragmented) forms in a terrace series represent the oldest river levels, and lower terraces reflect successively younger stages. In broad outline this relationship seems to hold, but it is now apparent that the sequences in many river valleys are far more complicated than this and stratigraphic evidence is usually required if the fluvial history of a river valley is to be reconstructed. For example, older terrace surfaces may be buried beneath younger alluvial fills and other important palaeoenvironmental indicators such as buried soils, peats, fossils and datable materials are usually found only in sections or in boreholes. This is illustrated in Figure 2.43. In the Main and Regnitz River valleys in Germany, a series of nine alluvial fills have each been cut and aggraded against an earlier alluvial unit. The upper surfaces of the units (representing a staircase of terrace surfaces in a descending

Figure 2.41 A suite of terraces formed by river incision into glaciofluvial gravels, Roy Valley, near Fort William, Scotland (photo: John Lowe).

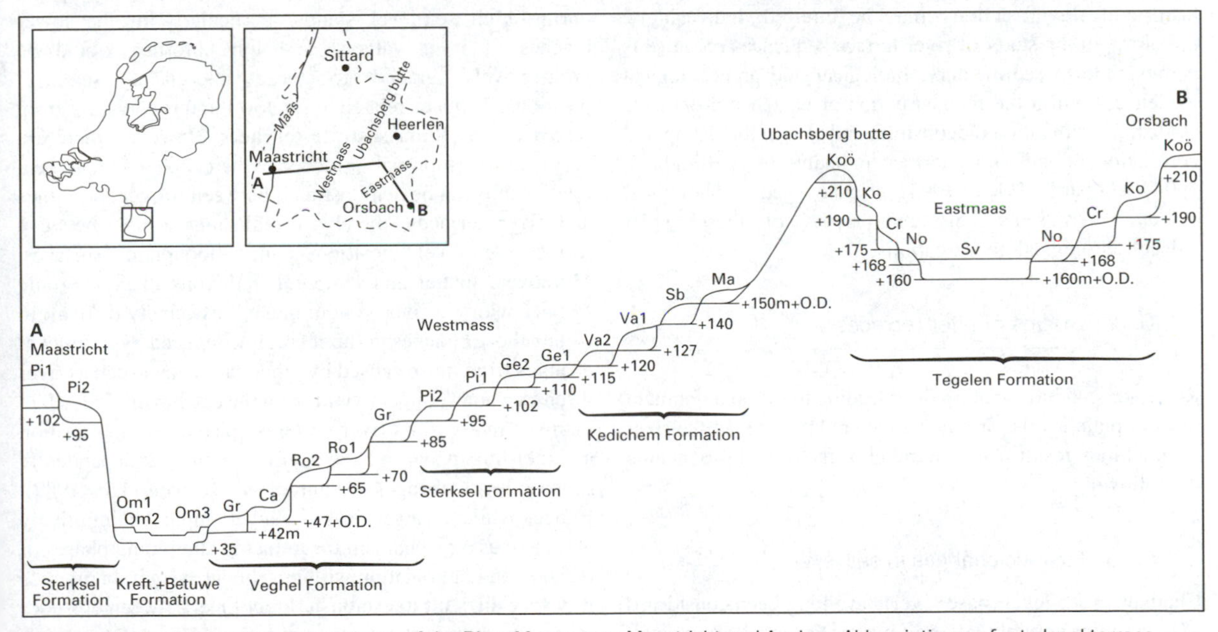

Figure 2.42 Quaternary terrace sequence of the River Meuse near Maastricht and Aachen. Abbreviations refer to local terrace names. Length of profile is 29 km (after Ruegg, 1994).

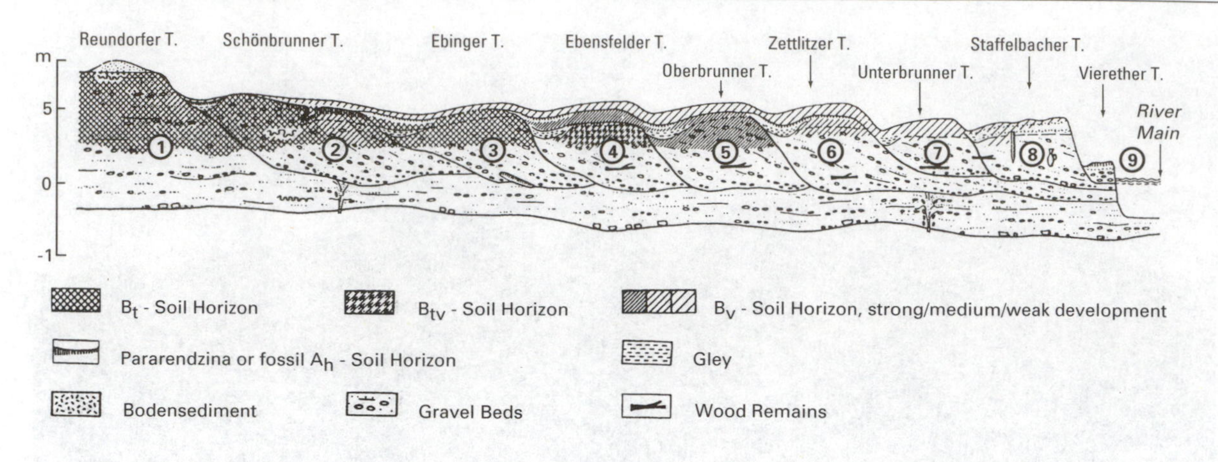

1 - Würm-Pleniglacial, **2** - post-Pleniglacial, pre-Allerød, **3** - Younger Dryas, **4** - Atlantic: 5850-4300 BC, or 7000-5400 y BP$^+$,

5 - Subboreal: 3250-1750 BC, or 4500-3500 y BP$^+$, **6** - Subatlantic: 200 BC - 250 AD, **7** - 550-850 AD, (4-7 = dendro ages; $^+$ = ^{14}C age),

8 - 15th - 17th century, **9** - early 19th century, T. = terrace.

Figure 2.43 Schematic diagram showing a section through the Würm and Holocene valley fill of the upper Main and Regnitz Rivers, Germany. The cross-profile shows how cut and fill processes have led to the truncation of older deposits and aggradation by progressively younger sedimentary units. The complex fluvial history contained within this sequence could not easily be reconstructed on the basis of geomorphological evidence alone (after Schirmer, 1988).

series towards the River Main) have been covered by a veneer of younger sediment which masks the underlying topography (Schirmer, 1988). Where exposures are not available and river terraces are recognised purely on the basis of geomorphological characteristics, a greatly oversimplified history of fluvial activity may be inferred. Increasingly, therefore, in the study of river terrace sequences recourse is being made to sedimentary, biological and archaeological evidence, both in the reconstruction of palaeoenvironments associated with alluvial deposition and also in the dating and correlation of individual terrace remnants (e.g. Bridgland, 1994; Gibbard, 1985, 1994). Hence, geomorphological evidence represents only one aspect of these wider palaeohydrological investigations.

2.6.1 Origins of river terraces

River incision into a valley floor leading to the abandonment of floodplain levels, or renewed aggradation along the river channel may result from a number of factors. These include the following

2.6.1.1 Eustatic changes in sea level

Changes in sea level (base level) have long been considered to be a major variable governing river terrace development. Eustatic sea levels appear to have fluctuated through a range of perhaps 150 m during the glacial and interglacial cycles of the Quaternary (section 2.5.2). The traditional model envisages a situation in which active downcutting and river incision would be expected to have accompanied falling sea levels, while increased aggradation might be anticipated during high sea-level stands, particularly in the lower reaches of river valleys, and long profiles would be progressively graded to these interglacial sea-level positions. Terraces related to sea level in this way have been referred to as **thalassostatic terraces**. However, there are very few empirical studies where an unequivocal relationship can be demonstrated between former shorelines and river terraces, thereby establishing a link between former sea-level positions and floodplain surfaces. Moreover, spatial and temporal variations in geomorphic activity within a river system make it extremely difficult to isolate those changes in fluvial régime induced by base-level changes from those caused by other parameters such as river discharge and sediment yield from the catchment. Indeed, in many of the world's river systems episodes of aggradation and incision appear to have operated entirely independently of base-level changes (Schumm & Brackenridge, 1987). Hence, while changes in base level might be intuitively attractive as a mechanism for inducing alternating phases of incision and aggradation within a fluvial system, in practice it is very difficult to establish direct links between episodes of terrace formation and sea-level changes (Dawson & Gardiner, 1987).

2.6.1.2 Climatic change

The influence of climate on river valley evolution has long been recognised, and is still regarded as a major influence on river terrace development in many parts of the world (Knox, 1984; Bohncke & Vandenberghe, 1991; Starkel, 1991). The relationship between river behaviour and climate depends on the nature of climatic change and on the effects of such changes on discharge and sediment load. In semi-arid and arid regions, river incision was common during pluvial episodes (section 2.7.1), when discharge was more constant and sediment yield was reduced by an increase in vegetation cover around drainage basin catchments. Phases of aggradation, on the other hand, were more characteristic of arid periods, when discharge was reduced and sediment yields increased during periods of more restricted vegetation cover. In temperate regions increased aggradation occurred during colder periods when rivers were more heavily debris-laden. Present arctic nival discharge régimes are marked by a major flood event during the spring resulting from the rapid melt of winter snow (Bryant, 1983). Hence, areas affected by periglacial conditions in the past would have experienced seasonal flooding due to snowmelt, and increases in sediment yield due to combination of a relatively sparse vegetation cover and to ground disturbance by periglacial processes. During warmer periods sediment yield would have been reduced and discharge variations less marked. Terraces related primarily to climatic changes have been referred to as '**climatic terraces**', but clearly some geographical variation in the relationship of terrace formation to climatic factors is to be expected. Moreover, it must be stressed that climate is only one of a number of factors affecting discharge, sediment load and sedimentation (Starkel, 1991).

2.6.1.3 Effects of glaciation

In areas formerly covered by glacier ice, deglaciation was typically associated with the accumulation of glaciofluvial sediments (mainly outwash gravels), followed by alternating periods of incision and aggradation or channel stability, which resulted in the formation of terrace sequences in most river valleys (Maizels, 1987; Maizels & Aitken, 1991). These are usually described as **outwash terraces**, and are frequently preserved as complex terrace systems comprising several surface levels along the margins of upland river valleys (Figure 2.41). Throughout the mid-latitudes, moreover, many major drainage basins were not directly affected by glacier ice, but lay sufficiently close to the ice margin to be indirectly influenced by glaciofluvial activity. Consequently, the geomorphology of these proglacial drainage basins may have been affected by the increased discharge, and also by the increased sediment load reflecting the large quantities of debris released from the ice. During cold stages, therefore, considerable aggradation might be expected in the upper reaches of such rivers, and terraces would remain as evidence of these aggraded surfaces following river incision in the subsequent warmer phase (Starkel, 1987, 1991). However, while it is likely that glacier activity in peripheral parts of drainage basins would find some local expression in alluvial aggradation, the role of proglacial aggradation in the development of river terrace sequences on the regional scale has almost certainly been overestimated (Clayton, 1977).

2.6.1.4 Tectonic changes

In general terms, tectonic uplift leads to rejuvenation of rivers and therefore to accelerated incision. The effects are best seen in the active orogenic areas of, for example, New Guinea and New Zealand, where relatively rapid uplift has produced sequences of widely separated terrace levels in many river valleys (e.g. Pullar *et al.*, 1967). Indeed, in the Tien Shan region of China, river terraces themselves have been warped by anticlinal movements associated with uplift during the Quaternary (Molnar *et al.*, 1994). Those areas affected by glacio-isostasy would also have experienced gradient changes as a result of differential land upift. Hence, in southern Finland, rapid land uplift has led to the development of abrupt local changes in gradient along the major southward-flowing rivers, each marked by a distinctive 'knickpoint' (Mansikkaniemi, 1991). Moreover, despite the abundance of glaciofluvial and related materials released by the melting ice sheet, few terraces have formed or been preserved in Finnish river valleys because of downcutting resulting from rapid glacio-isostatic recovery (Koutaniemi, 1991; Mansikkaniemi, 1991). In some large river valleys, only part of the basin may have been affected by uplift or tectonic downwarping, so that the original valley long-profiles indicated by contemporaneous terrace fragments have become warped. This particular factor complicates the investigation of terraces of the River Rhine, since downwarping characterises that part of the drainage basin adjacent to the North Sea, whereas significant uplift has occurred in the upper parts of the drainage basin (Brunnacker, 1975; Ruegg, 1994). The same problem is encountered in the Danube system where neotectonic processes associated with the continuing rise of the Carpathian Mountain system have led to deep incision in the middle/upper reaches of the river, while subsidence around the western margins of the Black Sea has resulted in sinking of the low terraces formed on the eastern Romanian Plain (Ghenea & Mihailescu, 1991). Clearly crustal warping must be taken into consideration when interpreting height–

distance diagrams of river terrace sequences, and when correlations are being made between terrace fragments on the basis of geomorphological evidence.

2.6.1.5 Effects of human activity

It is now apparent that in many river basins, hydrological régimes, sediment yield, and consequent changes in floodplain levels, have been affected by anthropogenic activity. Throughout the Mediterranean region, for example, archaeological evidence suggests that major alluvial fills post-date the main phase of Roman agricultural development (Butzer, 1980), and it now appears that many terraces formerly interpreted as 'climatic' can be related instead to human activity (Wagstaff, 1981; van Andel *et al.*, 1990). Forest clearance and cultivation practices remove the natural vegetation cover; this, in turn, reduces the water retentive capacity of the soil and leads to increased runoff. Overgrazing and overcultivation also result in substantially increased erosion and therefore higher stream loads. In the Po valley in northern Italy, for example, it has been estimated that 60 per cent of the plain was cleared and cultivated during the Roman period, but the field systems have been buried beneath great spreads of alluvial deposits. Large areas of cultivable land were subsequently lost because of increased alluviation that was itself a consequence of human interference with the river régime and catchment (Cremaschi *et al.*, 1994). Similar episodes of alluviation associated with prehistoric woodland clearance, and with prehistoric and historic farming activity, have been noted in upland Britain (Harvey & Renwick, 1987; Macklin *et al.*, 1991), and throughout northern and central Europe (Schirmer, 1988; Starkel, 1988). It is apparent, therefore, that over much of the northern temperate zone, Late Holocene alluviation and associated terrace development is as likely to be the product of human activity as of natural processes.

2.6.2 River terraces and palaeoenvironments

River terraces and their underlying sediments have long attracted the interest of those concerned with Quaternary environments. Not only do they provide evidence of former river régimes which, in turn, can provide insights into both environmental change and aspects of human history, but they are often developed upon sediments rich in floral and faunal remains. Fossils are preserved, for example, in abandoned meander scrolls, in oxbow lakes or in backswamp deposits, and from these the nature of the environment under which the terraces evolved can often be established. Indeed, in both western Europe and North America, some of the most important fossiliferous sites of both cold and warm phases are found in deposits associated with river terraces. In Britain, the terraces of the major rivers of central and southern England, particularly the Severn–Avon, the Trent and the Thames (see below), have been resolved into full glacial, interstadial and interglacial sequences partly on the basis of contained fossil assemblages (e.g. Maddy *et al.*, 1991a; Bridgland, 1994; Gibbard, 1994). Moreover, a combination of geomorphological, sedimentological, biological and geochronological evidence has allowed these individual terrace systems to be correlated throughout lowland England (Jones & Keen, 1993; see Table 2.2 below). In providing links between sequences in different areas, therefore, river terraces become valuable datum surfaces and offer a potentially useful framework for Quaternary chronology.

2.6.3 The terraces of the River Thames

One of the most intensively studied river terrace sequences is that of the River Thames in southern England. The Thames basin contains a long record of fluvial activity and river terrace genesis that reflects the influence of a number of the factors discussed above. Moreover, the Thames terrace system also exemplifies many of the difficulties that are encountered in the analysis and interpretation of river terrace sequences.

Research on the terraces of the River Thames has a long history extending back well over a hundred years (e.g. Whitaker, 1889; Prestwich, 1890). Work during the first half of the present century focused particularly on the geomorphology of the Thames Valley, and on the differentiation between different terrace levels (Bromehead, 1925; King & Oakley, 1936; Wooldridge, 1938; Hare, 1947). The results of this research were synthesised into the terrace sequence shown in Figure 2.44 (Wooldridge & Linton, 1955). In recent years, however, a new approach to the history of the River Thames, involving not only geomorphological mapping but also detailed lithostratigraphic and biostratigraphic investigations of the terrace gravels and their included organic deposits, has provided new insights into the temporal and spatial relationships of the Thames terrace sequence (Gibbard, 1985, 1994; Bridgland, 1994; Bridgland *et al.*, 1995). This new work has demonstrated that the terrace sequence and associated sediment stratigraphy of the Thames Valley are considerably more complicated than envisaged even twenty years ago (e.g. Clayton, 1977). Nevertheless, many of the terrace names introduced by Wooldridge & Linton have been retained, as these refer to surfaces that have been found consistently to represent important stratigraphic marker

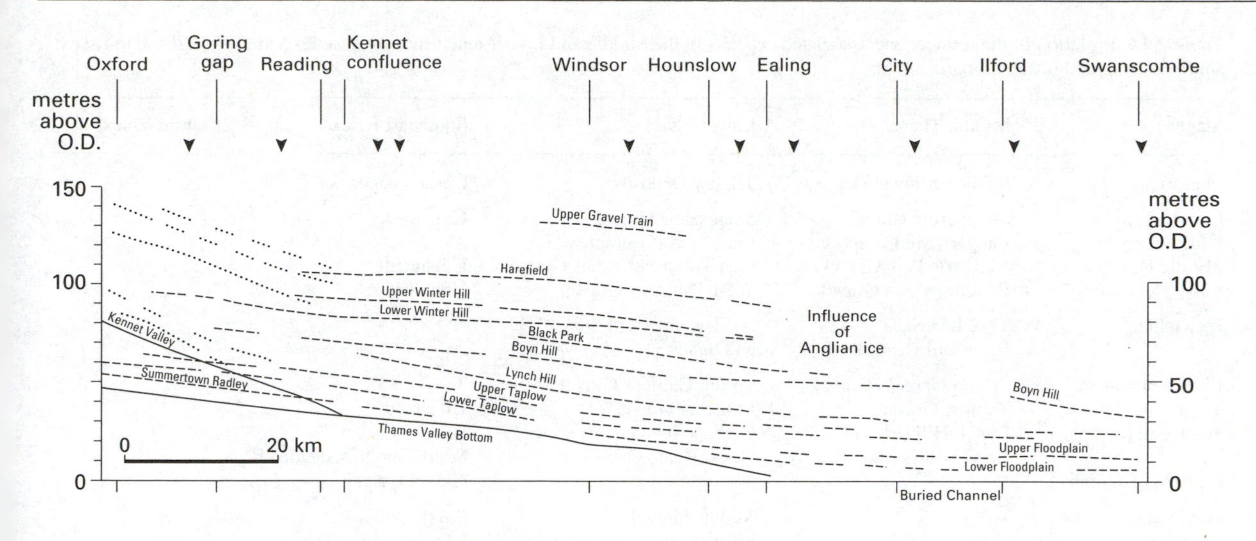

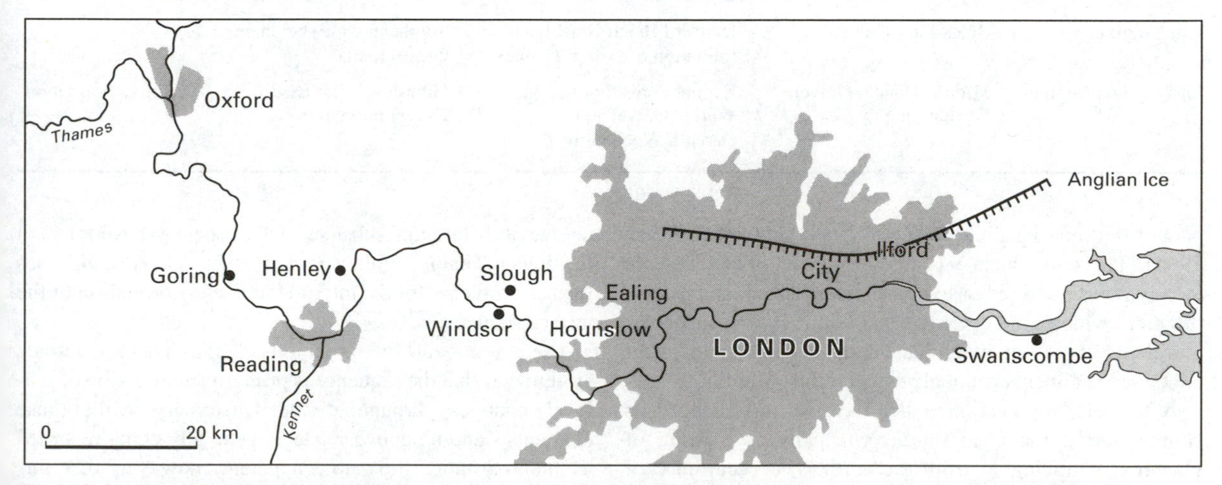

Figure 2.44 Reconstructed long-profiles of the terraces of the River Thames (dashed lines) and one of its tributaries, the River Kennet (dotted lines), along an approximate axis from Oxford to Swanscombe in the lower reaches of the Thames (after various sources).

horizons throughout the Thames Valley, although the terrace fragments are now named after the sedimentary units of which they are predominantly composed (e.g. Boyn Hill *Gravel*; Lynch Hill *Gravel;* see Table 2.1) rather than after the particular terrace levels (i.e. Boyn Hill *Terrace*; Lynch Hill *Terrace*). In accordance with modern lithostratigraphic procedures (section 6.2.3.1), these gravel units have, in turn, been classified as *members* (e.g. Gibbard, 1995) or *formations* (e.g. Bridgland, 1995). Hence, in Table 2.2, the Boyn Hill and Lynch Hill Gravels are referred to as the Boyn Hill *Formation* and the Lynch Hill *Formation*.

Despite the fact that much has been learned about the history of the Thames from these recent studies, a definitive history of the evolution of the Thames remains to be established. One reason for this is the nature of the evidence itself and, in particular, the fact that the period of Quaternary time represented by preserved deposits may be relatively short. Evidence from the Devensian part of the sequence, for example, suggests that the deposits reflect accumulation within a time span of no more than 30–35 ka, i.e. around 30 per cent of the timespan of the Devensian Stage. In reality, much less time is actually represented in the sedimentary record at any given locality, for there are numerous non-sequences or minor unconformities within sequences, arising from the nature of the depositional process. Downcutting between aggradations must clearly account for a substantial

Table 2.1 Correlation of the terraces and associated deposits in the Middle and Lower reaches of the Thames Valley with those in Essex, southeast England (after Gibbard, 1995).

Stage	Middle Thames	Lower Thames	Southeast Essex	Central Essex
Flandrian	Staines Alluvial Deposits	Tilbury Deposits	Coastal deposits	—
Late Devensian	Shepperton Gravel	Shepperton Gravel	Unnamed	—
Late Devensian	Langley Silt Complex	Langley Silt Complex	?	—
Middle Devensian	Kempton Park Gravel	East Tilbury Marshes Gravel	Unnamed	—
Early Devensian	Reading Town Gravel	West Thurrock Gravel	Unnamed	
Ipswichian	Trafalgar Square/ Brentford Deposits	Aveley, Crayford Silts and Sands	?	—
Late Wolstonian	Spring Gardens Gravel	Spring Gardens Gravel	?	—
	Taplow Gravel	Mucking Gravel	Unnamed	—
Wolstonian	Lynch Hill Gravel	Corbets Tey Gravel	Barling Gravel	—
	Boyn Hill Gravel	Orsett Heath Gravel	Southchurch–Asheldham	
Early Wolstonian			Gravel (part)	
Hoxnian	—	Middle Gravel	Southeast Essex	—
		Swanscombe Lower Loam	Channel Deposits	—
		Lower Gravel		
Late Anglian	Black Park Gravel	Dartford Heath Gravel (initiation of Lower Thames)	Southchurch–Asheldham Gravel (part)	—
Early to pre-Anglian	Middle Thames Gravel Formation	Epping Forest Formation, Warley Gravel and Darenth Wood Gravel	High-level East Essex Gravel Formation	Kesgrove Formation

part of the missing time (Gibbard, 1989; Gibbard & Allen, 1994). The consequence is that a number of terraces are locally poorly defined or extremely fragmented, and despite the more widespread employment of stratigraphic evidence, correlation between individual terrace remnants and their contained sediments is not always straightforward.

A further complication is that the river has changed its course during the Quaternary, principally as a result of glacier ice moving in from the north. The occurrence of buried valleys, spreads of fluvial gravels, and the distribution and gradient of certain terraces show that, prior to the Anglian Glaciation (generally regarded as the equivalent of OI Stage 12 in the marine sequence), the Thames flowed northeastwards immediately downstream from Reading, through the Vale of St Albans, to cross the present-day coastline near Colchester in Essex (Figure 2.45). There are also indications that minor tectonic deformation has occurred in and around the Thames basin during the Quaternary, including downwarping along the margins of the North Sea basin, glacio-isostatic deformation involving both crustal depression beneath the Anglian ice sheet and also forebulging beyond its margin (see above), and cambering and solution of chalk bedrock (Gibbard, 1985; Whiteman, 1992). While these effects are perhaps likely to have had less of an impact than was at one time considered (Bridgland, 1988; Bridgland *et al.*, 1993), some degree of differential subsidence does seem to have affected the lower Thames Valley (e.g. Devoy, 1979), and poses further problems for the interpretation and correlation of the terrace sequence.

The traditional interpretation of the Thames terrace system was that the sequence is primarily a reflection of sea-level changes throughout the Quaternary, with those elements standing above sea level generally being regarded as thalassostatic. It is now apparent, however, that this interpretation is no longer appropriate. For example, biostratigraphic evidence suggests that the majority of the terrace deposits aggraded under cold conditions (Briggs & Gilbertson, 1980; Gibbard, 1985, 1994), when sea levels were well below those of the present. In the upper parts of the drainage system, in particular, most terrace forms appear to relate to increased aggradation during deglaciation, when meltwaters were feeding large quantities of debris into the Thames system from the northwest (Whiteman & Rose, 1992).

Dating of the Thames terraces has proved to be particularly problematical. The traditional approach has been to develop a chronology of terrace development on the basis of elevation of the terrace fragments, stratigraphic position, and contained biological and archaeological evidence (Gibbard, 1985; Table 2.1). More recently numerical geochronological techniques have been

Table 2.2 Correlation of the terraces and associated deposits in the Thames Valley with those in other lowland valleys of England, based on amino-acid geochronological evidence (after Bowen *et al.*, 1995). The left-hand columns show amino-acid ratios obtained from Mollusca contained within the terrace deposits (AAR) and oxygen isotope stages from the deep-ocean sequence ($\delta^{18}O$).

AAR	$\delta^{18}O$	Lower Thames	Middle Thames	Upper Thames	Avon Valley	Severn Valley	Nene & Ouse
0.09	5e		Trafalgar Square	Cassington	New Inn Bed		
	6	Mucking Formation	Kempton Park Formation Taplow Formation	Stanton Harcourt Gravel Member	Ailstone Member	Ridgacre Form Kidderminster Station Member	Terrace No. 2
0.17	7	Aveley, Ilford (Uphall), Bakers Hole, Crayford, Purfleet	Redlands (Reading)	Stanton Harcourt	Ailstone Bed Froghall Strensham	Strensham	Stoke Goldington March Gravels at Somersham
	8	Mucking Formation Corbets Tay Formation	Taplow Formation Lynch Hill Formation	Summetown-Radley L.Gr Wolvercote Formation	Pershore Member	Bushley Green Member	Terrace No. 3
0.25	9	Purfleet, Ilford (Cauliflower), Belhus Park, Little Thurrock		Wolvercote Channel		Bushley Green Bed	Biddenham Woodston Beds
	10	Corbets Tey Formation Orsett Heath Formation	Boyn Hill Formation	Wolvercote Formation Hanborough Formation		Spring Hill Member	
0.3	11	Swanscombe, Clacton					
	12	Orsett Heath Formation Hornchurch Formation	Boyn Hill Formation Black Park Formation Winter Hill Formation Westmill Formation	Hanborough Formation Freeland Formation Moreton Formation Freeland Formation	Wolston Formation	Nurseries Formation Woolridge Member	

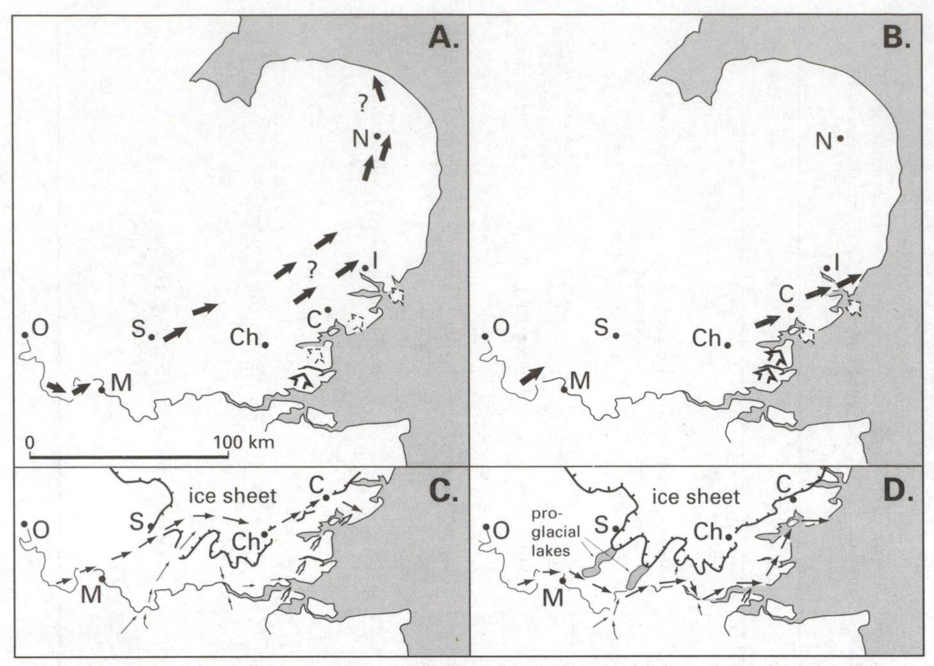

Figure 2.45 The evolution of the Thames drainage system up to, and including, the diversion of the river by Anglian ice. A. Westland Green gravel. B. Waldringfield gravel. C. Immediately prior to the arrival of Anglian ice. D. Anglian glaciation and diverted course of the river. Place names: C – Colchester; Ch – Chelmsford; I – Ipswich; M – Maidenhead; N – Norwich; O – Oxford; S – St Albans (after Bridgland, 1988).

employed, most notably amino-acid geochronology (Bowen *et al*., 1989, 1995; section 5.6.1; Table 2.2), but also radiocarbon (Gibbard *et al*., 1982; Briggs *et al*., 1985) and thermoluminescence dating (Gibbard *et al*., 1987). In some cases, there is agreement between these different approaches. For example, in both Tables 2.1 and 2.2, the Winter Hill and Black Park Gravels of the Middle Thames region are classified as Anglian (~ OI Stage 12) in age, while the Trafalgar Square deposits are assigned to the Ipswichian Interglacial (~ OI Substage 5e). In other cases, however, there are conflicts. For instance, in the lower Thames region, the Aveley Silts and Sands (Aveley, Ilford, Crayford) have usually been considered, on palaeobotanical grounds, to be of Ipswichian Interglacial age, whereas on the basis of mammalian and amino-acid geochronological data, the deposits have been assigned to the penultimate (~ OI Stage 7) warm stage (Bridgland, 1994; Table 2.2). These differences of interpretation reflect two problems in particular. First, it is apparent that successive glacial/interglacial oscillations produced similar environments at different times, and it becomes increasingly difficult to establish correlations between, and relative ages of, terrace deposits on the basis of sedimentological and biological evidence alone (Jones & Keen, 1993). Second, the deep-ocean oxygen isotope sequence indicates that warm and cold stages additional to those shown in Table 2.1 should be represented in the Thames terrace sequence, and these are perhaps reflected in the amino-acid geochronological correlations shown in Table 2.2. It should be noted, however, that limitations of the amino-acid geochronological method (section 5.6.1.2) mean that this approach to the dating of the Thames terraces is also not without its critics (Gibbard, 1995).

Despite over a century of research, therefore, many aspects of the Thames terraces remain enigmatic, and a secure chronology remains frustratingly elusive. Considerable progress has been made in recent years, however, with the recognition that a proper understanding of the history of the River Thames can only be achieved through interdisciplinary and multidisciplinary investigations involving, in addition to the geomorphological relationships between terrace fragments, stratigraphic, biological and archaeological evidence supported throughout by an independent dating framework. Such integrated approaches are leading to significant advances in our understanding of Quaternary fluvial landforms and deposits, not only in the Thames basin but in river systems throughout the world.

2.7 Quaternary landforms in low latitudes

There is abundant geomorphological and lithological evidence to show that major climatic changes have affected the tropical, subtropical and warm temperate regions of the world during the Quaternary. The periodic expansions and contractions of the great ice sheets in the high and mid-latitudes were accompanied by shifts in the main climatic zones of the low latitudes, and these produced marked spatial and temporal variations in regional rainfall values, seasonal distribution of rainfall, annual temperatures, and wind directions and strengths. Although these climatic changes were experienced throughout the tropics and subtropics, their effects were most pronounced in the desert and savanna margins of, for example, the Sahara, northwest India and parts of Australia (Goudie, 1992; Williams *et al.*, 1993a). In many of these areas, fossil landforms and

A.

Figure 2.46 A. Abandoned shorelines (platforms etched into the bedrock) of ancient Lake Bonneville, Utah, USA (photo: Andrew Goudie).

B.

B. Lake Chew Bahir in the southern rift valley of Ethiopia. The prominent abandoned spit (centre) was formed by wave action when the lake was deeper than today. The cliffs and former islands seen in the background are mantled by outcrops of carbonates formed by freshwater algae (stromatolites). Desiccation cracks can be seen on the former lake bed to left of person in photograph (photo: Andrew Goudie).

deposits are preserved that can be used to infer climatic changes during the Late Quaternary. Of particular importance are lacustrine features that provide evidence for wetter conditions at times in the past, and sand-dune complexes indicating former episodes of increased aridity.

2.7.1 Pluvial lakes

In many arid and semi-arid regions of the world, there are indications that **saline lakes** and ephemeral water bodies (**playa lakes**) have experienced phases of expansion and contraction during the Quaternary. Marked fluctuations have also been detected in the former levels of a number of present-day lakes in tropical and subtropical regions. These lakes and fossil lake features are common in those areas where geology has produced large basins and depressions in which drainage is predominantly internal and where outflow has been minimal. Lakes that show evidence of expansion and contraction unrelated to worldwide changes in base level have been termed **pluvial lakes**, as their high-water stages have been attributed to wetter climatic phases known as **pluvials** or **pluvial episodes**. Low water levels or periods of complete desiccation were, in turn, assumed to reflect **interpluvials** or **interpluvial episodes**. Because of the very close relationships that appear to exist between 'pluvial' lakes and precipitation and/or evaporation around the lake catchments, fluctuations in water level in closed-basin lakes are potentially useful indicators of continental palaeoclimates during the Late Quaternary.

Geomorphological evidence for the existence of pluvial lakes and for oscillations in water level includes abandoned clifflines, shorelines (Figure 2.46), beaches, bars and deltas, as well as abandoned watercourses that acted as overflows at times of high lake-level. These features can be mapped in the field (e.g. Mifflin & Wheat, 1979) and, where dating is possible, they enable the histories of pluvial lakes to be established (e.g. Oviatt, 1990). Although in the majority of cases former lake shorelines are horizontal features, care is needed in their correlation, for the differential effects of hydro-isostatic loading and unloading can, in certain situations, lead to shorelines of a similar age being found at different altitudes. In the Great Basin of Utah, for example, the weight of the water in pluvial Lake Bonneville caused isostatic depression of the crust to the extent that, following disappearance of the lake around 10 ka BP, the once level shoreline has rebounded isostatically so that shorelines on islands near the middle of the former lake are now *c.* 65 m above the elevation of contemporaneous shorelines near the lake margin (Smith & Street-Perrott, 1983). In heavily vegetated areas and regions of difficult terrain, increasing use is being made of remote sensing techniques for the identification and mapping of lake shorelines. Lake shoreline features can be readily identified on Landsat Thematic Mapper images where vegetational contrasts around abandoned shorelines show up well on bands 6 and 7 of the electromagnetic spectrum (Ebert & Hitchcock, 1978).

Strandline evidence is of particular value in establishing the former extent of pluvial lakes, some of which occupied considerable areas. In the southwest United States, for example, where one of the greatest concentrations of pluvial lake features in the world is to be found (Figure 2.47), Lake Bonneville, an ancestor of the Great Salt Lake, had a surface

Figure 2.47 The major pluvial lakes of the southwestern United States. Existing water bodies are solid, and stippling indicates the maximum exent of lakes during the Late Wisconsinan (after Spaulding *et al.*, 1983).

area at its maximum of 51 300 km^2 (the present-day Great Salt Lake is *c.* 4000 km^2) and a volume of *c.* 9500 km^3 and was over 370 m deep. At its maximum around 13 ka BP, Lake Lahontan in Nevada covered an area of 22 300 km^2

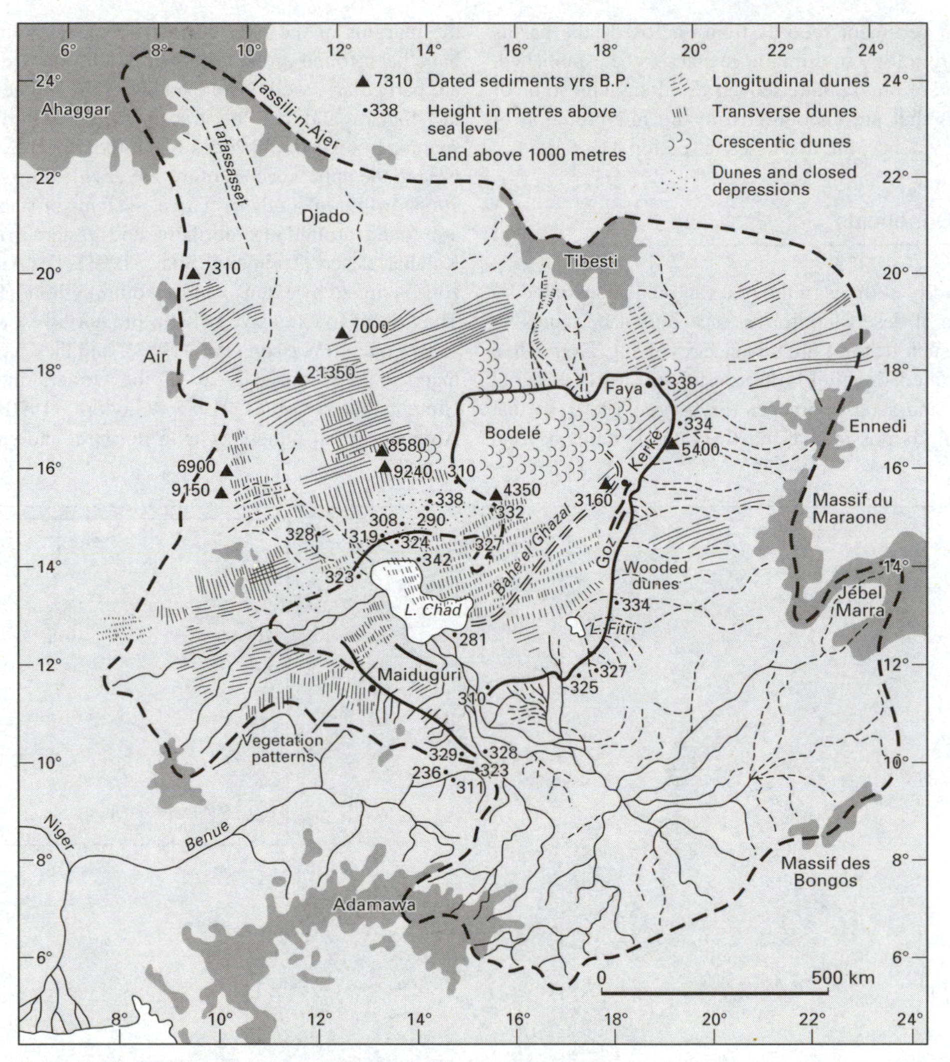

Figure 2.48 The Chad Basin, west Africa, showing the dune systems and the shoreline of 'Mega-Chad' (bold dashed line) at about 320 m (after Grove & Warren, 1968).

14.4 times larger than that of today (Benson & Thompson, 1987). In North Africa, Lake Chad (Figure 2.48) extended over 300 000 km^2 at its maximum during the Late Quaternary (the present area of the lake fluctuates between 10 000 and 25 000 km^2), yet was relatively shallow, being no more than 50 m deep (Grove & Warren, 1968). The most extensive former lake so far recorded, however, was the Aral and Caspian Sea systems, which may have reached maximum dimensions during the early part of the last cold stage (Chepalyga, 1984). This enormous water body, which covered an area in excess of 1.1 million km^2, stood more than 75 m above the level of the present Caspian Sea, and extended 1300 km up the Volga River from its present mouth (Goudie, 1992). The lake was pluvial only in part, however, for it was fed by glacial meltwaters via the Volga, Ural and Oxus drainage systems.

Abandoned lake shorelines in the low latitude regions of the world therefore provide clear evidence of both the former extent of pluvial lakes, and of fluctuations in lake-water level. Analysis of these data offers valuable insights into changes in atmospheric moisture balance during the Late Quaternary (e.g. Street-Perrott *et al.*, 1989; Hostetler *et al.*, 1994). More detailed climatic reconstructions can be made on the basis of stratigraphic evidence, however, while the dating of sediment records from enclosed lake basins enables a chronology of climatic changes to be established. These aspects of pluvial lake sequences, along with some of the problems that are encountered in the interpretation of this type of evidence, are discussed in section 3.6.

2.7.2 Dunefields

Increased aridity at times in the past can be demonstrated by the existence of desert landforms, especially sand dunes, in areas where such features are no longer evolving. Even when heavily vegetated, dunefields often stand out clearly on aerial photographs and satellite images (Fryberger, 1980), so that rapid mapping is possible of inactive dune systems beyond the margins of the present desert regions. Stabilised dunes have been found around the margins of many deserts. They are particularly well developed along the southern fringes of the Sahara and can be traced in a discontinuous band spanning some five degrees of latitude from Senegal in the west to the upper reaches of the River Nile (Grove & Warren, 1968), while in southern Africa, systems of stabilised dunes are found around the northern and eastern fringes of the Kalahari desert (Thomas & Shaw, 1991). The Australian arid zone is ringed by extensive fossil dune systems (Figure 2.49). They occur in extensive belts in the northwest of the Indian subcontinent (Wasson *et al.*, 1983), and they have also been mapped across large areas of the western interior of the United States (Muhs, 1985; Gaylord, 1990). In South America, fossil dunes occur in northern and eastern Brazil

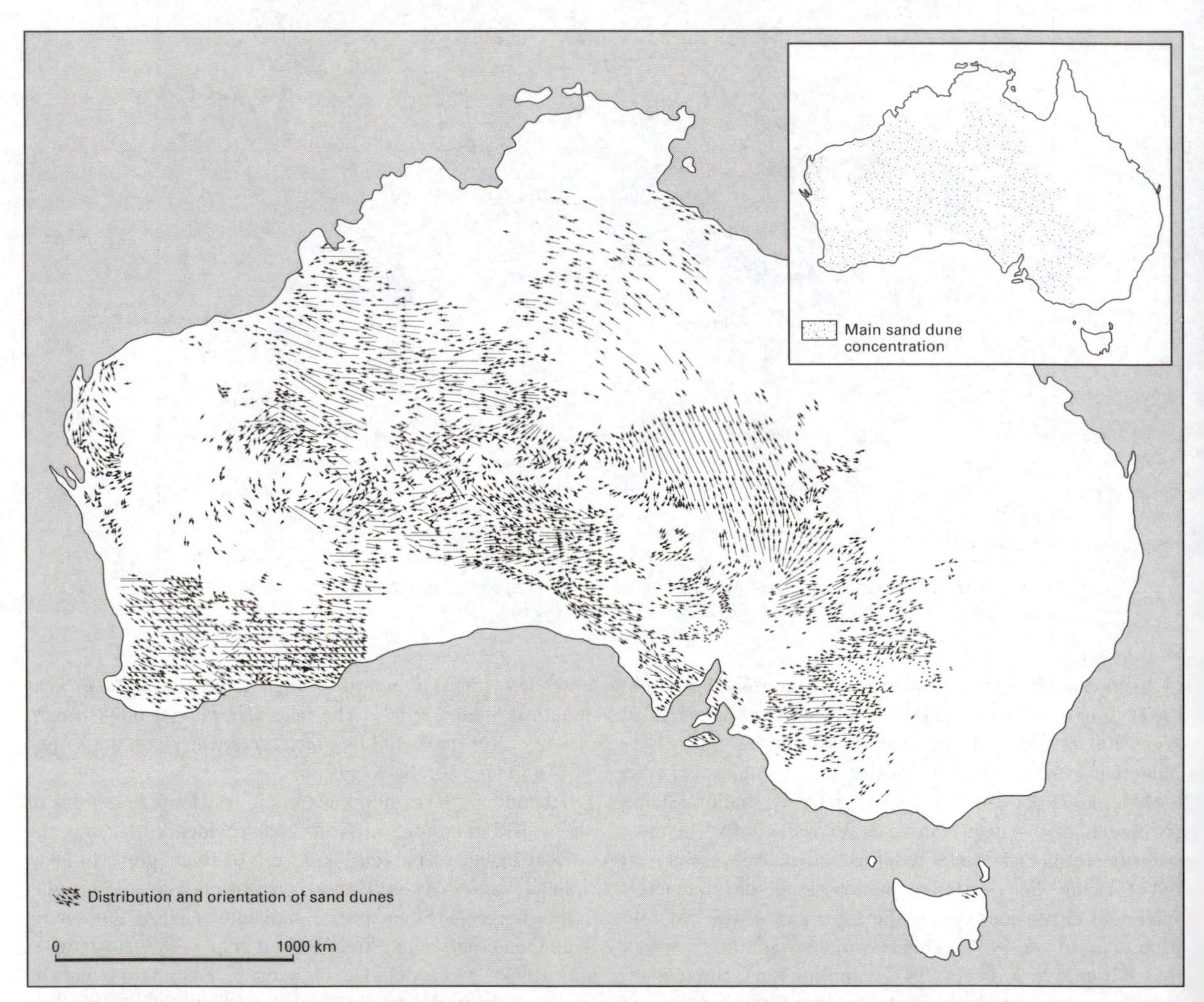

Figure 2.49 The distribution and orientation of sand dunes in Australia. The arrows show the orientation of dunes, while the shading in the inset shows the main concentration of dunes (after Wasson *et al.*, 1988).

Table 2.3 Sequence of climatic changes during the Late Quaternary in the Niger area of the southern Sahara, based on evidence of fossil dunes, lake sediments and soils (modified after Völkel & Grunert, 1990).

Years BP	Character of climate	East Niger (Great Erg of Bilma)	Central and Southern Niger (S of the 300 mm isohyet)
>40 000	Arid phase	Dune formation?	'Erg ancien'
40 000	Humid phase	Soil formation?	Weathering of the 'Erg ancien', Luvic Arenosols
30 000			
20 000	Arid phase with summit at about 18 000 BP	First dune generation	Dune formation, desert climate
	Change to semi-arid		
15 000	conditions	Dunes get inactive, first weathering, soil formation	Dunes get active, first weathering, soil formation
	More humid		
10 000	Humid phase with summit at about 8000 BP	Deep weathering of the first dune generation; Chromic Arenosols	Weathering with deep rubefying on the ancient dunes, Chromic and Cambic Arenosols
6000	Drier	Partly dune formation,	
5500	Arid phase (?)	second dune generation	Dunes mostly stable, soil degradation by deflation
4500	Humid phase	Soil formation	Soil formation
4000		Eutric Regosols	
3000	Drier,		Soil formation, during dry
2500	up to present-	Third dune generation	years deflation of soil
2000	day climatic		material
1000	conditions		In modern times widespread
0		Desert	degradation of soils

and on the pampas of Argentina (Thomas & Shaw, 1991). The geomorphological evidence in all of these regions points to phases of significantly increased aridity on numerous occasions during the Late Quaternary.

In some areas, it may be difficult to make a clear distinction between active and fossil dunes, particularly where sand has been remobilised and secondary dune patterns have become superimposed on an older primary set. Most fossil dunes have been distinguished on the basis of features indicative of a period of stability under more humid climatic conditions. These include the evidence of gullying and truncation by fluvial erosion; the presence of features produced by pedogenesis and chemical weathering, including chemical alteration of clay minerals, decalcification and staining by iron oxide; discordance with currently prevailing wind directions; and, in some cases, archaeological evidence of prolonged history of cultivation (Goudie, 1992). The most widely used criterion for inferring relict status is the presence of a dune vegetation cover. Care is needed, however, in the use of this form of evidence, as some dunes may be preferential sites for plant growth in arid environments, and hence vegetation cover may simply reflect better water-retaining properties and deep-rooting opportunities offered in such areas (Thomas & Shaw, 1991). This means that not all vegetated desert dunes are inactive features which have palaeoenvironmental significance.

Some fossil dunefields are found on the beds of former pluvial lakes where the presence of lacustrine deposits blanketing the inter-dunal depressions, and lake shorelines etched into dune flanks, provide convincing evidence for the alternation of arid and pluvial phases (e.g. Cooke, 1984). In several localities around the southern Sahara and Sahel, for example, dune generations of different ages are found associated with pluvial lake deposits, and the sequence of environmental changes has been dated. In the Erg of Bilma in eastern Niger, a major episode of dune formation in the period 20–16 ka BP was followed by conditions of increasing humidity which culminated during the early Holocene (*c.* 8 ka BP) in the development of an extensive pluvial lake system (Völkel & Grunert, 1990). Precipitation estimates for this pluvial phase are in the range 400–500 mm compared with present values of 30–50 mm. A drier interval around 6.5 ka BP led to a second phase of dune construction and, following disappearance of the lakes, a third episode of dune generation began around 2.5 ka BP (Table 2.3). The first and second generation of dunes show evidence of weathering, but the third generation is still active at the present day. Similar sequences of lake sediments and dune-building episodes have been noted in other regions of the southern Sahara (Petit-Maire & Riser, 1983), while extensive dune systems of two different ages have been traced across Mali, Niger and Mauretania (Fryberger, 1980). Multiple dune systems have also been found in the Sudan. The most recent of these appear to be Holocene in age and indicates a southward movement of the wind belts by about 200 km, while an earlier period of aridity represents a shift of some 450 km. In the valley of the White Nile, this early dune-building phase coincides with a major episode of alluviation dated to between 20 and 12.5 ka BP, during which time the river itself may have been blocked by migrating dune sands (Adamson *et al.*, 1980). Collectively, the evidence from the southern margins of the Sahara suggests that during the arid phases of the Late Quaternary, the major wind and precipitation belts may have shifted southwards through some 5° of latitude (Figure 2.50).

In addition to providing evidence on the extent of arid zones at times in the past, palaeowind directions can also be inferred from fossil sand dunes. In the Lake Chad basin, for example, systems of dunes formed during one or more Late Quaternary arid phases under the influence of winds from between the northeast and east (Figure 2.50) while a wind direction from the northeast is also implied by the orientation of dunefields in the 'empty quarter' of the western Sahara (Grove & Warren, 1968). Fossil longitudinal dunes in Australia show a clear counterclockwise swirl about the arid interior (Wasson, 1984), and the evidence suggests a summer anticyclonic system during the major dune-building episodes at least 5° further north than at present (Mabbutt, 1977). Thermoluminescence dating (Chapter 5) of relict dune systems in northern Australia suggests three major episodes of dune building during the Late Quaternary: 24–18 ka BP, 8.5–7.0 ka BP and 2.6–1.8 ka BP (Lees *et al.*, 1990).

Although fossil dunefields are therefore potentially extremely valuable indicators of the former extent of arid

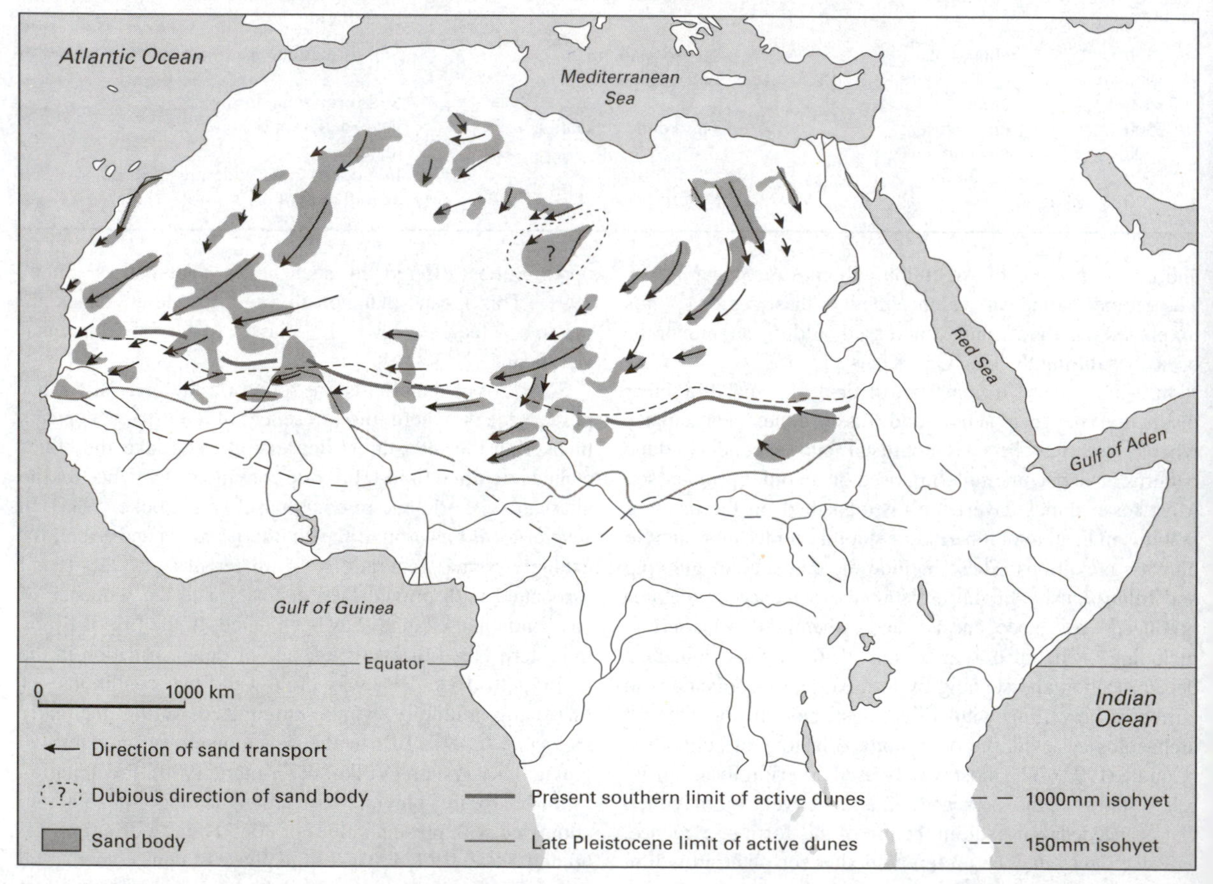

Figure 2.50 The southern limits of Late Pleistocene and contemporary desert dunes in the Sahara, and inferred directions of sand transport (after Williams *et al.*, 1993a).

environments, care must be exercised in their use as palaeoclimatic indicators. Desert conditions result from a variety of climatic factors, the most important clearly being temperature, precipitation and wind. A change in any of these variables may be sufficient to alter the balance between adequate groundwater retention for plant growth and excessive evaporation leading to severe drought. Hence, a change towards an episode of dune construction may result from higher temperatures, lower precipitation levels, increased wind strength, or a combination of all three. Similarly it cannot be assumed that higher precipitation levels constitute the only environmental factor leading to sand-dune stability (Thomas & Tsoar, 1990). For transitional periods, therefore, additional lithological and biological evidence may be required if correct palaeoclimatic inferences are to be made. A second problem arises out of the activities of human groups on the desert and savannah margins. It is clear that, over recent millennia, people have played an increasingly important role in the expansion of the desert margins (Goudie, 1994), and human-induced desertification has to be considered when palaeoenvironmental inferences are being made for the late Holocene on the basis of dunefield evidence. Finally, where fossil sand dunes are being employed as indicators of palaeowind directions, the problems encountered in the interpretation of fossil dunes found at high latitudes, which are discussed in Chapter 3, must also be considered.

2.7.3 Fluvial landforms

Other geomorphological evidence that has been used to infer changing climatic conditions in low latitudes includes fluvial landforms and weathering crusts. Relict fluvial features are found throughout the tropics and subtropics. In the Sahel of West Africa, for example, Landsat imagery has revealed the presence of integrated drainage networks of the River Niger that extended far northwards into the present arid zone (Talbot, 1980). These drainage channels were believed to be active in the early Holocene but are relict features at the present day, and provide further evidence of a change from wetter to more arid conditions during the mid- and late Holocene (see above). Relict drainage networks have also been detected by Landsat and SLAR imagery in the Amazon lowlands (Tricart, 1975). In parts of the interior of eastern Australia, complex systems of palaeochannels, many of which are very different in form from contemporary alluvial channels, appear to have developed under climatic régimes different from those of the present day (Rust & Nanson, 1986; Williams *et al.*, 1993a). Similarly, in northern India, changes in river régime from a phase of aggradation to one of marked incision reflect the climatic shift from the arid phase of the last cold stage to wetter conditions during the early Holocene (Williams & Clarke, 1984).

In reconstructing palaeoenvironments of the arid and semi-arid zone from relict fluvial landforms, however, a range of non-climatic factors must be taken into account before drawing palaeoenvironmental inferences (Rognon, 1980). Fluvial processes are governed by a range of environmental variables which include, in addition to climate, geology, relief, soil, vegetation cover and the influence of people. All of these, in combination, will affect the rate at which fluvial activity proceeds as well as the geomorphological response, and climate may not always be the dominant factor. For example, in highly arid regions, increased sediment load in rivers may result from increased rainfall, whereas in the semi-arid zone, lower rainfall levels may lead to increased sediment yield as a result of reduced vegetation cover (Mabbutt, 1977).

2.7.4 Weathering crusts

Weathering crusts, or **duricrusts**, are resistant surface mantles or cappings commonly found as protective layers at the surface of eroded bedrock or sediments in low latitudes (Goudie, 1973). They originate through the concentration in soils, sediments or permeable rocks of certain chemical constituents displaced through solution or translocation. These concentrations develop as hardpans and when exposed form cemented layers which are more durable than adjacent layers. There is a wide range of weathering crusts of different chemical constituents, but in general they can be divided into those that originate as weathering layers in humid tropical environments, and those that develop under arid or semi-arid conditions. **Laterites**, for example, evolve through the accumulation of hydroxides of aluminium and iron in humid tropical soils, and when these layers are exposed they form an extremely hard cemented horizon (Ollier & Galloway, 1990). Laterite or **ferricrete** crusts are typically found as caprocks in desert and savanna regions where their presence on plateau surfaces reflects inversion of relief (Ollier, 1991a). Their occurrence indicates a major change in climatic conditions, for they are believed to form where both temperature and precipitation are high. Climatic conditions presently occurring between 30°N and S of the equator are considered to be those conducive to lateritization (McFarlane, 1983), but precise climatic parameters governing the formation of laterites are difficult to establish. Moreover, as deep-weathering profiles and ferricretes are usually of considerable antiquity (Ollier, 1991a), they may be of relatively limited value in inferring former climatic conditions during the course of the Quaternary.

Weathering crusts that formed under more arid conditions include **calcrete**, also known as **caliche**, which is composed of cemented calcareous horizons and which often forms the hard rim of exposed escarpments in deserts and savanna regions (Goudie, 1983), **gypcrete** or gypsum cement (Watson, 1983), and silicious crusts known as **silcrete** (Summerfield, 1983). These weathering crusts and related structures in soils have been found in many low-latitude regions, and have been interpreted as reflecting Quaternary climatic change (Watson, 1988; Kapur *et al.*, 1990). The crusts are often degraded, which suggests that they may be relict forms from previous arid phases (Ollier, 1991b). Again the climatic conditions governing their formation are difficult to quantify, although annual rainfall limits (e.g. 250–300 mm in the Kalahari; 250–700 mm in Australia) have been suggested for the formation of silcrete and calcrete (Summerfield, 1983). Attempts have been made to date calcrete using both radiocarbon and uranium series dating (Goudie, 1983) but, in general, establishing the age of these weathering crusts is not straightforward and, as with laterites, many may date from early or even pre-Quaternary times.

2.8 Conclusions

Geomorphological evidence provides a useful starting point in the investigation of Quaternary environments. By using modern landforms as analogues, aspects of former glacial, periglacial, fluvial, marine and aeolian environments can be inferred. In many cases, only fairly general conclusions are possible, but in certain instances specific climatic parameters can be deduced. Throughout this chapter, however, it has been stressed that geomorphology is but one of the lines of evidence used in the reconstruction of Quaternary environments and that, wherever possible, landform evidence should be employed in conjunction with other forms of evidence. A second major data source lies immediately beneath the earth's surface, and it is to the stratigraphic record that we now turn our attention.

Notes

1. The use of aircraft requires survey by wide-angle camera lenses in order to cover the maximum possible area. This results in increasing distortion towards the margins of each photograph. Satellite images, on the other hand, are based on narrow-angle 'lenses'; distortion is therefore minimal and a single exposure from these higher altitudes can also cover a much larger area.
2. **Sandar** is an Icelandic term to describe the outwash plains that form in lowland areas in front of valley glaciers and ice sheets. The singular is **sandur**.
3. More recent work suggests that the Halland Coastal Moraines are not recessional features, but are push moraines formed during a readvance of the ice front. Moreover they may be younger than 12.4 ka BP in age (Fernlund, 1993).
4. **Drumlins** are low hills with an oval outline and are usually less than 60 m in height. They are formed mainly of till, although some contain stratified material or a bedrock core (see Menzies & Rose, 1987).
5. In glaciology, the term 'steady-state' refers to the condition whereby a glacier or ice sheet maintains equilibrium over a finite period as a result of a balance between the major controlling variables such as seasonal climatic variations, albedo, rate of flow, etc.
6. In some glaciers, particularly in the High Arctic, there may be a complex zone of transition between accumulation and ablation areas. This is usually referred to as the **superimposed ice zone** in which ice will have formed as a result, for example, of the refreezing of meltwater runoff or avalanche material from upglacier. The lower boundary of the superimposed ice zone is the equilibrium line (upglacier from which there is a gain in mass), whereas the firn line, i.e. the line dividing fresh snow from ice, forms the upper boundary. On most maritime temperate glaciers, where snowfall is heavy, there is frequently no superimposed ice zone, and therefore the firn line and the equilibrium line coincide (see Embleton & King, 1975; Paterson, 1994).
7. Permafrost has been defined by Muller (1947, p. 3) as 'a thickness of soil or other superficial deposit, or even of bedrock, at a variable depth beneath the surface of the earth, in which a temperature below freezing has existed continually for a long time' (from two to tens of thousands of years).
8. Strictly speaking, the term 'raised beach' relates to beach deposits that have been raised by land uplift above the level at which they were formed. However, the term is conventionally used to describe all beach features found above present sea level, irrespective of whether their position results from actual uplift of the land or a fall in sea level.
9. In fact, a small remnant of the Laurentide Ice Sheet still exists in the form of the Barnes Ice Cap on Baffin Island.
10. The term **Neotectonics** refers to late Cenozoic deformation of the earth's crust. The study of Neotectonics involves the study of such earth movements, their mechanisms, their geological origin, their geomorphological consequences and their future extrapolations (Vita-Finzi, 1986; Owen *et al.*, 1993).
11. The **marine limit** is the farthest point of marine influence inland.
12. **Rheology** is the branch of Newtonian mechanics that deals with the **deformation** and flow of materials that are neither solid nor completely liquid, such as semi-molten (ductile) rock or glacier ice. The term **deformation** refers to the process by which a substance changes shape without breaching continuity.

CHAPTER 3 Lithological evidence

3.1 Introduction

Although geomorphological evidence can provide useful insights into former climatic régimes and environmental conditions, a more detailed impression of events during the Quaternary can often be gained from the sedimentary record. Not only can valuable data on Quaternary environments be obtained from the sediments themselves, by relating observations on present depositional environments to features preserved in the recent stratigraphic record, but since many deposits are fossiliferous, inferences based on lithological changes can often be supported directly by those based on fossil evidence. Furthermore, because sedimentary sequences frequently reflect sediment accumulation over an extended time period, some appreciation can be gained of both spatial and temporal aspects of environmental change. Finally, while geomorphological evidence is restricted largely (although not wholly) to the terrestrial environment, sedimentary data can be obtained from beneath the waters of present-day lakes, from the world's ice caps and, perhaps most important of all, from the deep-ocean floors where lengthy sequences of virtually undisturbed deposits are preserved.

Quaternary sediments are generally unconsolidated and are of two principal types: inorganic (**clastic**) deposits, consisting of mineral particles (termed **clasts**) ranging in size from large boulders to very fine clays, and **biogenic** sediments, consisting of the remains of plants and animals. Biogenic sediments can, in turn, be divided into an organic component of humus and the decayed remains of plants and animals, and an inorganic component of such elements as mollusc shells and diatom frustules. In this chapter we are concerned primarily with inorganic sediments such as tills, aeolian and cave sediments, and fossil soils, although some of the properties of biogenic sediments are also considered. A full discussion of the fossil record contained in Quaternary sediments can be found in Chapter 4.

3.2 Field and laboratory methods

3.2.1 Sediment sections

Wherever possible, lithological investigations should be carried out on open sections so that variations in stratigraphy, both vertically and horizontally, can be carefully recorded. Before commencing fieldwork, the section should be cleaned of slumped material, and a 'fresh' face revealed by cutting back into the exposure. On large sections, steps can be cut on slumped material or directly into the face to provide temporary working surfaces, but a ladder may be necessary to reach the less accessible parts. Careful drawing of the exposure is the first stage in analysis, and this should be supported wherever possible by a photographic record. It may be useful to grid the face with measuring tapes as this will provide an accurate scale for section drawing. Detailed notes should be taken on all aspects of exposed stratigraphy, using Munsell colour charts to obtain a relatively precise description of colour changes between and within lithostratigraphic units. Where necessary, instrumental levelling from a benchmark will enable the various stratigraphic features to be related to a common datum, for purposes of altitudinal comparison both within and between sites.

The type of sampling framework employed will depend on the nature and purpose of the investigation. For certain types of study, for example the analysis of soil or loess profiles (both discussed in this chapter), a sequence of samples may be required from a representative vertical section of the exposure. Sometimes a set of monoliths measuring perhaps 25 cm square may be cut from the face for subsequent laboratory analysis, using either a sharp spade or metal boxes specially designed for the purpose that can be hammered into the face. Alternatively, bulk samples may be taken at a set vertical interval or, depending on the aims of the investigation, small samples of only 1 cm^3 may be all that are required. In all of these cases, however, sampling horizons should be carefully related to a measuring tape attached to the free face at the side of the sampling line; the trowels, spades, knives and spatulas should be cleaned between the extraction of each sample, and care must be taken over the packaging, sealing and labelling of the sediment samples. Detailed notes should be made throughout the sampling process. In other types of investigation, it may be necessary to take a number of samples from points scattered across the face of the section. These can be selected subjectively, but random sampling of a face that has been gridded will provide a more objective sampling framework. The same method can be applied where measurement of sedimentary properties needs to be undertaken in the field, such as the recording of the orientation and dip of pebbles or other clasts (termed sediment fabrics – see section 3.3.6.2).

3.2.2 Coring

Although section work is preferable to coring, relatively few natural sections are to be found, and there are many situations where it is impossible to excavate exposures for various reasons, such as problems of time, expense, sediment thickness or the likelihood of waterlogging. Hence, recourse must be made to coring. A range of coring equipment is now available, with different operating mechanisms and different levels of success in core recovery (see, e.g., Aaby & Digerfeldt, 1986; Faegri & Iversen, 1989). Hand-operated corers of either the side-sampling (e.g. *'Russian'*) or piston (e.g. *Livingstone*) type are widely used for sampling peat and lake sediments, although motor-driven equipment (e.g. *Rotary* or *Percussion* corers) is required for more cohesive sediment such as gravels and tills. Specialised motor-driven or hydraulically operated machines are also used for coring into deep lakes, marine sediments and ice sheets. Numerous problems are encountered in coring operations, including the fact that (i) cores may be distorted during recovery or extrusion from the sampling chamber, a problem that becomes particularly acute when cores are being taken from poorly consolidated sediments (e.g. some types of peat and lake muds); (ii) unless overlapping cores are taken, stratigraphic units that are either thin or of limited lateral extent may be missed during sampling; and (iii) it is often difficult to transport heavy coring equipment to remote areas where other logistical problems may also be encountered, such as a lack of water for hydraulic coring operations. Despite such difficulties, however, the development of increasingly sophisticated coring machinery in recent years has enabled the successful drilling of marine sediments, of the polar ice sheets and of deep lake sequences, all of which have revolutionised our understanding of the Quaternary environmental record (section 1.4).

Since most Quaternary sedimentary sequences, when traced laterally, vary in both thickness and complexity, the sedimentary history of a site is often difficult to ascertain when only a single core is obtained. It is common practice, therefore, to obtain several borehole records, usually arranged in transects, so that a two-dimensional, or even three-dimensional, schematic model of lithological variability can be constructed. The number of boreholes employed in such an exercise will depend upon the area to be surveyed, the complexity of the lithological sequence, the time available for fieldwork and other logistical considerations, such as costs. Remote sensing techniques, such as seismic and sonar sounding methods, may also be employed to assist in such an exercise (section 2.2.2).

3.2.3 Laboratory methods

Laboratory analysis of Quaternary sediments is an integral part of palaeoenvironmental investigations. Both the physical and chemical properties of sediments can provide valuable data on the nature of former depositional environments, and are often useful indices of climatic and other environmental changes. A very wide range of laboratory methods is now available, and a detailed account is beyond the scope of this book. In this section, however, we provide a brief introduction to the methods that are most commonly used in the description and analysis of Quaternary sediments. For further details the reader is referred to Bridgland (1986), Berglund (1986), Goudie (1990) and Gale & Hoare (1991).

3.2.3.1 Particle size measurements

Particle size distribution is a most important diagnostic property of a body of sediment, for even very subtle variations in average grain size or in the range of sizes of clasts may reflect important changes in sedimentary environment. The particle size distribution of coarser grades of sediment (sand size and above – Table 3.1) can be established by sieve analysis, but sedimentation methods are required for finer materials. Different grades of sediment suspended in a liquid will settle out at different rates and these can be established either by using a hydrometer to record changes in the density of the suspension over time, or by extracting subsamples from the suspension by means of a pipette and then measuring the changing concentrations of suspended matter over time by successive weighings (Gale & Hoare, 1991). An electrical sensory method has also been

Table 3.1 The Wentworth scale of particle size fractions and the equivalent Φ (phi) units. The phi units are obtained by conversion from the millimetre scale, where phi is $-\log_2$ of the diameter in millimetres. The phi scale has the advantage of using integer numbers only, and also makes the statistical description of sediments more straightforward.

Name	mm scale	Φ units
Boulder	more than 256	more than −8.0
Cobble	256 to 64	−8.0 to −6.0
Pebble	64 to 4	−6.0 to −2.0
Granule	4 to 2	−2.0 to −1.0
Very coarse sand	2 to 1	−1.0 to 0.0
Coarse sand	1 to 0.5	0.0 to 1.0
Medium sand	0.5 to 0.25	1.0 to 2.0
Fine sand	0.25 to 0.125	2.0 to 3.0
Very fine sand	0.125 to 0.0625	3.0 to 4.0
Coarse silt	0.0625 to 0.0312	4.0 to 5.0
Medium silt	0.0312 to 0.0156	5.0 to 6.0
Fine silt	0.0156 to 0.0078	6.0 to 7.0
Very fine silt	0.0078 to 0.0039	7.0 to 8.0
Coarse clay	0.0039 to 0.00195	8.0 to 9.0
Medium clay	0.00195 to 0.00098	9.0 to 10.0

developed for particle size work using a **Coulter counter** (Whalley, 1990). Sediments are suspended in an electrolyte and passed through an electrode-flanked aperture. Voltage pulses proportional to the volumetric size of the particles can then be counted and the particle size distribution established. More recently, more sophisticated equipment has been developed which uses laser or X-ray beams for rapid, automated measurement of the fines (silts and clays) suspended in a fluid medium (Swyitski, 1991). Particle size data are usually presented in the form of sigmoidal curves on probability graph paper (section 3.3.5.1, Figure 3.7) or in ternary diagrams (triangular graphs – section 3.3.5.1, Figure 3.6) or histograms.

3.2.3.2 Particle shape

Particle shape has been used to distinguish between sediments that have accumulated in different depositional environments, and has been employed particularly effectively in the analysis of glacial, glaciofluvial and fluvial sediments. Techniques include the purely visual assessment of particle shape, the use of prepared charts as a basis for the division of pebbles into classes ranging from angular to rounded, and the direct measurement of particles themselves. Shape measurements are then based on the axial ratios of the particles and include a variety of indices of elongation, roundness, flatness and sphericity (to derive classes of, for example, blades, rods, spheres and discs). These methods are described in Bridgland (1986) and Gale & Hoare (1991). More sophisticated automated methods of particle form analysis (e.g. image analysis) are discussed in Whalley (1990).

3.2.3.3 Surface textures of quartz particles

Different sedimentary environments (e.g. glacial, marine, aeolian) give rise to particular textural features on the surfaces of quartz and sand grains and these can be analysed using an **electron microscope.**[1] Characteristic textural features on quartz grains that can be detected by scanning electron microscopy (SEM) include fracture patterns, scratches, grooves, chatter marks, solution pits and cleavage flakes (Whalley, 1990). Moreover, it is often possible to identify superimposed features and hence more than one palaeoenvironment of modification can sometimes be inferred.

3.2.3.4 Organic carbon content

The determination of organic carbon content is of considerable importance in palaeolimnology where it provides an index of biological productivity in former lake basins. It is also useful in establishing the amount of organic material that is likely to be required for the radiocarbon dating of a sample (section 5.3.2). The most widely used method is **loss on ignition**, in which the amount of carbon in a sample is indicated by the weight loss following combustion in a furnace (Bengtsson & Enell, 1986). More accurate results can usually be obtained using standard titration methods or colorimetry techniques.

3.2.3.5 Metallic elements

The variations in proportions of metallic ions of, for example, calcium, potassium, sodium or magnesium in late Quaternary lake sediments are now regarded as important indicators of the changing erosional history of lake catchments (e.g. Mackereth, 1965; Bengtsson & Enell, 1986; see section 3.9.3.1). The concentration of such elements in a sediment sample can be determined using either a **flame photometer** or an **atomic absorption spectrophotometer (AAS).** The former operates on the principle that a metallic salt drawn into a non-luminous flame ionises and emits light of a characteristic wavelength, while the AAS measures the concentration of an element by its capacity to absorb light of its characteristic resonance while in an atomic state. In both cases, the light emissions are recorded photoelectrically (Whalley, 1990). More rapid methods, which measure the proportions of metallic ions present in minute quantities of sediment, use mass spectrometers to analyse atomic emissions with a high degree of precision. The **inductively coupled plasma atomic emission spectrometry (ICP)** method, for example, involves the creation of an aerosol (or plasma) of ionised gas from a homogenised sediment sample, which is then analysed for elemental content using mass (ICP-MS), atomic emission (ICP-AES) or atomic fluorescence (ICP-AFS) spectrometry (Thompson & Walsh, 1983).

3.2.3.6 Heavy minerals

Heavy-mineral assemblages often reflect the derivation or provenance of Quaternary deposits. They have been used in a number of Quaternary studies including the investigation of weathering profiles, the differentiation of tills (section 3.3.6.3), and the analysis of loess deposits. Heavy minerals are those with a specific gravity (SG) greater than 2.85 and are usually separated from the lighter mineral fraction in a sample by settling in a heavy liquid such as bromoform (SG 2.89). The heavy mineral assemblage is then dried and mounted on a slide, and the percentage of individual types can be determined using a petrological microscope. Full details of the method can be found in Whalley (1990).

3.2.3.7 Clay mineralogy

The clay mineralogy of a sediment can provide information both on the origins of the material and on any chemical changes that have occurred owing, for example, to the effects of different weathering processes since deposition. In Quaternary research, clay mineral analysis has also been employed in the differentiation of tills (e.g. Haldorsen *et al.*, 1989). The most widely used method in clay mineral analysis is **X-ray diffraction (XRD)**, which involves the rotation of a sample in a stream of directed electrons. The clay minerals present (e.g. illite, chlorite, montmorillonite) are identified by observing and comparing the spacing and intensity of peaks on diffractometer traces. A short account of the method is provided by Yatsu and Shimoda (1990).

3.2.3.8 Mineral magnetic analysis

Because sediments vary enormously in the quantity, size and type of magnetic minerals that they contain, they can often be readily characterised on the basis of their magnetic properties. A range of magnetic signals can be measured rapidly, both in the field and in the laboratory, and these enable the concentration and types of magnetic minerals present in sediments to be determined (see, e.g., Oldfield, 1991; Thompson, 1986). Variations in the magnetic properties of sediments can then be used to correlate sediment units as well as to establish changes in rates of sedimentation. In lake sediment records, environmental changes around the catchments can be inferred on the basis of mineral magnetic properties (e.g. Dearing, 1986). The application of magnetic measurements in the analysis of sediment sequences is discussed in more detail in section 5.5.1.

3.2.3.9 Stable isotope measurements

Several common elements (e.g. oxygen, carbon, hydrogen) exist in nature in different isotopic states, reflecting variations in the number of neutrons in the nuclei (see section 5.3). This results in slight, but very important, differences in their physico-chemical behaviour, and a selective separation between molecules of different atomic mass (see, e.g., sections 3.8.5 and 3.10.2) often occurs as a result of crystallisation, evaporation, precipitation, osmosis, metabolism, etc. (Lajtha & Michener, 1994). Changes in the ratios of isotopes of different atomic mass can subsequently become locked into fossils or precipitates, and hence into the sedimentary record. For example, where lake water levels fall during drought conditions, the 'environmental stress' that results may be reflected in a shift in isotopic ratios and this will be registered in the tissues and skeletons of biota inhabiting the lake as well as in chemical precipitates. The isotopic composition of meteoric water may also change due to climatic variations, and these can also lead to changes in isotopic ratios in lake sediments, ice sheets and cave speleothem. Variations in isotope ratios in sediments can be measured using mass spectrometers (Siegenthaler & Eicher, 1986), and the results often provide insights into the environmental conditions under which the sediments accumulated. In some cases, global climatic changes are reflected in the isotopic signals contained within sediment sequences. Isotope measurements of lake sediments, deep-sea sediments and ice cores, in particular, are described in more detail later in this chapter.

3.3 Glacial sediments

3.3.1 Introduction

Glacial sediments of Quaternary age cover large areas of the earth's surface, particularly in the mid-latitude regions. In Europe, for example, glacially derived **diamictons**[2] form an intermittent blanket over at least one-third of the land area, while in North America, such deposits are spread over half the continent. It has already been shown (section 2.3.3) that the moulding of these deposits into characteristic glacial landforms can be used to establish former glacier extent and direction of ice movement, and can form the basis for glacier modelling and climatic reconstruction. However, equally important palaeoenvironmental data can be derived from an analysis of the deposits themselves, for the distinctive properties of many glacial sediments allow inferences to be made about former glacier types, the mode of sediment deposition, ice-flow directions and the sources of sediment supply. An understanding of these properties can also have economic benefits as, for example, in tracing the sources of ore bodies or modelling groundwater flow in glacigenic deposits. Indeed, in view of the widespread nature of glacial deposits by contrast with well-defined landforms, such as moraines and drumlins, it could be argued that the lithological evidence has a more important role to play in the reconstruction of Quaternary environments, although the most secure palaeoenvironmental inferences are likely to be drawn from situations where geomorphological and sedimentological evidence are employed together.

3.3.2 The nature of glacial sediments

The nomenclature of glacial deposits can be very confusing. The superficial sediments that blanket the landscape in many parts of Europe were originally believed to have been derived from a great flood and were termed **diluvium**, although Sir Charles Lyell referred to them as **drift** because, along with many of his contemporaries in the early years of the nineteenth century, he believed that the deposits had been derived primarily from the melting of icebergs that had drifted in during a marine inundation. Curiously, the latter term has survived in the literature to the present day and is still used to refer to deposits formed by, or in association with, glacier ice or by ice-melt. The fundamental division of glacial 'drift' into **stratified** and **unstratified drift** was first proposed by T.C. Chamberlin at the turn of the century and has formed the basis for the classification of glacigenic sediments until relatively recently. Unstratified drift is usually referred to as **till** or (inappropriately) **'boulder clay'**. The former term was first used by Geikie in 1863 to describe coarse stony soil commonly found on the glacigenic deposits of northern Britain. It is preferable to, and now more widely used than, the term 'boulder clay' since many non-glacigenic diamictons contain boulders and clay whereas till frequently does not. The term 'till' is also more satisfactory than '**ground moraine**' which has often been employed to describe glacigenic deposits. 'Moraine' is a geomorphological term, and hence should not be used for lithological classification. All diamictons formed directly from glacier ice without the intervention of flowing water should be termed **till** (Eyles *et al.*, 1983).

Till is one of the most variable types of sediment to be found on the earth's surface. According to Goldthwait (1971), the principal characteristics that collectively distinguish till from other diamictons are:

(a) a lack of complete sorting which usually means the presence of some pebbles or boulders much larger than the dominant clay, silt or sand matrix;
(b) a homogeneous mix lacking any smooth laminations or regular graded bedding, combined with
(c) a mixture of mineral and rock types, some of which may be of distant provenance and not represented in the local strata.

A number of additional identifying features frequently found associated with till are:

(d) at least a small proportion of striated stones and microstriated grains;
(e) common orientation of the long axes of elongated grains and pebbles;
(f) relative compactness or close packing of sediment;
(g) presence of a striated surface on the underlying rock in certain cases;
(h) subangularity in clasts of all sizes due to frequent breakage in transport coupled with particle smoothing by abrasion (stones and boulders commonly have a 'bullet' shape as a result of these processes);
(i) either large-scale deformation structures (thrust planes, folding, etc.), which usually indicate a dominant direction of lateral compression and shear, or microfabrics indicative of compressional deformation (see below).

Stratified drift, on the other hand, is characterised by sorting of material by the action of glacial meltwater. These deposits show many affinities with fluvial sediments and are therefore referred to as **glaciofluvial**. The term covers a range of sedimentary environments, including **ice-contact deposits** (sediments formed adjacent to, or in contact with, glacier ice), **proglacial** or **outwash deposits** (sediments accumulating close to the frontal margin of an ice sheet or terminus of a glacier), and **glaciolacustrine deposits** (sediments accumulating in lakes located on, within, or adjacent to, glacier ice, or fed directly by glacial meltwaters). The sediments found in these various contexts range from coarse gravels and sands formed in braided streams to laminated clays and silts formed in lakes. Where the latter are seasonally frozen, **varves**[3] may develop.

The nature and composition of glacigenic sediment depend on a number of factors, including, *inter alia*, the thickness of the ice, the rate of melting, the topographic context (e.g. whether the ice is confined within a narrow valley, or spreads out into a piedmont lobe), the concentration and type of clasts contained within the ice, and the zone or part of the ice in which the sediments were initially entrained (see Figure 3.1). Sediments may accumulate in a **supraglacial** (on or close to the ice surface), **englacial** (within the ice body) or **subglacial** (at the base of the ice) position. Short-term oscillations of the ice margin may lead to the superimposition of sediments from each of these contexts, as well as the incorporation of sediments laid down during an earlier depositional phase. As a consequence, glacigenic sequences frequently reveal intercalations of tills and stratified sediments, with complex vertical and spatial **facies**[4] variations.

The analysis of glacigenic facies in contemporary ice-marginal environments reveals diagnostic features that can be employed in the interpretation of older Quaternary sequences (the 'analogue' approach). In recent years, a number of classificatory schemes of glacigenic sediment sequences have been developed on the basis of this approach (e.g. Eyles *et al.*, 1983; Brodzikowski & Van Loon, 1987, 1991; Eyles, 1993). Figure 3.2 shows some examples of the 'subenvironments' that might develop in ice-marginal areas, with characteristic lithofacies sequences (Table 3.2) for each of these. In each subenvironment, a number of distinctive facies types and associations can be identified, depending on the local style of ice melting, as well as on whether the sediments accumulate against, above or beneath active or inactive ice. In the continental ice-marginal environment (Figure 3.2A), for example, the type of sediment that is deposited is determined by distance from the ice margin and position of the material within the ice mass. Both of these factors influence hydrological conditions as well as the physical and mechanical properties of the accumulating sediment which, in turn, will determine the rate of sedimentation and the extent of sediment **deformation**.[5] Examples of facies variations in the near terminal and marine ice-marginal environments are illustrated in Figure 3.2B and C.

3.3.3 The classification of tills

Prior to the 1970s, till classification and genesis were based mainly on the analysis of glacigenic sequences that accumulated during the last, or earlier, glacial stages in Europe and North America. It was not until the later 1960s and the early 1970s that detailed observations of the active processes operating today in ice-marginal zones began to appear regularly in the literature. A number of important new approaches to the study of glacigenic sequences emerged from these publications. One example was the classification of tills and of the factors influencing their formation introduced by G.S. Boulton (1972, 1975; Boulton & Eyles, 1979). On the basis of observations of deposits associated with contemporary glaciers in Spitsbergen, three

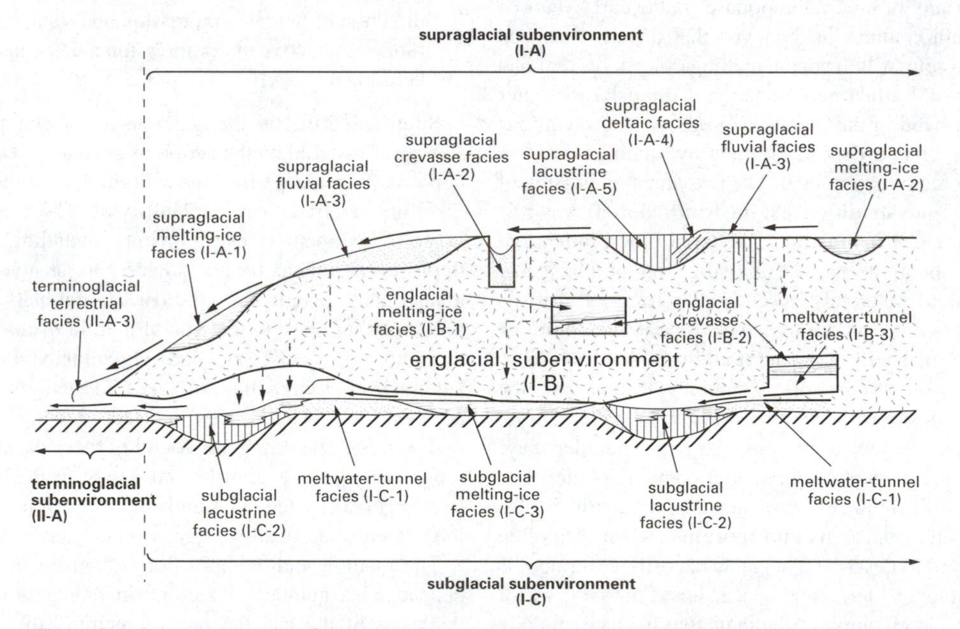

Figure 3.1 Schematic model of the the supra-, en- and subglacial 'subenvironments' within the marginal zone of a continental ice mass. A classification of the principal sediment facies in ice-marginal zones is shown in Table 3.2 (after Brodzikowski & Van Loon, 1987).

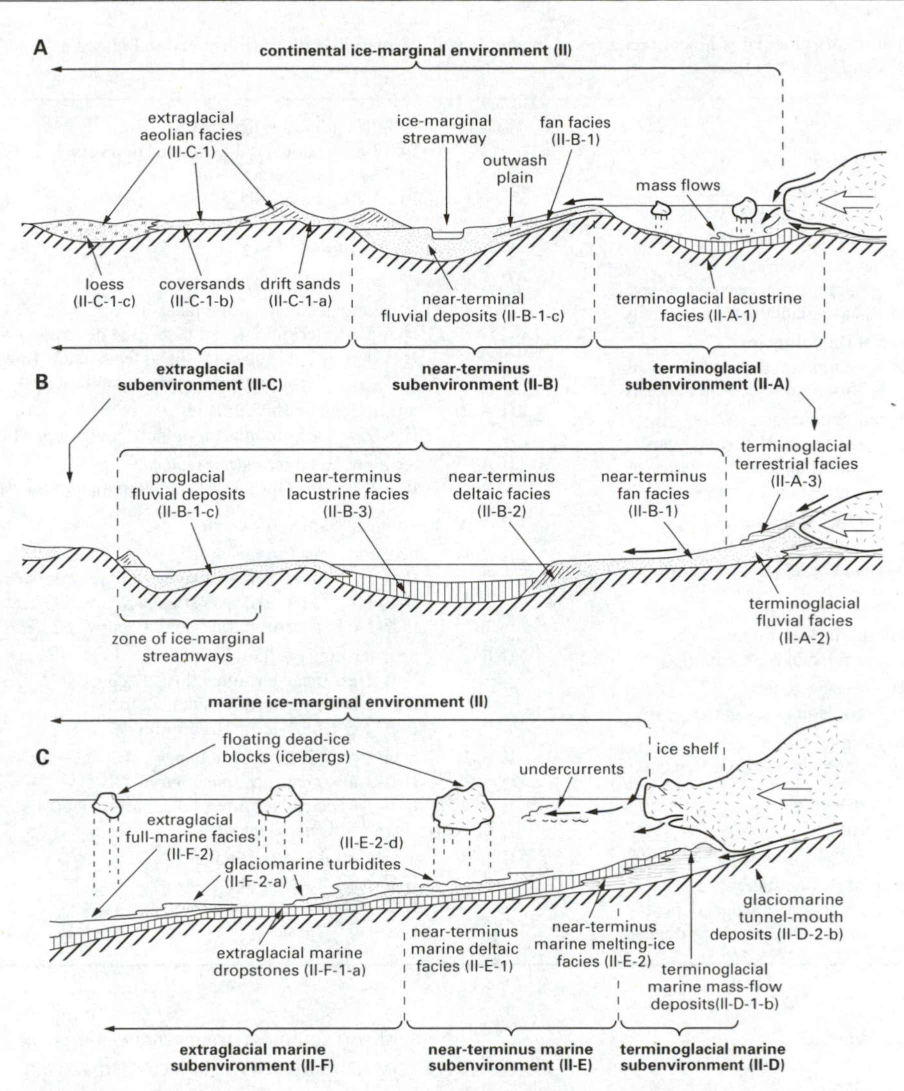

Figure 3.2 Schematic model of modes of deposition associated with a glacier margin: A and B illustrate the variety of sediment types associated with continental ice margins; C illustrates the sediment types associated with the marine ice-marginal environment. A classification of the principal sediment facies in ice-marginal zones is provided in Table 3.2 (after Brodzikowski & Van Loon, 1987).

main types of till were recognised: **flow till, melt-out till** and **lodgement till** (Table 3.3). **Flow till** is released as a water-saturated fluid mass from the downwasting glacier surface. Flowage of successive generations of this material leads to interbedding of flow deposits, not only with outwash and lacustrine sediments deposited on the ice surface, but also with spreads of outwash beyond the glacier terminus. As a consequence, complex stratigraphic sequences of till, outwash and lacustrine sediments can result from a single phase of ice wastage. **Lodgement till** (Figure 3.3A) is deposited at the ice base and accumulates on the subglacial floor either through pressure against bedrock protuberances or against patches of stagnant ice underneath the moving glacier body. **Meltout till** consists of englacial debris released from melting ice either above the glacier sole or at the glacier surface. In the former situation it is confined beneath the overlying ice body and the glacier bed, while at the surface it is trapped beneath the overburden of flow till. Hence, melt-out till can be either subglacial or supraglacial in origin (Boulton, 1980).

Table 3.2 Classification of principal sediment facies found in continental ice-marginal zones (illustrated in Figures 3.1 and 3.2) according to Brodzikowski & Van Loon (1987).

Code	Sub-code	Description
GLACIAL ENVIRONMENT		
I–A Supraglacial Subenvironment		
I-A-1		supraglacial melting ice facies
	I-A-1-a	supraglacial flow-tills
	I-A-1-b	supraglacial ablation tills
I-A-2		supraglacial crevasse facies
	I-A-2-a	supraglacial crevasse deposits
	I-A-2-b	supraglacial stream deposits
I-A-3		supraglacial fluvial facies
	I-A-3-a	supraglacial stream deposits
	I-A-3-b	supraglacial stream deposits
I-A-4		supraglacial deltaic facies
	I-A-4-a	supraglacial stream deposits
	I-A-4-b	supraglacial deltaic deposits
	I-A-4-c	supraglacial bottom-sets
I-A-5		supraglacial lacustrine facies
	I-A-5-a	supraglacial bottom-sets
	I-A-5-b	supraglacial lake-margin deposits
I–B Englacial Subenvironment		
I-B-1		englacial melting-ice facies
	I-B-1-a	englacial melt-out tills
I-B-2		englacial crevasse facies
	I-B-2-a	englacial crevasse deposits
I-B-3		meltwater-tunnel facies
	I-B-3-a	englacial crevasse deposits
I-C Subglacial Subenvironment		
I-C-1		meltwater-tunnel facies
	I-C-1-a	subglacial channel deposits
I-C-2		subglacial lacustrine facies
	I-C-2-a	subglacial channel deposits
	I-C-2-b	subglacial lacustrine deposits
I-C-3		subglacial melting-ice facies
	I-C-3-a	subglacial lacustrine deposits
	I-C-3-b	lodgement tills
	I-C-3-c	basal tills
EXTRAGLACIAL ENVIRONMENT		
II-A Terminoglacial Subenvironment		
II-A-1		terminoglacial lacustrine facies
	II-A-1-a	terminoglacial lacustrine deposits
	II-A-1-b	terminoglacial subaqueous mass-flow deposits
	II-A-1-c	terminoglacial tunnel-mouth deposits
II-A-2		terminoglacial fluvial facies
	II-A-2-a	terminoglacial tunnel-mouth deposits
II-A-3		terminoglacial terrestrial facies
	II-A-3-a	terminoglacial subaerial mass-flow deposits
II-B Near-terminus Subenvironment		
II-B-1		near-terminus fan facies
	II-B-1-a	near-terminus subaerial mass-flow deposits
	II-B-1-b	near-terminus stream-flood and sheet-flood deposits
	II-B-1-c	near-terminus fluvial deposits
II-B-2		near-terminus deltaic facies
	II-B-2-a	near-terminus fluvial deposits
	II-B-2-b	near-terminus deltaic foresets
	II-B-2-c	near-terminus bottom-sets
II-B-3		near-terminus lacustrine facies
	II-B-3-a	near-terminus bottom-sets
	II-B-3-b	near-terminus lake margin deposits
II-C Extraglacial Subenvironment		
II-C-1		extraglacial aeolian facies
	II-C-1-a	drift sands
	II-C-1-b	coversands
	II-C-1-c	loess

Table 3.3 Classification of tills.

Traditional classification	Modern classification	
Ablation till	Supraglacial till	Flow till Melt-out till
Lodgement till	Subglacial till	Melt-out till Lodgement till Deformation till

More recently, a fourth category of till, termed **deformation till** (Boulton, 1987; Dreimanis, 1993), has been recognised. Indeed, most lodgement tills seem likely to have been affected by subglacial deformational processes (Hart, 1995b). It now appears that many subglacial tills are not, as previously thought, built up gradually in layers by the release of material from the base of the ice (sometimes referred to as a 'plastering-on' process), but rather that a body of soft sediment, tranported *en masse* at the base of the ice, is continuously subjected to shear stresses and remoulding by deformation. The concept of a **'deforming bed'** is based on observations of modern glaciers which indicate that most of the movement of glaciers is achieved by continuous deformation of the subglacial debris (Sharp, 1988; Humphrey *et al.*, 1993). The confining, compressional stresses generate a variety of deformational features, including, for example, isoclinal recumbent folds, shears (Figure 3.3A and B), boudins, hook folds, diapirs, tension fractures, sediment wedges, brecciated beds and conjugate fracture patterns (see Boulton, 1987; Dreimanis, 1993; Hart, 1995b). However, deformational structures, mainly folds, can also form in flow tills, as a result of sliding or slumping under saturated conditions, and are also found in push moraines along an actively retreating ice margin (Hart, 1990). Care also needs to be taken to distinguish between intra-formational deformation structures (those generated

Figure 3.3 A. Typical lodgement till from a site at Müntschemier, Switzerland, showing widely separated cobbles and pebbles in a fine-sediment matrix which contains oblique, straight shear planes (scale in centimetres; photo: Jaap van der Meer).
B. Deformation at the base of a basal till at San Martin de los Andes, Argentina. Streamlined, striated and imbricately lodged boulders overlie a shear zone developed in rhythmites: note the sheared clastic dyke underneath the large boulder (20 cm knife for scale; photo: Jaap van der Meer.)

during the initial settling of the till) and those resulting from post-depositional deformation, by the readvance of ice over older, unconsolidated glacigenic sediments (see, e.g., Meer, 1987; Croot, 1988; Kluiving, 1994). In addition to macroform features, thin sections of tills (very thin slices of the fine matrix examined under a microscope) can also reveal signs of deformation, which are produced by rotational and horizontal displacements of the finer-grained components of the till matrix. These include microshears, fractured clasts, intraclasts and anisotropic fabrics (localised orientation of fine particles; Meer, 1993). Scanning electron micrographs are now commonly used to reveal these microstructures (Figure 3.4).

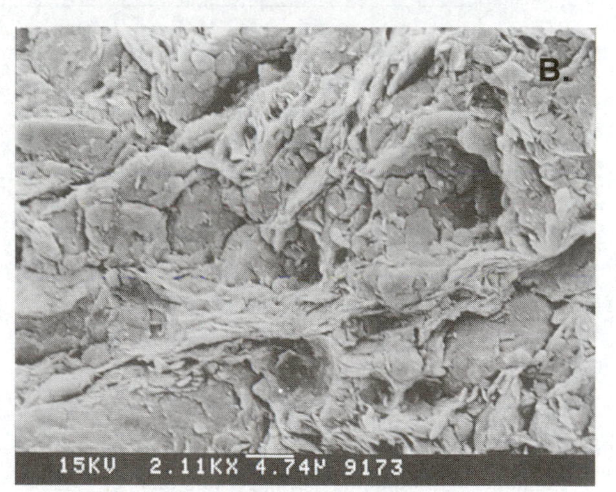

Figure 3.4 Scanning electron photomicrographs of vertical sections through the fine matrix of tills. A. Slate-rich till, showing anisotropy with elongated clasts dipping towards the right, indicating ice movement direction towards the left. B. Shears in silty, deformation till indicating movement of ice towards the right (photos: Lewis Owen).

Deformational features are ubiquitous in subglacial tills. They have been observed in numerous exposures in Britain (e.g. Hart & Boulton, 1991; Benn & Evans, 1996), continental Europe (Meer, 1993), North America (e.g. Hicock & Dreimanis, 1992), the Himalayas (Owen & Derbyshire, 1988) and Antarctica (Alley *et al.*, 1986). Indeed, the deformation of subglacial till is now thought to be the primary mechanism maintaining the high rates of ice movement associated with ice streams in Antarctica, and may also be responsible for the generation of glacier surges (Benn & Evans, 1996). Eyles (1993, p. 13) has concluded that '... it is probably true to say that subglacial deformation has been the most important till-forming mechanism in Pleistocene glaciated terrains', while Meer (1993) has suggested that, since almost all basal or lodgement tills show evidence of subglacial deformation, the sediments should be described in tectonic rather than sedimentary terms. Hence, many researchers are now using the term **glacitectonite** to refer to subglacially deformed material, although others have maintained a distinction between *deformation till,* homogenised diamict material formed by glacially induced shear of subglacial material, and *glacitectonite*, materials that have undergone subglacial shear but retain some of the structural characteristics of the parent material (e.g. Benn & Evans, 1995).

3.3.4 The influence of the thermal régime of glacier ice

A fundamental factor in the determination of the type of till that will be deposited at any one locality is the position at which debris is transported within the ice and this, in turn, is governed largely by the thermal régime of the glacier (Boulton, 1972). The thermal régime is determined by ice thickness, mass balance and, above all, climate, and can be used to define four boundary conditions at the glacier sole. These are:

A. a zone of net basal melting where more heat is provided to the glacier sole than can be conducted through the glacier;
B. a zone in which a balance exists between melting and freezing where the heat provided at the glacier sole is approximately equal to the amount that can be conducted through the glacier per unit time;
C. a zone of net basal freezing but where sufficient meltwater may still be present to raise the temperature and maintain parts of the sole at the melting point;
D. a zone in which the amount of heat provided at the sole is insufficient to prevent freezing throughout.

In zone A the glacier slips over its bed and material entrained in the basal ice layers will subsequently be deposited where frictional retardation against the bed is high. Similar processes operate in zone B although lodgement of material will tend to be greater with lower amounts of meltwater present. In zone C plucking of subglacial material occurs as the glacier slides over its bed, little lodgement till is deposited and material tends to be carried up into the ice through shearing action. Subsequent deposition therefore tends to be in the form of melt-out and flow tills. In zone D, the bed is frozen, no basal sliding occurs and glacier movement is entirely a result of internal shearing. Again, melt-out and flow tills are the dominant depositional types. Zones A and B tend to be associated with 'warm-based' or temperate glaciers, while zones C and D are found principally in 'cold-based' or polar glaciers. Hence in Spitsbergen the great thicknesses of till which have been released during retreat of the primarily cold-based glaciers are dominated by flow and melt-out tills, while in areas such as Iceland, Norway and the Alps where temperate glaciers predominate, most of the till deposited is of the lodgement type (Boulton, 1972).

There are several aspects of this work that have profound implications for the interpretation of Quaternary glacigenic sequences. First, if the relationships between glacier types and the form of deposition are applicable to older Quaternary glaciers, then it should be possible to make inferences about the thermal régime of former glaciers on the basis of their deposits. Boulton (1972) has suggested that many of the late Quaternary tills that have been described at localities in both Europe and North America, and which had been interpreted as subglacial in origin, bear striking resemblances to the supraglacial tills of Spitsbergen, and are, therefore, more likely to be flow tills than lodgement tills. This led to speculation that the last British ice sheet, for example, was largely cold-based, at least in its outer parts where it adjoined the permafrost zone beyond the ice margins (Boulton *et al.*, 1977). Secondly, observations of processes operating today in ice-marginal zones have shown how multiple till sequences can often be the product of a single retreating ice mass releasing flow tills at its margin. This has led to a fundamental reinterpretation of sequences containing several tills which were originally interpreted as the product of two or more glacial advances. Many of these sequences are now considered to be the product of vertical accretion, during a single phase of ice stagnation, of several diamictons and interbedded stratified units (e.g. Eyles *et al.*, 1982; Kluiving *et al.*, 1991; McCarroll & Harris, 1992). In the majority of cases, the diamictons are interpreted as flow tills which show clear evidence of soft-sediment deformation caused by flow under saturated conditions (e.g. Evans *et al.*, 1995).

3.3.5 Analysis of glacigenic sequences

In view of the complexities of till genesis, the analysis of glacigenic sequences requires careful application of a range of field and laboratory techniques in order to distinguish between different types of glacigenic sediment and to establish their field relationships. The aim should be to gain an understanding of both the internal structure and constituents of each sedimentary unit, as well as the three-dimensional geometry of the glacigenic sequence under investigation. A detailed description should be obtained of each lithological unit, including changes in grain size, shapes of particles, and the presence or absence of bedding and deformational structures such as shears and folds (Chapter 6, Table 6.1). Evidence of flowage of material, for example current bedding or ripple structures in stratified sediments or clast fabrics (section 3.3.6.2) in unstratified deposits, should also be recorded. It is particularly important to note the facies relationships between lithological units, including details of the nature of the contacts between them, any lateral variations both within and between beds, and any evidence for superimposition of sediments as well as the degree of deformation or of other post-depositional modifications (Figure 3.5). The analysis of facies relationships and facies variations can provide greater insights into the nature of the former depositional environment than the more traditional analysis of individual sedimentary units (Eyles, 1993). Some of these aspects of glacial sediment analysis are now considered in more detail.

3.3.5.1 Particle size and shape analysis

Individual till units often have characteristic grain-size variations and particle shapes. Particle size distributions are a function of a number of factors, including rock and mineral types of which the clasts are composed, transportational processes, transport distance and mode of deposition. In lodgement tills, for example, the particles tend to be closely packed and little winnowing by drainage takes place, so that if the sediment has a high clay content, this will tend to be preserved. Melt-out tills, on the other hand, may lose much of the finer matrix through rapid drainage which removes clay and silt particles, a process known as **illuviation**. Flow tills are much more variable – if the depositional process is passive, much of the fine matrix is preserved, whereas active slides followed by free water escape may reduce the amount of fine particles during settling. Laboratory measurement of particle size distributions, in the form of ternary plots (Figure 3.6) or cumulative frequency curves (Figure 3.7) can sometimes separate tills of different genesis, or isolate tills

Figure 3.5 Till/non-till sequence at Lunteren, The Netherlands, showing Saalian glaciofluvial deposits (A), Saalian till (B), Saalian glaciofluvial deposits (C) and Weichselian periglacial deposits (D). Older Pleistocene fluviatile deposits have been contorted (at E) during ice advance. Person at centre-right for scale (photo: Jaap van der Meer).

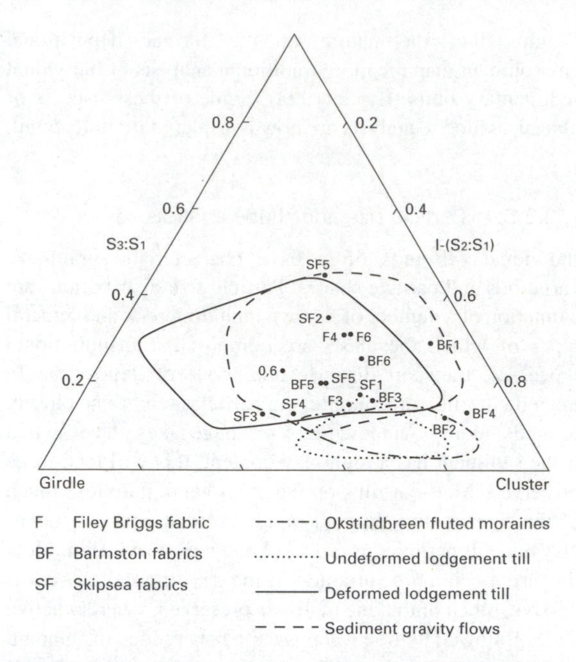

Figure 3.6 Ternary diagram using a statistical clustering method (eigenvalues) to characterise clast fabrics obtained from different types of diamicton in eastern England (from Evans *et al.*, 1995).

from other diamictons, and can even reveal subtle differences in lithology between tills of similar origin (see, e.g., Owen, 1994; Evans *et al.*, 1995). An analysis of particle shape and surface textures, along with signs of micro-wear, may also be important for establishing the mode of genesis of tills. There will be a tendency for particles to be more strongly affected by abrasion, faceting and crushing in lodgement than in supraglacial tills, for example, and so this evidence, in combination with particle size data, may help to differentiate between tills of different genesis (e.g. Bouchard & Salonen, 1990).

3.3.5.2 Lithofacies interpretations

Most interpretations of glacigenic sequences are based on the synthesis of a range of lithological data, involving detailed vertical logs of sediment exposures, and the analysis of clast lithology, fabric (section 3.3.6.2) and grain-size variations from as many beds as possible. These are plotted to provide an overview of the three-dimensional geometry of lithofacies represented at a site, and this forms the basis for the correlation of beds and reconstructions of their genesis (e.g. Figure 3.8; McCarroll & Harris, 1992). The number of logs and associated field measurements depends on the local

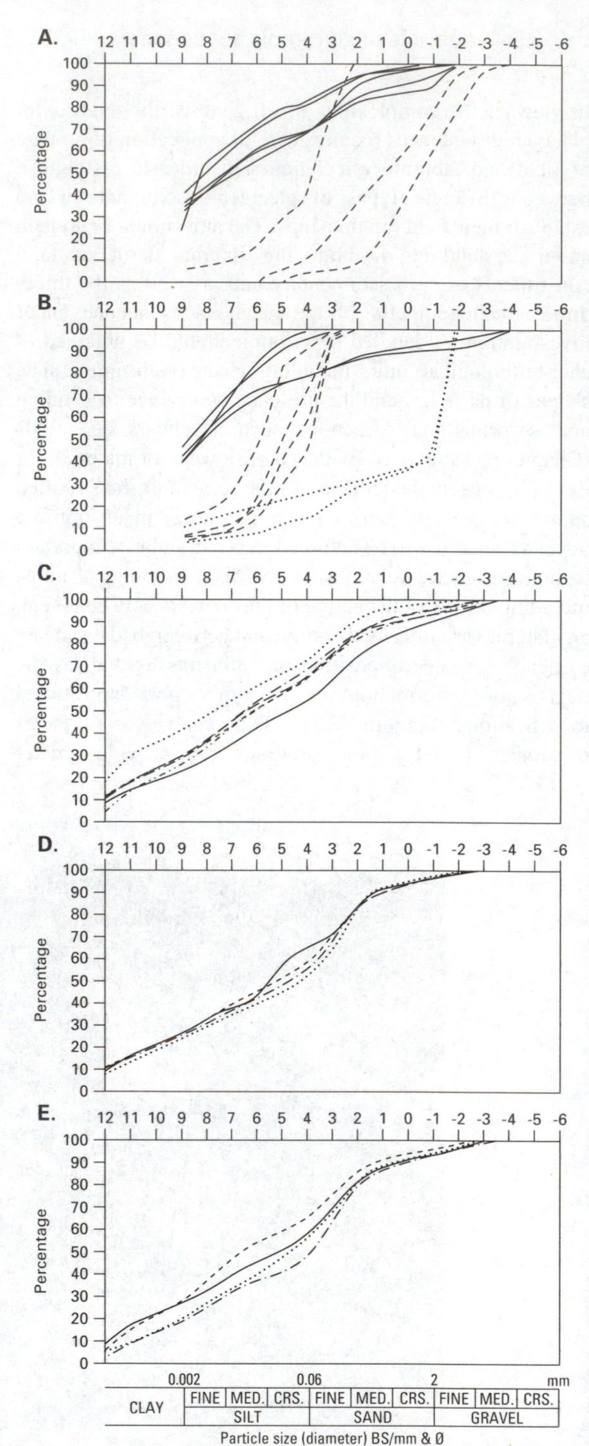

Figure 3.7 Particle size distributions (cumulative frequency curves) on the mm and Φ scales showing the distinctive curves obtained from diamictons and sorted sediments (A, B) and the way in which the technique may distinguish between different diamictons (C, D, E); note the more consistent frequency curves obtained from diamicton D compared with those of C and E (A, B from McCarroll & Harris, 1992; C, D, E from Evans *et al.*, 1995).

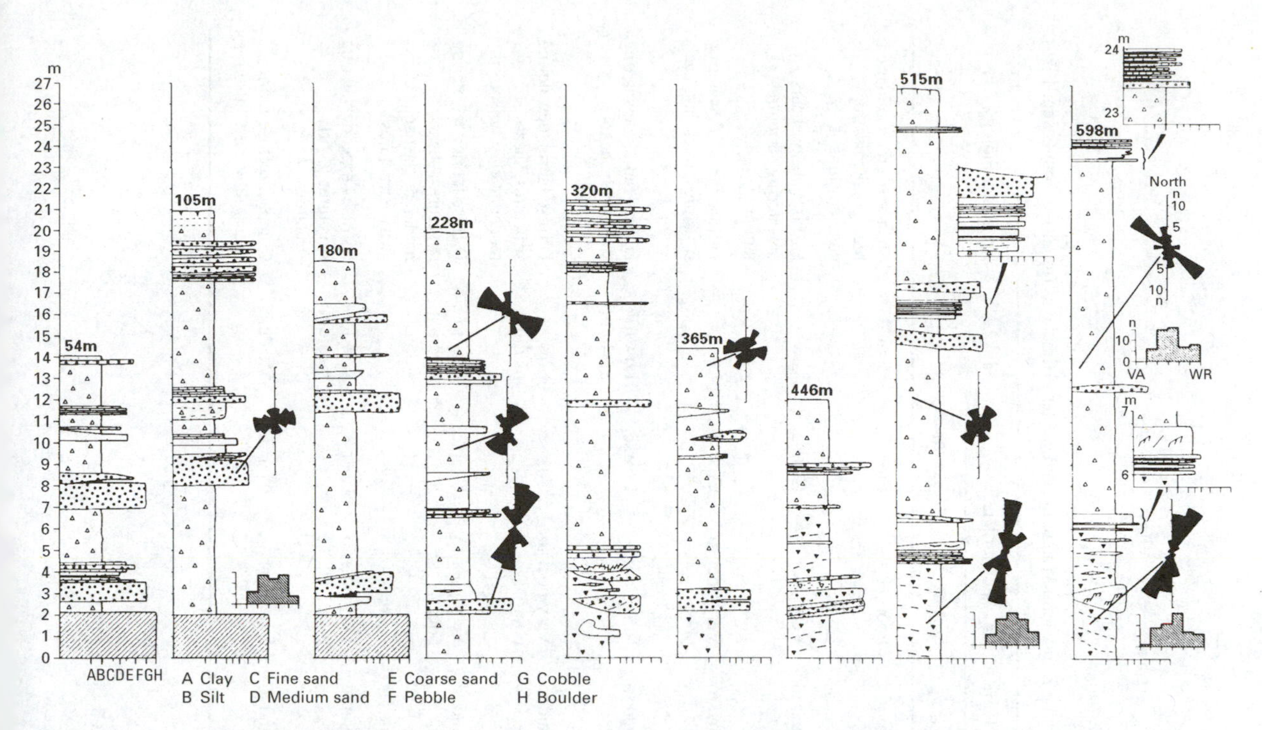

Figure 3.8 Schematic facies logs of glacigenic sediments in Wales, including rose diagrams of preferred orientations of long axes of clasts and clast roundness data (after McCarroll & Harris, 1992).

stratigraphic complexity. The more sedimentary characteristics that can be established for each bed, the more confidently can the mode of deposition be inferred. Table 3.4 gives an example of the sort of characteristics that can be logged, and how the combined information may help to differentiate between particular types of till and between tills and other diamictons (Owen, 1994).

Despite the meticulous and painstaking nature of this type of work, the evidence often remains frustratingly equivocal. This can be exemplified by the differences of view that currently exist over the interpretation of the sedimentary record of ice wastage in the Irish Sea basin in western Britain. Some have interpreted the stratigraphic evidence as indicating deposition within a glaciomarine environment (e.g. Eyles & McCabe, 1989), while others conclude that the coastal deposits around the Irish Sea were derived largely from wasting land-based glaciers (e.g. McCarroll & Harris, 1992). There is no single set of sedimentary features that can conclusively resolve such differences of opinion; interpretation depends on an evaluation of all of the lithological (and geomorphological) evidence that is available, and the matching of this information with what are considered to be best-analogue facies sequences from contemporary glacial contexts. Much depends, therefore, upon having an adequate knowledge of the links between modern glacier, glaciomarine and glaciofluvial processes, and of the lithofacies variations they produce. Our understanding of Pleistocene glacigenic sequences is therefore an iterative and on-going process, with constant refinement of the facies models upon which environmental reconstructions are based (Warren & Croot, 1994).

3.3.6 Ice-directional indicators

It has already been shown how certain landforms, such as drumlins and *roches moutonnées*, can be used to infer the direction of ice movement across a formerly glaciated area. However, some characteristics of glacial sediments can also yield valuable ice-directional information. The most widely used are indicator erratics, till fabrics and certain properties of the till matrix.

3.3.6.1 Erratics

The far-travelled particles found within, or on the surface of, a body of glacial sediment are known as **erratics**. The term is derived from the phrase *terrain erratique*, and was used initially by the French geologist de Saussure in the late eighteenth century to describe areas where material of foreign origin overlay local bedrock. The shortened term is

Table 3.4 The dominant sedimentary characteristics of diamictons of different genetic origin in the Karakoram Mountains and western Himalayas (modified from Owen, 1994).

Origin	Sedimentary structures	Clast fabric	Clast angularity	Particle size characteristic			Bulk density	Micromorphology
				Mean/phi	Sorting	Skewness		
Subglacial lodgement till	Sheared; overconsolidation structures	Strongly oriented up-valley	Edge rounding; angular	Fine sand to coarse silt	Poor	Fine	High	Shears; overconsolidation; high grain anisotropy; edge crushing
Subglacial and englacial melt-out till	Massive; sandy lenses; shears	Low anisotropy	Slight edge rounding	Medium sand to coarse silt	Very poor	Coarse	Low	Diffuse fabric; shears; dewatering structures; laminae
Supraglacial melt-out till	Massive; slumped; intercalations; dewatering structures	Low anisotropy	Very angular to angular	Medium sand to coarse silt	Very poor to poor	Coarse	Low	Diffuse fabric; dewatering structures; microlaminae
Supraglacial slide till	Massive; crude downslope stratification	Low anisotropy	Very angular to angular	Medium sand to coarse silt	Very poor	Coarse	Low to medium	Diffuse fabric; dewatering structures; shears
Supraglacial flow tills	Massive; downslope stratification; levéed channels	Weak down-valley	Very angular to angular	Medium sand to coarse silt	Very poor to poor	Coarse	Low to medium	Diffuse fabric; dewatering structures; shears; moderate clast anisotropy
Debris flow sediments	Massive; downslope stratification; levéed channels; shears	Weak down-valley	Very angular to angular	Medium sand to coarse silt	Very poor to poor	Coarse	Low to medium	Diffuse fabric; dewatering structures; shears; moderate clast anisotropy; swirly structures
Flowside sediments	Massive; crude sub-horizontal stratification; shears	Low anisotropy	Very angular to angular	Medium sand to coarse silt	Very poor to poor	Coarse	Low to medium	Diffuse fabric; dewatering structures; swirly structures
Rockslide sediments	Massive; shears	Very low anisotropy	Very angular to angular	Medium sand to coarse silt	Very poor	Coarse	Medium	Compact-diffuse fabric; shears; low clast anisotropy

Figure 3.9 The giant quartzite erratic near Okotoks, Alberta, Canada (photo: Neville Alley).

now generally applied to a particle of any size that is not indigenous to the area in which it is currently found. The most valuable types of erratic are those for which the sources are known, which are resistant to erosion, and which have distinctive appearances, unique mineral assemblages or unique fossil contents, thereby allowing unequivocal identification. These are usually termed **indicator erratics** and range in size from finely comminuted fragments to large blocks weighing several hundred tons. The famous Okotoks erratic (Figure 3.9) in southwest Alberta, for example, is a block of quartzite weighing over 18 000 tonnes which appears to have travelled over 100 km from its source in the Rocky Mountains to the northwest of Calgary. At the other end of the scale, distinctive **indicator minerals** can also be used to infer till origin or *provenance* (e.g. Peuraniemi, 1990). Erratic distributions can provide information on both local and regional ice-flow directions (Figure 3.10). They have also been used to reconstruct patterns of ice-sheet deglaciation, where changing flow patterns and migrations of ice divides can be inferred from erratic distributions (Bouchard & Salonen, 1990). In this respect they are also an important data source for the development of ice-sheet models (section 2.3.4).

One problem with using erratics as ice-directional indicators, however, is that their presence in a glacial deposit may not always reflect primary derivation (i.e. an erratic could have been removed during a previous glacial episode and reincorporated into a younger till) and this can lead to erroneous interpretations of former patterns of ice movement. Hence, although erratics can frequently provide useful ice-directional information, they are perhaps best used in conjunction with other independent sources of evidence.

3.3.6.2 Till fabrics

The arrangement of particles in a till is termed the **till fabric**. It was observed at a very early stage in the development of glacial geological studies that stones (clasts) within a till often displayed a preferred orientation, although it was somewhat later that a quantitative relationship was established between stone orientations in till and patterns of ice movement. Over the last few decades till fabric analysis, involving the measurement of orientation and dip of particles within a till matrix, has become one of the most widely used techniques for reconstructing former ice-flow directions (Warren & Croot, 1994). The technique rests on the assumption that, within the constantly deforming layer at the base of the ice, stones will become orientated to adopt the line of minimal resistance to flow, i.e. with their long axes parallel to the flow direction. Subsequent deposition of the subglacial debris in the form of lodgement till therefore preserves a record of the former direction of ice movement, and this can be established by measuring the orientation of pebbles in lodgement till exposures using a compass. Measurement of the dip by means of a clinometer or a similar instrument may also provide useful ice-directional information, as a tendency has been observed for pebbles in a lodgement till to dip up-glacier.

Till fabric data are usually presented in the form of polar graphs. Where two-dimensional data only have been obtained, a rose diagram is constructed showing the number or proportion of pebbles in different azimuthal classes. However, because each measured stone is represented by two opposite azimuthal values (e.g. 30° and 210°), the rose diagram actually consists of two reflected halves or mirror images. The data can be shown by a line through the middle of each sector (Figure. 3.11A), by the linking of such lines to form the typical rose diagram (Figure 3.11B), or by the shading of each azimuthal class to the extent of the line marking the outer limit of each class (Figure 3.11C). Where the dip of the pebbles has been recorded, orientation measurements are taken in the down-dip direction, and therefore each pebble is represented by a single dip and orientation value. In this case the diagram will show a full 360° distribution (Figure 3.11D). Orientation and dip can be plotted together in the form of a scattergram with the radius divided into degrees showing the angle of dip and the circumference divided into degrees showing the orientation of the pebble (Figure 3.11E). This is one of the most commonly used methods for depicting till fabric data, and in some cases the visual effect is enhanced by contouring the diagram as shown in Figure 3.11F.

The interpretation of till fabric data requires a knowledge of the statistical basis of analysing azimuthal data as well as of the variability in fabric pattern that characterises each lithological unit (Briggs, 1977; Dowdeswell & Sharp, 1986). As noted above, till fabric is generated by the action of flow. Under lodgement processes, the proportion of stones that become aligned parallel to the flow direction depends upon clast concentration, speed of flow, particle shape, flow mechanics, and so on. The first thing to establish, therefore,

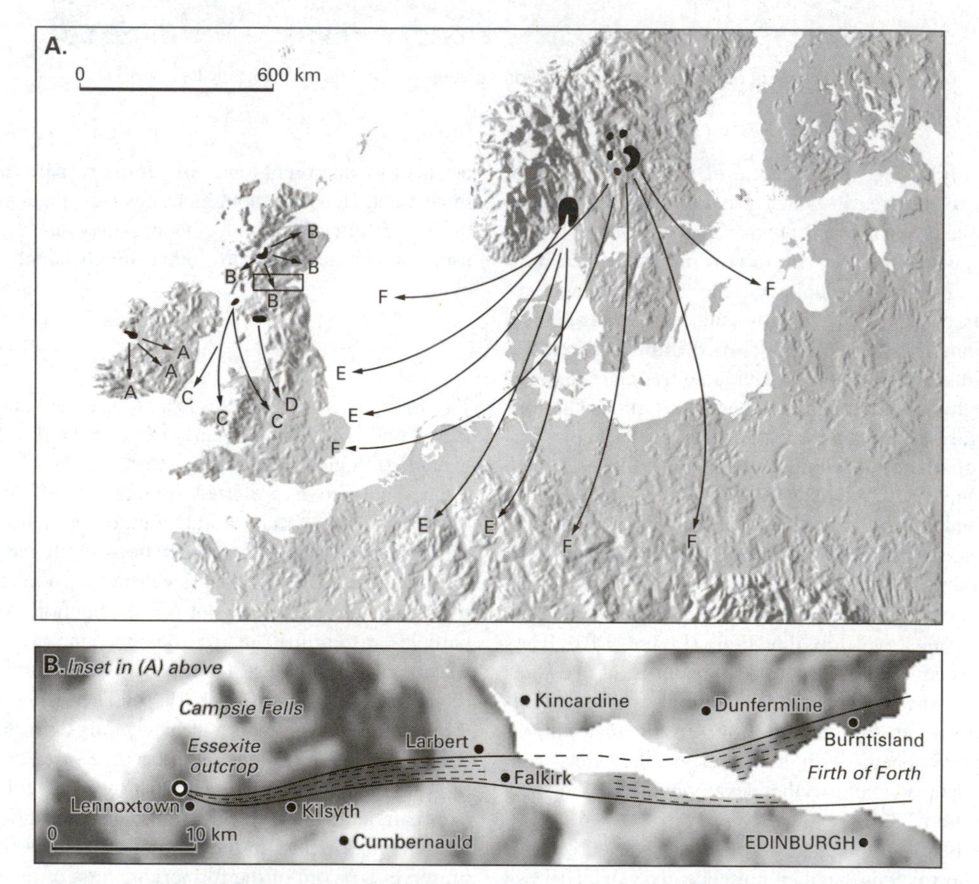

Figure 3.10 **A.** Distribution of some indicator erratics by ice in Britain and northwest Europe: A Galway granite, B Rannoch granite, C Ailsa Craig riebeckite–eurite, D Criffell granite, E Oslo rhomb porphyry, F Dala porphyries. **B.** The Lennoxtown boulder train in the Forth Valley lowlands of central Scotland. Based on diagrams in Sissons (1967) and West (1977).

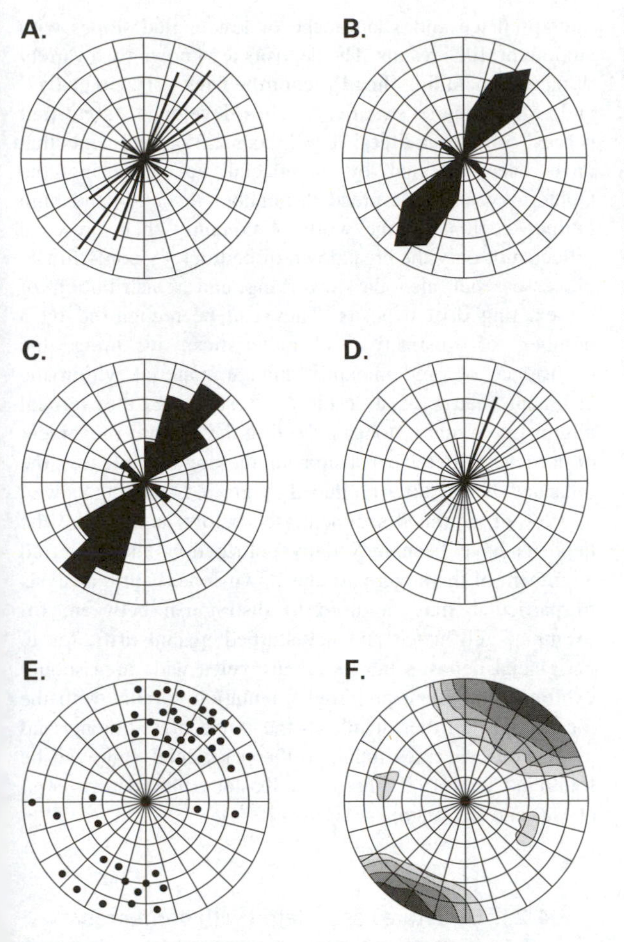

Figure 3.11 Different methods for representing till fabric data. For explanation see text.

is the relative strength of alignment of particles and how the degree of alignment varies throughout a lithological unit. Flow tills, for example, which were generated under saturated conditions and flowed down a steep ice surface before settling, tend to show a very strong preferred orientation throughout the sediment body. However, fabrics from flow tills invariably indicate only very localised conditions of flowage with neighbouring till units often exhibiting markedly different flow directions (Boulton, 1975). Melt-out tills, on the other hand, frequently show little preferred orientation where clast concentration was low and melting from within the ice was slow. Melt-out tills may retain their englacial fabric (Johnson *et al.*, 1995), but as this is easily lost during the final melting phase, these too can be unreliable indicators of regional ice movement.

Where the objective is to determine the regional ice-flow direction, fabrics should be obtained from lodgement tills. However, lodgement tills can also display considerable spatial variability, even within a single quarry (Rose, 1974). The fabrics of lodgement tills are affected by the morphology of the glacier bed, especially where basal ice is forced around prominent bedrock protuberances (e.g. Catto, 1990). They may also show marked local variability where the tills were affected by changes in the direction of regional ice movement during deposition (e.g. Hicock, 1992). A further complicating factor is that subsequent ice advances can cause reorientation of the fabric in previously deposited tills. In order to reconstruct regional ice-flow directions from till fabric data, therefore, a number of basic rules must be observed (Hirvas & Nenonen, 1990):

1. Samples should only be taken from lodgement till. This, however, is often easier said than done since some flow tills can resemble lodgement tills and it may not always be possible to differentiate between the two, particularly in limited exposures. Lodgement tills may also be affected by deformational stresses that are not always operating parallel to ice-flow direction. A large number of till fabrics will always be required from an area, therefore, in order to eliminate local anomalies. Moreover, stratigraphic and sedimentological characteristics of the tills should always be considered in the interpretation of fabric data.
2. In view of the above, it is, of course, axiomatic that individual fabric measurements should always be taken from what is known to be the same till unit.
3. Fabrics should be obtained, wherever possible, from areas of low relief so that the effects of local flow deflections are minimised. In undulating terrain, flat summits are preferred to valley-side and valley-bottom sites.
4. Fabrics should be used, as far as possible, with other indicators of ice-flow direction.

3.3.6.3 Properties of the till matrix

Certain physical and chemical properties of the till matrix can be used to make inferences about regional ice movement, and also to differentiate between tills from different source areas. This approach rests on the assumption that tills will inherit certain textural and chemical characteristics from bedrock over which the glacier has travelled. In certain cases, where a particularly distinctive lithological type lies up-glacier, it may be possible to detect the influence of that rock outcrop on the till matrix. In a sense, therefore, certain properties of tills can be used in the same way as indicator erratics. For example in Norway, Jørgensen (1977) showed how particle-size contrasts between tills from Pre-Cambrian bedrock differed from those derived from sedimentary Cambro-Silurian rocks, and then augmented these data with the results of chemical and mineralogical analyses to determine transport distance and direction of the till material. Further south, analysis of the clay mineral composition (particularly smectite and illite abundances) of the Saalian tills of The Netherlands revealed evidence of a source region in the Pre-Cambrian and Cambro-Silurian areas of the Baltic some 1000 km to the north (Haldorsen *et al.*, 1989). These data not only provided ice-directional information, but also suggested transport of some of the till material at relatively high englacial levels within the ice, and over considerable distances.

Other properties of the till matrix that have been used to infer ice direction include mineral magnetic data (Walden *et al.*, 1987, 1992), mineral geochemistry and heavy minerals (Peuraniemi, 1990), lead isotope ratios (Bell & Murton, 1995) and microfossil content (e.g. Dreimanis *et al.*, 1989; Meer & Wicander, 1992). In addition to providing insights into former patterns of ice movement, the analysis of indicator minerals in tills also has applications in economic geology, for ore bodies can be located using evidence of glacially transported debris (Peuraniemi, 1990).

In conclusion, careful study of glacial sediments, particularly tills, can provide a considerable amount of information about Quaternary glacial environments. A range of techniques, involving both geomorphological and sedimentological evidence, is now available for the establishment of former ice-flow directions, and it is the judicious application of these methods, particularly in combination, which makes possible the types of reconstructions shown in Figures 2.12 to 2.14. In addition, considerable advances have been made in recent years in understanding the patterns and processes of glacial sedimentation, and the results of these studies now hold out the prospect of being able to infer the type of glacier, and perhaps also the climatic controls, responsible for the deposition of a particular sediment body.

3.4 Periglacial sediments

3.4.1 Introduction

In the 'periglacial domain' (section 2.4), freeze–thaw activity causes fracture of the country rock and the accumulation of coarse, angular debris. This material moves downslope through the combined processes of flowage (**gelifluction**) and creep induced by the growth and melt of interstitial ice and a landscape of low-angled slopes with smooth profiles results. The deposits are known by a variety of names, including **'head'**, **'coombe rock'**, **'tjaele gravel'** and, where coarse stratification has developed, **'stratified screes'** or **'grèzes litées'** (e.g. Figure 2.25). Sediments that have been affected by periglacial action during the Quaternary are widespread throughout the mid- and high latitude regions of the world. Frequently their presence reflects not only the breakdown of bedrock by cold-climate processes, but also the reworking and redistribution of pre-existing drift deposits. They can be recognised by a number of distinctive characteristics, including the occurrence of predominantly angular material within the sediment matrix as a result of frost-riving, the vertical alignment of many stones reflecting the upward movement of particles with the expansion and melting of ice lenses, the presence of structures produced by ground cracking as well as flow of saturated sediment (see below) and, where the deposits have been moved by gelifluction, the preferred alignment of the larger particles downslope. Fabric analysis in particular may be used to distinguish between, for example, gelifluctеd and undisturbed glacial drift, for in periglacial deposits fabrics taken over a wide area should exhibit a consistent preferred orientation parallel with the local slope. Further details of the range of sediments and associated processes found in the periglacial zone can be found in Clark (1988), French & Koster (1988), Harris *et al.* (1988) and Ballantyne & Harris (1994).

3.4.2 Structures associated with permafrost

Although the presence of frost-shattered bedrock and extensive spreads of gelifluction deposits is indicative of a former periglacial climatic régime within a particular area, it normally provides only the most generalised of information about former environmental conditions. However, where sediments show evidence of ground-ice activity, more precise palaeoclimatic inferences are possible. This section deals with two types of structure that result from the deep penetration of ground ice: **ice-wedge casts** and **involutions**.

Ice wedges are considered to be diagnostic structures of permafrost (Harry & Gozdzik, 1988) and form where thermal contraction in winter opens vertical cracks in the permafrost table into which water seeps and subsequently freezes. The incremental accumulation of ice in veins and wedges along the axes of the contraction cracks leads to the growth of wedges. They can be up to 3 m in width and 10 m in depth, but mature wedges are typically 1–1.5 m in width and 4 m in depth. If they develop some time after the accumulation of the sediments in which they form, they are referred to as **epigenetic ice wedges**. Occasionally, however, cracks form in sediments that are still accumulating, and the wedges extend upwards to keep pace with sediment aggradation; these are referred to as **syngenetic** (or **synsedimentary**) ice wedges. The latter can grow to exceptional depths as, for example, the 10 m long forms reported from Poland (Klatkowa, 1990). Upon melting, the ice is replaced by material falling into the cracks from above and from the sides (Murton & French, 1993). In this way a cast or **pseudomorph** of the original form of the ice wedge is preserved (Figure 3.12). Ice wedges typically form as part of a network of thermal contraction cracks which appear as interconnected polygons on the ground surface, although in sections they are often found as single features.

Studies of active ice wedges in present-day cold regions show that they can develop in a variety of sediments and soils (e.g. Jethick & Allard, 1990) and that they occur only in the zone of **continuous permafrost** (e.g. Ballantyne & Harris, 1994). They therefore provide unambiguous evidence of the former existence of perennially frozen ground. In the **discontinuous permafrost zone**, most ice wedges appear to be inactive. In arid and semi-arid periglacial regions and in some localities that are free-draining, frost fissures that develop from thermal contraction of the ground are frequently filled with wind-blown sediment and are termed **sand wedges** (Figure 3.13). Occasionally, though, they are filled with other materials,

Figure 3.12 Ice-wedge pseudomorph, Westdorp, The Netherlands (photo: John Lowe).

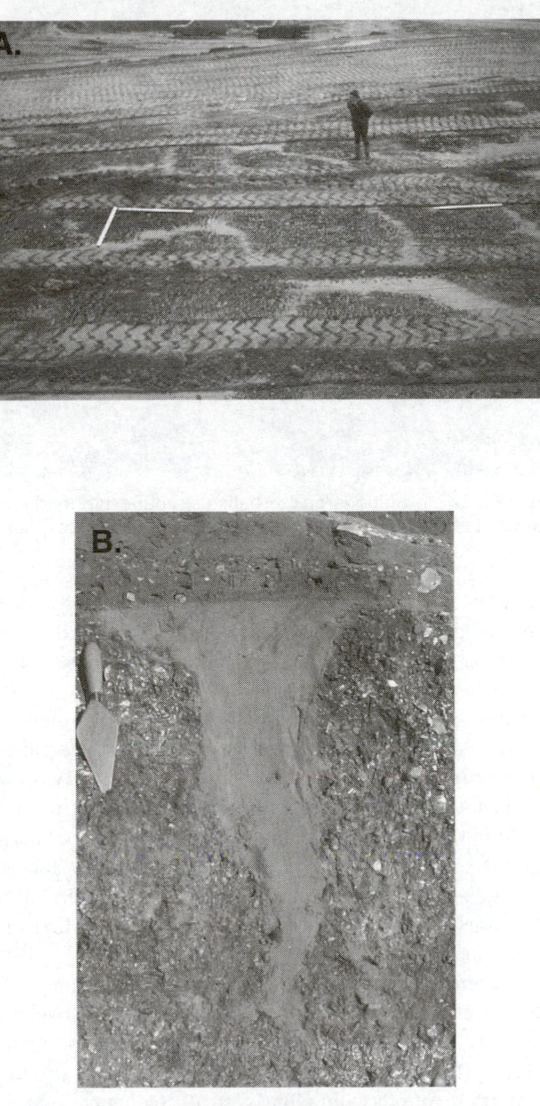

Figure 3.13 A. Early Anglian stage sand wedges forming patterned ground, exposed during quarry operations at Newney Green. B. Early Anglian sand wedge in profile, Broomfield. Both sites in Essex, England (photos: Peter Allen).

Figure 3.14 Involutions/cryoturbation structures, formed in Late Weichselian fluvial sediments during the Younger Dryas Stadial, exposed at the site of Bosscherheide, The Netherlands (photo: Jef Vandenberghe).

including soil, so that in eastern Europe the more general term of '**ground wedge**' is used (Harry & Gozdzik, 1988), while the term '**soil wedge**' is used to refer to those features that have a high soil content.

Sand-wedge pseudomorphs can sometimes be distinguished from ice-wedge casts by the characteristics of the sediment infill, but a clear differentiation between the two is not always possible in the field and the two forms can be found in close association in present-day permafrost regions (Worsley, 1984). Care also needs to be taken to distinguish ice wedge casts from other deformational structures, such as water escape features which, at first sight, often appear very similar (e.g. Burbidge *et al.,* 1988). Characteristic features of wedges that can assist identification include: (i) depth:width ratios of between 3:1 and 6:1, which conform with the dimensions of modern wedges (Ballantyne & Harris, 1994); (ii) slump structures and stratification within the cast-fill sediment, which is usually concave downwards ('sag' structures); (iii) the presence of large joints and normal faults within the wedge structure and of associated micro-joint patterns in adjacent sediments, produced by freezing during the development of patterned ground (Mol *et al.,* 1993); and (iv) if the wedge cast has not been truncated by erosion, the top of the cast may merge with other evidence indicating the position of the former permafrost table, such as a stone pavement or cryoturbated horizon (e.g. Vandenberghe, 1985; Ran *et al.*, 1990).

In many areas where periglacial conditions prevailed, unconsolidated sediments in open sections frequently display contortions in bedding, the interpenetration of one layer by another, and pockets which resemble **load structures**.[6] Such features are termed **involutions, cryoturbations** or **festoons** (Figure 3.14) and are widely interpreted as reflecting differential pressures induced by freezing within the active layer above the permafrost table (Vandenberghe & van den Broek, 1982). The convoluted structures sometimes appear to be irregular in spacing and heterogeneous in form, but a close inspection often reveals a degree of order. Vandenberghe (1988) has recognised six different types of involution on the basis of their morphology (Figure 3.15): (1) isolated folds of small amplitude (i.e. depth of structure) but large wavelength; (2) regular, symmetrical and well-developed forms with amplitudes of 0.6–2 m; (3) smaller-scale versions of type 2; (4) solitary features of teardrop (4a) or diapiric (4b) form; (5) sediment injected upwards into polygonal cracks; and (6) irregular deformation structures.

Three main modes of formation have been proposed for involution structures (Ballantyne & Harris, 1994). **Periglacial loading** involves the density inversion of sediments during the thaw of frozen ground. Melting may produce pockets or layers of mobile saturated sediment that become liquefied and injected into overlying deposits to be replaced by denser, less fluid material which sinks. This process probably accounts for most of the features classified as types 2, 3 and 4 in Figure 3.15. The second process involves the movement of material under **cryohydrostatic pressure**, as the 'freezing front' descends from the surface each autumn towards the upper surface of the permafrost table at the base of the active layer. This leads to a build-up of high pore-water pressures in the material trapped within the unfrozen part of the active layer, and subsequent deformation of the liquefied sediments. The third mechanism is **differential frost heave**, which results from the differential rate of freezing in sediments of varying composition. Frost will penetrate coarser sediment more quickly than finer deposits, and the water in the latter has a lower freezing point. This results in differential pressures as the ground freezes, leading to mass displacement (heaving) of sediment.

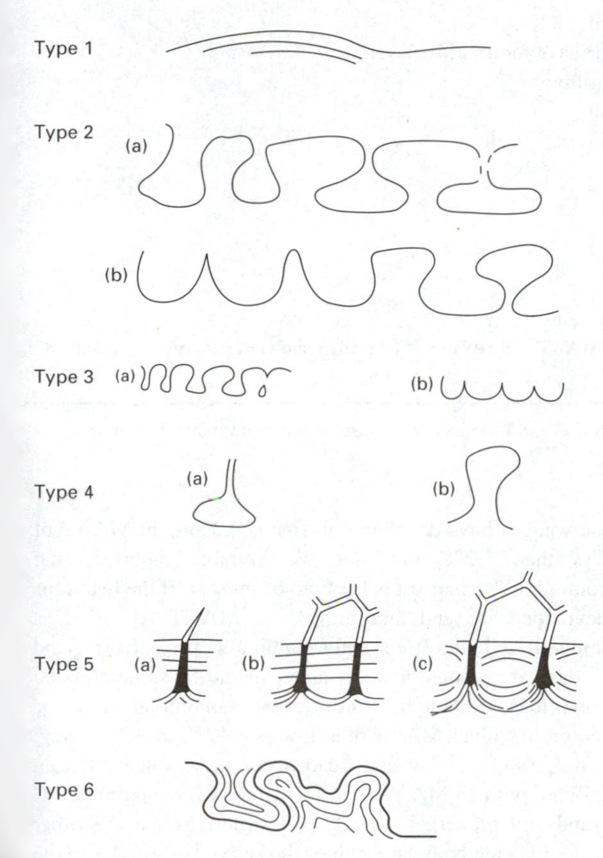

Figure 3.15 Classification of cryoturbation structures according to their form (symmetry, amplitude, wavelength and pattern of occurrence) (based on Vandenberghe, 1988, after Ballantyne & Harris, 1994).

As is the case in the interpretation of ice-wedge casts, however, care has to be taken before attributing a periglacial origin to involutions, since very similar structures can also be produced under non-freezing conditions where, for example, high-pore water pressures build up and escape routes are created in overlying layers by tectonic effects, or where density variations lead to the injection of liquefied sediments into overlying sediment bodies. However, there is often a close field relationship between cryoturbations and other undoubted periglacial phenomena, such as ice wedges, stone pavements and upturned stones, and thus where the depositional context is clearly periglacial, the involutions can be interpreted more confidently as frozen ground phenomena (Ran *et al.*, 1990; Vandenberghe, 1992).

3.4.3 Palaeoclimatic significance of periglacial structures

As ice wedges and, in some cases, involutions are associated with permafrost, the former extent of permanently frozen ground can be deduced from the distribution of ice-wedge casts and involution structures. Thus, if the conditions under which permafrost is generated can be ascertained, then the fossil forms will enable palaeoenvironmental reconstructions to be made (Table 3.5). However, permafrost covers very large parts of the Northern Hemisphere, extending southwards into Alaska, Canada, Siberia and other parts of Asia. It occupies some 2.15 million km^2 (22.4 per cent of the territory) of China today (Qui & Cheng, 1995) and some 25 per cent of the southern circumpolar region (Bockheim, 1995). Clearly, with such an areal extent, there can be no single definition of what constitutes a 'permafrost climate'. It is generally accepted, however, that permafrost will only occur where the **mean annual air temperature (MAAT)** drops below 0°C, although an upper MAAT limit for permafrost development of −2°C, especially for mountain regions, has also been suggested (e.g. Haeberli, 1985). The development of a thick permafrost layer requires centuries of sustained freezing. In both Siberia and North America, the present southern limit of continuous permafrost coincides with a MAAT isotherm of −8°C (Brown *et al.*, 1981), while the

Table 3.5 Environmental implications of some relict periglacial phenomena (modified after Ballantyne & Harris, 1994).

Feature	Environmental implications
1. NEAR-SURFACE CRYOGENIC STRUCTURES IN UNCONSOLIDATED SEDIMENTS	
Ice-wedge casts	Severe winter ground cooling. In fine sediments: continuous or discontinuous (?) permafrost; MAAT < −3°C to −4°C. In sand and gravel: continuous permafrost; MAAT < −6°C
Tundra polygons	As above
Sand wedges	As above. Abundant supply of aeolian sand may indicate regional aridity
Soil wedges	Severe winter cooling of seasonally frozen ground. In fine sediments: MAAT < +1°C. In sand and gravel: MAAT < −1°C
Pingo scars and related depressions	Shallow discontinuous permafrost. MAAT probably < −4°C to −5°C, but not markedly lower
Thermokarst depressions	Continuous (?) ice-rich permafrost
Large-scale non-sorted patterns	Probably indicative of permafrost
Large-scale sorted patterns	Strong but not conclusive indicator of permafrost: MAAT < 0°C to −2°C. May indicate active layer depth
Cryoturbation structures	Depth of annual freeze–thaw; some indicate depth of active layer above permafrost
2. MASS-WASTING LANDFORMS AND STRUCTURES	
Solifluction sheets and lobes	Seasonal freeze–thaw of soil
Active layer detachment slides	Permafrost. May indicate depth of former active layer
Ground-ice slumps	Rapid thaw of ice-rich permafrost
Granular head deposits	Seasonal freeze–thaw of soil
Clayey head deposits	Active layer sliding (see above): permafrost
3. AEOLIAN AND NIVEO-AEOLIAN FEATURES	
Loess deposits, coversands and sand dunes	Dominant wind direction
4. TALUS-RELATED LANDFORMS	
Protalus ramparts	Former regional snowfall patterns
Protalus rock glaciers	Discontinuous permafrost: MAAT < −1°C to −2°C; former snowfall patterns; mean annual precipitation

N.B. Evidence for continuous permafrost implies MAAT less than c. −6°C to −8°C; evidence for discontinuous permafrost implies MAAT in the range *c.* −1°C to −8°C.

southern limit of discontinuous permafrost in North America coincides with the −1°C isotherm. These relationships are not always consistent, however, for isolated but locally continuous areas of permafrost occur south of both of these isotherms. Altitudinal limits for permafrost have also been identified in some mountainous regions, such as Norway, where continuous permafrost occurs at MAATs that fall below −6°C and discontinuous permafrost below −1.5°C (King, 1983). In the Rocky Mountains of North America, discontinuous permafrost occurs where MAATs fall below −1°C.

More important in the context of palaeoclimatic reconstructions, however, are the specific climatic conditions required for the generation of ice-wedge casts and cryoturbations (Ballantyne & Harris, 1994). Where the ice wedges have developed in fine sediment, an MAAT of less than −3°C to −4°C is probably required, and continuous permafrost is likely to be present. If the feature is developed in sand and gravel, an MAAT of −6°C is implied, and almost certainly continuous permafrost. Sand wedges are thought to form under much the same climatic conditions, although more severe conditions may be required, with an MAAT of as low as −12°C to −20°C (e.g. Karte, 1983). Soil wedges can occur in seasonally frozen ground, with an MAAT of +1°C (fine sediments) to −1°C (sand and gravel). Cryoturbation structures, on the other hand, indicate both the depth of the active layer and also the likely MAAT, with large-scale structures indicating permafrost conditions and an MAAT of, at the most, −6°C (Vandenberghe, 1988).

In addition to palaeotemperature estimates, some indication may also be gained from periglacial sedimentary evidence of former precipitation gradients. It was noted above, for example, that sand wedges are often characteristic features of arid periglacial environments, i.e. polar deserts (Kolstrup, 1987; Vandenberghe & Pissart, 1993). The low levels of ground moisture mean that they tend to be well preserved, for in the absence of large amounts of ground ice, thawing will not lead to deformation. Ice-wedge casts, on the other hand, are indicative of more humid conditions, although it would seem that true ice wedges are also unlikely to form where precipitation levels are high. Evidence from present-day Arctic areas suggests that low winter snowfall is as vital a condition for the growth of ice wedges as severe cold, for as snow is an excellent insulator the ground will not crack if there is a snow cover in excess of 15–25 cm (Williams, 1975). A 25 cm snow cover is equivalent to a little over 100 mm of winter rainfall. Since about 60 per cent of total precipitation in many Arctic regions falls in the summer months, an annual precipitation of 250 mm (in rainfall equivalent) would appear to be the maximum possible if ice wedges are to form. An annual precipitation of 250 mm is markedly less than that found in many mid-latitude regions where ice-wedge pseudomorphs occur, implying significantly drier conditions in these areas at the time that the ice wedges were forming (Vandenberghe, 1992).

The distribution of ice and sand wedges and cryoturbations, in conjunction with other periglacial indicators (e.g. landforms), forms the basis for reconstructions of the periglacial environment of western Europe during the last cold stage. For example, Williams (1969) used the distribution of periglacial phenomena to establish the approximate limit of permafrost in England and Wales during the Last Glacial Maximum (Figure 3.16A) while similar data have been employed to establish the southern margin of permafrost in France, Belgium and The Netherlands (Figure 3.16B). The latter reconstruction suggests that the continuous permafrost limit may have been much further west than envisaged by Williams (Van Vliet-Lanöe, 1988). A synthesis of periglacial data from various parts of Europe shows the development of marked west–east gradients, with 'wet-aeolian facies' and ice-wedge casts predominating in the west, but sand wedges and 'dry aeolian facies' dominating in the east (Böse, 1991). The data also indicate much more severe and widespread permafrost in the earlier 'Pleniglacial' period than during the Late Weichselian Lateglacial.

The most detailed regional palaeoenvironment reconstructions based upon periglacial phenomena are those from The Netherlands. This area lay just to the south of the Fennoscandian ice sheet margins during successive cold stages, and periglacial features are abundant (and very well preserved) in the fluvial and aeolian sediments. Indeed, they are found in every cold stage deposit from the early Pleistocene through to the last cold stage (Vandenberghe & Kasse, 1989; Vandenberghe, 1992). On the basis of variations in the type and scale of periglacial features in different stratigraphic horizons, the intensity of former periglacial conditions can be deduced. This is illustrated in Figure 3.17, where the data suggest that the most intense cold episodes during the last cold stage were during the early Pleniglacial (*c*. 72–62 ka BP) and especially during the Late Pleniglacial (*c*. 26–13 ka BP) at the time of the last glacial maximum. Wetter and less severe conditions are inferred for the long Middle Pleniglacial period, between *c*. 62 and 26 ka BP (Ran *et al*., 1990; Vandenberghe, 1992; Vandenberghe & Pissart, 1993). Climatic reconstructions for the last cold stage based upon periglacial phenomena have also been suggested for Poland, another area of Europe that lay just beyond the limits of the last ice sheets. The evidence indicates that the Middle Pleniglacial was relatively mild and dry, but this was succeeded first by a subpolar desert with a semi-arid régime, and eventually by full polar desert conditions of extreme aridity during the Last Glacial Maximum (Krzyszkowski, 1990). These data show how former periglacial climatic zones, and the climatic gradients associated with them, can be established at both the regional and continental scales.

Fossil periglacial phenomena, therefore, appear to be a useful data source for the reconstruction of Quaternary environments. However, a number of workers have urged caution over the use of this type of evidence in palaeoclimatic reconstruction. Present-day Arctic areas are not necessarily good analogues for the periglacial régimes that prevailed in the mid-latitudes during former cold stages, since the latter would have experienced quite different solar radiation cycles from those that prevail at the poles today. As a consequence, many areas of the mid-latitudes would have experienced considerably more diurnal freeze–thaw cycles than the present High Arctic where the seasonal periodicity of daylight and darkness favours longer, more severe cycles and deeper ground freezing (Washburn, 1979). Certainly, mid-latitude areas did not experience the extremes of freezing that now characterise, for example, interior Siberia and Canada and, in view of the proximity of large glacier masses, probably had a climatic régime which differed in a number of respects from that of the present-day high latitudes, particularly in terms of precipitation, wind direction and wind intensity (French, 1996). Difficulties may also arise from the sometimes ambiguous nature of periglacial evidence, and from the fact that, in periglacial regions, geomorphological processes are affected as much by local site factors as by prevailing climatic conditions. Finally, there is the problem of dating. Ice wedges and

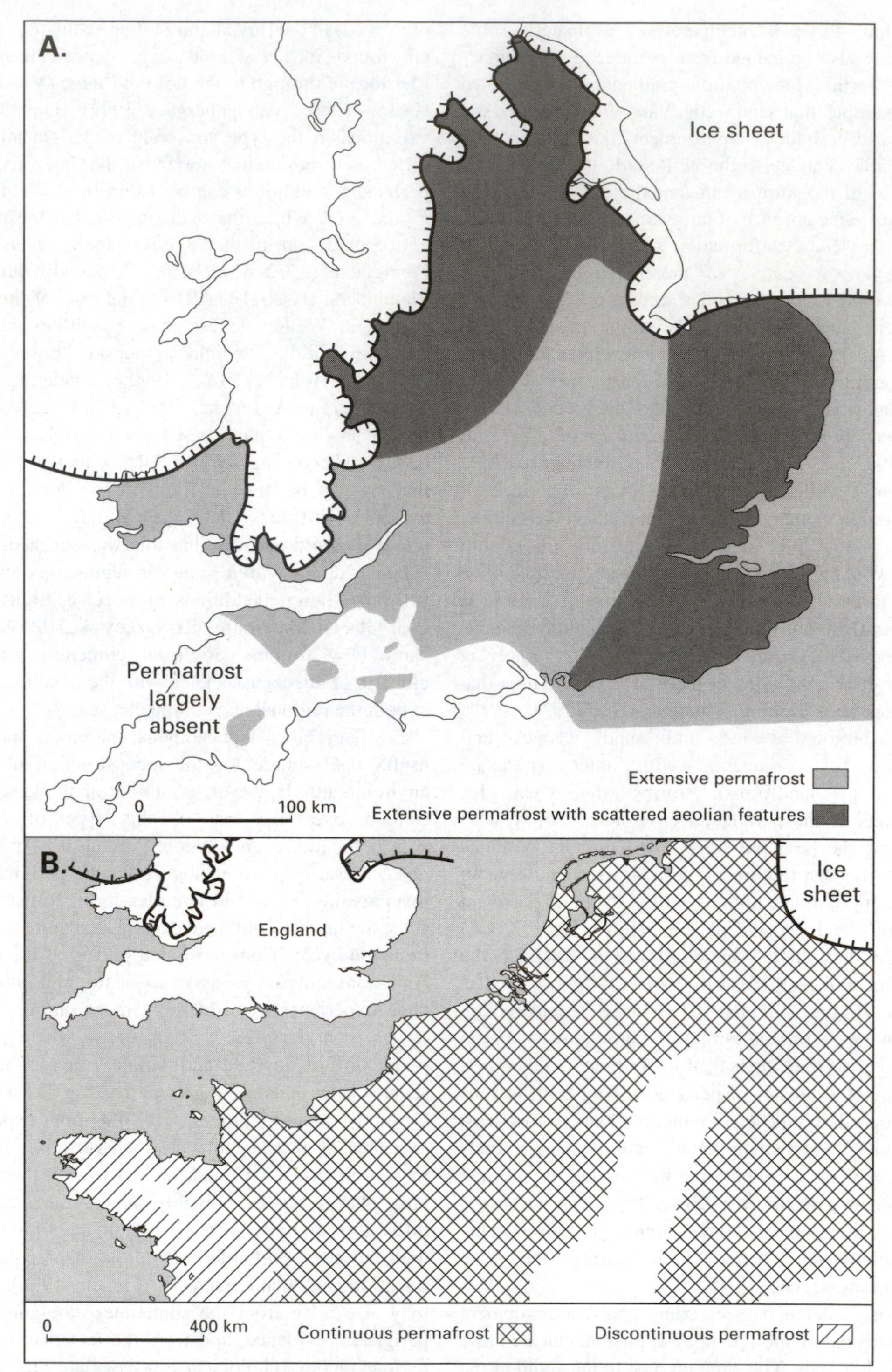

Figure 3.16 Reconstructions of the distribution of permafrost in England (A) and northern France, Belgium and The Netherlands (B) at the time of the last glacial maximum, based on the distribution of periglacial sediments and landforms (from Ballantyne & Harris, 1994).

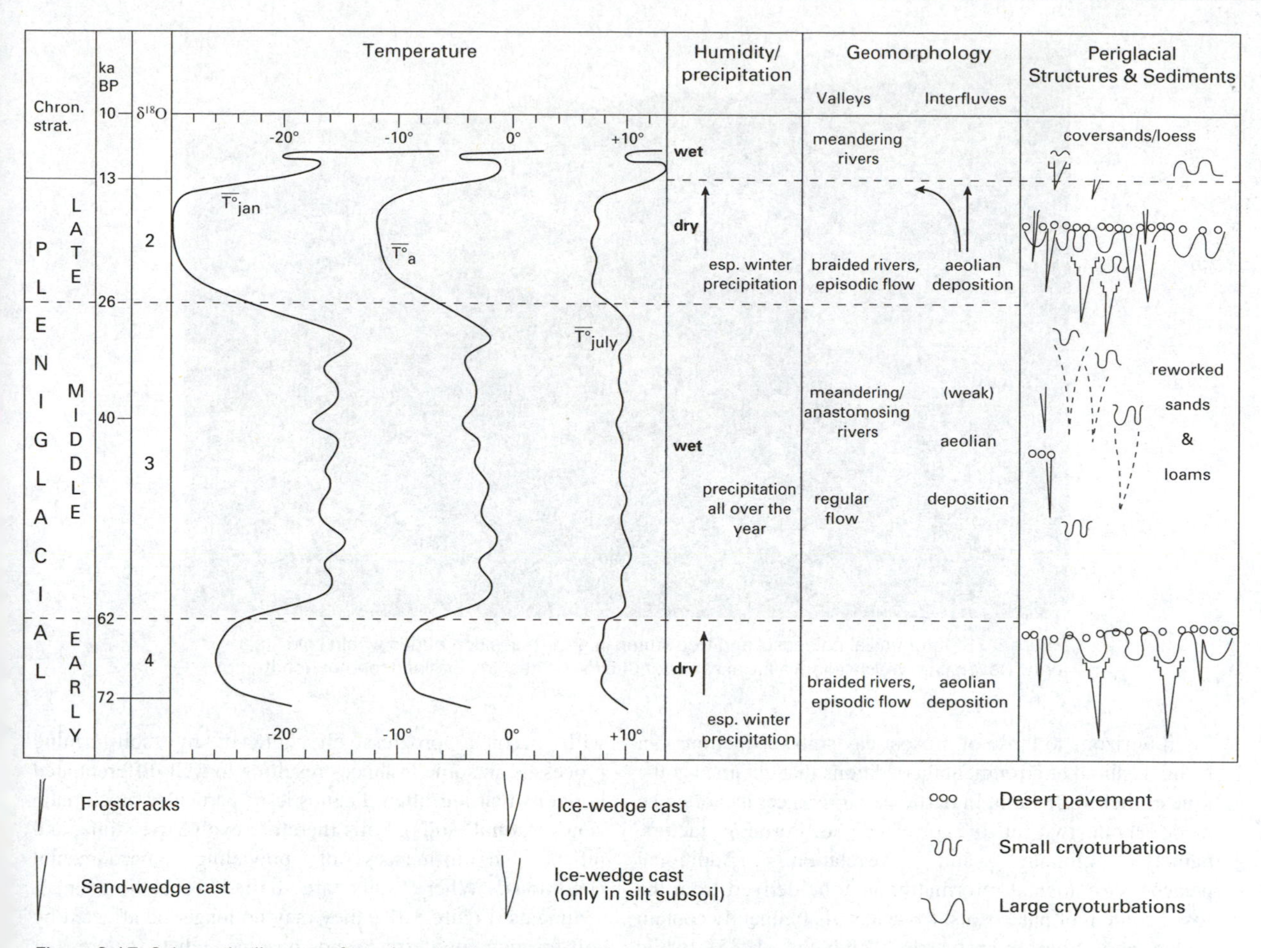

Figure 3.17 Schematic diagram of stratigraphical occurrence of periglacial features and inferred climatic conditions during the last cold stage ($\overline{T°}_a$ = mean annual temperature; $\overline{T°}_{jan}$ = mean winter temperature; $\overline{T°}_{july}$ = mean summer temperature) (from Vandenberghe, 1992).

cryoturbations cannot be dated directly, but only by interpolation using ages from materials contained within the host sediments. In both Europe and North America, relatively few sites have been found where fossil periglacial phenomena can be dated with any degree of precision. Despite these limitations, however, palaeoclimatic inferences based on periglacial phenomena are often supported by climatic estimates based upon other independent lines of evidence (section 4.5 and Chapter 7). This suggests that relict periglacial features can provide important insights into Quaternary climatic conditions, particularly during cold stages, but they are perhaps best employed as a supplementary rather than as a primary data source.

3.5 Palaeosols

3.5.1 Introduction

The term **'palaeosol'** was defined by Ruhe (1965) as any soil that has developed on a land surface of the past, and, while there are some difficulties with this definition (see below), it is a term that has been widely used in Quaternary literature and it has therefore been retained for convenience (see Catt, 1990). Palaeosols frequently formed under environmental conditions that differ markedly from those currently prevailing at a site, and therefore by relating the

Figure 3.18 Interstadial palaeosol and tree stump in growth position buried within cold stage (early Devensian/Weichselian) alluvial sands at Chelford, Cheshire, England (photo: Rob Kemp).

fossil horizons to those of present-day soils, deductions can be made about environmental conditions that obtained at the time of their formation. In particular, inferences can often be made about two of the principal soil-forming factors, namely climate and vegetation. Additional palaeoenvironmental information may be derived from the fossil content of palaeosols, for acid soils frequently contain pollen and other microfossils (Dimbleby, 1985), while molluscan remains are often found in calcareous soils (Carter, 1990). Palaeosols have also been used as a basis for the relative dating of landforms and sediments (Birkeland, 1992) and, insofar as they form time-parallel marker horizons, they can be employed in the subdivision and correlation of Quaternary successions (section 6.2.3.4). In this respect, they have proved to be especially valuable as markers in loess sequences (section 3.7), where the interbedded palaeosols constitute key reference horizons for correlation at the local and regional scales (section 6.3.2.3). Finally, palaeosols are potentially a rich data source for the measurement of rates and processes of mineral weathering (Colman & Dethier, 1986; Woodward *et al*., 1994).

3.5.2 The nature of palaeosols

Soils are formed by chemical, physical and biological processes operating in combination at the earth's surface. Where the surface remains stable, near-surface layers will become progressively altered by soil-forming processes, in some instances resulting in well-differentiated horizons that are often diagnostic of particular bioclimatic zones (**'zonal' soils**). Soils therefore evolve over time, and reflect the influences of prevailing environmental conditions. Where soils are buried beneath younger sediments (Figure 3.18), they may no longer be affected by soil-forming processes, and become relict features or **palaeosols**.

The distinction between a soil and a palaeosol, however, is not always straightforward (Catt, 1990). Most soils are, in fact, **polygenetic** or **'polycyclic'**, for they can evolve over such long periods that environmental conditions may change significantly during the period of soil formation, and thus soils often contain **'relict features'** (e.g. red coloration or cryoturbation structures) inherited from a former climatic régime (see, e.g., Kemp *et al*., 1993). In arid to semi-arid environments, for example, long-term changes in regional climate may lead to alternation between episodes of clay enrichment under more humid conditions and episodes of carbonate accumulation and surface salt deposition under arid conditions (Birkeland, 1984, 1992). Where relict features are particularly well developed in soils, therefore, these are referred to as **'relict palaeosols'**.

When buried by sediment, a soil may be modified physically (e.g. by root penetration) or chemically (e.g. by solute percolation) through the action of subsequent soil-forming processes at the new ground surface. A **'welded**

soil' may develop if the younger soil profile is superimposed upon, or merges with, the older one. Welded soils are common in loess, alluvial and colluvial sequences. In these 'pedocomplexes', it often becomes difficult to distinguish the separate effects of the different phases of soil formation (Dahms, 1994). A further problem arises in the case of **accretionary soils**. These form where the deposition of colluvial, alluvial or aeolian sediment over a soil is so slow that soil formation can keep pace with aggradation at the surface (Catt, 1990). The soil profile develops upwards and often generates very thick organic-rich horizons (A horizons) as a result. It is possible, of course, that environmental conditions may change during the accretionary process, so that lower horizons contain relict features, and, where exceptional rates of accumulation are experienced, the lowest horizons may become isolated from soil-forming processes. Clearly the distinction between active soil and palaeosol units may be difficult to make in these circumstances.

Distinguishing between relict, welded and accretionary soils is far from straightforward (Bronger & Catt, 1989), and hence these types of pedocomplexes are more problematic for the student of Quaternary environments than deeply buried soils. Equally difficult are **exhumed palaeosols**, which are those soils that were formerly deeply buried but have subsequently become exposed by erosion (Whiteman & Kemp, 1990). Exhumed palaeosols will be affected by contemporary pedogenic processes and they may therefore grade laterally into buried palaeosols and welded soils. Exhumed palaeosols have not been widely recognised, however, for in only a few areas, such as the loess landscape of the American Midwest, can they be shown to have emerged from beneath an eroding overburden and thus the exhumed nature of the palaeosol can be clearly demonstrated (Ruhe, 1965). In the absence of such stratigraphic evidence, it is often impossible to differentiate between relict and exhumed palaeosols, and thus the value of exhumed palaeosols in Quaternary research tends to be limited.

Buried soils may also be affected by **diagenesis**[7] as a result of burial and other post-formational influences. These include disturbance by periglacial processes, compaction due to the weight of any overburden (especially where glaciers advance over the sediments in which the palaeosols are developed), or the effects of changing groundwater levels and conditions. All of these can lead to changes in the physical and chemical properties of soils, such as microfabric structures and iron oxide, carbonate and organic matter content (Kemp *et al.*, 1994b). Where these changes have been particularly marked, it may be difficult to distinguish true palaeosols from **'pseudosoils'**, which are distinctive, coloured horizons in sediment sequences, caused by the mobilisation and subsequent accumulation of iron, manganese and other elements during diagenesis (Martini & Chesworth, 1992; Robinson & Williams, 1994).

Where a palaeosol is exposed across a site or region, it may be possible to establish the extent to which the nature of the profile varies laterally as well as vertically, to form a **palaeocatena**. Soil catenas, or **'toposequences'**, describe the gradational, lateral changes in soil profiles as a result of variations in surface gradient and the topographic position of the profile (Berry, 1990; Gerrard, 1990). Few examples of palaeocatenas have been published, but the classic study of Valentine and Dalrymple (1975) serves to demonstrate their value in distinguishing true soil horizons from diagenetic or weathering phenomena, for the former should show lateral variations in development which bear a logical relationship to original topography (e.g. variations in clay content and mineral or salt concentrations) whereas the latter will not. Not surprisingly, therefore, greater attention is now being paid to the spatial variations in modern soils in order to develop models of catenas that can assist in the identification and interpretation of palaeocatenas (e.g. Harrison *et al.*, 1990; Birkeland *et al.*, 1991).

3.5.3 Analysis of palaeosols

The recognition of buried palaeosols is not always easy, for a wide range of weathered materials will have been buried in Quaternary landscapes which may or may not be soils. The most important property of a soil, and that which distinguishes it from other sediments, is that it has developed distinctive, vertically differentiated layers or horizons in response to variations in physical, chemical and biological weathering, and the subsequent movement of weathering products up and down the profile (Catt, 1986). The often abrupt nature of the horizon boundaries, the truncation of underlying geological structures, and the areal or lateral continuity of soil are further characteristics that aid in the recognition of buried palaeosols.

The soil profile forms the upper part of the **weathering profile**, although some confusion has arisen, particularly in the American literature where the two terms have sometimes been used interchangeably. In certain cases, however, the weathering and soil profiles may be indistinguishable where, for example, erosion has removed the A and B horizons leaving only the weathered subsurface materials exposed in sections. In many palaeosols the organic content of the A horizon is not retained after burial (it may be lost by decomposition or as a result of erosion), although the mineral part of the A horizon may still be present and may be recognised by a clay content that is markedly different from the underlying B horizon (Birkeland, 1984; Catt, 1988). Generally, however, it is the B horizon which is of greatest importance in the identification of buried palaeosols. Some important diagnostic features of the B horizon include colour, texture variations (e.g. clay enriched or depleted horizons), weathered minerals, and enrichment (e.g. soils in semi-arid areas) or depletion (acid soils) in carbonate content. These properties can be used either singly or, more preferably, in combination, to demonstrate evidence of pedogenesis. A careful analysis of the properties of the B horizon may also establish the type of environmental conditions under which the soil evolved (Woodward *et al.,* 1994). Features that should be carefully recorded in the analysis of palaeosols, therefore, include colour changes (using Munsell colour charts), particle size distributions, clay mineral composition, organic matter content, evidence of soil macrostructures (e.g. peds, pans, nodules) and variations in calcium carbonate content. Further details on the methods involved are described in Catt (1990).

In recent years increasing emphasis has been placed on **soil micromorphology** as both a descriptive and a diagnostic tool in palaeopedology (Kemp, 1985a, 1996; Catt, 1990; FitzPatrick, 1993). Soil micromorphology is the term used to describe the distinctive arrangement of particles and voids making up a **soil fabric,** which can be established by an examination of soil thin sections under a microscope (Figure 3.19). It is widely regarded as one of the most reliable methods for detecting evidence of pedogenesis, distinguishing sequential phases of soil formation and inferring changes in environmental conditions (Kemp *et al.,* 1994b; Kemp, 1996). Soil micromorphological analysis can reveal evidence of weathering alteration of minerals, packing or orientation of clay particles, concentration and arrangements of voids, presence and type of calcite crystal growths, animal excrements, clay coatings, rootlet

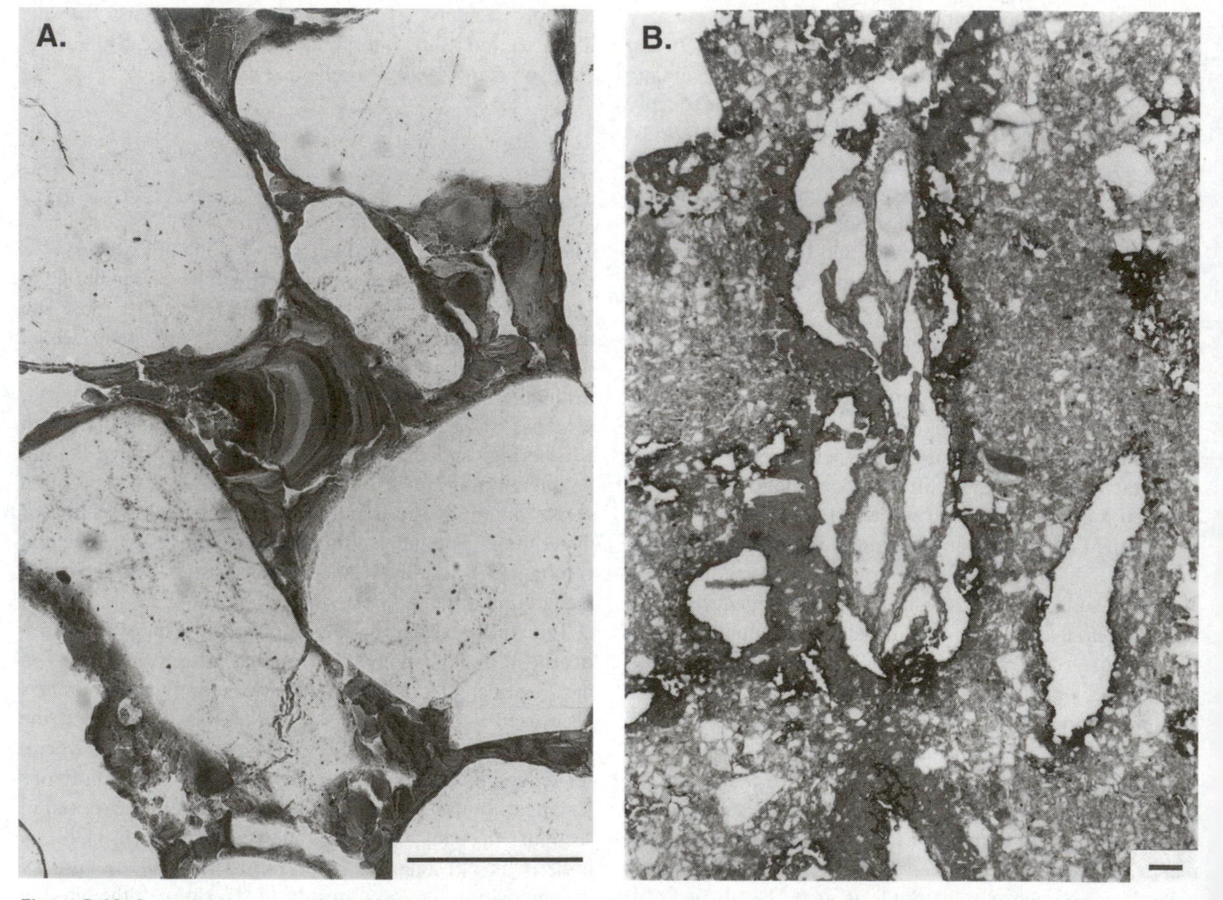

Figure 3.19 Some micromorphological features of assumed pedogenic origin within buried soils under plane polarised light. Scale bar = 0.25 mm. A. Laminated illuvial clay filling (centre). B. Calcitic root pseudomorph (vertically orientated structure at centre) (from Kemp, 1995).

pseudomorphs and a range of other features (Figure 3.19). Careful analysis of such evidence helps to unravel the history of soil formation and the changes in environmental conditions that may have occurred during pedogenesis.

A second technique that is being increasingly widely used in the investigation of palaeosols is **mineral magnetic analysis** (section 5.5.1). This involves the measurement of the magnetic properties of the mineral constituents of soils in order to identify both the types and concentrations of minerals present (Thompson & Oldfield, 1986). The method offers a rapid way of analysing complex soil-stratigraphic sequences, for magnetic minerals produced or enriched through pedogenic processes generate diagnostic mineral magnetic signals. Further details can be found in section 3.7.

3.5.4 Palaeosols and Quaternary environments

The interpretation of Quaternary environments on the basis of palaeosol evidence rests, as in other fields of Quaternary investigation, on the uniformitarian principle of inferring past conditions from the observed relationships between present-day soils and environments. Although this approach seems to work reasonably well with most types of biological evidence (Chapter 4), it is, perhaps, less satisfactory in geomorphological and pedological contexts where the dangers of **equifinality** (different processes leading to the production of similar forms) are always present. Soils with very similar physical and chemical characteristics can develop through a variety of genetic pathways, and these may be impossible to differentiate in the fossil soil profile. Thus, although buried palaeosols may be similar in morphology and in other characteristics to soils forming at the present day, they may not necessarily be analogous in terms of palaeoenvironment and regional conditions. A further difficulty with the use of modern soils as analogues for Quaternary pedogenesis is that it is seldom possible to separate the influences of environmental variables (e.g. climatic régime) from those of other soil-forming factors (e.g. parent material). Moreover, some soils may not be in equilibrium with present-day environmental conditions. Finally, fossil soils have proved very difficult to date, particularly using radiocarbon and uranium series methods (section 5.3), although a range of new techniques, including luminescence and palaeomagnetic measurements, and the analysis of cosmogenic nuclides in soils, offer the prospect of more precise dating of palaeosols (Catt, 1990).

Despite these problems, a number of inferences have been made about former environmental conditions on the basis of palaeosol evidence. For example, in southeast England, the widely developed Early and Middle Pleistocene palaeosol known as the *Valley Farm Soil* is clay enriched and distinctly reddish in colour (**'rubified'**), characteristics that are normally associated with a temperate or even Mediterranean environment (Rose & Allen, 1977). Hence it has been inferred that the soil formed under climatic conditions that were at least as warm as, or possibly warmer than, those prevailing in the region at the present time (Kemp, 1985b, 1987). This contrasts with the *Barham Soil*, a Middle Pleistocene palaeosol which is frequently superimposed on the Valley Farm Soil unit, but which displays both macromorphological characteristics (incorporated aeolian sediments; ground-ice structures) and micromorphological features characteristic of severe arctic conditions (Rose *et al.*, 1985).

This type of analogue approach has been adopted in other areas. In the loess region of central Europe, for example, Bronger and Heinkele (1989) noted that the palaeosol that developed during the last interglacial resembles a chernozem, or brown forest chernozem, which is characteristic of the area at the present day. They therefore suggested that the climate of the Eemian was similar to that of today. The penultimate palaeosol in the sequence, on the other hand, appeared to be a degraded chernozem, perhaps implying slightly wetter conditions than those of the present. Palaeosols from earlier interglacials, however, are distinctly rubified and were interpreted as indicating a significantly warmer, Mediterranean-type climate. In both east–central Europe and China, warm intervals of differing duration and intensity have been inferred on the basis of palaeosol evidence, with well-developed soil units being equated with interglacial periods, and less well-developed palaeosols reflecting interstadial episodes. In other areas of Europe, interglacial soils (e.g. Cromerian, Holsteinian, Eemian palaeosols) exhibit mature profiles that can be readily distinguished from the less well-developed interstadial soils (Catt, 1986).

Palaeosols have also proved to be valuable sources of palaeoenvironmental information where pedological data are augmented by palaeoecological evidence. For example, in northern Canada the migration of the forest/tundra boundary during the Holocene has been reconstructed using a combination of pedological, palynological and radiocarbon dating evidence (Sorensen, 1977). Podzols are associated today with spruce vegetation, whereas Arctic Brown soils are found mainly in tundra regions. Arctic Brown palaeosols may subsequently be altered by podsolisation, but polygonal patterns generated by ground-ice activity during the arctic phase may persist. The evidence suggests that, during the mid- and late Holocene, the position of the forest border ranged from 280 km north to at least 50 km south of the present tree-line in southwest Keewatin (Figure 3.20). These changes were frequently rapid, and appear to have occurred in response to variations

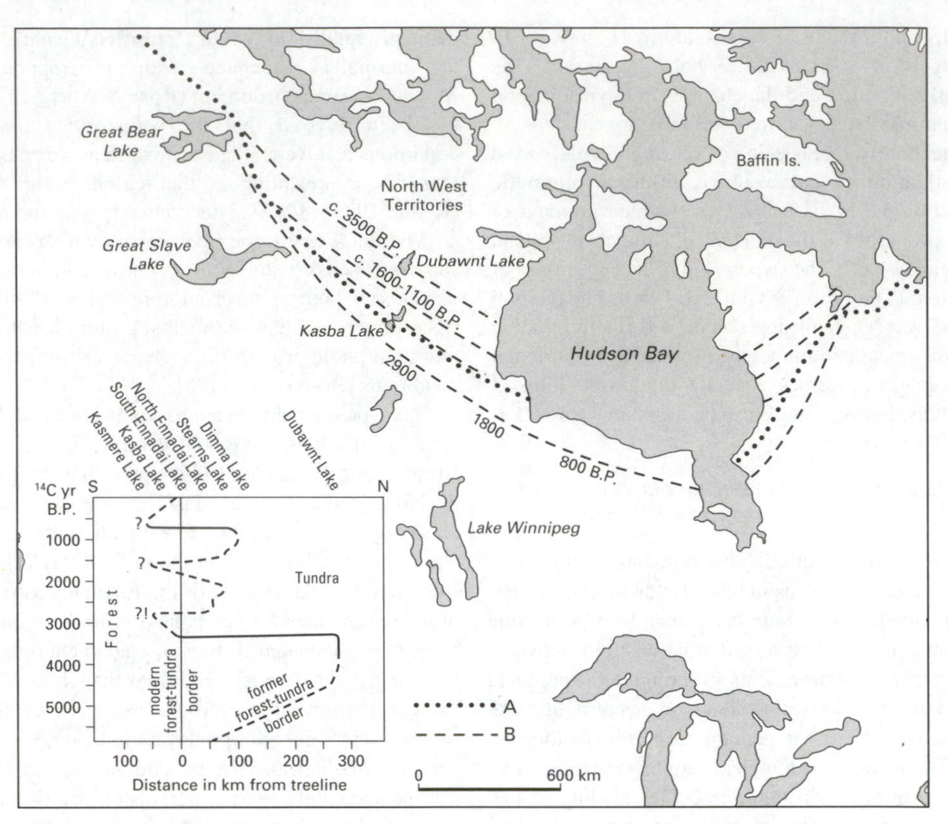

Figure 3.20 Holocene migrations of the forest/tundra border in southwest Keewatin, NWT, Canada, based on radiocarbon dates and soil morphology. A: present forest border; B: estimated palaeoforest border location (after Sorensen, 1977).

in the incidence of Arctic and Pacific air masses. The presence of arctic air during the summer months seems to be the most important vegetational and pedological control, and this implies that the position of the forest border is thermally determined.

Palaeosols, therefore, are of considerable value in Quaternary research. They constitute important stratigraphic markers, they provide a basis for correlation, they can be used as a means of relative dating and, above all, they can provide palaeoenvironmental information that is additional to, and often independent of, that derived from other sources such as fossil assemblages. Indeed, in some regions they may provide the only source of palaeoenvironmental information for periods during which little or no sediment has accumulated. While it is true that there is still much to be learned about modern soils that is crucial to the interpretation of palaeosols, it is equally the case that much can be learned by pedologists from the Quaternary archive of palaeosols, about the ways in which modern soils have evolved, and about the rates and effects of weathering and related pedogenic processes.

3.6 Lake level records from low latitude regions

3.6.1 Introduction

The geomorphological evidence for environmental changes in low latitude regions during the Quaternary was reviewed in section 2.7. However, much longer and more detailed lake level histories can be reconstructed using sedimentological data (Dohrenwend *et al.*, 1986). The records can be broadly divided into those from humid, low latitude regions, and those more typical of arid and semi-arid areas. In some lakes of the lowland equatorial and upland forest belts, sediment sequences span the whole of the Quaternary and extend back into the Pliocene (Hooghiemstra & Sarmiento, 1991) while exposures in alluvial sequences provide additional palaeoenvironmental information (e.g. Colinvaux *et al.*, 1985; Thomas & Thorp, 1995). Most attention has focused, however, on the 'pluvial lake' sequences (section 2.7.1) of the arid and semi-arid regions, for these records not only provide some of the most dramatic evidence of late Quaternary environmental change, but the data from these lakes are being increasingly widely used in the construction of palaeoclimatic models at both the continental and global scales (Street-Perrott & Roberts, 1983; Street-Perrott & Perrott, 1993; Williams *et al.*, 1993a).

3.6.2 'Pluvial' lake sediment sequences

Cores from saline lakes or from salt-encrusted pans of former playa lakes often reveal complex sequences of lake sediments (silts, marls, clays and organic muds), interbedded with units of alluvium, colluvium, aeolian sediments and occasional soil horizons. Unconformities are also common. In some places, for example in parts of Egypt and northern Kenya, the upper parts of the sequences are exposed and the intercalated lacustrine and terrestrial sediments can be examined in section (Nyamweru & Bowman, 1989; Brookes, 1993). The sedimentary records reflect episodes of lake expansion interrupted by periods of contraction (Figure 3.21) and, in some instances, complete desiccation (e.g. Benson & Thompson, 1987; De Deckker, 1988a; Teller *et al.*, 1990). At Searles Lake, California, for example, over 275 m of sediments have been recorded, while in a borehole from ancient Lake Bonneville (the expanded precursor of the Great Salt Lake, Utah), 28 phases of lake expansion and contraction are thought to be represented in a succession extending over 800 ka (Eardley *et al.*, 1973). In most cases, however, the borehole records span only the last glacial–interglacial cycle or even shorter time intervals.

The sedimentary record in closed basin (**endoreic**) lakes is closely related to changes in water balance which, in turn, reflect changes in regional climatic régime. If the links

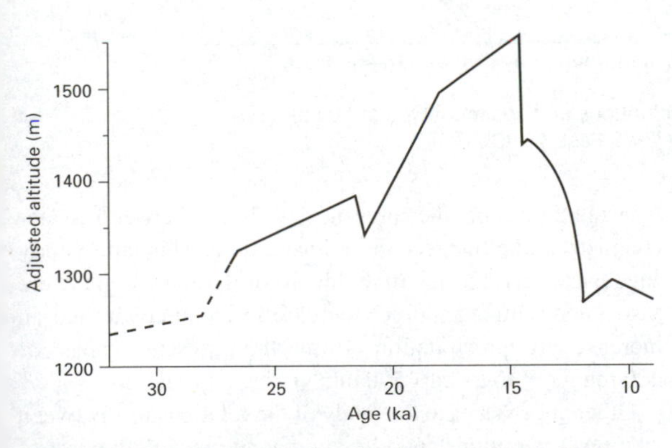

Figure 3.21 Time–altitude plot of variations in level of Lake Bonneville during the past 30 ka (modified after Oviatt *et al.*, 1992).

between lithological variations and water balance can be quantified, it is possible to reconstruct the former climatic conditions under which the lake sediments accumulated. The principal variables affecting water balance within a catchment have been summarised as:

$$P + R + G_I = E + O + G_O \pm \Delta S$$

where P is the precipitation onto the lake surface, R is the runoff from the lake catchment, G_I is the groundwater inflow, E is the evaporation from the lake, O is surface outflow, G_O is the groundwater outflow and ΔS is the change in storage (Street, 1981; Street-Perrott *et al.*, 1985). Changes in lake area or volume could, therefore, reflect the influence of any of these variables. However, not all causes of lake level changes are necessarily related to climate. For example, overspill from adjacent lakes (e.g. Benson & Paillet, 1989; Benson, 1994), influx of glacial meltwaters during cold stages, changes in basin configuration due to tectonic activity (e.g. Shaw, 1988), or the creation of dams by avalanche debris, talus cones or lava flows, can all affect lake volume and lead to fluctuations in the lake-water level. Where sedimentary (and indeed geomorphological) evidence can be obtained from a number of lake basins within a particular region, and where these show a consistent trend in water balance changes for a given time period, a regional climatic signal may reasonably be inferred, and hence the lake sediment record can be used as a proxy for climate change.

As lakes change in volume, the sedimentary and geochemical environment also changes. In salt lakes, for example, salinity declines as water volumes increase, and hence clastic sediments of relatively low salinity tend to accumulate when lakes are expanded. By contrast when lakes decrease in size, salinity levels rise, until primary carbonates are deposited first, followed by gypsum and then by halite (Figure 3.22; Teller & Last, 1990). Algal limestone growths (**stromatolites**) may also develop at this stage (Hillaire-Marcel *et al.*, 1986a). With seasonal drying of the lake, secondary gypsum deposition occurs on the exposed sediment surface and clay pellets also form. Prolonged desiccation will result eventually in soil formation. A re-expansion of the lake will reverse these trends, the precise nature of the sedimentary sequence depending upon the rate and magnitude of the water level changes. The sedimentary and geochemical responses to lake level fluctuations will vary considerably between individual lake basins, however (Figure 3.23), and each sedimentary sequence must therefore be interpreted in the context of local sediment chemistry, detrital input and hydrogeology (De Deckker, 1988b; Teller & Last, 1990).

Quantifying the linkages between water balance and climate on the one hand, and between water chemistry and water budgets on the other, rests heavily on reference to

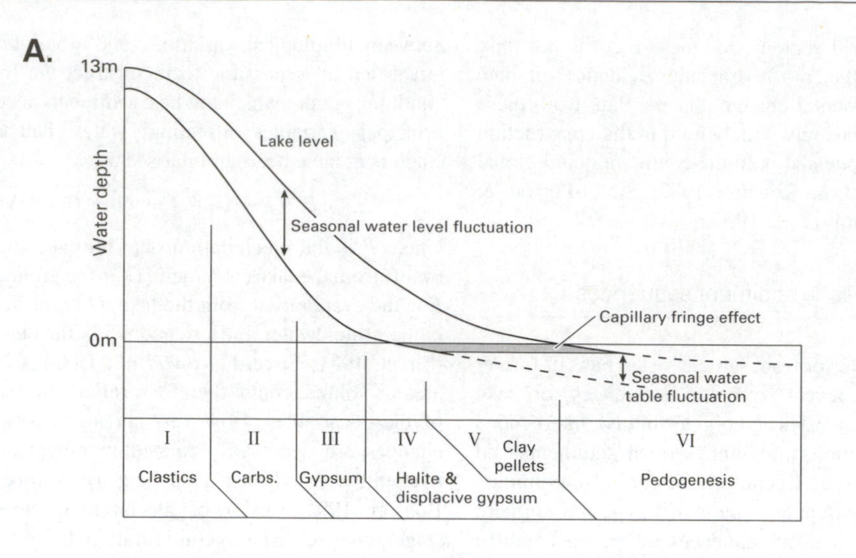

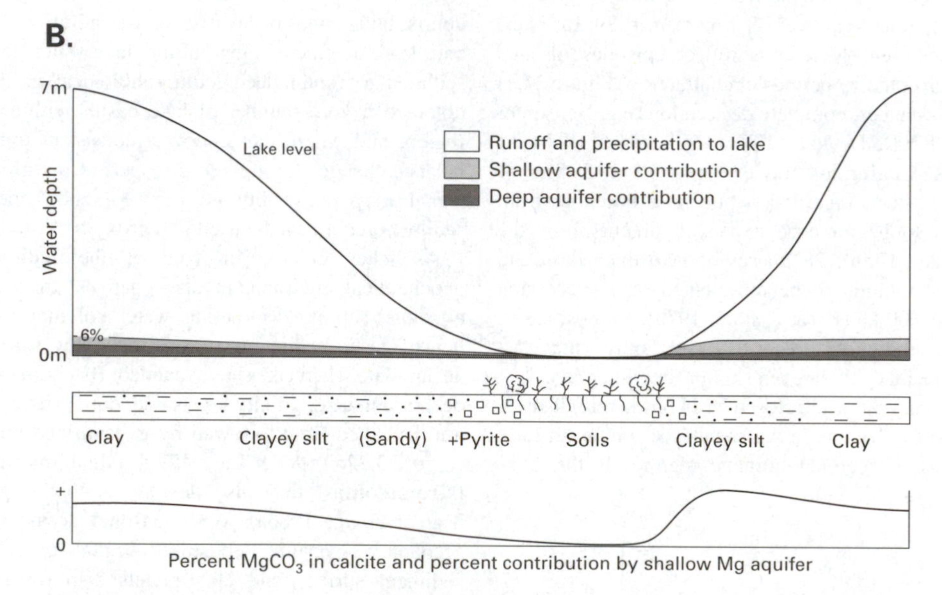

Figure 3.22 Relationships between lake level fluctuations and sediment type at (A) Lake Tyrell, Australia and (B) Lake Manitoba, Canada (after Teller & Last, 1990).

modern analogues. Where present-day equilibrium conditions can be established for a lake basin, modelling techniques can be used to simulate the conditions that give rise to changes in either water budget or water chemistry. Examples of this approach include the work of Hastenrath & Kutzbach (1985), who developed a water-balance model for Lake Titicaca (Bolivia) using modern climatic and radiation data, and Grosjean (1994), who reconstructed water level changes in Laguna Lejia (Chile) based on an index of past changes in water chemistry which was, in turn, derived from measurements of the present-day links between water chemistry, lake budgets and climatic data. This latter study suggested that former high levels of Laguna Lejia were associated with a 2.5 times increase in cloud cover, and an increase in precipitation from the present value of 200 mm yr^{-1} to at least 500 mm yr^{-1}.

Other approaches to the study of the relationships between lake level variations and climatic change include the use of biological indicators of lake salinity, particularly diatom and ostracod assemblages (sections 4.3 and 4.9), or the chemical

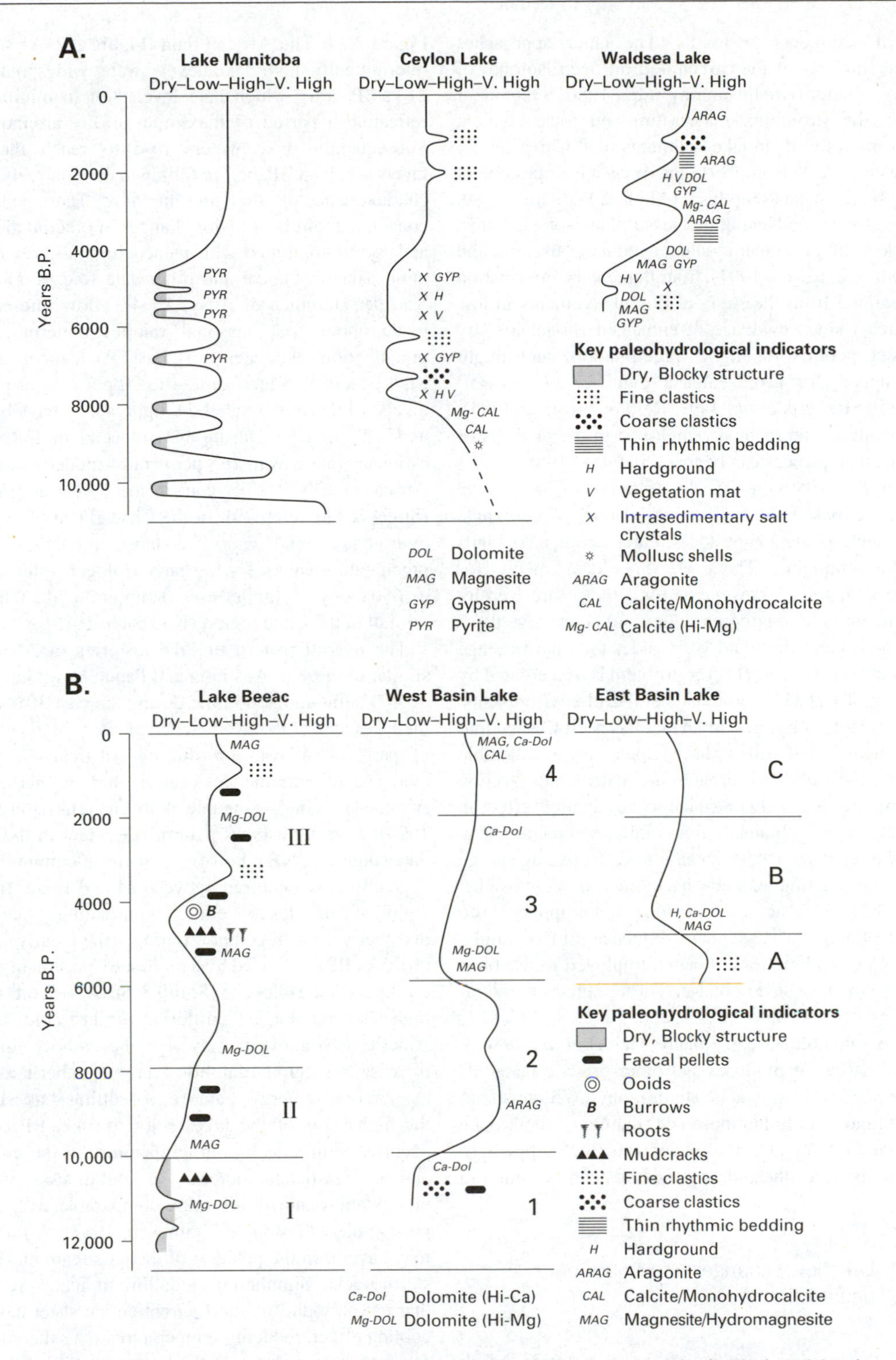

Figure 3.23 Relationships between lake level and various palaeohydrological proxies for (A) three lakes in the Canadian Prairies and (B) three lakes in Australia (after Teller & Last, 1990).

analysis of sediments or fossils. The latter approaches include the analysis of the carbon and nitrogen isotopes of lake organic matter (Krishnamurthy *et al.*, 1986; Sukumar *et al.*, 1993), the strontium, magnesium and trace element content of microfossils in lake sediments (e.g. Chivas *et al.*, 1986; Engstrom & Nelson, 1991), the oxygen isotope content of calcite, sediments and molluscs (Abell & Williams, 1989; McKenzie, 1993), and variations in sulphur isotope ratios, which reflect the presence of sulphur-reducing bacteria and water depth (e.g. Rosen, 1991). Complementary information can be obtained from the study of fluvial sediments in low latitude areas, since evidence of enhanced fluvial activity may reflect periods of higher precipitation, and might provide support for lake sediment evidence of 'pluvial' episodes (e.g. Roberts *et al.*, 1993; Reid & Frostick, 1993). Further details of the methods employed in the analysis of lake sediment sequences can be found in Gray (1988).

Prior to the 1980s, most chronologies of lake level changes were based on radiocarbon dating of carbonate materials, such as algal limestones, calcite-cemented sands and mollusc remains. These are not ideal media for radiocarbon dating, however, for they are easily contaminated by older carbon residues and hence give dates that may be several thousand years older than the true age (e.g. Williamson *et al.*, 1991). The problem is exacerbated by the mobility of $CaCO_3$ in arid and semi-arid environments, and by the fact that organic carbon that has been washed into lake sediments will often have spent a considerable 'residence time' in soils around the catchment (section 5.3.2.4). Again, this will contribute to an 'ageing' effect in radiocarbon dates obtained from lake sediments (e.g. Hillaire-Marcel *et al.*, 1989). Finally, the effective age range of radiocarbon dating extends back only to *c.* 45 ka BP (section 5.3.2) and hence the technique is not applicable to the dating of long lake sediment sequences. Increasingly, therefore, other techniques are being employed to establish chronologies of lake level change. These include uranium-series dating of carbonates (e.g. Szabo, 1990; Szabo *et al.*, 1995) and stromatolites (e.g. Hillaire-Marcel *et al.*, 1986a), amino-acid dating of molluscs and other organic materials (Magee *et al.*, 1995), magnetic stratigraphy (Williamson *et al.*, 1991) and thermoluminescence dating of both lake sediments and interbedded aeolian deposits (see section 3.7). Further details of all these dating methods can be found in Chapter 5.

3.6.3 Lake level changes and Quaternary palaeoclimates

Lake level fluctuations over the last 30 ka in Africa, North America and Australia/Papua New Guinea are shown in Figure 3.24. The African data (Figure 3.24A) suggest that intermittently high lake levels were widespread prior to 21 ka BP, after which lake levels fell to minimum levels reflecting a period of maximum aridity around 13 ka BP. Subsequently, lake waters rose to reach their highest levels at *c.* 9 ka BP, before falling to the relatively low levels characteristic of the present day. These changes are considered to reflect major changes in precipitation régimes and, when combined with palaeoclimatic inferences based upon palaeobotanical and other data (e.g. Kershaw, 1994; Van der Hammen & Absy, 1994), allow inferences to be made about past seasonal rainfall patterns. Maximum precipitation estimates for East Africa for the period 21–12.5 ka BP range from 54 to 90 per cent of present-day levels, while for the Sahel the figures may have been as low as 15–20 per cent. During the Holocene, on the other hand, estimates range from 165 per cent of modern values in East Africa, to 200–400 per cent in the Sudan and Mauretania (Street & Grove, 1979). In the Chad Basin of West Africa, hydrological and energy balance modelling suggested precipitation values for the early Holocene pluvial phase of *c.* 650 mm yr^{-1}, at least 300 mm more than the current rainfall in the Chad region (Kutzbach, 1980).

The overall pattern of lake histories in Africa is very similar to those in Australia and Papua New Guinea (Figure 3.24C), although at no time during the past 30 000 years do the lakes appear to have reached the very high levels typical of parts of Africa, nor do the Australasian data show evidence of extreme desiccation that is characteristic of African lakes during the late Holocene (Harrison & Dodson, 1993). A very different pattern is evident in the American data (Figure 3.24B), however, where maximum lake levels appear to have occurred between 24 and 14 ka BP, with the majority of lakes either contracting markedly or disappearing altogether during the early Holocene (10–5 ka BP), followed by a modest expansion in some lakes during the late Holocene (Smith & Street-Perrott, 1983). The high lake status achieved during the last cold stage in the American southwest may not necessarily indicate an increase in precipitation, however, but rather a reduction in evaporation. Energy balance modelling has shown that the highstands of the lakes prior to 14 ka BP could have occurred with a decline in temperature of the order of 7°C and a significant increase in cloudiness, but with a precipitation and runoff value comparable with that of the present day (Benson & Thompson, 1987). A further factor may have been the position of the jet stream and associated storm tracks. Simulation modelling indicates, for example that not only did the the Laurentide ice sheet have a direct cooling effect, reducing temperatures over the whole of the continent, but it also split the jet stream so that the southern branch was displaced to a latitude south of its present

A. Tropical Africa

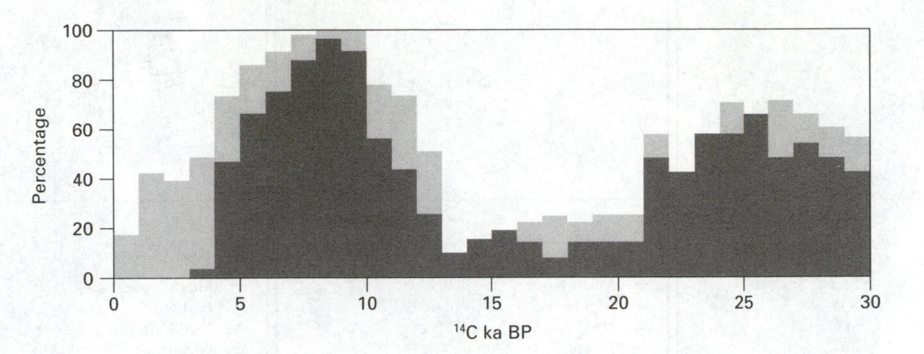

B. South West USA

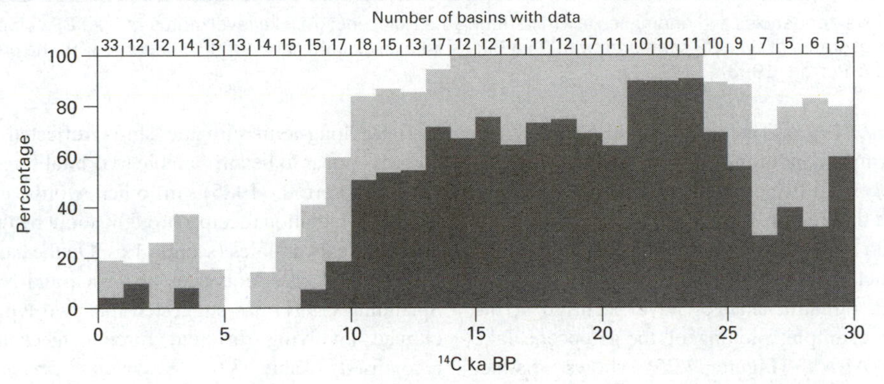

C. Australia and Papua New Guinea

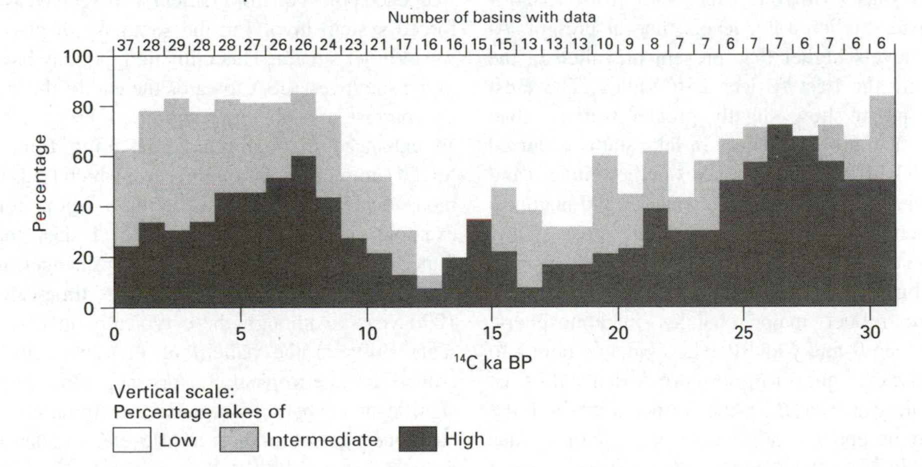

Vertical scale:
Percentage lakes of
Low Intermediate High

Figure 3.24 Histograms of lake level status for 1000-year time periods from 30 ka BP to present for (A) Africa (after Street & Grove, 1979), (B) United States (from Smith & Street-Perrott, 1983) and (C) Australia (from Harrison & Dodson, 1993).

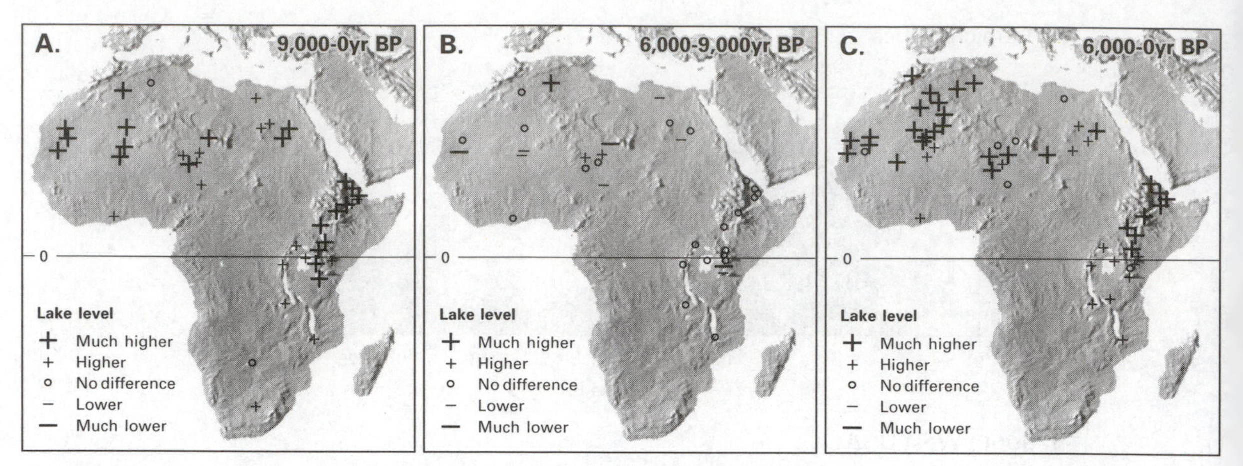

Figure 3.25 Lake level tendencies and anomalies in Africa during the Holocene: (A) lake level status at 9 ka BP compared with the present; (B) lake level status at 6 ka BP compared with 9 ka BP; (C) lake level status at 6 ka BP compared with the present time (from Street-Perrott & Perrott, 1993).

track (Wright *et al.*, 1993). The effect would have been to bring wetter, stormier conditions into the southwest of the USA, and this may be reflected in the relatively high status of pluvial lakes in that region during the last cold stage.

Lake level data not only form the basis for regional climatic reconstructions, but they also provide insights into the way in which climatic changes have occurred at the global scale. For example, plotting of the Holocene lake level data for Africa (Figure 3.25) shows distinct continental-scale patterns (Street-Perrott & Perrott, 1993). At 9 ka BP, all sites (with one exception) from 15°30'S northward register higher water levels than at present. At 6 ka BP, water levels higher than present prevailed in the Sahara and along the East African Rift Valley. The most northerly sites again show slightly greater wetness than today. Some clear regional changes in lake status occurred between 9 and 6 ka BP. In general, lakes in the southern and northeastern Sahara showed no net change or had begun to desiccate. By contrast, lake levels generally rose in the northwestern Sahara. These large-scale trends in lake status cannot be attributed merely to local influences, but rather they appear to reflect major changes in atmospheric circulation between 9 and 6 ka BP. The evidence points to strong monsoon rains right across northern Africa to the East African Rift Valley at 9 ka BP, while further south both the summer monsoons and the adjacent oceanic anticyclones were weaker. Although the Northern Hemisphere summer monsoon at 6 ka BP remained stronger than today, its northern limit had retreated southward in the central and eastern Sahara. Wetter conditions in the northwestern Sahara and Atlas Mountains were associated with decreased upwelling and warmer sea-surface temperatures offshore. The interior of South Africa, the Cape region and Madagascar became wetter after 6 ka BP, whereas the opposite trend may have occurred in the Namib Desert.

These long-term climatic shifts reflected in lake level records appear to be attributable to orbital forcing (Kutzbach & Street-Perrott, 1985), in other words to changes in terrestrial radiation receipts brought about by the operation of astronomical variables (section 1.6). On the basis of evidence from pluvial lake sequences in Africa and North America, Spaulding (1991) has suggested that two types of 'pluvial' climate involving different forcing mechanisms can be recognised (Table 3.6). A *'winter precipitation type'*, associated with expanded continental ice sheets and a steepened pole–equator gradient, is perceived as an internally forced system involving the southerly displacement of the westerly jet stream. This climatic type may have dominated in the southwest USA towards the end of the last cold stage. By contrast, a *'summer precipitation type'* is associated with an enhanced thermal régime resulting from an orbitally forced summer insolation increase, which leads to intensified monsoonal circulation in low latitude regions as reflected, for example, in the African lake level data for 9 ka BP. Superimposed upon these long-term changes, however, are short-term climatic oscillations over timescales of 100 to 1000 years. Although these typically involve very small-scale shifts in the pattern of pressure cells, particularly around the intertropical convergence zone, they can lead to significant changes in regional precipitation patterns, and have been linked to short-term cycles of solar activity (e.g. Faure & Leroux, 1990).

Fluctuations in water level in closed-basin lakes therefore provide a valuable means of investigating late Quaternary climatic change. 'Pluvial' lakes appear to respond rapidly to climate variations and are more direct indicators of water balance than vegetation or soils. Moreover, many pluvial lake deposits are found in areas where biological and other indicators of former environmental conditions are scarce. Not only do they provide direct evidence of former

Table 3.6 Characteristics of two principal types of pluvial climates based on model simulations and palaeoenvironmental data from southwest North America and North Africa. P_s, P_w and P_a are, respectively, summer, winter and average annual precipitation; T_a is average annual temperature (from Spaulding, 1991).

	Winter precipitation type	Summer precipitation type
Correlation	Glacial and/or stadial climates	Interglacial and/or interstadial climates
Dominant circulation type	Zonal	Meridional
N latitudes affected (approx.)	25–≤40°	10–35°
Last episode	~23 to 14 (12) ka BP	~11 to 6 ka BP
Representative regions (! = most pronounced)	SW North America! Northernmost N Africa	North Africa exc. Maghreb! SW North America below ~35°N
Vegetation/floristic elements at optimum distribution	*Artemisia* steppe, subalpine and pygmy conifer woodlands, boreal elements	Thorn-scrub, grasslands, savanna, subtropical (thermophilous) floristic elements
Climatic parameters		
ΔT_a	−7 to −4°C	~0°C(?)
Precipitation seasonality	Winter	Summer
% P_s (relative to present)	<−50	>100
% P_w (relative to present)	70–100	?
% P_a (relative to present)	40–100 (150)	>200 (in Saharo-Sahelian region)
Proximal forcing	Southward deflection of westerlies Intensification of westerlies Increased cyclonic disturbances Decreased *T* and, consequently, *P/E*	Poleward migration of ITCZ Robust subtropical high-pressure systems Enhanced land-surface heating and advective flow Intensified convection
Ultimate cause(s)	Expansion of continental ice sheets Steepened pole–equator gradient	Enhanced planetary thermal régime Orbitally forced summer insolation increase

precipitation régimes, but they also provide the basis for modelling climatic change and for reconstructing atmospheric circulation patterns in the low latitude regions of the world.

3.7 Wind-blown sediments

3.7.1 Introduction

Large areas of wind-blown sediment (**sand seas**) are currently forming in arid and hyper-arid regions of the world, while blankets of sandy material (grain size 64 μm to 2 mm) or **coversand** were deposited in many mid-latitude temperate regions during the last cold stage. Elsewhere, silt particles (2 to 64 μm) form the bulk of wind-blown sediment, material which is referred to as **loess**. Spreads of loess are found throughout the world, and these often accumulate to great thicknesses. Analysis of these loess sequences has shown that not only are they frequently of considerable antiquity (some extend back into the Pliocene), but they also contain evidence of cyclical climatic change, with phases of aeolian sedimentation (reflecting cold stages) interspersed with episodes of soil formation during warm stages. Wind-blown sediments, both loess and coversands, are also valuable indicators of former wind directions. In addition, dust of terrestrial origin can be identified in marine and ice cores, and these archives enable both the former transport paths and the relative flux of air-transported dust to be reconstructed at a global scale.

3.7.2 Loess stratigraphy

Loess deposits cover about 10 per cent of the earth's land surface, blanketing hilltops, plateau surfaces and valley floors alike (Derbyshire *et al.*, 1988; Pécsi, 1987). The most extensive spreads of loess are to be found in the 'Loess Plateau' of North China (Figure 3.26), where they cover an estimated area of 275 600 km^2 (Derbyshire *et al.*, 1995), large areas of central Asia (Dodonov, 1991; Bronger *et al.*, 1995), central Europe, where loess extends as a discontinuous belt from northern France to the Ukraine (Bronger & Heinkele, 1989; Velichko, 1990), the Great Plains of North America (Martin, 1993; Winspear & Pye, 1995) and the Pampas of Argentina and Uruguay (Iriondo & Garcia, 1993; Clapperton, 1993b). The loess of North China frequently exceeds 100 m in thickness, and near the city of Lanzhou reaches 300 m

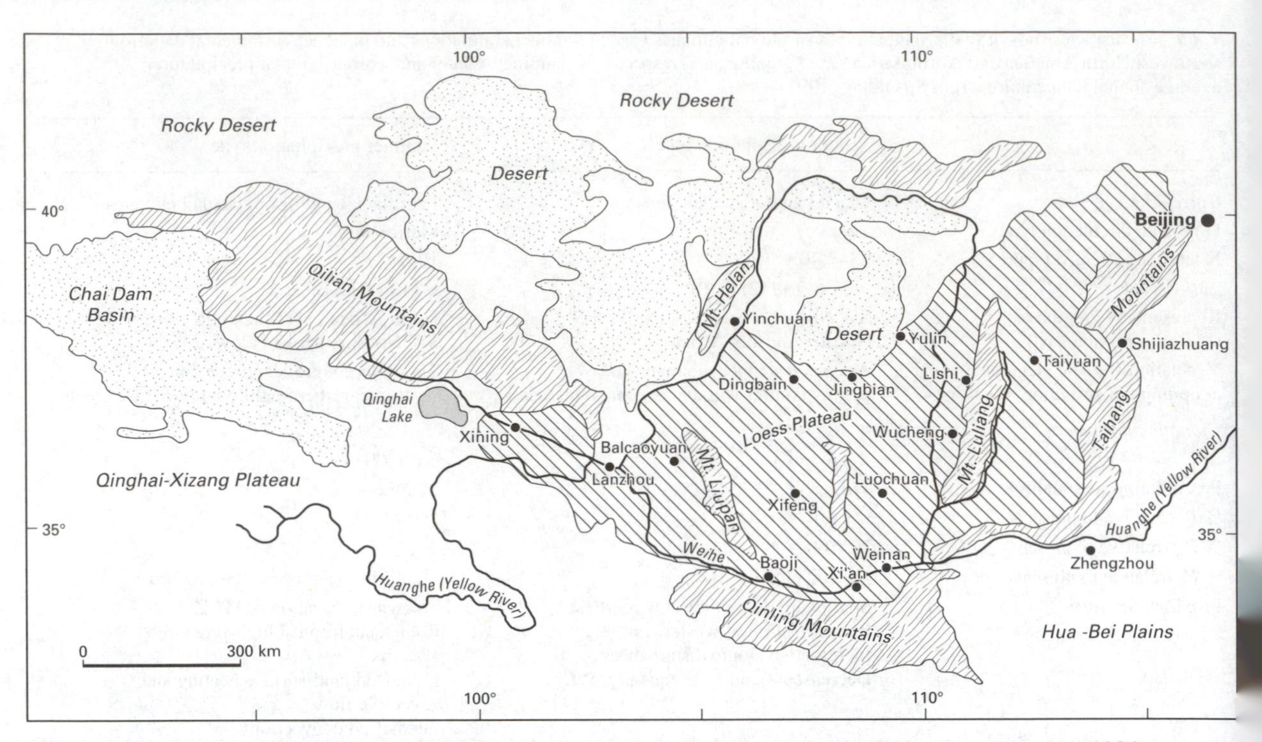

Figure 3.26 Map of the Loess Plateau, major deserts and mountain regions of north–central China (from Ding *et al.*, 1994).

(Derbyshire *et al.*, 1995). In central Asia, loess accumulations of about 90 m are common, but locally can exceed 200 m, whereas in Europe and North America thicknesses of about 30 m are more usual. More restricted spreads of loess occur in many other parts of the world, including the Mediterranean region, the Middle East, India and Pakistan, the western USA, New Zealand and Alaska.

Loess has a high carbonate content, sometimes exceeding 40 per cent by weight, and it frequently possesses a distinctive range of sedimentary properties, including heavy mineral and clay mineral suites, sediment magnetic variations and micromorphological features, all of which can be used to characterise and correlate loess sequences (e.g. Derbyshire, 1995). The fine-grained and friable nature of loess deposits gives rise to unstable landscapes into which rivers are rapidly incised (Dijkstra *et al.*, 1993), and this creates numerous exposures for detailed study (Figure 3.27). In section, the sediments usually display little visual evidence of stratification, although careful examination often reveals faint bedding. Microscopic and SEM examination of the particle fabric (micromorphology) can distinguish differences in texture which are important for palaeoenvironmental reconstruction (e.g. Feng *et al.*, 1994; Kemp *et al.*, 1995). These analyses have shown that there are two types of loess, one that is typical of areas such as China which have experienced arid conditions during glacial periods and humid conditions in interglacials, and a second type that has accumulated in areas with persistently high humidities, such as western Europe (Derbyshire *et al.*, 1988).

The most distinctive features in loess successions, however, are palaeosols which often show up in section as darker units (Figure 3.27). In China, magnetic susceptibility analysis of loess deposits has shown that certain soil units are more strongly marked than others, and can be traced in a number of exposures throughout a region, thereby providing a basis for correlation between individual loess successions (e.g. Rutter *et al.*, 1991; Liu *et al.*, 1994). For example, in parts of the Chinese Loess Plateau, up to 37 soil units have formed during the last 2.5 Ma (Figure 3.28). Some of these are very distinctive and widespread, such as the S_1 (last interglacial) and S_5 (oxygen isotope stages 13, 14, 15) palaeosols (Kukla *et al.*, 1990; Rutter *et al.*, 1991). However, a number of these palaeosols have proved to be complex polygenetic units, consisting of several soil horizons separated by thin **'second order' loess beds** with fainter **'second order' pedogenic features** in some of the more distinctive soil units (Kemp *et al.*, 1995). These variations appear to reflect climatic change, the well-developed palaeosols probably indicating relatively long periods of pedogenesis during interglacial stages, while the less prominent soil units reflect much shorter episodes of soil formation during interstadial periods (Kukla *et al.*, 1990).

Numerous palaeosols, possibly comparable in age with those of China, have also been discovered in central Asia

Figure 3.27 A. The loess–palaeosol succession at Luochuan, Shaanxi Province in the heart of the Loess Plateau of China. The prominent soil horizon visible in the main section to the right of centre is the S5 palaeosol complex (photo: Edward Derbyshire).

B. Loess–palaeosol sequence near Tongchuan, Shaanxi Province, Loess Plateau, China. The prominent soil in the upper part of the section is the last interglacial (S1) palaeosol. The angular unconformity and steeply dipping S2 and S3 palaeosols at the centre of the photograph indicate the degree of active landscape evolution in the region. The very thick (>5 m) red soil at the base of the sequence is the S5 palaeosol complex (photo: Edward Derbyshire).

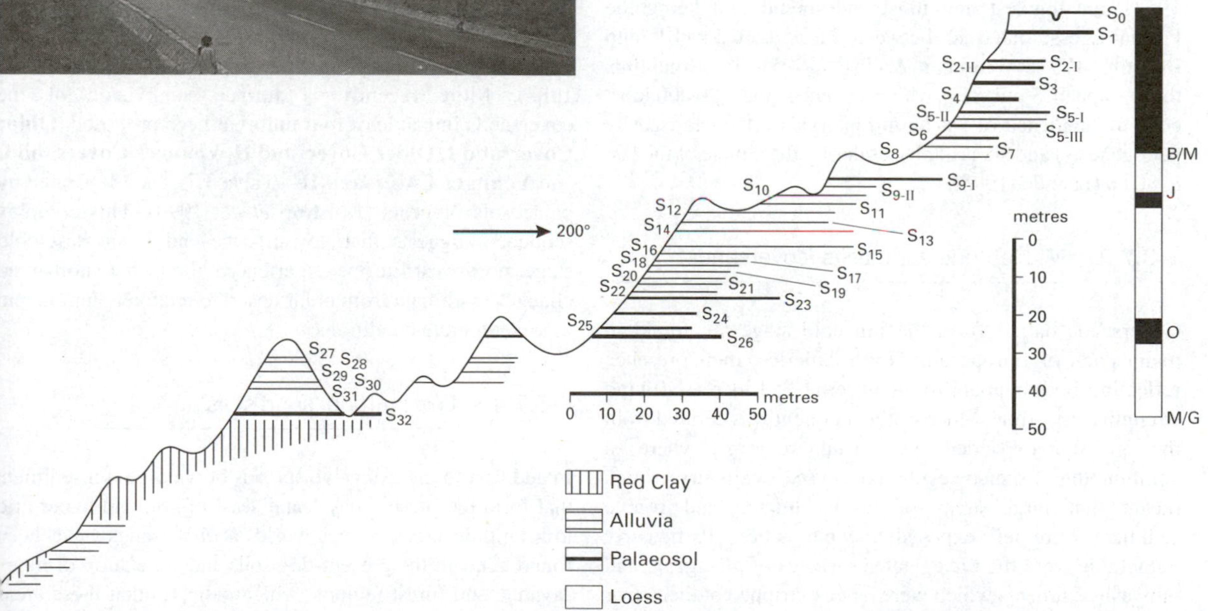

Figure 3.28 Loess–palaeosol succession at Baoji, north-central China (for location see Figure 3.26), showing 32 palaeosol units spanning the last 2 Ma. The palaeomagnetic timescale (see Figure 5.26) is shown on the right. B/M = Brunhes/Matuyama boundary *c.* 0.78 Ma BP, J = Jaramillo event, 0.99–1.07 Ma BP, O = Olduvai event, 1.79–1.956 Ma BP and M/G = Matuyama/Gauss boundary, *c.* 2.6 Ma BP (from Ding *et al.*, 1994).

though these have not yet been studied in detail (Bronger *et al.*, 1995). However, they extend back at least to *c.* 730 ka BP (the Brunhes–Matuyama magnetic boundary – Figure 5.26), and probably beyond (Dodonov, 1991). Some of the soil horizons are thicker and more strongly developed than in the present-day soils in the region, suggesting prolonged interglacial conditions. The central Asian sequences may eventually provide even more detailed palaeoclimatic records than those obtained from China (Bronger *et al.*, 1995). In Europe and North America, on the other hand, the loess sequences do not appear to extend as far back into the Quaternary as those in Asia, with the notable exception of that of the Saint Vallier in the Rhône Valley, which extends back over some 2 Ma (Billard *et al.*, 1987). Up to seven palaeosol units (each reflecting warm-stage conditions) have been described from loess sections from eastern Europe, whereas in western Europe the deposits are usually thinner and appear to span a more limited age range (Bronger & Heinkele, 1989). Indeed, in southeast England, which lies at the western extremity of the central European belt of wind-blown sediments, loess-like deposits, locally termed **'brickearths',** date from only the last cold stage (e.g. Catt *et al.*, 1987; Gibbard *et al.*, 1987). In North America, complex successions of loess, sand units and at least four cycles of pedogenesis, dating approximately from the last 400–500 ka, have been reported from the central Great Plains (Feng *et al.*, 1994), but most loess sequences in the USA only span the period from the last interglacial (e.g. Forman *et al.*, 1992; Leigh & Knox, 1994), the thickest and most widespread unit being the Peoria Loess, dated to between 21 and 10.5 ka BP ago (Martin, 1993; Winspear & Pye, 1995). In Argentina, the Pampean Sand Sea, which covers some 200 000 km^2, contains evidence of six humid intervals (characterised by pedogenesis) and six arid phases during the course of the last *c.* 80 ka (Iriondo, 1995).

3.7.3 Mid-latitude sand belts (coversands)

Coversands dating from the last cold stage are found in many parts of Europe and North America, their presence reflecting both availability of material and increased wind strength at that time. Much of the sediment was derived from the greatly expanded periglacial regions, where a combination of sparse vegetation and seasonally dry ground meant that large areas of unconsolidated and friable sediment were left exposed to wind action. Particularly vulnerable were the unvegetated surfaces of glacigenic and outwash sediments which were rapidly stripped of their finer components. Considerable quantities of sand were also removed from the continental shelves which had been exposed by a eustatic fall in sea-level of over 100 m, while unvegetated floodplains of large rivers provided a further source of material (e.g. Lea & Waythomas, 1990; Madole, 1995). A more vigorous atmospheric circulation around the ice sheets, coupled with strong katabatic winds blowing off the glaciers and ice sheets, resulted in the stripping of these surfaces and the deposition of the coversand and loess belts of the northern mid-latitude regions.

The coversands form a featureless surface, occasionally fashioned into undulating dunes, over large parts of the northern Great Plains of the USA and Canada. In lowland Europe, they stretch from northern France, The Netherlands and Belgium, eastwards through Denmark, northern Germany and Poland into Russia. This mantle of yellow to grey sand is often several metres in thickness, and individual sand units are relatively homogeneous, with little or no stratification (Koster, 1988). The sediments are frequently deformed by folding and tension cracks, reflecting **niveo-aeolian deposition**, where both sand and snow are driven by the wind to form interdigitating layers (Ballantyne & Harris, 1994). Subsequent melting of the snow destroys any bedding and leads to deformation and cracking of the sediment layers, collectively termed **denivation features** (Koster & Dijkmans, 1988).

The European coversands typically comprise a number of distinctive sand units, which are interbedded with finer sand and loamy layers. Early studies of these deposits in The Netherlands revealed two widely developed coversand units (the **'Older'** and **'Younger Coversands'**) which are frequently separated by a soil or peat unit of Bølling age (Table 3.7). These two units have subsequently been recognised in other parts of Europe (e.g. Niessen *et al.*, 1984). More recently, a further subdivision of the coversands into at least four units has been proposed (**Older Coversand I, Older Coversand II, Younger Coversand I,** and **Younger Coversand II** – Table 3.7), each separated by palaeosols or peats (Kolstrup *et al.*, 1990). This complex sequence suggests that, towards the end of the last cold stage, northwest Europe experienced abrupt but short-lived changes in climate from polar desert to relatively humid and more temperate conditions.

3.7.4 Low latitude 'sand seas'

In addition to the extensive spreads of wind-blown sediment that form the present-day 'sand seas' of arid and hyper-arid low latitude areas of the world, aeolian sands have been found beneath the present-day soils and vegetation of many savanna and forest regions, which indicates that these areas have experienced more arid conditions in the recent past (Lancaster, 1990). In South America, for example, some 25

Table 3.7 Lateglacial and Upper Pleniglacial stratigraphy in The Netherlands (from Kolstrup *et al.*, 1990, p. 208).

^{14}C years BP	Chronostratigraphy		Lithostratigraphic units and soils of coversand area
10 000	Holocene		
11 000	Late Glacial	Late Dryas	Younger Coversand II
11 800		Allerød	Peat or Usselo Soil
12 000		Early Dryas	Younger Coversand I
		Bølling	Peat or loam band
	Pleniglacial	Upper Pleniglacial	Older Coversand II
			Beuningen Gravel Bed and arctic soil
			Older Coversand I
30 000		Middle Pleniglacial	

per cent of the continent is covered with palaeo-aeolian features, which is attributed to a stronger atmospheric circulation régime at the last glacial maximum (Clapperton, 1993a). Similarly in many areas of North Africa and northern India, there are enormous spreads of aeolian material, much of which has been fashioned into dunes (section 2.7.2), and which provides evidence of the former extension of sand seas during the last cold stage (Goudie, 1992). By contrast, many present-day arid or hyper-arid regions contain evidence of earlier, wetter phases. Interbedded within, or buried beneath, the surficial sand seas can often be found lacustrine sediments from former playa lakes, or palaeo-drainage systems, both of which testify to the former existence of more pluvial conditions. Such evidence has now been described from the desert regions of Australia (e.g. Magee *et al.*, 1995), northern India (e.g. Agrawal *et al.*, 1990) and Africa (e.g. Grosjean, 1994). Other indicators of a recent climatic change from wetter to more arid conditions in many desert regions include archaeological evidence in the form of abandoned settlements, along with fossil remains which suggest a former, more extensive vegetation cover and a fauna characteristic of more humid environments (McKenzie, 1993; Szabo *et al.*, 1995).

3.7.5 Wind-blown sediments and palaeoenvironmental reconstructions

Former wind directions can be reconstructed on the basis of grain-size or mineralogical variations of wind-blown sediments. For example, grain-size analysis of the Peoria Loess on the Great Plains of the USA showed that the deposit becomes progressively finer towards the Mississippi River (Mason *et al.*, 1994). It is unlikely, therefore, that the loess was derived from river alluvium in the Mississippi Valley, as had been previously supposed, and more likely sources are the ice-marginal sediments deposited by the Wisconsinan ice sheet. In Britain, analysis of aeolian deposits at sites from the Kent coast to south Devon revealed a westward decrease in coarser particles and a complementary increase in light, flaky minerals (mica and chlorite), a trend which can most readily be interpreted as reflecting transport of loess by northeast or easterly winds (Catt, 1977). An easterly wind direction is also indicated by the close similarities between the mineralogy of the silt-size fraction of Late Devensian glacial deposits and loesses in eastern England, suggesting that the source of the loessic material was glaciofluvial sediments in the North Sea basin (Catt, 1987). Dating by thermoluminescence (section 5.3.6) confirms a Late Devensian age for the majority of these loessic deposits (Gibbard *et al.*, 1987; Parks & Rendell, 1992). A predominantly easterly wind direction implies the development of a major anticyclonic circulatory system centred over northwest Europe during the Late Devensian/Late Weichselian.

In eastern Asia, a range of palaeoclimatic data has been obtained from the analysis of wind-blown sediments and related deposits. In north–central China, particle-size, geochemical and soil-micromorphological evidence from loess–palaeosol sequences revealed a series of arid phases, characterised by increased dust deposition, separated by semi-arid episodes of soil formation (Kemp *et al.*, 1995). On the Chinese Loess Plateau, grain-size variations reflected in mineral magnetic signals (e.g. Maher *et al.*, 1994) have been used to demonstrate changes in both the strength and position of the Asian monsoon cell during the late Quaternary. Horizons characterised by ultra-fine grains (high magnetic susceptibility) are typical of episodes of reduced aeolian deposition and pedogenesis (Maher & Thompson, 1992), while periods of increased influx of wind-blown dust are indicated by lower magnetic susceptibility measurements (Rolph *et al.*, 1993; Figure 3.29). The latter represent periods when the winter monsoon (associated with dust transport from Siberia) was dominant, while evidence of pedogenic development reflects a reduction in the strength of northerly winds and an increase in intensity of the summer monsoon from the south (Ding *et al.*, 1994; Xiao *et al.*, 1995). Changes in the strength of the Asian monsoon on the Chinese Loess Plateau have also been inferred from variations in the concentration of a range of chemical indicators (Si, Ca, Fe, K and Mn) in the loess–palaeosol sequences (Zhang *et al.*, 1994). Finally, magnetic parameters have been used to obtain quantitative estimates

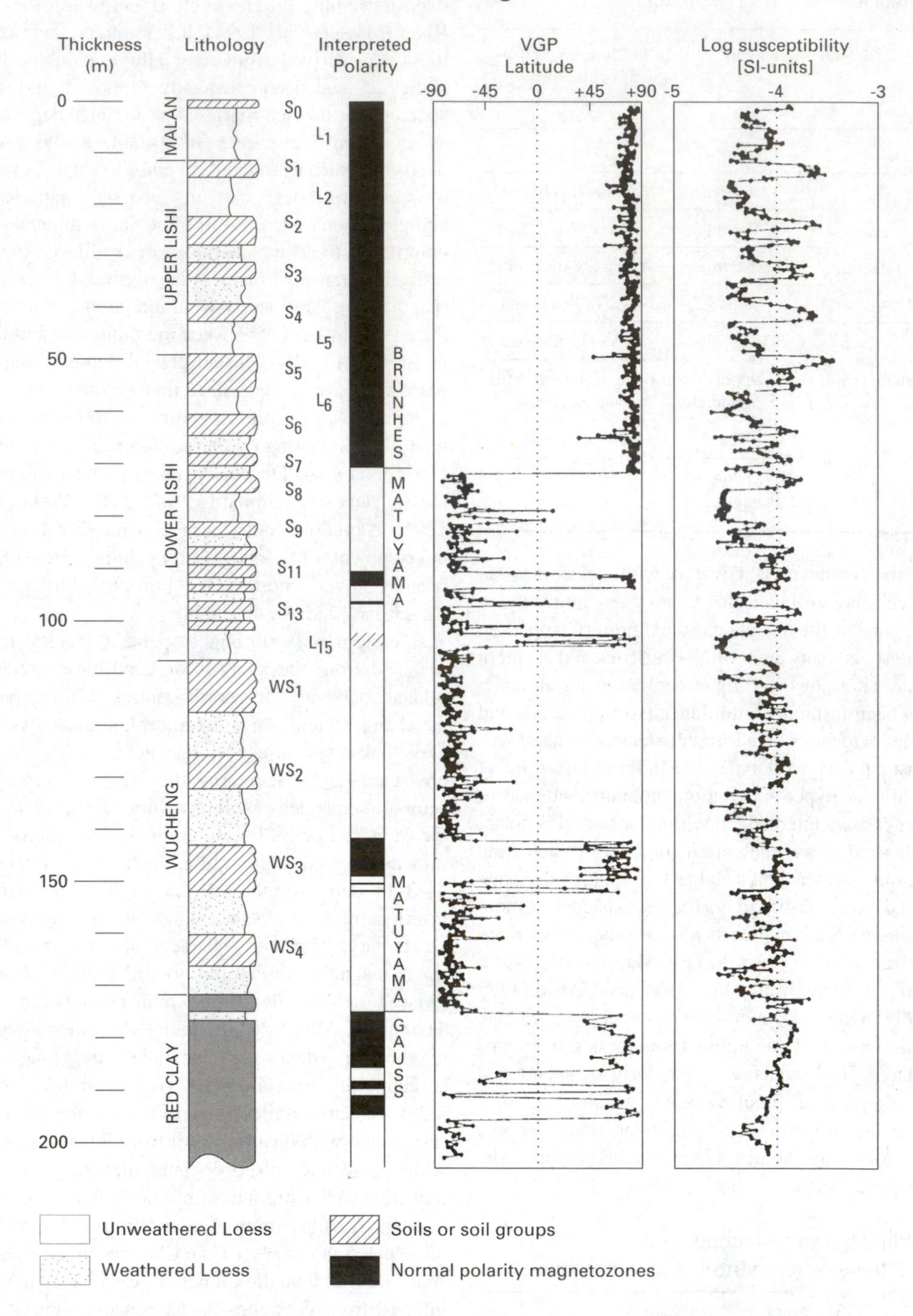

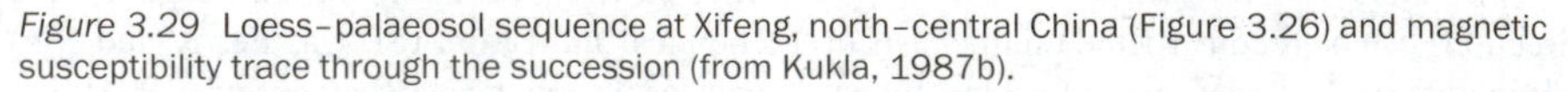

Figure 3.29 Loess–palaeosol sequence at Xifeng, north-central China (Figure 3.26) and magnetic susceptibility trace through the succession (from Kukla, 1987b).

of dust deposition on the Loess Plateau over the course of the past 140 ka, from which palaeoprecipitation estimates have been obtained and which, in turn, enable the varying strength of the monsoon to be inferred (Liu *et al.*, 1995).

Dust transport fluxes to the polar ice sheets can also be reconstructed using geochemical and isotopic tracers. Petit *et al.* (1990), for example, have found that the dust content of ice layers that accumulated in Antarctica during the last cold stage was up to 50 times higher than that recorded in modern ice layers. Martin & Fitzwater (1990) have suggested that increased fluxes of Fe-rich dust to the Antarctic are responsible for phytoplankton blooms in the Southern Ocean which, in turn, may have an effect on the amounts of CO_2 released to the atmosphere, and hence can influence global climates. Continental dust can also be detected in deep-marine cores using geochemical and magnetic methods, and these have enabled inferences to be drawn about large-scale air mass circulation during the last glacial–interglacial cycle (e.g. Grousset *et al.*, 1992; Rea, 1994).

Quaternary wind-blown sediments are therefore valuable data sources for palaeoenvironmental reconstruction. At the local and regional scales they provide evidence of sediment source and transport, and hence of palaeowind directions. At the global scale, these data can be used to reconstruct former atmospheric pressure systems, and also to infer those times when atmospheric dust transport was greater than that of the present time. The loess–palaeosol sequences are especially valuable archives of palaeoenvironmental data, and constitute some of the longest and most detailed records of Quaternary climatic change. They also provide one of the principal sources of evidence for the correlation of terrestrial and marine sequences, a topic that is discussed in greater detail in section 6.3.3.

3.8 Cave sediments and carbonate deposits

3.8.1 Introduction

Caves form natural sediment traps in which the deposits are largely protected from the effects of sub-aerial weathering agencies and erosion. In karstic regions especially, the sediments that have survived in caves frequently cover a much longer time interval than those on the neighbouring land surface (Ford & Williams, 1989). Three main types of material contribute to cave sediment sequences: clastic detritus, organic detritus and precipitated carbonates, the relative proportions of each depending upon rock type, size of fissure, groundwater régime, topographical or geological context and geographical location (Trudgill, 1985). Clastic detritus includes rock rubble, cave earth and water-lain sediments. Organic detritus may consist of skeletal parts of animals that occupied the caves and those of their prey, while many caves acted as occupation sites for humans and so cave successions may contain a rich legacy of organic, artefactual and cultural material of anthropogenic origin (Sutcliffe, 1985). In limestone regions, reprecipitated carbonates, collectively known as **speleothem**, constitute a third important component of the sedimentary fill. Precipitation of carbonate from flowing or dripping water in caves can generate a variety of forms of **dripstone**, the best known of which are stalactites and stalagmites, or **flowstone**. In many caves, dripstone and flowstone production appears to have been periodic or even cyclic, resulting in inter-stratification of speleothem and detrital layers (Figure 3.30), and artefacts, bones and clastic material can often become wholly embedded (and very well preserved) within the precipitated carbonate.

In this section, the palaeoenvironmental potential of detrital and carbonate deposits in caves will be examined. Of particular importance is speleothem, since this material (a) provides the longest and most detailed palaeoenvironmental records from cave contexts (Gascoyne, 1992), (b) can be dated by radiometric means (particularly by uranium-series dating – section 5.3.4) to provide a chronology of cave deposits (Schwarcz, 1989), (c) contains palaeoclimatic data that form a basis for the correlation of continental and oceanic records (Kashiwaya *et al.*, 1991; Baker *et al.*, 1995) and (d) offers valuable insights into the evolution of the landscape around the caves themselves (Lauritzen, 1993). Reference will also be made to other types of carbonate material that are found outside the cave environment (e.g. *travertine* and *tufa*), since the processes of formation, and the analytical techniques employed in their study, are often similar to those associated with speleothem (section 3.8.6).

3.8.2 Detrital sediment in caves

A major factor governing the processes of sedimentation in caves is the shape of the cave itself. A distinction is often made between **exogene** caves, which are shallow niches in the hillside, referred to by archaeologists and anthropologists as **rock shelters**, and **endogene** caves, which penetrate deep into the ground as chambers or passages. In the interior passages of endogene caves, relatively equable conditions prevail, for the air is protected from the temperature extremes that occur at the ground surface. Daily and seasonal changes in weather and climate therefore rarely penetrate, at least not directly, and therefore only the major and long-term climatic changes can affect the mode of sedimentation. Unless endogene passageways are

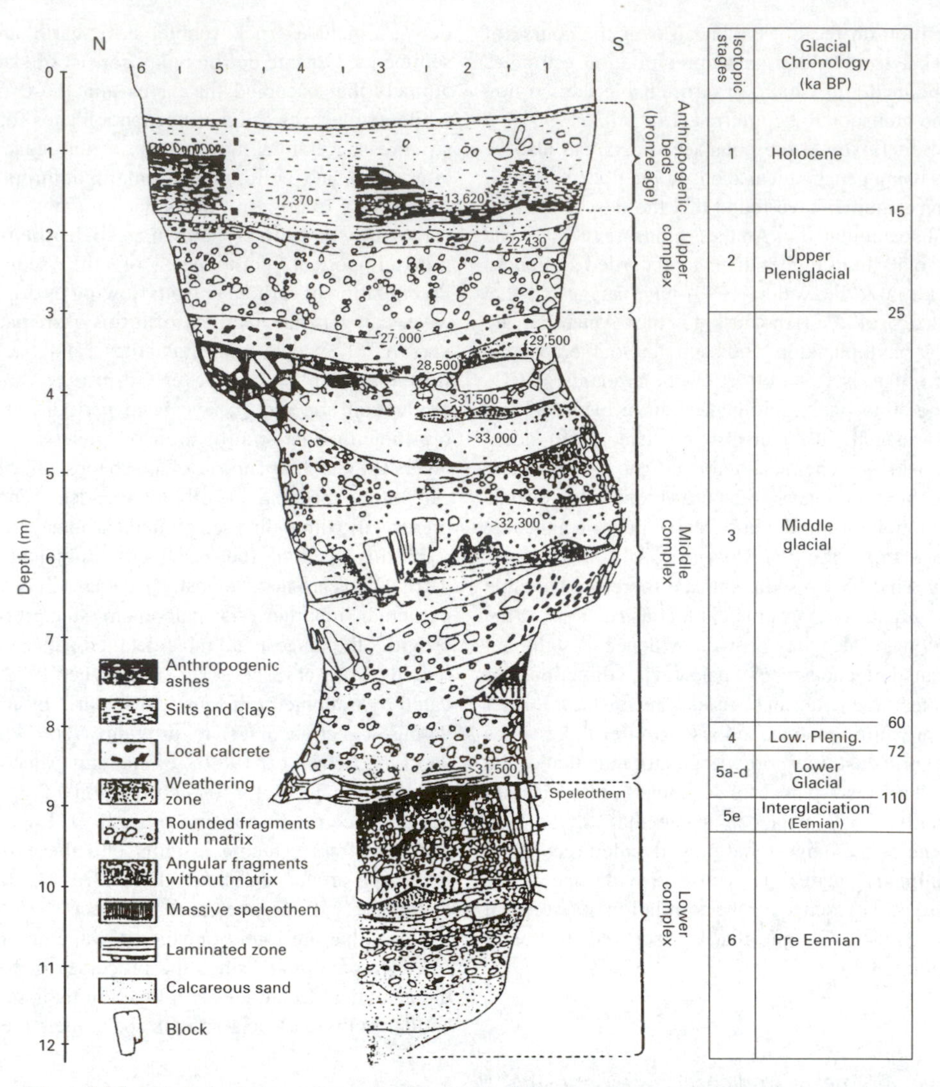

Figure 3.30 Lithological succession in the Baume de Gigny cave, France, showing alternations of clastic fill, laminated deposits, speleothem and anthropogenic layers spanning the last *c.* 145 000 years (from Campy & Chaline, 1993).

near the ground surface, in which case exotic material may be introduced into the system through clefts in the cave roof, the majority of sediments are derived from the bedrock within which the fissures have formed, the principal exception being water-lain sediments from conduits connected to the surface.

In exogene caverns, and near the entrance to endogene caves, sedimentation is more directly influenced by prevailing weather conditions outside. Moreover, as the deposits are derived both from within the cave (**autochthonous** component) and from the surrounding area (**allochthonous** component), the stratigraphy in these situations is often complex. As a rule, exogene caves wil[l] tend to have a higher allochthonous component tha[n] endogene caves. The entrances of exogene caves in formerl[y] glaciated terrain may contain till or other glacigeni[c] sediments, while coastal caves may contain beach gravels o[r] (in the inner reaches) finer marine sediments (e.g. Proctor & Smart, 1991). Wind-blown material is also commonl[y] encountered in exogene cave sediments as well as colluvia[l] and soliflucted sediment.

Autochthonous sediments consist principally of roc[k] rubble and cave earth. Angular fragments of rock are commo[n] deposits in many caves and have been weathered from th[e] roofs and walls to form **thermoclastic scree.** In the outer parts of caves, some rock fragments will have been derived from insolation weathering (expansion and contraction at the rock surface due to temperature changes), while in limestone areas, solutional weakening by percolating groundwaters will result in pieces of rock breaking off from the walls and roofs of a cave. **Cave earth** is composed of much finer materials (sand size and less) and may have a variety of origins. Near the cave entrance it is often largely allochthonous, being composed of wind-blown or water-lain sand or silt, or even inwashed colluvial sediment. In the deeper parts of limestone caves, however, cave earths are formed either from the acid-insoluble residues left by the solutional breakdown of the country rock, or from the secondary weathering of angular rock fragments that have accumulated on the cave floor. Cave earths are frequently red or brown in colour, due partly to the presence of oxides of iron and aluminium, but in some cases reflecting the influence of phosphate derived from fossilised faecal matter.

Few detailed studies have been made of clastic sediments in caves, at least for the purposes of reconstructing Quaternary environments, since most interest has been focused on the fossil, archaeological or speleothem content of cave successions. Indeed, many caves were excavated in the late nineteenth and early twentieth centuries primarily to recover archaeological remains, with scant regard for the stratigraphical context of this material. The situation has changed in recent decades, especially in western Europe, with the increased protection afforded caves because of the important rock art they contain (Sutcliffe, 1985), and following the development of dating methods that enabled the history of human occupation of caves to be established more firmly (Lumley, 1976).

Cave sediment sequences are normally complex and characterised by numerous hiatuses in sedimentation. Furthermore, the facies variations found in cave sediments are not as well understood as those associated with surface processes. The reconstruction of environmental conditions from clastic cave sediments is therefore not without its problems. However, Campy and Chaline (1993) have shown that in a number of well-dated cave sequences in France (e.g. Figure 3.30) there is a clearly marked regional pattern in the types of sedimentary evidence that they contain as well as in the relative completeness of the cave records. Sites in the northern Alps contain only relatively recent sediments, the oldest being of Older Dryas age (*c.* 12 ka BP) or even younger. Sequences spanning much longer time ranges are found in the caves of central and southern France. In the Franche-Comté (near Dijon), for example, the sequences extend back to the last interglacial or beyond, although there are major gaps in the sedimentary record of the last cold stage. The caves in the south of the country, in the Périgord for example, have the most complete sequences, with no significant depositional hiatuses. Campy and Chaline attribute these regional differences to the proximity of the cave systems to the last ice sheets, with increased rates of erosion by glacial meltwaters accounting for the short sequences in the Alps and the major gaps in the records elsewhere. They also observed a close similarity between the major lithological variations preserved in the caves of central and southern France, which suggests a common external climatic control over the processes operating in these caves.

A wide range of organic materials can be found in cave sediments, incorporated within or interstratified with clastic material. The most obvious organic components are usually the skeletal remains of natural cave dwellers and of their prey, while many caves contain rich snail assemblages, which, in karstic regions, may often be of considerable antiquity (e.g. Goodfriend & Mitterer, 1993). Other materials may be transported to caves by animals and people, including the decayed parts of plants carried in for bedding, litter or food, animal excreta and carcasses, charcoal deposits from former hearths, various artefact remains, and pollen and other microfossils (e.g. **phytoliths**[8]) derived from the vegetation carried in by animals and humans. The fossil content (biostratigraphy) of caves is discussed in Chapter 4.

Caves have frequently been used in prehistory for stabling of animals, especially in upland areas that provided rich pasture in the summer months. Evidence of the use of caves for this purpose can be found in sediment layers that are rich in phosphates, coprolites (animal excreta), charcoal and concentrations of calcium oxalate crystals that are derived from burnt coprolites (Courty *et al.*, 1991). Soil micromorphological techniques have been used for the identification of 'stabling layers' (horizons representing periods during which the caves were used for stabling purposes) which are often well preserved in cave sediment sequences (Courty *et al.*, 1989). Other micromorphological methods have been developed specifically for the analysis of burned organomineral deposits in cave successions (Wattez *et al.*, 1989). These 'anthropogenic' horizons have now been found in many European cave sequences, reflecting the widespread practice of transhumance in mountain regions during the Neolithic and Bronze Ages (Maggi *et al.*, 1991).

3.8.3 Speleothem

Speleothem is a secondary mineral deposit formed in caves in karst regions as **dripstones** or **flowstones.** Dripstones are deposits of calcium carbonate (although some may be composed of varying quantities of aragonite, gypsum or halite – Goede *et al.*, 1990) formed by water dripping from the ceilings or walls of a cave, or from the overhanging edge

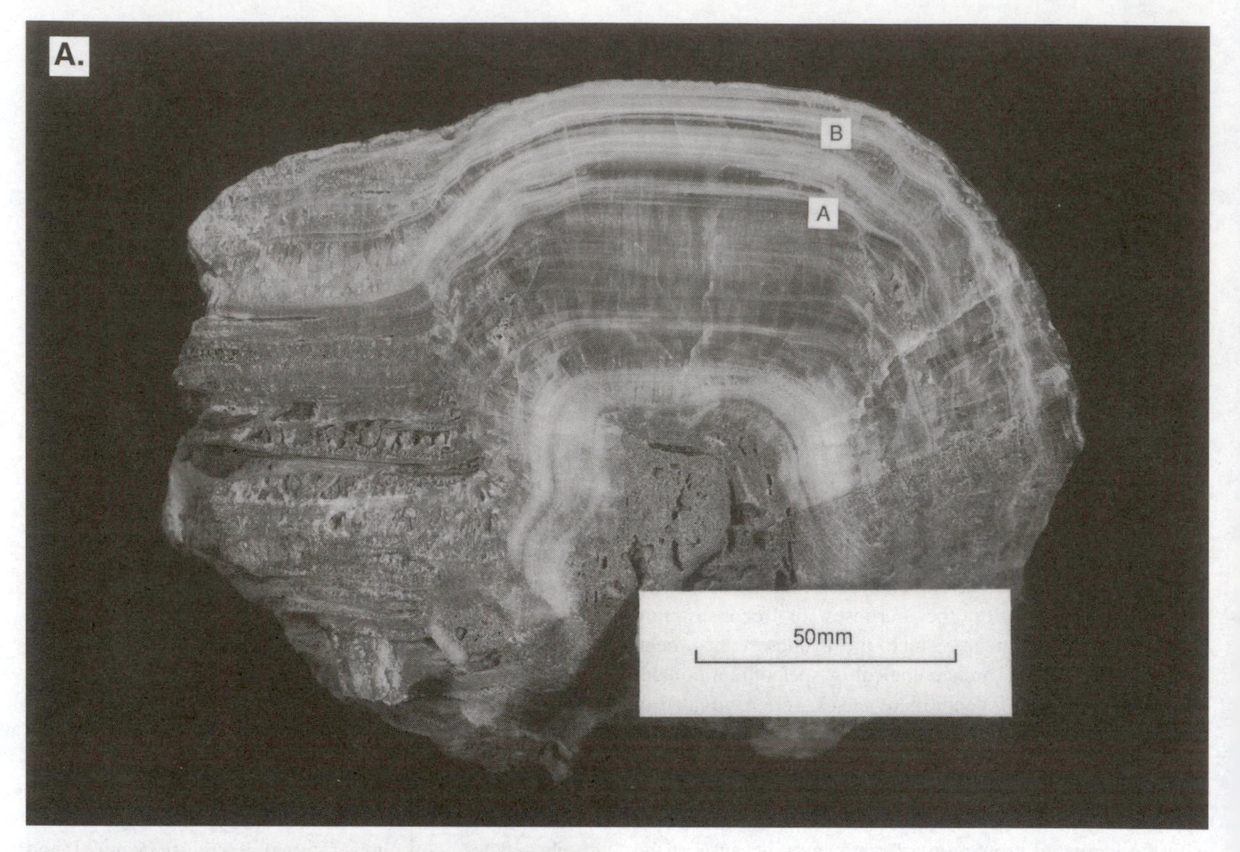

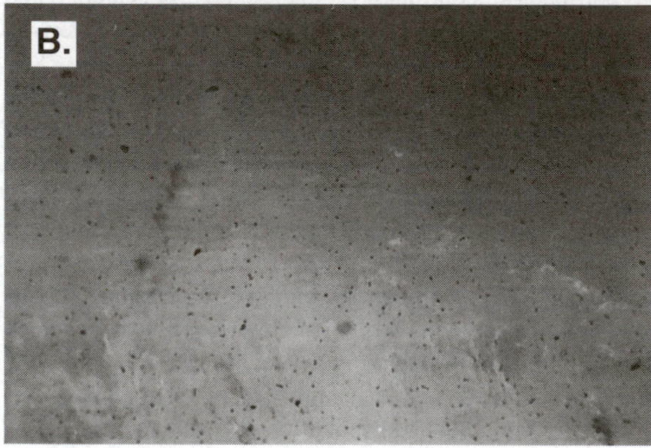

Figure 3.31 A. A luminescent banded sample of stalagmite (drops impacted from top surface) from the Mendip Hills, southwest England. The section between A and B has been dated to 277^{+44}_{-32} ka BP. B. Close up of A showing annually laminated speleothem bands. Average band width (= annual growth rate) is 0.026 ± 0.01 mm y^{-1} (photos: Andy Baker).

of a rock shelter. The most common features that develop in this way are **stalactites** and **stalagmites**. Flowstones are deposits of calcium carbonate, gypsum or other mineral matter that have accumulated on the walls or floors of caves in places where water trickles or flows over the rock. Upon reaching the cave floor, water may percolate into the interstices of clastic sediments, cementing them into a coherent, often very hard porous rock known as **cave breccia.** In some cases a complete cover of precipitated calcium carbonate blankets the floor of the cave, where it may become interbedded with the screes, breccias and cave earths. Such a flowstone cover has been referred to by some workers as a **stalagmite floor.** Other precipitated calcium carbonate deposits occasionally found in caves in karst regions include **travertine**, a light, compact and generally concretionary substance, extremely porous or cellular varieties of which are known as calcareous **tufa,** calcareous **sinter** or **spring deposit.** Compact banded varieties are sometimes referred to as **'cave marble'** or **'cave onyx'**. Many of these carbonate features are also common in subaerial environments in karstic regions, around streams and springs (see below).

The precipitation of calcium carbonate is caused by degassing of CO_2. Water that drips or flows into the cave chamber has usually originated on the surface where it has acquired a high concentration of CO_2 from biogenic production in soils. Due to ventilation effects, the air in the cave chamber has a lower partial pressure of CO_2 than the incoming water and so automatic degassing of CO_2 from the water takes place, leading to supersaturation and then precipitation of carbonate. With a continuous supply of water over long periods, large speleothem structures can build up by successive **growth layers** of carbonate (Figure 3.31). It is this characteristic that makes speleothems so valuable for the reconstruction of Quaternary palaeoenvironments, for they often preserve a continuous sedimentary record capable of analysis at a very high resolution (Lauritzen, 1993). In some sites the growth banding in speleothem may be annual in nature (Baker *et al.*, 1993), while speleothem can also be dated using the uranium-series dating method (section 5.3.4). Once a chronology of speleothem growth has been established, palaeoenvironmental inferences can be made on the basis of (a) the rate and abundance of speleothem formation and (b) temporal variations in the isotope ratios contained within the carbonate structure. These aspects will now be discussed in more detail.

3.8.4 Speleothem growth and environmental reconstruction

Precise dating of speleothem has demonstrated that the development of carbonate growth bands in cave systems may be cyclic or intermittent. This largely reflects environmental controls on cave hydrology, and so an analysis of the growth frequency of speleothem within cave systems can be used to reconstruct a number of different aspects of former environmental conditions, as the following examples demonstrate.

3.8.4.1 Speleothem growth and sea-level variations

Submarine cave systems in coastal karst areas often contain speleothem structures, indicating that sea level must have been below the level of the caves at the time the speleothem formed. Dating of speleothem structures in submerged caves around reef islands, such as the Bahamas and Bermuda, indicates that speleothem formation was controlled largely by global sea-level variations. Dating the onset and cessation of speleothem growth throughout the cave systems has enabled the amplitude and rate of global sea-level variations during successive glacial–interglacial cycles to be reconstructed (Harmon *et al.*, 1983; Mylroie & Carew, 1988; Ford & Williams, 1989). These data are valuable for providing independent tests of the models of global ice-volume based on oxygen isotope ratios from marine microfossils (section 3.10), since the changes in sea level inferred from the speleothem evidence should match the ice volume changes deduced from the marine isotope records (Harmon *et al.*, 1983).

3.8.4.2 Speleothem formation and tectonic activity

Speleothem evidence also provides useful information on local and regional tectonic histories. In the Bahamian archipelago, for example, while speleothem growth is abundant in many submerged caves, it is less well developed in caves above sea level, being confined to a very narrow zone between *c.* +1 and +7 m. The restricted elevations of emerged caves, and the fact that speleothem older than 100 ka is absent within them, led Carew & Mylroie (1995) to conclude that the Bahamian banks have been tectonically stable throughout the late Quaternary. By implication, therefore, it should be possible to derive a late Quaternary eustatic sea-level curve from Bahamian shoreline evidence (section 2.5.2). Speleothem growths may also provide insights into local tectonic activity. In the Devil's Hole cave in Nevada which is a massive fissure extending some 130 m below the local water table, the submerged part is lined with a layer of calcite that is more than 30 cm thick and which was precipitated over a period of more than 500 ka. Since there are no dissolutional features within the fissure, it does not seem to have formed by normal karstic processes. Riggs *et al.* (1994) therefore argue that a tectonic origin for the fissure is more likely (it is, in fact, situated within a zone of extensional faulting) and they use the term '*tectonic speleogenesis*' to refer to this type of fissure formation.

3.8.4.3 Speleothem formation and climatic change

Carbonate precipitation is strongly influenced by prevailing climatic conditions. It is either strongly reduced or arrested during cold episodes and increases to a maximum during warm intervals. In permafrost environments, groundwater percolation is at a minimum, while CO_2 concentrations in these groundwaters are also reduced because of the restricted biogenic activity associated with skeletal periglacial soils. There may, however, be some limited or sporadic speleothem growth, depending upon local conditions. Under a glacial ice cover, however, cave galleries become flooded, carbonate precipitation is not possible, and hence speleothem formation ceases completely (Lauritzen, 1993). During interglacial conditions, by contrast, higher precipitation levels result in greater water penetration, biological productivity increases, **vadose**[9] conditions prevail in cave systems, and maximum speleothem formation occurs. In mid- and high latitude regions, therefore, speleothem formation reflects the sequence of climatic conditions experienced during the course of a glacial–interglacial cycle. In Britain, for example, speleothem formation was relatively rare in the period 40–26 ka BP, ceased altogether between 26 and 15 ka BP (during the last glacial maximum) but became increasingly common from 15 ka BP onwards during the present interglacial (Atkinson *et al.*, 1986b). Records extending back beyond the last interglacial show ten distinct episodes of increased speleothem growth in British cave sites (Figure 3.32), the six most recent growth maxima occurring at *c.* 128.8 ka BP (the last interglacial; marine OI substage 5e), *c.* 103.1 (OI substage 5c?), *c.* 84.7 (OI substage 5a?), *c.* 57.9, *c.* 49.6 and 36.9 ka BP (Gordon *et al.*, 1989). The last five maxima may correspond to short-lived interstadial episodes during the course of the last cold stage, which have been recognised in Greenland ice cores (section 3.11 and Chapter 7). Similar relationships between speleothem formation and climatic change have been observed in other parts of northern Europe (Lauritzen, 1990), in North America (Sturchio *et al.*, 1994) and in Tasmania (Goede & Harmon, 1983). These data suggest that speleothem growth is strongly influenced by external climatic forcing (Baker *et al.*, 1995), and indeed evidence of the 41 ka obliquity and 21 ka precessional Milankovitch cycles (section 1.6) have been identified in speleothem carbonate (Kashiwaya *et al.*, 1991).

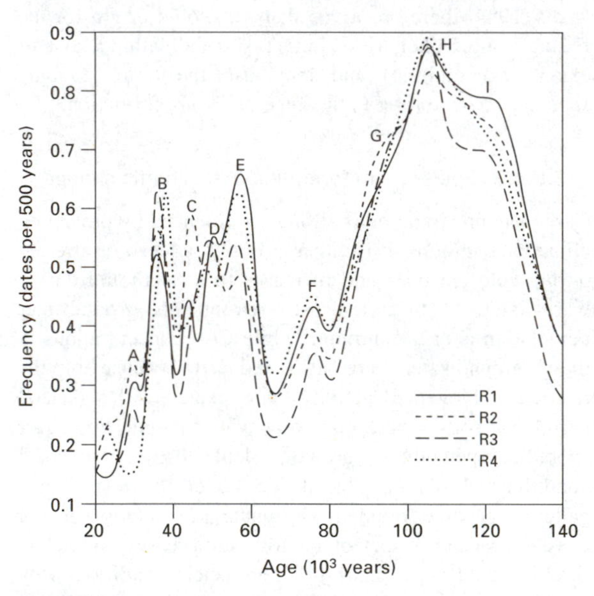

Figure 3.32 Comparison of growth frequency curves of speleothem during the period 140–20 ka BP obtained from four cave sites in the UK. The data suggest cyclical variations in the rate of speleothem formation, with maximum growth during the period 130–90 ka BP (peak H coincides with the last interglacial), and a series of oscillations in growth rates during the last cold stage (growth peaks A to F) (from Gordon *et al.*, 1989).

Speleothem records may also indicate periods of more arid conditions. In caves in low latitude regions, where variations in groundwater flow reflect regional changes in precipitation régime, speleothem growth is often episodic. A change to drier climatic conditions will lead to a reduction in the amount of water percolating from the surface while soil biogenic activity will also be reduced (Gascoyne, 1992). Goede *et al.* (1990) have suggested that this may explain why caves in southern Australia contain abundant speleothem older than 400 ka BP, but no significant amounts of carbonate deposit of a younger age. They attribute this contrast to a change to a prolonged period of aridity in the region during the late Quaternary.

3.8.4.4 Speleothem formation and rates of denudation

Dating of speleothem in mountain regions has demonstrated that the oldest material is frequently found in the highest caves, while the onset of speleothem formation is progressively later in caves at lower altitudes (Trudgill, 1985; Ford & Williams, 1989). This may reflect long-term lowering of the water table through valley incision, which results in the higher caves being perched above the vadose zone so that carbonate precipitation will then cease (see Atkinson *et al.*, 1978; Gascoyne *et al.*, 1983a, 1983b). As the water table falls, so new caves are formed or come within the vadose zone, and speleothem formation is initiated. Rates of valley incision implied by these data typically range between 0.3 and 1.2 m ka^{-1} (Lauritzen, 1993), although in glaciated regions downcutting was more episodic, as a result of increased rates of erosion (by ice or meltwaters) during cold stages compared with much lower rates of incision during warm intervals.

3.8.5 Oxygen isotope ratios in cave speleothem

In deep caves, where there is no direct contact with the external atmosphere and there is little or no air circulation, the temperature remains more or less constant and is in equilibrium with the mean annual surface temperature in the vicinity of the site. Any speleothem produced under such conditions tends to form in **isotopic equilibrium**[10] with the water from which it is precipitated (Gascoyne, 1992). A change in mean annual surface temperature can therefore change the isotopic composition of percolating water and this will be registered in the isotopic ratios locked into the carbonate crystal or in fluid inclusions trapped within it. By measuring the ratio of heavy (^{18}O) to light (^{16}O) oxygen isotopes (measured as a deviation, δ, from a standard – section 3.10.2) at regular intervals along the axis of

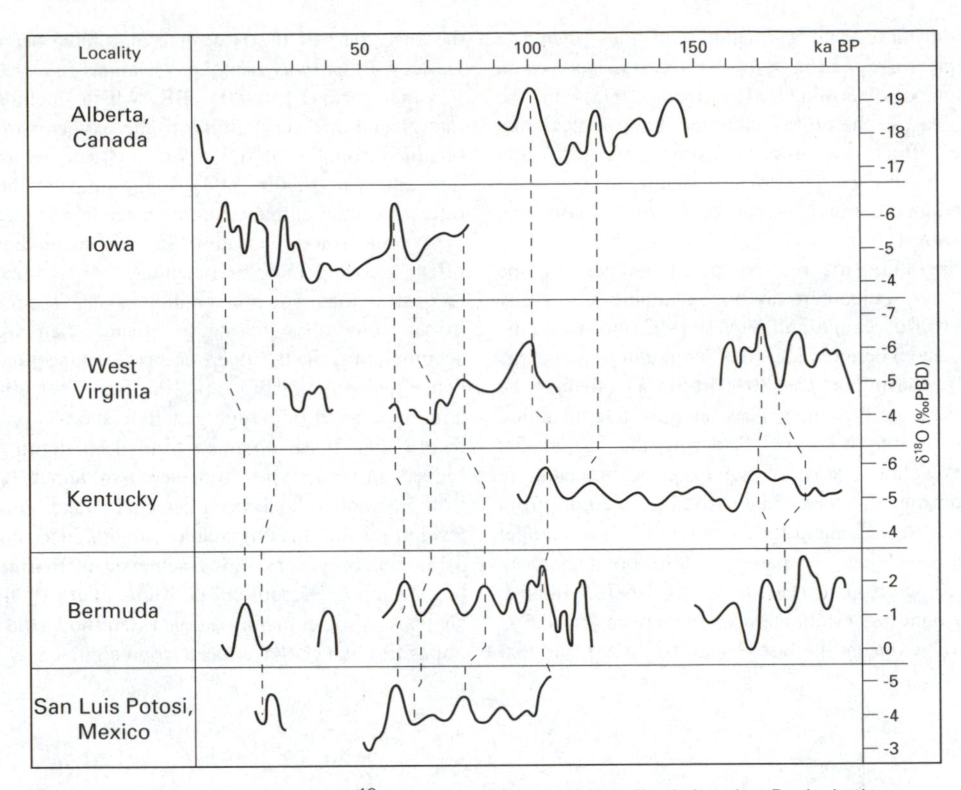

Figure 3.33 Generalised record of $\delta^{18}O$ variations from six sites in North America. Peaks in the curves reflect warm episodes and troughs colder intervals. Tentative correlations between the curves are shown by dashed lines (after Harmon *et al.*, 1978b).

speleothem growth bands, a record of isotope variations over time can be obtained. Such data have shown that the oxygen isotope ratios often vary in a cyclic fashion (Figure 3.33), and dating of these records by the uranium-series method allows direct correlation both with marine oxygen isotope cycles and also with other records of global climatic change (Schwarcz, 1986). Indeed, the match between the speleothem and ocean isotope data is so close that speleothem can be regarded as a 'geothermometer', the variations in oxygen isotope ratios predominantly reflecting variations in mean annual surface temperature (e.g. Harmon *et al.*, 1978b; Schwarcz, 1986). Oxygen isotope ratios are discussed in more detail in section 3.10.

Not all speleothem is suitable for this type of research, however, for a number of conditions must be satisfied before the isotopic ratios can be regarded as providing reliable palaeotemperature estimates. These include: (1) the carbonate precipitate must be in isotopic equilibrium throughout the period represented; (2) the oxygen isotope content of the water from which the carbonate was precipitated must be known; (3) no diagenetic alteration should have occurred; and (4) the deposit must be sufficiently free of detrital contamination for satisfactory dating of the sequence (Talma & Vogel, 1992) . Given the complexity of cave sediment sequences, it is not always easy to satisfy all four of these conditions. Perhaps the most difficult is (2), for isotope variations in cave carbonate can be the result of the combined influences of temperature, the **fractionation**[11] which occurs when carbonate precipitates from dripwater (a decrease in cave temperature will lead to higher ^{18}O concentration in the carbonate precipitate), and changes in the isotopic ratios of dripwater before entry to the cave system. This last factor relates to the effects on the global hydrological cycle of changes in oxygen isotope ratios in sea water during glacial–interglacial cycles (section 3.10) which feed through to atmospheric vapour, rainfall, soil/ground water recharge and hence ultimately to dripwater in caves. Estimates can be made of the magnitude of this effect through models of temporal changes of the isotopic ratio of water evaporated from the oceans, using marine micropalaeontological data (Gascoyne *et al.*, 1981; Talma & Vogel, 1992). The fractionation effects that occur during carbonate precipitation can also be estimated by analysing hydrogen isotope variations, since these are unaffected during crystal formation and they therefore enable the pre-fractionation ratios of oxygen isotopes in dripwater to be determined (Lauritzen, 1993). In the majority of cases, therefore, both the hydrological and fractionation effects in oxygen isotope records from speleothems can be estimated, and hence the palaeotemperature component can be isolated (Gascoyne, 1992; Lauritzen, 1995).

Palaeotemperature records based on oxygen isotope traces from cave speleothem are now available from many parts of the world. Talma and Vogel (1992), for example, have constructed a detailed record of temperature variations in South Africa during the past 30 ka from speleothem in the Cango Caves, Cape Province. These suggest a temperature reduction of around 6°C to 7°C during the last glacial maximum, as well as a series of high frequency temperature variations during the last 5 ka. Isotopic records from speleothems in New Zealand for the last 25 ka show parallel trends which are partly corroborated by dendroclimatological data (Goede *et al.*, 1990; Williams, 1991). These data suggest that temperatures were 4°C to 5°C lower than today during the last glacial maximum, and that the early part of the Holocene may have been up to 2°C warmer. In Norway, isotopic variations in cave speleothem for the period 150–80 ka BP, which includes the last interglacial, are very similar to the oxygen isotope records obtained from the GRIP ice core in Greenland over the same time interval (Figure 3.34), suggesting that both datasets reflect external climatic forcing mechanisms (section 3.11).

Oxygen isotope variations in cave speleothem therefore offer considerable potential for palaeoclimatic reconstructions. The precipitation of $CaCO_3$ per unit volume in speleothem is markedly greater than in sediments accumulating on the deep ocean floor (section 3.10), and hence the oxygen isotope profiles from speleothem provide a more detailed (although generally shorter) palaeoclimatic record than those obtained from deep marine sequences. Indeed, in some types of speleothem, annual variations in isotopic content have been detected (Baker *et al.*, 1993), a level of resolution comparable with that in ice cores (section 3.11) and only very rarely achieved in marine sequences (e.g. Kemp & Baldauf, 1993; Kemp *et al.*, 1994a). Detailed analysis of speleothem material often shows the presence of impurities which have been coprecipitated with the host

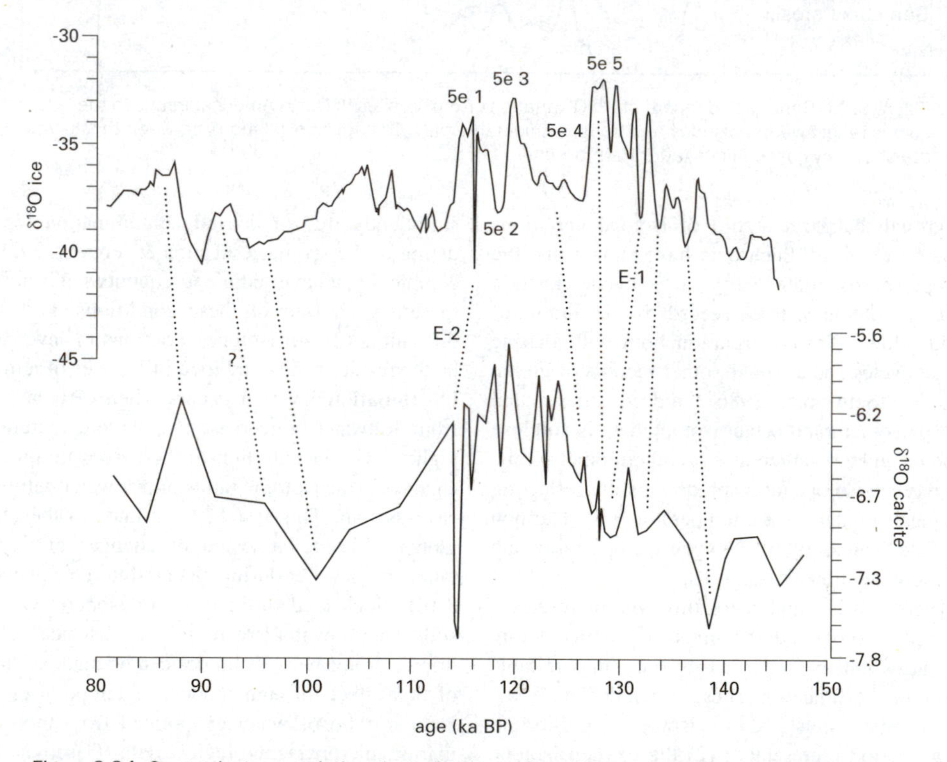

Figure 3.34 Oxygen isotope variations in speleothems from Norway (lower curve) compared with the GRIP ice core (section 3.11) isotope record (upper curve). The correlation is particularly strong during the period 140–110 ka BP, with the fivefold subdivision of the last interglacial (substages 5e-1 to 5e-5: section 7.3) clearly identified in both records (from Lauritzen, 1995).

carbonate, including trace elements and organic compounds derived from the soils through which the dripwater has percolated (Bull, 1983). The latter include pollen grains, amino acids, humic acids, fulvic acids and peptides (Lauritzen *et al.*, 1994). When examined under ultraviolet light, these impurities, especially when humic acid staining is present, display luminescence and very fine growth lines can often be detected (Shopov *et al.*, 1994). Some of these may be annual features, but others represent intra-annual bands. This technique, therefore, offers the prospect for the development of very high resolution (perhaps even down to a few days!) palaeoclimatic records from cave speleothem carbonate.

3.8.6 Other carbonate deposits

Other forms of carbonate accumulate in or around surface streams and springs in karstic regions. Some, such as **travertine** (carbonate coatings on bedrock), are produced in much the same manner as flowstone in caves. Groundwater saturated in $CaCO_3$ undergoes CO_2 degassing where it emerges into the open and carbonate is precipitated at the surface. In arid and semi-arid regions, the rate of production of CO_2-charged water depends upon local soil and microbial activity, and so the formation of travertine may reflect times of increased biological productivity associated, perhaps, with wetter climatic conditions (e.g. Livnat & Kronfeld, 1985). In some semi-arid regions, isolated patches of travertine may mark the levels at which higher, perched water tables developed during wetter periods. Dating of such features in the Grand Canyon, Arizona, for example, suggests that wetter episodes occurred at 338, 171, 71 and 15 ka BP, dates which correspond with those obtained for high lake level stands in that region (Szabo, 1990). Travertine precipitation from hyperalkaline springs in Northern Oman indicate wet conditions prior to 19 ka BP, a period of extreme aridity between 16.3 and 13 ka BP, a pluvial episode between 12.5 and 6.5 ka BP, and a late Holocene phase of hyperaridity which continues to the present day (Clark & Fontes, 1990). Again, this is very similar to the climatic sequence obtained from lake level records. It is also possible to derive carbonate accumulation rates for some sites that reflect seasonal or annual carbonate precipitation events. Where the carbonate was deposited at the lake margin, these data may enable the precise duration of lake still-stands to be estimated (e.g. Benson, 1993).

Tufa is a calcareous crust that develops around lake margins, stream edges or springs. It can take a variety of forms and can merge imperceptibly with other carbonate deposits, such as travertine, calcareous cements and stromatolites (see below). Tufas commonly mark the margins of high lake stands in arid and semi-arid regions, and the chronology of lake level variations in such areas is frequently based upon the uranium-series dating of tufa deposits (see, e.g., Lao & Benson, 1988). Oxygen isotope ratios in freshwater tufa also appear to be a sensitive indicator of past climatic changes, enabling, in particular, records of palaeotemperatures to be established (e.g. Pazdur *et al.*, 1988).

A third category of carbonate growths is **stromatolites**, which are carbonate encrustations around algae, bacteria or other organisms. These can take a wide variety of forms, depending upon the range of organisms present, the chemical composition of the water and the nature of the climatic régime (see, e.g., Cohen & Thouin, 1987). Stromatolites commonly accumulate to form reefs at the edges of lakes or in the intertidal zone, and the microbial structures of which they are composed can provide information on water depth and energy of the depositional environment during reef construction (Rasmussen *et al.*, 1993).

Finally, carbonate deposition is also an important pedogenic process, particularly in arid and semi-arid regions where **pedogenic carbonate** is deposited as nodules, root casings, pebble coatings and irregular concretions. Precipitation is largely of inorganic carbonate, but biomineralisation of carbonate in soils by soil fungi and other organisms also occurs (Monger *et al.*, 1991). The analysis of pedogenic carbonate may provide evidence of climatic change, ground surface stability and substrate age, for the amount, type and depth of carbonate production in soils is climatically determined (Pendall *et al.*, 1994). Analysis of oxygen isotope ratios of pedogenic carbonate can also provide useful palaeoclimatic information, since these reflect changes in humidity and frequency of soil wetting (Pendall *et al.*, 1994).

3.9 Lake, mire and bog sediments

3.9.1 Introduction

Preserved within sediments that have accumulated in lakes, mires and peat bogs is a diverse and often detailed record of environmental change. Given sufficient time, all lakes become infilled with sediment to form mires and bogs. Thus lake, mire and bog sediments are genetically related and frequently grade into one another. For this reason, the sediments are considered under a single heading. Lake, mire and bog deposits are important in a number of respects. First, the contained fossil flora and fauna provide evidence of both local and regional ecological changes. Second, the character of the sediments offers clues about former environmental conditions. This is particularly true in the case of lake sediments where variations in the physical and chemical properties reflect developments in the lake ecosystem, and also changes in the rates at which processes operated around the lake catchment. In both cases, the observed variations may be interpreted in terms of environmental change. Third, fossil lacustrine sediments and associated shoreline features often reveal a record of fluctuations in lake levels in response to climatic changes during the later part of the Quaternary. This topic was discussed in sections 2.7.1 and 3.6 in relation to lakes in arid and semi-arid regions, where prolonged episodes of drought frequently result in desiccation of the lakes. Here, however, we focus on lake sediments accumulating in the more humid regions of the world, where drought is less of a hazard and sediment supply tends to be more continuous.

Lakes can have a variety of origins. Håkanson and Jansson (1983), for example, recognise 11 main types (Table 3.8) of which three – tectonic, volcanic and 'glacial' – are of most interest to the Quaternary scientist, since it is within these that the most detailed sedimentary records have been found. **Tectonic lakes** form in areas of subsidence caused by folding or faulting, and include some of the largest lakes in the world, including Lakes Baikal (Siberia), Titicaca (Bolivia), Tanganyika, Victoria (East Africa) and Biwa (Japan), and the Black, Dead and Caspian 'Seas'. Some of these contain sediment sequences that span the whole of Quaternary time. **Volcanic lakes** (or **maars**) occur in calderas and craters, the length of the sediment record depending upon the time at which volcanic activity ceased. In Europe, a number of the maars in the French Massif Central, in central Italy and in Greece contain sediments that extend over several glacial–interglacial cycles. **'Glacial lakes'** are the many hundreds of small lakes that typically form in glaciated regions. They include **kettle lakes** (which develop in hollows created by the melting of buried ice) and lakes dammed behind or between glacial landforms (e.g. terminal moraines or drumlins) as a result of the blocking of drainage outlets. In some respects the term is ambiguous, however, as it can also be used to refer to lakes that form on, or within, glacier ice (section 3.3) or to ice-marginal lakes where glaciers advance to block a drainage outlet. Other types of lake listed in Table 3.8 tend to be more ephemeral or to contain sediment records that span only short periods of time.

Table 3.8 Classification of lakes according to Håkanson and Jansson (1983).

1. Tectonic lakes	a. Lakes formed by epeirogenic movements (e.g. Caspian Sea) b. Lakes formed by tilting, folding or warping (e.g. East African Rift system)
2. Volcanic lakes	a. Maars, calderas and crater lakes (e.g. Crater Lake, Oregon) b. Lakes formed by damming of drainage by lava or volcanic debris (e.g. Lake Kivu, Sea of Galilee)
3. Landslide lakes	Lakes formed by rockslides, mudflows and screes
4. Glacial lakes	Large variety of lakes formed in glacigenic sediments by ice-melting ('kettle' lakes) or through impeded drainage behind or between glacigenic constructional forms (e.g. moraines, drumlins)
5. Solution lakes	Ground solution of limestone, other calcareous rocks, gypsum and rock salt
6. Fluvial lakes	Plunge pools, delta lakes, meander lakes (oxbows and levées)
7. Aeolian lakes	Deflation basins; lakes dammed by wind-blown sediment
8. Shoreline lakes	Damming of material transported by longshore currents; tombolos and spit-lakes
9. Organic lakes	'Phytogenic dams' (blocking by vegetation; beaver dams); coral lakes
10. Anthropogenic lakes	Dams and excavation fills (e.g. Lake Mead, Arizona)
11. Meteorite lakes	Meteorite craters

In the glaciated regions of Europe and North America, present-day lakes are no older than *c.* 15 000 years, as they formed following the retreat of the last ice sheets, and hence the sediments in mires and bogs in these regions are usually considerably younger. In those areas that lay beyond the ice sheets, however, sediments continued to accumulate in lake basins throughout the last cold stage, and these sites often contain records extending back into the last interglacial and beyond. Coring of such sites is technically difficult and relatively expensive, and it is only in recent years, with the development of sophisticated hydraulic coring equipment and the establishment of international research teams to provide the necessary logistical and financial support, that long records have begun to be obtained from these deep lake basins (e.g. Creer 1991; Thouveny *et al.*, 1994).

The importance of these long lake sequences is that they provide a continuous record of changes in ecosystems, lake sediment processes and climate, sometimes over several glacial–interglacial cycles. Hence they offer a means of correlating terrestrial records with those obtained from deep-ocean sequences and from ice cores (section 6.3.3). In addition to these long lake records, fragmentary remains of lake and peat deposits dating to before the last glacial maximum are also found in many localities. In higher latitudes these are often intercalated between tills and other terrestrial sediments and usually reflect limnic accumulation during interglacials and interstadials or, in areas dominated by coarse-grained sediments that are free-draining, during

Figure 3.35 The Cromer Forest Bed on the Norfolk coast, eastern England. This peat, of Cromerian Interglacial age, underlies glacial sediments deposited during the Anglian cold stage (~OI Stage 12; *c.* 520–480 ka BP) (photo: Mike Walker).

intermittent periods of higher water tables during which temporary lakes or mires developed (Figure 3.35). Although they are useful in stratigraphic subdivision, the palaeoenvironmental significance of these deposits usually lies more in the fossils they contain than in the nature of the sediments themselves. The faunal and floral remains that are commonly found in lake and bog sediments are discussed in Chapter 4.

In this section we are concerned with those *lithological* characteristics of lake, mire and bog deposits that can be used as a basis for palaeoenvironmental reconstruction. The emphasis is placed on sediments that have accumulated during the late Quaternary (particularly during the Lateglacial and Holocene periods) in North America and northwest Europe, for these are often relatively accessible and can be sampled either in section or with hand-operated corers. As a consequence, they are frequently known in much greater detail than older deposits.

3.9.2 The nature of lake and bog sediments

Lake sediments are both allochthonous and autochthonous in origin, being derived partly from organic production within the lake ecosystem, and partly from the inwash of both organic and inorganic material from around the lake catchment (Aaby & Berglund, 1986). If the lake is rich in mineral nutrients, organic productivity will be high and the conditions are described as **eutrophic**. The typical deposit will be a green–brown organic-rich sediment known as **nekron mud** or by the Swedish name **gyttja**. In deeper waters, this will be extremely fine in texture and will consist of comminuted and largely unrecognisable plant material, but will grade into **detritus gyttja** with recognisable plant macrofossils (fruits, seeds, leaves, etc.) in shallow waters. Where the lake substrate is calcareous, lime may be precipitated from the water by aquatic plants (e.g. by some of the pondweeds, such as *Potamogeton*, and algae such as *Chara*) and other organisms, and a fine cream–white clay-rich sediment known as **marl** will accumulate. In general, sediments deposited under eutrophic conditions are predominantly autochthonous. Where the lake is poor in nutrients and organic productivity is low (**oligotrophic**), allochthonous sediments will often predominate. If the inwashed materials are low in organic content, clastic sediments (sands, silts and clays) will dominate, the finer grades of sediment being encountered in deeper waters. Where organic productivity is low, but the inwashed materials are dominated by humic substances from, for example, peats around the lake catchment, the lake waters are typically brown in colour due to the dissolved humic acids. In some lakes a dark brown **gel-mud** composed largely of colloidal precipitates will accumulate. Such conditions are often described as **dystrophic** and the deposit is known by the Swedish term **dy**. Finally, where the lake waters support a rich diatom flora the sediments are sometimes characterised by a white silicious mud composed almost entirely of diatom frustules (section 4.3) which is termed **diatomite**. These deposits may form under either eutrophic or oligotrophic conditions depending on the ecological affinities of the diatom species.

The nature of lake sediments is also partly related to lake size and the temperature and density variations experienced by lake waters. Lakes can be classified according to whether they are **chemically** or **thermally stratified**, a distinction which may be important for the interpretation of lake sediment sequences as well as their fossil content (Burgis & Morris, 1987; Behrensmeyer *et al.*, 1992). The density structure of lake water is controlled by temperature, by solids in the water column, and by dissolved compounds. In fresh water the temperature of maximum density is about 4°C at the surface and declines to about 3.4°C at a depth of 500 m due to increased pressure. In lakes where summer water temperatures exceed 4°C the surface waters become less dense than bottom waters and a thermal stratification evolves. The warm surface waters (**epilimnion**) are separated from deeper, cooler water (**hypolimnion**) by a marked thermal gradient, termed the **thermocline**. Stratification may also occur in winter due to freezing at the surface. Most lakes in temperate regions therefore experience vertical mixing (breakdown of thermal stratification) twice a year (**diamict** lakes), during what are called the spring and autumn **'overturns'**. In high altitude and high latitude lakes, however, where warming does not exceed 4°C, mixing may occur only once per year (**monomictic** lakes). Similarly, lakes in low latitude regions may not cool to below 4°C, and are also usually monomictic. Some lakes (**meromictic**) remain stable throughout the year, and do not overturn, usually because of a strong density gradient as a result of light freshwater overlying denser water at depth, the latter containing high concentrations of solids or dissolved salts. In very general terms, large ($>10\ km^2$) and deep (>10 m) lakes tend to be stratified, whereas shallow or small lakes tend to be unstratified (Behrensmeyer *et al.*, 1992).

The importance of density/thermal stratification lies in its effects on oxygen levels in the water column, on the preservation of organic detritus and on the influences these have on lithological composition. The epilimnion is usually well oxygenated, while the hypolimnion is **anoxic** (anaerobic), and this leads to important differences in lake chemistry, lake biota and the types of sediment that accumulate (Håkanson & Jansson, 1983). Organic remains, for example, are more likely to survive degradation under anoxic conditions. On the other hand, anoxic conditions may limit the variety and abundance of organisms inhabiting the lake, for they can lead to the production of hydrogen sulphide (H_2S) which, in high quantities, is toxic for most organisms. Exceptions include reducing bacteria that can thrive under these conditions. Heavy metals are reactive with sulphides and form compounds that are typically black or blue–black in colour. These are usually very obvious in lake sediments, as they contrast markedly with the lighter coloured oxidised clastic sediments or the brown colour of oxidised organogenic sediments. The former also often emit a pungent sulphurous odour and have a distinctive chemical composition. Therefore changes in lake stratification and the former degree of oxygenation of lake waters can be inferred from an analysis of sediment chemistry. In diamictic lakes which experience marked stratification, seasonal turnover may cause sudden and large-scale 'die-off' of aquatic organisms, such as algae (notably diatoms), and in some cases this can result in the formation of annual laminations (**organic varves** – Figure 5.18B) in the sediments accumulating on the lake floor (Saarnisto, 1986). A seasonal alternation between oxygenated and anoxic bottom waters may also lead to seasonal production of calcareous deposits, iron oxides or sulphur-rich sediments and hence **chemical varves**.

Over time, lakes silt up, plants encroach from the marginal zones, and areas of open water are progressively eliminated. The succession from open water to mire and bog is known as a **hydrosere** and the sediments gradually change in character from muds to peats. Three broad categories of peat can be identified and each is characteristic of a particular stage in the hydroseral succession. These are: (i) **limnic peats** which form beneath the regional water table and which are composed partly of transported plant debris and partly of decayed vegetation formerly growing *in situ*; (ii) **telmatic peats** which form in the swamp zone between high and low water levels and which are largely autochthonous in origin; and (iii) **terrestrial peats** which accumulate at, or above, the high water mark and which are entirely autochthonous in derivation. Each of these peat types will be composed of the remains of particular peat-forming plants, depending on the stage in the hydrosere and the trophic status of the lake water (Figure 3.36). The rate at which hydroseral succession progresses, and lakes become infilled, depends upon a number of factors, including the size of the lake, the size of the catchment, the rate of sediment supply and productivity within the lake. Many smaller lake basins in northwest Europe and North America have become completely infilled since the end of the last glacial stage, and are currently covered by peat deposits (Figure 3.37), whereas a number of larger lakes are still accumulating limnic sediments, except perhaps in shallow marginal areas where telmatic and terrestrial deposits have formed.

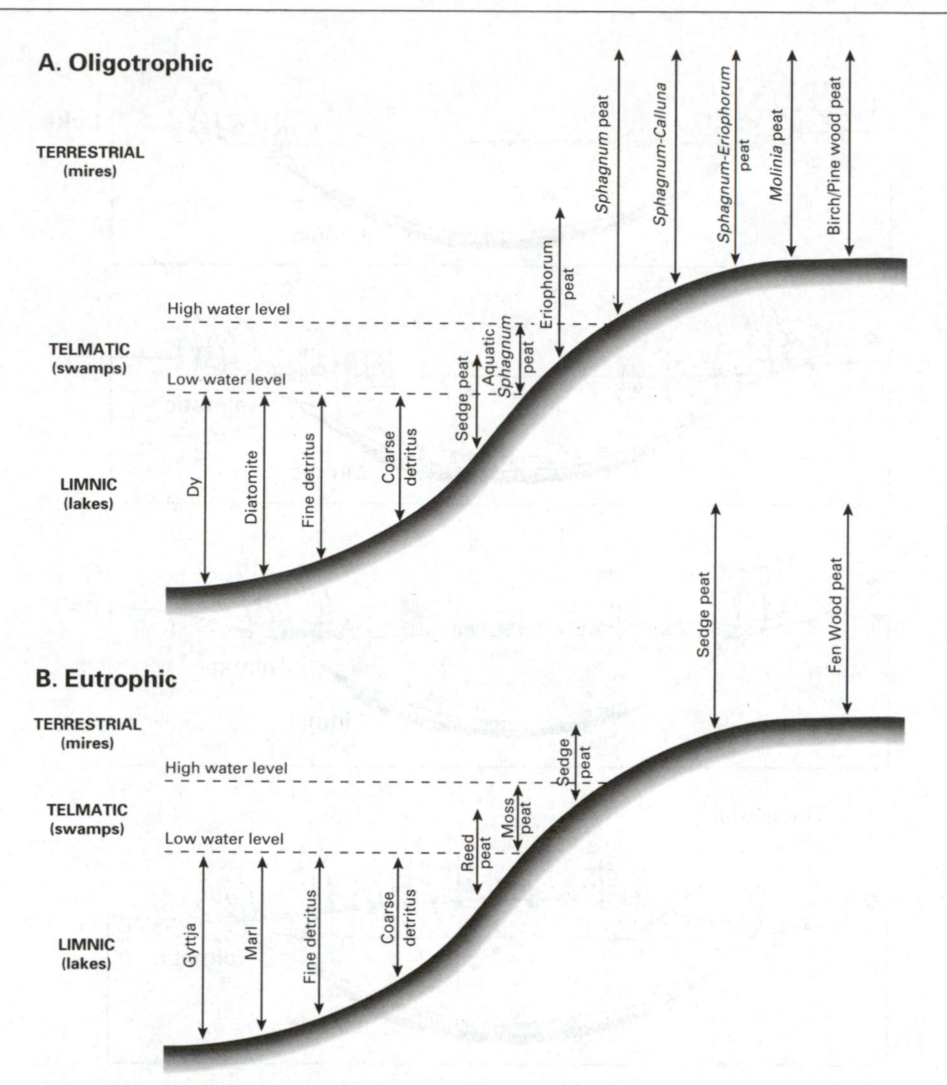

Figure 3.36 Some sediment types deposited with increasing depth of water under oligotrophic (A) and eutrophic (B) conditions (after Birks & Birks, 1980).

The terminology applied to peat-forming environments can be confusing. All waterlogged areas where peat develops as a result of reduced vegetal decay under anaerobic (anoxic) conditions are termed **mires** (Moore, 1986). Mires can be divided into those in which the high water table that induces peat formation is a consequence of groundwater conditions, either where drainage is impeded (**soligenous mires**) or where water accumulates in enclosed basins (**topogenous mires**), and those in which the water table is maintained by high atmospheric moisture levels (**ombrogenous** mires). Topogenous mires are, of course, part of the hydroseral sequence and are usually referred to as **fens** if eutrophic and **valley bogs** if oligotrophic. Soligenous mires are almost always oligotrophic. Ombrogenous mires (usually termed **bogs**) can be subdivided into raised bogs and blanket bogs. Raised bogs develop mainly in lowland areas where the peat-forming plants, principally *Sphagnum* mosses, produce a domed surface above the level of the surrounding ground. **Raised bogs** form the final stage of topogenous hydroseral successions (Figure 3.37) whereas **blanket bogs** are typical of upland areas and develop as a continuous cover over the landscape where rainfall is high (Barber, 1981, 1993; Moore, 1993).

Further details on sediments in lakes, mires and bogs can be found in Birks and Birks (1980) and various papers in Berglund (1986) and Chambers (1993).

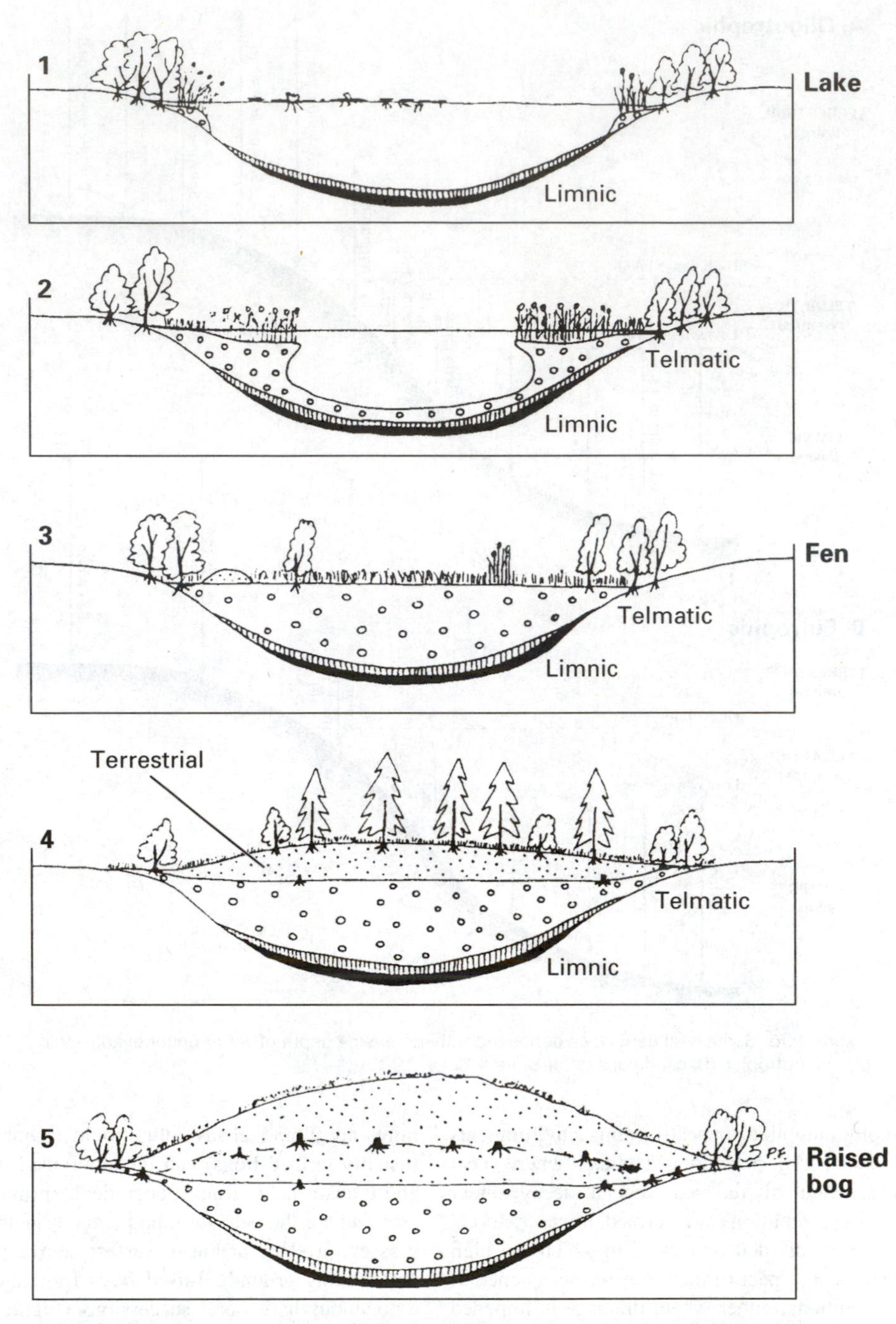

Figure 3.37 Schematic sequence of lake, fen and bog sediments in a small depression in northwest Europe (after Foss, 1987).

3.9.3 Palaeoenvironmental evidence from lake sediments

Like caves, lakes form natural sediment traps, and analysis of the sedimentary record in lake basins allows inferences to be made about environmental changes around the lake catchment. These include changes in the composition of the vegetation cover, which may reflect both natural (climatically induced) or anthropogenic processes, and changes in the rates of operation of geomorphological processes on catchment slopes (solifluction, gelifluction, etc.). In addition, both the levels of lake waters and the pattern of sedimentation within the lake will be influenced by changes in precipitation, while in calcareous lakes the isotopic record of the sediments will be determined, at least in part, by former temperature régimes. Hence it may be possible to make inferences about regional climatic change from the analysis of lake sediment records. These aspects of **palaeolimnology** (the study of ancient lake sediments) are considered in the following section.

3.9.3.1 Lake sediments and landscape changes

It has long been customary for those working on the pollen content of lake sediments to draw parallels between inferred vegetation changes and variations in sediment stratigraphy. In northwest Europe, for example, Lateglacial lake deposits typically consist of a threefold sequence of organic-rich lake muds (often gyttja or clay–gyttja) which overlie and are underlain by mineral sediments with a very low organic content (see Figure 3.38). The whole sequence often rests upon glacial gravels or sands (deposited during the last cold stage) and commonly lies beneath Holocene peats and organic muds. Palaeobotanical evidence suggests that the two minerogenic horizons accumulated during periods of reduced vegetation cover, the former during the pioneer phase immediately following local deglaciation, while the latter represents the cold phase of the Loch Lomond Stadial or Younger Dryas when a periglacial régime prevailed. Minerogenic material was therefore transferred from the catchment to the lakes, especially during the Younger Dryas phase, when surrounding slopes were affected as much by freeze–thaw activity and gelifluction as by overland flow. The organic-rich sediments, however, contain a fossil record reflecting a vegetation of shrub or woodland which developed under more stable conditions, as indicated, in turn, by a substantial reduction in the inwash of minerogenic material.

More sophisticated approaches to the investigation of lake deposits include analyses of the chemistry and magnetic properties of lake sediments. In Britain, pioneering research on Lateglacial sediments by Mackereth (1965) and Pennington *et al.* (1972) demonstrated how changes in the chemical composition of lake sediments could most readily be explained if the sediments were regarded as sequences of soils derived from the catchment. Increased soil erosion results in the transfer of large amounts of relatively unweathered material into lake basins, and the mineral fraction of the sediments that accumulates under such conditions is therefore characterised by higher proportions of metal elements, most notably sodium (Na), potassium (K) and magnesium (Mg) and, in certain cases, calcium (Ca), iron (Fe) and manganese (Mn), derived from exposed sediment or bedrock. During periods of reduced erosive activity and soil maturation under a vegetation cover, the mineral material transported into the lakes would be leached of its content of potassium, magnesium and sodium in particular, and during such 'stable' phases the lake sediments record lower concentrations of those elements. Periods of reduced erosion also coincide with higher values for organic carbon, which result partly from increased aquatic productivity within the lake basins, and partly from the increase in organic matter washed in from the catchment. Where acid soils and peats develop in the catchment, this may result in high iodine:carbon ratios or a high Fe and Mn content in lake sediments (Pennington *et al.*, 1972).

Chemical data from lake sediments can therefore be used to augment pollen-stratigraphic and other palaeoenvironmental data in the reconstruction of lake catchment histories. Figure 3.39 shows chemical profiles from two lake sequences on the Isle of Skye, western Scotland, that date from the Lateglacial and early Holocene periods (*c.* 13–9 ka BP). The lowermost sediments accumulated in the basins following the wastage of the last ice sheet from Skye and contain relatively high quantities of Na, K, Ca and Mg, and a low organic carbon content. These data reflect the inwash into the lakes of unweathered, finely comminuted clastic material derived from freshly exposed glacigenic sediments. Subsequently, an increase in carbon content reflects the gradual stabilisation of slopes around the catchments and maturation of soils as the vegetation cover expanded in response to climatic amelioration. The reduction of mineral inwash is marked by the decline in base elements in the sediment profile. A return to cold conditions during the Loch Lomond (Younger Dryas) Stadial (11–10 ka BP) led to the break-up of the interstadial vegetation cover, the destruction of soils by freeze–thaw and frost heave, and the renewed inwash of mineral material into the basins. These environmental changes are reflected in the reduced organic carbon content of the lake sediments and the significant increase in base element content. Finally, the abrupt climatic amelioration at the beginning of the Holocene is marked by a sharp upturn in the organic carbon curve and an equally dramatic decline in the curves for Na,

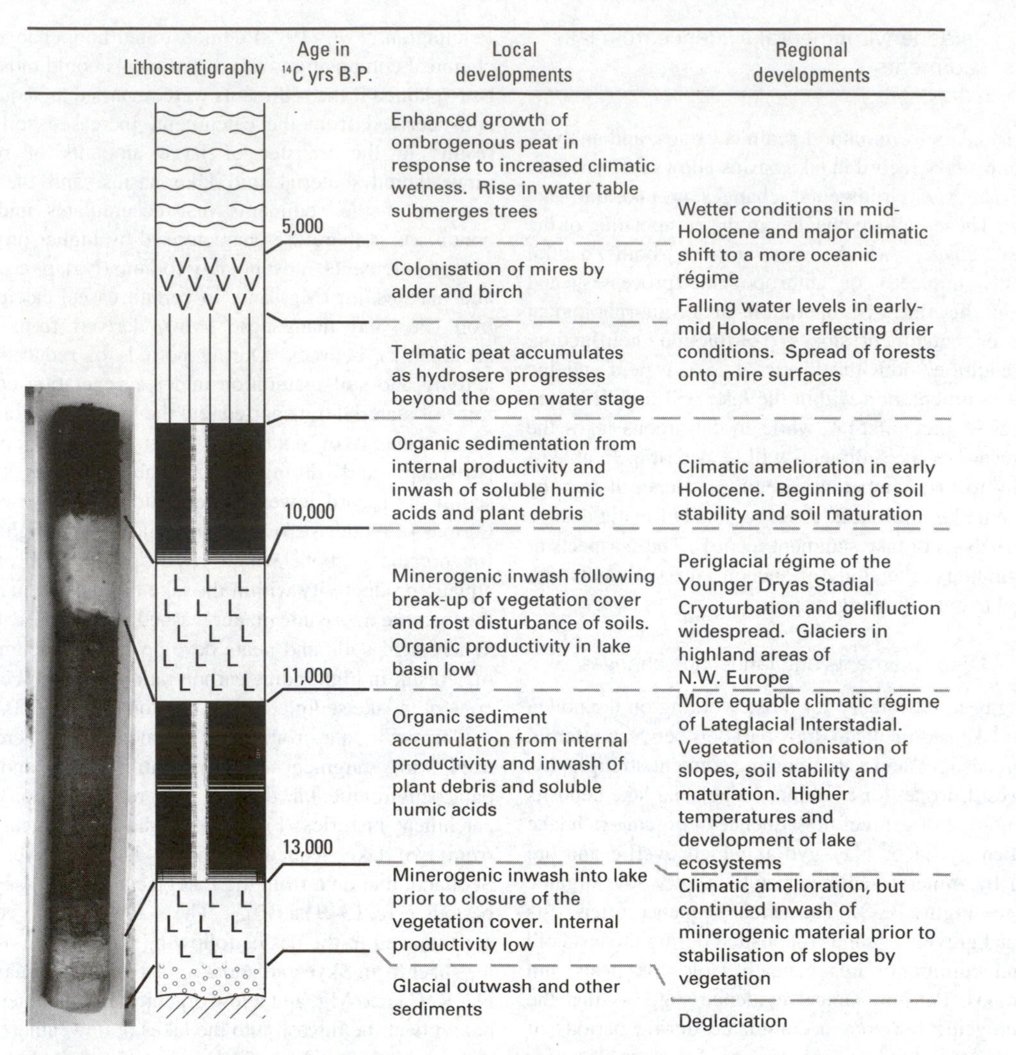

Figure 3.38 Lateglacial and early Holocene environmental changes inferred from lithostratigraphy of a typical northwest European lake and mire sequence (photo: Mike Walker).

K, Ca and Mg. In general terms, therefore, the chemical record from these lake sediments can be read as a proxy for regional climatic change (Walker & Lowe, 1990).

Anthropogenic effects on the landscape during the Flandrian can also be inferred from the chemical record of lake sediments. Mackereth (1965) noted the increase in concentration of base elements in several lake cores in northern England following the decrease in woodland cover recorded in pollen records at *c*. 5000 BP. This, he suggested, was a reflection of increased soil erosion around the lake catchments following forest clearance by Neolithic people. Accelerated soil erosion arising from human activity during the mid- and late Holocene is also reflected in the significantly increased sediment influx into many lake basins (Roberts, 1989). For example, at Braeroddach Loch near Aberdeen in northeast Scotland the onset of pastoral activity after *c*. 5390 BP resulted in a threefold increase in sediment accumulation, while during the period of modern agricultural practices (dated to 370 ± 250 BP), about 25 per cent of the sediment deposited during the past 10 600 years appears to have accumulated (Figure 3.40). The increase in the curve for sodium at these two horizons is particularly significant, suggesting inwash of mineral particles from exposed soils or substrates (Edwards & Rowntree, 1980). A Frains Lake in south Michigan, Davis (1976) demonstrated the contrast in erosion before and after deforestation of the catchment at around AD 1830. Prior to that date, the average annual sediment yield was estimated to be abou

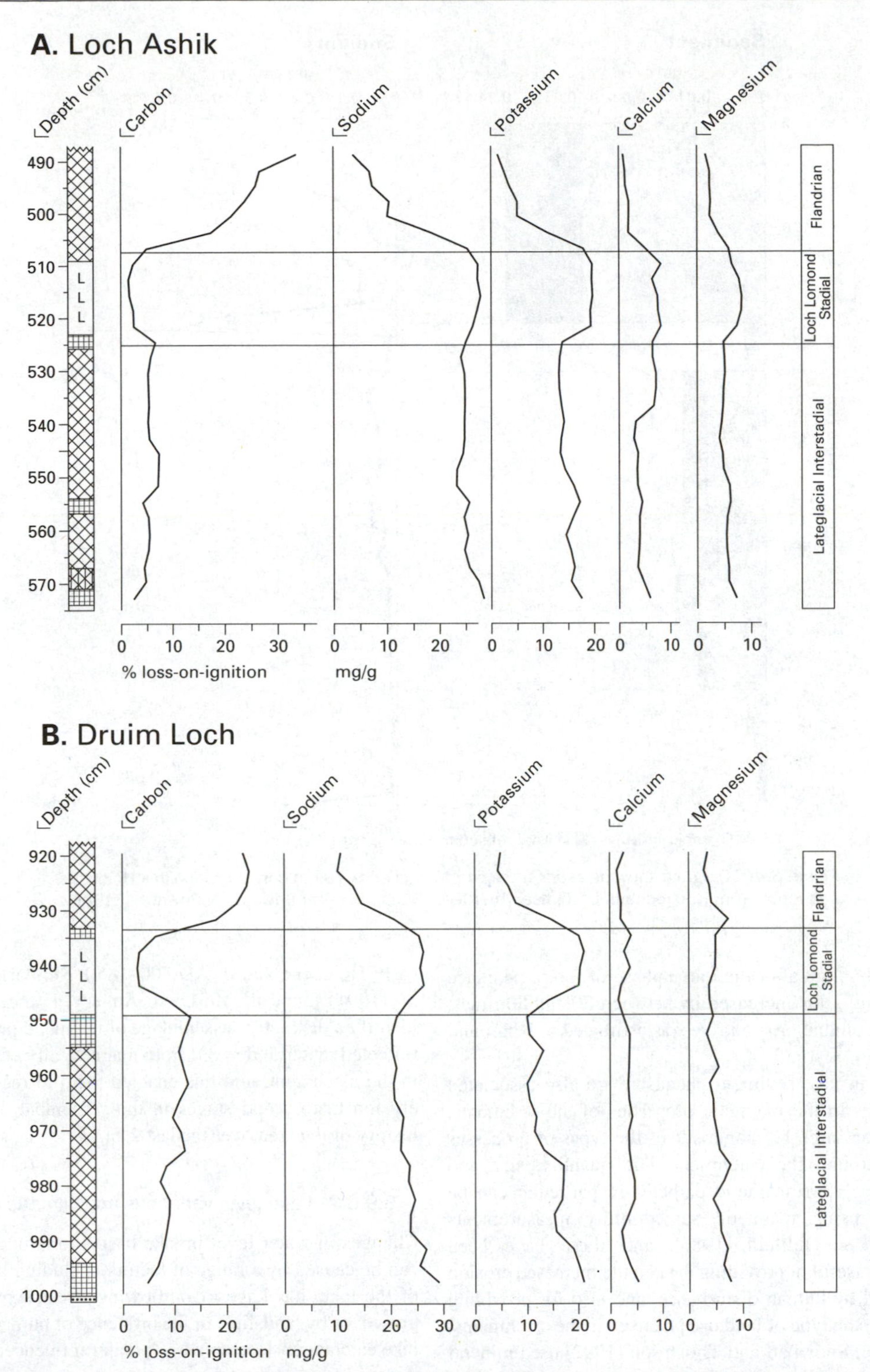

Figure 3.39 Variations in the abundance of selected chemical elements in Lateglacial and early Holocene sediments from two sites in the Isle of Skye, Inner Hebrides, Scotland: A , Loch Ashik; B, Druim Loch (from Walker & Lowe, 1990).

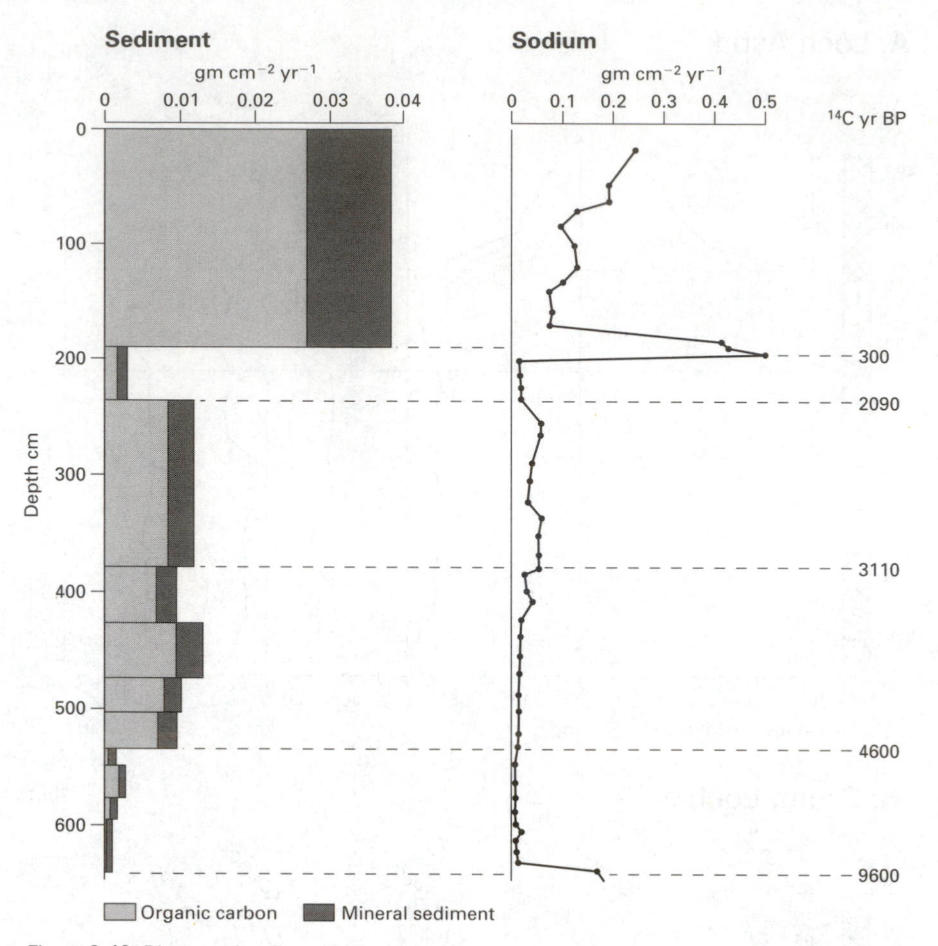

Figure 3.40 Diagram showing sediment accumulation and sodium concentration in a Holocene sequence from Braeroddach Loch, near Aberdeen, Scotland (after Edwards & Rowntree, 1980).

9 tonnes km^{-2}, but after the initial phase of forest clearance and ploughing, this increased by between 30 and 80 times. The present annual erosion rate was estimated to be around 90 tonnes km^{-2}.

Changes in lake sediment chemistry are also associated with changes in the magnetic properties of the sediments, and these can often be diagnostic of the types of processes operating around the catchment. The quantity, size and mineralogy of ferrimagnetic particles in particular can be established using magnetic susceptibility measurements (Thompson & Oldfield, 1986), and these have been particularly useful in providing data on the increased erosion rates caused by human disturbance, and also for providing insights into the type of land use practised in the catchments. For example, Snowball and Thompson (1992) used mineral magnetic measurements to show that sediment yields in a lake site in North Wales increased threefold between the early Holocene and *c.* AD 500–1850. Similarly, Higgitt *et al.* (1991) found that in Lac d'Annecy, France, they could identify a distinctive assemblage of magnetic properties that reflected material derived from magnetically enhanced soils in the catchment, and this enabled them to reconstruct soil erosion history and stages in the settlement and land use history of the area over the last 2 ka.

3.9.3.2 Lake level variations and climatic change

Changes in water level in lake basins in temperate regions can be caused by a range of factors, including the silting up of the lake, blockage of outflowing streams by vegetation growth or by landslips, or the influence of human activity on lake catchments (see above). A major influence, however, is the regional precipitation régime, changes in which can affect catchment hydrology, groundwater levels and, hence

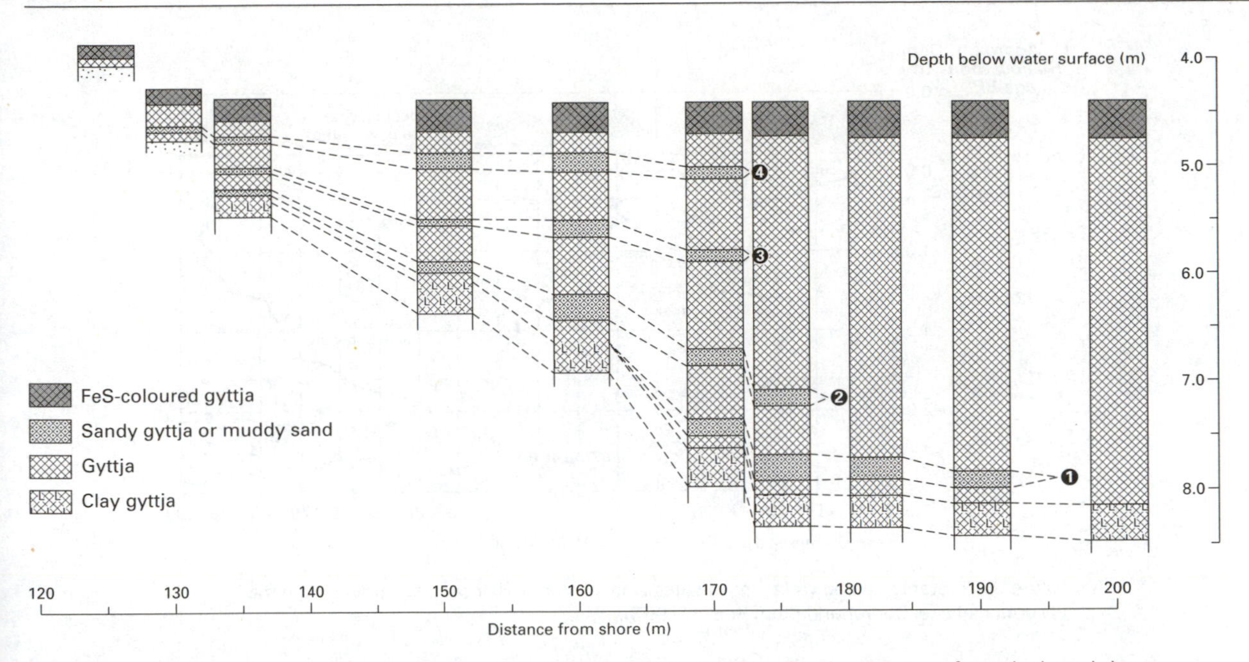

Figure 3.41 A transect of core records from Lake Växjösjön, southern Sweden, showing the presence of marginal sandy layers interpreted as indicating episodes of lower lake level (from Digerfeldt, 1975).

the depth of water in lakes. Evidence of past variations in lake levels is often preserved in the lake sediment record. Lower lake levels may be indicated by episodes of minerogenic sediment accumulation (Figure 3.41), by the occurrence of horizons of semi-terrestrial or terrestrial peats interbedded with limnic deposits, and by evidence for increased oxidation or disturbance of lake sediments (Digerfeldt, 1986, 1988). Where lake level fluctuations can be shown to be broadly synchronous at the regional scale, a climatic control might reasonably be inferred, and hence the lake sediment records can be used as a basis for palaeoclimatic reconstruction (Harrison & Digerfeldt, 1993). During the Lateglacial period in Europe, for example, lower lake levels in the mid- to late Allerød (*c*. 12–11 ka BP) are found throughout The Netherlands, northern Germany and southern Sweden, and a similar fall in lake levels is recorded towards the end of the Younger Dryas cold phase (Berglund *et al*., 1984; Bohncke *et al*., 1988). This suggests markedly drier conditions over a considerable area of northwest Europe during these two periods. For the Holocene, continental-scale compilations of lake-level data point to significant differences in precipitation régime between southern and northern Europe. Lake level records from the Mediterranean suggest a gradual trend towards wetter conditions during the early Holocene, which culminated around 6 ka BP, and was succeeded by increasing aridity. In southern Sweden, by contrast, drier conditions during the early Holocene were followed by a significant increase in effective moisture between 8 and 6 ka BP, a more arid régime between 5.5 and 3.5 ka BP, and a marked increase in effective moisture over the course of the last 3 ka. These regional precipitation contrasts, derived from lake level records, appear to reflect changes in the direct effects of insolation, and in the strength and position of the westerlies and of the subtropical high pressure cell (Harrison & Digerfeldt, 1993).

3.9.3.3 Lake sediments and palaeotemperatures

In calcareous lake sediments, a direct relationship has been found between stable isotope variations in lake sediments and palaeotemperatures. For example, oxygen isotope variations obtained from Lateglacial lake sediments in a site in Switzerland (Figure 3.42) suggest the highest temperatures during the last glacial–interglacial transition occurred during Bølling times (13–12.5 ka BP). This was followed by a gradual cooling, with perhaps minor climatic oscillations, during Allerød times (12–11 ka BP) and very low temperatures during the Younger Dryas. This climatic reconstruction is remarkably similar to that based upon beetle data obtained from Lateglacial records (sections 4.5 and 7.6.2). The investigation of stable isotope ratios in lake sediments therefore seems to offer considerable potential for deriving palaeoenvironmental information from lake sediments (Siegenthaler & Eicher, 1986; Hammarlund & Lemdahl, 1994).

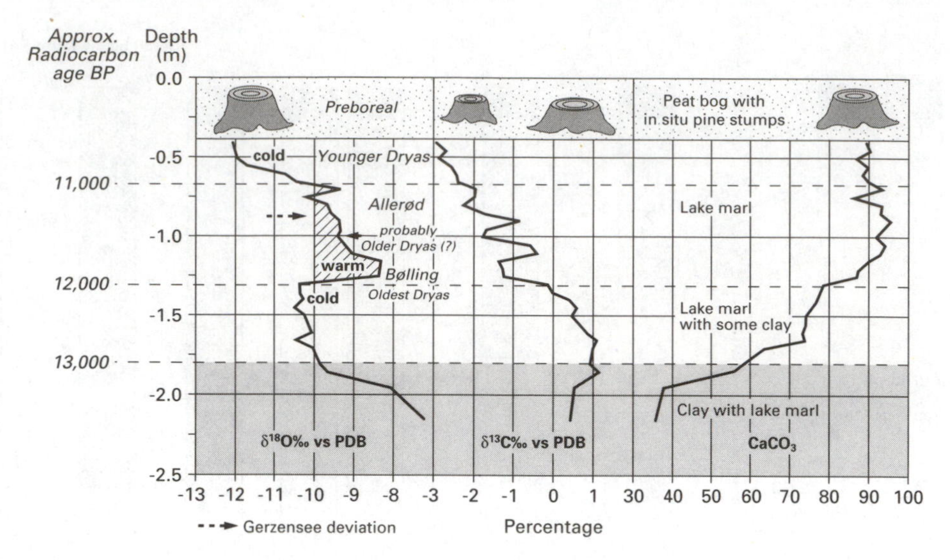

Figure 3.42 Stable isotope variations in Lateglacial and early Holocene sediments from the Wylermoos Lake, Switzerland (from Kaiser, 1993).

3.9.4 Palaeoenvironmental evidence from mire and bog sediments

In topogenous mires and bogs that develop over former lake deposits, the different types of peat may also provide indications of former climatic conditions. The rate at which hydroseral succession has proceeded cannot usually be interpreted simply in terms of climatic change, as local site factors will often exert a degree of control (Moore, 1986), but in certain circumstances it may be possible to make inferences about variations in the height of former water tables and hence about former precipitation levels. A good example would be where there is evidence for disturbance of the hydrosere (e.g. a transition from bog peat to reedswamp peat) reflecting a major change in local environmental conditions, such as a rise in silty water, possibly resulting from a change to a wetter climatic régime, or the occurrence of tree roots in a raised mire sequence which may indicate invasion of the bog surface by trees during a drier climatic period (Figure 3.37).

Ombrotrophic peats can also provide valuable palaeoclimatic information. A number of factors determine the rate of peat formation and the type of peat formed, including temperature (which affects metabolic and decomposition rates), local topography and groundwater pH (Moore, 1986; Charman, 1992). The most important control, however, is the height of the local water table (Clymo, 1984). Above the water table, decomposition can take place relatively rapidly because of increased microbial activity under aerobic conditions. The height of the water table fluctuates throughout the year due to seasonal variations in regional drainage conditions but, when examined over a timescale of decades, it will tend to have a minimum level, below which little decomposition can take place. This lower permanently waterlogged zone is termed the **catotelm**. Above this a shallow near-surface zone (the **acrotelm**) is regularly exposed for at least a brief period each year, and perhaps experiences prolonged dryness during the summer. The thickness of the acrotelm will be determined by a combination of water supply and summer drying, especially in ombrotrophic peats which are totally dependent upon precipitation for their water supply. Ombrotrophic peats therefore contain a record of past climatic régimes and a range of palaeoclimatic data can be obtained from the careful analysis of peat stratigraphy (Barber, 1993).

In the early years of the present century, data from Scandinavian peat bogs enabled the Scandinavian botanists Blytt and Sernander to divide the Holocene into five periods on the basis of marked changes in peat bog stratigraphy (Table 3.9), each of which were considered to reflect regional climatic changes. Subsequently, the **Blytt–Sernander climatic sequence** was related to regional pollen zones by von Post and other workers (Table 3.9), although this practice has now largely been discontinued, as it is apparent that the relationships between peat stratigraphy, pollen assemblage zones and climatic change are rather less straightforward than was envisaged in the 1920s (Blackford, 1993; Lowe, 1993a). Nevertheless, in many ombrogenous mires in northwest Europe, particularly in raised bogs,

Table 3.9 The Blytt–Sernander scheme of peat bog stratigraphy.

Period	Climate	Evidence
Sub-Atlantic	cold and wet	poorly humified *Sphagnum* peat
Sub-Boreal	warm and dry	pine stumps in humified peat
Atlantic	warm and wet	poorly humified *Sphagnum* peat
Boreal	warm and dry	pine stumps in humified peat
Pre-Boreal	subarctic	macrofossils of subarctic plants in peat

distinctive horizons are found separating dark, well-humified peats from overlying light-coloured, less humified *Sphagnum* peats, and these appear to indicate changes in former precipitation régimes (see Svensson, 1988a). Five such boundaries, which were termed **recurrence surfaces**, were described by E. Granlund in southern Sweden in 1932. Each was considered to reflect a change from drier to wetter conditions, and these were dated to *c*. 2300 BC, 1200 BC, 600 BC, AD 400 and AD 1200. Similar horizons have been found in peat profiles from other areas of western Europe, the most widely discussed of which is the **grenzhorizont** (boundary horizon) of northern Germany, first described by Weber more than 100 years ago. Radiocarbon dates place the age of the *grenzhorizont* at *c*. 500 BC, the transition from the Bronze to the Iron Age on the archaeological timescale. In many British raised bogs, a prominent recurrence surface has been dated to about the same time (Godwin, 1975). In Denmark, more than 20 stratigraphic changes from dark to light peat formation have been identified in bog sequences dating from the last 5500 years, and have been interpreted as reflecting cyclic shifts (periodicity *c*. 260 years) to cooler and wetter conditions (Aaby, 1976).

Although recurrence surfaces are potentially valuable in palaeoclimatic work, questions have been raised about their origin, about the dating of the features, and about the extent to which they reflect regional shifts in climate (e.g. Blackford, 1993). Radiocarbon dating has shown that not only may the ages of prominent recurrence surfaces from nearby sites differ by several hundred years, but there may be considerable within-site variations in the age of recurrence surfaces. This may partly reflect changing rates of peat growth on different parts of a mire surface, but it might also indicate mixing of carbon residues between peat horizons during oxidation and decomposition of the peat-forming materials. Where this has occurred, it can lead to major discrepancies between radiocarbon age determinations from the same stratigraphic horizon. A second problem relates to the extent to which recurrence surfaces develop in response to regional climatic (i.e. precipitation) changes, or whether they may also form as a result of site-specific factors, such as local changes in mire hydrology, particularly where human activity has been a factor (Moore, 1986, 1993).

In recent years, and partly in response to these problems, more sophisticated analyses of bog sediments have been undertaken in an attempt to explain the nature of the peat-forming process, and to establish the relationship between peat stratigraphy and climate, particularly as reflected in variations in bog-surface wetness. Peat humification changes can be detected using colorimetric methods (Blackford & Chambers, 1991), or on the basis of variations in the types and degree of preservation of plant macrofossil remains found within the peats (Barber *et al*., 1994 – see Figure 4.17). In addition, the ratio of inorganic to organic materials may provide a basis for fine-resolution stratigraphy, as the drying of the former peat surface, for example, may be reflected in higher concentrations of mineral particles derived by aeolian transport. Useful climatic data can also be obtained from the chemistry of bog sediments, while analyses of oxygen isotope and hydrogen/deuterium ratios have shown that these vary systematically in peats, with larger deviations corresponding to climatic changes inferred from other data sources (Brenningkmeier *et al*., 1982). In particular, there appears to be a close relationship between stable isotope variations in peats and long-term variations in both annual temperature and atmospheric moisture content. There is also a close relationship between peat humification changes and the biochemistry of organic detritus, reflecting the fact that the chemical pathways through which decomposition proceeds are species-dependent. Insofar as the species represented in the organic detritus reflect drainage conditions at the time of peat formation, analysis of the biochemical composition of the peats may allow inferences to be made about former hydrological conditions and, by implication, about past climate (Moers *et al*., 1990).

Application of these and other techniques has provided new insights into the palaeoclimatic record contained in peat profiles. In upland Britain, for example, episodes of increased bog-surface wetness point to significant shifts in climate during the late Holocene, involving higher rainfall and/or lower temperatures. There are, moreover, indications of a periodicity or cyclicity in the observed climatic pattern (Barber *et al*., 1993, 1994; section 4.4). Detailed analyses of peat profiles have also resulted in more consistent dating of recurrence surfaces, with many sites in northwest Europe now indicating shifts to wetter climate at *c*. 3850, 3500, 2800, 2200, 2050 and 1400 BP (Blackford & Chambers, 1991).

In addition to providing evidence on climatic change, bog sediments may also contain other palaeoenvironmental information. For example, anthropogenic disturbance of the vegetation cover may be indicated by the presence of minerogenic horizons reflecting inwash of eroded material (Edwards *et al.*, 1991), while tephra (volcanic ash) layers in peat profiles have been used to make inferences about the possible effects of volcanic ashfalls on Holocene vegetation (e.g. Blackford *et al.*, 1992) and perhaps also on prehistoric communities (e.g. Baillie, 1989; Burgess, 1989). Tephras also provide a possible basis for correlating between individual peat sequences.

3.10 Stable oxygen isotope stratigraphy of deep-sea sediments

3.10.1 Introduction

On the deep-ocean floors, sediments have been accumulating in a relatively undisturbed manner for thousands, or even millions, of years. They consist partly of terrigenous deposits, i.e. detrital material derived from erosion of the land masses surrounding the ocean basins, and partly of biogenic sediments composed largely of accumulations of the calcareous and siliceous skeletal remains of micro-organisms that formerly lived in the ocean waters. Terrigenous detritus (ranging in size from fine sand to clay) arrives on the ocean floor by a number of different pathways, but the principal transporting agencies are turbidity currents, bottom currents, wind and ice. In the mid- and high latitudes, both coarse and fine terrigenous detritus appears to have been delivered to the ocean floors mainly during glacial periods (Ruddiman *et al.*, 1989), reflecting in particular the ice-rafting of glacially eroded debris (see below) and, to a lesser extent, the transport of aeolian sediments from the greatly expanded periglacial regions (section 3.7). Indeed, wind-blown sediment may have constituted a major proportion of the fine detrital input in low latitudes during glacial times (Rea, 1994). Sea-level lowering of over 100 m would also have resulted in the discharge of large quantities of terrestrial debris from the major rivers as they flowed across the continental shelves, and this material would subsequently have been spread down the continental slopes and across the abyssal plains in gravity-controlled sediment flow (Andrews, 1990).

In many ocean sediment sequences, therefore, a broad correlation can be detected between the deposition of terrigenous material and former glacial episodes, and this is reflected most clearly in the large volumes of **ice-rafted debris** (**IRD**) that are found in deep-ocean sediments. In the North Atlantic, for example, IRD deposition appears to have been by far the most important mechanism for supplying terrigenous sediment to the ocean floor and it has been estimated that IRD may make up as much as 40 per cent of the total amount of sediment deposited in Quaternary cold stages (Robinson *et al.*, 1995). Cycles of IRD deposition in the North Atlantic during the course of the last cold stage are reflected in distinctive layers of glacially derived material (Figure 3.43) in the ocean sediments around 45°N (Heinrich, 1988; Grousset *et al.*, 1993). These have been termed **Heinrich layers** and reflect episodic deposition of IRD from icebergs drifting eastwards from the margins of the Laurentide ice sheet (Alley & MacAyeal, 1994). The significance of these deposits in the context of the history of the North Atlantic during the course of the last interglacial–glacial cycle is discussed in Chapter 7.

In the deeper oceans, the sediments tend to be finer grained and are often dominated by biogenic material consisting of the accumulations of the carbonaceous and siliceous remains of micro-organisms that formerly lived in the ocean waters. Such sediments are known as **marine oozes** (section 4.11) and are frequently characteristic of interglacial or warmer episodes. They contain recognisable fossil remains which provide a record of ocean circulation, ocean water temperature and, by implication, atmospheric temperatures throughout the Quaternary. The use of fossil assemblage data to reconstruct changes in oceanic circulation is considered in more detail in Chapter 4.

The most detailed evidence for environmental change in the oceans during the course of the last 2 million years, however, comes not from the sediments, but rather from the chemical and isotope content of the marine organisms contained within them. Variations in the ratios between, for example, aluminium, barium, calcium and cadmium, and between the isotopes of carbon, oxygen and uranium, reflect the combined influences of circulation, nutrient supply and water temperature, and therefore provide a basis for the reconstruction of environmental change within the oceans (see section 4.11). Of particular importance in Quaternary research, however, are variations in the ratios of oxygen isotopes in marine sediments. Oxygen isotope analysis of cave speleothems has already been discussed (section 3.8), but the technique is also applicable to other depositional contexts, including ice cores (section 3.11), lake carbonates (Siegenthaler & Eicher, 1986) and deposits containing molluscan remains (Goodfriend, 1992). However, it is the application of oxygen isotope analysis to deep-ocean sediments that has had by far the greatest impact in

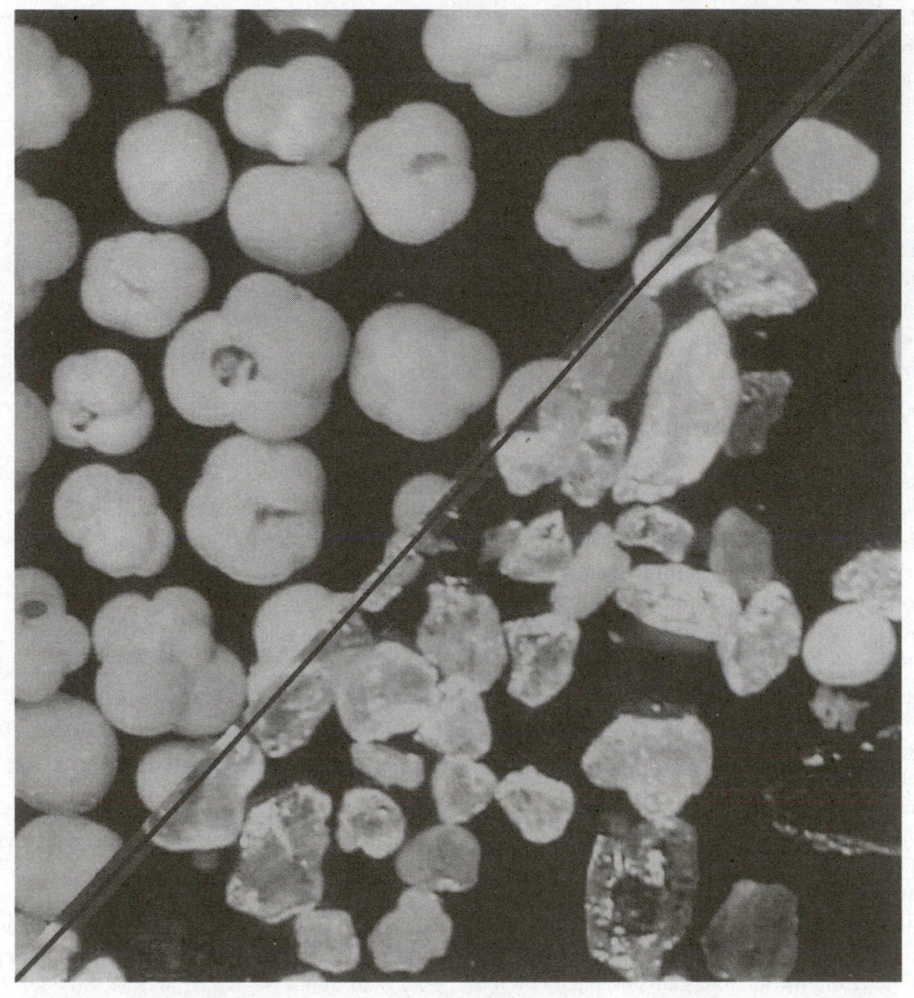

Figure 3.43 Photographs illustrating the marked contrast between clast-dominated Heinrich layers (lower right) and the normally microfossil-rich sediments (Foraminifera in this image; upper left) that accumulated on the floor of large areas of the North Atlantic during the last cold stage cycle (photo: Elsa Cortijo).

Quaternary science. The method not only provides one of the principal indices of global environmental change during the Quaternary, but also serves as a basis for global stratigraphic subdivision and correlation (Chapter 6). Moreover, it was on the basis of marine oxygen isotope evidence that the influence of the Milankovitch radiation cycles on the earth's climatic history was first convincingly demonstrated (Imbrie & Imbrie, 1979). The remainder of this section is devoted, therefore, to a consideration of the principles and applications of oxygen isotope analysis of deep marine sediments, although there is also a brief discussion of a complementary method, namely carbon isotope analysis of marine microfossils.

3.10.2 Oxygen isotope ratios and the ocean sediment record

Oxygen can exist in three isotopic forms (^{16}O, ^{17}O and ^{18}O) but only two (^{16}O and ^{18}O) are of importance in oxygen isotope analysis of marine deposits. ^{18}O:^{16}O ratios in the natural environment vary between about 1:495 and 1:515, with an average of approximately 1:500. This means that only about 0.2 per cent of oxygen in natural circulation is ^{18}O. Ratios of oxygen isotopes are measured not in absolute terms but as relative deviations ($\delta^{18}O$ per mil) from a laboratory standard value. The standards normally employed are **PDB** (belemnite shell – see section 5.3.2) for the analysis of carbonates and **SMOW (Standard Mean Ocean Water)** for the analysis of water, ice and snow (Craig, 1961). The latter is used in the isotopic analysis of glacier ice cores (section 3.11). PDB is +0.2 per mil in relation to SMOW. The standard materials are monitored and distributed to laboratories by the International Atomic Energy Agency in Vienna.[12] Mass spectrometric analyses are carried out on CO_2 gas prepared from the fossil material and on the standard reference material. Oxygen isotope ratios are then expressed as positive or negative values relative to the standard (δ = zero), thus:

$$\delta^{18}O = 1000 \times \frac{^{18}O/^{16}O \text{ sample} - {}^{18}O/^{16}O \text{ standard}}{^{18}O/^{16}O \text{ standard}}$$

A $\delta^{18}O$ value of −3‰ indicates that the sample is 0.3 per cent or 3.0‰ deficient in ^{18}O relative to the standard. A $\delta^{18}O$ value of −10‰ is even more deficient in $\delta^{18}O$ and is therefore described as **isotopically lighter** than a value of −3‰. It is important to understand this because reference is commonly made in the literature to isotopically 'light' and isotopically 'heavy' segments of oxygen isotope records (see below).

The variation in the isotopic composition of ocean waters over time can be reconstructed from the $\delta^{18}O$ values of carbonate shells and skeletons preserved in deep-sea sediments. Many marine organisms secrete (or build) carbonate structures and oxygen is abstracted from sea waters for this purpose. Thus, the oxygen isotope ratios in fossil carbonates buried in sediments on the ocean floors should reflect the ratios prevailing in the oceans at the time of their secretion. Analyses have been carried out on the remains of a range of marine micro-organisms, but by far the most widely used fossils are the tests of the planktonic and benthic Foraminifera (section 4.10). There is now a considerable body of evidence to show that the ratios of ^{18}O:^{16}O in ocean waters varied in a quasi-cyclic fashion during glacial and interglacial cycles as a result of a natural **fractionation** of oxygen isotopes during evaporation of water from the sea surface. Evaporation of water at the sea surface leads to fractionation of oxygen isotopes because the lighter $H_2{}^{16}O$ molecule is drawn into the atmosphere in preference to the heavier $H_2{}^{18}O$ molecule. This process is temperature dependent and so it is particularly marked at higher latitudes where colder air masses are increasingly less able to absorb the heavier isotope. Thus the moisture-bearing winds which nourish the polar glaciers contain relatively higher quantities of the lighter ^{16}O, and this, in turn, is reflected in the isotopic composition of glacier ice. A generalised gradient of modern $\delta^{18}O$ values over the Northern Hemisphere would typically show δ values close to 0‰ for surface ocean water and, in atmospheric water, −10‰ for low latitude regions, −20 to −30‰ at about 50–60°N, and as low as −60‰ over the polar ice sheet (Robin, 1983).

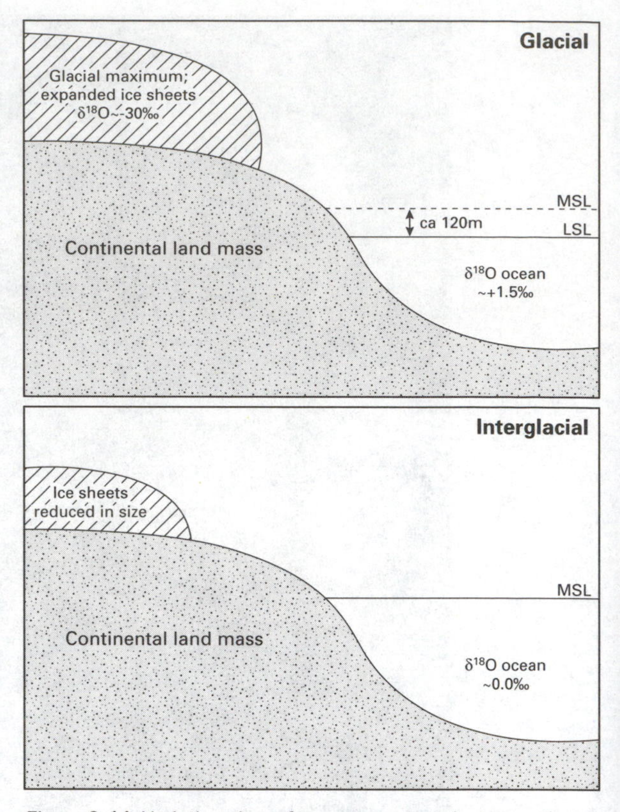

Figure 3.44 Variations in surface water oxygen isotope ratios during times of glacial maxima and interglacial high sea-level stands (minimal ice cover).

During the cold phases of the Quaternary, with markedly expanded ice masses in both Northern and Southern Hemispheres, large quantities of $H_2{}^{16}O$ were trapped in the ice sheets leaving the oceans relatively enriched in $H_2{}^{18}O$ (i.e. isotopically more **positive**, or **heavier**). Conversely, the melting of the ice masses during interglacial periods liberated large volumes of water enriched in $H_2{}^{16}O$ back into the oceans (Figure 3.44), resulting in lighter ratios, generally close to a δ value of 0‰ at times of minimal global ice cover. Analysis of ^{18}O:^{16}O ratios in Foraminifera has revealed that the overall glacial–interglacial variation in isotopic composition of ocean waters was, in fact, extremely small. For example, in one of the best-known standard oxygen isotope profiles, that obtained from marine core V28-238 (Figure 6.5), the isotopic differences between successive glacial and interglacial stages range from 0.47 to 1.37‰ (the extremes are −0.9 and −2.4‰) relative to the standard. Some parts of the ocean surfaces were, it seems, only about 1 to 1.5‰ more positive than at the present day according to data obtained from planktonic (surface-dwelling) species (Shackleton & Opdyke, 1973). Isotopic data obtained from benthic (ocean bottom-dwelling) species have much more positive $\delta^{18}O$ values than surface species, reflecting the higher $\delta^{18}O$ content of deep ocean water (see Pisias *et al*., 1984; Jansen, 1989). Nevertheless, the glacial–interglacial oscillations reflected in isotope profiles from benthic species match very closely those obtained from planktonic species, although the $\delta^{18}O$ ratios in the former traces generally vary within the range +2.5 to +5.0‰ (Shackleton *et al.*, 1990).

Oxygen isotope traces through cores of deep-ocean sediment reveal records of glacial–interglacial changes spanning, in many instances, the whole of the Quaternary. There is, moreover, a remarkable similarity between the isotopic profiles from different parts of the world's oceans, in terms of both the number and amplitude of oxygen isotopic cycles (Patience & Kroon, 1991). This suggests that the oceans as a whole have been responding in a coherent fashion to a common forcing mechanism (the Milankovitch radiation cycles), and hence variations in the isotopic signals are broadly synchronous worldwide. The common peaks and troughs in the curves can therefore be designated as **isotopic stages**, and these form the basis for stratigraphic subdivision of the individual profiles and for correlation at the global scale (Jansen, 1989). These aspects of oxygen isotope stratigraphy are considered in more detail in sections 6.2.3.5 and 6.3.3. Here we are concerned with the palaeoenvironmental applications of marine oxygen isotope records.

3.10.3 Environmental influences on $^{18}O/^{16}O$ ratios in marine sediments

The $\delta^{18}O$ of carbonate in marine organisms is dependent upon two main factors, the temperature and isotopic composition of sea water during secretion. Initially the dominant influence was thought to be water temperature, for as the fractionation that occurs when carbonate is precipitated slowly in sea water is at least partly temperature-dependent, it was anticipated that former temperatures could be established by measuring the degree of isotopic fractionation in marine carbonate fossils. Following pioneering work by H.C. Urey in 1947, Epstein and others developed an empirical equation for this relationship in 1953, which gave a value of around 0.23‰ per 1°C (see Jansen, 1989). In order to use this equation to obtain palaeotemperatures, however, it was necessary to know the former isotopic composition of sea water, and as this cannot be established directly, it had to be estimated. An important factor in these calculations, however, was the extent to which the changing isotopic balance of the world's oceans has been affected by fluctuations in land ice volumes. In other words, it was necessary to know the **glacial ice storage component**. C. Emiliani, who pioneered the stratigraphic application of oxygen isotope analysis in 1955, based his estimates of the glacial storage component in former changes in isotopic composition of sea water on models of the volume of global ice and its assumed average isotopic composition. These suggested that only about 0.4‰ of the average 1.7‰ $\delta^{18}O$ variation between glacial and interglacial stages was caused by ice storage variations, and that the more important factor was therefore water temperature. Using Epstein *et al.*'s equation, he therefore interpreted the fluctuations in isotopic content of planktonic Foraminifera in cores from the Caribbean and Equatorial Atlantic as reflecting palaeotemperature changes of 6°C between the last glacial and the postglacial periods.

Emiliani's ideas were challenged, however, following the development of more sophisticated models of the isotopic composition of the polar ice sheets and of revised estimates of the volumes of ice that existed in the past (Dansgaard & Tauber, 1969). When Shackleton (1967) and others (e.g. Duplessy, 1978) began to publish isotopic data obtained from benthic Foraminifera, it became clear that the temperature effect is small compared with that of glacial ice storage. Deep water should not be affected by temperature variations to the same degree as surface water, yet the isotopic signal in benthic and planktonic isotope traces is broadly comparable. Today, therefore, there is a general consensus that most of the isotopic variation between glacial and interglacial stages can be accounted for not by temperature, as Emiliani and his contemporaries believed, but rather by ice storage effects (Williams *et al.*, 1988). In the case of the last deglaciation, for example, it has been estimated that the ice volume effect accounts for 1.1‰ of the total isotopic shift recorded in planktonic Foraminifera (Bard *et al.*, 1987a).

Oxygen isotope profiles from marine deposits are, therefore, more readily interpreted as a record of global **palaeoglaciation.** In Figure 3.45, the marine isotope signal has been drawn to show patterns of global ice accumulation and wastage over the course of the last 600 ka (the scale is the reverse of that normally depicted, with isotopically heavier values to the right). The isotopic trace is asymmetrical, reflecting a gradual increase in ice volume during the course of a cold stage, followed by a sudden decline indicating rapid ice wastage (Broecker & Denton, 1990b). These latter points on the isotope curve are referred to as **terminations** (I–VII in Figure 3–45) and are discussed in more detail in section 6.2.3.5. Small amplitude variations

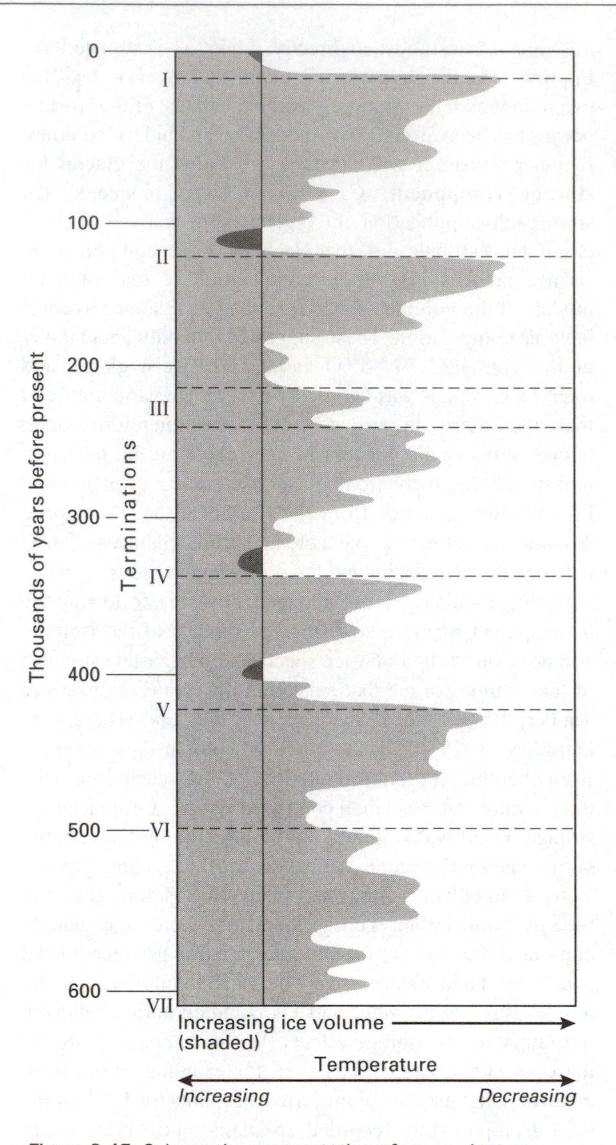

Figure 3.45 Schematic representation of oxygen isotope variations for the past 600 ka. Isotopically light values to the left, and isotopically heavier values to the right. The isotopic signal is interpreted as a proxy for extent of global palaeoglaciation. I to VII are termination events (section 6.2.3.5). The vertical line represents the isotope ratio that corresponds to the limited ice cover typical of the late Holocene. Only three short periods during the preceding 600 ka appear to have experienced comparable climatic conditions (after Broecker & Denton, 1990b).

in isotope values between successive terminations suggest that the gradual build-up of land ice was interrupted by short-lived episodes of glacier retreat. The overall shape of the curve is common to all marine isotope profiles, and hence the clearly defined terminations form marker horizons that can be used as a basis for inter-core correlation (section 6.2.3.5). The diagram also shows that, for much of the last 600 ka, global land ice cover was far greater and, by implication, the climate was cooler than at present. Hence, the current relatively equable climate is atypical when viewed in the context of the isotope records for the last 600 ka. Moreover, none of the last eight or nine interglacial periods shown in Figure 3.45 appears to have had a duration of more than 10–12 ka.

If changes in the isotopic composition of benthic Foraminifera can be taken as an index of land-ice volumes, then oxygen isotope data can also be regarded as an indicator of eustatic sea-level changes, since an expansion of global ice volume must mean a concomitant reduction in eustatic sea level. It has been suggested that a sea-level change of *c.* 10 m will be represented by a 0.1‰ shift in the oxygen isotope signal (Shackleton and Opdyke, 1973). At the height of the last glacial, deep-ocean waters were enriched in ^{18}O by about 1.6‰, which is equivalent to a sea-level lowering of *c.* 165 m at the last glacial maximum by comparison with the present day (Shackleton, 1977). However, it now appears that sea-level estimates based on benthic isotopic data tend to overestimate the scale of eustatic sea-level fall during the last glacial cycle by about 50 m (Figure 3.46), and it is only for the last and present interglacials that there is general agreement between the marine terrace and marine isotopic data over former sea levels (Shackleton, 1987; Jansen, 1989). Nevertheless, the oxygen isotope signal from the deep ocean floors remains a potentially invaluable independent monitor of glacio-eustatic sea level.

Although isotopic measurements from marine microfossils are still occasionally used to derive estimates of former ocean temperatures (e.g. Duplessy *et al.*, 1992), overall, the temperature effect remains frustratingly difficult to isolate. In planktonic Foraminifera, the isotopic signal is influenced by a range of factors including temperature as well as local salinity and other variations caused by upwelling effects (Patience & Kroon, 1991). A further complicating factor is that seasonality and living depth during foraminiferal growth are still poorly known (Bard *et al.*, 1987a). Isotopic records from planktonic Foraminifera are often characterised by abrupt ('spiky') fluctuations, making distinction between the different components of the $\delta^{18}O$ data extremely difficult. More consistent isotopic traces tend to be obtained from benthic species, as these generally inhabit waters of more constant temperature, salinity and chemistry. However, these records provide little information on ocean surface temperature changes and, moreover, even in the deep oceans some temperature variations occur which can influence the isotopic signal (Shackleton, 1987). Estimating former water temperatures from marine isotopic data, therefore, remains extremely problematical, and much still remains to be learned about the various factors that influence particular isotopic data sets.

3.10.4 Limitations in oxygen isotope analysis

There are a number of limitations affecting the interpretation of oxygen isotope data. These include the following

3.10.4.1 Resolution

Sedimentation rates vary markedly throughout the oceans. In deeper waters, where the terrigenous influx is often minimal and biogenic production is dominant, the rate of sediment accumulation is usually very low. Although long records, sometimes spanning the whole of the Quaternary, can occasionally be obtained from these deposits, an average sample for oxygen isotope analysis may span a time interval of several thousand years. Much higher temporal resolution can be obtained in sequences that have accumulated more rapidly, such as those on or close to the continental shelves (e.g. Austin & Kroon, 1996) or from bottom deposits dominated by IRD (e.g. Robinson & McCave, 1994; McCave *et al.*, 1995). However, these sedimentary records are often very much shorter, and are usually restricted to the last glacial–interglacial cycle.

3.10.4.2 Sediment mixing

Where mixing has occurred, as a result of bottom-dwelling burrowing organisms, or through the action of turbidity currents, the clarity of the oxygen isotope record will tend to be blurred (Jansen, 1989). In particular, the peak-to-peak amplitude in oxygen isotope values may be significantly reduced (Shackleton & Opdyke, 1976). Benthic organisms typically affect sediments to a depth of 20 cm as they feed and burrow (Patience & Kroon, 1991), although in some cases the depth of bioturbation may reach 50 cm (Bard *et al.*, 1987a). Reworking caused by the scouring of ocean bottom currents can also result in depositional hiatuses, while distortions of the sediment record may occur during sampling (Ruddiman *et al.*, 1987).

3.10.4.3 Isotopic equilibrium between test carbonate and ocean water

Because of fractionation effects, some species of benthic Foraminifera do not secrete carbonate that is in isotopic equilibrium with the ocean water that they inhabit (Grossman, 1984). Hence, when benthic tests are being measured, more importance is attached to species such as *Uvigerina senticosa* and *Globocassidulina subglobosa* which are known to deposit carbonate in isotopic equilibrium with deep ocean waters (Shackleton and Opdyke, 1977). A number of planktonic species also calcify at different depths during their life cycle (e.g. some species of *Globorotalia*), and this can lead to differences in isotopic ratios between adults and juveniles of the same species. Some knowledge of the species-dependence of such effects is therefore required and appropriate corrections can then be applied. Certain species are regarded as particularly good indicators of the original isotopic ratios of surface waters (e.g. *Globigerinoides sacculifer* and *Globigerina bulloides*), while others, such as *Globorotalia menardii* and *Neogloboquadrina pachyderma*, are selected because they are thought to provide reliable records of isotopic ratios in deeper waters (Jansen, 1989).

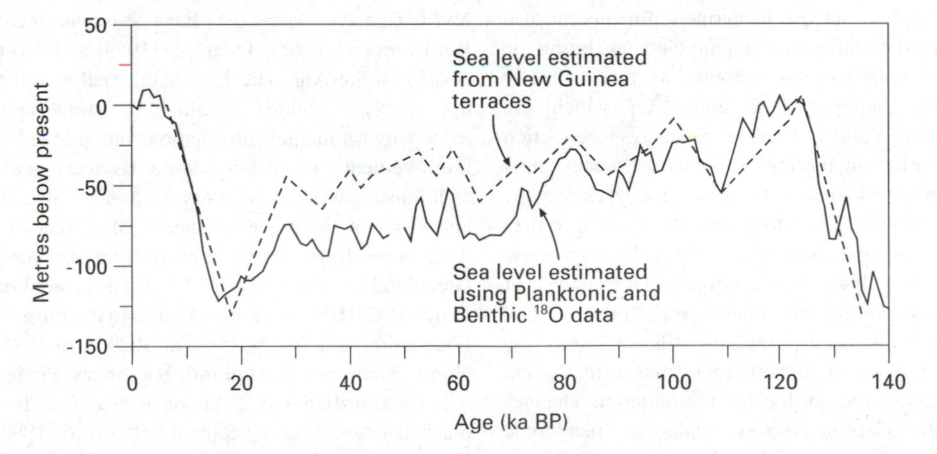

Figure 3.46 Sea-level variations based on benthic oxygen isotope record obtained from East Pacific core V19-30 and sea-level records, based on marine terraces in New Guinea (after Shackleton, 1987).

3.10.4.4 Carbonate dissolution and diagenesis

After death, carbonate microfossils sink within the water column, and many will become dissolved or disaggregated before reaching the ocean floor. Some foraminiferal species are more susceptible to dissolution than others, so that during settling there is a selective removal, usually of the species that lived closer to the surface. Between 3 and 5 km depth, $CaCO_3$ solution equals $CaCO_3$ supply, the level defined as the **carbonate compensation depth (CCD)**. Below that depth only the most robust microfossils arrive on the sea bed. Even where complete dissolution does not take place, the water depth plus depth of sediment burial may be sufficient to lead to diagenesis in the form of carbonate recrystallisation (Killingly, 1983). Stratigraphic studies of foraminiferal assemblages are, therefore, usually restricted to sediments that have accumulated in waters that are shallower than the CCD, and where evidence of diagenesis is absent or negligible.

In spite of these limitations, however, it is apparent that oxygen isotope analysis of deep-sea sediments is a technique of great importance in Quaternary research. Although perhaps not as secure an indicator of palaeotemperatures as was at one time believed, oxygen isotopes do, nevertheless, provide a unique record of glacial/interglacial cycles, of changing global ice volumes, and of glacio-isostatic oscillations of sea level. Moreover, the recognition of comparable isotope stages from different areas of the world's oceans provides a means of correlating environmental changes on a global scale (Chapter 6).

3.10.5 Carbon isotopes in marine sediments

The analysis of carbon isotopes in marine sediments can also provide valuable data on oceanographic changes during the Quaternary. As with oxygen, carbon has two naturally occurring stable isotopes (^{13}C and ^{12}C) which are fractionated during a range of natural processes (see section 5.3.2). $\delta^{13}C$ profiles in marine sediment sequences show cyclic variations, very similar to those in oxygen isotope traces, and it appears, therefore, that these also reflect important environmental changes. Similar problems to those of oxygen isotope analysis are experienced in the analysis of carbon isotopes, including, for example, variations in carbon isotope content between microfossils caused by fractionation as a result of different vital effects, and regional variations due to localised circulation changes (Jansen, 1989). Carbon isotope data in planktonic Foraminifera, however, provide information on former productivity changes in the upper layers of the oceans and on the flux of ^{12}C in surface waters (Shackleton & Pisias, 1985). This evidence also offers insights into former atmospheric variations in CO_2 (Shackleton *et al.*, 1992), and therefore forms a complementary source of information to the ice cores (section 3.11). Benthic Foraminifera record deep-water circulation changes in the oceans. In the North Atlantic, for example, vertical circulation brings oxygenated waters into the deeper parts of the ocean, a process referred to as **ventilation**. At times of reduced vertical mixing, oxygen levels fall, productivity is reduced and this will be reflected in the $\delta^{13}C$ signatures obtained from the fossil records. Hence, a history of deep-water circulation changes can be reconstructed from the analysis of carbon isotope ratios in the remains of benthic Foraminifera (Sarnthein *et al.*, 1994).

3.11 Ice-core stratigraphy

3.11.1 Introduction

Since the 1960s, work initiated by groups such as CRREL (United States Army Cold Regions Research Engineering Laboratory) and the British Antarctic Survey has led to the development of rotary drilling rigs capable of raising undisturbed cores from deep within the world's ice sheets. Analysis of deep cores from Greenland and Antarctica, and of shallower ones obtained from the smaller ice caps and glaciers, has revealed a record of annual increments of snow and ice accumulation extending back to beyond the last interglacial. The first cores were obtained from the Arctic: from Camp Century (1966), Dye 3 (1981) and Renland (1988) in Greenland, and from Devon Island (1976) in NWT, Canada. Deep cores have also been recovered from Byrd Station (1968), Dome C (1979) and Vostok Station (1985) on the Antarctic Ice Sheet. Analyses of these cores not only yielded valuable stratigraphical and palaeoenvironmental information, but also led to technical improvements in drilling, core recovery and analytical techniques, as well as providing new insights into the behaviour of the great ice sheets. More recently, two new deep cores have been obtained from the summit of the Greenland Ice Sheet (Figure 3.47). The **Greenland Ice-core Project (GRIP)**, coordinated through the European Science Foundation, reached bedrock at 3029 m in 1992, while the North American **Greenland Ice Sheet Project (GISP)**, which was drilled only 30 km or so away from the GRIP site, reached bedrock at a depth of 3053 m in 1993. The data obtained from these investigations have revolutionised our understanding of the patterns and rates of past global climatic change, and of the linkages between the ocean–atmosphere–terrestrial systems.

3.11.2 Ice masses as palaeoenvironmental archives

Glacier ice accumulates in a sequence of annual layers, and analysis of cores from the high latitude ice sheets has shown that these contain a wealth of palaeoclimatic evidence. The annual increments of ice reflect the balance between accumulation and ablation over the course of a year, and hence the variation in thickness of the annual layers may provide useful information on, for example, amount of winter snowfall, degree of melting (determined by summer temperature régimes), etc. More detailed palaeoclimatic information can be obtained, however, from aerosol particles and other exotic material that has settled on the glacier surface, and which subsequently become incorporated into the ice layers. These include dust particles from volcanic or desert sources, a variety of trace substances, and microbial or other biological materials (e.g. pollen grains and fungal spores), their relative abundance reflecting the vigour of atmospheric circulation, wind direction and aerosol flux. Other elements also become incorporated into the annual increments of ice sheets and constitute valuable sources of palaeoenvironmental data. These include trace gases (e.g. carbon dioxide and methane) that become trapped in minute air bubbles within the ice crystals, and which provide evidence of both short- and long-term changes in atmospheric gas composition, and stable isotopes (particularly isotopes of oxygen), which not only act as a proxy for climate change but also offer a basis for correlation between marine and terrestrial records (section 6.3.3). Finally, natural and artificial radioactive isotopes contained within the ice layers provide an independent means of dating ice cores. Ice cores, therefore, contain a range of evidence that can be used to reconstruct past climatic conditions and are widely regarded as one of the most important archives of Late Quaternary palaeoenvironmental data (section 3.11.4).

3.11.3 Analysis of ice cores

3.11.3.1 Annual ice increments

Near the surface of an ice sheet, the annual increments of ice comprise a darker and a lighter component. The winter layers are lighter in colour whereas those of the summer melt season are darker (Figure 3.48), the lower rate of accumulation and partial melting in the summer leading to a higher concentration of impurities and hence a darker colour by comparison with the winter layers. At depth, however, the annual layers are less obvious because they become thinner and distorted through pressure from the accumulating overburden, and as a result of flow deformation. They can be detected, however, using light transmission or X-ray methods, or on the basis of changes in physical or chemical properties of the annual ice increments (Figure 3.49). Further details of these methods can be found in Langway (1970), Robin (1983) and Oeschger & Langway (1989).

Once a record of incremental layers has been established, each layer of ice can be assigned an age in **ice-accumulation years** (number of annual layers below present surface). For

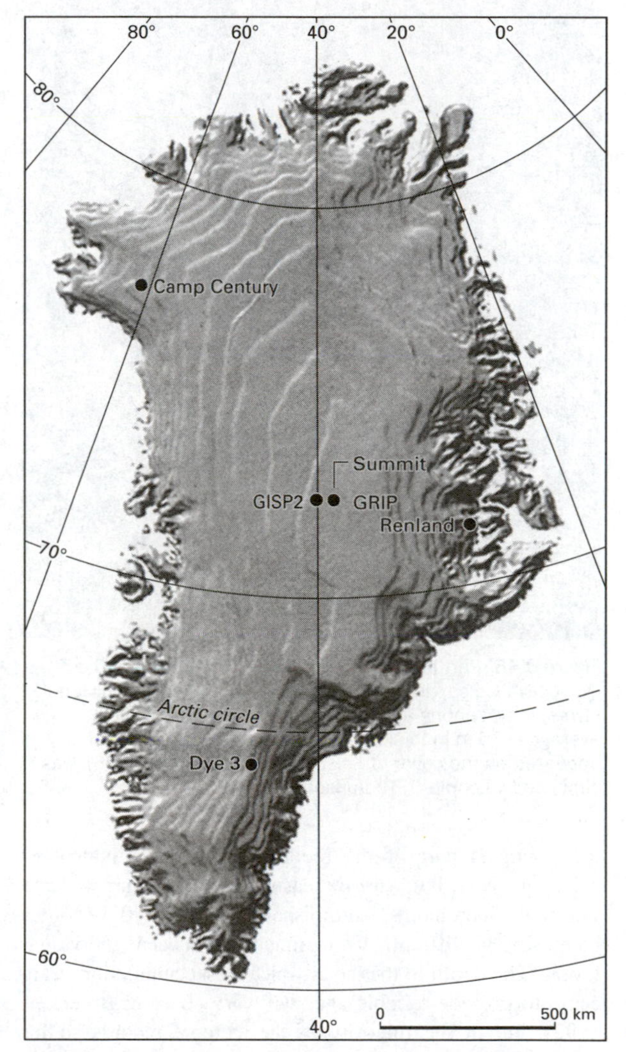

Figure 3.47 Location of importatnt drilling sites in Greenland from which deep ice cores have been obtained.

Figure 3.48 Part of a 50 m ice cliff near the margin of the Quelccaya ice cap in the topical SE Peruvian Andes. Individual layers, representing annual increments of accumulation, average 0.75 m in thickness. The photograph, which first appeared on the cover of *Science* (1979, **203**, no. 4386), was supplied by Lonnie G. Thompson.

the youngest part of the record, these are equivalent to calendar years, but with increasing depth the clarity of the seasonal increments diminishes, so that it becomes increasingly difficult to distinguish between individual layers. The depth in the ice at which ice-accumulation years cease to provide reliable ages will vary (Bard & Broecker, 1992), but in the majority of the ice-core records so far obtained, ice-accumulation years have been established for most of the Holocene. Dating of the older section of ice cores relies on other methods, and these are discussed in section 5.4.4.

3.11.3.2 Dust content

Wind-blown dust particles in glacier ice can be detected using light-scattering methods. They provide one of the most useful techniques for identifying incremental layers from deep ice, because dust peaks are not significantly affected by diffusive processes. They are therefore commonly used at the site drilling camp, in order to give rapid first-order assessments of the age and stratigraphy of ice cores. Measurement of the concentration of dust particles, however, is more difficult. The majority of insoluble aerosols in polar ice have a grain size of 0.1 to 2 μm, which makes them very difficult to measure using conventional laboratory procedures. However, with the development of the Coulter counter (section 3.2), the rapid measurement of the concentration of particles with a grain size as small as 0.3 μm is possible at a very high level of stratigraphic resolution (i.e. at the mm scale). This has been an important breakthrough in the analysis of ice cores, for not only do variations in the concentration of dust particles provide evidence of past aeolian activity (including wind strength and direction), but they also constitute marker horizons within the cores. They are therefore important in both a palaeoenvironmental and a stratigraphical context.

3.11.3.3 Chemical content

The chemical content of the ice can be established by means of **Electrical Conductivity Measurements (ECM)**. This technique involves the transmission of an electric DC current through the ice, the strength of the electrical current being directly proportional to the balance of acids and bases present in the ice. The current increases with higher concentrations of strong acids, especially sulphuric and nitric acids, but a decrease occurs where the acids are neutralised, for example as a result of the presence of alkaline dust (Taylor *et al.*, 1993b). A new high-frequency electrical conductivity method for measuring the **dielectrical properties (DEP)** of ice is a refinement of the ECM technique. ECM and DEP measurements provide an independent method for establishing variations in dust content in polar ice, with high concentrations of alkaline dust, usually associated with cold climate episodes (Mayewski *et al.*, 1993), being reflected in reduced conductivity values (Figure 7.21). High concentrations of HNO_3 and H_2SO_4 in ice layers can be readily identified since they give high conductivity values. The acid content is mostly derived from volcanic aerosols, and therefore

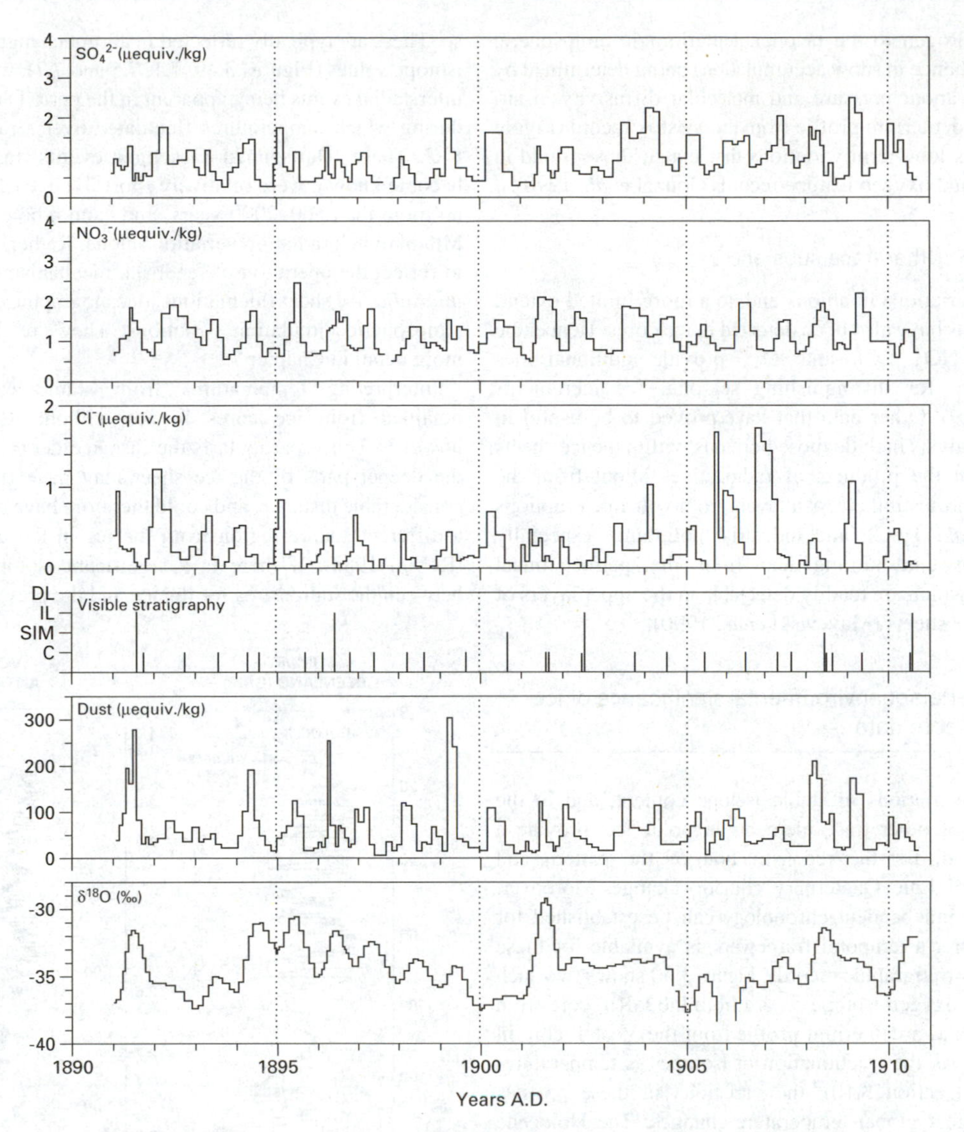

Figure 3.49 Seasonal variations in chemistry, dust content and stable oxygen isotope ratios in ice layers dating between AD 1890 and AD 1910 in a core from central Greenland (after Steffensen, 1988).

provides a record of past volcanic activity. Finally, ECM measures also reveal relatively high levels of ammonia in ice cores, which may indicate episodes of biomass burning.

3.11.3.4 Stable isotope records

The oxygen isotopic composition of ice is measured by mass spectrometry and is expressed as deviations ($\delta^{18}O$) per mil from a standard, which is SMOW (section 3.10; but see footnote 12). During evaporation from the ocean surface, atmospheric water becomes depleted in the heavier ^{18}O isotope by, on average, about 10‰ (Robin, 1983). However, there is seasonal variability in this value, so that the range of $\delta^{18}O$ between summer and winter precipitation over the ice sheets is commonly about 15‰. Thus seasonal changes can be detected in ice cores by precise measurements of oxygen isotope variations (Figure 3.49). Although the amplitude of the $\delta^{18}O$ signal decreases with depth due to diffusion effects, significant variations are still detectable over at least the last 160 ka (Figure 3.50), and these predominantly reflect changes in global climate (see below). Hydrogen isotopes behave in much the same way as oxygen isotopes, the ratio of normal hydrogen to the heavier deuterium in atmospheric water (and hence in snow accumulation) being determined by saturation vapour pressure and molecular diffusivity in air. Hence, the deuterium profile from the Vostok record (Figure 3.50) shows long-term variations that match those found in the Greenland oxygen isotope records (Jouzel *et al.,* 1990).

3.11.3.5 Other trace substances

Seasonal variations in anions and, to a more limited extent, some cations have also been detected in ice cores. Hence, the analysis of NO_3^-, Cl^- and SO_4^{2-} provide additional lines of evidence for distinguishing seasonal ice increments (Figure 3.49). Other data that have proved to be useful in ice-core analysis include those horizons within the ice sheets that contain the products of radioactive fallout from the atomic bomb-testing era, ash layers from volcanic eruptions (Palais *et al.*, 1992), and industrial pollutants, especially those from coal-burning and from the petrochemical industries, which are readily detectable in the upper layers of the polar ice sheets (Mayewski *et al.,* 1990).

3.11.4 Palaeoenvironmental significance of ice-core data

Downcore variations in stable isotope content, and in the abundance of other trace elements in polar ice, provide a powerful tool for the reconstruction of the pattern and amplitude of Late Quaternary climatic change. Moreover, because an independent chronology can be established for each ice core, a temporal framework is available for these continuous stratigraphic records. Figure 3.50 shows two such profiles, an oxygen isotope record from the GRIP core from Greenland and a deuterium profile from the Vostok core in Antarctica. As the fractionation of isotopes is temperature-dependent (section 3.10), the variations in these profiles broadly reflect global temperature changes. The Holocene and Eemian Interglacials are readily discernible in the records, as are the last cold stage and the cold stage that preceded the last interglacial. The shapes of the curves are also very similar to those derived from the isotopic analysis of deep-ocean cores (Figure 7.1). This broad measure of agreeement between two independent datasets from the North and South Polar regions, and from marine isotopic sequences, is difficult to explain other than by the operation of a common external forcing mechanism. As such it provides strong confirmatory evidence for the Milankovitch hypothesis of long-term climate change described in section 1.6.

An important feature of the GRIP and GISP2 data, however, is the evidence for numerous high-frequency climatic oscillations during the course of the last 120 ka or so. These are typically reflected in abrupt changes in oxygen isotope values (Figures 3.49, 7.1, 7.5 and 7.7), with some 20 interstadial events being apparent in the period 80–20 ka BP, during which temperatures fluctuated over a range of 5 to 8°C. These **'Dansgaard–Oeschger events'**, as they have become known, were relatively short-lived, each lasting for no more than 500–2000 years, and cannot be explained by Milankovitch radiation variations alone. Rather, they appear to reflect the operation of feedback mechanisms involving, *inter alia*, ice sheet fluctuations, oceanographic changes and atmospheric circulation variations. They are discussed in more detail in Chapter 7.

Interpreting temperatures from stable isotope data obtained from ice cores is not without its problems, however. This is particularly the case in older records, where the deeper parts of the ice sheet may have flowed some considerable distance, and could therefore have originated in a different source region from the ice in the upper layers (Robin, 1983). If there is a significant isotopic gradient between the source area for the ice and the present drill site,

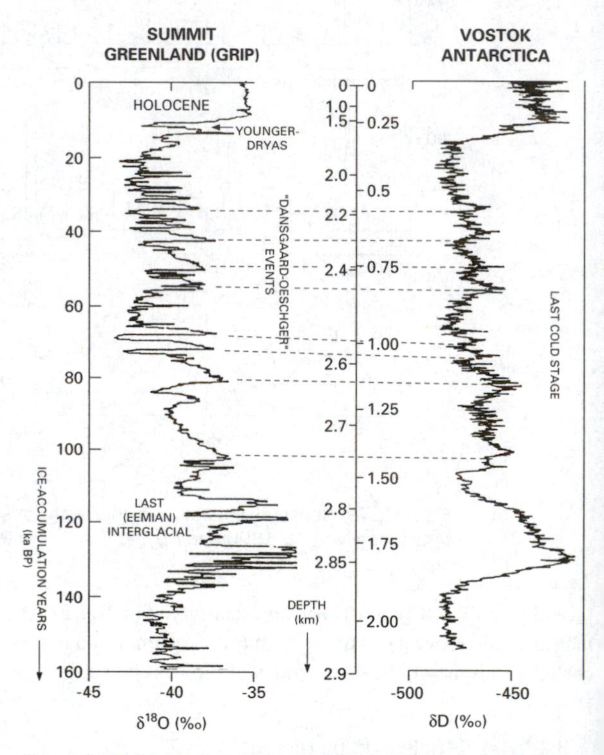

Figure 3.50 Stable isotope variations during the last 160 ka recorded in the GRIP Greenland Summit core (oxygen isotopes) and the Vostok core from Antarctica (deuterium ratios). Note that the vertical scale is in ice-accumulation years; conversion to a depth scale would lead to a marked transformation of the plots, since the Holocene sequence in both cores is much thicker than older parts of the sequence (modified from Peel, 1994).

this could be reflected in down-core isotopic variability which is independent of temperature. Corrections may have to be made for this effect, depending upon the regional isotopic variability of the ice (Dansgaard & Oeschger, 1989). One advantage of the GRIP and GISP2 records is that the drill sites were close to the summit of the Greenland Ice Sheet and hence, providing that the ice divide has not migrated significantly, flow within the ice sheet at these locations would have been minimal. A second problem concerns the range of factors that may affect isotopic ratios in glacier ice, and the fact that there is, as yet, no simple way of quantifying the relationship between oxygen isotopic ratios in precipitation and regional climatic conditions (Lorius *et al.*, 1989). One way in which this difficulty might be overcome is to compare isotopic values in seasonal snow/ice layers with meteorological data for the same time period, and the data can then be used to calibrate the isotopic signal for earlier episodes (e.g. Dansgaard *et al.*, 1975). An alternative approach involves the application of statistical methods to correlate borehole temperature records and isotopic data with a view to developing calibration models for the conversion of oxygen isotope ratios to palaeotemperatures (Cuffey *et al.*, 1992).

Ice-core records also provide evidence of long- and short-term changes in atmospheric gas content. For example, data from the Vostok core show that over the course of the past 160 ka, similar variations have occurred in CO_2 and CH_4 and, moreover, these parallel very closely the oxygen isotope-derived temperature record for the same time period (Figure 3.51). Methane records have also been obtained from the Greenland ice cores (Dansgaard *et al.*, 1993; Blunier *et al.*, 1995), and these also show a close correlation with reconstructed palaeotemperature records. This apparent relationship between atmospheric gas content and global temperatures has led to the suggestion that fluctuation in these 'Greenhouse gases' may have exerted an influence on Late Quaternary global temperatures (Chappellaz *et al.*, 1990). However, despite the similarities between the records, a direct cause and effect relationship remains to be established, for both atmospheric gas content and temperature may both have co-varied with other environmental influences (section 7.8).

Over shorter timescales, the gas content in the upper layers of polar ice sheets constitutes an important record of recent human activity. For example, the gradual increase in CO_2 in Greenland ice from the mid-eighteenth century onwards, and the more rapid increase during the course of the last 50 years, provide clear evidence of the extent of human impact on the atmospheric environment (Neftel *et al.*, 1985). The data indicate an increase in atmospheric CO_2 of 20–30 per cent in less than 200 years, due mainly to an increase in fossil fuel combustion and, to a lesser extent, to

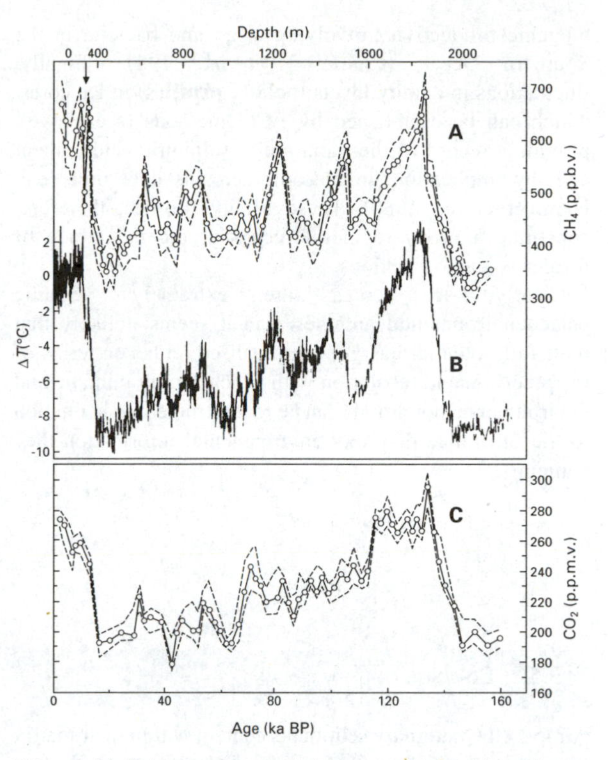

Figure 3.51 Palaeoenvironmental records obtained from the Vostok ice core, Antarctica, spanning the last 160 ka: (A) methane content, showing mean values (circles) and 2σ uncertainty ranges; (B) surface palaeotemperatures inferred from oxygen isotope variations; (C) variations in CO_2 content (mean values and 2σ uncertainty ranges; from Chappellaz *et al.*, 1990).

clearance of forests and conversion of biomass to CO_2. There are parallel trends in the curves for CH_4 (Stauffer *et al.*, 1988). The anthropogenic signal in ice cores is also reflected in dramatic recent increases in the concentration of a range of chemical pollutants, including lead, soot, nitrates, etc. (Mayewski *et al.*, 1990).

Ice-core data have also contributed to an understanding of many other aspects of environmental change. For example, the dust content of ice provides evidence of changes in atmospheric aerosol loadings and of the former extent of arid or poorly vegetated landscapes (Jouzel *et al.*, 1990). Thus, data obtained from two ice cores in the north–central Andes of Peru suggest that during the last glacial stage the atmosphere may have contained 200 times as much dust as today (Thompson *et al.*, 1995). Episodes of increased storminess may be reflected in variations in the concentration of sea-salt particles in ice cores (Shaw, 1989), while variations in sulphate or biosulphates in the Antarctic Ice Sheet have been used to infer changes in levels of biogenic productivity involving algae and bacteria in the Southern Ocean (Charleson *et al.*, 1987). Finally, fluctuations in acidity levels (**acidity profiles**) in ice cores, which can be established by ECM methods (see above), provide a record of the variation in sulphuric acid content and, by implication, in volcanic aerosols over time (e.g. Hammer *et al.*, 1980). These acidity profiles, therefore, constitute a proxy temporal record of the magnitude of former volcanic eruptions.

Clearly, ice cores are extremely versatile palaeoenvironmental archives, and it seems unlikely that their full potential has yet been realised, either in terms of the precision and resolution with which former climatic and environmental conditions can be reconstructed, or in relation to the full range of proxy environmental information they contain.

3.12 Conclusions

Almost all Quaternary sediments contain within their matrix important clues about their mode of deposition, and often about the climatic régime under which the sediments accumulated. This information can be extracted from the stratigraphic record by the application of the various physical and chemical methods described at the beginning of this chapter, and by analogy with sedimentological processes that can be observed in operation at the present day. Technological developments in field and laboratory methods, perhaps best exemplified by the extraction and analysis of cores from the world's ice sheets, have led to remarkable advances in our understanding of the operation of the earth's climate system over timescales ranging from a few years to thousands of millennia. No less impressive are the data that have been obtained from loess–palaeosol sequences, where the application of a range of analytical techniques has provided a high resolution environmental record, in some cases spanning the whole of the Quaternary. It must be emphasised, however, that the sedimentary evidence can seldom be properly evaluated in isolation. In the analysis of, for example, glacigenic deposits, periglacial sediments or pluvial lake sequences, the stratigraphic record should, wherever possible, be integrated with geomorphological evidence to produce a synthesis of landscape or climatic change. The same applies equally to the fossil record, the third category of evidence used in the reconstruction of Quaternary environments, and which forms the subject matter of the following chapter.

Notes

1. An **electron microscope** consists of a cathode-ray tube through which a beam of electrons is passed. The electrons are concentrated on, and pass through, the specimen to produce a magnified image on a photographic plate. After development, the electron photomicrograph shows the structure of the object in terms of its electron density.
2. The term **diamicton** refers to non-sorted terrigenous sediments and rocks containing a wide range of particle sizes, regardless of genesis.
3. **Varves** are laminated sediments, consisting of annual layers (usually couplets) produced by the seasonal delivery of clastic material, or seasonal precipitation or biological activity (see section 5.4.2).
4. **Facies**: a body of sediment that is characterised by a combination of lithological, physical or biological properties and can be distinguished from adjacent bodies of sediment by a well-defined geometry and structure; facies, or combinations of facies (**facies associations**) reflect certain types of sedimentary process or a particular sedimentary environment.
5. **Deformation**: alteration of the primary (original) bedding or attitude of lithological units or their components through stress forces (e.g. compression, extension or shear caused by traction) resulting in folding, faulting or alteration of internal structures and fabric. See also note 2.12
6. **Load structures** often form where sand is deposited over a hydroplastic or fluid mud layer. Under the weight of the sand, the mud layer becomes distorted and bent downwards. In some cases, the sand layer sinks and forms lobes; in others, the mud layer becomes pushed up to form tongues (see Reineck & Singh, 1973).
7. **Diagenesis**: the alteration of minerals and sediments by the influences of oxidation and reduction, hydrolysis, solution, biological changes (e.g. induced by anaeorobic bacteria), compaction, cementation, recrystallisation, and the alteration of the lattice structure of clays by expulsion of water and ion exchange.
8. Most plants secrete opaline silica bodies and these assume the shape of the cell in which they are deposited. These forms are known as **phytoliths** and many are characteristic of the plants in which they are found. They may, therefore, provide important clues to past vegetation.
9. Water-filled passages in bedrock, which are below the groundwater table, are termed **phreatic** passages. Water flow is very slow, and little, if any, downward erosion occurs. Immediately above the long-term water table, passages are only filled with water on a seasonal basis, or they may contain streams flowing down to the water table. This is termed the **vadose zone**, in which significant downcutting can take place.

10. **Isotopes** are atoms of an element that are chemically similar but have different atomic weights (section 5.3.1). Oxygen, for example, consists principally of two isotopes, the heavier ^{18}O isotope and the lighter ^{16}O. ^{18}O:^{16}O ratios in the atmosphere, seas, groundwater and ice are often controlled by climatic conditions.

11. **Fractionation** is the selective separation of chemical elements or isotopes during natural physical, chemical or biochemical processes, such as during evaporation, condensation, transpiration and metabolism.

12. A problem has recently arisen because the supply of PDB (Peedee belemnite) standard has been exhausted and because SMOW (Standard Mean Ocean Water) does not have a unique definition. This means that, because laboratories do not use the same reference material to establish their isotope ratio scales, laboratories are reporting different values for the same material. In order to eliminate confusion in the reporting of stable isotope ratios, therefore, the Commission of Atomic Weights and Isotopic Abundances of the International Union of Pure and Applied Chemistry has recommended that the use of SMOW and PDB be discontinued, and that isotope abundances of oxygen- (and also hydrogen- and carbon-) bearing materials should be reported relative to the reference water VSMOW (Vienna Standard Mean Ocean Water) and VPDB (Vienna Peedee belemnite). These are defined by adopting a $\delta^{18}O$ value of $-2.2‰$ and a $\delta^{13}C$ value of $+1.95‰$ for NBS (National Bureau of Standards) carbonate relative to VPDB. SLAP (Standard Light Antarctic Precipitation) isotopic abundance scales should be normalised so that values for $\delta^{2}H$ (deuterium) and $\delta^{18}O$ are $-428‰$ and $-55.5‰$ respectively relative to SMOW (Coplen, 1995).

CHAPTER 4 Biological evidence

4.1 Introduction

Biological evidence, in the form of plant and animal remains, has always been a cornerstone in the reconstruction of Quaternary environments. The analysis of fossil evidence employs **uniformitarian principles**, namely that a knowledge of the factors that influence the abundance and distribution of contemporary organisms enables inferences to be made about environmental controls on plant and animal populations in the past. Applying this approach to the interpretation of Quaternary fossil assemblages, therefore, the majority of which have living counterparts, it should be possible to reconstruct former environmental conditions with a reasonable degree of confidence (Rymer, 1978). The use of modern ecological information in this way is an essential element of **palaeoecology**, the study of the interrelationships of organisms in the past, both with their physical environment and with other plants and animals (Birks & Birks, 1980; Warner, 1990a; Delcourt & Delcourt, 1991). Plant and animal remains are also used to subdivide the geological record, a field of study known as **biostratigraphy**. This chapter is concerned almost entirely with the palaeoecological aspects of biological evidence; the principles and practices of biostratigraphy are considered more fully in Chapter 6.

4.1.1 The nature of the Quaternary fossil record

The fossil evidence upon which Quaternary palaeoecological studies are based falls into two major categories: **macrofossil evidence**, which ranges from whole skeletons of large vertebrates to small fragments of plant or animal remains, all of which can be identified by eye or with low power magnification (up to *c.* ×40); and **microfossil evidence**, consisting of minute remains of former biota that can be identified only by using microscopes. Microfossils are generally less than 1 mm in size, and include pollen, diatoms, other algae, fungal spores and zooplankton (e.g. Cladocera). When sediments are disaggregated and examined, it is often surprising how abundant and diverse both the macro- and microfossil components can be, and even sediments that appear on first sight to be 'barren' may contain a number of tiny, but recognisable, plant and animal remains. These might include seeds, fruits, leaves, pieces of wood, charcoal, insect remains, molluscs, fish scales and so on. Occasionally spectacular fossils, such as skeletons of large vertebrates and tree trunks or stumps (Figure 3.18), are preserved. The range of biota that can be identified in Quaternary fossil material is therefore extensive, although so far only a few types have been studied in detail. These include pollen, diatoms, plant macrofossils, insects, molluscs, ostracods, foraminifers and vertebrates, all of which are discussed in this chapter. Other organisms for which only limited palaeoecological data are available, including chironomids (non-biting midges), Cladocera, fungal spores and certain types of marine plankton, are also briefly considered.

4.1.2 The taphonomy of Quaternary fossil assemblages

The study of the processes that lead to the formation of a fossil assemblage is referred to as **taphonomy**, and understanding the taphonomy of an assemblage is an essential prerequisite for palaeoenvironmental interpretation. Fossils are best preserved where deterioration, resulting from the activities of micro-organisms and the operation of chemical agencies, is minimal, and hence anaerobic environments are more likely to lead to survival rather than those where oxidation has occurred (Allison & Briggs, 1991). Some organisms may be found in the position of growth (***in situ* fossils**), such as tree stumps buried by the rapid accumulation of peat or wind-blown sand. The majority of fossils, however, appear to have been transported from their growth or life position by, for example, wind (e.g. pollen and spores), water (a great range of fossils) and animals (e.g. prey of carnivorous taxa). These processes can introduce bias into the eventual fossil assemblage through the **selective transport** of fossils, or by **differential destruction,** in which only the more robust specimens survive. Some fossils may subsequently be removed from their initial place of deposition and redeposited in a new locality, perhaps in an entirely different depositional context. These are referred to as **secondary** or **derived fossils**. Clearly it is important to distinguish such secondary components from the **primary fossils** when palaeoenvironmental reconstructions are being made on the basis of this form of evidence.

4.1.3 The interpretation of Quaternary fossil assemblages

In addition to the taphonomical problems outlined above, other difficulties arise in the interpretation of the fossil record. First, fossil remains have to be identified to a sufficiently low taxonomic level in order that uniformitarian principles can be applied. Many types of fossil (e.g. Coleoptera, diatoms) can be identified to species level, but in other cases (pollen, for example), it is often only possible to make identifications to the genus or family level. This can pose problems when different species within a genus have contrasting ecological affinities (see section 4.2). Secondly, if the uniformitarian approach is to be employed, a number of assumptions have to be made about contemporary plant and animal populations. It must be assumed, for example, that we adequately understand and are able to isolate the environmental parameters that govern present-day plant and animal distributions, and that present plant and animal populations are in equilibrium with those controlling variables. Thirdly, when examining the fossil record, it has to be assumed that former plant and animal distributions had reached equilibrium with their environmental controls, that former plant and animal assemblages have analogues in the modern biota, and that the ecological affinities of plants and animals have not changed through time.

The extent to which these assumptions can be met varies both with the type of fossil evidence (pollen grains, animal bones, shells, etc.) and with the nature of the assemblage, but it is rare indeed for all of the above conditions to be satisfied. In this chapter, different types of biological evidence that have been used in the analysis of Quaternary environments are evaluated in the light of the qualifications set out above. It must be stressed, however, that while the different forms of evidence are discussed individually, in reality organisms do not exist as such but are, of course, components of **ecosystems**. In the reconstruction of past environments, therefore, it is important to understand the interrelationships between a range of organisms, i.e. the **palaeobiology** of fossil assemblages. Hence, in recent years, Quaternary scientists have increasingly been developing **multi-proxy** (as opposed to **single proxy**) investigations, in which environmental reconstructions are based on the integration of evidence from several different data sources (pollen, plant macrofossils and fossil insects, for example). Moreover, while it is clear that palaeoecologists must have a knowledge of the principles of modern ecology in order to make reasonable interpretations of fossil assemblages, it is equally true that a sound knowledge of Quaternary fossil records is essential for an understanding of the origins of the modern biota. Hence, while the present might reasonably be regarded as the key to the past, the reverse is equally true in that the development of contemporary ecological theory relies heavily on a secure foundation in Quaternary palaeoecology. These points are explored in more detail towards the end of this chapter.

4.2 Pollen analysis

4.2.1 Introduction

Of all the methods currently employed in the reconstruction of Quaternary environments, undoubtedly the most widely adopted and arguably the most versatile is the technique of pollen analysis (or, more correctly, pollen stratigraphy[1]). The method has been used since the 1920s as a means of correlating Quaternary stratigraphic units, to reconstruct vegetational history (Huntley & Birks, 1983; Delcourt & Delcourt, 1991) and to investigate the impact of human activities on late Quaternary vegetation and landscape in many parts of the world (e.g. Behre, 1986). An extensive literature on the principles and applications of pollen analysis is now available, and it is not possible to do full justice to the method within the space available in this volume. The aim here, therefore, is to provide a general introduction to pollen analysis as a palaeoenvironmental technique, and to outline the principal strengths and weaknesses of the method. Further details can be found in Birks & Birks (1980), Faegri & Iversen (1989), Moore *et al.* (1991) and in the summaries provided by Berglund & Ralska-Jasiewiczowa (1986) and MacDonald (1990).

4.2.2 The nature of pollen and spores

Pollen grains (Figure 4.1) are formed in the anthers of the seed-producing plants (**angiosperms** or **gymnosperms**). They contain the **male gamete** of the plant and aim to reach the stigma of the female part of the flower where fertilisation can take place. Spores, which are commonly included in pollen diagrams, represent the **sporophyte** stage of lower plants (**cryptogams**), such as ferns (**Pteridophyta**) and mosses (**Bryophyta**). The sporophyte is dispersed to suitable habitats where the second stage in plant generation, the **gametophyte**, can grow. Pollen grains and spores are frequently dispersed in very large numbers in order to maximise the opportunites for successful pollination or gametophyte growth, and many accumulate on the ground surface or in water bodies. Some will subsequently become incorporated and fossilised in sediments and it is the extraction, identification and counting of these preserved fossil grains which forms the basis of the technique of pollen analysis.

Most pollen grains and spores are extremely small, few exceeding 80 to 100 μm in diameter, with the majority falling in the size range 25 to 35 μm. A typical pollen grain consists of three elements (Faegri & Iversen, 1989). The central portion is the living cell which is surrounded by a covering of cellulose known as the **intine.** Neither of these survives in the fossil form. The outer layer or **exine** consists of a remarkably resistant, waxy coat of material called **sporopollenin**, a substance whose composition and structure is not yet fully understood, but which appears to consist predominantly of monomers and polymers of carotenes and carotene esters. The prime function of this outer, resistant wall is to protect the young gametophyte from desiccation and microbial attack, but it also has the effect of preserving pollen grains in sediments when almost all other organic constituents are reduced to structureless and unrecognisable components.[2] The exine is characterised by a variety of morphological and structural features which, along with the number and distribution of germinal apertures, and the overall size and shape of the grain, form the basis for pollen and spore identification (Figure 4.1).

Pollen grains and spores are disseminated by a variety of means. Spores are usually dispersed by wind, but pollen grains can be spread by water, by insects, by birds and by animals (including humans). Those plants that liberate wind-

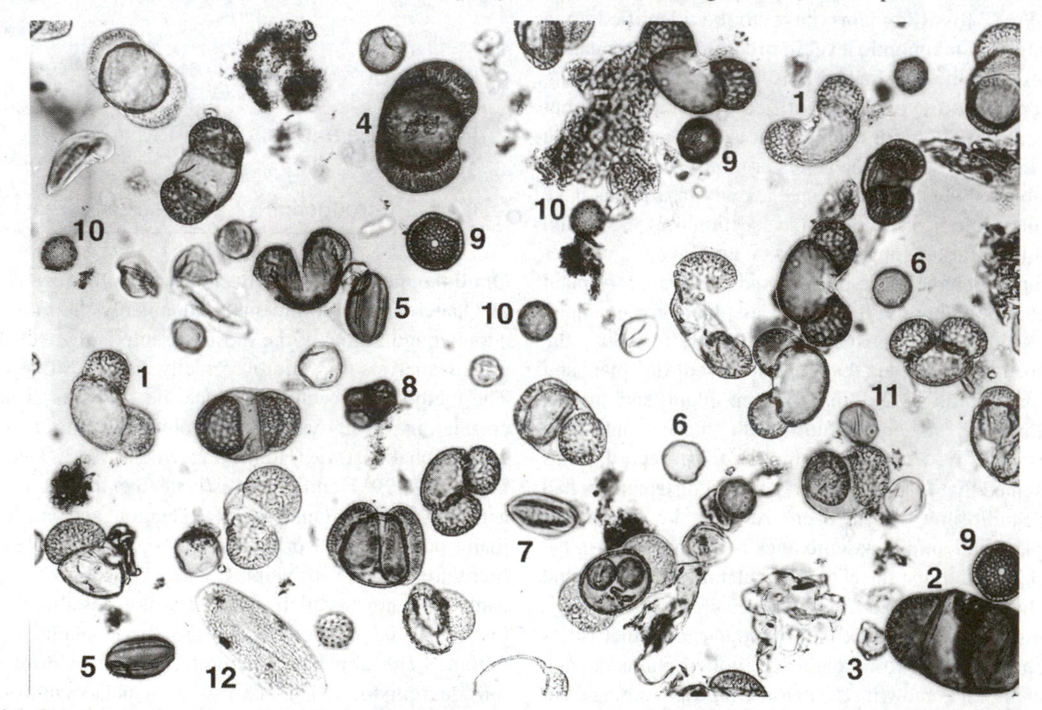

Figure 4.1 Subsample of a pollen assemblage typical of Würmian cold stage deposits from the site of St Front, France. The arboreal component is dominated by several types of *Pinus* (1), while *Picea* (2) and *Betula* (3) are also recorded in smaller numbers. Pollen of *Cedrus* (4) are atypical of this assemblage, and probably indicate long-distance transport. A wide variety of non-arboreal pollen is represented, including, for example, species of *Helianthemum* (5), *Plantago* (6), *Ephedra* (7), *Calluna* (8) and several genera within the families Caryophyllaceae (9), Chenopodiaceae (10), Poaceae/Gramineae (11) and Liliaceae (12). The long axis of the *Helianthemum* grain (5) is *c.* 45 μm. Photo montage courtesy of Maurice Reille and Valerie Andrieu.

borne pollen are termed **anemophilous**, and generally produce far greater numbers of grains than the **entomophilous** taxa, which rely on insects or other zoological vectors for transfer. Wind dispersal is facilitated by the small size, smooth surface features and low specific gravity of the grains, while in the gymnosperms (e.g. pine *Pinus*, spruce *Picea*) air bladders or sacs have evolved enabling pollens of these taxa to stay airborne for very long periods and also to travel considerable distances (Moore *et al.*, 1991). The entomophilous grains possess a hardy, armoured surface which often has prominent spines and a coat of sticky material that causes them to adhere to each other and to the body of the animal. Generally, these grains are large (in excess of 60 μm) or very small (less than 15 μm) and are usually less well represented in the fossil record than the wind-dispersed types. The whole field of pollen production and dispersal is extremely complex and is considered in more detail in section 4.2.5.

4.2.3 Field and laboratory work

Pollen grains and spores are usually well preserved in lake and pond sediments and in peats, and it is these deposits that have been most widely investigated in pollen analytical work. They are also found in soils (Dimbleby, 1985), cave earths (Hunt, 1989) and ocean floor deposits (Dupont, 1989), though in these sediments they are often lower in concentration and less well preserved, and they have also been recovered from polar ice cores (Fredskild & Wagner, 1974) and from speleothems (Burney *et al.*, 1994). The degree of preservation of pollens and spores depends upon a range of factors, the most important of which are the grain size of the sediment matrix, the extent to which anaerobic conditions have persisted since deposition, and the thickness and structure of the exine. Pollen grains are less well preserved where loosely compacted peats allow aerobic or microbial attack and they can be damaged or destroyed by desiccation or by mechanical abrasion in coarse-grained sediments on river floodplains, for example, or in lakes close to a point of stream inflow.

Samples containing fossil pollens and spores can be taken from sections exposed in river banks, cliffs, road cuttings or building excavations, or by digging pits. Alternatively, samples can be obtained by means of coring (Aaby & Digerfeldt, 1986). Samples must be sealed air-tight and are usually kept in a cool store (at *c.* 1–3°C) to prevent desiccation and/or microbial attack. This also protects them from contamination by pollen circulating in the atmosphere, especially during the pollen and spore production season. In the laboratory, following sediment dispersal, sieving and/or chemical flotation (density separation), samples are chemically treated in a variety of ways to remove as much of the sediment matrix as possible. Lignins and cellulose can be reduced in volume, if not entirely removed, by oxidation and acetolysis (Berglund & Ralska-Jasiewiczowa, 1986), while minerogenic sediments may either be removed by digestion in hydrofluoric acid, or be separated off by differential centrifugation or by floating the organic detritus (including pollens) out of the matrix using a 'heavy liquid'. Carbonates and calcareous sediments are treated with hydrochloric acid. The residues containing the pollens and spores may be stained with an organic dye such as safranin which enhances the surface detail of some grains (although this stage may be omitted as some dyes limit the quality of photomicroscopy) and then mounted on to glass slides in a suitable medium such as glycerine jelly or silicon oil. Counting is then performed at magnifications of 100× to 1000×, depending on the detail required for identification purposes. By traversing the slide in a systematic way, a count can be made of all of the identifiable pollens and spores until a predetermined number (the **pollen sum**) has been reached. This should be high enough (e.g. 300–500 grains) to account for most of the variability in the spectrum (see Birks & Birks, 1980). Identifications, which can generally be made easily to family level, commonly to genus level but less frequently to species level, are based on distinctive exine characteristics using pollen keys, collections of photographs and laboratory reference material made up from modern pollen samples. Details of the laboratory procedures and examples of keys and photographs can be found in Faegri & Iversen (1989), Moore *et al.* (1991) and Reille (1992).

4.2.4 Pollen diagrams

Where samples have been taken from a stratified sequence of sediments, such as a lake or peat sequence, an analysis of the pollen content of a single horizon will reveal a mixture of pollen types, collectively termed the **pollen assemblage** or **pollen spectrum**. Analysis of a series of horizons (or spectra) may show changes in pollen content which may, in turn, be interpreted as indicating temporal changes in vegetation cover in the area adjacent to the site. These changes are usually depicted graphically in the form of pollen diagrams which are based either on percentage values or on pollen concentration data, the latter sometimes being termed 'absolute pollen diagrams'. **Percentage** (or **relative**) **pollen diagrams** usually take two forms. In some cases a pollen sum is selected for each level, and individual pollen and spore types are then expressed as percentages of that sum (Figure 4.2). It is usual practice to exclude pollen of obligate aquatic taxa (e.g. pondweeds and water lilies) and spores

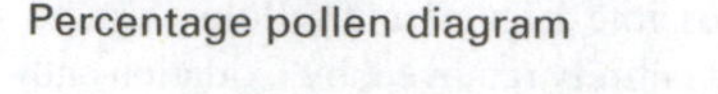

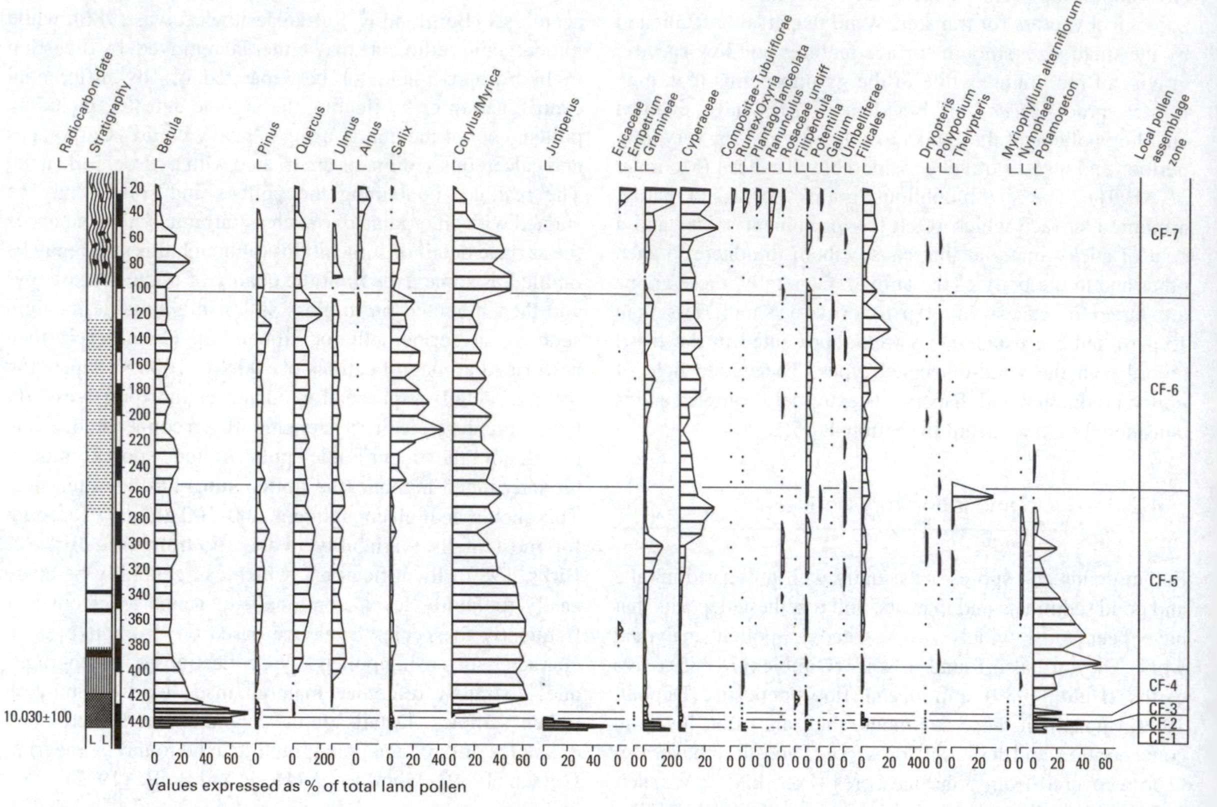

Figure 4.2 Percentage pollen diagram spanning the Early and Middle Holocene from Craig-y-Fro, Brecon Beacons, South Wales (modified after Walker, 1982b).

from the pollen sum, the former because they are produced in a totally different environment from the terrestrial taxa which are usually the main concern of the analyst, and spores because they are formed in a different way (and have a different function) from pollen grains. An alternative form of percentage diagram is one based on a sum of pollens of tree taxa only (an **arboreal pollen diagram**). In this type of diagram, all pollen and spore types are counted until a specified number of tree taxa have been identified, and then each pollen and spore type is expressed as a percentage of the tree pollen sum (Figure 4.3). In the early stages of pollen analysis, when researchers were especially concerned with forest history, the latter type of diagram was invariably employed, and indeed it is still widely used in archaeological investigations and in studies of Holocene landscape change, where the focus is often on the changing composition of forest communities (Behre, 1986). Where sediments of the cold stages of the Quaternary are being investigated, however, much lower frequencies of tree pollen are usually recorded, and a '**land pollen diagram**' (based on total pollen recorded, excluding those of obligate aquatic taxa) is considered more appropriate. A list of generic and common English names of taxa most frequently depicted in NW European pollen diagrams is shown in Table 4.1.

One problem arising from the representation of pollen data in percentage form is that the curves for individual taxa are, of necessity, *interdependent*. In other words, an increase in the influx of, for example, *Betula* (birch) pollen to a site will automatically lead to a suppression of the *percentages* of other taxa represented in the pollen diagram. Thus, statistical fluctuations will be reflected in the pollen diagram which do not necessarily represent ecological changes (Prentice & Webb, 1986). One way of overcoming this problem is by the use of '**absolute pollen diagrams**', as these are based not on percentage changes but on changes in the *total number* of pollen grains per unit volume of sediment. A number of techniques are now available for establishing the amount of pollen (*pollen concentration*) in a unit volume of sediment (Peck, 1974), but the one most commonly used involves the addition of a known quantity of **exotic pollen grains** (sometimes referred to as the '**spike**') to the fossil sediment during the laboratory preparation. In northwest Europe, for example, pollen of *Eucalyptus* is commonly used for this purpose since the plant is exotic to the region and the pollen can be easily distinguished from fossil grains. If the quantity of exotic pollen added to the sample is known, then the observed ratio of exotic to fossil pollen enables the total number of fossil pollen in a sample to be estimated, and thus changes in pollen concentrations at different levels in a profile can be depicted diagrammatically (Figure 4.4). If the chronology of sediment accumulation can be established by, for example, radiocarbon dating a series of samples from the same profile, then it may be possible to calculate the rate of sediment accumulation. In this case, the **pollen concentration diagram** can be converted into a **pollen influx diagram** which expresses the data in the form of pollen grains per square centimetre of surface of sediment per year or for some other unit of time.

Although this technique will overcome the problem of statistical interdependence of taxon curves, and is therefore potentially of considerable value in palaeoenvironmental reconstruction, it too has problems of application. Variations in rates of deposition within a body of sediment will tend to distort pollen concentration values, and great care must be taken in the comparative analysis of pollen concentration data. This is particularly true when comparisons are being made between different types of sediment as, for example, between lake muds and telmatic or terrestrial peats. Even within a seemingly uniform deposit, however, it cannot be shown unequivocally that the sedimentation rate has been uniform. In late Quaternary deposits, radiocarbon dates may help overcome the problem, particularly where AMS determinations are made on plant macrofossil material (section 5.3.2). Hitherto, however, the majority of radiocarbon dates from peat and lake sediment profiles have been obtained from bulk samples of material, and here difficulties can arise over the often thick lenses of sediment (up to 10 cm) which are commonly extracted from cores in order to provide sufficient material for dating purposes. Moreover the fairly large standard deviations on some radiocarbon dates may render the age determination almost meaningless in a stratigraphic context. Although potentially valuable, therefore, pollen concentration diagrams must be used with caution and are best employed in conjunction

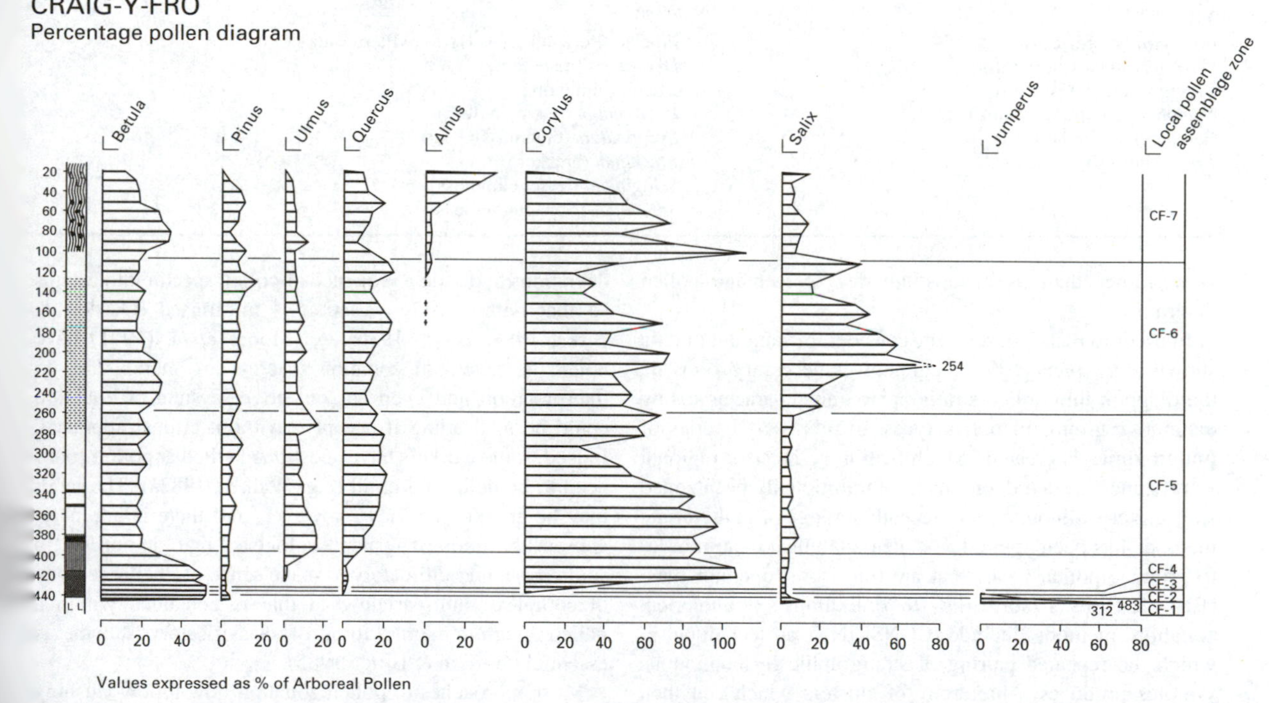

Figure 4.3 Pollen diagram from Craig-y-Fro, Brecon Beacons, South Wales, based on arboreal pollen percentages (after Walker, 1982b).

Table 4.1 Some common taxa found in late Quaternary pollen assemblages from northwest Europe (• new *Flora Europaea* terminology, where appropriate).

TREES			
Betula (birch)	*Alnus* (alder)	*Pinus* (pine)	*Picea* (spruce)
Ulmus (elm)	*Quercus* (oak)	*Fagus* (beech)	*Fraxinus* (ash)
Tilia (lime)	*Populus* (poplar)	*Carpinus* (hornbeam)	*Sorbus* (rowan)
Abies (fir)	*Taxus* (yew)		
SHRUBS			
Corylus (hazel)	*Salix* (willow)	*Juniperus* (juniper)	*Hedera* (ivy)
Calluna (ling)	*Ericaceae* (heather fam.)	*Empetrum* (crowberry)	*Ilex* (holly)
Hippophäe (buckthorn)			
HERBS			
Gramineae (• Poaceae; grass family)		*Rumex* (sorrel/dock)	
Cyperaceae (sedge family)		*Ranunculus* (buttercups)	
Caryophyllaceae (pink family)		*Thalictrum* (meadow rue)	
Chenopodiaceae (goosefoot fam.)		Rosaceae (rose fam.)	
Cruciferae (cabbage fam.; • Brassicaceae)		*Filipendula* (meadow sweet)	
		Dryas (mountain avens)	
Compositae Tubuliflorae (tubular florets = daisy fam.); • Asteroideae			
Compositae Liguliflorae (toothed and ligulate florets= dandelion fam.) • Cichorioideae or Lactucoideae (tribes Lactucae and Carduease)			
Artemisia (mugwort)		*Potentilla* (tormentil)	
Galium (bedstraw)		*Saxifraga* (saxifrage)	
Epilobium (willow-herb)		Scrophulariaceae (figwort fam.)	
Helianthemum (rock-rose)		*Succisa* (scabious)	
Labiatae (labiate fam.)		Umbelliferae (carrot fam.; • Apiaceae)	
Leguminosae (pea/vetch fam.); • Fabaceae		*Urtica* (nettle)	
Plantago (plantain)		*Valeriana* (valerian)	
AQUATICS		SPORES	
Littorella (shoreweed)		Filicales/Polypodiaceae (ferns) • Pteropsida	
Myriophyllum (water milfoil)		*Dryopteris* (male fern)	
Nuphar (yellow water-lily)		*Isoetes* (quillwort)	
Nymphaea (white water-lily)		*Polypodium* (polypody fern)	
Potamogeton (pondweed)		*Lycopodium* (clubmoss)	
Typha (bulrush)		*Pteridium* (bracken)	
		Selaginella (lesser clubmoss)	
		Sphagnum (Sphagnum moss)	

with, rather than as a substitute for, percentage pollen diagrams.

In order to make sense of the considerable amount of data shown in a typical pollen diagram, it is necessary to divide the diagram into pollen-stratigraphic units characterised by distinctive groups of pollen types. In this way, a series of **pollen zones** is created, which may have local or regional significance. Pollen diagrams have traditionally been zoned subjectively, although more recently a range of multivariate methods has been applied to pollen analytical data in order to produce pollen zones that are free from 'operator bias' (Birks & Birks, 1980; Birks, 1986). Examples of numerical zonation methods include CONSLINK, a technique in which the repeated pairing of stratigraphically continuous samples produces a hierarchy of clusters which can then form the basis for pollen zonation, and SPLITSQ which is a hierarchical partitioning technique which uses least-squares deviations to divide a sequence of pollen spectra into groups so that within-group variation is minimised (Gordon & Birks, 1985; Birks, 1986). As Moore *et al.* (1991) have noted, however, all zonation systems are merely aids to interpretation and even an 'objective' system of zonation could be misleading if accepted without critical appraisal. Indeed, some workers have questioned whether pollen zones need to be delimited at all (e.g. Walker, 1982a). The latter may be an extreme view, however, and there is a general consensus amongst palaeoecologists that in order to summarise, and ultimately to make sense of, the large body of complex, multivariate data that is contained within a pollen diagram, some form of classificatory scheme is essential (Gordon & Birks, 1985).

Most approaches to pollen zonation now follow Cushing (1967a) in that the pollen diagram is first divided into **local pollen assemblage zones**, usually on the basis of the

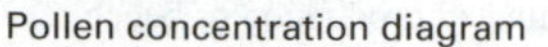

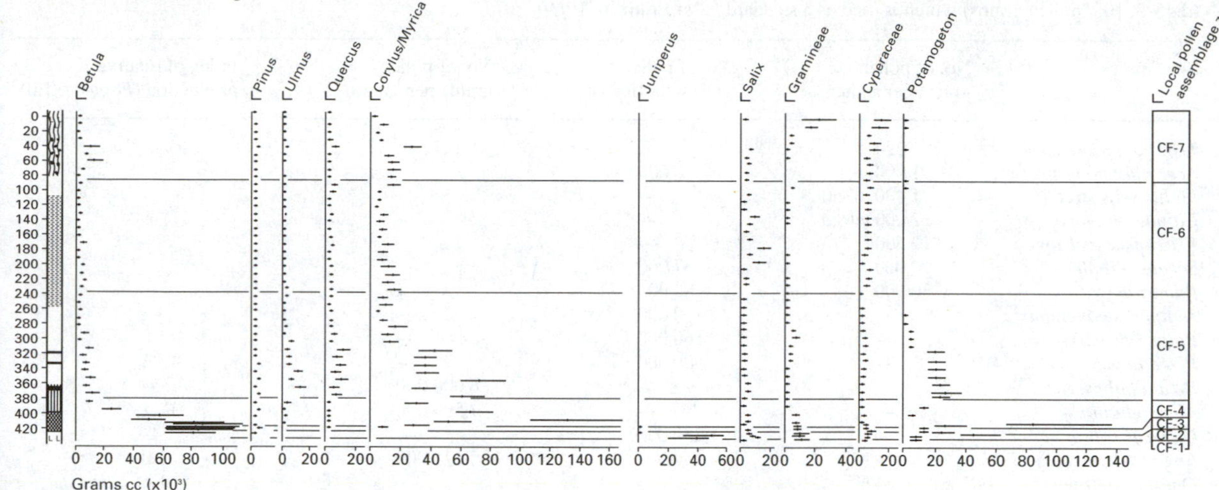

Figure 4.4 Pollen concentration diagram from Craig-y-Fro, Brecon Beacons, South Wales (after Walker, 1982b).

principal terrestrial taxa. **Regional pollen assemblage zones** can then be established on the basis of comparisons between individual pollen diagrams. Within relatively restricted geographical areas, vegetation changes are likely to have been more or less synchronous, and hence the boundaries between regional pollen assemblage zones will be effectively time-parallel. Regional pollen zonation schemes can therefore be constructed which summarise the principal changes in vegetation cover, especially changes in forest composition, over given time periods. Examples of such schemes are those developed for western Scotland (Lowe & Walker, 1986a), the French Massif Central (Beaulieu *et al.*, 1982), the Pyrenees (Reille & Lowe, 1993) and the northern Apennines (Lowe & Watson, 1993).

4.2.5 The interpretation of pollen diagrams

The interpretation of a pollen diagram is undoubtedly the most difficult part of pollen analysis, for it requires a knowledge of pollen production and dispersal, pollen source and deposition, pollen preservation and the relationship between fossil pollen and former plant communities. Only when these factors have been carefully evaluated can inferences be made about former vegetation cover and, by implication, about former climates and environments.

First, it is important to appreciate the fact that not all plants produce the same quantities of pollen. It has already been noted that entomophilous species usually produce much less pollen than anemophilous plants and will therefore be under-represented by comparison in modern surface samples and in the fossil record. Less well represented are the **autogamous** plants such as wheat (*Triticum*) which are self-pollinating and which liberate very few pollen grains into the atmosphere. Even more extreme in this context are the **cleistogamous** plants (e.g. *Viola*), the flowers of which never open; thus pollen is very rarely released. Within each of these plant types, however, there is considerable variation. The lime (*Tilia cordata*) and common heather or ling (*Calluna vulgaris*), for example, are both insect-pollinated yet they usually liberate large quantities of pollen. On the other hand, beech (*Fagus sylvatica*) and oak (*Quercus petraea*) are both wind-pollinated, yet are often relatively low pollen producers. Table 4.2 gives some indication of the general variability in pollen production of some taxa that are commonly encountered in pollen diagrams from Europe. Although there are marked differences in the absolute quantities of pollen produced by these plants in different areas or vegetation associations, the rank order between them is often similar (Moore *et al.*, 1991). From a knowledge of such rankings, some analysts have attempted to derive correction factors (or **R-values**) for tree and shrub taxa. The implication is that taxa with high R-values are likely to be over-represented in the pollen record, while those with low values will be under-represented. By applying scaling or weighting factors to the original pollen counts, corrected pollen diagrams may be constructed, which are considered to reflect more closely the relative importance of each taxon in the contemporary vegetation (Figure 4.5).

Table 4.2 Pollen production of various plant species. The index of relative pollen production in the final column is based upon estimates of the pollen production of an individual plant over a period of 50 years and is expressed relative to the estimated production of beech (2.45×10^{10} pollen grains) which is used as a standard (after Erdtman, 1969).

Species	No. of pollen grains per anther	No. of pollen grains per flower	No. of pollen grains per catkin	Index of relative production (*Fagus* = 1.0)
Trifolium pratense	220	—	—	—
Acer platanoides	1 000	8 000	—	—
Malus sylvestris	1 400–6250	—	—	—
Calluna vulgaris	2 000 tetrads	—	—	—
Fraxinus excelsior	12 500	—	—	—
Secale cereale	19 000	57 000	—	—
Rumex acetosa	30 000	180 000	—	—
Juniperus communis	—	400 000	—	—
Pinus sylvestris	—	160 000	—	15.8
Picea abies	—	600 000	—	13.4
Betula pubescens	—	—	6 000 000	—
Alnus glutinosa	—	—	4 500 000	17.7
Quercus robur	—	1,250 000	—	—
Fagus sylvatica	—	—	175 000	1.0
Quercus petraea	—	—	—	1.6
Carpinus betulus	—	—	—	7.7
Betula pendula	—	—	—	13.6
Corylus avellana	—	—	—	13.7
Tilia cordata	—	—	—	13.7

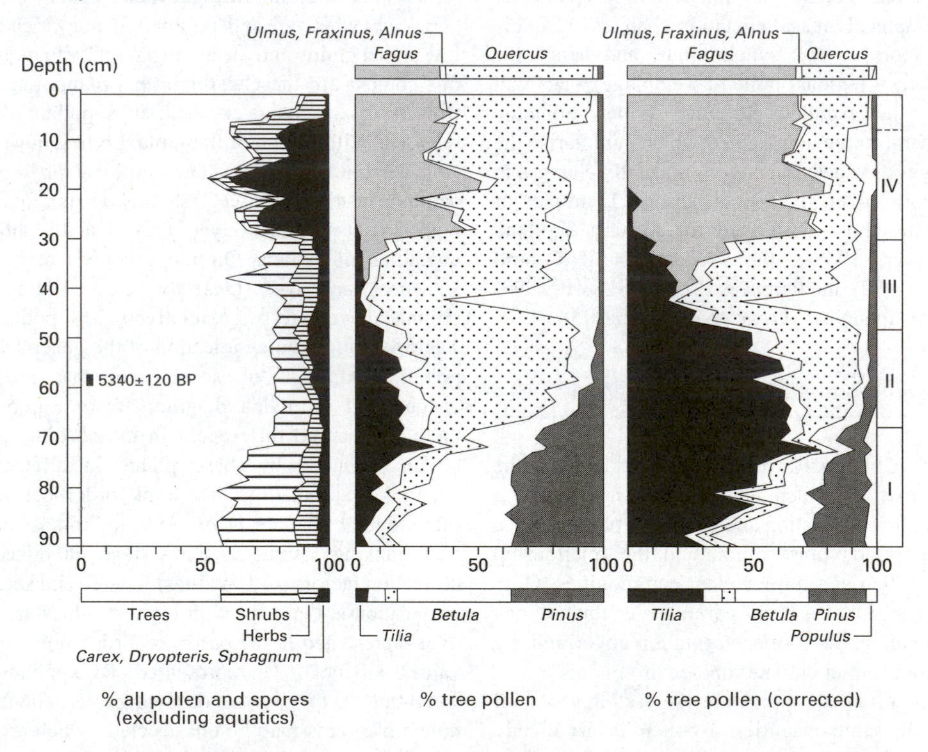

Figure 4.5 Pollen diagrams from a small hollow in *Fagus–Quercus* forest in Denmark. The tree pollen percentages have been corrected in the right-hand diagram using 'R-factors' to convert raw pollen counts (after Andersen, 1973).

Secondly, it is necessary to know something about the source of fossil pollen in a body of sediment. It is important to establish whether plants were growing on the bog surface or within the lake basin, around the margins of the site, in the immediate vicinity, or some distance away. Moreover, it is necessary to know something of the mechanisms involved in the transport of pollen from its source to the eventual point of deposition. Initially, attention was focused on airborne transport of pollen. Tauber (1965), for example, suggested that in a forested region airborne pollen arriving at a bog or lake surface would have travelled by one of three pathways; either through the trunk space, or through the forest canopy, or by raindrop impact from the air above. Factors such as wind speed through the trunk space and canopy, the density of woodland cover, thickness of foliage, time of pollination of the trees, and the size, shape and proximity to the source of the first major bog or lake will all play a part in determining the pollen composition of the final assemblage. The mix of airborne pollen deposited on a site will also depend upon the surface area of a site in relation to surrounding vegetation. Pollen influx to a small lake basin surrounded by a dense tree cover will be dominated by pollen from the immediate surroundings, whereas a large open lake or bog surface will receive a higher proportion of airborne pollen derived from a larger regional catchment or transported over long distances (Jacobsen & Bradshaw, 1981). Analysis of sediment samples from the former site will, therefore, provide evidence of local vegetation cover, while data from the latter will be more likely to reflect the regional vegetation cover. Other research has highlighted the importance of the transport of pollen grains by streams. Investigations of modern pollen recruitment to lake sediments has shown that in some lakes up to 90 per cent of the pollen accumulating on the lake bottom may be derived from inflowing streams and groundwater (e.g. Bonny, 1980). In addition, there will be a local pollen input from aquatic plants growing in the lake or from mire plants growing on a bog surface or at the edges of a lake, while there may also be a secondary component from pollen which has been deposited around the site catchment and has subsequently been remobilised and incorporated into lake, fluviatile or estuarine sediments at a later date (Moore *et al.*, 1991; see below).

In order to establish how far these various components contribute towards the development of a pollen assemblage in a lake or bog site, studies have been undertaken of modern pollen dispersal, and these data have subsequently been compared to the species composition of the vegetation around the sampling site. Data on present-day pollen rain can be obtained from moss polsters on bog surfaces (e.g. Bradshaw, 1981; Heide & Bradshaw, 1982) or from specially designed pollen traps for collecting both atmospheric pollen and pollen settling in lake waters (e.g. Bonny, 1980; Bonny & Allen, 1984). This type of study can provide valuable quantitative data on the relationship between local, regional and long-distance components in the pollen rain and the modern vegetation cover.

In general, it would seem that most wind-borne pollen is deposited within a few kilometres of its source, and only a very small proportion of the pollen grains liberated into the atmosphere are likely to travel very far. Nevertheless, studies have shown that far-travelled pollen may, in certain circumstances, constitute an important element of the atmospheric pollen rain. Tyldesley (1973), for example, recorded tree pollen, largely of birch and pine, over the Shetland Islands off the north coast of Scotland in densities of up to 30 grains per cubic metre of air. The Shetland Islands today are treeless and the source of this exotic pollen is believed to be Scandinavia. Similarly, Fredskild (1973) reported significant concentrations of arboreal pollen (principally *Alnus, Betula, Pinus* and *Picea*) at sites in eastern Greenland whose source area is believed to have been northeast North America. Many similar studies have demonstrated that upper air currents throughout the world often contain small amounts of pollen, which represent a sort of constant 'background' component that can potentially contribute to local pollen accumulation. The relative importance of this component in the resulting pollen assemblages will depend mainly upon the influx of pollen generated by local and regional sources. It will be insignificant (perhaps undetectable) where there is a dense local vegetation cover providing a high local pollen influx, but in barren or poorly vegetated polar or sub-polar regions, for example, where the local pollen influx may be exceedingly low, this component may assume *relatively* high values. It may, therefore, have affected pollen assemblages in many Quaternary cold stage deposits in mid- and high latitude regions.

A third factor to be considered when pollen diagrams are being interpreted concerns the nature of pollen deposition. Differential settling velocities of pollen in lakes and ponds, coupled with the disturbance of sediment on the lake floor, either by currents or by burrowing organisms, can lead to complications in the fossil record. Equally misleading can be the occurrence of redeposited or secondary pollen that has been washed into the lake by stream flow, overland flow, solifluction or collapse of the basin edge sediments, and the subsequent redistribution of material across the lake floor. These grains will clearly be of a different age from those arriving at the lake surface from the atmospheric pollen rain, and although they can often be distinguished from the primary pollen by signs of exine deterioration (see below), they are potential sources of confusion in the interpretation of the biostratigraphic record.

In general, fewer uncertainties are caused where terrestrial sites are used in preference to lakes, although here too complications may arise. Studies have shown that when pollen arrives on the surface of a bog, there may be a tendency for both lateral and vertical mixing to occur, with the larger grains remaining on the surface while smaller grains may migrate downwards into the peat (Birks & Birks, 1980). However, these movements are believed to be relatively insignificant when set against the timescales usually involved in peat accumulation. More problematical is the behaviour of pollens and spores in soils. A major difficulty with soil pollen analysis is that the processes of leaching and capillary action will have the effect of moving fossil grains up and down the profiles (Dimbleby, 1985). Mixing by earthworms and other soil organisms further exacerbates the problem, although discrete pollen assemblages have been detected where earthworm populations have burrowed to progressively shallower depths as soil profiles have developed (Keatinge, 1983). Further problems relating specifically to soil pollen analysis are discussed by Dimbleby (1985) and Moore *et al.* (1991).

Fourth, many fossil pollen grains show signs of deterioration resulting from physical, chemical and biological attack on the exine. Experimental work by Havinga (1985), for example, has shown that pollen and spores vary in their susceptibility to such processes as oxidation and corrosion (Table 4.3) and that this variability may be partly attributable to the nature of the depositional environment. Some grains, for example spores of the clubmosses (*Lycopodium*) and certain ferns (*Polypodium*), show remarkable resistance to deterioration, while others such as the more delicate grains of nettle (*Urtica*) and poplar (*Populus*) may be destroyed altogether. As a consequence, some pollen types tend to be under-represented in the fossil record while others may be over-represented. Cushing (1967b) has described four categories of deterioration in pollen grains:

(a) **corrosion:** where the exine is pitted or etched;
(b) **broken:** where the grains are ruptured or split, or pieces have completely broken away;
(c) **crumpled:** where the grains are folded, twisted or collapsed;
(d) **degraded:** where the structural elements are fused together presenting a 'solid' or 'fossilised' (waxy) appearance to the grain.

These different categories of deterioration reflect very closely the nature of the depositional environment. Corroded grains usually indicate oxidation in poorly compacted peats, while degraded grains frequently reflect secondary deposition, the exine surfaces having undergone structural modification through reworking. As such, deteriorated pollen diagrams may provide useful corroborative information on local environmental conditions or of redeposited pollen (Lowe, 1982; Walker & Lowe, 1990). Soil pollens frequently show all of the above characteristics, and recognition is consequently made more difficult by the often poor state of preservation of these grains.

Table 4.3 Corrosion and oxidation susceptibility of selected pollens and spores (after Havinga, 1964).

A. Sequence of increasing corrosion susceptibility of selected pollens and spores	
Lycopodium	(*low*)
Conifers	↓
Tilia	↓
Corylus	↓
Alnus, Betula	↓
Quercus	↓
Fagus	(*high*)
B. Sequence of increasing oxidation susceptibility of selected pollens and spores	
Lycopodium clavatum	(*low*)
Polypodium vulgare	↓
Pinus sylvestris	↓
Tilia spp.	↓
Alnus glutinosa, Corylus avellana	↓
Betula spp.	↓
Carpinus betulus	↓
Populus spp.; *Quercus* spp.; *Ulmus* spp.	↓
Fagus sylvatica, Fraxinus excelsior	↓
Acer pseudo-platanus	↓
Salix spp.	(*high*)

Finally, there is the vexed question of how far it is possible to relate pollen assemblages to plant communities, and how far we are justified in making inferences about former climatic and environmental conditions on the basis of pollen analytical data. It is now generally accepted that many former plant communities, especially those dominated by herbaceous taxa, which were characteristic of large areas of western Europe and North America during the cold phases of the Quaternary, have no analogues in the modern flora. One such example is the *Artemisia*-dominated association that appears in many pollen diagrams from northwest Europe during the Younger Dryas or Loch Lomond Stadial of the Lateglacial period (Lowe, 1994). Although Pennington (1980) has described an *Artemisia borealis*- and *Silene*-rich community in Greenland which produces pollen rain superficially resembling that of the Lateglacial cold phase, latitudinal and altitudinal variations, in association with seasonal and diurnal temperature fluctuations, would probably have combined to produce significantly different plant communities during past cold phases in western Europe from those existing in present-day tundra regions. A similar problem has arisen in interior

Canada where there appears to be no contemporary analogue for the plant communities reflected in a number of early Holocene pollen records (MacDonald & Ritchie, 1986; MacDonald, 1987).

These difficulties are further compounded by the limitations imposed on palaeoecological inferences by the taxonomic imprecision of pollen identification. Under normal microscopy, it is occasionally possible to identify pollen grains to the species level; distinctions can usually be made, for example, between species of plantain (*Plantago*), saxifrage (*Saxifraga*) or clubmoss (*Lycopodium*). More frequently, identifications are made only to the generic level, for example in identifying pollen of birch (*Betula*), willow (*Salix*) or mugwort (*Artemisia*), while in other cases it is often difficult to subdivide beyond the family level. Grass (Poaceae) and sedge (Cyperaceae), for example, are rarely taken to the generic or specific level. However, considerable progress has been made in recent years in the quality of instrumentation (SEM analysis and optics of higher resolution in standard microscopy, for example) while detailed taxonomic studies have provided clearer diagnostic criteria for identifications to lower taxonomic levels (e.g. Punt & Clarke, 1976;[3] Reille, 1992). Despite these technical and methodological advances, however, pollen diagrams invariably consist of a data bank at a variety of taxonomic levels, and this obviously imposes major constraints upon the reconstruction of former plant communities, particularly as some plant families and genera contain elements with markedly contrasting ecological affinities.

The reader may be forgiven for thinking that, in the light of what has gone before, difficulties in the interpretation of pollen data render the technique of dubious value to the analysis of Quaternary environments. That this is clearly not the case is demonstrated by the remarkable degree of consistency in the large number of pollen-based research publications that have appeared in the Quaternary literature in recent years. Some of the many applications of pollen analysis are briefly considered in the following section.

4.2.6 Applications of pollen stratigraphy

4.2.6.1 Local vegetation reconstructions

Tracing the course of local vegetation developments has been, and still remains, a central theme of pollen analysis. Pollen records from peat and lake cores enable inferences to be made about the history of a particular peat bog or lake ecosystem (e.g. Aaby, 1986; Charman, 1994), while in those sites where pollen diagrams have been obtained from a number of profiles ('three-dimensional pollen analysis'), changes in the local vegetation cover can be mapped through time (e.g. Smith & Cloutman, 1988). In coastal regions, pollen analysis can help to elucidate the history of sea-level change, as marine 'transgressions' and 'regressions' will be reflected in local pollen records by changes between saltmarsh and terrestrial or freshwater plant communities (Tooley, 1978; Shennan *et al.*, 1994). Understanding local vegetation changes and what has influenced them are prerequisites to establishing the scale and pattern of regional changes in vegetation, as well as the impact of human influences (section 4.2.6.4).

4.2.6.2 Regional vegetation reconstructions

For many years, pollen data have constituted one of the principal lines of evidence for reconstructing vegetational history at the regional and extra-regional scales (Godwin, 1975; Huntley & Webb, 1988). Not only does the technique enable large-scale vegetation patterns to be established (see below), but it allows the history of both individual species and entire vegetational assemblages to be traced through time (e.g. Bennett, 1983; Webb, 1987). Regional biotic catastrophes may also be reflected in pollen records including, for example, the *Tsuga* decline at *c.* 4.8 ka BP in western North America and the *Ulmus* decline at around 5 ka BP in western Europe. Both events have been attributed to the effects of disease (Davis, 1981; Huntley & Birks, 1983; Allison *et al.*, 1986), although in the case of the elm decline, human activity may also have been a contributory factor (Peglar & Birks, 1993).

4.2.6.3 Space–time reconstructions

Where a network of pollen sites exists with the records securely dated, maps can be compiled showing changes in the vegetation pattern through time. These may be based upon **isopollen maps**, which provide 'snapshots' of regional vegetation cover for selected time periods (Huntley & Birks, 1983). Pollen percentages of selected taxa are obtained from all sites within a region for a particular time interval; these are then plotted on maps which are contoured to reflect spatial variations in taxon abundance. Data can be displayed for consecutive periods showing the changing distributions of selected taxa over time. By synthesising evidence relating to a number of taxa, **palaeovegetation maps** can be produced to show the overall change in vegetation composition and distribution through time (Figure 4.6). Isopollen and palaeovegetation maps for the Lateglacial and Holocene periods have been constructed for both Europe (Huntley & Birks, 1983; Huntley, 1990; Huntley & Prentice, 1993) and North America (Webb, 1987; Webb *et al.*, 1993a). Such maps are valuable palaeoenvironmental models, for they provide a basis for

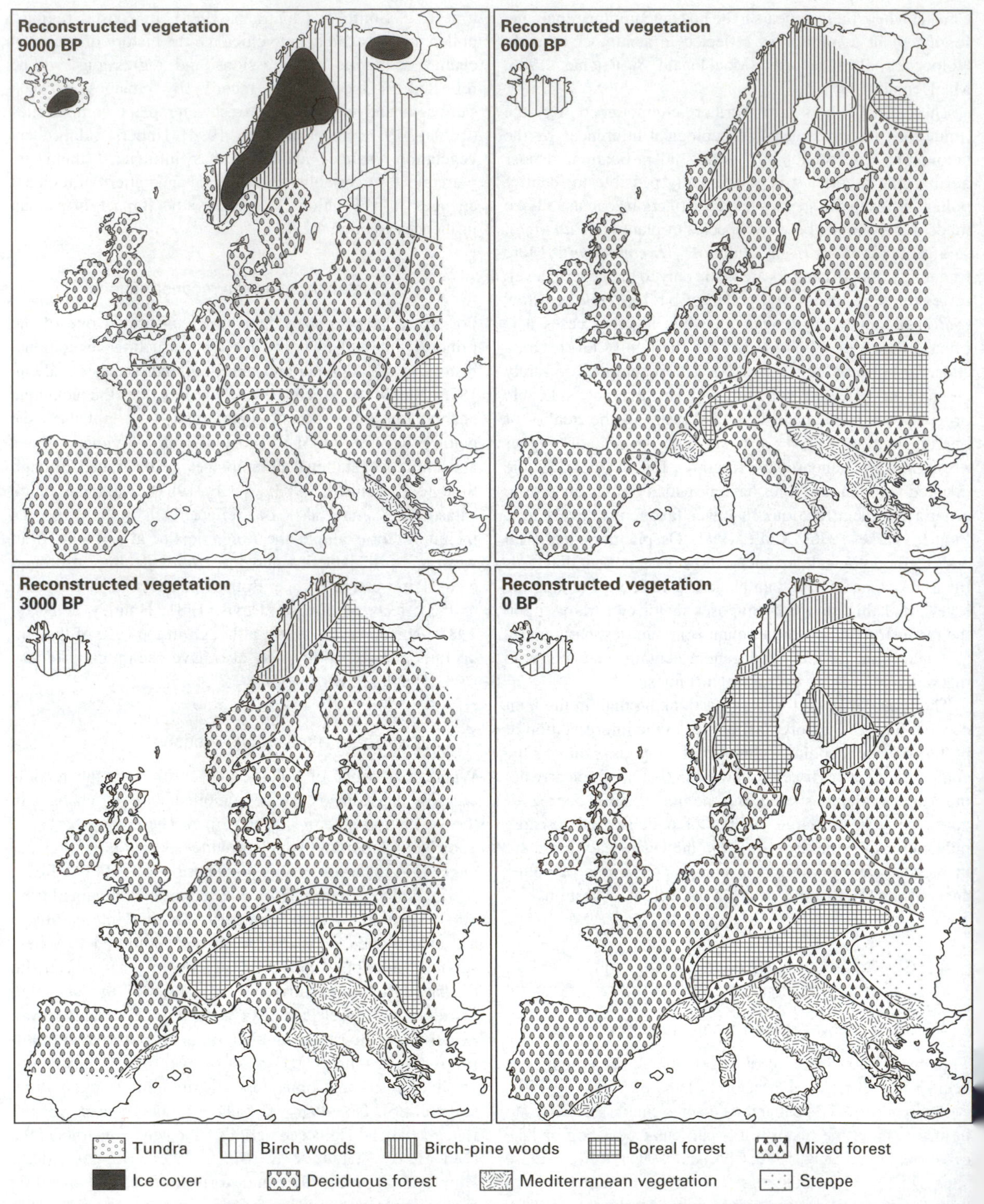

Figure 4.6 Reconstructions of dominant vegetation types in western Europe at 9 ka, 6 ka, 3 ka BP and the present day, based on pollen data (modified after Huntley & Prentice, 1993).

testing hypotheses relating, for example, to the responses of individual taxa or vegetation as a whole to global climatic changes or other environmental influences. The maps may also have applications in the studies of faunal history, of soils, and of former **biomass** (e.g. Prentice *et al.*, 1992), the last-named being of particular importance in the development of climatic models and in contributing to our understanding of the global carbon cycle (see section 4.15.1).

4.2.6.4 Human impact on vegetation cover

Human impact has been the dominant factor affecting the vegetation cover of Europe since the beginning of the Neolithic period (approximately 7000 years ago), and the effects of widespread woodland clearance, farming practices and the introduction of new plant species into regions has left clear imprints in pollen records, especially in sites closely associated with human activities. Significant alteration of woodlands through burning (e.g. Caseldine & Hatton, 1993) or clearance (e.g. Walker, 1993) are reflected in reduced arboreal pollen percentages, while episodes of pastoral or cereal cultivation may be recorded by increased representations of pollen of **ruderal** taxa ('weed' invaders of tilled or pastoral land) and, in the case of cultivation phases, by the appearance of pollen grains of cereal taxa. The presence of these so-called 'anthropogenic indicators' in pollen diagrams (Behre, 1986) provides insights into land management techniques (Kaland, 1986), prehistoric and historic farming practices (Gaillard *et al.*, 1992), activities on and around settlement sites (Dimbleby, 1985) the environmental impact of early mining operations (Mighall & Chambers, 1993) and military invasion and occupation (Dumayne & Barber, 1994). In combination with indirect evidence of human modification of the landscape, such as increased sediment accumulation in lakes, the influx of charcoal into peat and lake sediments and changes in soil characteristics, pollen-stratigraphic evidence provides an important archive of the evolution of the **'cultural landscape'** (Birks *et al.*, 1988; Berglund, 1991; Edwards & MacDonald, 1991; Peglar & Birks, 1993).

4.2.6.5 Pollen data and climatic reconstructions

Pollen data have long been used to reconstruct Quaternary climates. Indeed, some of the earliest systematic attempts to derive climatic parameters from fossil evidence employed pollen records (Faegri & Iversen, 1989). These and subsequent approaches have relied largely on **indicator species** in the pollen records, i.e. plants with known climatic affinities (Zagwijn, 1994). However, problems relating to the taphonomy of pollen assemblages (see above) and also the differential migrational response of plants, especially trees, to climatic change, may lead to erroneous palaeoclimatic inferences. More recent approaches to climatic reconstructions using pollen data have involved the development of **pollen response surfaces**, which measure in a quantitative way the dependence of broad-scale vegetation patterns on climate (Bartlein *et al.*, 1986; Prentice *et al.*, 1991). Modern pollen data are calibrated to contemporary climatic parameters using **transfer functions**,[4] and these relationships can then be used to infer past climatic conditions from isopollen data (Huntley, 1992, 1993b; Guiot *et al.*, 1993). However, transfer functions assume equilibria between biota and the environment and they may produce spurious results at times of disequilibria (see below), since the total fossil assemblage at such times could be a mélange of species not normally found in close association. Nevertheless, pollen response surfaces are considered a powerful new tool for climatic reconstruction, and are being increasingly widely used in the development of integrated models of past global climates (Huntley & Prentice, 1993; Wright *et al.*, 1993).

4.3 Diatom analysis

4.3.1 Introduction

Although several members of the algal kingdom have been studied by Quaternary palaeoecologists, including the green algae, blue-green algae and chrysophytes (Cronberg, 1986; Van Geel, 1986; Smol, 1990), it is **diatoms** that have attracted most attention, since their classification and ecological preferences are much better understood. Diatoms have been studied for over two centuries and the analysis of the diatom content of Quaternary sediments actually predates pollen analysis (Battarbee, 1986). By the end of the nineteenth century, a considerable amount of work had been undertaken on diatom remains, and while most of these studies had little stratigraphic or palaeoecological value, they laid the foundations for later research on diatom taxonomy. Quaternary diatom remains have proved extremely useful as indicators of local habitat changes, particularly in lake sediments, but also in both shallow and deep marine deposits. The analysis of diatom floras has provided new insights into a wide range of palaeoenvironmental issues, such as the reconstruction of past lake level changes, water chemistry variations, sea-level variations and the disturbance of lake ecosystems by human activities.

4.3.2 The nature and ecology of diatoms

Diatoms are microscopic, unicellular members of the Bacillariophyta of the algal kingdom (Figure 4.7). They secrete a siliceous shell or structure, known as a **frustule** which can range in length from 5 μm to *c.* 2 mm, depending on the species (Brasier, 1980; Round *et al.*, 1990). The frustule is often compared to a pill-box, as it consists of two overlapping valves or **thecae**, the larger one (**epitheca**) fitting over the smaller one (**hypotheca**) in a box fashion to enclose the living (protoplasmic) mass. The valves are linked together by connecting or girdle bands (**copulae**). The wall of the frustule may be a single layer of silica or it may be more complex, consisting of a double silica wall separated by vertical silica slats. Frustules are commonly circular (**centric**) or elliptical to rod-like (**pennate**) in shape and are perforated by intricate patterns of tiny apertures (**punctae** or **areolae**). The arrangement of the perforations is one of the most important diagnostic characteristics of diatoms, although other structural details, including reticulations, canals and ribs, are important for classification below the generic level, so that very careful scrutiny of the valves under a high-powered microscope is necessary for species identifications. The frustules are composed of an amorphous hydrated silica, similar to opal, which enhances their preservation potential in a range of sedimentary environments.

Diatoms are found in a wide range of aqueous to sub-aqueous environments and make up about 80 per cent of the world's primary producers. They exist in bottom-dwelling (**benthic**), attached (**epiphytic** – attached to plants; **epilithic** – attached to stones) and free-floating (**planktonic**) forms, and while all species require light and are therefore limited to the **photic zone** (usually less than 200 m water depth), they occupy a large number of ecological niches. In the sea they are found in lagoons, shelf seas and deep oceans, they are common in the intertidal zone in estuaries and salt marshes, and they are often abundant in ponds, lakes and rivers. Certain species even live on wetted rocks, in the soil or attached to trees. The freshwater, soil or epiphytic niches are dominated by pennate diatoms, while the centric forms tend to be more common as plankton in marine waters, especially in the sub-polar and temperate latitudes. Marine benthic habitats, however, are characterised by pennate forms.

Figure 4.7 Light microscope images (differential interference contrast) of diatom frustules of the species *Cymbella pusilla* Grunow (the shorter, more ornamented form) and *Nitzschia palea* var. *debilis* (Kütz.) Grun. in samples obtained from a salty swamp at Zaafrane–Douz Road, Tunisia (collected by Leila Ben Khalifa, Paris; photo: Laurence Carvalho). Length of *Cymbella* frustules = 14 μm.

The distribution of diatom species is determined by a number of variables, including water acidity and salinity, oxygen availability, nutrient content and water temperature. Freshwater diatoms, for example, are controlled largely by salinity, pH and trophic status, while sea-surface temperatures, oceanic frontal contrasts and nutrient up-welling influence the distribution of many marine taxa (Sancetta *et al.*, 1991; Villareal *et al.*, 1993). Changes in any of these parameters can have a major effect on the structure and composition of the diatom community. The autecology of modern diatoms has been extensively studied (see Battarbee, 1984, 1986, 1988) and a range of quantitative methods has been developed to relate modern diatom assemblages to contemporary habitat and environmental conditions (see below).

4.3.3 Field and laboratory methods

Diatom valves, like pollen grains, are best preserved in fine-grained sediments since they can be easily damaged or destroyed in coarse-grained deposits. Samples for analysis can be obtained from vertical exposures in shallow water marine or estuarine deposits, though more frequently they are extracted from sediment cores obtained from lake, shelf seas or the deep ocean floor (e.g. Jones *et al.*, 1989; Sancetta & Silvestri, 1986; Koç Karpuz & Jansen, 1992). Diatoms are not susceptible to oxidation or microbial degradation, but cores for diatom analysis are usually sealed air-tight to prevent drying out of the sediment which can lead to fracturing of the valves.

Diatom frustules may be separated from the sediment matrix by a variety of laboratory procedures. Organic matter is removed by oxidation, the most common methods being by digestion in H_2O_2 or in a mixture of potassium dichromate and sulphuric acid, while carbonates and certain other salts can be dissolved by heating gently in dilute hydrochloric acid. Much more difficult is the removal of ninerogenic matter, since diatoms are soluble in some acids, nd especially in HF (hydrofluoric acid) which is commonly sed in the preparation of pollen samples. This means that iatom and pollen counts cannot be carried out on the same lides since the samples have to be prepared separately. Coarse mineral particles (>500 μm) can be removed by ieving or by gentle swilling in a beaker, but finer particles equire either some form of flotation using heavy liquids, or he less efficient differential centrifugation method Battarbee, 1986, 1988). The residues are mounted on slides nd counted under a microscope using phase-contrast llumination and magnification of up to ×1000. Diatoms are often more abundant than pollen in sediment samples, and a ount of 500 or 1000 valves may form a statistically ignificant total (the 'diatom sum'). As with pollen grains, dentifications are based upon type collections, keys and photographs in diatom manuals and catalogues (listed in Battarbee, 1986, 1988; Round *et al.*, 1990). Nevertheless, liatoms are often difficult to classify and recent inter-comparison exercises have revealed discrepancies in axonomic conventions between different countries, a problem that is being addressed by the development of nternational taxonomic quality control standards (e.g. Munro *et al.*, 1990).

Three principal methods have been developed for the neasurement of diatom concentrations in sediment samples. The **aliquot method** uses a measured volume of sample suspension which is pipetted on to a circular coverslip and he total number of diatoms in the sample volume is estimated. The **evaporation tray method** uses a measured volume of suspension which is added to a coverslip and from which the water is allowed to evaporate at room emperature. The diatoms are then counted and since the original volume is known, the concentration can be neasured. The use of **microsphere markers** is the third nethod. These are tiny spherules made of glass or plastic hat are added in known numbers to an aliquot of sample suspension, and concentrations of diatoms can be calculated from the ratio of diatoms to spherules. This third method is similar in principle to the estimation of pollen concentrations using exotic pollen markers, and is perhaps he most reliable (Battarbee, 1986).

Diatom counts based on samples selected from a stratified sediment sequence are normally presented in the form of a percentage diagram using bar histograms (e.g. Figure 4.8). This can be subdivided in a similar way to pollen diagrams, either subjectively on the basis of visual inspection of the data or by various forms of numerical analysis (Gordon & Birks, 1972). Indeed, zonation of diatom records often employs procedures used in pollen analysis (see above). Other methods of data presentation that have been employed by diatom analysts include the tabular format, where counts are listed in percentages, and the **composite** or **ratio** diagram, whereby groups of diatoms with particular environmental affinities are added together and the ratio variations between the different groups are depicted (e.g. Figure 4.9B).

4.3.4 The interpretation of Quaternary diatom records

The difficulties that arise in the interpretation of diatom assemblages are, in many ways, similar to those discussed above for fossil pollen. Diatom valves are light and easily transported, and thus in estuarine sediments, for example, there is frequently a complex admixture of marine, brackish and freshwater forms, while lake muds may contain diatoms derived not only from the lake ecosystem itself, but also from inflowing streams and catchment soils. Freshwater diatoms often occur in marine sediments, having been blown in by the wind, while in the deep oceans, diatom remains have been found that have been transported many hundreds of kilometres from their source. Selective destruction of diatoms is another potential error source, with complete or partial dissolution of the frustules under pressure at depth in the oceans, while in brackish and freshwater contexts, the less robust and weakly silicified forms will tend to dissolve where conditions are very alkaline. In these environments the diatom death assemblage will be biased in favour of the stronger and more heavily silicified forms. Grazing by herbivores may also affect the composition of fossil assemblages. Reworked diatom frustules can sometimes be detected where valves are broken or partially dissolved, or demonstrate signs of mechanical abrasion, although secondary diatoms within a body of sediment may not always be easily recognisable.

Despite these problems, diatom analysis has proved to be a particularly valuable technique for environmental reconstructions (Battarbee, 1991). In recent years, the technique has become widely used in studies of eutrophication, acidification and salinity changes (Battarbee *et al.*, 1990). These and other applications of diatom analysis are discussed in the following section. In addition, diatoms in deep-ocean sediments constitute a valuable source of palaeoecological data and this aspect of diatom analysis is also considered in this chapter (section 4.11).

4.3.5 Applications of diatom analysis

4.3.5.1 Diatoms as salinity indicators

A major control on diatom distributions is salinity, and hence individual diatom species can be classified on the basis of their salinity preferences. The **halobian system** of

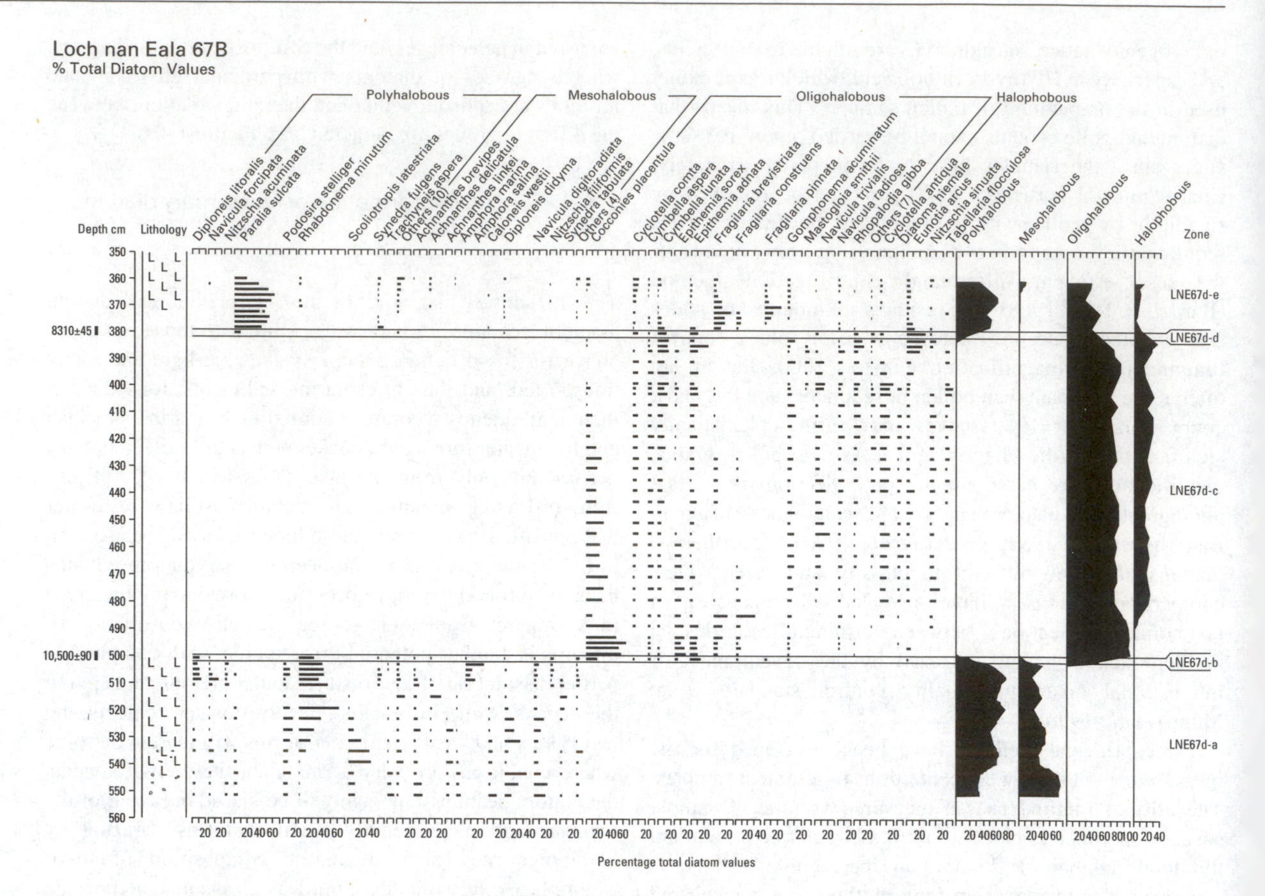

Figure 4.8 Diatom diagram, calculated as percentages of total diatom valves, from lake sediments in northwest Scotland (after Shennan *et al.*, 1994). The diatom assemblages reflect marine sedimentation in units LNE67d-a and LNE67d-e, with freshwater conditions in LNE67d-c.

classification, introduced by Kolbe in 1927 for application to marine and estuarine assemblages, and subsequently modified by Hustedt (1957), has four main groupings:

(a) **polyhalobous** diatoms: those that thrive in salt concentrations of >30 per cent;
(b) **mesohalobous** diatoms: those that thrive in salt concentrations of 0.2–30 per cent;
(c) **oligohalobous** diatoms: those that generally require salt concentrations of <0.2 per cent;
(d) **halophobous** diatoms: those that cannot tolerate even slightly salty water.

The sensitivity of diatoms to salinity changes is reflected in two particular aspects of Quaternary research: sea-level change, and the environmental record in closed-basin lakes.

(a) Sea-level variations Diatoms have long been employed as indicators of changing sea level, the earliest accounts being from Scandinavia in the 1920s, where evidence of marine or brackish water conditions in lakes now well above sea level formed an important line of evidence in the reconstruction of the extent of glacio-isostatic recovery following the wastage of the last ice sheet (Battarbee, 1986). Subsequently, the analysis of diatom assemblages from coastal localities has become, as with pollen data (see above), a standard technique for identifying marine 'transgressions' and 'regressions' in littoral sediment sequences (e.g. Tooley & Shennan, 1987; Shennan, 1989). Regressive contacts are characterised by the replacement of polyhalobous diatoms first by mesohalobous and/or oligohalobous species, and ultimately by halophobous forms, whereas the reverse is the case at transgressive contacts (Figures 4.8 and 4.9). In this way, diatom assemblages can reflect precisely the highest point of marine influence, and therefore provide important stratigraphic markers for the reconstruction of shoreline displacement

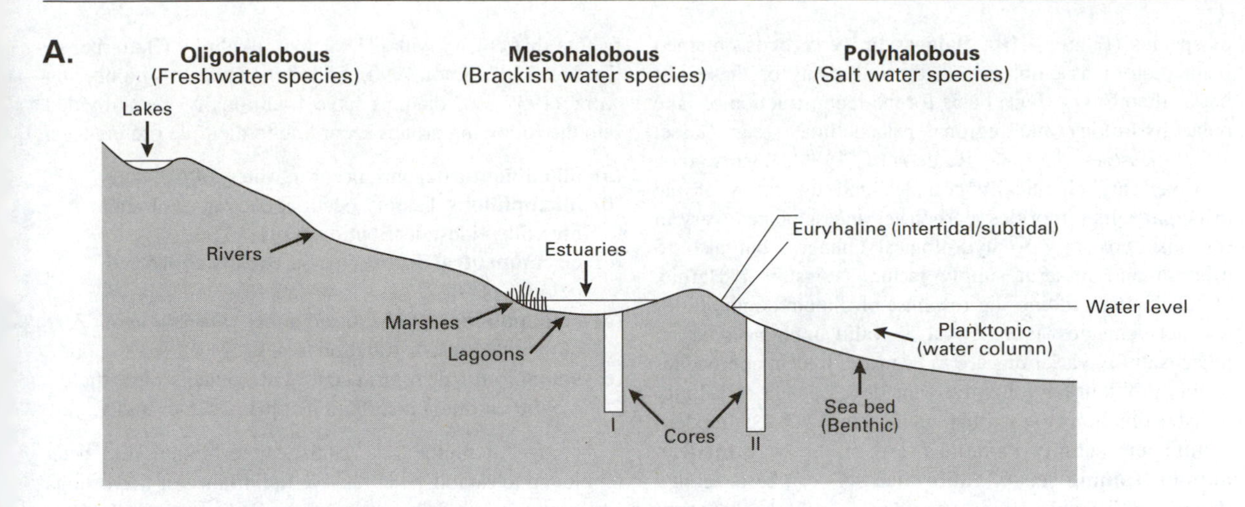

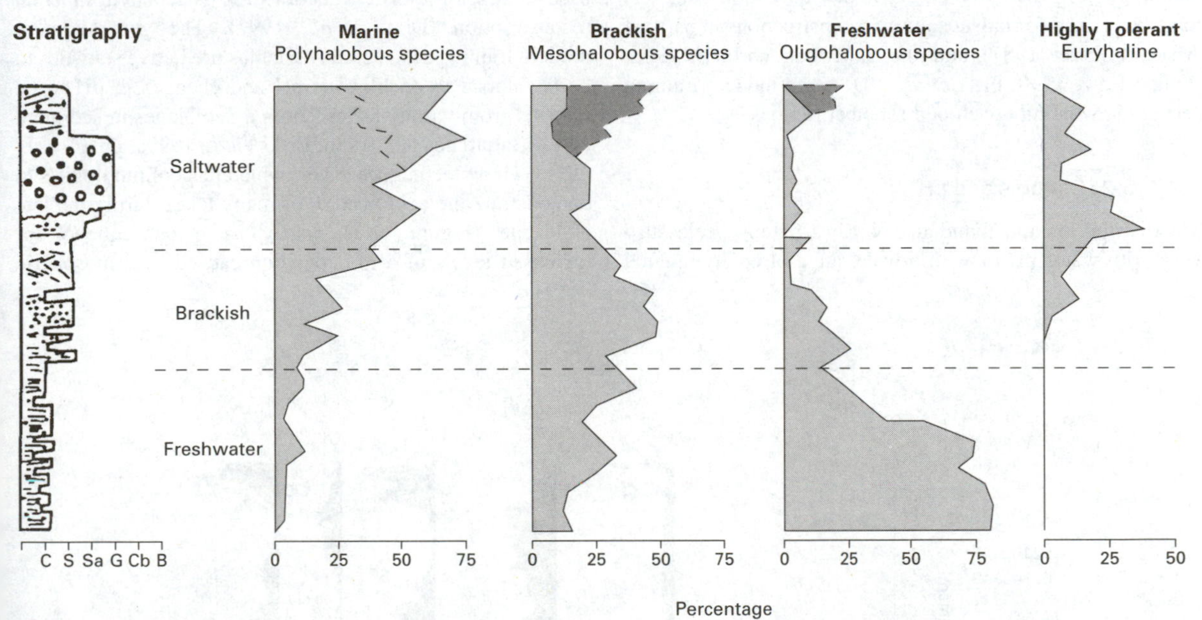

Figure 4.9 A. Isolation lake basins close to shore and affected by variations in sea-level change. During times of relatively high sea level, the lake basins may be submerged or periodically inundated by the sea (euryhaline; polyhalobous diatoms dominate the assemblages). B. Schematic representation of diatom assemblage changes reflecting a marine incursion (transgressive phase) (after Carter, 1992).

curves and patterns of isostatic rebound (e.g. Stabell, 1985; Long & Shennan, 1993; see also section 2.5).

(b) Closed-basin lakes As was shown in sections 2.7.1 and 3.6, lakes in low-latitude regions are frequently sensitive to changes in effective moisture, especially where this alters the salt content of the lake water (**'salt lakes'**), and hence past variations in salinity can be used as a basis for reconstructing climatic change (Teller & Last, 1990). Fluctuations in lake volume and salt concentrations in lake waters will, in turn, be reflected in changing compositions in diatom assemblages because of the different salt tolerances

of species (Figure 4.10). **Palaeosalinity records** obtained from diatom assemblages in the sediments of these salt lakes, therefore, offer a basis for the reconstruction of lake palaeohydrology and regional palaeoclimate (e.g. Gasse, 1987; Gasse *et al.*, 1987; Radle *et al.*, 1989). Lake waters, however, are chemically complex and the ratios of the principal salts (chlorides, carbonates and sulphates) vary in response not only to hydrological changes, but also to dilution and mineral supply factors (Eugster & Hardie, 1978). Understanding the response of diatom communities to such changes is essential if valid inferences about palaeosalinity variations are to be made. Recent approaches to this problem have involved the development of salinity transfer functions to establish relationships between modern diatom and salinity variations, and in this way **modern diatom training sets** or **calibration sets** can be generated (Gasse & Tekaia, 1983; Fritz *et al.*, 1991). Statistical calibration procedures are then employed to establish predictive measures between diatom assemblages and water chemistry (e.g. Birks *et al.*, 1990a). Calibration data sets have been used for palaeosalinity reconstructions in parts of Africa (Gasse, 1987) and the American northern Great Plains (Fritz, 1990; Fritz *et al.*, 1991), and modern training sets are now being developed for other regions.

4.3.5.2 Diatoms and pH

The distribution and abundance of many diatom species also vary with water pH or with a wide range of environmental factors that covary with pH, such as alkalinity (Battarbee & Charles, 1987; Smol, 1990). Following Hustedt's pioneering work (1937-39), diatoms have traditionally been divided into the following groups according to their pH preferences:

(a) **alkalibiontic** diatoms: occur in waters of pH values >7;
(b) **alkaliphilous** diatoms: occur at pH values of about 7, but with widest distribution at pH >7;
(c) **circumneutral** diatoms: occur equally above and below a pH of 7;
(d) **acidophilous** diatoms: occur at pH values of about 7, but with widest distribution at pH <7;
(e) **acidobiontic** diatoms: occur at pH values of less than 7, with optimum distribution at pH of 5.5 or under.

A range of multivariate statistical techniques has been employed to establish the relative importance of the various environmental variables that influence diatom distributions in lake ecosystems (Battarbee, 1984; Stevenson *et al.*, 1989a; Dixit *et al.*, 1991) and, once identified, these can be used to develop transfer functions for palaeoenvironmental reconstruction (Dixit *et al.*, 1993). The generation of modern training sets therefore enables predictive statistics to be developed to enable past pH and changes in pH to be inferred from variations in diatom assemblages in sediment cores (Battarbee *et al.*, 1986; Birks *et al.*, 1990a; Dixit *et al.*, 1991). These techniques have enabled palaeolimnologists to demonstrate the acidification of many lakes during the late Holocene (Figure 4.11), partly as a consequence of increased levels of acid deposition caused by atmospheric

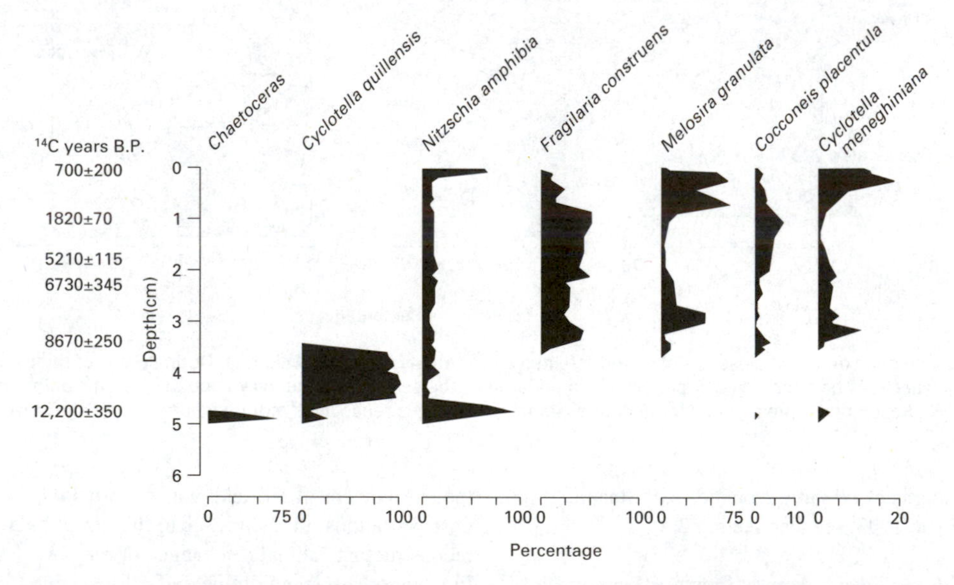

Figure 4.10 Diatom diagram from Lake Valencia, Venezuela. The saline lake planktonic diatom *Cyclotella quillensis*, which dominates the basal assemblages, is replaced by freshwater diatoms at *c.* 8500 BP, indicating a rise in water levels and dilution of dissolved salt levels (after Bradbury *et al.*, 1981).

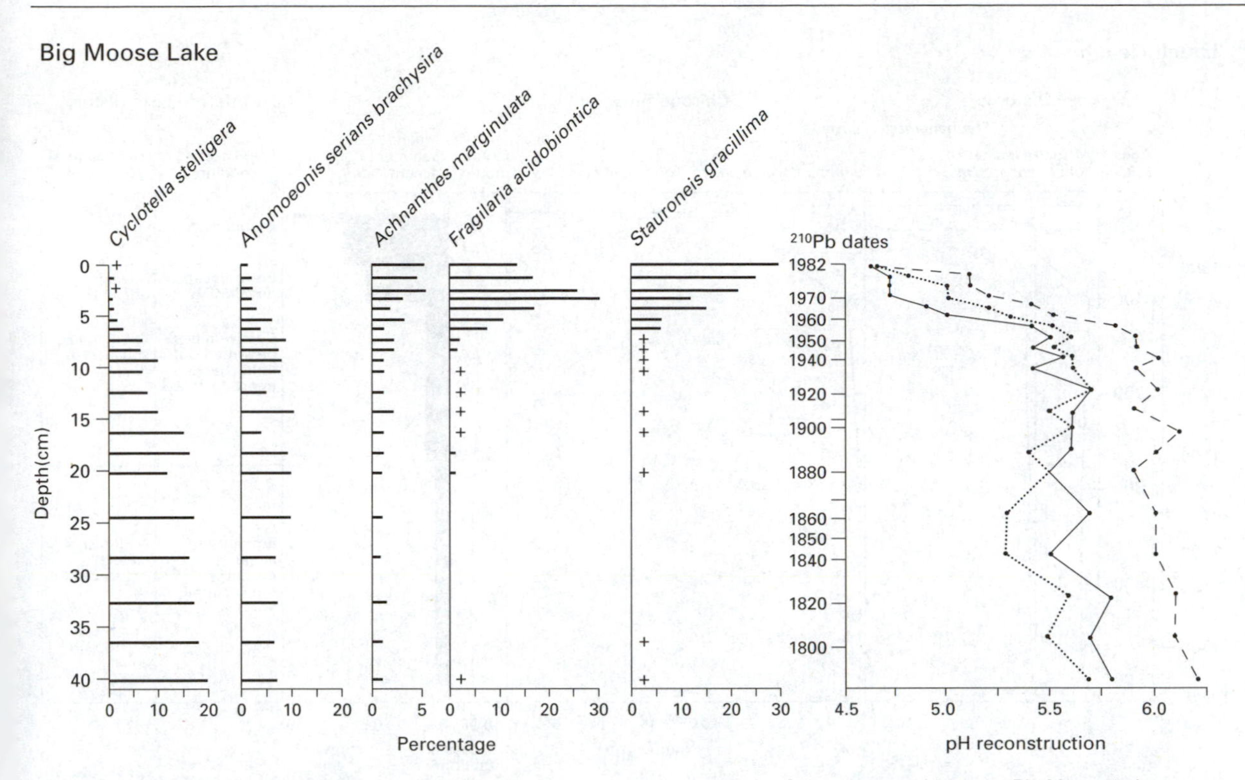

Figure 4.11 Evidence for pH changes in recent lake sediments, based on analysis of diatom assemblages at Big Moose Lake, Adirondack Mountains, New York State: the data suggest that pH has reduced by between 1 and 1.5 units since AD 1950 (after Battarbee, 1984).

pollution (Battarbee, 1984; Battarbee & Charles, 1987; Jones *et al.*, 1989). Models of lake acidification processes based on fossil diatom records can then be used to formulate strategies for the future management of lakes and lake catchments in areas affected by acid deposition (Sullivan *et al.*, 1992).

4.3.5.3 Diatoms and trophic status

Diatom communities are also very sensitive to changes in nutrients and are therefore good indicators of trophic status. Many lakes have experienced a trophic stability (see section 3.9) through most of the Holocene. However, for some lowland lakes, deforestation and the spread of farming in the mid- to late Holocene has led to marked changes in nutrient balance (Fritz, 1989), whilst the recent increase in sewage effluents, phosphorus-rich detergents and intensive agricultural methods has resulted in a sudden nutrient enrichment of many lakes throughout the world. Such **cultural eutrophication** is reflected in diatom records from lake sediments (Battarbee, 1978), particularly in those lakes that were naturally oligotrophic (Figure 4.12), by increases in overall diatom productivity, in the relative abundance of planktonic taxa (Smol *et al.*, 1983; Battarbee, 1986), or in diatom-inferred phosphorus concentrations (Anderson *et al.*, 1993; Bennion, 1994).

4.3.5.4 Diatoms and the archaeological record

Diatom assemblages are becoming increasingly widely used as indicators of past human activity (Battarbee, 1988). Archaeological applications include the locating of occupational sites in proximity to coastlines and brackish-water inlets (Miller, 1982), reconstructing water quality changes in former settlement sites that were located close to waterfronts (Juggins, 1988), and establishing the source of clays used in pottery making (Alhonen *et al.*, 1980). Fossil diatom assemblages from lake sediments have been used to reconstruct historic and prehistoric land-use changes (Renberg *et al.*, 1993), while applications of diatom analysis in recent 'industrial archaeology' include the evaluation of the effects of strip mining on landscapes of parts of the American Midwest (Fritz & Carlson, 1982) and the environmental effects of uranium mining and milling in Ontario (McKee *et al.*, 1987).

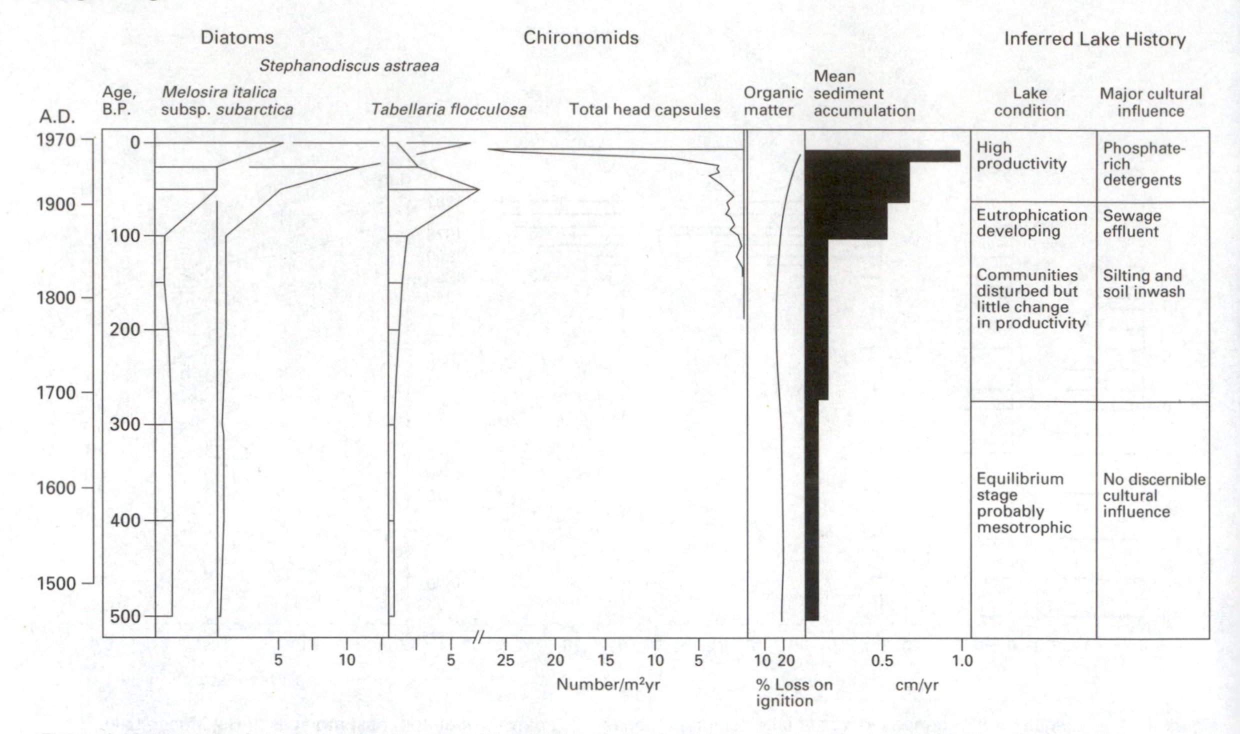

Figure 4.12 Evidence for recent eutrophication of Lough Neagh, Northern Ireland, based on increased abundance of diatom species and chironomids (after Battarbee, 1978).

4.4 Plant macrofossil analysis

4.4.1 Introduction

The study of fossil plant remains is one of the earliest branches of Quaternary studies, for research was underway into the Quaternary floras of the British Isles as long ago as the 1840s. Similar studies were being undertaken in Denmark and Germany and the results of many of these investigations were synthesised in Clement Reid's remarkable volume *The Origin of the British Flora* published in 1899. This book contained the first clear statement of Quaternary vegetation changes in western Europe, and appeared almost twenty years before the development of pollen analysis as a palaeoenvironmental technique. The analysis of plant macrofossils can provide valuable complementary information to microfossil data, but it can also provide an independent approach to the reconstruction of environmental conditions.

4.4.2 The nature of plant macrofossils

Plant macrofossils range in size from minute fragments of plant tissue to pieces of wood (even whole trees – Figures 3.18 and 4.13) that can be measured in cubic metres. They include recognisable remains of vascular plants, such as fruits, seeds, stamens, buds, scales, cuticle fragments and, very occasionally, bryophyte remains (Dickson, 1986) and oospores of algae. Carbonised plant macrofossils, usually in the form of wood or seeds, are also encountered, especially in archaeological contexts (Tolonen, 1986).

Plant macrofossils are found in a variety of depositional environments, but most commonly in lacustrine and fluviatile sediments (especially in fine alluvium), and in acid peats. Occasionally, rich assemblages of plant remains are recovered from soils or sediments on archaeological sites where the fossil remains will include cultivated plants, weeds of cultivation and uncultivated species collected for food. Fossil dung or animal middens may also preserve plant remains in good condition, and in some regions these have

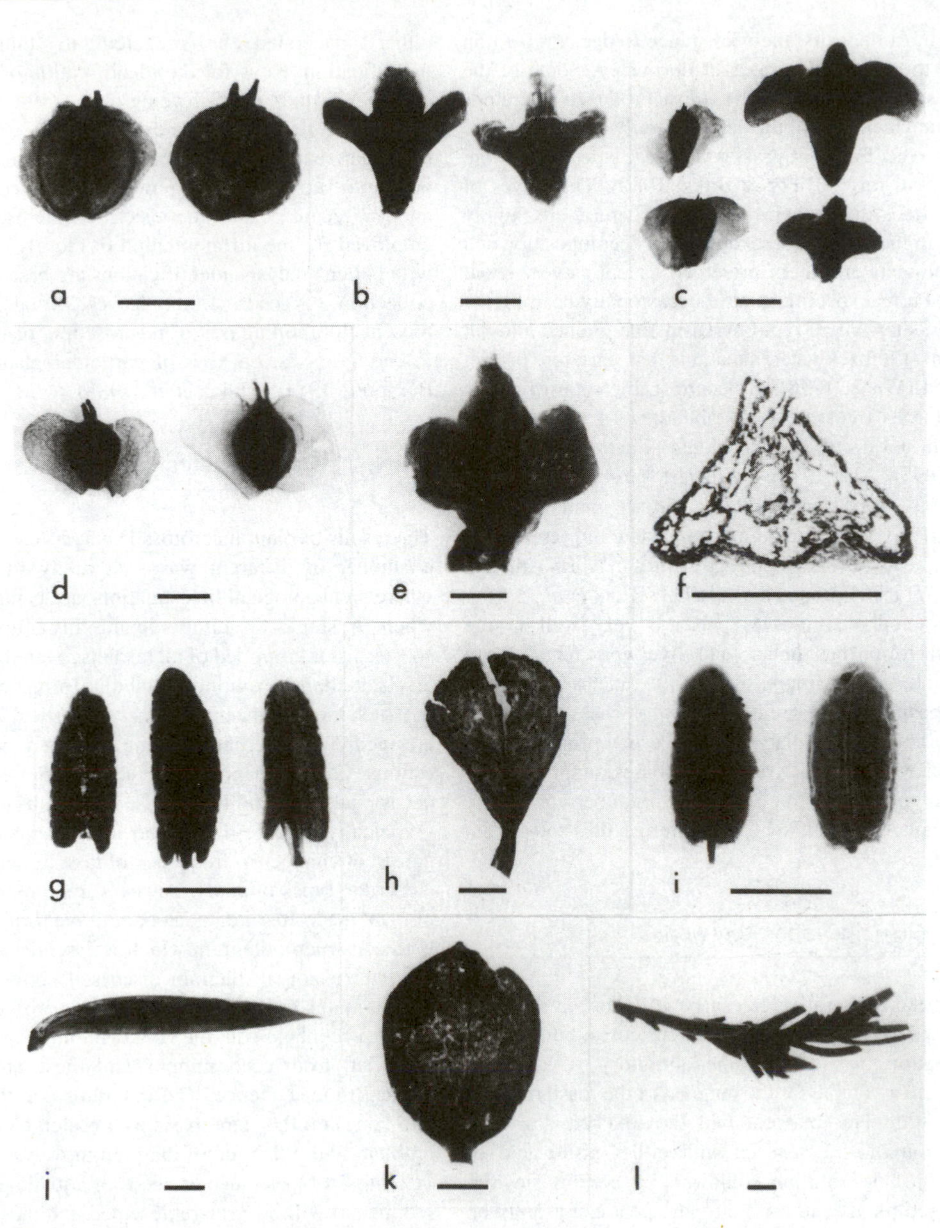

Figure 4.13 Selection of plant macrofossils of ecologically important boreal, arctic/boreal and arctic/alpine taxa. The bar scale represents 1 mm, except for (f) where it represents 0.5 mm. a. *Betula glandulosa* fruits; b. *B. glandulosa* catkin scales; c. *B. papyrifera* catkin scale; d. *B. papyrifera* fruits; e. *B. papyrifera* catkin scale; f. *Pinus strobus* needle; g. *Dryas integrifolia* leaves; h. *Salix herbacea* leaf; i. *Empetrum* leaves; j. *Juniperus communis* needle; k. *Vaccinium uliginosum* leaf; l. *Cassiope hypnoides* twig (after Mayle & Cwynar, 1995; photo by courtesy of Francis Mayle).

provided valuable records of vegetational and environmental change (e.g. Betancourt *et al.,* 1990). However, it is in acid peat deposits that the remains are often best preserved, as the fossils are often found in growth position and have been protected from oxidation.

The preservation of fossil plant material is very variable. Wood may survive in recognisable condition for many thousands of years, either in waterlogged sites or in very dry soils in arid environments, but in other situations decomposition may be extremely rapid. Seeds and fruits will survive in most deposits, their resistance to decay reflecting adaptation to withstand periods of dormancy. Some of the very small seeds, however, such as those produced by orchids and by certain members of the heather family (Ericaceae), are rarely preserved. Seeds of grasses (Poaceae) are also seldom found in fossil form (Birks & Birks, 1980). The leaves of deciduous trees with their delicate structure are highly vulnerable to mechanical breakdown and decomposition and in lake sediments are rarely preserved except as very small fragments. Perfect specimens have occasionally been found, however, in sites where rapid burial in fine-grained alluvial deposits in still backwater situations has ensured perfect preservation (Watts, 1978). By contrast, the robust needles from coniferous trees are often abundant as macrofossils, occurring in a variety of depositional situations (Mayle & Cwynar, 1995). Also, leaves of dwarf shrubs from tundra environments such as willow (*Salix*) and bilberry (*Vaccinium*) have been found in a good state of preservation in lake sediments of Lateglacial age at sites in North America (Watts, 1967) and Europe (Birks, 1993; Birks *et al.,* 1993). Of the lower plants, mosses preserve very well in the macrofossil form, but lichen and liverwort remains are seldom found, though impressions of the latter are common in tufa deposits. In the great majority of cases, therefore, Quaternary macrofossil analysis is concerned primarily with the study of wood, seeds, fruits and mosses, augmented by information provided by a limited number of easily identifiable plant remains such as conifer needles and certain leaves.

4.4.3 Field and laboratory work

The larger plant macrofossils can be collected in the field from exposed sections or from sediment cores, but in most cases extraction takes place in the laboratory. A variety of techniques are available for the removal of the fossil remains from the sediment matrix, but the majority involve disaggregation of the material with either nitric acid or sodium hydroxide solution, followed by careful sieving. Finer lake sediments and ombrotrophic peat can usually be broken down simply with the aid of a jet of water, washing through a sieve typically with a 250 μm mesh. During disaggregation, some fruits, seeds or leaves will rise to the surface of the liquid and these can be picked off with a fine paintbrush, while macrofossils can be removed in a similar fashion from the mesh of the sieves. Some fossils, such as fruitstones, can be kept dry, but others need to be stored in alcohol, glycerine or other preserving fluids. The more delicate structures such as translucent leaves and seeds are best kept mounted on microscope slides. Wood, except in the case of some very obvious species (e.g. birch) must be either macerated or prepared in thin section for identification. Keys for the identification of wood sections can be found in Schweingrüber (1990). Most plant macrofossil remains can be examined on a white plate under binocular scanners or low-powered stereo-microscopes. On occasions, however, high-powered microscopy is required, and in recent years, the electron microscope has been employed for the differentiation of closely similar taxa. As with pollen analysis, identifications are based on a reference collection of seeds, fruits, leaves, wood, etc. from the present flora, on atlases of macroscopic plant remains and, in certain cases, on keys of particular plant families (e.g. Berggren, 1981; Schoch *et al.,* 1988).

4.4.4 Data presentation

The results of plant macrofossil analysis can be presented in a number of different ways. At many sites, particularly where archaeological investigations are being carried out or where a single stratum is being investigated, a simple species list is compiled of all taxa discovered. Where several levels are being examined, a tabular format may be adopted in which the presence or absence of particular plant remains are recorded. Alternatively, the data may be expressed as estimates of abundance using such descriptive terms as rare, occasional, frequent, In this case, the results may be depicted graphically (Figure 4.14), and thus an impression can be gained of changes in frequency of taxa through time.

Certain types of study, however, require the presentation of plant macrofossil data in quantitative form. One approach is to construct a diagram which is essentially similar to the percentage pollen diagram discussed above, in which the plant remains from different levels in a profile are expressed as a percentage of the total number of macrofossils identified from each sample, and these are plotted on a vertical time-sequence. The diagrams can then be divided into zones on the same basis as a pollen diagram. A major problem with this type of diagram, however, is that certain macrofossil types (such as seeds of aquatic plants in limnic sediments) will be over-represented and thus the curves for other taxa will be suppressed. Equally, it may be difficult to arrive at a satisfactory macrofossil sum, particularly when a range of different types of plant macrofossil material is present. An alternative approach is to construct a concentration diagram showing the occurrence of total numbers of plant macrofossils per unit volume of sediment (e.g. Warner, 1990a; Mayle & Cwynar, 1995) at different levels in the profile (Figure 4.15), and this may be converted to a macrofossil influx diagram showing accumulation per year if a dating framework can be established (e.g. Birks & Mathewes, 1978). As with pollen concentration diagrams

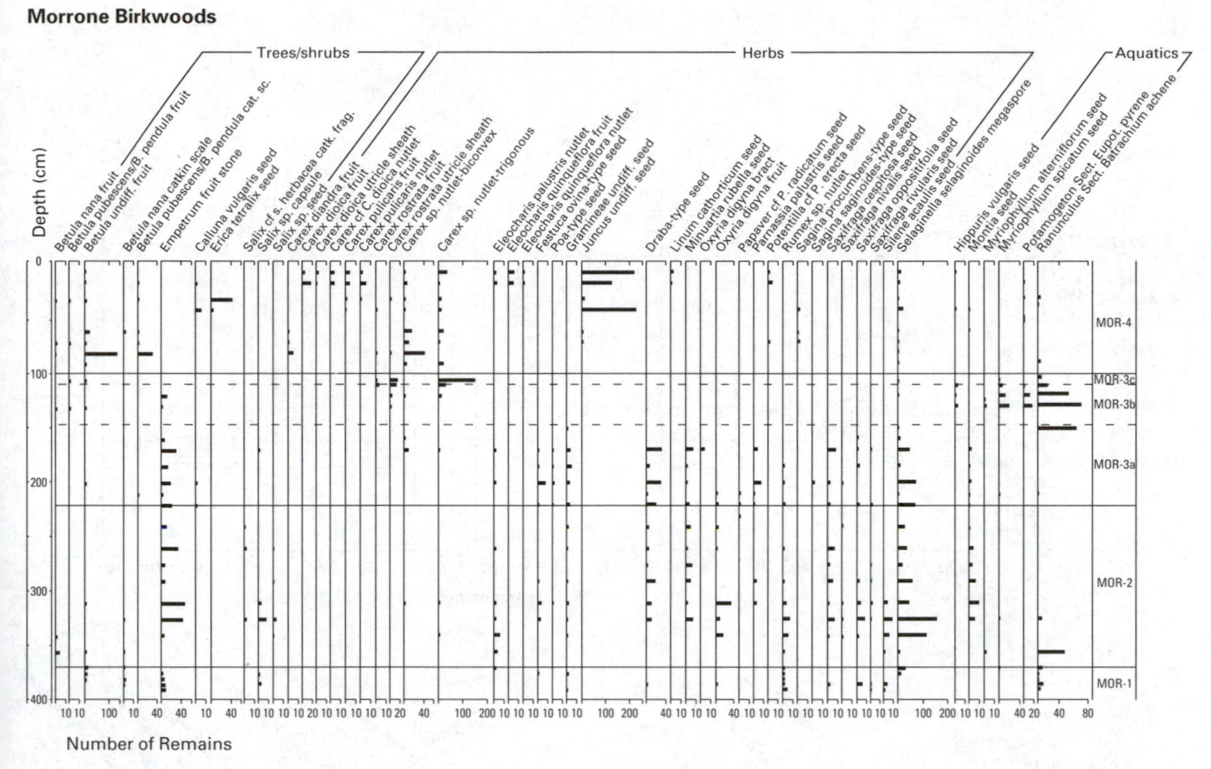

Figure 4.14 Vascular plant macrofossil diagram from a site in the Morrone Birkwoods, eastern Highlands of Scotland (from Huntley, 1994).

however, fluctuations in sedimentation rates and an insufficiently sensitive timescale can pose serious interpretative problems and, as yet, very few influx diagrams have been constructed for plant macrofossil data. Finally, a square grid graticule can be used for the quantitative estimation of macrofossil remains examined under a low power microscope, and the resulting data can be analysed statistically using, for example, weighted averages (e.g. Barber *et al.*, 1994).

4.4.5 The interpretation of plant macrofossil data

The majority of plant macrofossils found in a sediment body are derived locally (autochthonous) and are therefore of limited value in the reconstruction of regional vegetational patterns. They do, however, provide data on the composition of former plant communities growing in and around the site of deposition. Important information can often be gained, therefore, on local hydroseral developments and also on changes in the trophic status of lake waters and associated fens (see, e.g., Birks & Birks, 1980). Interpretation of the record is aided by the fact that, unlike pollen grains, a very large number of plant macrofossil remains can be identified to the species level, and therefore many of the problems arising from taxonomic imprecision which often prove so frustrating in pollen analysis are not encountered in the study of plant macrofossils. Moreover, there are some plants such as the rushes (Juncaceae) and poplar (*Populus*) whose pollen seldom survives in the fossil form, and which are therefore only occasionally represented in the pollen record. The former presence of these types may, however, be detected by their macrofossil remains. The same applies equally to those plants that are very low pollen producers.

Unfortunately, however, the occurrence of plant macrofossils in most Quaternary sediments is sporadic. Many deposits, while rich in fossil pollen, are entirely devoid of recognisable plant remains (and vice versa). Other sediments may contain plant macrofossils, but a large quantity of material may be needed in order to produce relatively few fragments of fossil vegetable matter. Moreover, although identification is frequently possible to the specific level, the abundance of diagnostic detail in different fossil remains is very variable, and while there are seeds and fruits that are relatively easy to recognise, the identification of small fragments of achene or epidermis, for

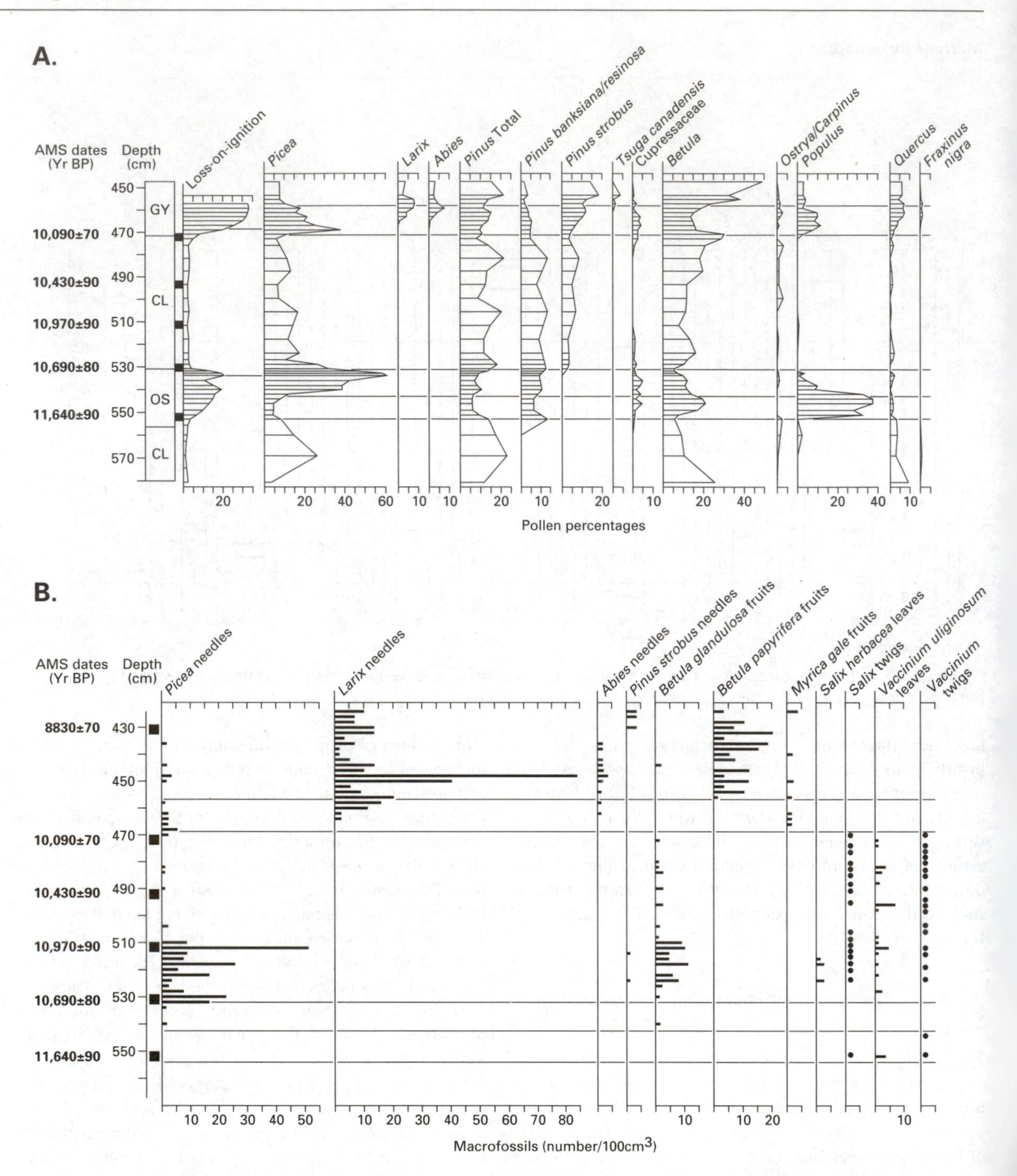

Figure 4.15 Relative pollen diagram (A) and plant macrofossil records (concentrations) (B) from a sediment sequence at Splan Pond, New Brunswick, maritime Canada. Note that the pollen of *Betula* are not differentiated in the pollen diagram, but that *Betula glandulosa* could be distinguished from *B. papyrifera* in the macrofossil record (from Mayle & Cwynar, 1995).

example, may require a great deal of work. For these reasons, plant macrofossils may only be worth studying if they are abundant, well preserved and easily extracted from the sediments in which they occur, or where they make up most, if not all, of the sediments, such as in ombrotrophic peats. In some cases, however, it may be worth examining the macrofossil content of a site where relatively few remains are preserved, either because it might help to solve a particular ecological problem encountered in pollen analysis, or because the macrofossil remains are required for radiocarbon dating (section 5.3.2).

As with pollen analysis, a proper understanding of the origins of a plant macrofossil assemblage is required before palaeoecological inferences can be attempted (West *et al.*, 1992). In fen or bog sites, macroremains tend to be almost entirely of local origin, apart from a few with very good wind dispersal such as fruits of birch (*Betula*) or seeds of sycamore (*Acer).* Hence, these records will usually be dominated by the remains of the peat-forming plants such as *Sphagnum* mosses and the cotton sedge (*Eriophorum vaginatum*), accompanied by the remains of species such as the bog myrtle (*Myrica gale*) and common heather (*Calluna vulgaris*) which are often found growing on bog surfaces. In lake sediments, however, the plant macrofossil assemblage is more diverse, for although locally derived fossils (particularly those of aquatic plants) will tend to predominate, exotic elements from outside the lacustrine ecosystem may also be present. The proportion of autochthonous to allochthonous fossil material will be determined by such factors as production and dispersal of seeds, fruits, leaves, etc., and mode of sedimentation within the lake basin.

Vegetational productivity varies considerably between species, and even between individuals of the same species, depending upon reproductive strategies (in the case of fruits and seeds) and vegetational response to environmental conditions. Watts and Winter (1966) noted, for example, that woodland shrubs and herbs tended to produce relatively few seeds, while certain trees and annual weedy plants, especially those found on the mud surfaces of lakes with markedly fluctuating water levels, may have a very high seed production. Other factors will also come into play. Many seeds and fruits will not find their way into the fossil record because they are taken for food by birds and animals, while others may be subject to attack by fungal parasites. Thus the seed population available for dispersal is what remains after such predation (Watts, 1978).

Plant macrofossil material arrives at the lake surface by a variety of pathways. Wind clearly plays a major role in moving seeds and leaves but, as with the dispersal of pollen, the nature of the vegetation surrounding the lake may have an important limiting effect. Dispersal through a woodland stand, for example, is less efficient than wind transport over bare or open ground. Sheetwash, solifluction and snowbed melts on slopes around the basin catchment may be important processes (West *et al.*, 1992), particularly during periods of reduced vegetation cover and soil instability. Birds and animals may be instrumental in transporting plant material to the lake surface either on or within their bodies. Transport of vegetative remains by inflowing streams may also be important (Holyoak, 1984), especially following periods of heavy rain, and will be particularly instrumental in carrying into the lake the seeds and fruits of such waterside species as alder (*Alnus glutinosa*) which are largely dependent on running water for their dispersal. Finally, but most importantly, there will be the input from vegetation growing around the basin littoral, and from both floating and submerged aquatics within the lake itself. This is the component that usually dominates in plant macrofossil records, together with remains of the plants growing immediately adjacent to the lake edge (see, e.g., Warner, 1990a), and studies of recruitment in present-day lakes show that the richest macrofossil accumulations tend to occur close to the lake littoral and to be dominated by the local flora.

Once the plant material arrives at the lake surface, its incorporation into lacustrine sediments is not necessarily straightforward. Some fruits and seeds will become waterlogged and sink immediately, while others may remain afloat for a considerable period of time. Seeds of aquatic species are particularly noted for their flotation characteristic with certain types deriving their buoyancy from a covering of corky tissue, while others such as the white water-lily (*Nymphaea alba*) have an aril consisting of a thin cellular bag containing many air bubbles. Those seeds which float for long periods may be washed up on the lake shore and thus disappear from the fossil record. Even those that sink to the lake floor do not necessarily become incorporated into the sediments immediately, but may be moved by turbulence or by bottom-living creatures some distance from their original point of deposition before becoming incorporated into the lake floor.

Watts (1978) has proposed the following general model for the recruitment of vegetative material to lake sediments (Figure 4.16) which, although designed for seeds, also has application to other types of plant macroremains.

1. After predation and parasitism has been accounted for, the surviving seed population is dispersed.
2. Wind, particularly strong or violent winds, carries seeds to lake surfaces. This process is most effective in the case of trees where seeds are launched from a height. Alternatively, some seeds are carried to lakes by streams and surface runoff during periods of high rainfall. Birds and animals play a minor role in transporting soft fruits.

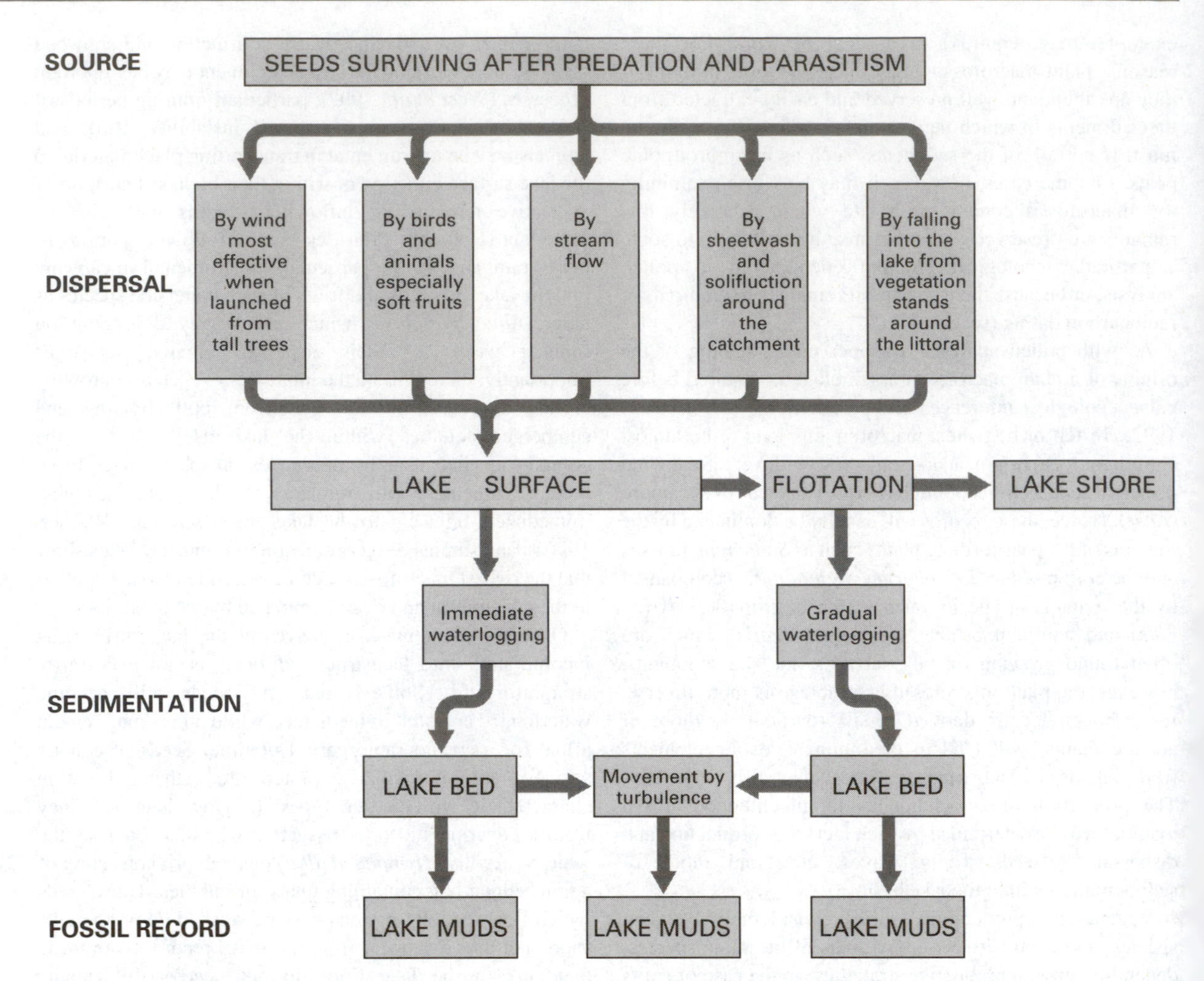

Figure 4.16 General model of seed recruitment to lake sediments (modified after Watts, 1978).

3. Seeds that are not waterlogged remain afloat and are carried into shallow water by wind and wave action.
4. Seeds of some species become waterlogged preferentially and sink, while others succeed in resisting waterlogging altogether and are washed ashore.
5. The waterlogged seeds are moved along the lake bed by turbulence until they settle out by gravity in the coarser marginal sediments or are trapped near the margin by submerged plants.

The model is speculative and was designed to stimulate experimental work involving trapping such as that carried out in fluviatile sediments in tundra regions (Holyoak, 1984; West *et al.*, 1992). A number of studies have demonstrated that macroremains in lake sediments were deposited predominantly within the zone of rooted aquatics, and that the fossils consisted almost entirely of plant material from the aquatics themselves (e.g. Warner, 1990a). The occurrence of upland plants from the basin catchment was both accidental and unpredictable. Clearly, further studies of this type are required if we are to gain a proper appreciation of the origins of the macrofossil assemblages in Quaternary lake sediments.

Although the foregoing discussion has been concerned with lake sites, many of the points apply equally to plant macroremains in riverine sediments. In fluvial deposits, however, a further complicating factor is the often frequent occurrence of plant macrofossil material from different time periods that has been incorporated into a single assemblage. This is a problem common to all fossil assemblages and has already been considered with reference to reworked or secondary pollen. Mixed assemblages of plant fossil materi are occasionally encountered in lake sediments where, for example, there have been marked fluctuations in the levels of lake waters and intermittent erosion of the shoreline, but they are especially common in river terrace deposits where they can often be highly misleading to the unwitting palaeoecologist. A good example is provided by the discovery of nuts of the hornbeam (*Carpinus betulus*) at the base of terrace gravels of the last glacial period in the Fens of eastern England. Careful analysis of the site revealed that the fossils were of secondary origin and had been derived from sediments deposited at the height of the preceding interglacial (Sparks & West, 1972). Unlike secondary pollen, it is not always possible to detect reworked plant macrofossil material by signs of physical and chemical deterioration, and often a meticulous evaluation of the assemblage is necessary in order to isolate such exotic components.

Overall, the detailed investigation of plant macrofossils requires many more hours of work than does pollen analysis and, as the results tend to be largely site-specific, markedly fewer plant macrofossil studies have been undertaken since the widespread adoption of pollen analysis as a palaeoenvironmental tool. This is unfortunate, for the analysis of plant macrofossils should be seen not as an alternative to pollen analysis in the reconstruction of former vegetation, but as an adjunct. The two techniques can be mutually supportive and can provide new insights into flora, vegetation and landscapes of the Quaternary, as well as providing information on local site developments. When used in conjunction, therefore, plant macrofossil analysis and pollen analysis offer a more secure basis for palaeoecological inference than either technique used in isolation.

4.4.6 Applications of plant macrofossil studies

4.4.6.1 Resolving taxonomic limitations of pollen records

Since pollen can travel considerable distances, it may be difficult to establish the percentage value of a taxon that represents local presence of the plant. This is particularly problematic when considering the records of pollen that are known for their ability to travel considerable distances, such as *Pinus* (pine) or *Olea* (olive). Local presence can therefore only be established on the basis of plant macrofossil evidence. For example, Younger Dryas sediment records from the Pyrenees are invariably characterised by relatively high values of *Pinus* pollen, but it is only in the lower altitude sites that pine stomatal guard cells are found (Reille & Lowe, 1993). This suggests that the higher sites lay above the tree-line at that time, and received pollen by wind transport from pine stands on the lower slopes. A second area where plant macrofossil evidence has proved a useful adjunct to pollen analysis is in the differentiation between plant types where pollen grains may be almost identical. Palynologists have encountered considerable difficulties, for example, in distinguishing pollen of tree birch from that of dwarf birch (*Betula nana*), a distinction which is crucial for understanding the pattern of vegetation colonisation following the last glaciation. Data from sites in Norway, the British Isles and Canada (e.g. Figure 4.15) have enabled separate episodes of birch succession to be established, an initial phase dominated by dwarf birch, followed by a later phase of tree-birch expansion (e.g. Birks, 1993; Huntley, 1994; Mayle & Cwynar, 1995).

4.4.6.2 Leaf stomatal density and environmental change

Controlled laboratory experiments on leaves up to 200 years in age have established that the **stomatal density** on leaves of certain plants varies with CO_2 concentration (Beerling & Woodward, 1993). Research on fossil leaves of *Salix herbacea* (dwarf willow) from the last cold stage and from the Lateglacial period has shown marked temporal variations in stomatal density which are considered to reflect atmospheric CO_2 variations (Beerling *et al.*, 1992). This represents an exciting new development in the use of plant macrofossils, for the results of this research suggest that physiological responses of leaves may be used to reconstruct patterns of past CO_2 variations which complement data derived from ice-core records (section 3.11 and Chapter 7), and hence contribute to global models of environmental change (Beerling *et al.*, 1993; Beerling & Chaloner, 1993).

4.4.6.3 Peat macrofossils and palaeoclimate

The stratigraphy of ombrotrophic bogs is closely linked to variations in local hydrological conditions (bog surface wetness) which, in turn, are strongly influenced by regional climatic conditions (see section 3.9). Although the relationship between peat stratigraphy and climate has been investigated in detail (e.g. Barber, 1981; Clymo, 1991), the data have often proved difficult to quantify. It has now been demonstrated, however, that the composition and state of preservation of plant macrofossils in ombrotrophic peats show high-frequency variations which appear to reflect short-term changes in climate wetness (Barber *et al.*, 1994). These may be reflected, for example, in changes in the proportions of several species of the bog moss *Sphagnum*, since each species has a different tolerance to bog surface

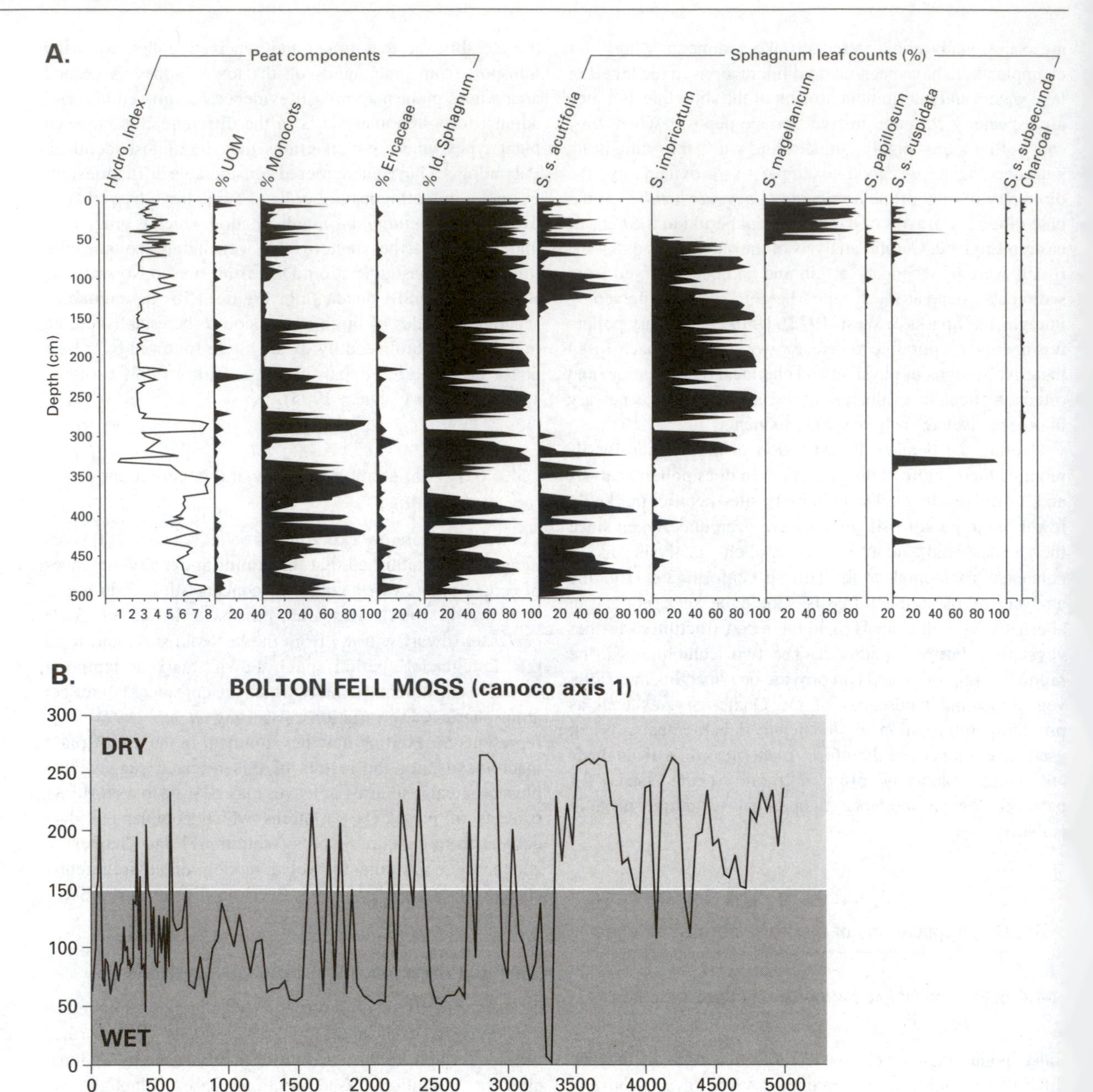

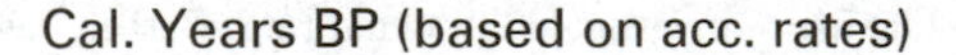

Figure 4.17 Analysis of *Sphagnum* leaves in a Holocene peat record from Bolton Fell Moss, northwest England (A) and spectral analysis of these records (B) showing the dominance of an 800-year cycle over the last 7000 years (from Barber *et al.*, 1994).

wetness (Figure 4.17) or in the relative importance of macrofossils of plants requiring drier bog conditions, such as *Erica*. These changes can be quantified using weighted averaging methods, and the results show a very strong correlation with known climatic changes in northern England over the last 1000 years. Moreover, spectral analysis similar data extending through much of the Holocer indicates a cyclic periodicity of about 800 years, which thought to reflect ocean-driven climatic influences (Barber *al.*, 1994).

4.5 Fossil insect remains

4.5.1 Introduction

Fossil insects are often abundant in a wide range of Quaternary deposits. Typically these include sediments that accumulated in ponds or near lake margins, in backwaters of rivers, in peats or indeed in any depositional environment conducive to the preservation of plant debris. Many different orders of fossil insects have been observed in Quaternary deposits, including bugs (Hemiptera–Homoptera), two-winged flies (Diptera), caddis flies (Trichoptera), stink bugs (Pentatomidae), seed bugs (Lygaeidae), leaf hoppers (Cicadelidae), bees, ichneumons (Hymenoptera), dragonflies (Odonata), non-biting midges (Chironomidae), water striders (Gerridae), shore bugs (Saldidae), water boatmen (Corixidae), backswimmers (Notonectidae) and beetles (Coleoptera) (Berglund, 1986), while members of the ant family (Formicidae) have also been investigated. Although not technically insects, the mites (Arachnida), especially oribatid mites, have been considered together with insect assemblages, and are particularly important components of the soil fauna of arctic and alpine ecosystems (Elias, 1994).

It is the remains of beetles, however, that are the best known of this array of fossils (Figure 4.18). They are the most commonly observed group because they are robust and commonly display brilliant and often iridescent colouring of blues, greens and golds, and consequently their presence in Quaternary sediments has long been a source of fascination to laymen, naturalists and entomologists alike. More recently, however, and largely through the work of Professor G.R. Coope, fossil Coleoptera have proved to be a powerful new tool for the investigation of Quaternary environments, particularly climatic change. Although fossils of the other insect groups have occasionally proved useful in palaeoecological studies (e.g. Crosskey & Taylor, 1986; Elias, 1994), in general they are more difficult to deal with. They are rarely well preserved due to their delicate structures, some of the groups are difficult to identify beyond generic level, and their present-day distributions are less well known than those of the Coleoptera. As a

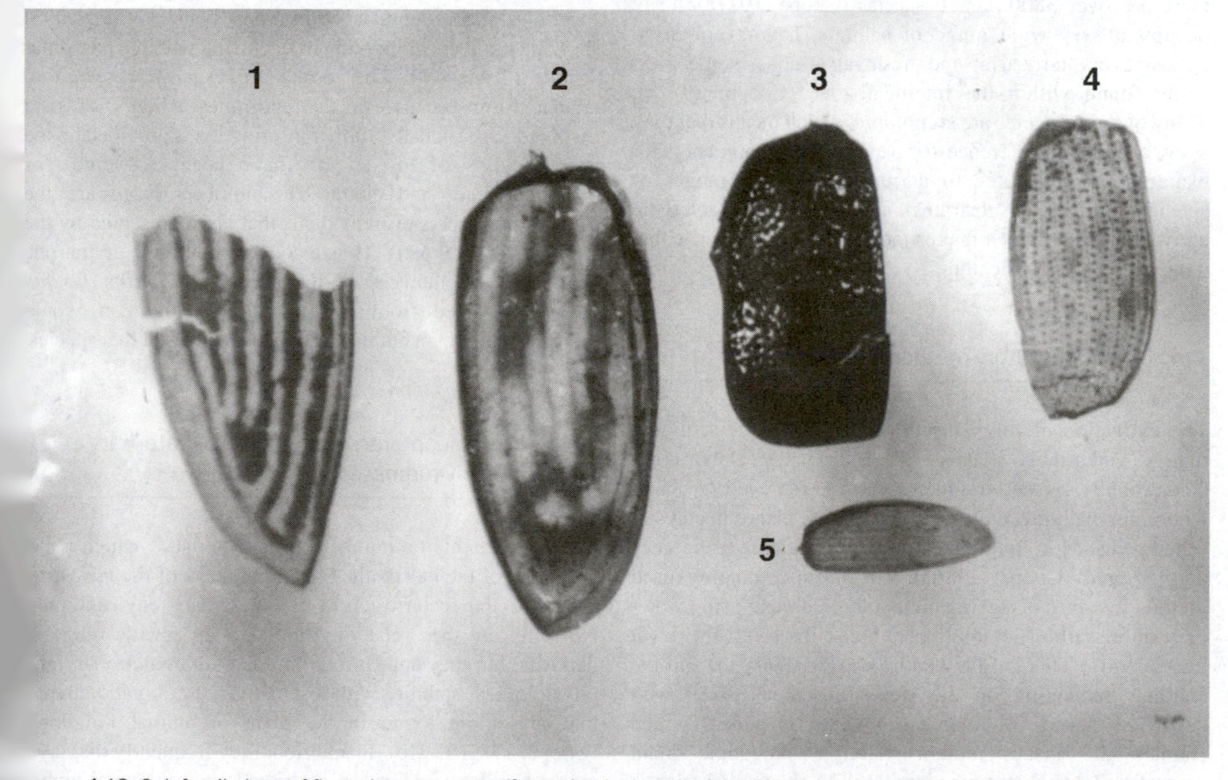

gure 4.18 Sub-fossil elytra of five coleopteran taxa (for scale, the length of the *Olophrum* elytron is c. 1.5 mm): (1) *Potamonectes* *riseostriatus* (DEG.), (2) *Aphodius* sp., (3) *Olophrum* sp., (4) *Helophorus* sp., (5) Hydrophilidae indet (photo: Adrian Walkling).

consequence, the remainder of this section is devoted mainly to the Coleoptera, although the Chironomidae are also discussed briefly in the next section, since their potential as palaeoclimatic indicators is now becoming apparent.

4.5.2 Coleoptera

Coleopteran remains are usually the most diverse group of insect fossils in Quaternary deposits and often they are the most abundant. The chitinous exoskeletons of which they are composed are highly robust and contain sufficient structural detail to permit many fossils to be identified to species level (Figure 4.19). They have been collected and studied by entomologists in many parts of the world and there is now a considerable body of knowledge in the form of atlases and monographs which summarise their distribution and ecological associations. The order Coleoptera is one of the largest in the animal kingdom, accounting for 25 per cent of all known species of organisms. They form the most important insect order, with more than 300 000 known species (Elias, 1994). About 1500 new species are described each year and within Britain alone there are over 3800 named species (Coope, 1977a). They occupy a very wide range of habitats, having colonised almost every terrestrial and freshwater niche, some even being found within the intertidal zone (Coope, 1977b). Many of these species are **stenotopic**, which means that they show a marked preference for particular environments (for example those adapted to narrow temperature ranges or specific habitats or substrates), and it is this characteristic above all others which makes the Coleoptera such valuable palaeoecological indicators.

4.5.3 Laboratory methods

The extraction of fossil insect remains from the sediment matrix invariably takes place in the laboratory. Occasionally, insect fragments can be removed by hand where, for example, they occur on bedding planes in clays or felted peats. More frequently, however, flotation techniques are required (Coope, 1986a). The most commonly used method involves disaggregation of the sediment using water or sodium carbonate solution to break the sediment down into a slurry. This is followed by sieving (300 μm) and the residues remaining on the sieves are then mixed with kerosene (paraffin), after which water is added, which enables the insect remains, along with some plant macrofossils, to float to the surface. The floating fraction is decanted, washed and sorted in alcohol under a low-power microscope. The insect fossils range in size from less than 1 mm to several centimetres and careful examination may be necessary at this stage in order to ensure that very small specimens are not overlooked and that the subsequent collection is not heavily biased in favour of the more conspicuous species.

The fossil remains are then gummed onto cards or stored in dilute (*c.* 20 per cent) alcohol and examined under a microscope. Electron microscopy may be necessary where examination of very fine structural detail is required for confident identification of some species. Since the entire insect fossil is rarely recovered from the sediment body (commonly identification is made on an elytron (wing cover) or thorax, or even a small fragment of an elytron), keys to identification are of limited value and confident identifications require careful comparisons with modern specimens. A comprehensive comparative collection of modern beetles is therefore essential for palaeoecological research. Some parts of the fossils possess few diagnostic features, but in many cases the heads, thoraces, elytra and genitalia (particularly in the male specimens) display a wealth of useful characters that enable specific determinations to be made (Figures 4.18 and 4.19).

Data from fossil insect analysis are usually presented in the form of a species abundance list (e.g. Walker *et al.*, 1993) showing numbers of individuals occurring within a particular sample (Table 4.4). The results are usually presented in tabular form because of the enormous amount of data involved. Occasionally, further information is provided on the specific parts of insects that have been recovered, for example, heads and elytra. The numbers listed are the minimum numbers of individuals that are represented by the recorded skeletal parts. Thus a collection of three elytra, one head and two thoraces of the species *Olophrum fuscum* (Grav.) obtained from one stratigraphic horizon would indicate a minimum number of two individuals of that species

4.5.4 Coleopteran analysis and Quaternary environments

Coleoptera exhibit a number of characteristics which make them one of the most valuable components of the terrestrial biota for the reconstruction of Quaternary environments. Not only are they relatively abundant in a wide range of deposits, but they appear to combine both evolutionary and physiological stability with a sensitivity to climatic change that is seldom found in the plant or animal kingdom (Coope, 1987, 1994b). In addition, they frequently display a marked preference for very restricted environmental niches or conditions.

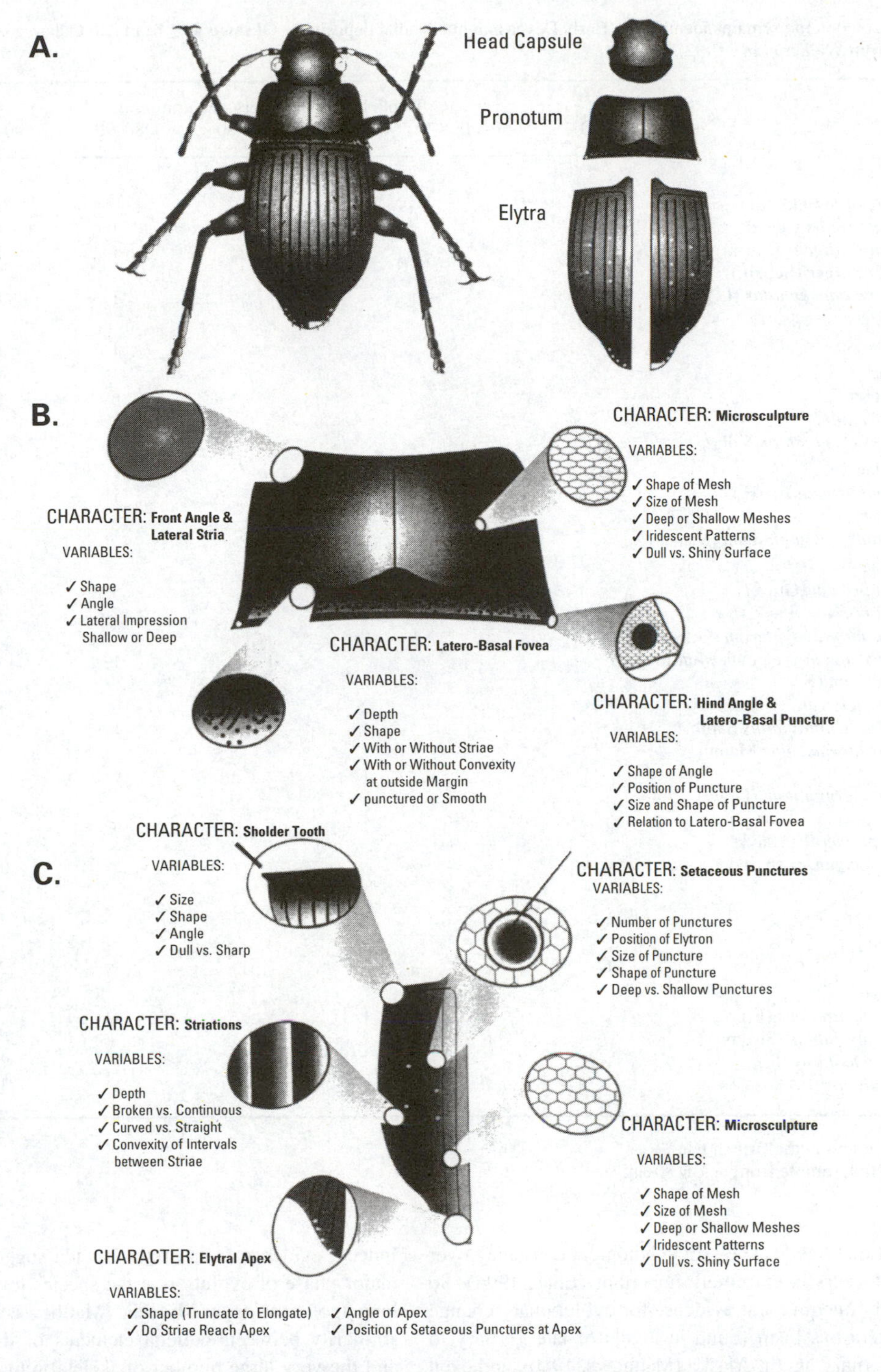

Figure 4.19 Generalised drawings of coleopteran sclerites frequently preserved as Quaternary fossils, showing a range of diagnostic features used in fossil identification: A: dorsal surface of a ground beetle (Carabidae); B: pronotum of ground beetle; C: ground beetle elytron (from Elias, 1994).

Table 4.4 Coleopteran remains identified in Early Devensian interstadial deposits (~OI stage 5c?) from Allt Odhar, a site in northern Scotland (from Walker *et al*., 1992).

	Depth below top of organic unit (cm)						
	46–50	50–60	60–70	70–80	80–90	90–95	COMP†
Carabidae							
Diacheila polita Fald.*						1	
Patrobas assimilis Chaud.					1	1	3
Pterostichus diligens (Sturm)					1	1	2
Pterostichus niger (Schall.)							1
Calathus melanocephalus (L.)					1	1	
Dytiscidae							
Hydroporus sp.					1	1	1
Hydraenidae							
Limnebius sp.					6	1	1
Helophorus nubilus F.					1		1
Helophorus cf. *glacialis* Villa*							1
Hydrophilidae							
Coelostoma orbiculare (F.)					1	1	1
Sphaeridium sp.							1
Megasternum boletophagum (Marsh.)							1
Staphylinidae							
Olophrum fuscum (Grav.)				1	5	1	7
Olophrum boreale (Payk.)*			1	1	17	1	15
Euconecosum brachypterum (Grav.) and/or *Euconecosum norvegicum* Munst.*			2		11		12
Acidota crenata (F.)					1		1
Lesteva longelytrata (Goeze)			2		5	1	7
Boreaphilus henningianus Sahlb.*			1		1		1
Euasthetus laeviusculus Mannh.							1
Stenus sp.			1		9	1	2
Lathrobium terminatum (Grav.)							1
Tachinus sp.			1				1
Gymnusa brevicollis (Payk.)					3		
Aleocharinae gen. et sp. indet.					5	1	1
Elateridae							
Hypnoidius sp.							1
Scarabaeidae							
Aphodius sp.	1		2	1		2	1
Curculionidae							
Otiorhynchus arcticus (F.)			1				
Otiorhynchus dubius (Strom.)			1				
Notaris bimaculatus (F.)							1
Notaris aethiops (F.)		1	2	4	1	1	2

* No longer found in the British Isles.
† COMP = Bulk sample from below 80 cm.

The fact that beetles show morphological constancy over millions of years is extremely important (Elias, 1994). So far, the only unequivocal evidence for evolutionary change in Coleoptera has been found in fossils of late Tertiary to early Quaternary age in Alaska (Matthews, 1980), and even in these cases the differences are so small as to be of a comparable order of magnitude to present-day racial differences between individual species (Coope, 1977b). Indeed, evidence from arctic Canada suggests that the last major phase of evolution at the species level took place as long ago as the upper Miocene (Matthews, 1976). The close similarity between skeletal elements of living Coleoptera and the very large number of skeletal remains that have so far been examined from British Quaternary sequences has led Coope (1977a, 1987, 1994a) to the conclusion that the Coleoptera demonstrate a remarkable morphological stability throughout at least the last million years and most probably during the entire Quaternary period. It would seem, therefore, that there are good grounds for believing that the British Quaternary fossil Coleoptera represent exactly the same species as those in living assemblages.

Equally significant is the fact that ecological preferences of most coleopteran species do not seem to have changed to any great extent during the Quaternary. This is obviously more difficult to establish, but the available evidence tends to suggest that, in the great majority of cases, species of beetles are found in similar associations in both fossil and living assemblages. In Britain, for example, the warm-adapted insect assemblages of one interglacial period are essentially similar to those of others, even though the interglacials may differ in age by hundreds of thousands of years (Jones & Keen, 1993). The beetle assemblages of various cold stages also have a great number of species in common. Moreover, independent palaeobotanical and geological evidence indicates that most fossil beetle species were associated with similar types of environment to those that they occupy today. It does seem, therefore, that physiological stability in Coleoptera accompanied morphological constancy throughout the greater part of the last two million years. There are, however, some exceptions to this. For example, *Hypnoides rivularis* (Gyll) is a species which now has its southern limit across Fennoscandia at about 60°N, and yet has been found at a number of sites in Britain, including the Lateglacial Interstadial of the Windermere profile, in association with species of a more temperate aspect (Coope, 1977a). *Timarchia goettingensis* (L.), on the other hand, is today widely distributed in Europe south of latitude 60°N, yet it is encountered in Middle Devensian deposits at Upton Warren in central England in association with many species now found in tundra environments. In both cases the beetle species may well have changed their ecological tolerances and could therefore be misleading as palaeoenvironmental indicators if found in isolation. In most Quaternary insect assemblages, however, careful examination of the total assemblage will normally serve to isolate such aberrant species.

4.5.4.1 Habitat preferences

Any assemblage of fossil insect remains will contain species from a variety of local habitats. Botanical factors, soil type, microclimatic environment, hydrological conditions and chemical variations will all restrict the distribution of insects at the local scale (Elias, 1994). To a palaeoecologist who is interested in the environmental history of a particular site, therefore, it is important that the range of habitats represented by the fossil assemblage is identified and, as far as is possible, quantified. Some beetle species are substrate dependent, such as *Bembidion obscurellum,* which lives on dry, sandy soils, *Dyschirius globosus* which requires moderately humid soils with clay, sand or peat and a sparse vegetation cover, and *Bembidion schueppeli,* which is restricted to river banks. A large number of beetles are associated with aquatic habitats. Thus actively flowing water is indicated by *Esolus, Limnius volckmari* and *Ochthebius pedicularius* while *Potamonectes depressus (elegans)* and *Halyplus obliquus* live in clear ponds with sandy or silty bottoms. Other beetles indicate the presence of particular plants or other animals upon which they depend for food. Most of the staphynilid beetles, for example, are predators, living on small arthropods and worms in leaf litter. A profusion of dung beetles in an assemblage would indicate the local presence of mammals. A number of phytophagous Coleoptera feed only on reeds, such as several species of the *Donacia* genus which live on *Carex, Scirpus, Sparganium* and tall marsh grasses. *Hydnobius puntatus* feeds on fungal hyphae, *Hypera postica* is dependent upon various leguminous plants, such as *Medicago, Melilotus* and *Trifolium*, and *Simplocaria semistriata* feeds exclusively on moss.

Beetle assemblages can therefore provide valuable information on a diverse range of contemporaneous habitats or co-dependent biota, and may provide environmental insights that are difficult to obtain from other lines of evidence. For example, large numbers of the genus *Bledius* were found among the earliest colonisers of the surface of the lower cover sands (deposited during the last cold stage) in The Netherlands in deposits containing numerous remains of the blue-green alga *Gleotrichia* (Van Geel *et al.,* 1989), an important nitrogen-fixer. The algae may have provided an important food source for these beetles at a time when few other organisms had succeeded in colonising the area. Large numbers of *Bledius* recovered from early Lateglacial sands in a site in eastern England may therefore indicate that nitrogen-fixing algae played an important role in the colonisation process at that site also (Walker *et al.,* 1993). Some beetle species are obligate dwellers in tree bark or leaf litter, and fossil remains of such beetles have been used to augment palynological evidence for the presence of trees in parts of North Italy during the last glacial–interglacial transition (Ponel & Lowe, 1992). Fossil insect assemblages preserved in packrat middens in the deserts of the southwest USA contain a mixture of temperate and desert species not found together in any part of North America today (Elias, 1994). This may indicate a late Pleistocene climatic régime for which there is no modern analogue. These examples show that a knowledge of the autecology of beetles can provide useful additional evidence for Quaternary palaeoenvironmental reconstructions.

4.5.4.2 Palaeoclimatic inferences based on coleopteran assemblages

The most important factor that has governed the distribution of most insect species during the Quaternary has been climate, particularly thermal conditions (Coope, 1990). Modern beetle distribution maps show that the geographical range of many species corresponds with well-defined climatic zones (Figure 4.20) and especially with summer temperature thresholds. Those insect species whose distributions are narrowly restricted are termed **stenotherms**, while those that can tolerate a broader range of climatic conditions are termed **eurytherms**. The former are much more important in palaeoclimatic research, since they enable precise inferences to be made about former temperature régimes. The acute sensitivity of beetles to temperature variations is reflected in data from The Netherlands which show that the relative abundances of species in the modern beetle fauna have changed in response to climate on a decadal timescale since 1890 (Hengeveld, 1985).

Clearly, however, there are problems in utilising fossil beetles as climatic proxies. It can never be established for certain, for example, that an insect species has colonised the entire climatic range to which it is suited, nor that past distributions were entirely in equilibrium with the prevailing climatic conditions. On the other hand, Coope (1977b) has argued that the enormous scale of the changes in geographical distribution of species in response to climatic change during the Quaternary, and the rapidity of many of these changes, indicate that Coleoptera must have been able to colonise new available habitats extremely quickly. In the majority of cases where the range limit of a coleopteran species coincides with a climatic boundary, this relationship has been used to derive quantitative palaeotemperature estimates since the early days of Quaternary palaeoentomological research (Coope, 1959).

The species in fossil coleopteran assemblages represent a variety of different climatic ranges. For example, assemblages from Britain may consist of species that fall into the following eight categories (Coope, 1987):

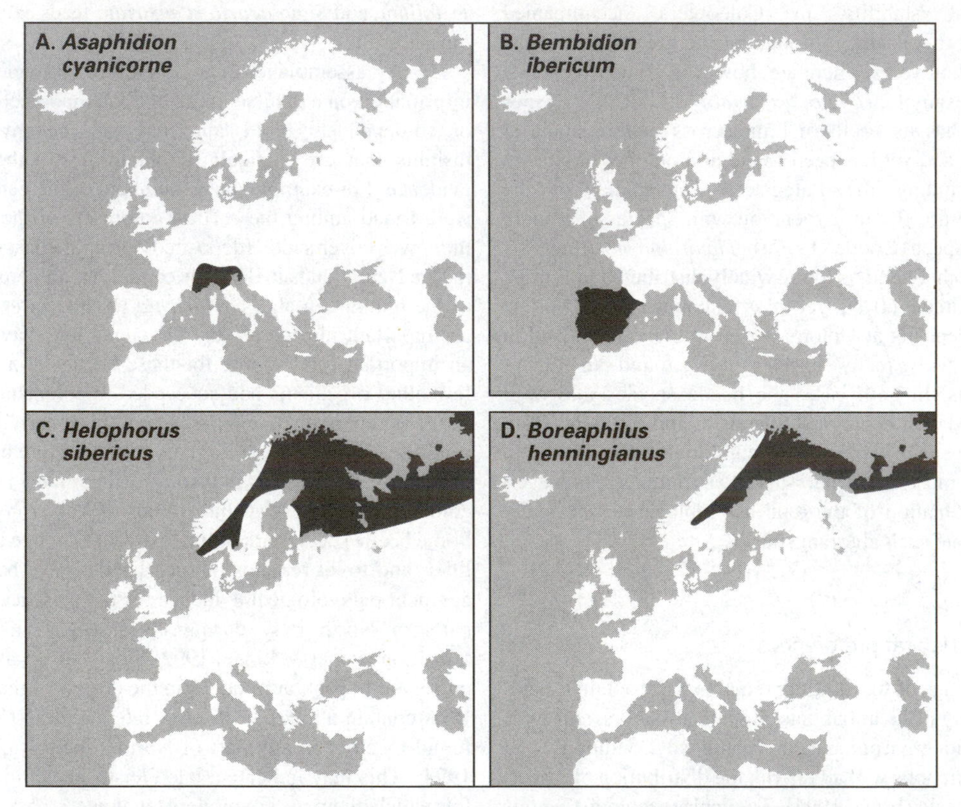

Figure 4.20 Present-day European distributions of Coleoptera found in Lateglacial deposits (*c.* 14.5–10 ka BP) at Glanllynnau, North Wales. A. *Asaphidion cyanicorne* Pand. B. *Bembidion ibericum* Pioch. C. *Helophorus sibericus* Mot. D. *Boreaphilus henningianus* Sahib (after Coope & Brophy, 1972).

A. Southern European species
B. Southern species whose normal ranges just fail to reach Britain
C. Southern species whose normal ranges are south of central Britain
D. Widespread species whose normal ranges are north of central Britain
E. Boreal and montane species whose normal ranges extend down into the upper part of the coniferous forest belt
F. Boreal and montane species whose normal ranges are above the tree line
G. Eastern Asiatic species, some of which also range into North America
H. Cosmopolitan species with very wide geographical ranges.

An interglacial assemblage may be dominated by fossils from categories A, B and C with representatives of group H, while cold stage faunas may have a high representation of categories D, E and F with some elements of H. The problem is how to convert such complex assemblage data into statistics representing the contemporaneous regional macroclimate.

Early attempts to derive macroclimatic data from fossil assemblages employed the **range overlap method**, in which the modern distributions of species represented in a fossil assemblage are plotted (Figure 4.20) and the zone of overlap of the ranges are identified (e.g. Coope, 1959). Modern climatic statistics obtained from meteorological stations that lie within the zone of overlap, such as mean annual temperature or the annual temperature range, can then be used to provide a quantitative estimate of the regional macroclimate that prevailed at the time the fossil taxa coexisted. In practice the method works best when a large number of stenothermic species are represented in an assemblage, which may enable fairly precise palaeoclimatic estimates to be inferred. The main difficulty with this approach, however, is that it is possible that a species does not occupy its full potential geographical range. Furthermore, some taxa may temporarily coexist during a transitional phase of adaptation to new climatic conditions, and the resulting mix of fossils may therefore be largely an ephemeral one, with no modern analogue (a **non-analogue assemblage**). This problem may be particularly acute in the interpretation of fossil insect assemblages derived from sediments that accumulated during an episode of abrupt climatic change, since insects appear to have responded much more rapidly than other biota to changing climatic conditions. One of the challenges of Quaternary palaeoecology, therefore, is to distinguish such temporary associations from those representing more stable climatic episodes, and, while meeting this challenge, to note that non-analogue faunas do not necessarily imply non-analogue environments.

In an attempt to avoid the errors that may arise from the use of the indicator species approach, the **Mutual Climatic Range (MCR)** method has been developed to obtain palaeotemperature estimates from beetle records (Atkinson *et al.*, 1986a, 1987). This is based on the pioneering work of Iversen (1944) and Grichuk (1969), and is an extension of the range overlap method, but it employs the ranges of *all* of the taxa included. Moreover, because of the considerable body of statistical information that is involved, a computer is necessary for the storage and manipulation of the data. Modern distribution maps are first obtained for as many as possible of the species in the fossil assemblage, and the climatic range of each beetle type is then established using contemporary meteorological data. The two most important variables govering beetle distributions appear to be the temperature of the warmest month (T_{MAX}) and the temperature range between the warmest and coldest months (T_{RANGE}), the latter providing an index of seasonality. By knowing the distribution in terms of T_{MAX} and T_{RANGE}, the geographical range of each species may be plotted in '*climate-space*', and for each species a '*climatic envelope*' is thus produced (Figure 4.21). For any fossil assemblage, therefore, the **mutual climatic range** can be determined from a computer-generated plot of the climatic parameters relating to each beetle in the assemblage. From these plots, the values of T_{MAX}, T_{RANGE} and T_{MIN} (temperature of the coldest month) can be obtained, and these constitute the 'best estimates' of the mutual climatic conditions within which the particular mix of fossils formerly coexisted. The method is most successful (i.e. produces the narrowest range estimates) where an assemblage contains a large number of species, and where a number of these are obvious stenotherms.

The advantages of the MCR method over the indicator species approach are that it avoids subjective interpretations and possible bias, as well as over-generalisation from the use of geographical overlays. Moreover, geographical range limits are often too broad, and cannot take into account such factors as altitude, oceanicity, microclimatic variations and so on. The MCR approach ignores geographical location, and focuses entirely on climatic parameters governing species distributions. Hence a complex geographical distribution may be reduced to a simple climate plot, reflecting the fact that the often diverse geographical locations in which a species occurs may, in fact, have common climatic characteristics. Also, and most importantly, it does not really matter if the species does not occupy its full (potential) geographical range, so long as it

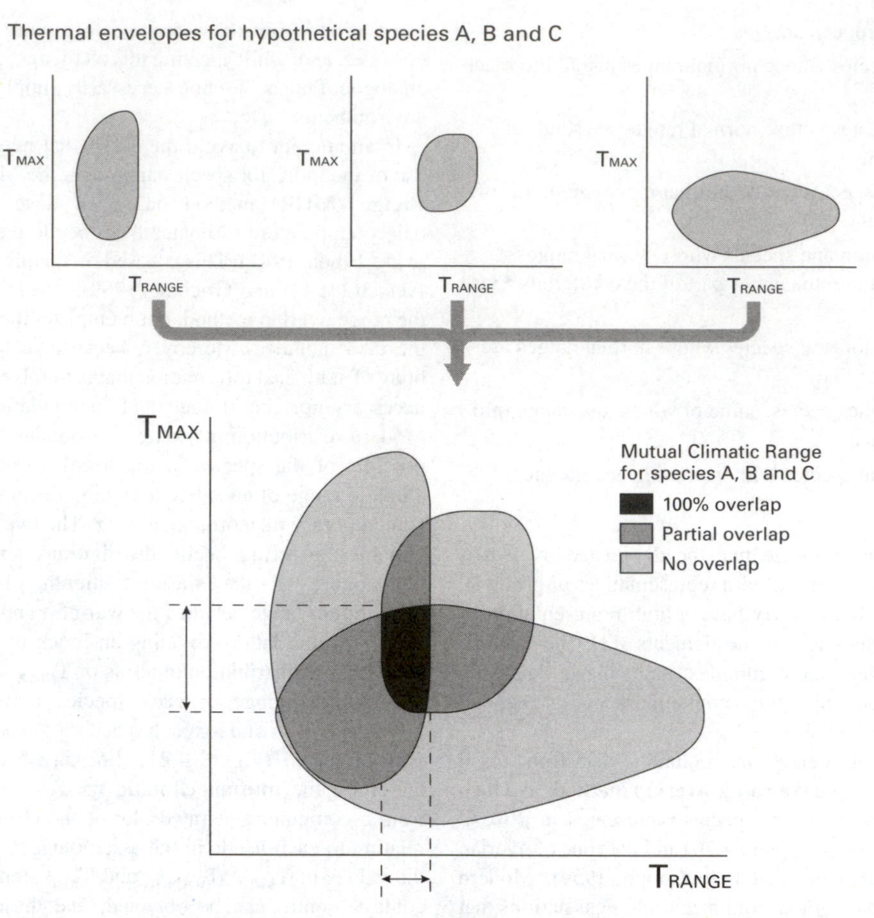

Figure 4.21 Schematic representation of the Mutual Climatic Range method of quantitative temperature reconstructions (courtesy of Adrian Walkling).

reaches potential boundaries in sufficient places. The MCR method has been widely used to reconstruct climatic conditions in Europe during the last glacial–interglacial transition (Ponel & Coope, 1990; Lemdahl, 1991; Guiot *et al.*, 1993; Walker *et al.*, 1993; Hammarlund & Lemdahl, 1994) and has also been employed to generate palaeoclimatic data from Quaternary herpetofaunal (reptile) and palaeobotanical records (Sinka, 1993).

Temperature reconstructions based upon beetle MCR data obtained from sites in northwest Europe have played a very important role in recent attempts to reconstruct the magnitudes, timing and rates of climatic changes during the last glacial–interglacial transition (*c.* 14–9 ka BP). Beetle MCR data (Figure 4.22) suggest an abrupt thermal rise at about 13 ka BP, with mean July temperatures in Britain possibly rising by as much as 1°C per decade, culminating in a 7°C rise. The corresponding rise in the mean temperature of the coldest month may have been as much as 20°C (Atkinson

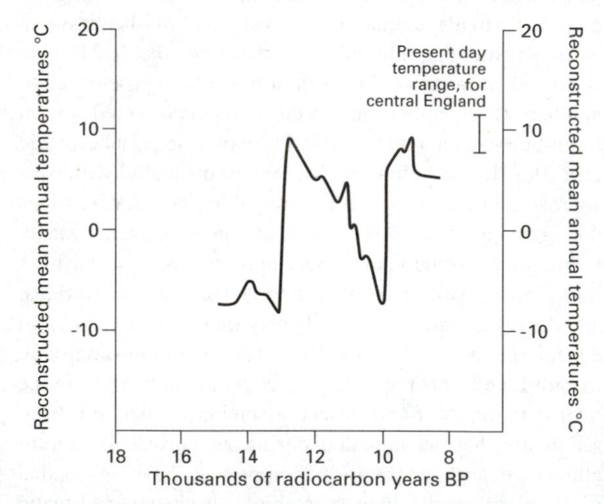

Figure 4.22 Reconstructed mean annual temperatures in England for the period 14.5–8 ka BP based upon MCR analysis of beetle remains (courtesy of G.R. Coope).

et al., 1987; Walker *et al.*, 1993) and a thermal maximum was achieved between 13 and 12.5 ka BP (Lowe *et al.*, 1994; Walker, 1995). This was followed by a period of gradual climatic cooling, possibly associated with minor climatic oscillations, between about 12.5 and 11 ka BP. A much sharper fall in temperature is inferred for 11 ka BP, at the start of the period of severe cold known as the Younger Dryas, which was terminated by an abrupt thermal increase at *c.* 10 ka BP. This interpretation, particularly with respect to the early part of the record (13–12 ka BP), was initially considered to be in conflict with climatic reconstructions based upon pollen data (see, e.g., Coope, 1977a; Coope & Pennington, 1977), since the maximum development of terrestrial plants was typically dated to about 12–11 ka BP. This discrepancy is now widely considered to reflect the different *response rates* of beetles and terrestrial plants to climatic warming at the close of the last cold stage (e.g. Walker *et al.*, 1993; Walker, 1995). The beetle MCR curve is now strongly supported by other climatic reconstructions based upon Greenland ice-core evidence, marine micropalaeontological records from the northeast Atlantic and isotopic evidence from lake sequences in Switzerland (e.g. Walker *et al.*, 1994; Walker, 1995; Lowe *et al.*, 1994; Lowe & NASP Members, 1995; Lowe *et al.*, 1995a), and beetle MCR data are being used increasingly to reconstruct temperature gradients over northwest Europe during the last glacial–interglacial transition (Coope & Lemdahl, 1995; Lowe *et al.*, 1995b). These aspects are discussed in more detail in section 7.6.

4.5.4.3 Insect fossils and archaeology

Coleopteran assemblages, in common with a number of other proxy biological indicators, clearly record the effects of human impact on the landscape of Europe during the mid-Holocene (Buckland, 1979). Extermination of some beetles followed the clearance of trees upon which they depended, while other species that are closely associated with cereals, pasture land or cattle dung provide a basis for inferring past agricultural and pastoral activities (e.g. Buckland, 1981; Panagiotakopulu & Buckland, 1991; Elias, 1994).

The analysis of coleopteran assemblages has proved particularly valuable in the interpretation of occupation sites (Kenward, 1982, 1985) since certain beetles are associated with stored products (grain and other foods), some invade tanneries, sewers, refuse pits, stables, etc., while others are human parasites or are predators on human and other corpses (Figure 4.23). The study of **archaeoentomology**, however, is

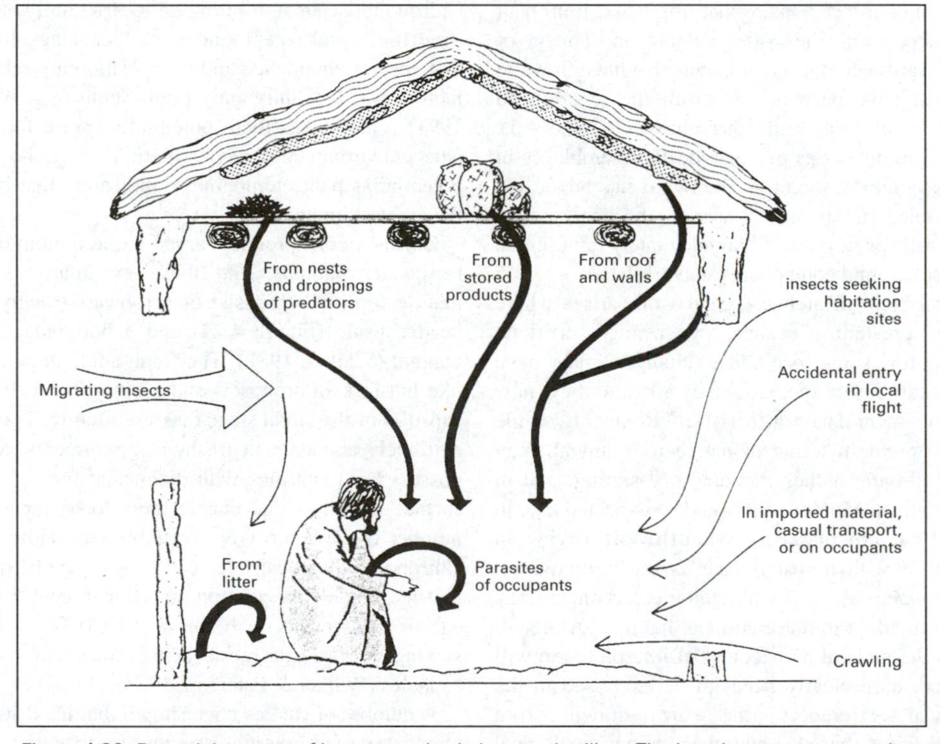

Figure 4.23 Potential sources of insect remains in human dwellings.The broad arrows represent insects originating from within the dwelling, and slim arrows those originating from outside (after Kenward, 1985).

Table 4.5 The abundance of Coleoptera and Heteroptera of various habitats in an assemblage from a modern drain sump, and the availability of the habitats in the immediate surrounding area (after Kenward, 1976).

Ecological grouping	Number of individuals	Percentage of total fauna	Availability of habitat
Aquatic and aquatic marginal	9	3.5	Not recorded within 250 m
Open ground	12	4.5	Some habitat for most species within 10 m
At roots of low vegetation	34	13.0	Scattered isolated plants present within 30 m
Phytophages	28	11.0	Hosts of some recorded within 100 m, but rare
Rotting plant matter			
Primarily	38	15.0	Certainly absent within 10 m; probably some accumulations within 250 m
Facultative	72	28.0	
Herbivore dung			
Primarily	4	1.5	Absent within 250 m; probably very rare within 1 km
Often in dung	21	8.0	
Dead wood	22	8.5	Present within 2 m
Synanthropic	6	2.5	Study area entirely dominated by man – no natural habitats
Often in association with man	57	22.0	
Often in areas disturbed by man	259	100.0	

far from straightforward. Kenward (1985) has discussed the difficulties involved with this type of work and has noted that, although most assemblages from occupation sites contain a large proportion of insect remains that originated from near the point of deposition, they often contain an element of **'background' fauna** composed of insects that have flown to the site, or that have been derived from the regurgitated pellets or faeces of birds and other animals (Table 4.5). Because of the complexities of insect death assemblages in archaeological contexts, therefore, Kenward has advocated that large samples (minimum 50 species) be analysed, in order that the full species diversity is adequately represented and that the background component can be isolated.

A further problem in archaeological work arises where people have created a range of wholly artificial environments, for some ecological changes may have occurred in certain insect species as they adapt to these new biotopes. Many natural and artificial environments, while being very different in terms of macroenvironment, may contain microclimatic niches that are very similar, and in some cases identical. Moreover, it would be expected that, in those insects that are markedly **synanthropic** (living in close association with humans) such as the grain weevil (*Sitophilus granaricus*), slight physiological changes may have occurred to adapt to these artificial habitats. Although in most cases the general archaeological interpretation will not be affected, care clearly needs to be exercised in the palaeoecological inferences that are drawn from archaeological sites based upon the known ecological affinities of modern insect species.

4.6 Chironomidae

Chironomidae are non-biting midges, a family of the Diptera (true flies), and recent studies of their autecology indicate that species abundance and composition are related to such factors as pH, salinity and trophic status (e.g. Walker *et al.*, 1991). This makes them potentially useful for Quaternary palaeoenvironmental reconstruction. It is, however, their potential as palaeotemperature indicators that has attracted the greatest interest.

Chironomidae produce larvae in the bottom of almost all freshwater habitats, and these eventually develop into mature forms that consist of a robust, strongly-sclerotised head capsule (Figure 4.24) and a body that resembles a maggot (Walker, 1987). The final, adult stage is mosquito-like but lacks a proboscis and a biting habit. It is the head capsules of the larval stage that are often well preserved and extremely abundant in freshwater sediments. Most genera possess head capsules with diagnostic forms, structures or surface markings that enable them to be identified, and a number of keys are now available (see Hofmann, 1986). Chironomid head capsules can be separated from sediment matrices by deflocculation in 10 per cent KOH or, for calcareous deposits, in 10 per cent HCl. This is followed by sieving, a 100 μm mesh being sufficient to trap most capsules (Walker & Paterson, 1985).

A number of studies have shown that the distribution and abundance of chironomid species are strongly influenced by summer surface water temperature, especially in arctic and

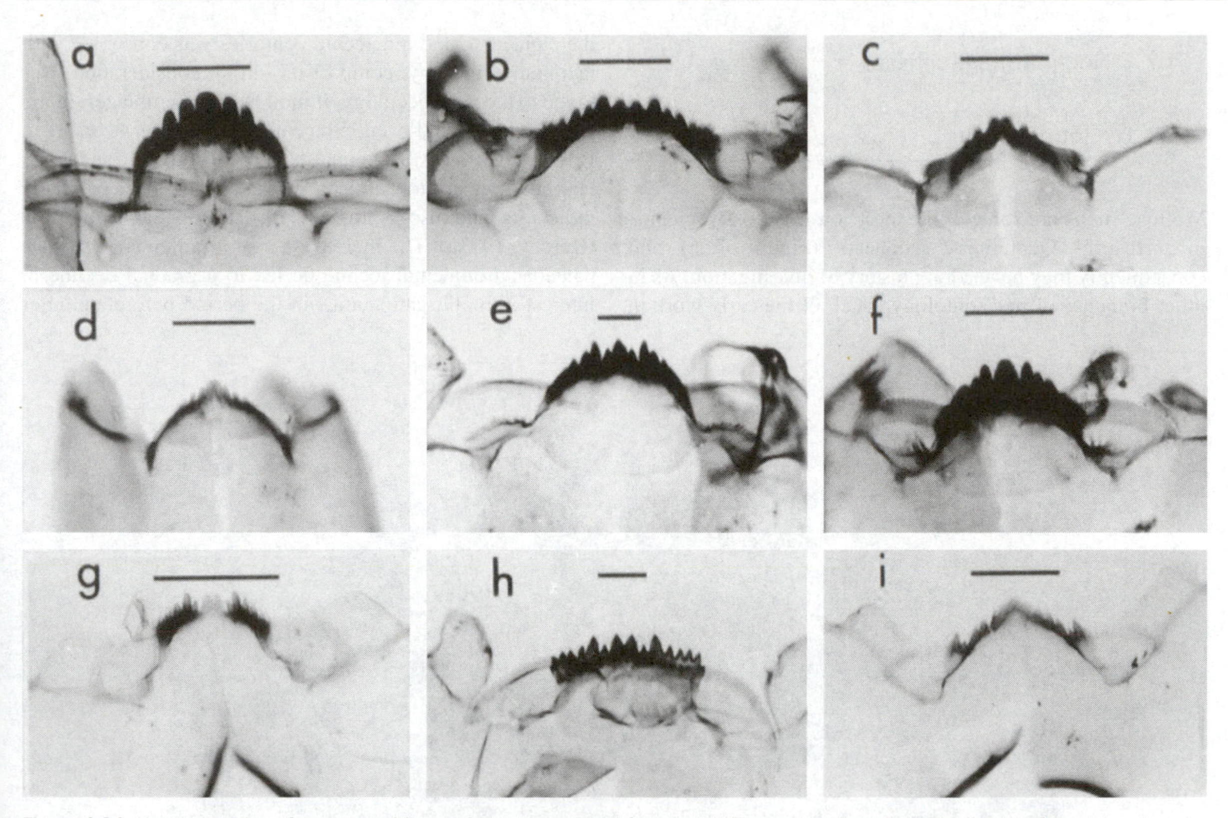

Figure 4.24 Head capsules of common chironomid taxa recovered from Late Wisconsinan Lateglacial sediments from lakes in eastern Canada (photo: André Levesque). (a) Subtribe Tanytarsina, (b) *Sergentia*, (c) *Heterotrissocladius*, (d) *Hydrobaenus/Oliveridia*, (e) *Chironomus*, (f) *Dicrotendipes*, (g) *Microtendipes*, (h) *Polypedilum*, (i) *Cladopelma*. Cold water taxa include *Sergentia*, *Heterotrissocladius* and *Hydrobaenus/Oliveridia*, and are most common in sediments deposited immediately following deglaciation, as well as sediments deposited during cooling events such as the Younger Dryas stadial. The other taxa illustrated are most common during warmer intervals. The scale bar represents 50 μm. vp = ventromental plates, the shape and surface texture of which are diagnostic for taxonomic classification (from Levesque *et al.*, 1994, p. 323).

alpine environments (Walker & Mathewes, 1987a, 1987b; Walker *et al.*, 1991). A close relationship has been established between *predicted* (i.e. chironomid-inferred, using weighted-average regression methods) and *measured* summer surface water temperatures for sites in Labrador. Using the data from Labrador, a strong correlation between predicted temperature and altitude has been obtained for sites in British Columbia. Fossil chironomids have subsequently been used to reconstruct temperature variations in New Brunswick, Canada during the Wisconsinan Lateglacial period (Wilson *et al.*, 1993; Cwynar *et al.*, 1994). This evidence suggests that summer surface-water temperatures declined rapidly by *c.* 6–7°C at around 10.8 ka BP, which coincides with the start of the Younger Dryas period in Europe. Recent work in Europe also suggests that chironomid assemblages have much potential for quantitative temperature reconstructions. Present-day chironomids in Alpine streams are sensitive to temperatures in the range 0–6°C (Milner & Petts, 1994) while fossil chironomids in sediment successions of the last glacial–interglacial age reveal marked assemblage variations which correlate strongly with temperature variations obtained using the beetle MCR method (S.J. Brooks, unpublished data).

The analysis of chironomid head capsules offers two important advantages over other methods of Quaternary palaeotemperature reconstructions. First, many hundreds of head capsules can be obtained from as little as 1 cm^3 of sediment. This should enable a higher resolution record than is possible with other methods, such as fossil Coleoptera, where far larger samples of sediment are required. Secondly, the method enables summer water-surface temperatures of small lakes to be reconstructed, and this provides valuable microclimatic information which can be compared with estimates of contemporaneous ambient air or ground temperatures derived using other approaches.

4.7 Non-marine Mollusca

4.7.1 Introduction

Mollusc shells are some of the most common fossil remains in terrestrial Quaternary sediments (Figure 4.25) and consequently they have a long history of investigation. As in other branches of palaeontology, much of the early work in the eighteenth and nineteenth centuries was concerned with taxonomy and, by comparison, little consideration was given to the palaeoecology of molluscan assemblages. In the late nineteenth and twentieth centuries, however, workers in Britain such as A.S. Kennard, B.B. Woodward and F.W. Harmer began to utilise molluscs as palaeoclimatic indicators, and also as a means of dating geological events (Kerney, 1977a). The increasing use of pollen analysis as a palaeoenvironmental technique led to a gradual decline in interest in molluscan studies in the period before and after

Figure 4.25 Assemblage of land and freshwater molluscs obtained from Hoxnian Interglacial deposits from a site near Peterborough in eastern England (photo: David Keen). 1. *Cepaea nemoralis* (1a – top view; 1b – underneath; 1c – broken specimen); 2. *Candidula crayfordensis*; 3. *Cochlicopa lubrica*; 4. *Bithynia tentaculata*; 5. *Pisidium* sp. 1–3 are land species and 4 and 5 are fluvial species. The mixed assemblage is interpreted as reflecting the action of a river sweeping over a floodplain. The *Cepaea* specimens are approximately 25 mm across.

the Second World War, and the modern phase of molluscan investigations did not begin until the 1950s with the introduction of a quantitative approach to the study of shell-bearing deposits (see Lozek, 1986). This approach, developed initially for the investigation of non-marine Mollusca, has since been applied to the study of marine shells (section 4.8) and over the last thirty years or so has led to the assembly of a considerable body of data on local environmental and climatic changes during the Quaternary period. In recent years, however, attention has also turned to other palaeoenvironmental aspects of molluscan studies, such as shell morphology and colour-banding on the shells of certain species, stable isotopic composition of shell organic matter and shell carbonate, and rates of amino-acid diagenesis (Goodfriend, 1992; Horton *et al.*, 1992).

Mollusca possess a number of advantages over other terrestrial and freshwater fossil groups that have been used in the reconstruction of Quaternary environments. First, specimens can nearly always be identified to species level and therefore more meaningful palaeoenvironmental conclusions can often be drawn. Second, shells are present in oxidised sediments (e.g. slopewashes, loess, tufa) which generally lack other fossil remains such as pollen or Coleoptera. Third, specimens are large enough to be recognised in the field (see Evans, 1972; Kerney & Cameron, 1979) and a very good idea of the general palaeoecology can generally be gained. This may help to determine where samples should be taken for other fossil groups. Fourth, much is known about their present-day ecology and geographical distributions (see, e.g., Ellis, 1978; Lozek, 1986), and since molluscs are frequently sensitive to changes in the local physical or chemical environment (Taylor, 1988; Goodfriend, 1992), fossil molluscs may provide useful insights into past changes in the local environment.

4.7.2 The nature and distribution of molluscs

Molluscs are invertebrates in which the soft parts of the body are generally enclosed within calcareous shells. They have a wide range of morphological characteristics, many of which are shared between marine and continental species. The two principal classes ar far as the Quaternary palaeoecologist is concerned, however, are the **Gastropoda** (snails) or **univalves** which usually possess a single spiral or conical shell (although in the case of the slugs the shell is reduced to an internal remnant), and the **Bivalvia** (mussels and clams) in which the animal possesses two valves. Other classes include the **Polyplacophora** (chitons or 'coat-of-mail' shells), **Scaphopoda** (tusk-shells) and the **Cephalopoda**. The latter are the most highly organised of the Mollusca and range from *Nautilus* which has a large calcareous, external, coiled shell, through the squids, which have only a thin horny vestige of a shell embedded in the mantle, to the octopuses which have no shell. Some gastropods breathe by gills (**prosobranchs**) and are mainly aquatic, while others such as snails and slugs breathe by a rudimentary lung (**pulmonates**) and, although they can live in water, are primarily terrestrial in habitat. The crystalline form of calcium carbonate in the shells of most molluscs is pure aragonite, although the internal shells of certain slugs are composed of calcite. In both cases, the shells are subject to little change in either their crystalline or their chemical composition following the death of the organism, and are therefore sometimes referred to as **subfossil** rather than fossil (e.g. Evans, 1972).

Land and freshwater Mollusca are extremely useful palaeoecological tools because of their wide distribution and preservation in a variety of deposits. They show a marked preference for habitats that contain sufficient lime for building their shells, although they are found not only in chalk and limestone regions, but also in calcareous drifts, in colluvial deposits, in cave earths, in loess and on coastal dunes and beaches. Molluscs do occur in non-calcareous environments, for some species (e.g. *Zonitoides excavatus*) are strongly calcifuge and are found only in non-calcareous regions. In such areas, however, the number of species is more restricted, shells are often thinner and less well preserved, and weathering and leaching in acid environments are more likely to lead to shell dissolution. In general, the richer the base status of the locality, whether it be land or freshwater, the richer the fauna, and molluscan remains would be expected to be discovered in a wide range of river, marsh, lake, woodland and open-land sediments (e.g. Preece, 1990; Keen, 1990; Shotton *et al.*, 1993; Preece & Day, 1994; De Rouffignac *et al.*, 1995). Terrestrial molluscs are also found in a variety of archaeological deposits including soils, ditch, pit and well sediments, ploughwash and other colluvial deposits, and occupation horizons and building debris (Evans, 1972).

4.7.3 Field and laboratory work

Although molluscan remains can be collected from open sections in the field by hand, they are best extracted under laboratory conditions (Lozek, 1986) because hand-picking of individual shells will bias the resulting samples towards the larger species. Bulk samples from sections or from cores are air dried and immersed in water, a small quantity of a dispersive agent such as H_2O_2 or NaOH being added if there is organic material present. The froth, which includes the snails, is then decanted through a 0.5 mm sieve. The process is repeated several times until no more snails are present in the froth and the residual slurry is then poured into a second 0.5 mm sieve. Both sieves are dried and the residues passed through another set of sieves (1 mm, 710 μm, 2411 μm) for ease of sorting. Molluscan remains can be either removed by hand or with the aid of a moistened brush under a low-powered binocular microscope or scanner. Identifications are always based on type collections of modern shell material aided by the numerous molluscan reference works that are now available. In many respects, identification of molluscan remains is not as difficult as in other branches of palaeontology. This is particularly true of land and freshwater molluscs, for which not only is there an extensive taxonomic literature but also, in Britain at least, the total number of species (both living and extinct) in late Quaternary assemblages does not exceed 200 (*cf.* the 3800 or more species of Coleoptera). Nevertheless, identifications can be complicated by the fact that many mollusc species vary markedly in morphology and in markings from juvenile to adult stage, while both colouring and fine sculpture can be affected by local site conditions. Some fossils may also be damaged by mechanical abrasion, some can have their surface markings masked by carbonate overgrowths, while others, particularly the fragile specimens, may be highly comminuted. In certain cases, specialised techniques allow even small fragments to be identified. For example, the marine genera *Mytilus, Modiolus* and *Pinna* have shells which possess a characteristic crystal structure that can be recognised with a high-powered microscope. Similarly, differences in shell microsculpture are often diagnostic of land Mollusca, so that specific determination of fragmentary remains can be made using a light microscope, augmented by the use of scanning electron photomicrographs (Preece, 1981).

The results of molluscan analyses can be presented in a variety of ways. Some workers prefer the use of species lists, perhaps using symbols to depict the frequency of occurrence (+ = presence; * = common; 0 = abundant, etc.). This type of tabular format is usually adopted where a limited number of samples have been taken from a profile and where a full quantitative assessment of the change in species composition over time is required. It is often employed in studies of marine Mollusca (e.g. Peacock, 1989). More frequently, however, the data are depicted graphically, either in histogram form for single samples or, where a sequence of sediments is being investigated, on a diagram which has the vertical axis showing the depth below ground datum and the horizontal axis the number of species plotted. The results can be presented in terms of relative abundance or absolute numbers (Figure 4.26), in both cases the histogram bars being drawn proportional to the thickness of the sample horizons. Once constructed, the diagram can be divided int **molluscan assemblage zones** which allow furthe palaeoecological or biostratigraphical inferences to be made These zones will initially be of local significance only, bu may be extended (as in pollen analysis) to form a zonatio scheme that has regional applicability, implying an orderl immigration sequence of molluscan species within th region (e.g. Kerney, 1977b; Kerney *et al.*, 1980; Preece & Robinson, 1984; Preece & Day, 1994).

4.7.4 Taphonomy of non-marine molluscan assemblages

The interpretation of molluscan assemblages is complicate by a number of taphonomic problems which should be take into account before palaeoenvironmental reconstructions ar attempted. Although the investigation of species bias i modern snail assemblages is a relatively recer development, a number of important points have alread emerged. Briggs *et al.* (1990), for example, found significant difference between mollusc assemblages o modern floodplains and those preserved in floodplai deposits. Selective species preservation is also characteristi of assemblages in some shell middens because c differential destruction of fragile shells by the action c predatory animals such as birds (Carter, 1990). Ther appears to be a closer correspondence between life and deat assemblages in lake (Cummins, 1994) and estuarin environments (Kidwell & Bosence, 1991), though the deat assemblages can be more species-rich because they ofte consist of admixtures of shells accumulated over a period c time. They may therefore reflect short-term variations i local mollusc communities. Less is known about th taphonomy of mollusc assemblages in cave sediments despite the fact that they are frequently rich in well preserved mollusc fossils because caves are often forme within calcareous rocks, that they have a uniqu microclimate (equable temperature and humidity) and tha the sediments within are protected from erosion an weathering. Mollusc assemblages within cave sediments ar surprisingly diverse, being derived from the litter an vegetation that accumulate close to the cave mouth, from th action of scavenger animals that inhabit the cave, fror washing in of species into cave fissures from surface soils c rock faces, and from the dropping of shells into cave fissure by birds (Hunt, 1993).

Mollusc assemblages are therefore very diverse an occupy a wide range of ecological niches, and the range c taphonomic pathways by which they become preserved i known only superficially at present.

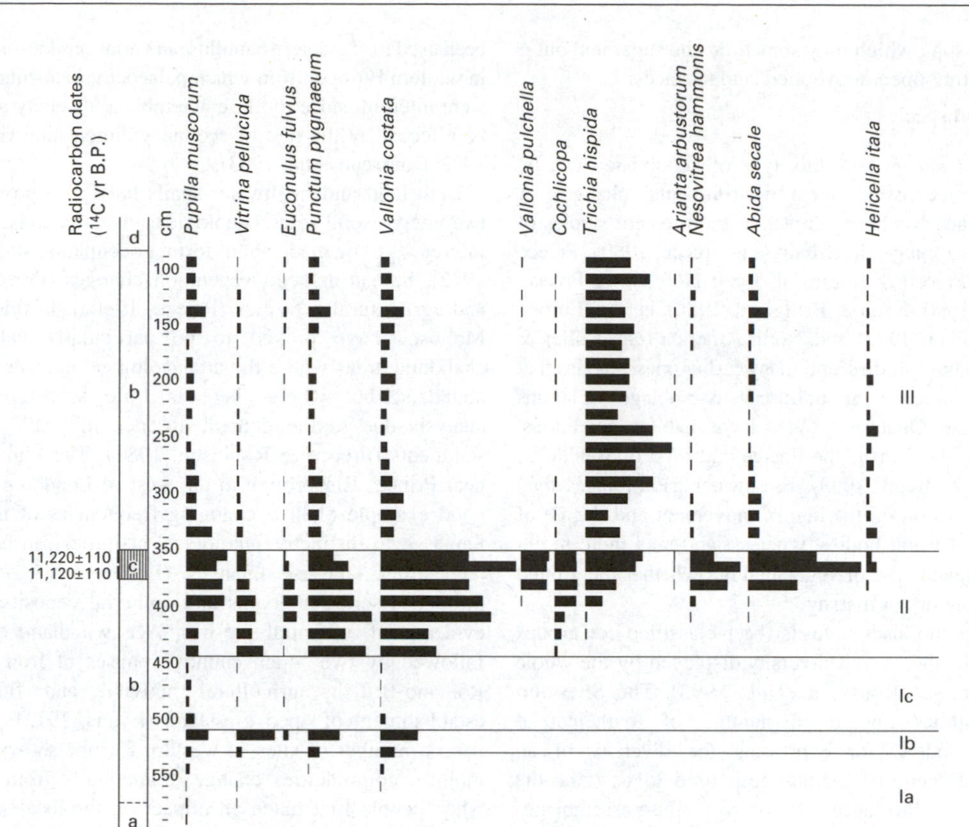

Figure 4.26 Molluscan histogram from Dover Hill, Folkestone, showing absolute abundances of species per 2 kg sample (after Preece, 1997).

4.7.5 Habitat preferences of non-marine Mollusca

Terrestrial Mollusca can be used in a number of ways to make inferences about former local habitats and subsequently about environmental change through time. One approach is to divide species into groups which have common habitat preferences. In modern British sites, for example, four groups of freshwater Mollusca are generally recognised, following the classification introduced by B.W. Sparks (see West, 1977):

1. A **'slum' group** composed of individuals tolerant of poor water conditions, ephemeral or stagnant pools with considerable changes in water temperature, for example the water snail *Lymnaea truncatula* and the small bivalve *Pisidium casertanum.*
2. A **'catholic' group** comprising Mollusca that will tolerate a wide range of habitats except the worst slums, for example *Lymnaea peregra* and *Pisidium milium.*
3. A **ditch group** that includes species often found in ditches with clean or slowly moving water and abundant growth of aquatic plants, for example *Valvata cristata* and *Planorbis planorbis.*
4. A **moving water group** composed of molluscs commonly found in slightly larger bodies of water such as moving streams and larger ponds where the water is stirred by currents and wind. Typical species are *Valvata piscinalis* and *Lymnaea stagnalis,* together with the larger freshwater bivalves.

Land Mollusca can similarly be divided into four groups:

(a) **marsh** and associated species, for example *Vallonia pulchella*;

(b) **dry land** species, for example *Vallonia costata* and *Pupilla muscorum;*

(c) *Vallonia* spp., which may sometimes be separated out as indicating open, unwooded land surfaces;

(d) **woodland** species.

Figure 4.27 shows how this type of subdivision can be employed to reconstruct local environmental changes. A similar approach has been adopted in more recent studies of molluscan assemblages in Britain (e.g. Preece, 1980a; Preece *et al.*, 1984; Preece & Robinson, 1984; Holyoak & Preece, 1985; Keen, 1990; Keen & Bridgland, 1986), central Europe (Kaiser & Eicher, 1987) and North America (e.g. Miller & Bajc, 1990). These studies, and many earlier ones, suggest that the primary influence on molluscan assemblage variations during the late Quaternary was local habitat conditions, especially (for land taxa) the degree and type of vegetation cover (Jones & Keen, 1993). Freshwater assemblages may provide information on the rate of movement and degree of oxygenation of water bodies, whereas land taxa indicate not only the dominant type of vegetation but whether the ground conditions were dry or marshy.

Non-marine molluscs have also been classified into groups on the basis of the overall diversity displayed by the whole fossil assemblage (Rousseau *et al.*, 1993). The **Shannon diversity index**,[5] one of a number of mathematical 'abundance models' for expressing the diversity of an assemblage (Engen, 1978), has been used to describe the relative diversity differences of modern mollusc assemblages (Rousseau, 1991, 1992), with high index values associated, for example, with the forests of Burgundy and low values characterising assemblages in the northern mountain environments of Sweden and Norway. This index has also been used to characterise molluscan variations loess sequences in western Europe, from which palaeoenvironmental changes were inferred, since mollusc assemblage diversity appears to be affected by the rate of aeolian sedimentation (Rousseau, 1992; Rousseau *et al.*, 1993).

Both land and freshwater snails have also proved to be extremely useful in archaeological investigations, allowing inferences to be made about former occupation sites (Evans, 1972), human-induced vegetation changes (Preece, 1993) and agricultural activities (Preece, 1980a). In this respect, Mollusca have proved to be particularly valuable in chalkland areas where the archaeological evidence is often abundant, but where there is little scope for pollen analysis due to the general absence of peats or limnic sediments (Preece & Robinson, 1984). The Pink Hill site near Princes Risborough to the west of London provides a good example of how changing frequencies of molluscan faunas with distinctive ecological affinities can be used to reconstruct land-use changes (Figure 4.28). A section through a series of loams and colluvial deposits revealed evidence of an initial pre-Iron Age woodland clearance, followed by two 'open country' phases of Iron Age and Romano-British agricultural activity, and finally the establishment of open grassland (Evans, 1972). Evidence from a number of sites in western Europe shows how the mollusc communities change dramatically from the time when people first made an impact on the local vegetation cover. Very low species diversity characterises the mollusc associations during the last 2000 years in particular, and this anthropogenic effect tends to mask other environmental controls on molluscan assemblages (Rousseau *et al.*, 1993).

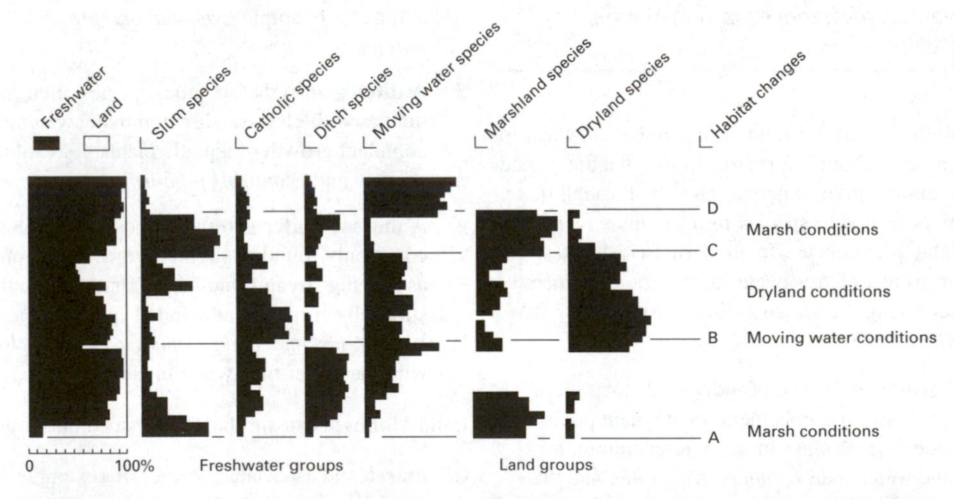

Figure 4.27 Local habitat changes shown by ecological groups of molluscan faunas in the Ipswichian (Last) Interglacial deposits at Histon Road, Cambridge. The diagram at the left-hand side shows the ratio of freshwater to land taxa. The changes most probably reflect a meandering stream shifting across the river floodplain (after Sparks, 1961).

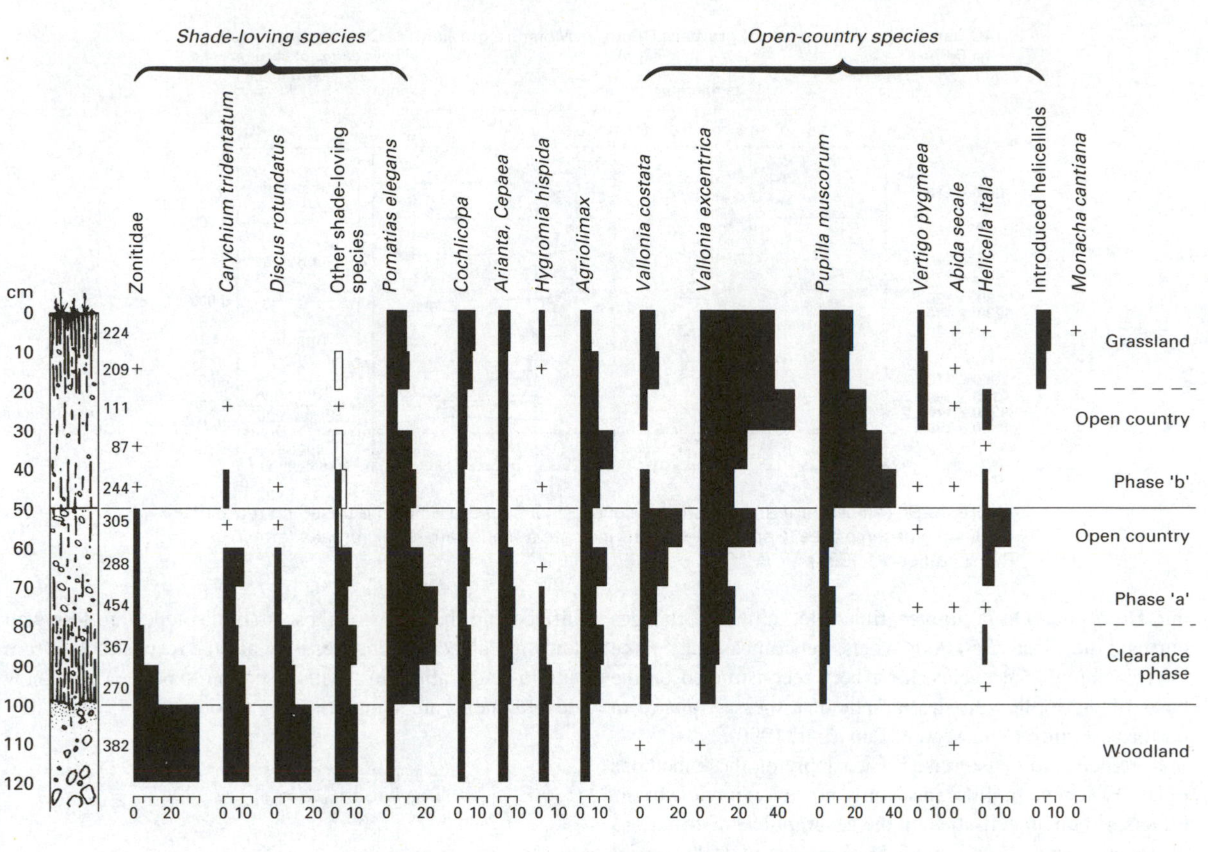

Figure 4.28 Molluscan assemblages of the Pink Hill (near Princes Risborough, Buckinghamshire, England) dry valley fill (after Evans, 1972).

4.7.6 Non-marine Mollusca and palaeoclimate reconstructions

Initial attempts to derive palaeoclimatic information from non-marine molluscan records relied upon the indicator species approach. Thus Kerney (1968), for example, noted that two species, *Pomatias elegans* and *Ena montana,* were common in Britain during the mid-Holocene (*c.* 8–5 ka BP) whereas both now have a more southerly distribution in Europe and are restricted to southern and western Britain. *Ena montana* was regarded as requiring summer warmth, while *Pomatias elegans* and *Lauria cylindracea* are essentially oceanic and cannot tolerate low winter temperatures. Kerney therefore suggested on the basis of this evidence that both summer and winter temperatures are lower now in Britain than during the mid-Holocene. In a similar fashion, Sparks and West (1970) used the occurrence of molluscan remains of *Vallonia enniensis, Clausilia pumila* and *Corbicula fluminalis* in the last interglacial deposits at Wretton in Norfolk to infer a more southern and continental climatic régime in Britain during that period, with summer temperatures perhaps 2°C higher than those of the present day.

In a more recent approach to palaeoclimatic reconstruction, Rousseau (1991) has developed a transfer function based upon a 'training set' of modern molluscan assemblages and their relationships with modern climatic variables. Correspondence analysis of a large number of datasets suggests that modern non-marine mollusc assemblages respond to two main influences, a temperature gradient and a moisture gradient, and these can be determined from characteristic groupings of species (Rousseau *et al.*, 1993). Using this approach, temperature and moisture variations in Burgundy during the Holocene have been reconstructed from fossil molluscan assemblages (Figure 4.29). The data suggest a two-step warming (similar to that inferred from oceanic evidence) at the beginning of

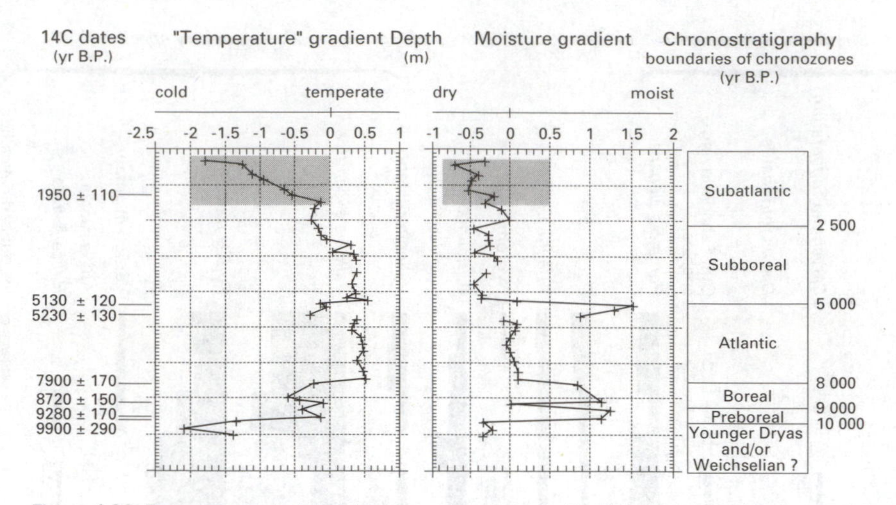

Figure 4.29 Temperature and moisture reconstructions for the Holocene based on terrestrial molluscan assemblages. Gaps in the graphs indicate major stratigraphic breaks (after Rousseau *et al.*, 1994).

the Holocene. On a longer timescale, climatic changes during the last 350 000 years, encompassing three interglacial–glacial cycles, have been reconstructed on the basis of the molluscan stratigraphy of a loess sequence in northeast France (Rousseau & Puissegur, 1990).

Care needs to be exercised when applying these methods to late Holocene molluscan records, however, in view of the effects of human activities on the geographical distributions of certain species (section 4.7.5). Horton *et al.* (1992) point to the records of six molluscan species obtained from interglacial deposits in the English Midlands, none of which are found in or near the region today. Preece and Robinson (1984) also argue that the distributions of several molluscan taxa contracted markedly in western Europe during the latter half of the Holocene. The extent to which these changes in species distributions reflect either climatic influences, or human pressures, or both factors operating together, is one of the most difficult questions facing Quaternary palaeoecologists today.

An alternative method of deriving palaeoclimatic information from non-marine molluscs, and one which may be independent of any human-induced influences, involves the analysis of isotopic ratios in shell carbonate matrices. As appears to be the case with cave speleothem (section 3.8), variations in oxygen isotope content in shell carbonate are thought to reflect changes in the temperature of rainwater. For example, variations in the $^{18}O/^{16}O$ ratios measured in species of *Vallonia* and in *Trichia plebeia* from a Lateglacial sequence in Switzerland were found to compare closely with the main temperature oscillations inferred for the sequence from other biostratigraphical data (Kaiser & Eicher, 1987). $^{13}C/^{12}C$ ratios also showed similar trends, but the factors influencing the ratios of these stable isotopes are less well known. Palaeotemperatures may also be reconstructed from amino-acid ratios in shells (section 5.6.1). These new developments are summarised by Goodfriend (1992).

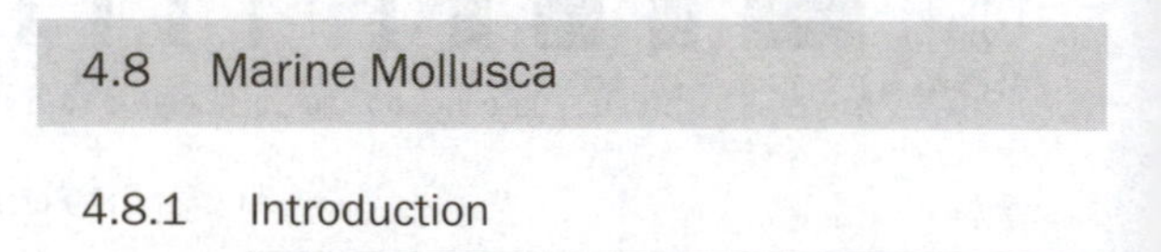

4.8 Marine Mollusca

4.8.1 Introduction

The shells of marine Mollusca have been found in a range of deposits in coastal areas. They often occur in beach gravels and estuarine clays now lying some distance above sea level, having been raised isostatically following the wastage of the last ice sheets (Sutherland, 1981), and dating of these marine fossils provides a chronology of sea-level change and deglaciation (e.g. Miller & Mangerud, 1985; Bowen *et al.*, 1985). Molluscan assemblages have been recovered from boreholes both onshore and from the sea bed, while shell remains, often highly fragmented, are found at localities inland, having been stripped from a former sea bed and transported to their present position by glacier ice (e.g. Merritt, 1992). Although perhaps less widely used in Quaternary studies than their freshwater counterparts, marine Mollusca are nonetheless an important additional source of palaeoenvironmental information. Indeed, the Pleistocene Epoch was originally defined on the basis of the composition of marine molluscan faunas.

Marine molluscs occupy a great range of ecological niches from pools and rock outcrops in the intertidal zone to the deeper waters off the edge of the continental shelf, although they are seldom found at depths greater than 1 km. The mode of life of marine molluscs varies considerably (Peacock, 1989), but many taxa are **infaunal**, burrowing with a muscular foot into soft sediments (e.g. *Turritella, Mya, Macoma, Arctica, Nucula)* or boring into bedrock *(Hiatella, Zirfaea).* Others are **epifaunal**, attaching themselves by threads to surfaces or other organisms (*Mytilus, Modiolus*) or by cementing one valve to the surface (*Ostrea edulis*). Only a few bivalves are free-swimming (*Chlamys*). The gastropods, scaphopods and bivalves are principally benthic and sedentary in life and, upon death of the individual, often become fossilised *in situ* to form autochthonous death assemblages. Marine Mollusca are often well preserved in Quaternary sediments (Figure 4.30), although those from the intertidal zone may be fragmented by wave action.

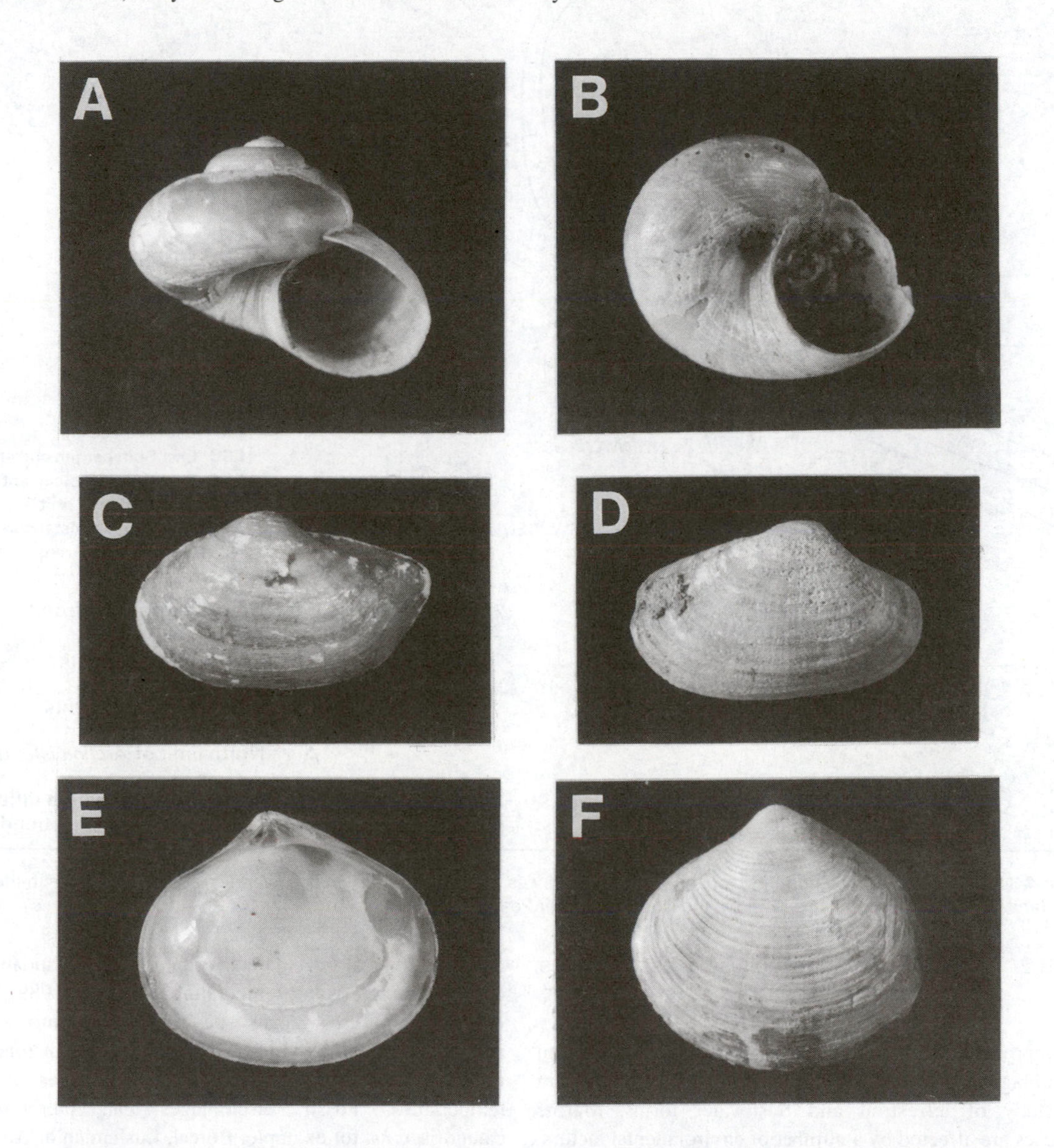

Figure 4.30 Arctic bivalve molluscs from 'arctic bed' of probable Loch Lomond (Younger Dryas) Stadial age at Greenock, Scotland. A. *Margarites groenlandicus* (side view ×3.3). B. *M. groenlandicus* (basal view ×3.3). C. *Portlandia arctica* (left valve outer ×2.8). D. *Lyonsia arenosa* (right valve outer ×2.1). E. *Tridonta striata* (right valve inner ×3.0). F. *T. striata* (left valve outer ×2.6). Nomenclature updated to 1995. Photos by Douglas Peacock; reproduced by kind permission of *Scottish Journal of Geology*.

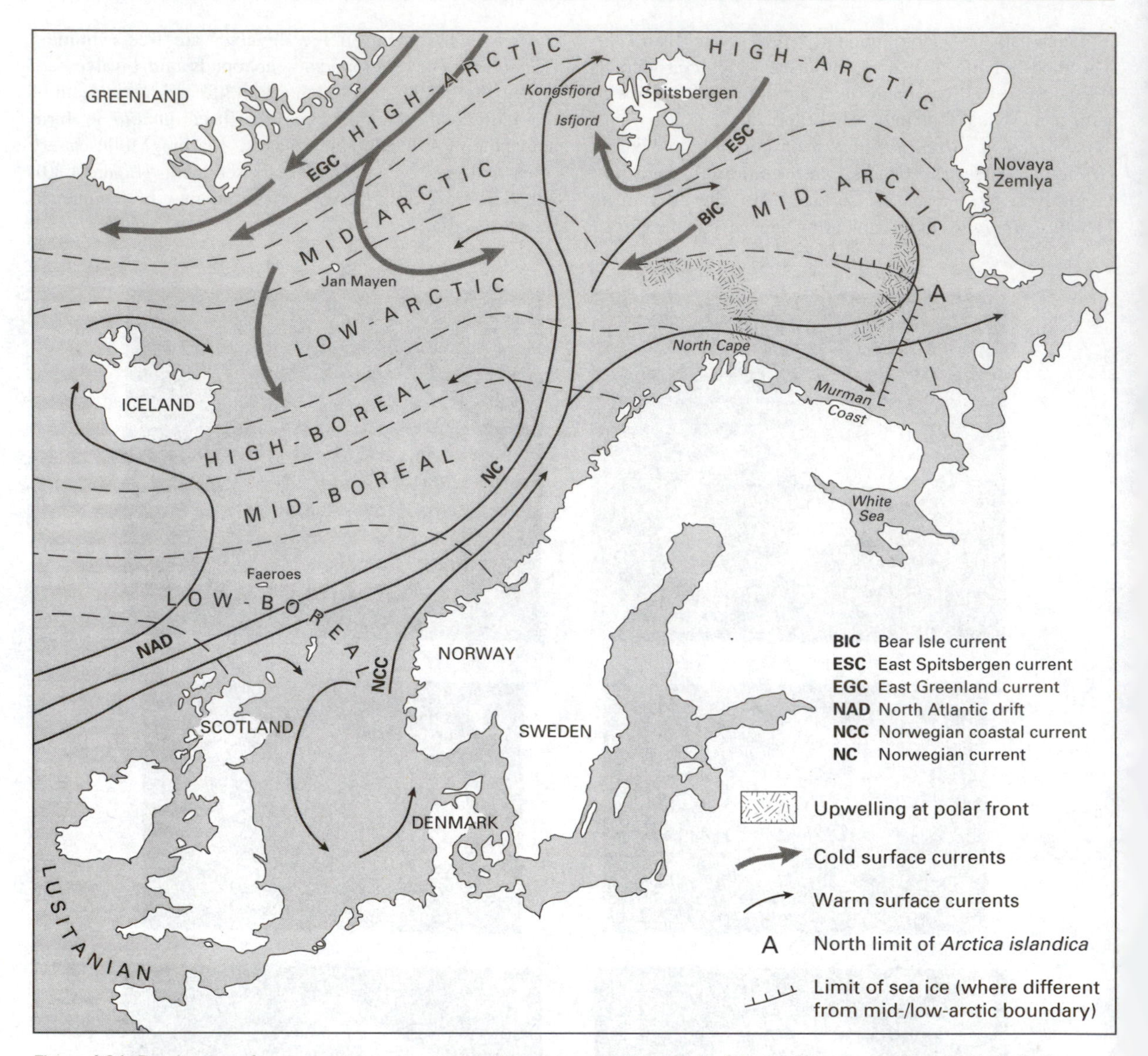

Figure 4.31 Dominant surface currents, zoogeographical provinces, northern limit of *Arctica islandica* and sea ice limits in the Greenland–Iceland–Norwegian Seas and in the North Sea (after Peacock, 1989).

4.8.2 Analysis of marine molluscan assemblages

The approaches adopted in the analysis of marine molluscan assemblages are essentially the same as those employed in the study of terrestrial and freshwater forms. Marine molluscs are affected by a number of environmental factors, such as substrate, food supply, temperature, salinity, oxygen level, nutrient availability, 'depth' (which affects, in combination, food supply, habitat, shelter, light and temperature), competition, predation and life strategy, but the main control over their distribution and abundance are current flux and water temperature (Peacock, 1989, 1993). Marine molluscs can be grouped into major **zoogeographical provinces** (Figure 4.31), ocean zones that reflect gradual, or sometimes sudden, changes in water temperature. Fossil assemblages can therefore be categorised as, for example, Boreal, Lusitanian or Arctic on the basis of the ecological affinities of modern molluscan species. Major currents (Figure 4.31) also have a role to play, as they affect nutrient supply as well as regulating water temperature, and they influence the dispersal of larva stages. Salinity variations may exercise an important control in shallow or enclosed seas, as indicated by carbon and oxygen isotope ratios obtained from subfossil mollusc shells in the Baltic (Punning *et al.*, 1988).

4.8.3 Marine Mollusca and palaeoclimatic inferences

There are three main approaches to deriving palaeoclimatic information (changes in sea-surface or water-column temperature) from marine molluscan assemblages. The first is based upon historical migrations of the principal zoogeographical provinces referred to above (Figure 4.31), and was typical of many of the early attempts to use molluscs for palaeoclimatic reconstructions. For example, in his discussion of the shell-bearing marine clays of Lateglacial age around the coasts of Scotland, Sissons (1967) distinguished between deposits of 'Arctic' affinity which were thicker and more widespread on the east coast of Scotland and which were characterised by abundant shells of species common in modern Arctic waters, and sediments from the west coast, which were designated as 'sub-Arctic' on the basis of their quite different fossil content. The former assemblages were believed to have been laid down in the seas around Scotland while glacier ice still covered much of the mainland, while the latter group was characteristic of the succeeding period when ice cover was markedly less extensive. More recently, Peacock (1989) has reviewed the much wider range of molluscan evidence that is now available and has produced a curve showing variation in summer sea-surface temperature around the British coasts during the last glacial–interglacial transition (Figure 4.32).

The second approach employs the indicator species concept. Mangerud (1977), for example, used the variation in occurrence through a sequence of marine deposits at Ågotnes, Norway, of three species (*Modiolus modiolus, Littorina littorea* and *Mytilus edulis*) whose present-day distribution and ecology are fairly well known, to plot the palaeo-positions of the **North Atlantic Polar Front**[6] off the west coast of Norway during the Lateglacial period. Hjort and Funder (1974) used the occurrence of the common

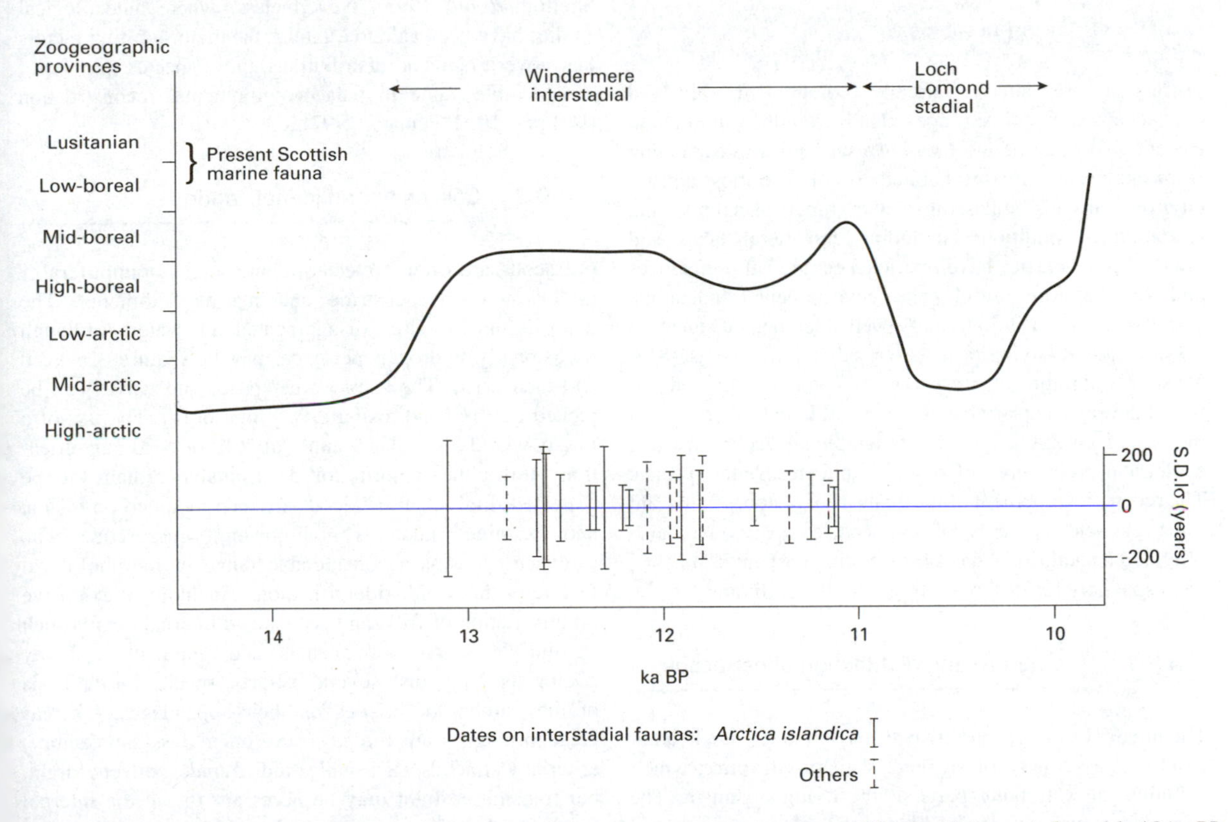

Figure 4.32 Schematic representation of variations in surface-water sea temperatures around the Scottish coasts from 14–10 ka BP, based on changes in dominant zoogeographical provinces represented in molluscan assemblages (after Peacock, 1989).

mussel (*Mytilus edulis*) to infer that temperatures along the coast of east Greenland were higher during the period 8–5.5 ka BP than at the present day. The North Atlantic Drift appears to be characterised by two dominant indicator species, *Arctica islandica* and *Modiolus modiolus*, in areas like Bear Island and Spitsbergen (Peacock, 1989), while the brachiopod *Macandrevia cranium* has been suggested as a marker species indicating the flow of Atlantic drift water along the Norwegian Shelf (Thomsen, 1990).

The third approach concerns the detailed analysis of growth structure or isotopic composition of certain marine mollusc species. There appears, for example, to be a relationship between the pattern of growth lines (growth increments) in the shells of *Cerastoderma edule* and length of growing season (Peacock, 1989), while isotopic ratios in some species obtained from the offshore region of the east coast of North America appear to correlate with seasonal temperature variations (Williams *et al.*, 1982).

4.9 Ostracod analysis

Ostracoda are small, laterally compressed, bivalved crustaceans whose time range extends from the Cambrian to Recent. Over that period they have succeeded in expanding from exclusively marine habitats to colonise most aquatic environments encompassing a wide range of salinity and temperature conditions, including ephemeral lakes and ponds. Some species have restricted ecological preferences and are therefore useful palaeoenvironmental indicators. The fossil record is extremely well documented, the first fossil ostracod having been described as long ago as 1813. Most stratigraphical research has been devoted to investigation of marine ostracods, although there is an increasing databank on the stratigraphical occurrence and ecological preferences of brackish and freshwater species. The recent development of the analysis of isotopic ratios and trace element contents of ostracod carapaces has also provided an additional basis for inferring past environments, and especially for differentiating salinity conditions.

4.9.1 The nature and distribution of ostracods

The majority of ostracods are between 0.6 to 2.0 mm in adult length. They consist of an outer shell or **carapace**, which contains the soft body parts of the living organism. The carapace is usually ovate, kidney-shaped or bean-shaped (Figure 4.33) and consists of two chitinous or calcitic valves that hinge above the dorsal region of the body. The biological classification of Recent ostracods rests very largely on the characteristics of the soft parts, but as these features are very rarely preserved in the fossil form, the taxonomic classifications have to be based on the nature of the carapace which fossilises relatively easily.

The majority of marine ostracods are bottom-dwelling forms, and only a small number occupy the planktonic realm. Moreover, pelagic species usually possess weakly calcified shells and are therefore relatively rare in fossil assemblages. The distribution patterns of living ostracod communities are governed by a wide range of factors (e.g. Angel, 1993; McKenzie & Jones, 1993). These include physical parameters such as water temperature, salinity and nature of the substrate, and biological factors such as food chains and natural associations. It is, however, difficult to cite any one control as universally dominant, for while many workers feel that, in the case of marine ostracods in particular, water temperature is the most important, others would argue that salinity is more fundamental, while in the freshwater situation the nature of the substrate may be the overriding influence (Engstrom & Nelson, 1991; Mitlehner, 1992; Shotton *et al.*, 1993). Nevertheless, where autecological studies have been able to establish the major limiting factors that govern ostracod distributions, those species may be of considerable value in palaeoenvironmental reconstruction (Holmes, 1992; Penney, 1993).

4.9.2 Collection and identification

Ostracods are often collected, along with Foraminifera or molluscs, from lacustrine and marine sediments. The deposits are usually disaggregated in water (although occasionally hydrogen peroxide may be required), sieved and then dried. The ostracod carapaces and valves can be picked out by hand, using a very fine brush. The use of a low-powered binocular scanner of ×40 or ×60 magnification allows the majority of determinable remains to be collected. Individual ostracods are then mounted on a slide and examined under a high-powered microscope. The carapaces possess a considerable range of morphological features that aid identification, including extensive ornamentation of frills and spines, and internal details such as muscle scars, pore canals and duplicature. Many zoologists distinguish several ostracod species on the basis of the number of bristles on their appendages, whereas palaeontologists are forced to use other (less satisfactory) criteria. Ostracods are usually studied under reflected light, but transmitted light may be necessary to see the internal features. Identifications are based on modern type collections, stereoscan photographs and descriptions in

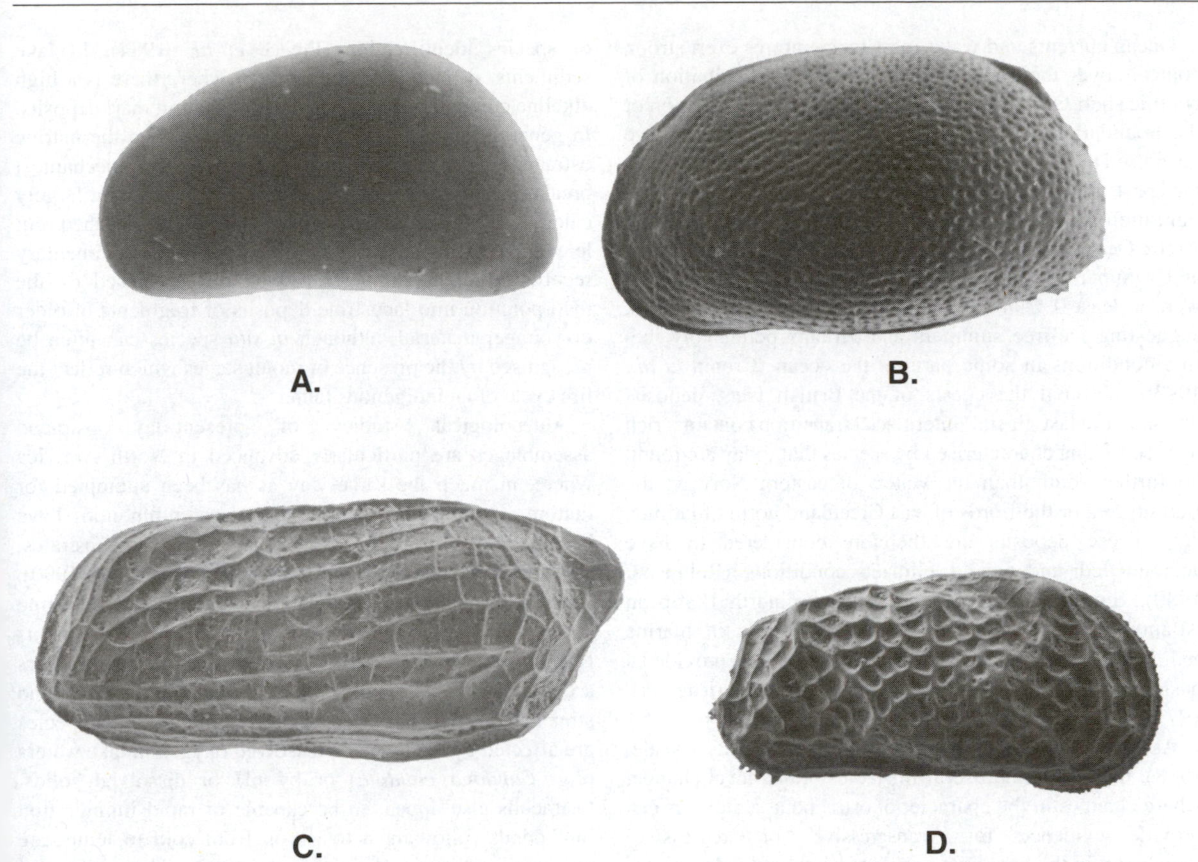

Figure 4.33 Ostracoda. A. *Candona candida* (Muller, 1785) (female left valve, *c.* 1.2 mm long). B. *Metacypris cordata* (Brady & Robertson, 1870) (female left valve, 0.6 mm long). C. *Semicytherura arcachonensis* Yassini, 1969 (male left valve, 0.61 mm long). D. *Baffinicythere howei* Hazel, 1967 (male right valve, 1.18 mm long). A and B are freshwater species. A change from a 'candida fauna' to a 'cordata fauna' usually reflects increasing productivity of a lake (change from oligotrophic to meso- or eutrophic conditions). C is an outer estuarine marine species, presently restricted in the eastern Atlantic to waters south of southwest France, but found in Hoxnian Interglacial sediments in Britain. D is a characteristic sublittoral Arctic marine species (photos: David Horne).

micropalaeontological manuals, and the data are expressed either as species lists or in diagrammatic form showing the change in frequency of occurrence through time. Further details on collection, preparation and study can be found in Brasier (1980) and Delorme (1990).

4.9.3 Ostracoda in Quaternary studies

Certain rapidly evolving ostracod lineages are useful markers in marine biostratigraphic sequences, especially where Foraminifera are absent (e.g. Whatley, 1993). However, as they lack planktonic larvae, many shallow and warm water species cannot cross physical barriers and are therefore restricted to particular geographical areas. Moreover, some of the problems already considered in the interpretation of terrestrial fossil assemblages are also found in ostracod analysis. At the generic level, the poor state of taxonomy often inhibits the comparison of fossil and recent forms. Also there is evidence to suggest that several ostracod species that are now benthic in character developed from shallow-water ancestors. Fortunately, it appears that migration in the opposite sense, in other words from deep and cold to shallow and warm water, seems unlikely to have occurred. Further, there are indications that the dominant elements in certain benthic fossil assemblages may be due less to environmental factors than to selective preservation of the more thick-shelled species. Providing that these difficulties can be overcome, however, marine ostracods can be extremely useful as palaeotemperature and palaeosalinity indicators, and they may also provide valuable data on palaeobathymetry.

Ocean currents and water mass temperatures exert strong controls over the geographical (latitudinal) distribution of benthic shelf Ostracoda today and changes in the position of the boundaries of these zoogeographical 'provinces' have been used to infer changes in the position of water masses in the past (Wood & Whatley, 1994). For example, the migration of temperate-water marine Ostracoda into the Arctic Ocean during the Pliocene is one indicator that winter and summer ocean temperatures along the Arctic margins were at least 0°C and 3°C respectively during that period, suggesting ice-free summers and perhaps perennially ice-free conditions in some parts of the ocean (Cronin *et al.*, 1993b). Around the coasts of the British Isles, deposits dating to the last glacial–interglacial transition contain a rich ostracod fauna characterised by species that today are found no further south than the waters of eastern Norway, the Barents Sea or the fiords of east Greenland north of latitude 76°. These deposits are therefore considered to have accumulated under cold climatic conditions (Robinson, 1980). Similarly, the presence of some north European Atlantic ostracod species at particular levels in marine sediments of the Italian Quaternary sequence has provided a basis for distinguishing cold-climate episodes (Ruggieri, 1971).

As ostracods can be good indicators of salinity (Neale, 1988), they are often useful in studies of sea-level change, where changes in the character of ostracod assemblages can provide evidence for 'transgressive' or 'regressive' sequences. In the Somerset Levels of southwest England, for example, Kidson *et al.* (1978) used ostracod evidence, along with foraminiferal, molluscan and plant macrofossil remains, to show that within the local Burtle Beds, an interglacial marine transgression reached a height of up to 12 m above the present (OD), at which time temperatures appeared to be similar to those of today. Moreover, biometrical analysis of the ostracod *Cyprideis torosa* found within the deposits suggested that the marine transgression was of last (Ipswichian) interglacial age. In similar integrated studies, Haynes *et al.* (1977) traced the course of the Holocene (Flandrian) marine transgression along the coast of Cardigan Bay in west Wales, where the transition from brackish to salt-water conditions was clearly represented in ostracod records at a large number of sites, while Mitlehner (1992) has reconstructed the depositional environment represented by ostracod-bearing clay deposits of Hoxnian interglacial age at a site in west Norfolk. The assemblages indicate an overall change from a low energy, mud-dominated sedimentary environment to a much higher energy régime, but with salinities indicating the prevalence of marine conditions throughout the accumulation of the succession.

Freshwater ostracods have been less extensively studied than their marine counterparts, due largely to the difficulty of species identification (Preece *et al.*, 1984). In lake sediments, ostracods are rare except where there is a high alkaline content, but they may be abundant in marl deposits. In general, they possess thinner carapaces than the marine ostracods and can be easily destroyed by mechanical breakdown and chemical corrosion. Because of their largely calcitic shell, the fossil remains are easily leached out leaving, in extreme cases, a complete gap in the sedimentary record. Fossil assemblages can be further biased by the incorporation into lacustrine deposits of fragments of older or younger material, although *in situ* species can often be recognised by the presence of moult stages which reflect the life cycle of an indigenous fauna.

Autecological studies of present-day ostracod assemblages are particularly advanced in North America where, in much the same way as has been attempted for diatom assemblages (section 4.3), transfer functions have been developed to relate species distributions to substrates, salinity, oxygen/anoxia and temperature (Delorme, 1990). This builds on earlier investigations that indicated that some species are restricted to still-water lake environments (*Candona subtriangulata, Cytherissa lacustris*) while others are indicative of moving water habitats in springs and streams (e.g. *Cypria obesa, Ilyocypris gibba*). Other species are affected by the levels of dissolved oxygen in lake waters (e.g. *Candona caudata*) or by pH or dissolved solids. Ostracods also appear to be capable of rapid immigration into ponds following a transition from cold to temperate climatic conditions (Keen *et al.*, 1988). Thus a careful analysis of ostracod assemblages can provide useful information not only on the nature of former freshwater habitats and on water quality, but also on palaeoclimates. For example, Mourguiart and Carbonel (1994) have reconstructed palaeolake levels in the Bolivian Altiplano on the basis of quantitative measures of changes in ostracod assemblages and related these to former precipitation variations.

A new development which appears to have considerable potential for reconstructing Quaternary palaeoenvironments is the analysis of carbon and oxygen isotope ratios (Talbot, 1990; Lister *et al.*, 1991) and of trace chemical element variations (Engstrom & Nelson, 1991) in ostracod carapaces. Technical developments in high-precision analysis, such as atomic mass spectrometry (AMS) and inductively coupled plasma–mass spectrometry (ICP-MS) have enabled high resolution studies of individual carapaces to be carried out. These suggest that isotopic and trace element variations in ostracod carapaces can be used as 'tracers' of changing temperature and salinity conditions (Holmes *et al.*, 1992). Lake palaeosalinity levels have also been reconstructed using measures of Mg and Sr ratios in ostracod carapace shells (Chivas *et al.*, 1986).

4.10 Foraminiferal analysis

Foraminifera are **protists** (or **prokaryotes** – single-celled organisms[7]) that possess a hard calcareous shell often distinctively coiled to resemble that of a gastropod or cephalopod. They were first described and illustrated in the sixteenth century, but were not studied systematically until the latter part of the nineteenth century following the remarkable voyage of *HMS Challenger* which began in 1872. The discovery during that expedition of living Foraminifera in deep-sea waters, and of fossil remains in sediments that were dredged from the sea floor, revolutionised marine micropalaeontology. Since then Foraminifera have become invaluable tools in Quaternary stratigraphy, palaeoceanography and palaeoclimatic reconstruction.

4.10.1 The nature and distribution of Foraminifera

Foraminifera consist of a soft body (protoplasm) enclosed within a shell or **test** secreted by the organism which is variously composed of organic matter, minerals (calcite or aragonite) or **agglutinated** (foreign particles held together by various cements) components. The tests may be single chambered, but more frequently consist of a number of chambers separated by **septae**. Connections between the chambers, through which cytoplasmic material can move, are formed by small holes in the septae known as **foramina**, from which the group derives its name. In many common species, the chambers are added in a spiral pattern, producing a coiled shell, while others develop far more complicated structures (Figure 4.34). They are classified on the basis of a number of characteristics, such as the **rhizopodia** (cytoplasmic extensions used in locomotion and feeding), degree and form of coiling, number of chambers, number and pattern of apertures, and surface ornamentation. Not all of these characteristics will be preserved, however, in fossil forms.

Foraminifera range in size from less than 0.40 mm (the planktonic forms) to some of the benthic species which may measure up to 10 cm in width (so-called larger Foraminifera). They are tolerant of a range of salinity and temperature, being found in saltmarshes, shallow brackish water in estuaries, on the continental shelf and in the waters of the deep oceans of the world. Most Foraminifera are marine and benthic, although a few genera are pelagic, while a very small number of species (**thecamoebids**) are adapted to freshwater environments (section 4.13). The marine planktonic and benthic forms have proved to be particularly useful in global correlation and climatic reconstruction and these will be considered in more detail in the next section of this chapter. The present discussion will be concerned principally with foraminiferal remains in shelf-seas and inshore waters. Further discussion on the nature, distribution and ecology of Foraminifera can be found in Brasier (1980), Bolli *et al.* (1985), Murray (1991) and Lipps (1993).

4.10.2 Collection and identification

Foraminifera can be extracted from sediments obtained from surface samples or from cores. The matrix is usually crushed and disaggregated using either water or hydrogen peroxide. The samples are then washed through sieves and the residues dried; the retained Foraminifera can be picked out by hand with the aid of a binocular microscope. Where large numbers of sand grains are present (in some shelf sediments, for example) the foraminiferal remains can be concentrated using a heavy liquid such as ethylene bromide/absolute alcohol solutions, the tests being 'floated' from the sand. The smaller Foraminifera are examined under a high-powered microscope using reflected light. Occasionally staining (with, e.g., malachite green or a similar food dye) is required to bring out the surface structures more clearly. Larger Foraminifera are often studied in thin section where wall and growth plan may be better seen under transmitted light. As with ostracods, identifications are based on type collections, descriptions and stereoscan photographs in Foraminiferal manuals. The data can be presented simply as species lists or, more commonly, as '**range charts**' expressed in percentage form or as abundance per unit volume of sediment (e.g. Knudsen & Sejrup, 1993). Further information on the collection and study of Foraminifera can be found in Brasier (1980) and Lipps (1993).

4.10.3 Foraminifera in Quaternary inshore and shelf sediments

Foraminiferal remains in sediments of most inshore waters and the shelf seas are dominated by benthic forms, in contrast with the deep-sea sediments in which planktonic Foraminifera are particularly abundant. Palaeoenvironmental inferences from these bottom-water assemblages are constrained by a number of factors (Lord, 1980) which include the following.

1. *Statistical reliability.* A complete assemblage of at least 300–400 individuals is required for statistical comparability between samples. The assemblage may still be biased, however, as the agglutinated Foraminifera are more susceptible to post-mortem disintegration than the calcareous or siliceous tests. Care is also required in the use of heavy liquids for the

Figure 4.34 Modern benthic Foraminifera from a shallow marine setting (14 m) in Loch Creran on the west coast of Scotland (photo: Heather Anne Austin): 1. *Cribrostomoides jeffreysii* (Williamson) (oblique side view ×130); 2. *Cibicides lobatulus* (Walker and Jacob (side view ×102); 3. *Oolina williamsoni* (Alcock) (side view ×167); 4. *Ammonia beccarii* (Linné) (ventral view ×140); 5. *Elphidium albiumbilicatum* (Weiss) (side view ×137); 6. *Brizalina pseudopunctata* (Höglund) (side view ×183).

concentration of foraminiferal tests, and the heavy residues must always be picked over carefully for the foraminiferal remains left behind, because of pyritisation or permineralisation of some tests.

2. *Reworking.* Reworking and redeposition of tests is a frequent occurrence, and while such factors as wear, poor preservation and unusual population structure may indicate a mixed assemblage, errors in interpretation may still occur.
3. *Sampling.* Isolated samples can be misleading, and several samples from sediment sequences should be taken whenever possible.
4. *Taxonomic comparability.* Although most of the Foraminifera of northwest Europe are well known, not all are easy to recognise and confusion over specific differentiation of, for example, *Elphidium* types and related forms continues to pose a problem. *Elphidium clavatum* is a commonly cited species with cool-water/subarctic connotations, but it is often misidentified.
5. *Correlation.* Some foraminiferal species have limited geographical ranges and, as very few sedimentary sequences containing foraminiferal remains are adequately dated, inter-area correlation of successions can therefore often be difficult to achieve.

These problems notwithstanding, foraminiferal data hav provided much useful information on the interpretation o inland and shelf-sea sequences, examples of which ar summarised below.

4.10.3.1 Sea-level change

Foraminifera may be preserved in sediments that accumulated in estuaries or shallow marine conditions and which are today elevated above sea level. These may, therefore, provide evidence of past sea-level variations, such as the foraminiferal-bearing sediments of Holocene (Flandrian) and last (Ipswichian) interglacial age preserved in places around the coasts of the British Isles (e.g. Haynes *et al.*, 1977; Kidson *et al.*, 1978). Sejrup (1987) used foraminiferal assemblages, in conjunction with molluscan data, from an abandoned clay-pit exposed close to the shore in Karmøy, southwest Norway, to reconstruct changes in sea-level during the last glacial stage, while variations in sea water offshore of Gothenburg, southwest Sweden, have been inferred from foraminiferal assemblages in shallow marine cores (Bergsten & Dennegård, 1988).

4.10.3.2 Water temperatures

The present-day distribution of planktonic Foraminifera is closely related to surface-water temperature (section 4.11), and therefore former changes in sea-surface temperatures will be reflected in the species composition of fossil foraminiferal assemblages. Foraminiferal records in North Sea sediments, for example, have enabled researchers to reconstruct variations between arctic and boreal (temperate) water surface conditions during the late Quaternary (e.g. Jensen & Knudsen, 1988; Knudsen & Sejrup, 1993). The climatic reconstructions have not been quantified but the data appear to indicate influx of temperate (Atlantic) waters during interglacial periods, and dominance of arctic waters during cold stages. The early Pleistocene 'Crag' deposits (mollusc-rich shelly deposits now raised above sea level) of Suffolk, eastern England, have yielded rich foraminiferal assemblages that record variations between boreal and arctic environments, and these data have been used to support palaeoclimatic inferences based upon palynological and mollusc records (Funnell & West, 1977; West, 1980). Lateglacial marine sequences around the Norwegian and Danish coasts also contain foraminiferal assemblages that provide data on both water temperatures and salinity (Nagy & Ofstad, 1980). The earliest faunas are arctic in character, reflecting both low water temperatures (much lower than the present day) and low salinity values, the latter resulting from meltwater influx from the wasting Late Weichselian ice sheet. The Foraminifera are primarily benthic, although a few arctic planktonic forms occur. Early Holocene foraminiferal remains are still primarily benthic, but contain fewer high-arctic forms and indicate rising water temperatures and a decreasing freshwater influence. Middle Holocene faunas are mainly boreal in character, and contain a high percentage of planktonic forms typical of the Foraminifera found in the northeast Atlantic waters at the present day. Both temperature and salinity characteristics of Norwegian coastal waters, therefore, appear to have changed little from the Middle Holocene to the present. Indirect evidence of climatic variation can also be deduced from, for example, foraminiferal assemblages indicative of pack-ice development in parts of the ocean that are presently ice-free (e.g. Haake & Pflaumann, 1989), or from assemblages commonly associated with glacier margins in presently non-glaciated regions (e.g. Sejrup *et al.*, 1987; Austin & McCarroll, 1992).

4.10.3.3 Other palaeoenvironmental applications

Foraminifera from shelf and nearshore environments can also provide useful data on other aspects of Quaternary marine environments. For example, foraminiferal assemblages in sediment successions that accumulated in the Skagerrak–Kattegat Straits during the last deglaciation indicate sudden shifts in local salinity levels between episodes when the straits were dominated by the inflow of warm, saline water from the Atlantic, and phases of low salinity during the drainage of the Baltic Ice Lake (Bergsten, 1994). Elsewhere in Scandinavia, Sejrup (1987) used foraminiferal evidence to reconstruct changes in marine energy conditions in successive interstadial and stadial episodes during the Eemian/early Weichselian transition, while in the Ionian Sea (Mediterranean), Rasmussen (1991) was able to reconstruct former oxygenation levels and associated water mass turnover during the early Holocene from foraminiferal assemblage data.

Foraminifera, therefore, can be used as indicators of both local and regional environmental conditions, although as with ostracod assemblages described earlier, they are best employed in conjunction with other fossil remains, particularly molluscs, ostracods, diatoms and pollen. Their most significant contributions to Quaternary environmental reconstruction, however, have undoubtedly been in the provision of oxygen isotope data (sections 3.10, 5.5.3 and 6.2.3.5) and in the information that they can provide about changes in the character and movement of ocean water masses and currents, a topic which is considered in more detail in the following section.

4.11 Micropalaeontology of deep-sea sediments

4.11.1 Introduction

It has already been shown (section 3.10) how the ratios of oxygen isotopes in marine microfossils can provide a record of the expansion and contraction of ice sheets during the Quaternary, and the applications of this technique in Quaternary stratigraphy will be considered in Chapter 6. However, the marine microfaunal and microfloral assemblages themselves can also provide valuable palaeoenvironmental data, since they commonly retain a record, at least in part, of former ocean-water temperatures that will, in turn, be a reflection of former climatic conditions. From an analysis of the biostratigraphy of the deep-sea sediments, therefore, it is possible to make reasoned inferences about climatic régimes that prevailed over large areas of the world's oceans at different times during the Quaternary period. In addition, marine microfossil records have been used to reconstruct changes in oceanic circulation (palaeoceanography) and in nutrient supply (palaeoproductivity), as well as in the dissolved oxygen content of ocean waters.

Some of the marine organisms found in the deposits of the deep-ocean floors that are employed in palaeoclimatic research have been discussed above in relation to fresh or brackish water situations, or to relatively shallow shelf seas. These include foraminifers, diatoms and ostracods. Of the remaining marine organisms, the most valuable in terms of their application to palaeoclimatic research have proved to be the Radiolaria and coccolithophores.

4.11.2 Radiolaria

Radiolaria are marine, amoebic protozoans that secrete elaborate skeletons composed largely of amorphous (opaline) silica, which is extracted from sea water in the same way that Foraminifera extract calcium carbonate (Anderson, 1983). The skeleton, which consists of a complex network of elements, is contained within the living protoplasm and thus the hard parts forming the fossil do not dissolve in sea water until the creature dies. The single-celled radiolarians are usually circular in shape and average between 100 and 2000 μm in diameter (Figure 4.35). There are now about 400 to 500 relatively common species and these are found in all ocean waters from the tropics to the sub-polar seas, in surface waters down to depths of over *c.* 4 km. Most of the taxa have specific, sometimes narrow, bathymetric preferences, though some can occur in a very wide range of water depths (**eurybathyal**) (Casey, 1993).

Although information on the Quaternary stratigraphy of radiolarians is currently limited, there is growing evidence that they have considerable potential for Quaternary research, for assemblage data have been employed to reconstruct temporal changes in water-mass conditions and sea-surface temperatures in, for example, the Pacific (Schramm, 1985; Morley & Dworetzky, 1991). As with the Foraminifera, they are tolerant of a range of temperature and salinity conditions, but studies of abundance variations have shown that radiolarians can be grouped into major ocean provinces, reflecting changes in near-surface water properties, including sea-surface temperatures (Morley & Dworetzky, 1991).

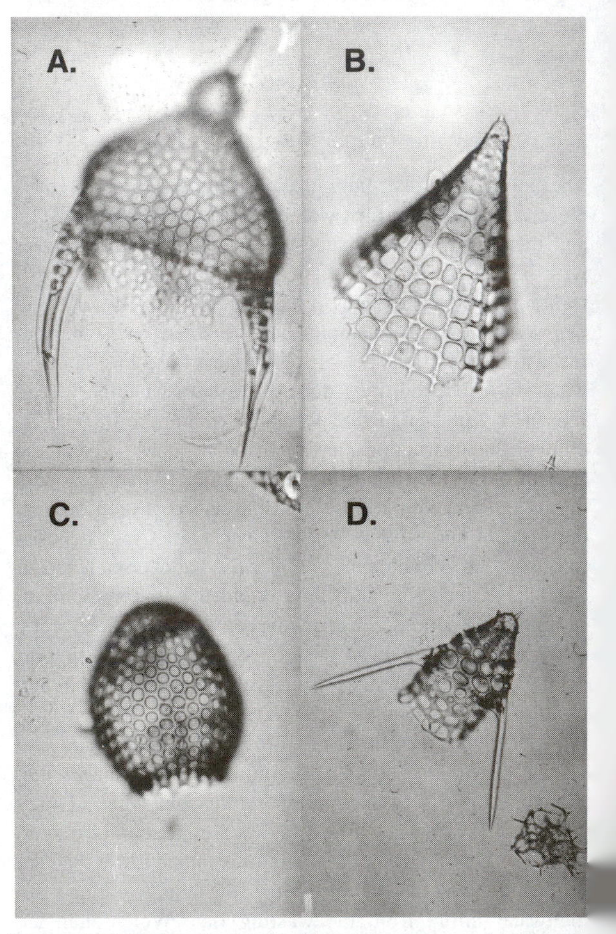

Figure 4.35 Radiolaria obtained from core-top samples of Holocene sediments from the Arabian Sea: A. cf. *Pterocanium* sp. (×200); B. *Peripyramis circumtexta* (×300); C. *Carpocanastrum* sp. (×295); D. *Lycnocanium* sp. (×480). Photos supplied by Alan Lord.

4.11.3 Coccolithophores

Coccolithophores are the most common members of a group of unicellular autotrophic marine algae known as **calcareous nannoplankton**. They are generally spherical or oval in shape, and are mostly less than 100 μm in diameter (Figure 4.36). The living organism is covered by a layer of organic scales upon which small calcite platelets, commonly ranging in size from 2 to 25 μm and called **coccoliths**, are secreted (Siesser, 1993). These may envelop the cell completely to form a hollow sphere (**coccosphere**) which eventually disintegrates and falls to the ocean bed. The individual button-shaped coccoliths are usually all that remain of the former living creature.

Like other marine flora, coccolithophores are autotrophic, possessing chloroplasts that are used to photosynthesise food. They possess whip-like threads (**flagellae**) to generate motion, and a few are known to ingest bacteria and small algae. They are therefore difficult to classify, possessing characteristics of both plants and animals. Although a few species are adapted to either fresh or brackish water, the majority of present-day coccolithophores are marine, but are much less common at salinities greater than 38‰ or less than 25‰. Being photosynthetic, they are confined to the photic zone of the water column and are rarely encountered below 200 m depth (Winter & Siesser, 1994). They are found in very large numbers in present-day ocean surface waters where they rival diatoms as the most abundant phytoplankton. Their abundance and species distributions are governed by a combination of light, salinity and temperature, but it is their temperature requirements that are best known, and the down-core abundances of key indicator species formed the basis of some of the early research on Quaternary palaeoclimatic and palaeoceanographic reconstructions (e.g. McIntyre & Ruddiman, 1972; McIntyre *et al.*, 1972), and this approach has also been adopted in more recent palaeoclimatic research (e.g. Weaver & Pujol, 1988; Molfino *et al.*, 1990).

4.11.4 Marine microfossils in ocean sediments

Planktonic Foraminifera are the major contributors to deep-sea sediments and, along with coccoliths, account for

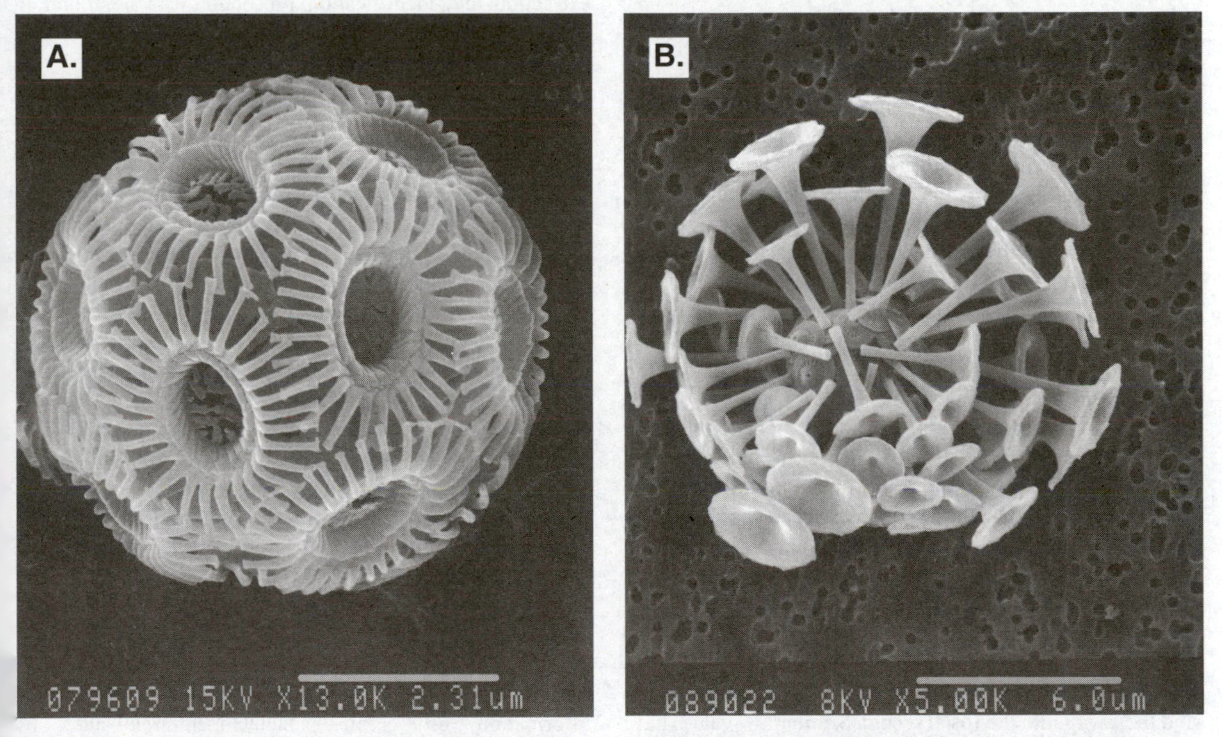

Figure 4.36 SEM micrographs of coccolithophores. A. *Emiliania huxleyi* coccosphere (white bar = 2.31 μm) – a composite exoskeleton formed of about 20 calcareous plate-like coccoliths and enclosing a single-celled planktonic alga. *E. huxleyi* evolved in the late Quaternary but is now the dominant coccolithophore in the world's oceans and one of the most abundant organisms on earth. B. *Discosphaera tubifera* coccosphere (white bar = 6.0 μm). SEM images supplied by Jeremy Young of the Natural History Museum, London.

more than 80 per cent of modern carbonate deposition in seas and oceans (Brasier, 1980). Most of the shells now being deposited are from planktonic species of *Globigerina* and it has been estimated that about 30 per cent of the present ocean floor (over 60 million km^2) is covered by the grey mud known as *Globigerina* **ooze**. These oozes are forming at depths up to 5 km in ocean waters between 50°N and 50°S. Coccolith oozes form principally in the tropical and subtropical regions, where the remains may average up to 30 per cent by weight of the sediments. In arctic regions, by comparison, the values may be as low as 1 per cent. By contrast with Foraminifera, however, coccolith remains settle much more slowly, and are therefore more susceptible to carbonate dissolution. Although some coccoliths may settle out more rapidly if they are contained within the faecal pellets of planktonic grazers, it has been estimated that less than 25 per cent of all coccolith species are actually preserved in the fossils of ocean sediments (Ruddiman & McIntyre, 1976). Below 3–5 km, the calcium carbonate compensation depth, nearly all $CaCO_3$ enters into solution, and thus only the most resistant calcareous fossils will be found. The sediments there will be dominated by siliceous remains, predominantly of Radiolaria.

Radiolaria accumulate in abundance in equatorial sediments where productivity is high in the water column above. However, as the productivity of calcareous organisms is also high, the radiolarian remains are often masked by foraminiferal and coccolith fragments. Only in areas such as the tropical northern Pacific, where large areas of the sea floor lie below the carbonate compensation depth, are radiolarian oozes encountered. Diatomaceous oozes are also found in the abyssal depths of the Pacific and Indian Oceans and in parts of the Atlantic (Kemp & Baldauf, 1993; Kemp *et al.*, 1994a). Some of these oozes are laminated and appear to have formed along oceanic temperature 'fronts' where upwelling leads to large concentrations of diatoms near the surface, and periodic die-off promotes sudden accumulation (**diatom mats**) on the sea floor (Sancetta *et al.*, 1991; Kemp *et al.*, 1994a). Siliceous oozes are most common in the high latitude areas of the north Pacific and around Antarctica. In these regions, calcareous fossils are rare and both radiolarian and diatom remains are abundant. As with carbonates, however, silica is soluble in sea water, dissolution being especially rapid in the upper levels of the water column. Only those radiolarian species with a solid opaline skeleton reach the sea floor and, overall, it has been estimated that perhaps as few as 10 per cent of all Radiolaria find their way into the fossil record. Similar low values have been suggested for diatoms. Both Radiolaria and diatoms are prone to exhumation and reburial in younger sediments and this poses further problems of interpretation for the marine biostratigrapher.

4.11.5 Laboratory separation of marine microfossils

Faunal and floral remains are extracted from deep-ocea cores in the laboratory by disaggregation of the sediment water or, where necessary, in hydrogen peroxic (Foraminifera, Radiolaria), nitric acid or hydrochloric ac (Radiolaria), or sodium hexametaphosphate (Calgon) in th case of coccoliths. The larger fossils can be hand-picke from the meshes of sieves, while for the smaller remain particularly the coccoliths, it is necessary to concentrate th microfossils into a liquid which can then be mounted on microscope slide. High-powered microscopy (up to ×160 may be necessary for ultra-detailed study using transmitte reflected and polarised light and, as in oth micropalaeontological work, increasing use is being made the electron microscope. Identifications may be mac difficult by the solution of diagnostic parts, by th mechanical wear of the skeletal remains, and by th tendency, especially in the case of carbonate fossils, fo calcite overgrowth and recrystallisation to obscure th morphology of the surface features. Further details extraction and identification procedures can be found i Brasier (1980) and Lipps (1993).

4.11.6 Marine palaeoclimatology

The distributions of Foraminifera, Radiolari coccolithophores and marine diatoms are partly determine by nutrient requirements. The planktonic forms are all foun in abundance in zones of upwelling, for example, pronounced vertical mixing, where mineral nutrients ar readily available. For this reason, large numbers of thes micro-organisms are frequently encountered just seaward the continental slope. In most cases, however, th fundamental determinant is water temperature, and detaile ecological studies have shown that many species ar associated with water masses that possess distinctiv thermal characteristics. From the present distribution marine plankton, and allowing for current circulation an Coriolis effects, it is possible to recognise distin equatorial, tropical, sub-arctic and arctic province (illustrated in Figure 4.37 for the present-day distributions c planktonic Foraminifera). Hence, the analysis of marin microfossil assemblages can provide a unique source information on ocean palaeotemperatures and, b implication, on former climatic conditions.

The initial approach to Quaternary temperatur investigations using data from deep-ocean cores was base simply on the presence or absence of certain key species i

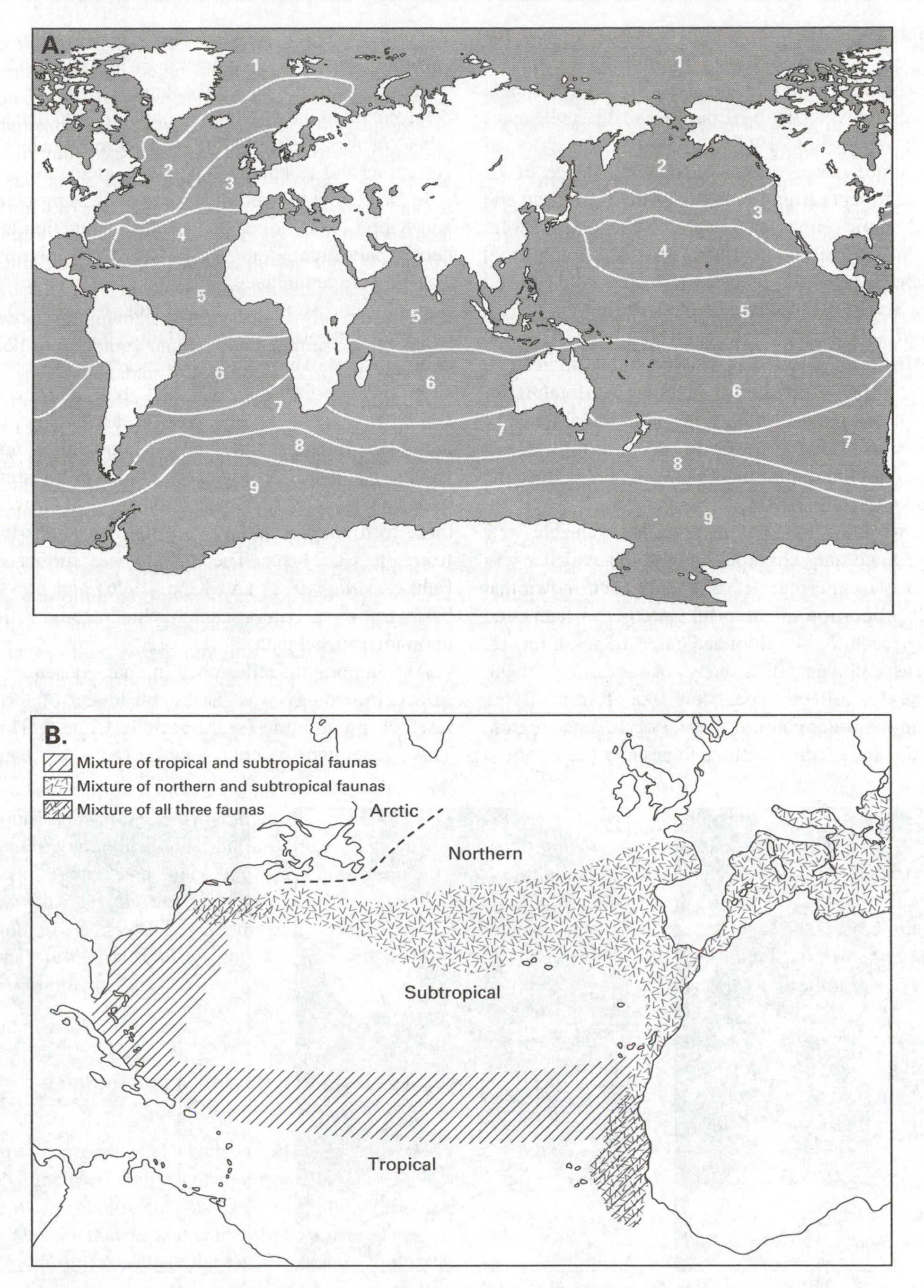

Figure 4.37 A. Modern planktonic foraminiferal provinces: 1 Arctic, 2 Subarctic, 3 Transitional, 4 Subtropical, 5 Tropical, 6 Subtropical, 7 Transitional, 8 Subantarctic, 9 Antarctic (after Bé and Tolderlund, 1971). B. Generalised distribution of planktonic foraminiferal faunas in the North Atlantic. Four distinct faunal provinces (Arctic, Northern, Subtropical and Tropical) are recognised; shading indicates zones of mixing between these groups. 1, mixture of tropical and subtropical faunas; 2, mixture of northern and subtropical faunas; 3, mixture of all three faunas (after Cifelli & Benier, 1976).

fossil assemblages. Early work demonstrated that the abundance of the planktonic foraminiferal species *Globorotalia menardii* could be used to infer climatic change, an idea developed by Ericson and his colleagues (e.g. Ericson & Wollin, 1968) to construct a series of palaeotemperature curves based on the abundance of *G. menardii* in sediments from the floors of the Caribbean and subtropical Atlantic. High percentages of *G. menardii* were interpreted as indicating warmer, possibly interglacial periods, while reduced frequencies reflected cold, glacial periods. In a series of influential papers during the 1970s, McIntyre *et al.* (1972), McIntyre and Ruddiman (1972) and Kellogg (1976) used selected planktonic faunal indicators, particularly the markedly polar foraminifer *Neogloboquadrina pachyderma*, along with the absence of coccolith remains at certain levels in cores from the North Atlantic, to record the migration of the Polar Front since the last interglacial (Figure 4.38).

Although work of this nature provided valuable new insights into Quaternary climatic changes, the evidence was not always easy to interpret. It has already been shown that only a small proportion of the planktonic ocean fauna and flora actually reach the sea floor and enter the fossil record. The death assemblage in a body of ocean sediment, therefore, rarely reflects accurately the former living assemblage in the water column above. For the same reason, however, indicator species in these invariably biased fossil assemblages are not always a reliable index of palaeotemperature change. Moreover, although temperatures are generally believed to be the major determinant in the distribution of planktonic fauna and flora, other factors need to be considered such as salinity variations and seasonal temperature fluctuations.

In an attempt to overcome some of these problems, Imbrie and Kipp (1971) developed a transfer function approach to derive palaeotemperature estimates from microfossil data. The relative abundances of some 26 species of planktonic Foraminifera were calculated for 61 core-top samples distributed throughout the Atlantic basin. Factor analysis distinguished five distinct ecological assemblages: tropical, subtropical, polar, subpolar and gyre margin, and these could be related to observed oceanographic parameters including summer and winter temperatures and salinity values. The palaeoecological equations or transfer functions based on least-squares regression were then used to relate these to fossil assemblages at different depths in the cores from the Caribbean. The method was further refined by Imbrie *et al.* (1973) and Kipp (1976) and now forms the basis for most palaeoceanographic reconstructions based upon microfossil data.

An important milestone in palaeoceanography and palaeoclimatology was the establishment of the CLIMAP research programme (see especially Cline & Hays, 1976). This generated a considerable body of data on the

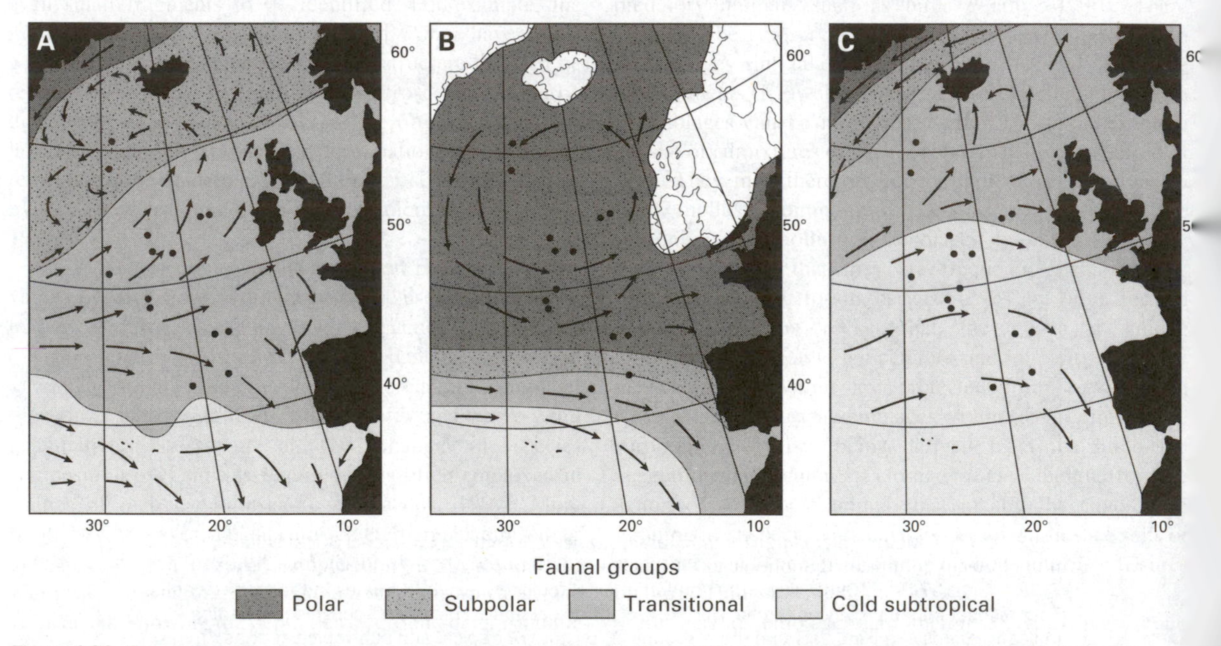

Figure 4.38 Oceanographic and palaeoceanographic maps of the northeast Atlantic Ocean based on ecological affinities of planktonic Foraminifera: A. present day; B. last glacial maximum (*c.* 18 ka BP); C. last interglacial (*c.* 125 ka BP) (after Ruddiman & McIntyre, 1976).

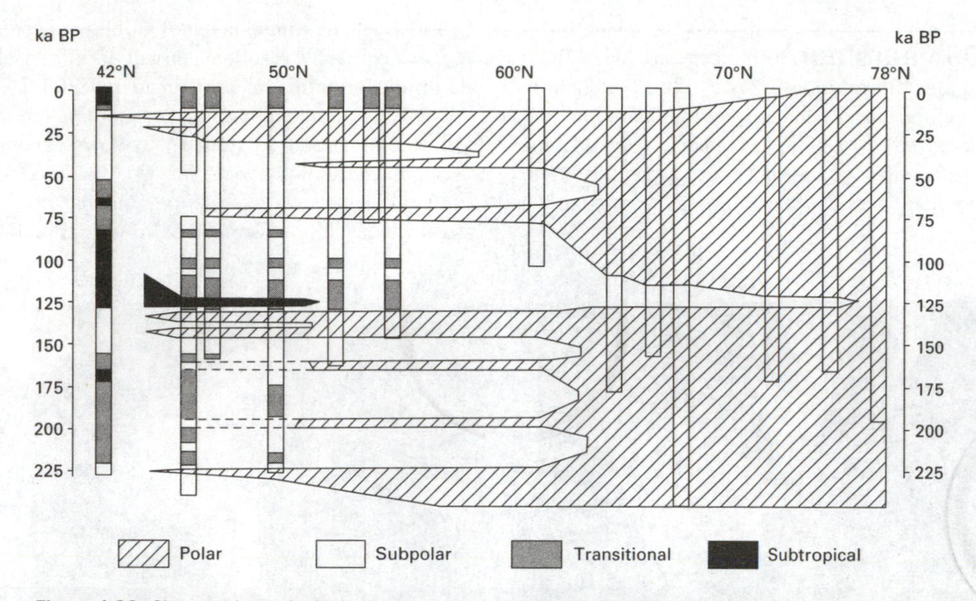

Figure 4.39 Changes in ecological water masses in the Norwegian Sea and the northern North Atlantic over the past 225 000 years; the reconstruction is based on analysis of foraminiferal assemblages in a transect of marine sediment cores obtained from the sea-bed (after McIntyre & Ruddiman, 1972).

Quaternary history of the world's oceans and atmosphere, and laid the foundations for the high-resolution palaeoceanographic reconstructions that form such an important component of contemporary Quaternary research (see Chapter 7). Within the North Atlantic, for example, Ruddiman and McIntyre (1976) and Kellogg (1976) were able to reconstruct time–space variations in water mass boundaries for the last 225 000 years based on the occurrence of distinctive foraminiferal and coccolith assemblages in the deep-ocean cores (Figure 4.39). These water masses appear to have migrated across more than 20° of latitude (some 2000 km) at rates of up to 200 m per year. From these data it would appear that glacial surface water temperatures in the area were up to 12.5°C lower in winter and 13.0°C lower in summer than at present, and within the last 600 ka alone, at least eleven separate southward movements of polar water occurred.

The CLIMAP Project (1981) produced maps showing inferred sea-surface temperatures (**SSTs**) for the world's oceans at the last glacial maximum, *c.* 18 ka BP, compared with the present day (Figure 4.40). Not only did these provide the first quantified model of the state of the oceans during full glacial conditions, but they also highlighted those parts of the oceans that experienced the largest temperature differences between the last glacial maximum and the present day (Figure 4.40), with the major change in the North Atlantic perhaps being the most distinctive feature.

Such reconstructions provide essential **boundary conditions**[8] for the development of Global Circulation Models (GCMs) which aim to simulate past climatic changes at a global scale (section 7.7). For example, models of the global climate at 18 ka BP developed by the COHMAP group employ the CLIMAP global SST reconstructions as boundary conditions for the ocean surface (COHMAP Members, 1988). For this purpose, the SST values were averaged for 2° grid squares (Kutzbach & Guetter, 1986). The COHMAP experiments were the first to simulate global palaeoclimate changes at 3 ka intervals from 18 ka BP to the present day. The CLIMAP SST reconstructions for 18 ka BP were also used to set the boundary conditions for a 15 ka BP simulation, since the oceanic records do not show any evidence of major warming between 18 and 15 ka BP (Kutzbach & Ruddiman, 1993). For later periods, SST estimates were interpolated between the CLIMAP 18 ka BP values and modern SSTs.

Both the CLIMAP and COHMAP experiments have been extremely valuable in providing a fuller understanding of the way the global climate system operates. They have enabled the following conditions to be modelled at a global scale for both January and July: surface temperatures, precipitation, air pressure at sea level, surface wind speeds and probable surface storm tracks (Kutzbach *et al.*, 1993). By comparing the reconstructions for 18, 12, 9 and 6 ka BP with modern surface conditions, the major differences in circulation

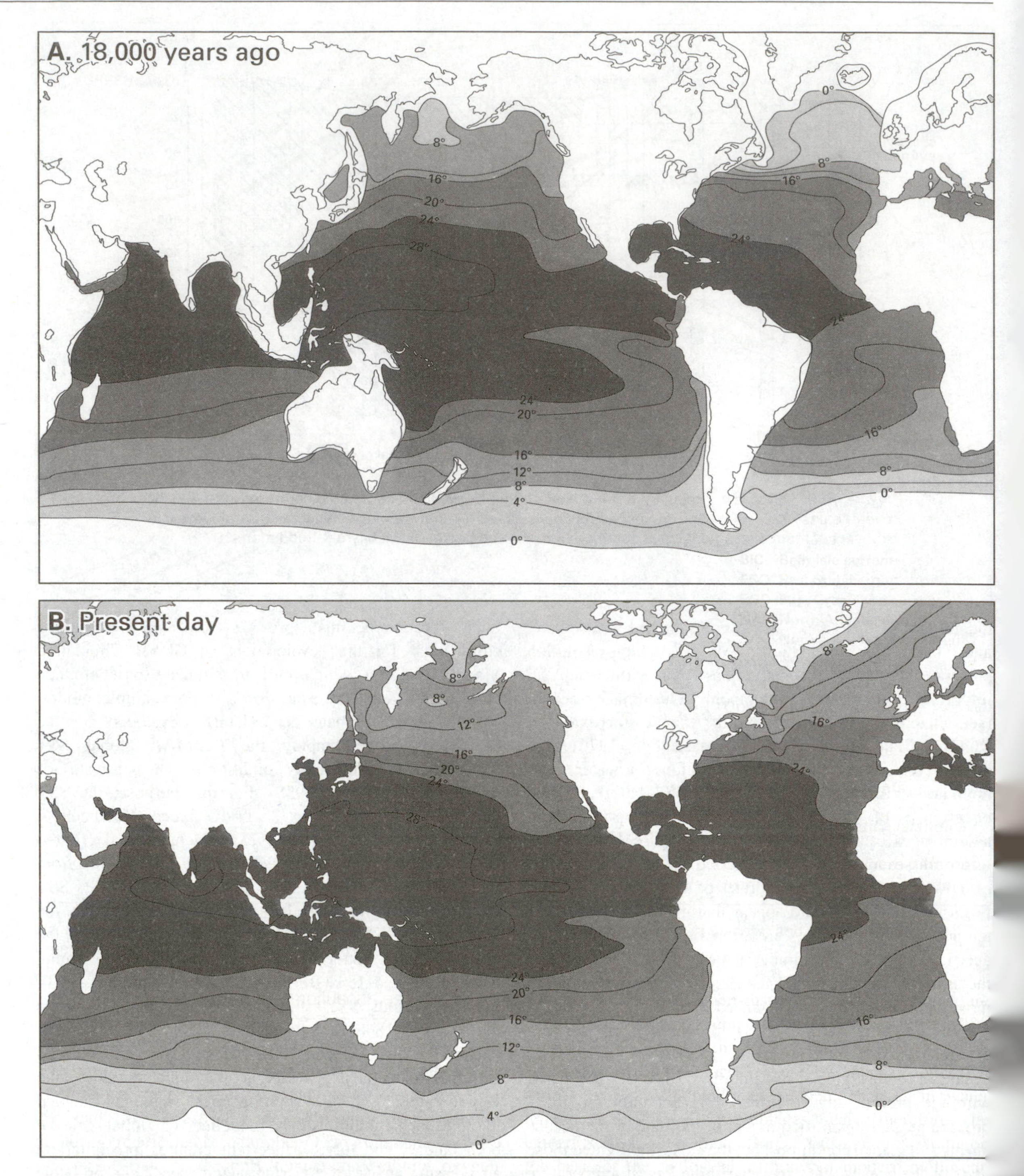

Figure 4.40 Reconstruction of surface ocean temperatures in August for (A) the Last Glacial Maximum (*c.* 18 ka BP) compared wit (B) the present day, based on the CLIMAP group's results (1981).

patterns can be established, and the mechanisms behind these circulation shifts can be assessed. SST reconstructions at improved temporal and spatial scales are therefore key elements in global circulation modelling (section 7.8). While much of this research has been concentrated in the Atlantic (e.g. Overpeck *et al.,* 1989; Ruddiman & Mix, 1993) and Pacific Oceans (Wright *et al.*, 1993), new SST data have also been generated for other, previously less well-studied marine areas, such as the Southern (Antarctic) Ocean (Crowley & Parkinson, 1988) and the Indian Ocean (Prell *et al.,* 1990). In addition, refinements have been sought in the transfer functions from which the SST estimates are derived (e.g. Mix *et al.,* 1986).

Ultimately it may be possible to use these data to predict future climatic changes, for realistic reconstructions of past rates of change of oceanic circulation may have implications for the way in which the oceans might respond in future to changing environmental conditions. Ruddiman and Mix (1993), for example, demonstrate how, on the basis of this type of evidence, the most thermally reactive parts of the oceans can be identified, and it is these zones that may be the most sensitive to future climatic perturbations. One such area is the northeast Atlantic, which experienced the highest amplitude of climatic change between 18 ka BP and the present day (Figure 4.40). In addition, rapid sedimentation on shelf areas in this region has taken place during the past 15 000 years or so, and these provide a basis for reconstructing SST variations at a high temporal resolution. Factor analysis of modern (core-top) diatom assemblages and their relationships to modern SSTs in the Greenland, Iceland and Norwegian (GIN) Seas has been used to infer down-core temperature variations over the past 14 ka (Koç Karpuz & Schrader, 1990; Koç Karpuz & Jansen, 1992). Comparison of a number of core records on this basis has enabled detailed reconstructions of sea-ice cover and temperature variations over the GIN seas during the past 13.5 ka (Koç *et al.*, 1993) from which inferences can be made about the location of the North Atlantic Polar and Arctic Fronts (see Chapter 7 for more detail).

A new approach to marine palaeoclimatology is the measurement of long-chain carbon compounds (C_{37-39} or U^{k}_{37} alkenones) in microfossil remains. Several of these compounds have been identified in the organic residues of coccolithophores (Marlowe *et al.,* 1990), and their abundance has been found to be closely correlated with the oxygen isotope record for the past 600 000 years (Eglinton *et al.*, 1992). It seems that the measurement of trace quantities of long-chain organic compounds, made possible by recent advances in chromatographic techniques, might offer a new, independent method of ocean palaeotemperature reconstruction.

4.11.7 Marine palaeoproductivity and palaeocirculation

Just as the present gyres and climate fronts (water mass boundaries) of the oceans are determined by the prevailing circulation pattern, so too are the distribution of nutrients and the rate of exchange between the ocean and the atmosphere on the one hand, and between shallow and deep ocean water on the other. Changes in any of these parameters will be reflected in chemical changes in the water column and in ocean sediments. It has recently been found that, in addition to an oxygen isotope signal, marine microfossils contain a record of trace element variations in ocean waters, and this constitutes a powerful new tool in palaeoceanography. For example, Cd/Ca ratios can be used to infer past variations in productivity. Cd resembles phosphate in terms of its behaviour in ocean water and so a measurement of Cd/Ca ratios in foraminiferal tests can be used as a proxy for nutrient (P) levels (Boyle, 1988). Keigwin *et al.* (1991) measured Cd/Ca ratios in benthic Foraminifera in the North Atlantic and identified four anomalies, dating between 14 500 and 10 500 years BP, that correspond with evidence for meltwater discharges. They suggested that variations in the Cd/Ca ratios reflect changes in nutrient supply which may, in turn, reflect changes in the rate of North Atlantic Deep Water formation. Other useful '**chemical tracers**' are carbon isotopes, ^{230}Th, opal, Ba/Al and Ti/Al ratios (Broecker, 1990; Francois *et al.,* 1990; Dymond *et al.,* 1992). By comparing ratios observed in benthic microfossils with those observed in contemporaneous planktonic specimens, the nutrient flux and degree of oxygenation in deeper water can be inferred from comparisons with shallower water. Where there is reduced nutrient supply and reduced oxygen content in benthics by comparison with planktonic specimens, this suggests a more sluggish ocean 'turnover' (see, e.g., Boyle & Keigwin, 1987; Oppo & Fairbanks, 1990).

4.12 VERTEBRATE REMAINS

4.12.1 Introduction

Fossil animal bones and teeth, particularly those of large vertebrates, have long been an attraction for enthusiastic collectors and, as a result, museums in many parts of the world are full of the skeletal parts of Pleistocene mammals. Many of these remains were removed from exposures in cliffs and in river valleys towards the end of the last century by well-meaning Victorian enthusiasts who, unfortunately, often paid scant regard to the stratigraphic context within which they lay, or indeed to the less spectacular but equally important collecting of smaller animal remains which together formed the total assemblage of the stratum. Thus, although the fossil remains indicate that extinct animals such as the mammoth (*Mammuthus primigenius*) and woolly rhinoceros (*Coelodonta antiquitatis*) were to be found in many areas of the Northern Hemisphere during the Quaternary (Haynes, 1993), and also that creatures such as the hippopotamus (*Hippopotamus amphibius*) and musk-ox (*Ovibus moschatus*) formerly occupied ranges that are very different from those of their present-day counterparts, albeit at different times, much less is known about the smaller vertebrates and their Quaternary distributions. In the last few decades, however, considerable progress has been made in understanding the spatial and temporal variations in both large and small animal populations during the Quaternary, and their relationships to environmental changes.

Vertebrate remains are found in Quaternary sediments in a variety of forms. Occasionally hair, muscle and hornsheaths are preserved, and, in exceptional locations such as the permafrost of arctic Siberia, tar pits and peat bogs, mummified carcasses have been found (Sutcliffe, 1985; Turner & Scaife, 1995). Other evidence of the former presence of animals includes nests and middens of rats, burrows, hyaena dens, **coprolites** (droppings – section 3.8), bird pellets, diagnostic teeth marks by predators on other bones, 'trace fossils' (prints and tracks) and skincasts (Churcher & Wilson, 1990). Overall, however, it is teeth and bones (and, occasionally, antlers) that make up the majority of fossil vertebrate remains (Figure 4.41), and these form the principal focus of the following discussion.

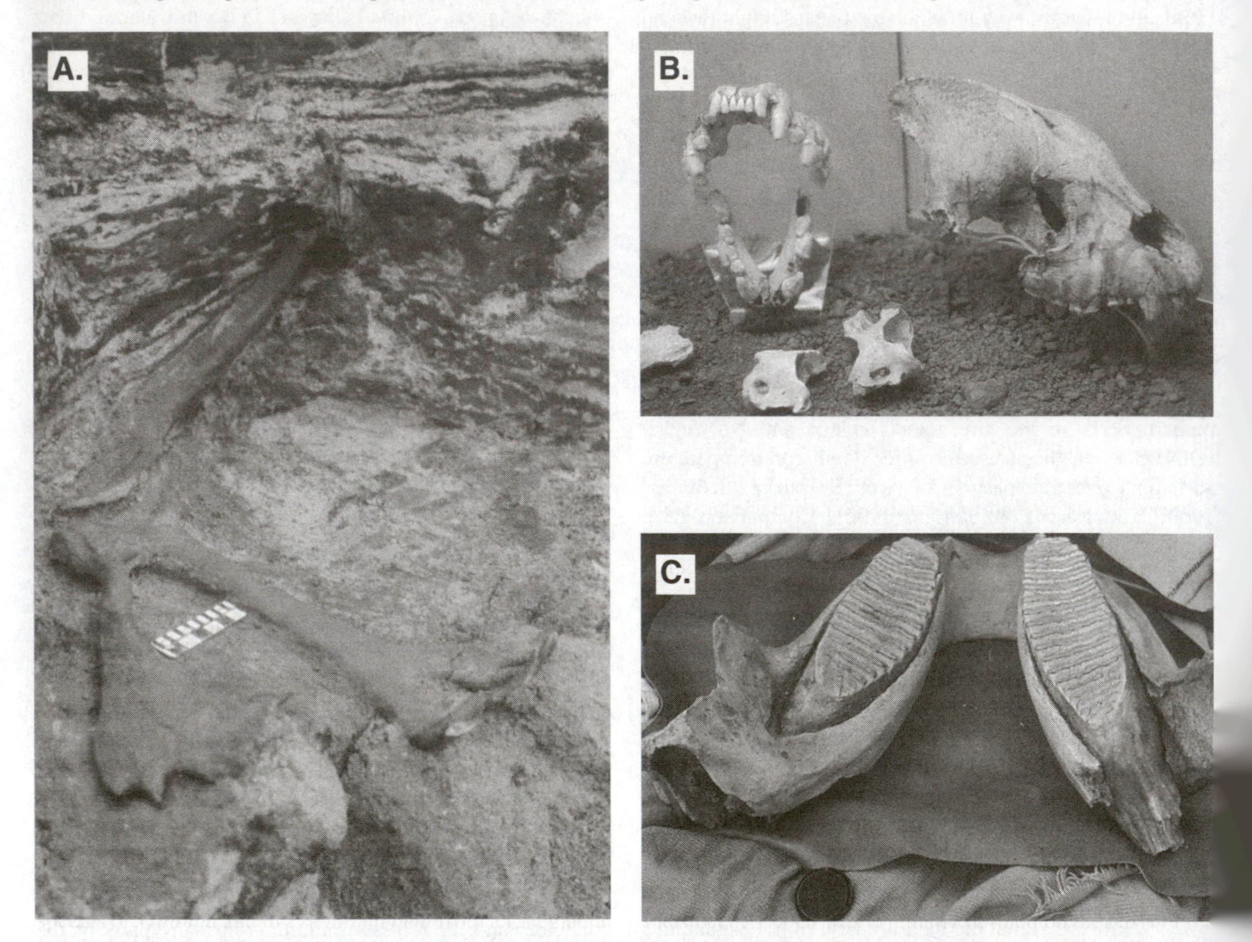

Figure 4.41 A. Reindeer (*Rangifer tarandus*) antler exposed within Devensian cold stage fluvial sands at the site of Shropham, Norfolk, England (photo: Adrian Lister). B. Jaws and skull of spotted hyaena (*Crocuta crocuta*) recovered from the Middle Devensian fill of Kent's Cavern, England (photo: Adrian Lister). C. Lower jaw of mammoth, complete with molars, recovered from fluvial gravels deposited during a temperate phase (probably late Oxygen Isotope Stage 5e) at Cassington, near Oxford, England (photo: Mike Walker).

4.12.2 The structure of teeth and bones

Teeth are extremely complex in structure but in most mammals consist of three distinct substances of differing hardness: the hard brittle outer casing (**enamel**), the softer **dentine** of which the greater part of the tooth is composed, and **cement** which covers the dentine of the roots and occasionally the valleys and folds of the main tooth body (Cornwall, 1974). In the fossil, the enamel provides the most durable element except where burning has affected the original dental material, in which case the dentine of the tooth roots may prove to be the most resistant. Teeth are of considerable importance in palaeoenvironmental work for not only do they provide data on the age (years of life) of the animal, but they also give an indication of dietary preferences (i.e. herbivore or carnivore). In more recent sediments, they tend to be outnumbered by fragmentary remains of bone, but occasionally in older Quaternary deposits, teeth are often proportionately more strongly represented than elements of the post-cranial skeleton (Stuart, 1982; Davis, 1987).

Fresh animal bone consists of both organic material and inorganic material in the approximate ratio by weight of 30:70. The organic fraction is contained within the shafts of long bones (e.g. femurs, tibias, vertebrae) and comprises cell tissue (fats, etc.) and a fibrous protein called **collagen**. The collagen is very resistant to decay and may survive for thousands of years following the death of the animal, while the remaining organic matter undergoes autolysis after death and is rapidly decomposed. Surrounding the collagen fibre is bone mineral material, the principal component of which is a phosphate of calcium, **hydroxyapatite** ($Ca_{10}OH(PO_4)_6$). The structure and composition of animal bones are of considerable interest to the palaeoenvironmentalist as they affect the way in which fossilisation takes place, and the chemical structure of bones in particular provides a means of dating the fossil material (see section 5.6.2).

4.12.3 Fossilisation of bone material

Quaternary vertebrate remains have been recovered from a wide range of deposits. These include cave and fissure sediments, lacustrine and marine deposits, fluvial sediments (especially river terraces), peat bogs, soils and a variety of situations associated with human activities such as middens, cess-pits and burial chambers. At some sites, whole skeletons have been found, but more frequently the fossil assemblages consist of disarticulated skeletons and an admixture of bones of varying sizes and in differing states of preservation. Animal bones are perhaps more susceptible to physical and chemical changes than any other biological remains encountered in Quaternary deposits. They are highly susceptible to 'weathering' by a variety of processes, if exposed to oxidation, temperature extremes or chemical exchange with ground solutions, although in some instances the extent of weathering has been used to advantage to estimate duration of exposure of bone assemblages (Behrensmeyer, 1984).

As soon as a bone becomes incorporated into a body of sediment, it begins to undergo chemical changes that vary considerably in nature and degree with the chemistry of the surrounding matrix. In most deposits where air is freely circulating, the mineral parts of the bone will tend to be more resistant to decay, while the organic substances will break down rapidly into simple compounds such as ammonia, carbon dioxide and water. Mineral salts in solution in the surrounding sediment, particularly calcium and iron, will be deposited in the vacant pore spaces and the bone may eventually become completely **permineralised**. This is one reason why deposits in caves in limestone regions are often comparatively rich in mammalian remains (Sutcliffe, 1985). As the process usually proceeds fairly slowly, a high degree of mineralisation will generally indicate considerable age. On the other hand, if the host deposit is acidic (e.g. peats or peaty-gleys) and depleted in bases, both the organic and mineral fractions will decompose and the bone will disappear completely, leaving no trace of its former existence. Thus while prehistoric burials on chalklands in areas such as southern England have often yielded well-preserved bones, those in adjacent regions where porous, sandy soils are found contain few bone remains.

In waterlogged situations, a completely different set of reactions occurs. In deep lakes where the substrate is of limestone and where bases are in plentiful supply, bones are not only well preserved but are often extremely hard. In some cases even the organic elements have been converted into a stable wax-like substance known as **adipocere** (Cornwall, 1974). At the other extreme, in peat bogs or in oligotrophic lakes, the anaerobic nature of the depositional environment often results in the organic portions of the bones being preserved, while attack by humic acids leads to complete destruction of the mineral fraction. Skeletal remains will, therefore, be found in a soft or pulpy state in advanced stages of decalcification (Turner & Scaife, 1995).

Finally, there are the effects of burial on bone that are purely physical. The seasonal drying of clay soils, for example, will result in the fissuring and eventual destruction of even the strongest bones. Bones may be similarly shattered by frost-heaving and by the action of ground ice. Soil creep, solifluction and mechanical abrasion in river gravels will have similar damaging effects and will result in the progressive fragmentation of bone remains.

4.12.4 Field and laboratory techniques

Because bones can be found in such a variety of conditions, great care must be exercised in the excavation of bone-bearing deposits. Mapping and surveying augmented by field description, sketches and photographs should precede the removal of bone fragments from the sediment matrix. In some cases, it may be possible to remove the larger bones by hand. These can be left to dry out and then cleaned with a brush or by gentle agitation in water. Many bone remains, however, even if heavily mineralised, are quite brittle and it may be necessary to treat these with a penetrative plastic solution (e.g. polyvinyl acetate in toluene) before removal from the matrix can be attempted (Leiggi *et al.,* 1994). If the bones are wet, especially those that are markedly decalcified, an emulsion of the plastic solution in water may be needed in order for the strengthening material to penetrate the bone fibres. Bones that are so treated, however, are effectively useless for any subsequent chemical analysis or for radiocarbon dating. The presence of very small bones or teeth (of rodents, for example) can usually only be detected by sieving the matrix following the removal of the larger faunal remains by hand (Davis, 1987).

Identification of bone remains is usually carried out in the laboratory and often proceeds in two stages. As most identifications are based upon fragmentary evidence, the first step is to identify the bone of which the fragment is a part. This is usually achieved by comparing the fragment with fresh skeletal material from a range of animals of different sizes. The second, and more difficult, stage is to track down the animal from which the unknown bone was derived. Here a reference collection of type material is essential, although the development of a type collection for the Quaternary vertebrates involves many more difficulties (and considerably more expense) than are encountered in the construction of a reference collection for Quaternary pollen grains or coleopteran remains. The Mammalia, for example, include a proportion of taxa that are now extinct, while evolution and speciation during the Quaternary pose additional complications. Moreover, as relatively few sites have so far been properly investigated, it is scarcely surprising that very few museums possess a good or reliable reference collection suitable for identification purposes. In spite of these difficulties, however, positive identifications of Quaternary vertebrate remains are steadily increasing and it is now proving possible to construct fairly detailed faunal lists for the major stages of the Quaternary (Stuart, 1982; Martin & Barnosky, 1993), while the palaeobiogeography and evolutionary trends in particular groups of animals can be reconstructed in detail (e.g. Haynes, 1993; Gee, 1993; MacFadden, 1994).

4.12.5 The taphonomy of fossil vertebrate assemblages

The first stage in the interpretation of fossil bones and teeth is to establish how a particular grouping of vertebrate remains (Figure 4.41) came to be associated together. The various factors that can result in a more or less distorted picture of the living community as represented by the fossil assemblage were reviewed by Stuart (1982). Three different depositional environments will serve to demonstrate the complexities of fossil vertebrate assemblages.

4.12.5.1 Cave and fissure deposits

Some of the richest vertebrate assemblages in the world are those found in cave sediments, particularly in limestone regions, yet the ecological history of cave faunas is frequently very difficult to interpret because of the multiple origins of the fossil material (Figure 4.42). Some bones, for example, may have been washed into the caves or fissures by streamflow and are therefore allochthonous to the site. Caves were often occupied during the Quaternary by carnivores including cave bear (*Ursus*), wolf (*Canis lupus*), red fox (*Vulpes vulpes*), sabre-toothed cats (e.g. *Smilodon*) and spotted hyaena (*Crocuta crocuta*), and hence many cave assemblages will be biased towards the prey of these animals. Small vertebrate remains could, for example, have been derived almost entirely from pellets dropped by owls roosting in the cave roofs, and could include either woodland or open-country rodents depending on the species of owl involved. Many of the large vertebrate bones will have been dragged into the cave by predators so that the resulting assemblage will give some indication of the original large vertebrate fauna of the vicinity. However, the cave assemblages will inevitably be biased towards the predators themselves as many would have eventually died in the caves and thus contributed their bones to the assemblage. In some instances, the caves seem to have acted as natural pitfall traps with animals having fallen in through holes in the cave roof, a good example being Joint Mitnor Cave in Devon (Sutcliffe, 1960), or where the configuration of a cave opening is such that animals can venture in but are unable to escape. The resulting bone assemblages, therefore, will be partly biased towards the scavenging animals, such as the hyaena which would have been attracted to the cave by dead and dying animals. The difficulties of interpretation of vertebrate assemblages in cave sites are further exacerbated by the often complex stratigraphy of cave sediments (Figure 4.42 and section 3.8).

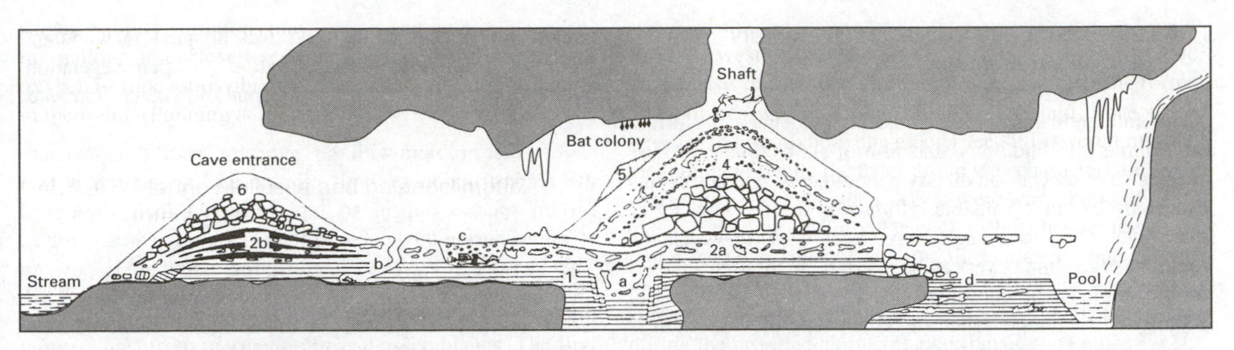

Figure 4.42 Vertical section through an imaginary cave sediment fill with preserved fossil bones. 1, Water-laid sands and clays; 2a, deposits of an animal lair; 2b, hearths of fires made by humans; 3, stalagmite floor; 4, talus cone with bones of animals which fell down a shaft; 5, bones and dung of bat; 6, second talus cone at cave mouth; showing disturbance by (a) collapse pit, (b) burrow, (c) human burial, (d) washing out and redeposition by a stream (after Sutcliffe, 1985).

4.12.5.2 Lacustrine sediments

Lake sediments often contain whole or partial skeletons of mammals, amphibians and fish. Remains of large mammals such as elk (*Alces alces*), reindeer (*Rangifer tarandus*) and mammoths (e.g. *Mammuthus primigenius*) found in lake deposits probably represent animals that died either by drowning after breaking through thin ice or after having become trapped in the soft mud on the lake floors in their efforts to drink, wallow or feed (Coope & Lister, 1987). Often the adjacent sediments will have been disturbed by the struggles of the animal to become free (Stuart, 1982). Many human occupation sites, such as the early Mesolithic hunting settlement at Star Carr in Yorkshire (Clark, 1954), were by lakes and rivers and therefore a proportion of the remains of animals that were hunted also found their way into the lake. Fish and amphibians clearly reflect the former presence of these animals in the lake waters, but again humans may have been responsible for the concentration of faunal remains in the littoral sediments.

4.12.5.3 Fluviatile sediments

The origins of the vertebrate remains in river sediments may be almost as diverse as those found in cave deposits. Large vertebrate remains become incorporated in riverine deposits in similar ways to those outlined above for lacustrine contexts, though allochthonous material is common because corpses can be floated downstream. From an analysis of the assemblage, it may be possible to gain some impression of relative population densities, of the lifespan of particular taxa, and of the distance of their habitats from the site of deposition. Fish and amphibian remains will be locally derived and will tend to be over-represented in the assemblage. The small animal remains are, however, more difficult to interpret. Some may be the remains of waterside creatures such as voles and rats, but Mayhew (1977) has emphasised the role of predatory birds as being responsible for the concentration of small animal bones in Quaternary fluviatile sediments. At West Runton in Norfolk, he was able to match corroded teeth and bone in river sediments of Cromerian age with specimens from the regurgitated pellets of modern kestrels and buzzards, and he therefore suggested that a major proportion of the fossil small mammal material from that site was transported to the point of deposition by avian predators. Stuart (1980) has suggested that carnivore droppings would provide an additional source of small mammal remains. Ironically, due to the extreme fragility of avian bones and their tendency to float, bird remains are seldom preserved in lacustrine or fluvial deposits. Consequently, the Quaternary history of the avian fauna is poorly understood.

A further difficulty that frequently arises over the interpretation of fossil assemblages from fluviatile deposits is that vertebrate remains of different ages are easily incorporated into the sediments as the river banks are eroded. Because similar (but not necessarily identical) animal populations existed during successive warm and cold stages of the Quaternary, the likelihood of erroneous ecological interpretations from mixed bone assemblages is very real, and care needs to be taken to establish the degree to which the mix of skeletal remains are contemporaneous. Sometimes it may be possible to recognise bones of different ages on the basis of varying degrees of physical deterioration. Alternatively, relative ages of bones may have to be established by chemical means, the most widely used techniques being the measurement of amino-acid ratios or of fluorine, uranium and nitrogen content, described in section 5.6.

4.12.6 Vertebrate fossils and Quaternary environments

Vertebrate remains provide useful evidence of former environmental conditions, and also of past climates. Many species have specialised dietary requirements, and herbivores especially can give valuable information on the character of the contemporaneous vegetation. Other ways in which the regional vegetation cover can be inferred are through analysis of the contents of coprolites and a knowledge of animal ecology, for example where a certain type of vegetation is required for camouflage, den or nest construction. Vertebrate fossils can be indicative of former ground conditions, for instance burrowing animals that require access to soft, uncompacted deposits, or hoof structures adapted to certain ground surfaces. Vertebrate remains can also greatly assist archaeological investigations, providing information on domesticated animals, diet and pests of sites of human habitation (e.g. Klein & Cruz-Uribe, 1984).

In terms of climatic reconstruction from faunal evidence, the traditional approach utilises the known distributions of present-day species as a basis for inferences about past climates. During the last interglacial, for example, hippopotamus (*Hippopotamus amphibius*), the pond tortoise (*Emys orbicularis*) and the lesser white-toothed shrew (*Crocidura* cf. *suaveolens*) were all found in southern Britain. The hippopotamus is now confined to tropical Africa, the pond tortoise is now found only in the Mediterranean and in southeast Europe with its northern breeding range apparently limited by the 18°C July isotherm (Stuart, 1979), while the lesser white-toothed shrew is also essentially southern European in its distribution. These data are, therefore, strongly suggestive of warmer summers and milder winters in the British Isles during the Last (Ipswichian) Interglacial by comparison with the present day, a hypothesis supported by both palynological and coleopteran evidence.

The use of vertebrate records for palaeoclimatic reconstruction is not without its problems, however. A straightforward relationship between animal distribution and climatic parameters should not be assumed, for, as pointed out by Zeuner in 1959, the primary adaptation of many vertebrate animals during the Quaternary may well have been to vegetation and only secondarily to climate. Climatic inferences may be particularly suspect during interglacials when both closed woodland and open grassland conditions often occurred in close proximity. In some last interglacial deposits in Britain, for example, the two mammoths *Mammuthus primigenius* and *Palaeoloxodon antiquus* were present shortly after the thermal maximum, yet the traditional interpretation is that the latter was the warm species while the former is indicative of cold conditions. A more likely explanation is that *P. antiquus* was adapted to woodland while *M. primigenius* was a native of open vegetation, perhaps akin to the warm loess steppe of Siberia (Sparks & West, 1972).

A further problem with the 'indicator species' approach to the reconstruction of both climate and vegetation is that certain species appear to have changed their ecological affinities during the course of the Quaternary (Stuart, 1982). The hamster (*Cricetus cricetus*), for example, is now an obligate steppe creature, yet it was present in the mixed oak woodland (albeit perhaps in more open habitats) of the Cromerian Interglacial in southern Britain; similarly, the musk-ox (*Ovibus moschatus*) which is now a tundra animal, occurred in open but hardly sub-arctic conditions at the end of the last interglacial, and was also present in the Early and Middle Devensian steppe faunas of central England. The fossil record of *Cervus elaphus* is even more intriguing, for red deer remains have been found associated with a variety of environmental conditions ranging from oakwoods (e.g. during the Cromerian Interglacial) to possibly full-glacial tundra environments of the Devensian (last) cold stage. These examples serve to reinforce the point that, in the absence of corroborating palaeobotanical data, assemblages are more reliable indicators of former environmental conditions than a single taxon.

While fossil vertebrate assemblages (as opposed to individual species) should be employed wherever possible in palaeoenvironmental reconstruction, this aim is frequently difficult to achieve in practice. Taxonomic imprecision as a consequence of the poor state of preservation of fossil material, or simply the lack of type material, often places constraints upon the reconstruction of whole animal communities from fossil assemblages. Further limitations arise from the manifold problems associated with evolution and extinction (see section 4.15) (Currant, 1989), and from the fact that many fossil animal assemblages appear to have no modern equivalents. The frequency of climatic and environmental change during the Quaternary may have resulted in an acceleration in rates of evolution and of morphological characteristics compared with preceding geological periods. Kurtén (1968) has suggested that, for the Pleistocene as a whole, the average **'turnover rate'** was 11 per cent per 200 000 years and that the mean longevity of species was *c.* 3 million years. However, from the penultimate interglacial to the close of the last glacial, a 9 per cent turnover rate per 75 000 years and a mean species longevity of 1.6 million years has been estimated. Clearly, therefore, the older the fossil assemblage under investigation, the more acute the difficulties of interpretation become, for not only does the proportion of extinct species increase with the age of the assemblage, but the degree of phyletic relation to the living form also decreases (Lundelius, 1976).

4.13 OTHER FOSSIL GROUPS

4.13.1 Chrysophytes

:hrysophytes, often referred to as 'scaled chrysophytes', re a group of planktonic algae that are covered by siliceous cales. They are potentially very useful as alaeoenvironmental indicators since (1) the scales may be xtremely abundant and well preserved in lake sediments; 2) they can be identified to species level; (3) fossil ssemblages often contain many different species, and pecies variation and abundance can be used to derive cological information; and (4) assemblage composition has een found to be strongly related to lake water chemistry Cumming *et al.,* 1992). This fossil group has attracted ncreasing interest in the last decade or so since it has been hown that assemblage variations provide a sensitive record f pH variations (Cumming *et al.,* 1991), that evidence from hrysophyte assemblages has been used in studies of lake cidification (e.g. Smol & Dixit, 1990). Chrysophyte species bundance and composition are also influenced by metal oncentrations (Dixit *et al.*, 1989) and may therefore be used s an indicator of lake pollution or changes in nutrient status e.g. Cumming *et al.*, 1992).

4.13.2 Cladocera

:ladocera are freshwater invertebrates, the skeletal ragments of which are often abundant in lake sediments Frey, 1986; Hann, 1990). They live mainly in the littoral one, but the skeletal fragments are usually transported to the leeper parts of lake basins. Although poorly studied by omparison with other lake fauna, they can provide valuable nformation on lake palaeoecology, especially population lynamics (e.g. Hann & Warner, 1987; Nilssen & Sandøy, 990). They can also be used to reconstruct lake histories. or example, Cladocera and diatom evidence from lakes in innish Lapland showed evidence of lower lake levels from : 8–4 ka BP, reflecting much drier conditions during the nid-Holocene (Hyvärinen & Alhonen, 1994).

4.13.3 Coral polyps

:oral polyps form a detailed and varied alaeoenvironmental archive that is becoming increasingly mportant in Quaternary research. **Coral reefs** are formed by number of interlocking elements, including autochthonous nd allochthonous remains of animals and plants, clastic sediment and chemical alteration processes. The main 'builders', however, are polyp colonies that are normally long-lived, spanning hundreds to thousands of years and, together with the other elements mentioned above, they produce massive calcareous structures close to mean sea level. Growth rates of about 10–30 mm yr^{-1} are common. Gradual changes in sea level have enabled reefs to build vertically, maintaining their shallow-water habitat over time, and thus provide crucial records of sea-level variations (see Chapter 2). As the reef grows, the growth rates and chemical composition of coral are affected by the physical and chemical condition of the surrounding sea water as well as by biological factors, and a record of variations in these parameters is therefore contained within the coral structure (Scoffin, 1987; Quinn *et al.,* 1993). The most important parameters governing rate of coral reef growth appear to be water temperature and water depth. Since these vary seasonally, many species of coral produce seasonal growth bands with clear density differences that are visible by X-radiography or fluorescence techniques (Patzold, 1984; Klein *et al.,* 1990). The rapid growth and annual banding of coral provides a basis for palaeoenvironmental reconstruction at high temporal resolution, potentially as detailed as records based upon tree rings (Chapter 5). Moreover, a number of chemical indices can be used in conjunction with 'coral stratigraphy' as a basis for palaeoenvironmental reconstructions. These include (1) stable isotopic composition, which reflects variations in sea-water temperature (Dunbar *et al.,* 1994) or in local precipitation or runoff (Tudhope, 1994), and (2) trace element composition, such as barium–calcium ratios, since increased precipitation or runoff is reflected in elevated barium levels in coral (Shen & Sandford, 1990). Strontium–calcium ratios in coral also appear to be controlled by sea-water temperature, and may provide a means of estimating sea-water temperatures to a resolution of better than 0.5°C (Beck *et al.,* 1992). Inter-annual variations in such indices have been used to reconstruct **ENSO ('El Niño' Southern Oscillation)** variations during recent decades (McGregor, 1992), and temperature and CO_2 variations in tropical surface waters over the past 120 years (Patzold, 1986).

4.13.4 Fungal remains

Fungal remains, especially hyphae and fruiting structures (which resemble spores), are very common in Quaternary deposits, and are often encountered during routine pollen analysis. They are especially common in lake sediments and peats. The fossil components are difficult to classify, even to genus level, although systematic recording and classification of the most common types found in late Quaternary deposits have led to the identification of key 'marker' fossils (e.g. Van Geel, 1986; Van Geel *et al.*, 1986). Fungal remains appear to offer considerable potential for Quaternary palaeoenvironmental reconstruction. For example, the identification in a sediment profile of saprophytic fungi known to invade peats and organic lake sediments during phases of desiccation could provide evidence of previous drier episodes, whereas mycorrhizal fungi, associated today with particular types of vascular plant, could indicate the former occurrence of such plants, despite the absence of pollen or plant macrofossil remains of those plants. The presence of parasitic fungi could explain the demise of vascular plants in pollen records, while fungi associated with animal dung or the plants associated with activities of humans (e.g. food production, medicines or ingredients of rituals) could provide valuable archaeological information (Pirozynski, 1990).

4.13.5 'Rhizopods' or testate amoebae

'Rhizopods' or **testate amoebae** (Protozoa) are freshwater foraminiferids, unicellular animals which have a thin shell (or **test**) enclosing the cytoplasm (section 4.10). They are also sometimes referred to as *'testaceans'* or *'thecamoebae'*. The tests, which are often abundant in fine-grained freshwater sediments and especially in peats, are extremely varied in form and are often identifiable to species level (Warner, 1990b). They are best preserved in *Sphagnum* peats and their most important application appears to be in the detection of hydrological variations affecting peat formation, although they may also provide information on variations in the chemical composition of mire water (Tolonen, 1986).

4.14 Multi-proxy palaeoecological studies

The majority of Quaternary palaeoecological research projects have involved the analysis of only one or two lines of biological evidence. This is a reflection both of the time-consuming nature of palaeoecological research and of the particular specialisms of most palaeoecologists. Increasingly, however, research in Quaternary science is following a **multi-proxy**, as opposed to a **single-proxy**, approach, partly because, as has been noted above, each individual strand of biological evidence is capable of providing misleading palaeoenvironmental information (due to taphonomic problems, for example), and partly because much more sophisticated (and secure) environmental reconstructions are likely when several different proxies provide converging and mutually supporting data. Moreover, a much wider range of biological evidence is now being investigated, and a great deal more is known about the ecological affinities of modern biota and, in particular, about their environmental controls. In an increasing number of cases, therefore, it is now possible to model environmental change using these different lines of evidence, but the models are, of course, much more powerful when they combine evidence from more than one climatic or environmental proxy. As a consequence, Quaternary research is now characterised by teams of scientists, each perhaps with a different expertise, cooperating in collaborative research programmes which are not only multidisciplinary but frequently interdisciplinary in nature. The first modern example of such collaboration was the CLIMAP project in which marine micro palaeontologists, marine geochemists and atmospheric scientists came together to develop models of ocean–atmosphere–biosphere–cryosphere interactions for the last glacial maximum (see sections 4.11.6 and 7.5).

Multi-proxy research programmes now characterise many Quaternary palaeoecological investigations. A reconstruction of the local and regional environment of southeast Iowa during the last cold stage, for example, involved the analysis of pollen, bryophytes, vascular plant remains, small mammals, molluscs and insects. These records reflected a mosaic of local habitat conditions, ranging from bare, gravelly soils to minerotrophic fens and shallow, possibly ephemeral, clear-water pools. At the regional scale, the different proxies provided a consistent picture of a climatic régime with mean July temperatures some 11–13°C colder than in present-day Iowa (Baker *et al.*, 1986). A second example concerns the investigation of Cromerian Interglacial (Middle Pleistocene) deposits in the English Midlands, based upon the analysis of pollen, plant macroremains, insects, ostracods, molluscs and mammals. Again, a range of local habitats could be identified, while at the regional scale, the different lines of evidence are consistent with a cool-temperate, but fluctuating, climatic régime at the end of the interglacial stage (Shotton *et al.*, 1993). In both of these cases, a high degree of confidence can be attached to the palaeoenvironmental inferences, because these are based upon different records and yet are mutually supportive.

The benefits to be derived from employing a multi-proxy approach are particularly well illustrated in a study of the Lateglacial and Holocene sequence of soils, colluvium and pond sediments from Folkestone, southeast England (Preece, 1997). Data have been obtained from plant macrofossils, pollen, molluscs, Coleoptera and vertebrates, as well as from palaeosols, other sediments and archaeological remains. At this particular site, however, with the exception of Mollusca, no single proxy spans the entire sequence, and hence palaeoenvironmental reconstruction is based upon a careful integration of the different biological records, along with lithological evidence relating to episodes of pedogenesis and colluviation.

Although time-consuming and resource-intensive, therefore, multi-proxy investigations are likely to become an increasing feature of Quaternary research. Indeed, they are essential if the multi-faceted nature of the Quaternary record is to be properly interpreted and reasonable inferences drawn about patterns, processes and causes of environmental change. The validity of the inferences that we make depends, however, on establishing the past distribution of biota (the **palaeobiogeography** of the period under investigation) as accurately as possible, and also improving still further our understanding of the **palaeoecology** of former organisms whose remains make up the fossil record. Just as the interpretation of modern distributions of individual species requires a knowledge of their mutual associations, so must be the case for **palaeoecosystems**. This, however, is a two-way process: studying the past is just as important to contemporary ecology as modern ecology is to understanding past conditions, and it is to this notion that we now turn our attention.

4.15 Quaternary palaeobiology and ecological theory

Quaternary palaeobiological studies provide a historical context which is of fundamental importance for understanding the present-day distributions of organisms, the timescales over which ecological changes operate and the responses of the modern biota to habitat modifications imposed directly or indirectly by human influences. In the words of Bennett (1988, p. 717): 'Quaternary palaeoecology has reached a stage and a precision where it can provide positive input to ecology by demonstrating what past distributions of plant taxa were like, how they have changed, and at what rates. The timescale of Holocene palaeoecology alone (10^4 years) is about 200 times longer than the longest of modern studies of temporal changes in plant communities.' In the following section, we examine some of the ways in which Quaternary biological data have been used to inform contemporary ecological theory.

4.15.1 Biomass and global climate change

In this chapter, climate has often been shown to be the major independent variable that determines the abundance and distribution of organisms. Just as important, however, is feedback between the total biomass that exists on the surface of the earth and in the oceans on the one hand and global climate change on the other. The nature and pattern of the vegetation cover on land affects global climate in a variety of ways, but principally by altering atmospheric gas ratios, surface albedo values and fluxes within the global hydrological cycle. Of particular importance is the way in which vegetation and soils together contribute to the regulation or modulation of atmospheric CO_2 levels, since increased biomass represents an increase in stored CO_2. In the same way, the ocean biomass also stores CO_2, and gas exchange between the oceans and atmosphere appears to be another major climatic feedback mechanism. Modelling of the atmospheric system (GCMs – section 7.7) suggests that global climate is extremely sensitive to changes in vegetation cover (Bonani *et al.*, 1992; Henderson-Sellers *et al.*, 1993; Shukla *et al.*, 1990) and that the northward extension of boreal forests about 6000 years ago (Ritchie, 1987), for example, was both a response to high latitude warming and possibly also a contributing factor to climate change through alteration of albedo (COHMAP Members, 1988; Wright *et al.*, 1993). Large positive feedbacks between climate and boreal forests may have occurred in the past, and this alone may have resulted in a 4°C warming in the spring season in high latitudes during the mid-Holocene (Foley *et al.*, 1994). An important priority for Quaternary palaeoecologists, therefore, is to establish past biomes and biomass variations and to begin to evaluate their role in global climatic changes.

4.15.2 Migration of biota and community structures

A number of established theories concerning modern species richness and community structures are currently being re-evaluated in the light of recent palaeobiological evidence. For example, it had long been assumed that during cold stages temperate deciduous forests in the Northern Hemisphere migrated south, maintained the species associations that are observable today, and then moved northwards once again during warm stages. Recent research on the rate and routes of tree migrations, locations of glacial refugia and forest development and composition during warm stages challenges these assumptions. The palaeobotanical data indicate that, at the beginning of an interglacial, forests migrated north from widely scattered refugia in mid-latitude mountain areas and individual tree species became extinct in the northern part of their ranges. In general, there was no migration south, but survival of trees depended upon scattered remnants of once widespread forest types that persisted through a cold stage to enable re-migration at the start of a subsequent interglacial (Bennett *et al.*, 1991).

The distribution and mix of trees at any particular stage appear to have evolved in a stochastic manner (Huntley, 1991; Birks & Line, 1993), with evolutionary adaptation playing a minor role during the Quaternary, and migration responses due to climatic change being dominant. Responses of taxa to climate change were 'individualistic', and the mix of trees in a particular region at any particular time is largely accidental (Huntley & Prentice, 1993). This raises fundamental questions about established concepts of species competition and community structures (Bennett & Lamb, 1988; Huntley & Webb, 1988). The conclusion reached by E.M. Reid in the 1930s, for example, that European temperate forests are less taxonomically diverse than their North American counterparts because of the different topographies of the two continents, has recently been questioned. North American trees have a high genus:species ratio, whereas European trees display a higher species:genus ratio and hence native European woodland can be regarded as just as taxonomically diverse (Huntley, 1993a). This appears to reflect extended phases of isolation in widely separated refugia during the cold stages of the Quaternary, and hence the modern woodland composition results from a climatic selection process rather than a topographical one.

Coope (1994a, 1994b) has consistently argued that the way in which beetles responded to the climatic fluctuations of the Quaternary was to change their geographical ranges extremely rapidly, often at the global scale. Hence, species that are endemic in parts of eastern Asia today were found in western Europe (including Britain) during earlier Quaternary episodes when the climatic régime was suitable. In Coope's view, beetles were able to survive the vicissitudes of climatic change during the last million years or so, and to do so without the need for evolutionary adaptation, simply because they were able to move quickly to areas where suitable climatic conditions could be found. In other words, by migrating rapidly when necessary, Coleoptera were able to exist in climatic conditions that remained essentially constant (from the beetle's point of view) throughout the Quaternary!

Quaternary palaeoecological records may also shed light on unusual biotic mixes that occur in certain regions, and that have puzzled ecologists for many years. The biota of the Shetland Islands, the Faroes, Iceland and Greenland, for example, are dominated by Eurasian species and contain no endemics. A number of hypotheses have been advanced to explain this situation, including human introductions, existence of land bridges in the past and glacial survival. A more likely explanation, however, is that the islands lost all of their biota during the intense cold of the Quaternary cold stages, but were quickly reinvaded at the start of interglacial periods (Coope, 1986b). This process may have been assisted by the re-establishment of northward-flowing Gulf Stream waters and associated air circulation in the North Atlantic at the beginning of each warm stage.

4.15.3 Extinctions

Plant and animal extinction has been a major theme of Quaternary research, but has received added impetus from the concern that human actions may accelerate the demise of many species in the future (see below). Most research has been concerned with faunal histories, however, as the evidence for both evolution and extinction is more apparent in faunal as opposed to floral records. In discussing evolution and extinction, distinction needs to be made between **faunal turnover**, i.e. the evolution of one genus from another (such as the mid-Pleistocene evolution of *Arvicola*, which survives today as the the water vole, *A. terrestris,* from the extinct rodent *Mimomys* – Sutcliffe, 1985) and true extinction, the extirpation of a species leaving no survivors. Quaternary palaeoecological research can provide important data for measuring both faunal turnover rates and for establishing the timing and scale of animal extinctions.

The most striking event in the Quaternary faunal record is the **mass extinction** which occurred during the last glacial–interglacial transition, and which saw the disappearance of many animal genera, especially of large mammals. During the late Pleistocene woolly mammoths, woolly rhinoceroses and giant deer were widespread in Europe and northern Asia; mammoths, mastodons, ground sloths and sabretooth cats were present in North America; mastodons, ground sloths, litopterns and glyptodonts roamed the South American continent and a range of giant marsupials occupied Australasia (Stuart, 1991). By the end of the Pleistocene (*c.* 10 ka BP) all of these large mammals, and many other genera, became extinct, while the ranges of other animals were significantly reduced (e.g. the loss of the lion – *Panthera leo* – from North America). This mass extinction event is different from those that occurred earlier in the geological record, at the Permian/Triassic and Cretaceous/Tertiary boundaries, for example, in that while the earlier ones affected a very wide range of biota, both terrestrial and marine, the late Pleistocene 'event' affected only large terrestrial mammals (Martin, 1984)

Invertebrates, plants and marine vertebrates survived almost unscathed.

The evidence suggests that extinctions were sudden and severe in North America, and it has been suggested that this was principally the result of **'overkill'** by humans who found these beasts easy prey (Martin & Klein, 1984; Mead & Meltzer, 1985). This **overkill hypothesis** has not been universally accepted, however, since there is often an absence of evidence for kill sites, in some areas of the world there was a long period of coexistence of humans and the affected genera before the 'event', and it is difficult to understand why the extinctions in Eurasia were much less severe and appear to have occurred over a longer timescale than in North America (Stuart, 1991). A number of other hypotheses have been advanced to explain the late Pleistocene extinctions, including climatic change, loss of habitats, disease, isolation, breeding strategies, or a combination of these with 'overkill' by humans. None provides a wholly satisfactory explanation for the phenomenon, however, although the favoured explanation is one of human predation aided by rapid climatic and other environmental changes.

The seemingly dramatic late Pleistocene mammal extinctions may be balanced against the evidence for species constancy within the beetle kingdom during the past million years or so. Although extinction of species may be difficult to establish in the geological record, Quaternary coleopteran data show no obvious high extinction rates but rather constancy of species and even of beetle communities during the last million years (Coope, 1994a, 1994b). The reason for this appears to be the ability of beetles to quickly adapt their geographical ranges to keep pace with rapid environmental changes. Because there was no need to evolve in order to adapt to environmental change, and so long as routes to suitable habitats remained open, species constancy could be maintained. To what extent other members of the invertebrate kingdom were able to develop a similar adaptive strategy remains to be established.

4.15.4 Conservation, biodiversity and habitat destruction

A major objective of conservation policies, until relatively recently, has been to preserve the landscape and selected ecosystems in their present-day state. The public desire has been to conserve what is apparently a long-term 'natural heritage'. Gradually, however, this perception has altered, largely as a result of Quaternary palaeoecological research which has shown just how much of the landscape and vegetation cover of the globe is, in fact, a reflection of recent human activity, and how the distributions and co-associations of species have altered dramatically throughout the Quaternary (e.g. Roberts, 1989; Bell & Walker, 1992).

Biodiversity is recognised today as a conservation issue of the highest priority for conservationists, since many consider that a significant number of the world's species will become extinct through habitat destruction and over-exploitation (Wilson, 1988; Morowitz, 1991; Williams *et al.*, 1993b). The scale of the impending changes can only be properly evaluated in the light of knowledge of how ecosystems have responded to changes in the past. Historical data are vital for the identification of priority ecosystems, areas and individual organisms that are most at threat and that are 'salvageable', for it may already be too late to protect some biological associations (La Salle & Gauld, 1993).

Habitat destruction is clearly a major factor in species extinctions (Ehrlich & Ehrlich, 1981; Wilson, 1988) and it is imperative to be able to predict the rate, pattern and effects of habitat destruction for environmental management purposes. Habitats have been destroyed (or at least relocated) many times in the past because of climatic and other environmental changes. The Quaternary record therefore provides a basis for measuring the rates and effects of habitat reorganisations at a range of spatial scales. There is a widespread belief that it is 'marginal' or rare species that will be predominantly affected by future habitat extinction. A more worrying scenario is painted by Tilman *et al.* (1994) who argue that dominant species may also be vulnerable and that the current dominant position held by some species may be fleeting. Coope (1994a, 1994b), for example, warns that although beetles survived major environmental changes in the past, largely through their ability to shift geographical ranges quickly, this does not mean that they are invulnerable to environmental change. Destruction of natural habitats by human activities, such as drainage of wetlands and forest destruction, may be so widespread that communities may become isolated, with no viable pathway for migration during times of stress.

Quaternary palaeoecological research therefore provides an essential perspective for evaluating these important issues, insofar as it can establish the rates and patterns of the shifts of distributions of organisms in the past, the links between these and environmental factors, and the various adaptive strategies employed by different organisms.

4.16 Conclusions

Palaeobiological evidence, in the form of fossil fauna and flora, is probably the most effective and direct means we have at our disposal for reconstructing past environmental conditions. The analysis of all forms of biological evidence, however, is time-consuming, often costly and requires a very high level of specialisation. These factors must be weighed against the type of information required, the level of sophistication in the data that are being sought and the importance of the research topic to which they are being applied. No single technique can provide all of the evidence that we need in order to understand fully the nature of Quaternary environmental changes. Each data source outlined in this chapter offers a slightly different perspective and the point has been made repeatedly that the most fruitful lines of enquiry are frequently those in which several techniques are employed in conjunction, or where biological evidence is supported by geomorphological, sedimentological or geochemical data. In these circumstances, the tools are available to enable Quaternary scientists to attempt reconstructions not only of environments that existed at specific times but also of the history of environmental changes and of specific responses of the biota over time. Before these steps can be taken, however, two further aspects of Quaternary research need to be examined, namely the establishment of a timescale for environmental change and the means whereby Quaternary sequences at widely separated localities can be correlated in both the spatial and temporal dimensions. These form the subject matter of the following two chapters.

Notes

1. Used literally, the term *pollen analysis* refers to the description and classification of pollen grains, whereas environmental reconstruction is based upon interpretation of pollen assemblages and their stratigraphical variations. However, the term is so widely used that it is retained here, though strictly the term *pollen stratigraphy* is perhaps more appropriate.
2. Apparently unaltered spore walls consisting of a **sporopollenin** substance similar to that found in pollen and spores today have been recovered from Palaeozoic and even older rocks (Faegri & Iversen, 1989).
3. The *Northwest European Pollen Flora* consists of a series of specialist volumes (keys) for the different vascular plant families, published by Elsevier under the general editorship of W. Punt and G.C.S. Clarke.
4. **Transfer functions** are essentially variants on multiple linear regression models. In palaeoecological studies they have been employed to establish quantitative relationships between biological data and environmental variables. If it is assumed that an assemblage of organisms is related to the environment by some complex function (the transfer function), and if this relationship can be determined for modern situations, then multivariate numerical analysis should allow that function to be applied to fossil assemblages, thereby enabling former environmental parameters to be reconstructed quantitatively (Birks & Birks, 1980).
5. **Shannon Diversity Index**: this is a mathematical model developed to describe the statistical regularity in or between statistical samples. In the case of ecology and palaeoecology, the model is used to develop measures of the mathematical (or species) diversity between samples (assemblages) (see Engen, 1978).
6. The **North Atlantic Polar Front** is a prominent hydrographic/oceanographic boundary, also termed the **Subarctic Convergence**, which separates warm water of high salinity flowing northwards from cold low-salinity water flowing from the Arctic. Weather patterns are usually 'trained' along the southern flank of this boundary, which is informally termed the 'Polar Front' in the literature.
7. **Prokaryotes** differ from all other organisms principally because the DNA within their cell is loosely organised and not bounded by a membrane into a nucleus. They lack chromosomes.
8. **Boundary conditions** are the assumed or measured surface conditions used to constrain global climate models, and include such factors as sea surface temperatures, albedo values, incoming solar radiation and atmospheric transparency. For further explanation see section 7.7.

CHAPTER 5 Dating methods

5.1 Introduction

Dating techniques in the Quaternary time range fall into three broad categories:

Methods that provide age estimates There are two types of dating technique that enable the age of fossils, sediments or rocks to be established directly in years before present (BP). These are **radiometric methods**, which are based on the radioactive decay of certain unstable chemical elements or related phenomena (such as damage to crystal lattices caused by radiation) and **incremental methods** which involve measurements of regular accumulations of sediment or biological materials through time.

Methods that establish age-equivalence These methods make use of contemporaneous horizons that can be identified in separate and often quite different stratigraphic sequences. Certain distinctive stratigraphic markers are regionally and, in some cases, globally synchronous, and where these can be traced laterally between sediment sequences, they can be taken to represent common time-planes in the stratigraphic records. If the age of the markers can be established in one locality by the application of any of the age-estimate methods (above) then equivalent horizons within other successions can be indirectly dated by correlation.

Relative age methods These techniques only establish the **relative order of antiquity** of fossils or stratigraphic units. The relative antiquity of geological materials is most obvious where superposition[1] can be established but, under certain circumstances, the relative age of Quaternary landforms and sedimentary units can be established from the degree of degradation or alteration resulting from the operation of chemical processes through time.

5.2 Precision and accuracy in Quaternary dating

Many of the dating techniques currently employed in Quaternary research can be applied only to restricted spans of Quaternary time (Figure 5.1), and each method has its own distinctive set of problems which lead to uncertainties in interpretation. In evaluating any age determination, particularly those obtained by radiometric means, it is important to make a distinction between **precision** and **accuracy**. The former refers to the **statistical uncertainty** that is attached to any physical or chemical measurement, while the latter relates to the degree of correspondence between true age and that obtained by the dating process (Figure 5.2). In other words, it refers to the extent of **bias** in an age determination. In considering accuracy and precision, it is useful to think of the analogy of a watch. A *precise* watch that tells the time to the nearest second may actually be inaccurate by 10 minutes; conversely, an imprecise watch with no second hand may still be *accurate* and tell exactly the correct time (Pilcher, 1991a). Both accuracy and precision determine the **reliability** of dates, but establishing whether or not a date can be regarded as reliable also requires a knowledge of other factors such as contexts of deposition, taphonomy of fossil material and post-depositional diagenetic processes. For example, precise measurements can be obtained from fossils containing contaminants, but if the contaminants are not recognised as such and corrected for, then the inferred age will be invalid.

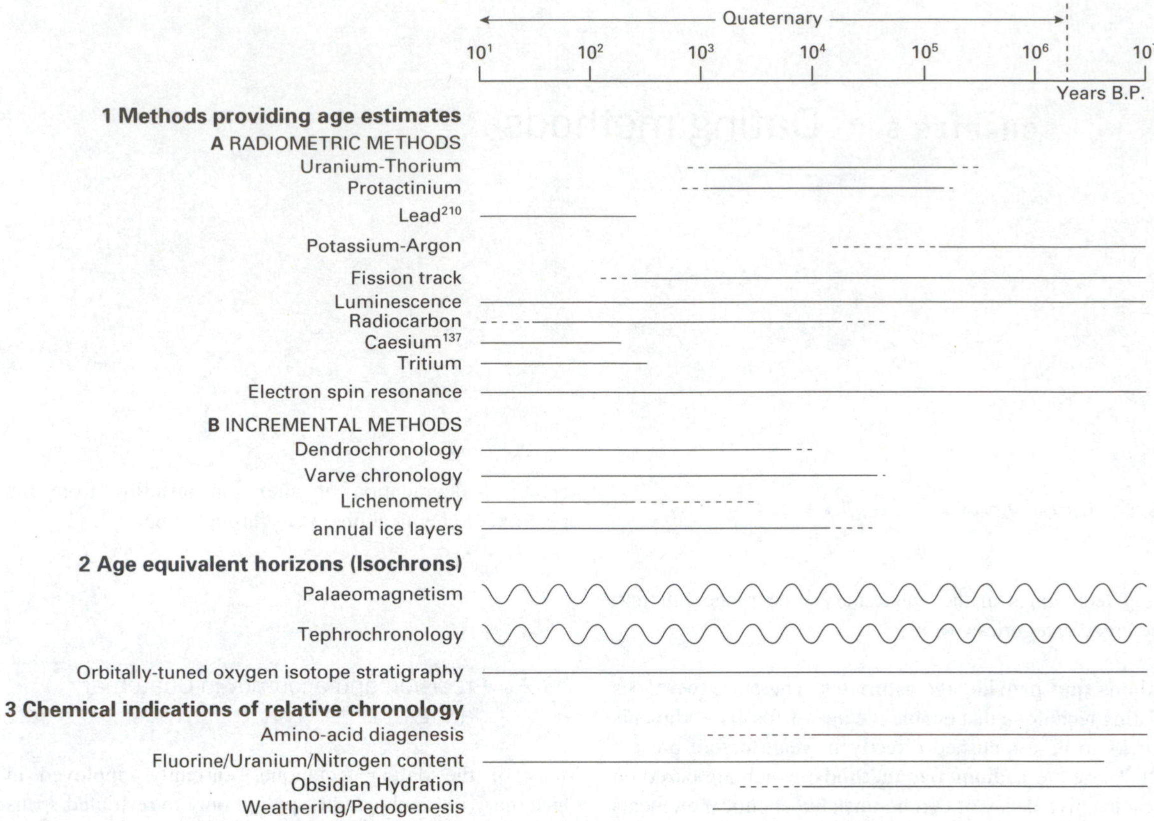

Quaternary

10^1 10^2 10^3 10^4 10^5 10^6 10^7

Years B.P.

1 Methods providing age estimates

A RADIOMETRIC METHODS

Uranium-Thorium

Protactinium

$Lead^{210}$

Potassium-Argon

Fission track

Luminescence

Radiocarbon

$Caesium^{137}$

Tritium

Electron spin resonance

B INCREMENTAL METHODS

Dendrochronology

Varve chronology

Lichenometry

annual ice layers

2 Age equivalent horizons (Isochrons)

Palaeomagnetism

Tephrochronology

Orbitally-tuned oxygen isotope stratigraphy

3 Chemical indications of relative chronology

Amino-acid diagenesis

Fluorine/Uranium/Nitrogen content

Obsidian Hydration

Weathering/Pedogenesis

Figure 5.1 Ranges of the various dating methods discussed in the text. Broken lines show possible extensions with further improvements in techniques; wavy lines indicate that dating is limited to specific time intervals within the Quaternary.

5.3 Radiometric dating techniques

Radiometric dating methods are based on the radioactive properties of certain unstable isotopes which undergo spontaneous changes in atomic organisation in order to achieve a more stable atomic form. Some radioactive elements, such as uranium, occur naturally and are commonly found in rocks, sediments and fossils. **Radioactive decay** (atomic transformation) is **time-dependent**, and if the rate of decay is known, the age of the host rocks or fossils can be established. Rates of radioactive transformations vary markedly; some elements decay in days or even seconds, whereas others transform gradually over millions of years. A number of radiometric dating methods have now been developed, but in this section only those techniques that are directly applicable to the Quaternary timescale are discussed.

5.3.1 The nucleus and radioactivity

The nucleus of an atom contains positively charged particles called **protons** and particles with no electrical charge known as **neutrons**. These are densely packed in the nucleus so that although the nucleus occupies only about 10^{-14} of the volume of an atom, it contains nearly all of the mass. The other major type of particle contributing to the structure of an atom is the **electron**. For practical purposes, electrons can be considered as tiny particles of negative charge, with negligible mass, spinning around the nucleus in orbits. Electrons vary in number for different chemical elements, and are arranged in electron shells (or orbitals) of different radial distance from the nucleus (see Figure 5.5 below). The analogy is often drawn between this arrangement and the planets orbiting the sun. Strictly speaking, this is not correct, as modern physics has shown that sub-atomic entities cannot

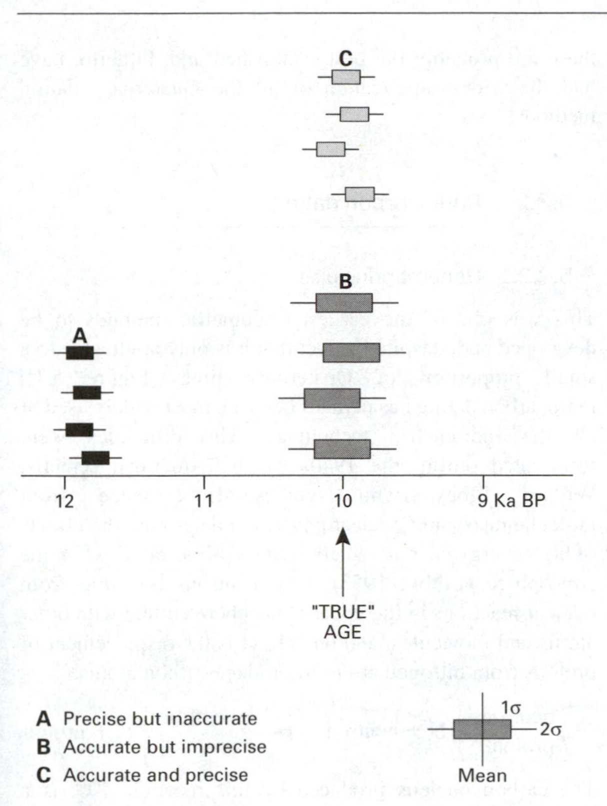

Figure 5.2 Precision and accuracy in Quaternary dating (courtesy F.M. Chambers).

be considered as discrete particles (Close *et al.*, 1987). However, for the purposes of this discussion, it will help if these nuclear units are regarded as nuclear particles.

Chemical elements are classified according to **atomic number (Z)**, which is the number of protons contained in the nucleus. Hydrogen has an atomic number of 1, oxygen 8 and uranium 92. The **atomic mass number (A)** of an element is the number of protons plus neutrons; that of hydrogen is 1 and of oxygen is 16. It is conventional to give the numerical value of A as a superscript and Z as a subscript on the left-hand side of the symbol for a chemical element, for example $^{238}_{92}U$ (uranium 238). The atomic mass number of elements can vary, since the number of neutrons in the nucleus is not always constant. Elements having the same number of protons, but a different number of neutrons (e.g. ^{16}O and ^{18}O; ^{12}C and ^{14}C) are known as **isotopes**. They have the same chemical properties, since the number of electrons remains constant for each element, but isotopes differ in mass. Each isotope of an element is called a **nuclide**. The particles that constitute the nucleus are bound together in a way that is not fully understood, but if a nucleus contains too many or too few neutrons, it becomes unstable and the repulsive forces between the similarly charged particles overcome the binding forces keeping them together. This results in spontaneous emission of particles or energy, which is the basis of radioactivity. Isotopes involved in such radioactive processes are known as **radioactive nuclides**.

Three types of energy emission occur during radioactive decay. **Alpha (α) particles** consist of two protons plus two neutrons and are the positively charged nuclei of helium atoms. They collide with surrounding atoms and acquire electrons to form helium gas. Nuclides that emit alpha particles lose mass and positive charge. By this process, the atomic number changes, and thus one chemical element can be formed by the 'decomposition' of others. **Beta (β) particles** are negatively charged electrons, and their emission does not alter mass, but changes atomic number. It is also possible for an electron to jump from one orbital into another, and in some rare cases they may even transfer into the nucleus. **Gamma (γ) rays** are powerful forms of radiation that occur during radioactive decay. They are not important in the calculation of decay constants but do contribute to the build-up of thermoluminescence properties in minerals (section 5.3.6). Moroever, the cosmic rays that constantly bombard the earth's upper atmosphere (section 5.3.2) consist largely of gamma rays.

The atom that undergoes atomic transformation is termed the **parent nuclide** (or '**mother nuclide**') and the product is the **daughter nuclide.** This single-stage transformation is known as **simple decay**. Many radioactive transformations, such as uranium series (see below), involve more complex pathways where the transformation of the nuclide with the highest atomic number to a stable nuclide involves the production of a number of intermediate unstable nuclides. This is known as **chain decay** (see Figure 5.6 below). Intermediate nuclides involved in such chains are both the product of previous transformations and the parents in subsequent radioactive decay, and such nuclides are termed **supported. Unsupported decay** involves the transformation of a parent nuclide that is not, in itself, the product of decay, or is separated from earlier nuclides in the chain as a result of physical, biogenic or sedimentary processes.

Radioactive decay processes are governed by atomic constants. The number of transformations per unit time is proportional to the number of atoms present, and for each decay scheme there is a **decay constant** (λ) which represents the probability that an atom will decay in a given period of time. The transformation of an individual atom occurs spontaneously and unpredictably, but where a large number of atoms of a particular nuclide are considered, there is a predictable time rate at which overall disintegration proceeds. The law of radioactive decay is given by:

$$\frac{-\delta N}{\delta t} = \lambda N$$

where N is the number of atoms, t is a time constant and λ is the decay constant for that nuclide. For all nuclides the decay is exponential (see Figure 5.3), and is best considered in terms of the **half-life** ($t_{0.5}$). This is the period of time required to reduce a given quantity of a parent nuclide to one half. For example, if 1 g of a parent nuclide is left to decay, after $t_{0.5}$ only 0.5 g of that parent will remain. It will then take the same period of time to reduce that 0.5 g to 0.25 g, and to reduce the 0.25 g to 0.125 g, and so on. The relation between the half-life and the decay constant is given as:

$$\text{Half-life } (t_{0.5}) = \frac{\log_e 2}{\lambda} = \frac{0.693}{\lambda}$$

The application of the principle of radioactivity to geological dating requires that certain fundamental conditions be met. If an event (such as the cooling of a magma, the formation of salt precipitates, the death of an animal and the burial of its bones, etc.) is associated with the incorporation of a radioactive nuclide, then providing (a) that none of the daughter nuclides is present in the initial stages and (b) that no parent or daughter nuclides are added to or lost from the materials to be dated (i.e. the radioactive process has proceeded within a **closed system**), then an estimate of the age of that event can be obtained if the ratio between parent and daughter nuclides can be established, and if the decay rate is known. All estimates of time derived by radioactive decay are termed **radiometric clocks**; some methods are based on measurements of the progressive disappearance of nuclides during disintegration, while others ('accumulation clocks') measure the increasing quantity of a particular nuclide through time.

In the following section a selection of the principal radiometric methods employed by Quaternary scientists is discussed. Radiocarbon dating and uranium-series disequilibrium dating are considered at greater length as these are probably the best established and, hitherto, have had the widest application of all the Quaternary dating methods.

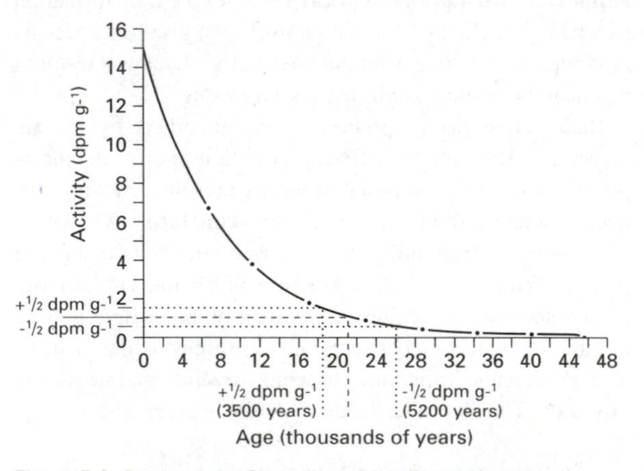

Figure 5.3 Decay curve for radiocarbon. For explanation see sections 5.3.2.1 and 5.3.2.2.

5.3.2 Radiocarbon dating

5.3.2.1 General principles

This was one of the earliest radiometric methods to be developed and, despite the fact that it is only applicable to a small proportion of Quaternary time (Figure 5.1), radiocarbon dating has perhaps been the most widely used of all the radiometric techniques. The principles were formulated during the 1940s by the American scientist Willard Libby who synthesised evidence from radiochemistry and nuclear physics to determine the effects of high energy cosmic radiation (the cosmic-ray flux) on the atmosphere (Libby, 1955). Free neutrons resulting from nuclear reactions in the upper atmosphere collide with other atoms and molecules, and one effect is the displacement of protons from nitrogen atoms to produce carbon atoms:

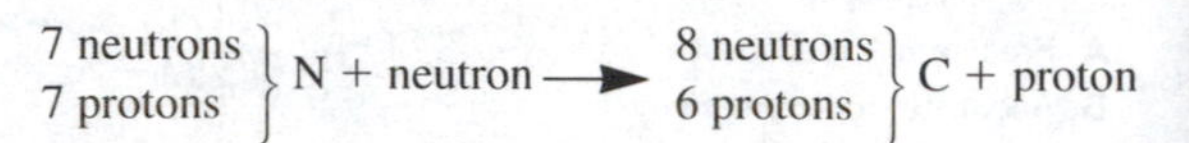

$$\left.\begin{matrix}\text{7 neutrons}\\\text{7 protons}\end{matrix}\right\}\text{N} + \text{neutron} \longrightarrow \left.\begin{matrix}\text{8 neutrons}\\\text{6 protons}\end{matrix}\right\}\text{C} + \text{proton}$$

The carbon nucleus produced by this reaction, ^{14}C, is a radioactive isotope of carbon which eventually decays to form the stable element ^{14}N:

$$\left.\begin{matrix}\text{8 neutrons}\\\text{6 protons}\end{matrix}\right\}\text{C} \longrightarrow \left.\begin{matrix}\text{7 neutrons}\\\text{7 protons}\end{matrix}\right\}\text{N} + \beta^-$$

Decay is by beta (β) transformation, i.e. the emission of β^- particles.

^{14}C atoms are rapidly oxidised to carbon dioxide and, along with other molecules of carbon dioxide ($^{12}CO_2$), become mixed throughout the atmosphere and absorbed by the oceans and by living organisms. In other words, ^{14}C, which is continually being produced in the upper atmosphere, becomes stored in various **global reservoirs** – the atmosphere, the biosphere and the hydrosphere.

Radiocarbon dating originally rested on four fundamental assumptions: (a) that the production of ^{14}C is constant over time; (b) that the ^{14}C:^{12}C ratio in the biosphere and hydrosphere is in equilibrium with the atmospheric ratio; (c) that the decay rate of ^{14}C can be established; (d) that a closed system has existed since the death of the organism. Although questions have subsequently arisen relating to each of these assumptions (see below), they can be accepted in general terms. All living matter absorbs carbon dioxide during tissue-building in a ratio that is *broadly* in equilibrium with atmospheric carbon dioxide. As long as the organism is alive, carbon used to build new tissues will be in isotopic equilibrium with (i.e. will exist in similar isotopic ratios to) those in the contemporaneous atmosphere. Upon death, ^{14}C within the organic tissues will continue to decay, but no replacement takes place. Hence, if the rate of decay of ^{14}C is known, date of death can be calculated from the measured residual ^{14}C activity.

The activity of ^{14}C in the atmosphere is approximately 15 dpm g^{-1} (15 disintegrations per minute per gram), and this activity is halved every 5700 years or so (Figure 5.3). The half-life of ^{14}C was originally calculated at 5568 ± 30 years (Libby, 1955), but subsequently this has been more accurately determined as 5730 ± 40 years (Godwin, 1962). However, because a large number of ^{14}C dates were published prior to the measurement of the new half-life, it has been conventional to base radiocarbon dates on the former half-life value, and 5570 + 30 years remains the internationally agreed fixed constant for all radiocarbon measurements (Mook, 1986). This avoids confusion as dates calculated using the same half-life, irrespective of value, are directly comparable. Conversion to the longer half-life can be made by multiplying radiocarbon ages based on the standard half-life by 1.03.

5.3.2.2 Measurement of ^{14}C activity

In order to detect ^{14}C activity in organic materials, extremely sensitive equipment is required. This is because the natural occurrence of ^{14}C is such that for every one million million atoms of ^{12}C in a living organism, there is only one atom of ^{14}C. Moreover, ^{14}C is a low-energy β particle emitter. Two approaches are used to measure the residual ^{14}C activity in a sample: (a) **Conventional radiocarbon dating**, which involves the detection and counting of β emissions from ^{14}C atoms over a period of time in order to determine the rate of emissions and hence the activity of the sample, and (b) **Accelerator mass spectrometry**, which uses particle accelerators as mass spectrometers to count the actual number of ^{14}C atoms (as opposed to their decay products) in a sample of material (Aitken, 1990; Bowman, 1990).

(a) Conventional radiocarbon dating Two methods are employed in conventional radiocarbon laboratories to detect emissions of beta particles: **gas proportional counting** and **liquid scintillation counting**. In the former, a suitable gas (usually carbon dioxide, ethylene or methane) is prepared from the carbon in the sample and collected in a chamber, down the centre of which runs a charged wire. This detects, and counts, pulses of current that flow through the gas when it is ionised by the radioactive decay. The current is *proportional* to the energy of the β particle (electron), and hence it is possible to discriminate between decays from different radioactive elements in a proportional counter. In liquid scintillation counting, samples are first combusted to CO_2, reacted with molten lithium metal to give lithium carbide, mixed with water to release acetylene and finally polymerised to benzene. A 'scintillator' is then added, usually a phosphoric substance which emits pulses of light (photons) in response to radioactive disintegrations, and these can be counted by photoelectrical means.

From the decay curve (Figure 5.3), it can be seen that material aged approximately 10 ka will have an activity of only 4 dpm g^{-1}, and older samples correspondingly lower values. The limit of practical counting using conventional methods is eight half-lives (about 45 ka), for beyond that age the curve becomes so flat and insensitive that it is difficult to separate samples of different activities with any statistical certainty. However, greater ages have been measured by the technique of **isotopic enrichment** where the amount of ^{14}C present in a sample is enhanced so that the frequency of decay can be more accurately determined by gas or liquid scintillation counters. This approach, which takes advantage of isotopic fractionation, uses either thermal diffusion or photodissociation by means of a laser beam (Bowman, 1990). With the former, finite ages in excess of 60 ka BP have been obtained (Grootes, 1978; Woillard & Mook, 1982), although the technique has not been widely applied, for it requires relatively large samples of material and, by comparison with standard dating procedures, is time-consuming and hence more costly.

In the calculation of radiocarbon dates obtained by conventional methods, laboratories compare sample activities to a **modern reference standard**. The internationally accepted reference standard for all ^{14}C dating is the modern activity level of NBS oxalic acid held by the American Bureau of Standards. Because of the relative scarcity of this material, however, some laboratories now use a secondary standard, such as Australian National University (ANU) sucrose or a paper cellulose sample provided by the International Atomic Energy Agency (IAEA). The reason why such standards are necessary is that there have been variations in ^{14}C production rates through time, and modern levels are artificially high (see below). Comparability between laboratories and between samples of different ages therefore requires reference to a standard. The time in years (R) since the death of an organism can be calculated from the following equation:

$$R = \frac{1}{\lambda} \log_e \left(\frac{A_0}{A} \right)$$

where λ is the decay constant of ^{14}C, A_0 is the ^{14}C activity of the modern reference standard, and A is the measured ^{14}C

activity of the sample of unknown age. In arriving at a measure of R, account has to be taken of a number of factors that affect the determination of the activity of a sample. These include sample volume or gas pressure, dilution ratio (especially for small samples), atmospheric pressure (which affects background radiation during measurement), and loss of sample during counting (which results in samples having to be reweighed at the end of counting and a correction applied). Because these cannot always be quantified, however, some uncertainty is always associated with the calculated age of a sample. One source of uncertainty that can be quantified is the probable effects of the randomness of radioactive decay on the counting statistics. As a consequence, radiocarbon dates along with other radiometric age determinations are always reported as mean determinations with a plus or minus value of one standard deviation about the mean. A radiocarbon date of 2000 ± 100 years should be interpreted as indicating that there is a 68 per cent probability that the true age of the sample lies between 1900 and 2100 years BP or, for a 95 per cent probability, the age is within the range 1800 to 2200 years BP. Even with two standard deviations on the 'date', there is still a 1 in 20 chance that the true age lies outside the range 1800 to 2200 years BP.

It is important to remember, however, that age is not the quantity that is being measured, but *activity of the sample* which, on the basis of a number of assumptions, is interpreted as indicating 'age'. The plus-or-minus refers to the uncertainty associated with determining activity, and this is why on older dates the plus value of the standard deviation is often quoted as being larger than the minus value. This can be seen in Figure 5.3 by considering activity of 1 dpm g^{-1} ± 0.5 dpm g^{-1} which, translated into 'age', gives a significantly higher plus than minus value. Because of the asymptotic decay curve, technically there is always a difference between the plus and minus value, but this is regarded as insignificant for younger material.

(b) Accelerator mass spectrometry (AMS dating) Mass spectrometers are widely used in physics to detect atoms of specific elements based on differences in atomic weights (section 3.10). Charged particles moving in a magnetic field will be deflected from a straight path by a factor that is in proportion to atomic weight the lighter the particle, the greater will be the amount of deflection. Normal mass spectrometers cannot discriminate, however, between ^{14}C and other elements of similar weight (e.g. ^{14}N), but if the particles are subjected to large voltage differences so that they travel at very high speeds, even the very small number of ^{14}C atoms in a sample can be detected. This is the principle of **accelerator mass spectrometry** (Linick *et al.*, 1989). The most commonly used system is a **tandem accelerator**, so called because there are two stages of acceleration. Samples are converted to graphite (although a number of laboratories use a CO_2 source) and mounted on a metal disc. Caesium ions are then fired at this 'target' and the negatively ionised carbon atoms (C^-) produced are accelerated towards the positive terminal. Nitrogen does not form negative ions, and hence almost all of the ^{14}N, which tends to mask the ^{14}C signal in a conventional mass spectrometer, is eliminated before it can reach the detector. During passage through the 'stripper', four electrons are lost from the C^- ions and they emerge with a triple positive charge (C^{3+}). Other molecules are lost at this stage. Repulsion from the positive terminal leads to a second acceleration of the carbon ions through focusing magnets where deflection occurs according to mass, and the concentration of ^{14}C and of the stable carbon isotopes ^{13}C and ^{12}C can therefore be measured (Bowman, 1990).

In order to obtain an age for the sample material, the $^{14}C/^{12}C$ ratio measured for the sample is compared to those for the targets in the same set which are made up from a material of known ^{14}C activity (usually oxalic acid – see above). This gives a sample/modern ratio from which, after corrections, a radiocarbon age can be calculated in years BP (Linick *et al.*, 1989). The plus or minus value that accompanies all AMS dates reflects, as in conventional dating, statistical uncertainties associated with the precise measurement of the ^{14}C decay curve, as well as random and systematic errors that inevitably occur during the measurement process. These can arise, for example, from contamination by modern carbon of samples, targets or the ion beam itself (Hedges, 1991)

(c) Evaluation of counting techniques There are two principal advantages of AMS dating over gas proportional or liquid scintillation counting. First, very small samples of material can be dated, with most AMS laboratories routinely counting samples containing 1 mg of organic carbon or less. This compares with a typical sample size of 5–10 g of organic carbon required by most conventional dating laboratories, although some laboratories are capable of handling small samples of carbon (10–100 mg) typically using gas proportional counters (Otlet *et al.*, 1986). Such systems have the disadvantage, however, of requiring long counting times (measured in months as opposed to days) and seem unlikely to rival AMS in the future dating of small samples of material (Pilcher, 1991b). The second advantage of AMS is one of time. The actual determination may take only a matter of hours whereas typical liquid scintillation or gas proportional counters are frequently occupied for days. Hence a typical AMS laboratory can perform more than

1000 analyses per year, far more than most decay counters can measure to a comparable precision (Linick *et al.*, 1989). The potential of AMS is clearly considerable, and its versatility is shown by a range of applications which include the dating of small plant macroremains from lake sediments (Ammann & Lotter, 1989), fossil pollen grains (Brown *et al.*, 1989), fossil insect remains (Elias & Toolin, 1990), microfauna from deep-ocean cores (Bard *et al.*, 1991) and tiny fragments of the Shroud of Turin (Damon *et al.*, 1989).

AMS dating does have certain disadvantages, however. The capital cost of establishing an AMS facility (in excess of £2m at 1995 prices) is an order of magnitude greater than that involved in developing a conventional radiocarbon laboratory, and running costs are also commensurately higher. Hence the cost of an AMS date is more than the cost of a conventional radiocarbon age determination. At present, there are relatively few AMS laboratories; in Britain, for example, there is only one AMS facility (at Oxford) compared with six conventional radiocarbon laboratories. A further problem with AMS dating concerns the level of precision. The error estimate is typically around 1 per cent, which is equivalent to a one standard deviation precision of ±80 years at around 5.5 ka BP (Pilcher, 1991a) and this may increase to about 100–120 years at 12 ka BP (Hedges, 1991). This compares with values of ±50 and ±80 for the equivalent time periods from most conventional laboratories. Indeed, some laboratories (Belfast, Groningen, Heidelberg, Pretoria and Seattle) have conventional facilities capable of producing very high levels of precision. Although relatively large samples of material are required (5–20 g carbon) and determinations are more expensive, these 'high precision' laboratories can produce dates with a standard deviation of less than 20 years, and sometimes as low as ±12 years (Pilcher, 1991a). Unfortunately, it seems, this level of accuracy cannot yet be attained using AMS and hence the ideal combination of high precision and small sample size remains to be achieved.

5.3.2.3 Sources of error in ^{14}C dating

(a) Temporal variations in ^{14}C production Discrepancies discovered more than thirty years ago between radiocarbon and calendar ages of recent wood samples (de Vries, 1958) first challenged the assumption that ^{14}C levels had not varied significantly over time. This has subsequently been confirmed by data from tree-ring series which show that atmospheric ^{14}C activity has fluctuated markedly throughout the Holocene (Pearson *et al.*, 1986, and see below), apparently in a quasi-periodic manner (Sonett & Finney, 1990). The record of long-term ^{14}C variations has since been extended into the Lateglacial where AMS dates from lake sediments in Switzerland show pronounced 'plateaux' of constant ^{14}C age (Figure 5.4) at around 12.7, 10 and 9.5 ka BP, each plateau being considered to reflect an episode of reduced atmospheric ^{14}C concentration (Ammann & Lotter, 1989). On a longer timescale, comparison between AMS determinations of ^{14}C and U–Th (section 5.3.4) on fossil corals suggests that radiocarbon assays underestimate true age by as much as 2.5 ka at 16 ka BP, and 3.5 ka at 20 ka BP (Bard *et al.*, 1990b; 1992). These discrepancies primarily reflect long-term variations in atmospheric ^{14}C production, although the coral dates may also have been affected by fluctuations in deep ocean ventilation (Broecker *et al.*, 1990).

The divergence between ^{14}C age determinations and those obtained by other dating methods means that it is necessary to make a clear distinction between '**^{14}C years**' and '**calendar years**' when discussing Late Quaternary chronologies. Attempts have been made, however, to reconcile the two timescales by comparing ^{14}C dates with those obtained from the same samples of material using independent dating methods. In this way it is possible to **calibrate the radiocarbon timescale** against a chronology of calendar years. The different approaches to radiocarbon calibration involving uranium series dating, dendrochronology and varve chronology are discussed later in the chapter in the context of each of those particular dating methods.

The causes of long-term atmospheric variations in ^{14}C remain to be established, but there is general agreement that a major component is periodic variation in the intensity of ^{14}C-producing radiation reaching the earth (Harvey, 1980). Factors influencing the cosmic ray flux include changes in the geomagnetic fields of both the earth and the sun, and changes taking place on the surface of the sun itself. There is now strong empirical evidence to suggest that most of the variation in the Holocene $^{14}C/^{12}C$ record can be attributed to the geomagnetic and solar influence on the flux of primary cosmic rays entering the atmosphere (Stuiver *et al.*, 1991), although instabilities in North Atlantic thermohaline circulation may also exert an influence (Stuiver & Brazunias, 1993). A further factor may have been changes in the carbon distribution in the ocean–atmosphere system, especially in the oceans which constitute one of the major global carbon reservoirs. This may have been particularly important during the transition from the last cold stage to the Holocene (Suess & Linick, 1990), where the plateaux in ^{14}C ages (see above) have been tentatively linked to rapid release of fossil CO_2 from the oceans following the onset of North Atlantic Deepwater formation (Sarnthein *et al.*, 1994; section 7.8.2). Over longer timescales, palaeomagnetic records suggest a close relationship between changes in geomagnetic field intensity and ^{14}C production (Mazaud *et*

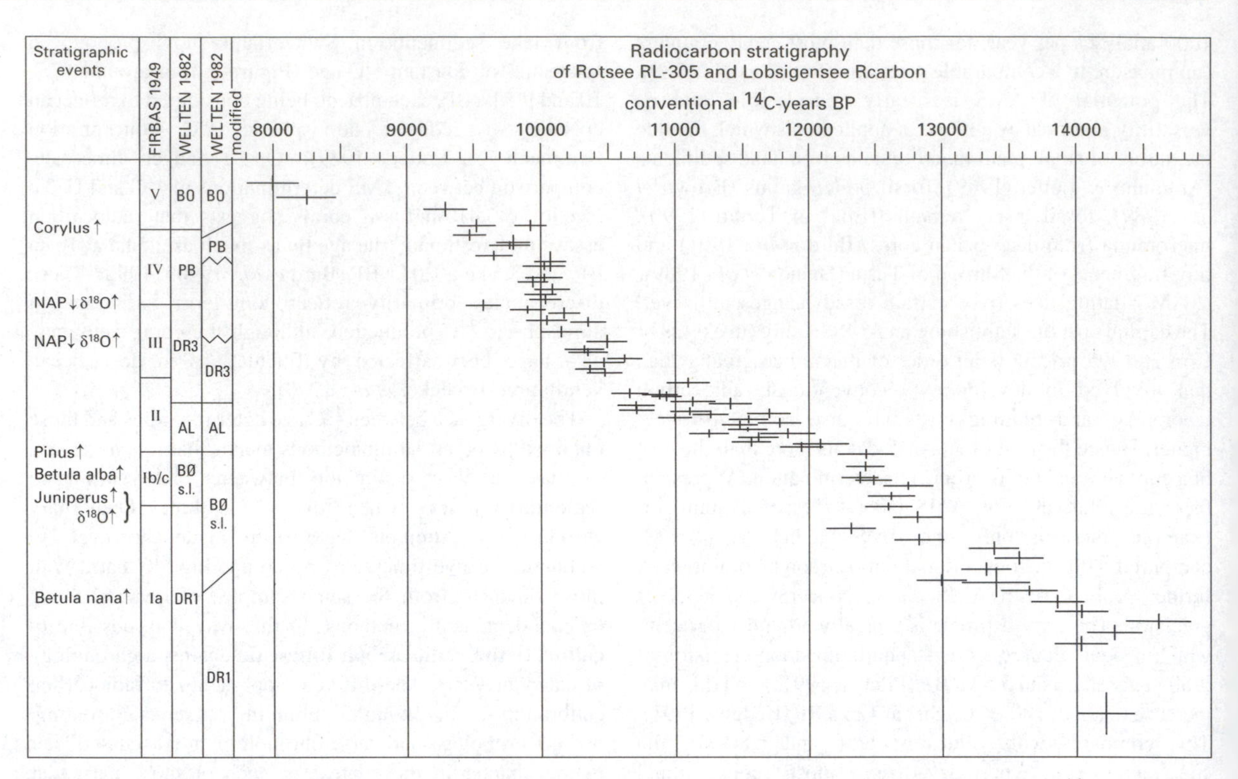

Figure 5.4 Radiocarbon stratigraphy of Swiss Lake sediments showing 'plateau' effects at around 12.7 and 10.0 k ^{14}C years BP (after Ammann & Lotter, 1989).

al., 1992). In the time interval from 10 to 40 ka BP, for example, field intensity was significantly reduced, falling to about one fifth of current levels at 40 ka BP (Salis & Bonhommet, 1992). This period coincides with the apparent 'younging effect' in ^{14}C ages in fossil corals referred to above, and implies higher rates of atmospheric ^{14}C production at times of markedly reduced geomagnetic field intensity.

In addition to natural variations, atmospheric ^{14}C levels have recently been affected by human activity. Over the past 200 years ^{14}C levels have been progressively diluted as a result of the combustion of fossil fuels, which has liberated large quantities of 'inert' ^{12}C into the atmosphere (Houghton *et al.*, 1990). In the last 45 years or so, however, this **industrial effect** has been offset by greatly increased production of ^{14}C resulting from the detonation of thermonuclear devices. The combined effects of industrial activity and atomic explosions means that modern organic samples are unsuitable as reference samples for radiocarbon activity. A value of 0.95 times the measured activity of the NBS standard is regarded as equivalent to the natural ^{14}C activity of AD 1890 wood (pre-industrial effect), and this is corrected to AD 1950, the reference year for all ages quoted in ^{14}C years BP.

(b) Isotopic fractionation Of the three naturally occurring isotopes of carbon about 98.9 per cent is ^{12}C, 1.1 per cent is ^{13}C and only 1 part in 10^{10} per cent is ^{14}C. In nature, however, a **fractionation** of this ratio commonly occurs. Photosynthesis, for example, results in an enrichment of ^{12}C relative to the other isotopes in most plant tissues, whereas ocean waters preferentially absorb ^{14}C. These effects are small, but can significantly affect radiocarbon dates where measurement to less than ± 1 per cent error is required (Harkness, 1979). In addition, fractionation can also occur in the laboratory induced by the conversion of sample carbon to the gas or liquid form. Most radiocarbon laboratories today make corrections for the probable effect of fractionation based on thermodynamic laws which show that the heavier isotope ^{14}C is twice as enriched as ^{13}C (Olsson, 1986). The latter can be measured in a small sub-sample of the material to be dated. The ^{13}C:^{12}C ratio is compared with a standard, PDB limestone

elemnite carbonate from the Cretaceous Peedee Formation ' South Carolina), and values are published as deviations om this standard:

$$\delta^{13}C\% = \frac{^{13}C/^{12}C\ \text{sample} - {}^{13}C/^{12}C\ \text{standard}}{^{13}C/^{12}C\ \text{standard}}$$

Most terrestrial samples have a negative $\delta^{13}C$ value ompared to the PDB standard. Different photosynthetic athways exist in plants so that very different levels of actionation occur (Table 5.1). However, it is practice to ormalise' ^{14}C activities during calculations of ^{14}C age, by eating each sample as if an average enrichment had taken ace. The normal value is taken to be −25‰, which is the ean isotopic composition of wood. If the $\delta^{13}C$ value of a mple is found to be −25‰, then no adjustment is made. 'ith a value of −30‰, however, a 5‰ depletion in the C:^{12}C ratio is implied, which in turn indicates a probable)‰ depletion in the ^{14}C:^{12}C ratio. Thus, the measured ^{14}C tivity would be increased by 10‰, which is equivalent to out 83 years (Harkness, 1979).

Table 5.1 Approximate $\delta^{13}C$ values for various materials. The ranges on these data are typically ±2 or 3‰, but substantially more variability is possible. With each per mil deviation from −25‰ representing *c.* 16 years, these data clearly illustrate the need for fractionation corrections to be applied to measured ^{14}C age results (from Bowman, 1990, with minor modifications by the author).

Material	$\delta^{13}C$ value
Wood, peat and many C_3 plants	−26‰
Bone collagen*	−19‰
Freshwater plants (very variable)	−16‰
Arid zone plants (C^4 plants)	−13‰
Marine plants	−15‰
Atmospheric CO_2	−8‰
Marine carbonates	−0‰

*For direct or indirect C_3 consumers.

(c) Circulation of marine carbon Because ^{14}C is ansferred from the atmosphere to the oceans only across e ocean surface, and because the mixing rate of surface nd deep waters is very slow, ^{14}C in deep ocean waters ecays without replenishment. Sea waters therefore have an **pparent age**. In the surface waters of the North Atlantic is is around 400 years (Bard *et al.*, 1987b; 1991), while in arts of the equatorial East Pacific the figure is *c.* 580 years Shackleton *et al.*, 1988). In the deep oceans, however, nger **residence time** means that sea water may have an pparent age in excess of 2000 years (Ostlund & Stuiver, 980). Hence ^{14}C dates on Foraminifera from deep ocean ores have to be corrected for the age of sea water, with fferent correction factors being applied to planktonic and enthic species. Similarly corrections have to be made to ates on fossils from coastal localities where upwelling of eep (and hence 'older') water has occurred. Measurements f apparent age on contemporary marine molluscs include)5 ± 40 years for coastal waters of the UK (Harkness, 983); 450 ± 40 years from the fjords of Norway; 365 ± 20 ears for coastal localities of Iceland (Håkansson, 1983); nd 788 ± 33 years for the northeast Pacific (Southon *et al.*, 992). A correction factor of 400 years has also been applied ^{14}C dates on samples of submerged corals from the waters round Barbados (Fairbanks, 1989). A major difficulty here, owever, is that the apparent age of present-day marine olluscs may not always represent an appropriate correction actor. Recent data from the North Atlantic, for example, uggest that during the Younger Dryas cold episode, the tmosphere–sea surface ^{14}C difference was 700–800 years by comparison with today's value of 400–500 years, perhaps reflecting reduced advection of surface waters to the North Atlantic and more extensive sea ice cover at that time (Bard *et al.*, 1994). If so, it may be necessary to use a correction factor which is significantly greater (in terms of years) than that obtained from modern Mollusca when radiocarbon dating marine fossils, both from the Younger Dryas and from the last cold stage.

(d) Contamination In the case of **organic sediments**, contamination can occur because younger or older carbon has been added to the sample material. The former can arise from root penetration through a profile, infiltration by younger humic acids through older peat or soil horizons, and the downward movement of younger sediments through bioturbation. The possible effects of contamination by younger carbon are shown in Table 5.2A. Because of the high activity of modern carbon in comparison to fossil material, relatively small amounts of contaminant can result in major errors in radiocarbon dates.

Contamination by older carbon can also take a number of forms. Inwash of older inorganic carbon residues (graphite, coal, chalk, etc.) into lake basins leads to a dilution of the local ^{14}C:^{12}C ratio, and hence an ageing factor (**mineral carbon error**) will affect ^{14}C dates from such sediments. In late Flandrian sediments, inwash of soils or sediments can arise as a consequence of anthropogenically induced erosion around lake catchments (Oldfield, 1978). Mineral carbon error in ^{14}C dates may not be restricted to areas of carbon-rich bedrock, however, but may also occur in recently deglaciated terrain where inert carbon may have been released from igneous and metamorphic rocks by glacial erosion and subsequently concentrated in lake sediments

Table 5.2 A. The effect of contamination by modern carbon on the true radiocarbon age of samples. B. The effect of contamination by inert carbon on the true radiocarbon age of samples (after Harkness, 1975).

A.

True age (years)	Measured age as a result of		
	1% contam.	5% contam.	10% contam.
600	540	160	Modern
1 000	910	545	160
5 000	4 870	4 230	3 630
10 000	9 730	8 710	7 620
Infinitely old	36 600	24 000	18 400

B.

Contamination (%)	Years older than true age
5	400
10	850
20	1 800
30	2 650
40	4 100

(Sutherland, 1980). Sub-aquatic photosynthesis, water uptake in carbonate-rich groundwaters, and carbonate secretion by freshwater or offshore organisms may also be affected by diluted ^{14}C levels. In these instances, the resulting age error is termed **hard water error,** and may add up to 1200 years to the apparent age of limnic material (Peglar *et al.*, 1989). In addition, contamination of lake sediments may arise from the inwashing of older *organic* carbon detritus, and this redeposited or allochthonous carbon will also produce an ageing effect in dated samples (Björck & Håkansson, 1982). The extent to which contamination by older carbon can affect ^{14}C dates is shown in Table 5.2B. In general, samples that are initially rich in organic carbon will only be seriously affected where the amount of contaminant is high. Many organic sediments such as lake gyttjas, however, typically consist of only 4 to 5 per cent (and sometimes as low as 1 per cent) organic carbon, and therefore relatively small amounts of inert carbon incorporated into the sediments can introduce significant errors.

A further problem in the dating of limnic sediments arises because the $^{14}C/^{12}C$ ratio in lake waters may be lower than that in the atmosphere (the **reservoir effect**). This is partly a consequence of the fact that the exchange rate at the lake surface is relatively slow and hence lake waters may have a somewhat lower activity than the atmosphere, but it may also reflect seepage into lakes of groundwater containing dissolved carbonates. In Swedish lakes, the reservoir effect can add 300–400 years to ^{14}C dates from lake sediments (Olsson, 1986).

Radiocarbon laboratories have standard pre-treatment procedures to guard against the more obvious effects of contamination by older and younger carbon, but these cannot counteract, nor indeed detect, all forms of contamination. Samples are first checked for the presence of rootlets and other obvious signs of 'foreign' matter, and they are then usually cut or ground into small portions, examined and sieved to remove dust. This is followed by treatment with hot dilute acid to remove carbonates and with alkali solutions to remove humic acids that may coat the samples. In addition to these routine laboratory procedures, attempts are now being made to isolate contaminants in organic sediments by dating individual components of the material (Walker & Harkness, 1990; Shore *et al.*, 1995). For example, the presence of non-contemporaneous carbon may be reflected in contrasts between ^{14}C dates from the alkali soluble ('humic') and alkali insoluble ('humin') fractions of a sample of organic sediment (Table 5.3). These laboratory procedures all rest on the assumption, however, that field sampling has been meticulously carried out and the *onus* remains on the collector to ensure the utmost care in selection, handling and despatch of samples to the radiocarbon laboratory.

Problems of contamination are often encountered where **shells** are used for dating, since carbon exchange takes place more readily in carbonate structures. Contamination may be by older or younger carbon, depending on the dissolved contaminants introduced, and can result from the gradual accumulation of particulates or solutional carbon in the interstices of the carbonate matrix, or from the recrystallisation of the carbonate matrix and an **exchange** of sample carbonate with contaminant carbon. The former type can be largely avoided, however, by choosing samples that possess a 'tight' shell matrix (Mangerud, 1972) .

Exchange usually affects the outer part of a shell more than the inner layers, and it has therefore become practice to remove the external portion prior to dating. Procedures vary between laboratories, but up to 25 per cent of the outer part is typically leached and discarded (Sutherland, 1986). The remainder is then treated to produce an 'outer' and an 'inner' fraction which are dated separately, the inner date being preferred where a noticeable difference between the two ages occurs (Peacock & Harkness, 1990). The development of AMS dating has resolved some of the difficulties that have arisen with the dating of shell, however, for age determinations can now be made on very small samples of

Table 5.3 Radiocarbon dates on Lateglacial and early Flandrian sediments from Llanilid, South Wales. The 'weighted mean' ages are those that would have been obtained had a single determination been made on the total acid insoluble organic carbon in each sample. Contamination by older carbon residues in SRR-3455 is reflected in the discrepancy between dates of the 'humin' (alkali insoluble) and 'humic' (alkali soluble) fractions, with the humic measurement providing the most reliable age estimate. By contrast, the discrepancy between the fraction ages in SRR-3459 to 3466 appears to be attributable to contamination by younger carbon and hence the 'humin' timescale is preferred for the age of these samples. In SRR-3456 to 3458, there is no statistically significant difference between the fraction ages, suggesting that in each of these samples contamination is minimal (after Walker & Harkness, 1990).

Laboratory number	Depth (cm)	'Humic' age (years BP)	Organic ^{13}C (‰)	'Humin' age (years BP)	Organic ^{13}C (‰)	Weighted mean (years BP)
SRR-3466	176	9 140 ± 65	−29.2	9 320 ± 60	−29.1	9 200 ± 45
SRR-3465	133	9 015 ± 65	−29.9	9 570 ± 60	−28.9	9 170 ± 45
SRR-3464	126	9 355 ± 60	−28.0	9 850 ± 65	−27.6	9 510 ± 45
SRR-3463	123	9 410 ± 65	−28.9	9 920 ± 65	−28.5	9 580 ± 50
SRR-3462	100	11 080 ± 70	−26.9	11 160 ± 70	−29.2	11 100 ± 50
SRR-3461	83	11 060 ± 70	−28.4	11 470 ± 80	−29.0	11 175 ± 55
SRR-3460	73	11 080 ± 70	−28.7	11 300 ± 70	−28.3	11 150 ± 50
SRR-3459	55	11 170 ± 70	−29.2	11 410 ± 70	−28.0	11 245 ± 50
SRR-3458	30	11 710 ± 75	−30.4	11 655 ± 70	−29.4	11 690 ± 55
SRR-3457	22	12 140 ± 70	−23.8	12 255 ± 70	−24.1	12 190 ± 50
SRR-3456	18	12 380 ± 70	−25.6	12 495 ± 70	−25.9	12 420 ± 50
SRR-3455	5	13 200 ± 70	−25.3	14 200 ± 75	−26.5	13 685 ± 55

inner shell material. Moreover, ^{14}C dates on Mollusca can now be obtained from hitherto unpromising contexts, such as marine boreholes, where only small amounts of shell carbonate are available (Housley, 1991).

In the case of **bone** material, carbon exchange after death is always likely to have occurred, and hence although in theory both elements of bone (calcium hydroxyapatite and collagen) can be dated, it is usually only the proteinaceous fraction which is used for ^{14}C measurements. Again, with the development of AMS dating, it has proved possible to obtain age determinations on individual amino acids within the collagen to check for consistency in dates and hence for the presence of contaminants (Bowman, 1990). The use of the amino acids themselves in establishing relative order of antiquity of protein-bearing materials is considered later in this chapter.

(e) Biogeochemistry of lake sediments AMS dating has enabled ^{14}C ages to be obtained on different biogeochemical components of lake sediments (humic acid fractions, lipids, chlorite treatment residues, HF/HCl treatment residues, etc.). Where this type of dating has been carried out, significant age differences have been found within a single sediment sample (Fowler *et al.*, 1986) and between samples from contemporaneous horizons (Lowe *et al.*, 1988). Moreover, marked discrepancies have also been recorded in ^{14}C activity between macrofossil cellulose and the sediments from which the plant macrofossils have been obtained, with macrofossils typically providing younger ^{14}C ages than the sediments within which they were contained (Andree *et al.*, 1986; Peteet *et al.*, 1990). These data suggest that more needs to be learned about the biogeochemistry of lake sediments, if reliable age estimates are to be obtained from these media by means of radiocarbon dating (Lowe, 1991).

5.3.2.4 Radiocarbon dating of soils

One of the most difficult materials to date by the ^{14}C method is soil (Matthews, 1985). All soils contain both organic and inorganic carbon and can, therefore, be dated by radiocarbon. However, soils are dynamic systems and receive organic matter over long time periods. Any radiocarbon date on a soil will therefore be heavily influenced by the **mean residence time** of the various organic fractions in the soil (Geyh *et al.*, 1971). When a soil is buried, addition of organic matter ceases and a radiocarbon age will reflect both the mean residence time and the time that has elapsed since burial. The date at which pedogenesis commenced, which is of primary interest to the stratigrapher, will be almost impossible to establish. Further complications are added by the constant recycling that takes place within the soil profile, notably by humic acid filtration and root penetration (Aaby, 1983). Some meaningful dates have been obtained from buried palaeosols by comparing radiocarbon ages of the different organic fractions (Matthews & Dresser, 1983; Matthews, 1991), or by using different materials in soils such as carbonised wood (Polach & Costin, 1971) or charcoal (Goh & Molloy, 1979), but in general soils continue to pose problems for ^{14}C dating (Geyh *et al.*, 1983). However, their ubiquity in the landscape, together with their importance as relict and buried palaeosols, suggest that if the technical difficulties can be resolved, the dating of soils has immense potential in Quaternary environmental reconstruction (Matthews, 1985).

Of equal complexity is the dating of inorganic carbon in calcium carbonate-enriched horizons of soils, due largely to the fact that as calcium carbonate is readily soluble, solution and reprecipitation can take place. Each time this occurs, new carbon is added to the system because of the carbon dioxide content of air (Birkeland, 1974). In humid regions, radiocarbon ages of porous carbonates are often too young (Bowler & Polach, 1971), while in arid regions the age of carbonates is sometimes greater than that of the soil from which they have been derived (Williams & Polach, 1969).

5.3.2.5 Quality assurance in radiocarbon dating

Because different practices have been adopted in ^{14}C laboratories, the international radiocarbon community has agreed to participate in regular quality assurance schemes designed to ensure reliability and comparability of results between laboratories. This had previously been done on an *ad hoc* basis, but now formal inter-laboratory calibration exercises have been devised in order to ensure satisfactory performance by the various ^{14}C laboratories. The most recent of these, which was concluded in 1989, involved over 50 laboratories worldwide and included liquid scintillation, gas counting and AMS laboratories (Scott *et al.*, 1990).[2] Because the results showed the existence of systematic laboratory biases and additional sources of variability not accounted for by the quoted errors in published ^{14}C dates, detailed proposals for quality control and assurance have been proposed. These include the introduction of a published protocol for internal laboratory procedures; the introduction of additional reference materials, provided independently by the International Atomic Energy Agency in Vienna; and a provision for regular international intercomparisons in the future (Scott *et al.*, 1991). The aim of the protocol is to establish uniformity of practice between ^{14}C laboratories and to ensure quality of performance in the service they provide to the wider scientific community.

5.3.3 Potassium–argon and argon–argon dating

Potassium–argon dating is a technique that allows the age of volcanic rocks to be established. ^{40}K is a radioactive nuclide that undergoes branching decay, leading to one of two daughter nuclides depending on the type of transformation (Figure 5.5). Most ^{40}K decays by β emission to produce ^{40}Ca, each particle emitted from the nucleus resulting in the conversion of a neutron to a proton. The atomic number is therefore increased by one, resulting in an element of different chemical properties but with a virtually unchanged atomic mass. Electron capture (from the surrounding electron shells) by the nucleus is the alternative radioactive process. This converts a proton into a neutron (Figure 5.5) to reduce the atomic number by one and yield argon, a gas. Only one of these branches, the $^{40}K/^{40}Ar$ pathway, is useful for dating, for ^{40}Ca is so ubiquitous in nature that it is not possible to separate ^{40}Ca atoms produced by the decay of ^{40}K from those already present in rocks at the time of formation.

$^{40}K/^{40}Ar$ transformation is an accumulation radiometric clock. Once a volcanic rock has cooled, ^{40}Ar from the decay of ^{40}K becomes trapped within the lattice, and its abundance increases with time. Hence measurement of the amount of ^{40}Ar in a volcanic rock enables an estimate to be made of the time that has elapsed since the formation of the rock. It has to be assumed, however, that no original ^{40}Ar is present and that the system remains closed to both ^{40}Ar and ^{40}K after crystallisation; hence dates will only be valid for host rocks from which argon gas cannot escape. Fortunately, many mineral lattices retain argon, and it is only if rocks become melted, recrystallised or heated to a critical temperature that substantial loss of argon will occur. In the case of a volcanic rock that has been reheated or metamorphosed, therefore, the method will determine the age at which the final phase of modification ceased and not the age of initial formation. It also has to be assumed that no non-radiogenic ^{40}Ar is present (i.e. argon from the atmosphere). In practice, some atmospheric ^{40}Ar will be included in rocks and minerals, but this can be corrected for by comparing the $^{36}Ar/^{40}Ar$ ratio in the host rock with the known atmospheric ratio for the two gases. Potassium concentrations are measured by means of atomic absorption spectrophotometry or flame photometry, while the abundance ratios of the three argon isotopes (^{36}Ar, ^{38}Ar and ^{40}Ar) are determined by gas isotope mass spectrometry (Richards & Smart, 1991).

Radiometric dating by the $^{40}K/^{40}Ar$ method is largely restricted to volcanic and metamorphic rocks, since sedimentary rocks do not retain argon. Yet not all volcanic and metamorphic rocks are suitable, for sufficient potassium must be present to make dating possible. The nature of the mineral lattice is also an important factor, for not all minerals retain argon over long time periods, particularly when under stress. Orthoclase and microcline, for example, do not retain argon well. Biotite, on the other hand, is one of the most suitable minerals, for not only does it retain argon but it is also rich in potassium. Although age estimates as low as 1.2 ka have been reported (Gillot *et al.*, 1982), the large standard errors on such dates (typically ± 100 per cent) mean that the determinations have little real value. In

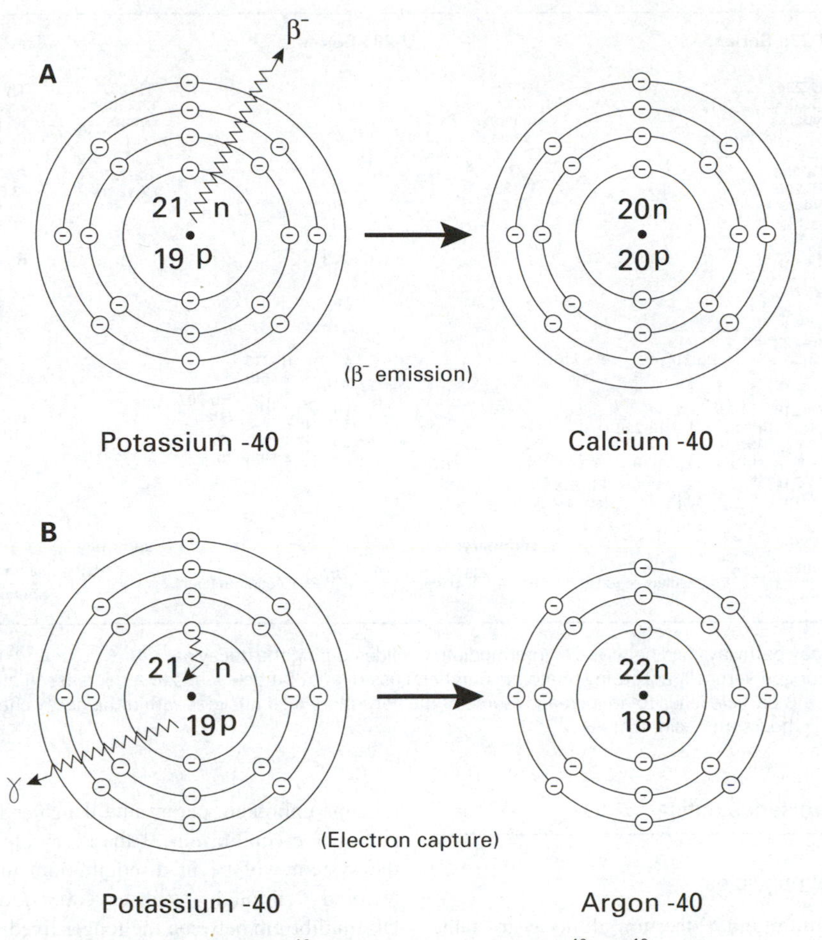

Figure 5.5 Branching decay of ^{40}K. A. Conversion of atoms of ^{40}K to ^{40}Ca through the emission of a β particle from the nucleus. B. Conversion of ^{40}K to ^{40}Ar through electron capture by the nucleus from one of the electron shells.

practice, meaningful age estimates have only been achieved on samples >100 ka.

A more recent development has been that of $^{40}Ar/^{39}Ar$ dating (Hall & York, 1984). This technique involves the irradiation of a sample with fast neutrons in a nuclear reactor to produce ^{39}Ar. The ^{39}Ar abundance is proportional to that of ^{39}K which is, in turn, proportional to ^{40}K. Hence a single mass spectrometric analysis can be employed to determine the $^{40}Ar/^{40}K$ ratio (Richards & Smart, 1991). The advantage of this approach is that dating can be carried out using very small samples, typically less than 10 g of material (tephra, for example). The technique offers the prospect of more reliable age estimates from a single rock sample, and also the possibility of determining whether 'closed system' conditions have been maintained following cooling of the rock (Richards & Smart, 1991). In addition, although it has so far been most widely employed in the dating of older volcanic materials (Bogaard *et al.*, 1989), it offers the potential for dating deposits younger than 30 ka BP (Gillot *et al.*, 1982).

The contributions of K–Ar dating to Quaternary research include a chronology for early hominid evolution in East Africa based on dates from lavas and related deposits (McDougall, 1981; Walter *et al.*, 1991), with the earliest find of the genus *Homo* being dated to *c.* 2.4 million years (Hill *et al.*, 1992); the development of an Early and Middle Pleistocene glacial chronology for western North America by dating tephras intercalated with tills (Richmond & Fullerton, 1986b); the dating of Middle Pleistocene events in central Europe (Bogaard *et al.*, 1989); and the provision of a timescale for the palaeomagnetic stratigraphic sequence (see below).

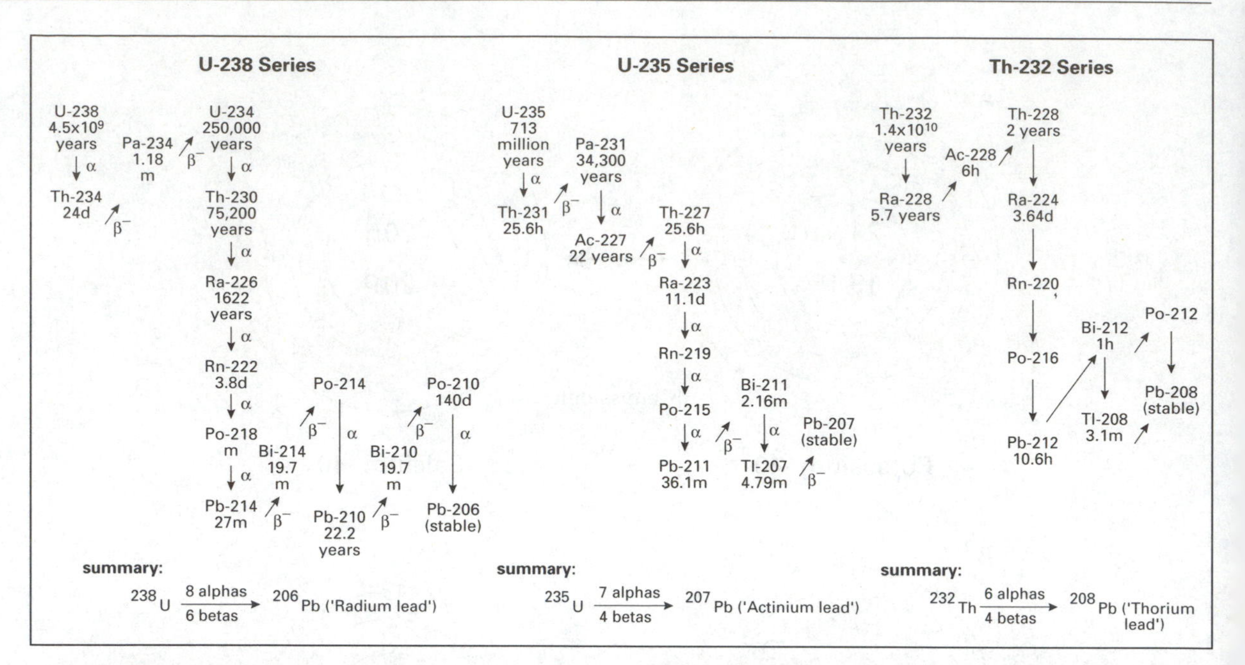

Figure 5.6 Chain decay pathways and half-lives of intermediate nuclides during the decay of ^{238}U, ^{235}U and ^{232}Th to stable lead. The elements are arranged vertically according to atomic number. Loss of an α particle leads to a decrease in atomic number, whereas emission of a β particle leads to an increase. Some of the very short-lived nuclides within the decay chain have been omitted (d = days; h = hours; m = minutes).

5.3.4 Uranium-series dating

5.3.4.1 General principles

^{238}Uranium, ^{235}uranium and ^{232}thorium all decay to stable lead isotopes through complex decay series of intermediate nuclides with widely differing half-lives (Figure 5.6). The helium gas formed by a particle emission may become trapped within host rocks, or may slowly diffuse out, ultimately to be liberated into the atmosphere. In theory, the age of a rock or mineral can be obtained from the amount of stable lead produced, or from the amount of helium liberated, but these measures are restricted to the dating of much older rocks (Jäger & Hunziker, 1979). Within the more limited timescale of the Quaternary, only those intermediate nuclides with relatively short half-lives can be employed. However, nuclides with half-lives of only a few years or less are impractical for dating and even the intermediate nuclides with half-lives of hundreds or thousands of years cannot be used for dating of materials where radioactive disintegration has proceeded in an undisturbed system. This is because in most rocks, an equilibrium state has been achieved in which nuclides formed by decay are disintegrating at rates similar to their rate of production by the parent nuclide. If the decay chain remains unbroken, parent and daughter isotopes remain in radioactive equilibrium. If the decay chain is broken, then the system will be in disequilibrium until equilibrium is restored through subsequent radioactive decay. Disequilibrium between the longer-lived isotopes can occur in natural systems for a variety of reasons. These include the loss of radon by gaseous diffusion through a porous rock matrix, and the separation of different chemical elements during weathering, transport and deposition in the hydrosphere (Smart, 1991a). Where the uranium decay is interrupted, and some decay products are selectively removed, the **uranium-series disequilibrium dating method** can be applied.

Disequilibrium methods are based on the following geochemical principles (see Schwarcz, 1989). Uranium and weathering products containing uranium are highly soluble, whereas other products of the U-series, such as thorium (^{230}Th) and protactinium (^{231}Pa), are readily absorbed or precipitated. Thus thorium and protactinium are co-precipitated with other salts to accumulate on the floors of lakes and on the sea bed, while uranium remains in solution. A selective separation, or **fractionation**, of these decay products therefore occurs. Accumulating sediments will contain quantities of thorium and protactinium but will be deficient in uranium, whilst organisms that secrete carbonate direct from ocean waters (such as molluscs and corals) will build a carbonate shell or skeleton that contains uranium, but very little thorium or protactinium. The same principles apply to carbonate precipitates such as speleothems and travertines, where fractionation results in the separation of uranium from decay products, and the age of precipitation can be measured from the renewed accumulation of radioactive decay products in the speleothem calcite. The age of lake or ocean floor sediments can be estimated by measuring the rate of decay of thorium or protactinium down a sediment profile, while the age of carbonate fossils, speleothems, teeth and bone can be derived from measurement of the accumulation of decay products of uranium within the carbonate matrix. The former has been referred to as the **daughter excess (DE)** type of uranium disequilibrium series dating method, i.e. the daughter nuclides are initially present in excess of concentration at secular equilibrium before decaying over time, while the latter is termed the **daughter deficient (DD)** group of methods, in which the daughter nuclide is initially absent but increases with time until equilibrium is achieved. DE methods have been most widely used in the dating of ocean sediments using the decay of thorium or protactinium, while the DD methods are principally based on measurement of $^{230}Th/^{234}U$ ratios in a variety of materials including tufa, speleothem, shell, bone and phosphates, all of which are initially deficient in thorium (Smart, 1991a).

5.3.4.2 Measurement, problems and age range

Conventional measurement of U-series ages is by means of alpha spectrometry following chemical extraction of thorium and uranium from the sample material. Subsequently, however, thermal ionisation mass spectrometry (TIMS) has been employed to determine U-isotope ratios (Edwards *et al.*, 1987; Li *et al.*, 1989). As in ^{14}C dating, this approach enables individual atoms to be counted directly as opposed to the monitoring of α particles emitted during radioactive decay. It is therefore more rapid and, as count rates are not restricted by the half-life of the isotope, there is the potential for extending the age range of the technique. In addition, the method offers a greater level of analytical precision enabling, for example, meaningful ages to be obtained from Lateglacial and Holocene materials (Bard *et al.*, 1990b). Sensitivity and precision in U-series dating has been further refined by the introduction of accelerator mass spectrometry (AMS). Not only does this approach permit the dating of very small samples of material, but it also enables both ^{14}C and U-series dates to be obtained from the same material (e.g. fossil coral), thereby allowing direct comparisons between the two timescales (see below).

A number of assumptions underlie the uranium series disequilibrium method (Smart, 1991a). It must be assumed that the decay coefficients have been accurately determined and that the activity ratio of daughter to parent nuclide can be measured to a high level of precision. These two requirements can largely be met. A third assumption, namely that there has been no loss or gain of nuclides since deposition, is more difficult to satisfy, and a number of materials (e.g. marine molluscs) show evidence of departure from **closed system** behaviour (Table 5.4). Dating of such material necessitates the use of correction factors which, in turn, requires a detailed knowledge of the geochemistry which gives rise to the isotope disequilibrium. Open-system behaviour may be reflected in reversals in isotopic ages in a stratigraphic sequence or the samples themselves may show evidence of open-system activity. For example, petrographic study of calcite may indicate recrystallisation and similar evidence may be detected in bones (Schwarcz, 1989). A fourth assumption is that the sample has not been contaminated at the time of formation by detrital materials that already contain daughter nuclides. Hence, in the measurement of the $^{234}U/^{230}Th$ ratio in speleothem calcite, for example, it is assumed that the ^{230}Th content was zero at the time of crystal formation, for any residual ^{230}Th in the sample would result in an age that was too old. Again, empirical evidence has shown that this assumption does not always hold, and that dated materials may contain varying amounts of detrital thorium. The presence of detrital ^{230}Th can be detected by measuring the amounts of ^{232}Th in a sample. This long-lived isotope occurs in water as a trace impurity and, where present, is indicative that contamination has occurred. The $^{232}Th/^{230}Th$ ratio may then be used as a basis for the correction of detrital contamination (Schwarcz & Latham, 1989). One way to identify the effects of detrital contamination and to correct for them is by the **isochron technique** (Smart, 1991a). This involves measuring the activity ratios of different minerals, or of whole rock fragments, obtained from the sample to be dated. If straight-line plots are obtained this indicates that the different mineral phases were formed simultaneously, and the best estimate of age is that ratio at which the isochron lines intersect. A large scatter in the data set indicates that detrital contamination is a major problem.

The age range of the uranium-series disequilibrium method varies with the nuclides employed. The practical dating range using conventional methods has been to five half-lives, and hence $^{230}Th/^{234}U$ has been employed to date samples in the range 5 to 350 ka, while $^{231}Pa/^{235}U$ and $^{231}Pa/^{230}Th$ have upper limits of around 200 ka and 250 ka respectively. Recent developments with mass spectrometric counting, however, have reduced the lower limits of the $^{230}Th/^{234}U$ method to less than 100 years (see above).

Table 5.4 Reliability of uranium-series dates for terrestrial materials due to deviations from closed-system behaviour and contamination by ^{230}Th and ^{234}U from detritus (after Smart, 1991a).

Reliability	Material	Closed system?	Contaminated
Reliable	Unaltered coral	Closed	Clean
	Clean speleothem		Clean
	Volcanic rocks		—
	Dirty speleothem		Contaminated
Possibly reliable	Ferruginous concretions	Possibly closed	Contaminated
	Tufa		Contaminated
	Mollusc shells		Contaminated
	Phosphates		Contaminated
Generally unreliable	Diagenetically altered corals	Open	Clean
	Bone		?
	Evaporites		Contaminated
	Caliche		Contaminated
	Stromatolites		Contaminated
	Peat and wood		?

Materials in the range 100 ka to 10 Ma may, in theory, be dated using the ^{4}H/U ratio, while the ^{234}U/^{238}U method has a potential age range of up to 1.5 Ma (Smart, 1991a).

5.3.4.3 Applications

Speleothems ^{230}Th/^{234}U dating has been widely applied to the dating of cave calcite precipitates – stalagmites, stalactites, flowstones, etc., the palaeoclimatic significance of which has already been considered (section 3.8). ^{234}U is precipitated from karst waters during the formation of speleothem carbonate, and almost all of the ^{230}Th subsequently found appears to be authigenic, i.e. it has originated from decay of ^{234}U that forms part of the speleothem chemistry. The dating of cave speleothems forms the basis for a chronology of, for example, climatic change (Gordon *et al.*, 1989), cave archaeology (Green, 1984), cave palaeontology (Stringer *et al.*, 1986), cave geomorphology (Gascoyne *et al.*, 1983a) and sea-level change (Proctor & Smart, 1991). In addition, application of the ^{234}U/^{238}U method has enabled ages in excess of 350 ka to be obtained from speleothem carbonate (Gascoyne *et al.*, 1983b).

Corals Corals appear to offer one of the most suitable media for uranium series determination for, after death, coral skeletons act as closed systems until the coral is dissolved or changes to calcite. Moreover, they contain sufficient uranium (typically 2–3 ppm) for the application of both the ^{230}Th/^{234}U and ^{231}Pa/^{235}U methods (Smart, 1991a). Since these are independent decay chains, they can provide an internal check on calculated ages. Uranium-series dating of raised coral reef complexes forms the basis for a chronology of Late Quaternary sea-level fluctuations, with high stands of sea level during the recent interglacials recorded in raised reef complexes (Pickett *et al.*, 1989), some of which have been influenced by tectonic uplift (Hoang & Taviani, 1991).

Carbonate deposits These include such diverse materials as travertines, calcretes, lake marls, stromatolites, phosphates and evaporites. All contain uranium precipitated at the time of deposition, but all tend to contain varying amounts of detrital thorium which need to be corrected for using the ^{232}Th/^{230}Th ratio. Some open-system behaviour may also occur with uranium loss by recrystallisation and leaching (Lao & Benson, 1988). Uranium-series dates on this type of material have proved especially useful in providing a chronology for pluvial lake-level changes (Thompson *et al.*, 1986), but other recent applications include the dating of subglacial calcite from beneath the Laurentide ice sheet (Hillaire-Marcel & Causse, 1989) and last cold-stage travertines in Israel (Kronfield *et al.*, 1988).

Molluscs Thus far, the use of fossil molluscs has been relatively unsuccessful, due partly to the fact that they contain initially only very small amounts of uranium (one-fiftieth of that contained in corals, for example), and also to the fact that they do not function as geochemically closed systems, for diagenetic uptake of uranium is common following death of the organism (Broecker & Bender, 1972). Hence, uranium-series dates on molluscs have tended to be regarded as unreliable (Kaufman *et al.*, 1971). On the other hand, apparently coherent chronologies have been obtained on, for example, Mollusca from localities around the Mediterranean basin (Hillaire-Marcel *et al.*, 1986b; Hearty, 1987). In view of the fact that many of the strandlines that need to be dated are not characterised by coral (a more suitable dating medium) but often contain abundant shells, attempts are likely to continue to find a means of obtaining reliable uranium-series dates from fossil molluscs.

Bone Following the death of an animal, uranium from groundwater enters and is trapped within the bone apatite. In theory, therefore, the reappearance of thorium and protactinium in bone apatite can be determined using either the $^{230}Th/^{234}U$ or $^{231}Pa/^{235}U$ ratios, and some coherent results have been obtained using the U-series method (Rae & Ivanovich, 1986; Rae *et al.*, 1987). However, discrepancies between $^{230}Th/^{234}U$ and $^{231}Pa/^{235}U$ dates on the same sample, and between ^{14}C dates and U-series dates on bone, suggest that problems of leaching (resulting in uranium depletion) and open-system behaviour (resulting in uranium enrichment) remain to be resolved before bone can be regarded as a suitable medium for this form of dating (Smart, 1991a).

Peat Peat, along with other organic material, takes up uranium from groundwaters and can become relatively enriched in the element (Vogel & Kronfeld, 1980). Since peat has a high adsorption capacity, any percolating groundwater will transfer its uranium content to the upper peat surface layer, thus protecting the older layers from further acquisition of uranium. As with bone, the incorporation of uranium-bearing inorganic detrital material into the peat deposits has proved to be a problem (van der Wijk *et al.*, 1986). However, some encouraging results have been obtained on both peat (Walker *et al.*, 1992; Heijnis *et al.*, 1993) and wood (de Vernal *et al.*, 1986), and refinement of the method offers the exciting prospect of dating interglacial peats and other organic deposits that currently lie beyond the range of the radiocarbon technique (Heijnis & van der Plicht, 1992).

Further details on the applications of uranium series dating in Quaternary and environmental science can be found in Ivanovich and Harmon (1995).

5.3.4.4 Calibration of the ^{14}C timescale

With the development of AMS techniques, it has become possible to obtain both U-series and ^{14}C dates from the same samples of material. This means that, for the first time, direct comparisons can be made between ^{14}C age determinations and those from an independent radiometric technique. As there are good grounds for believing that the production rate of daughter nuclides in the U-series decay chain is constant, U-series AMS dates can be used to calibrate ^{14}C age determinations which will inevitably have been influenced by the variations in atmospheric ^{14}C production described above.

The first steps towards ^{14}C calibration by means of U-series dating have been described by Bard *et al.* (1990b, 1992, 1993). ^{14}C and $^{230}Th/^{234}U$ ages were obtained on samples of fossil coral, U-series ages being determined initially by TIMS techniques and subsequently by AMS. The results (Figure 5.7A) show that ^{14}C ages are systematically younger than true ages during most of the radiocarbon range, except for a brief period between 500 and 2500 calendar years BP. The maximum divergence between the two timescales obtained by AMS occurs at around 16.0 ^{14}C ka BP when U-series dates indicate that ^{14}C determinations may be in the order of 2.5 ka too young. At around 20 ka BP, the TIMS U-series determinations suggest that the discrepancy reaches *c.* 3.5 ka, a deviation which has subsequently been confirmed by $^{230}Th/^{234}U$ and ^{14}C dating of cave speleothem (Holmgren *et al.*, 1994). The calendar duration for the Holocene ranges from 11.2 to 11.5 ka, an estimate in good agreement with data from the Greenland ice cores (Dansgaard *et al.*, 1993; Taylor *et al.*, 1993b), and German dendrochronological records (Becker, 1993). Indeed, the close agreement between the U-series and dendrochronological timescales between 6 and 11 ka BP (Figure 5.7B) lends further support to the accuracy and precision of measurement of time in calendar years by means of mass spectrometric $^{230}Th/^{234}U$ dating. Hence U-series dating offers a reliable way of calibrating the ^{14}C timescale over the entire dating range of ^{14}C, in other words well beyond the range of conventional dendrochronological calibration (section 5.4.1.5).

5.3.4.5 Uranium trend dating

This variant on the U-series method attempts to model open-system behaviour of the uranium decay series $^{238}U–^{234}U–^{230}Th$ in the weathering zone (Muhs *et al.*, 1989). Water movement through a body of sediment leads to fractionation of the parent and daughter isotopes, and by determining the range of activities of the various isotopes from a number of samples from the same material, it may be possible to obtain an estimate of the timing of the onset of leaching and, by implication, the date of deposition. In order to approximate age in years, however, the model has to be calibrated using materials of known age and, moreover, each model is site specific. Although still largely experimental, the method has been applied to a range of deposits including

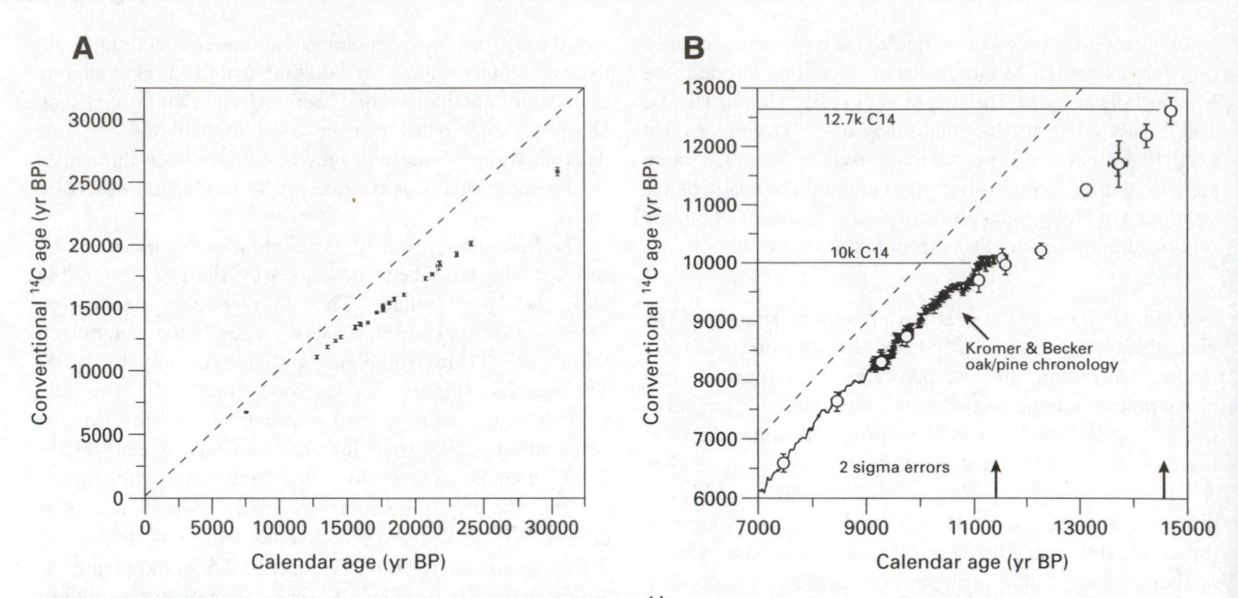

Figure 5.7 A. U–Th ages obtained by TIMS and plotted against AMS ^{14}C ages obtained on corals from Barbados and Mururoa. ^{14}C ages are expressed in years BP with statistical errors given at 2σ. For the interval between 8.5 and 20 ^{14}C ka, a simple linear calibration is: cal age BP = 1.24 (^{14}C age BP) − 840. B. An enlargement of A showing the German oak and pine dendrochronological dataset (after Kromer & Becker, 1992). Note the close correspondence between dendro-age and U–Th ages (after Bard *et al.*, 1993).

volcanic ash, alluvium, loess, till and soil, and appears to be applicable to a time range of 5 to 900 ka (Smart, 1991a).

5.3.5 Fission track dating

This method, which dates uranium-bearing crystals, is based on the **spontaneous fission** of ^{238}U: that is, the nucleus (of atomic number 92) divides to form elements of medium atomic number from about 30 to 65 (e.g. barium-56). An important consequence of spontaneous fission is that the energy released leads to high-speed collisions between fission fragments and neighbouring atoms. In rocks containing uranium, fission fragments cause **damage trails** or **tracks** in the wake of their movement through the host crystal lattice. The 'damage' induced is a result of ionisation of the atoms that come into contact with fission fragments. The positive charge acquired by adjacent atoms leads to mutual repulsion and therefore disorder in the crystal lattice. The tracks can be retained for millions of years and their number is a function of both uranium content and time. Although many minerals contain uranium, factors such as uranium abundance, and track retention and density mean that only zircon and glass are routinely dated in Quaternary rocks (Naeser & Naeser, 1988).

The age of a mineral or glass can be obtained by measuring the amount of uranium in the sample and the number of spontaneous tracks that it contains. The latter can be counted under a microscope after the surface has been polished and etched with a solvent to enlarge the fission track intersections. The sample is then irradiated in a nuclear reactor with thermal neutrons to induce fission in atoms of the less abundant uranium isotope ^{235}U; this produces a new set of fission tracks which can also be etched and counted. The induced fission track count provides a measure of ^{235}U abundance, from which the concentration of ^{238}U in the sample can be obtained from the known ^{235}U/^{238}U ratio in volcanic rocks (Hurford, 1991; Wagner & van den Haute, 1992).

The method, however, is not always straightforward. First, fission tracks may be less than 10 μm in size, and painstaking sub-microscopic examination of samples is required. Secondly, tracks can 'heal' or be erased (a process known as **annealing**) through the heating of the host materials or through spontaneous diffusion of ions. Indeed, because of the ease with which glasses anneal, the fission track ages of glass samples should always be regarded as minimum age estimates (Naeser & Naeser, 1984). Thirdly, since the density of fission tracks depends on uranium content and age, the method cannot be applied where the density of tracks is too low (typically in samples of <100 ka) or, at the other extreme, so high that counting becomes impossible. Ironically, rocks with a rich uranium content are not suitable because of the very high density of tracks, and in these instances the selection of minerals with a low uranium content, such as sphene, may form a more suitable counting medium.

Materials that have been dated by this method range from glass less than 140 years old (Wagner, 1978) to some of the oldest rocks on earth. The most significant contribution of the technique, however, has been in the field of tephrochronology (Walter, 1989), particularly in the age range 50 ka BP (the upper limit of radiocarbon) to 500 ka BP. Examples include the dating of the Rockland Tephra (*c.* 400 ka BP), a widespread pyroclastic layer and hence a key stratigraphic marker horizon in western North America (Meyer *et al.*, 1991), and the much older but equally significant Bishop Tuff (*c.* 700 ka BP) also from the western United States (Izett & Naeser, 1976). Other applications include the dating of microtektites both in terrestrial contexts (Storzer & Wagner, 1969) and in deep-ocean cores (Gentner *et al.*, 1970), the latter offering the means of establishing a chronology in sediments where other dating control is limited. Fission track dating has also been applied to archaeological materials, including the dating of early Pleistocene hominid remains in East Africa (Gleadow, 1980), the dating of more recent archaeological contexts in Mexico (Steen-McIntyre *et al.*, 1981) and the dating of obsidian artefacts and tracing of trade routes in South America (Miller & Wagner, 1981). In general, however, dating of recent archaeological materials is a minority application of the technique, because of the high analytical errors associated with such young and frequently uranium-poor material (Hurford, 1991).

5.3.6 Luminescence dating

Any material that contains uranium, thorium or potassium (all sediments and volcanic rocks contain all three), or lies in close proximity to other materials containing these radioactive substances, is subject to continuous bombardment by α, β and γ particles. This leads to ionisation in the host materials and the 'trapping' of metastable electrons within minerals. These electrons can be freed by heating, and under controlled conditions a characteristic emission of light occurs which is proportional to the number of electrons trapped within the crystal lattice. This is termed **thermoluminescence (TL)**, and the light emitted is additional to the normal incandescent light which would result from heating (Aitken, 1985). The latter will be emitted each time a sample is heated, whereas thermoluminescence, once liberated, can only reappear after further exposure to radiation. Thermoluminescent properties will accumulate progressively in a sample exposed to continuous radiation. Hence, thermoluminescent intensity is a product of the amount of radiation received per year (**radiation dose rate**) and time; if the former can be calculated by measurement of electron eviction, then an age can be assigned to the onset of electron trapping. In Quaternary materials this means that TL can be used to date the firing of objects (pottery, flint, etc.), or the burial of sediments which contain large quantities of quartz and feldspar material (e.g. loess). This is because both firing and prolonged exposure to sunlight (bleaching) effectively empty the electron traps, thereby resetting the TL clock to zero (Wintle & Huntley, 1982; Wintle, 1990).

Thermoluminescent properties (the **TL signal**) are reflected in the intensity of light emitted by the sample as it is heated from room temperature through to temperatures in excess of 500°C, and measured by a highly sensitive photomultiplier. This produces a **glow curve** (Figure 5.8), in which peaks are found at temperatures characteristic of the energies of the trapped electrons in the sample. The radiation dose received by the sample since the zeroing event, a measure that is referred to as the **Equivalent Dose (ED)** or **palaeodose**, is established by comparing the TL signal of the sample reflected in the glow curve with the TL signal induced in it by exposure to a calibrated laboratory radiation source. The latter is a measure of the TL sensitivity (Wintle, 1991).

In addition, an **annual dose** or **dose rate** has to be estimated.This is computed from the internal radioactive content of the sample (i.e. amounts of uranium, thorium and potassium), combined with a measurement of external radioactive contribution from the site of deposition. Where sediments are being dated, a further element that needs to be taken into account when estimating dose rate is water content, for a fraction of the energy in the sample will be

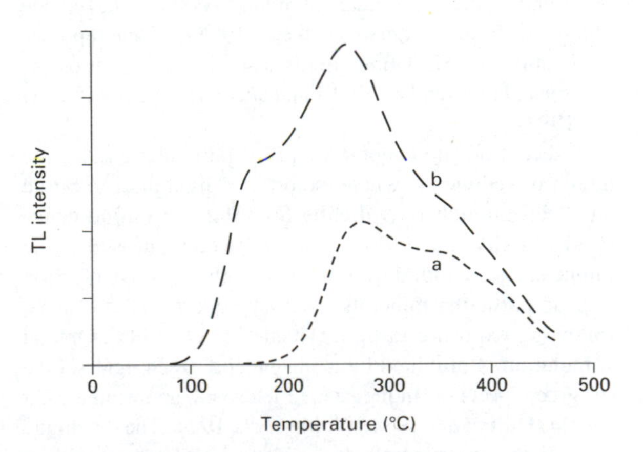

Figure 5.8 TL glow curves for a sample of Chinese loess. a, Natural TL; b, TL resulting from the addition of a 150 Gy radiation dose (after Wintle, 1991).

dissipated through water rather than through the mineral grains, and this can lead to a significant underestimate of age (Aitken, 1985). Once a value has been obtained for annual dose, age (in years) can be calculated from the following:

$$\text{TL age (years)} = \frac{\text{Equivalent dose}}{\text{Annual dose}}$$

Sources of error in TL dating include systematic errors associated with the calibration of laboratory radiation sources, which means that a precision of better than ±5 per cent (at the 68 per cent level of confidence) is unlikely to be achieved (Aitken, 1990); disequilibrium in the uranium decay chain (see above), migration of radioelements through surface sediments and estimation of past water content in sediments, all of which pose problems for the establishment of annual dose rate (Wintle, 1991); and problems arising from incomplete zeroing of the TL signal in sediments and the possibility of leakage from the electron traps (Aitken, 1985).

The lower age range of TL dating reflects the sensitivity of the sample and the efficiency of the zeroing mechanism, while at the upper end of the dating range **saturation** (the point at which the electron traps become completely filled) and the long-term stability of the TL signal are the principal constraints (Wintle, 1991). The upper practical limit for quartz, for example, is around 100 ka BP (Aitken, 1985), but ages up to 150 ka have been reported (Yanchou *et al.*, 1987). A significantly greater age range (>500 ka) may be obtained using feldspars (Mejdahl, 1988), although these minerals are subject to problems of thermal instability (Aitken, 1985). For speleothems, dates of up to 100 ka BP have been obtained (Debenham, 1983), while there are published dates on burnt flint in excess of 200 ka (Valladas, 1992). At the other end of the scale, dates of around 100 years have been obtained from pottery (Aitken, 1985). Contemporary applications of TL dating are discussed in three symposium volumes (Townsend *et al.*, 1988; Faïn *et al.*, 1992; Aitken *et al.*, 1994).

A recent development which is of particular use for the dating of sediments, where exposure to light prior to burial has been limited, is **Optically Stimulated Luminescence (OSL)** (Aitken, 1992). This technique measures the luminescence emitted from the most light-sensitive electron traps in particular minerals, especially quartz and feldspars, following exposure to light (Huntley *et al.*, 1985). Optical stimulation is provided by using either a green light source (or green laser) or, in the case of feldspars, an infrared light source (Hütt *et al.*, 1988; Wintle *et al.*, 1994).The technique is applicable to minerals of a wide range of ages (Smith *et al.*, 1990), although the lower practical limit appears to be around 1000 years (Aitken, 1990).

5.3.7 Electron spin resonance (ESR) dating

The basic principles of ESR dating are similar to those of TL. The technique involves the direct measurement of radiation-induced paramagnetic electrons trapped in crystal defects in a body of rock or other material (indeed, the technique is sometimes referred to as **electron paramagnetic resonance**). These 'free' electrons were generated by α, β and γ radiation from natural radioelements (e.g. U, Th and K) and have accumulated in the crystal lattices of minerals over time (Hennig & Grün, 1983; Grün, 1989). Exposure to high-frequency electromagnetic radiation in a strong magnetic field 'excites' the electrons and their resonance can be detected as the magnetic field is changed. This is because when the trapped electrons (or 'spins') are in resonance, electromagnetic power is absorbed in proportion to the number of electrons present; the greater the number of electrons, the greater the absorption (Aitken, 1990). Hence, the latter is a reflection of age and, as in TL dating, is known as the **palaeodose**, although in ESR dating, the terms **accumulated dose (AD), total dose (TD)** and **equivalent dose (ED)** are also used (Smart, 1991b). In order to obtain an age for the sample, a second element is required, namely the **annual dose** or the **ESR sensitivity** of the sample which reflects the average radiation flux or dose rate at the sample site. This includes both internal and external radiation, plus cosmic ray radiation. Additional doses of gamma radiation to the sample in the laboratory and reference to calibrated standards enable an estimate to be obtained for annual dose rate. The age of the sample (in years) is then obtained from:

$$\text{Age} = \frac{\text{Palaeodose}}{\text{Annual dose}}$$

A major problem with ESR dating is the difficulty encountered in estimating the annual dose, which typically produces an error at best of ±10 per cent (Hennig & Grün, 1983). It has to be assumed that the ESR signal has been zeroed on incorporation into the deposit, and also that the radiation flux at the sample site has remained constant over time; neither of these assumptions can always be confirmed (Smart, 1991b). In addition, although the method is, in theory, applicable to the entire time range of the Quaternary, in practice some materials have a shorter applicable timescale than others (Table 5.5). On the other hand, the method does have a number of advantages over TL. A greater range of materials can be dated by ESR including, for example, tooth enamel, mollusc shells and coral which are unsuitable for TL because of decomposition on heating (Aitken, 1990). In addition, insofar as the technique does not empty the electron traps, unlike TL, replicate measurements can be made on a single sample.

Table 5.5 Potential applicable ranges for ESR dating of various materials (after Smart, 1991b).

Material	Applicable range (ka)
Calcite – speleothems	0–800
Aragonite – molluscs	0–100
Aragonite – corals	0–250
Quartz – faults and sediments	0–>2000?
Hydroxyapatite – tooth enamel	0–2000
Gypsum	0–1000

The applications of ESR dating are considerable and are reflected in the papers of the three symposium volumes referred to above (Townsend *et al*., 1988; Faïn *et al*., 1992; Aitken *et al*., 1994). These discuss the dating of such diverse materials as speleothem calcite, travertines, corals, saline sediments, quartz grains, volcanic rocks, molluscs, Foraminifera in deep-sea cores, tooth enamel, bone and burnt flint artefacts. Although marked discrepancies are apparent between ESR and other radiometric dates on certain media such as teeth and corals (Blackwell *et al*., 1992; Grün *et al*., 1992), the relatively consistent results that have been obtained from other materials, notably speleothem calcite (Smart, 1991b), suggest that if the experimental problems can be resolved, ESR dating could become a technique of considerable significance in the development of a Quaternary timescale.

5.3.8 Other radiometric methods

5.3.8.1 Long-lived radioactive isotopes

A number of long-lived cosmogenically produced radioactive isotopes offer a basis for measuring time intervals within the Quaternary. These include ^{36}Cl (half-life 300 ka), ^{26}Al (730 ka), ^{10}Be (1.6 Ma) and ^{41}Ca (*c*. 100 ka) (Aitken, 1990).

Chlorine-36 Reaction of cosmic rays with certain elements in minerals leads to the formation and progressive accumulation of ^{36}Cl on exposed rock surfaces. Some ^{36}Cl is produced in terrestrial rocks by neutron activation of ^{35}Cl (derived from the decay of U- and Th-series nuclides) and from fission of ^{238}U, but the rate of production is very low, some 2–3 orders of magnitude below the levels of ^{36}Cl resulting from cosmic ray activity. As buried rocks only experience a low subsurface neutron flux, therefore, exposure of the rock leads to activation of the 'cosmic ray clock' and the onset of ^{36}Cl build-up. Hence, the amount of cosmogenic ^{36}Cl (measured relative to stable Cl by AMS) is proportional to the length of time that has elapsed since the exposure of the rock surface (Phillips *et al*., 1986; Zreda *et al*., 1991). Possible error sources include the presence of ^{36}Cl inherited from an earlier exposure episode, preferential leaching of ^{36}Cl relative to stable Cl, and erosion of rock surfaces (and hence disruption of the cosmic-ray clock) by spalling, shattering or disintegration (Phillips *et al*., 1990). The method has been used to date Pleistocene lacustrine sedimentation (Jannik *et al*., 1991), to establish the age of individual glacial features (Phillips *et al*., 1994), and in the development of Late Quaternary glacial chronologies (Figure 5.9). ^{36}Cl also been employed in the dating of groundwaters (Gove, 1987), and as an environmental and atmospheric tracer (e.g. Beasley *et al*., 1991; Wahlen *et al*., 1991).

Beryllium-10 and aluminium-26 The cosmogenic isotopes ^{26}Al and ^{10}Be in samples of quartz have been used to estimate erosion rates and exposure histories of surface materials (Middleton & Klein, 1987), but the greatest interest in beryllium has centred on ^{10}Be profiles from ice cores (e.g. Raisbeck *et al*., 1992), as these provide corroborative data on long-term atmospheric ^{14}C variability, and possible linkages with solar and geomagnetic changes (Stuiver *et al*., 1991; Stuiver & Brazunias, 1993). In addition, ^{10}Be peaks, such as those found at *c*. 35 ka and 60 ka BP in Antarctic ice cores, may constitute a basis for correlating southern and northern ice core records (Beer *et al*., 1992). Over a much longer timescale, a record has been obtained from loess–palaeosol profiles in China of ^{10}Be variations extending back to 5.4 Ma BP (Ning *et al*., 1994).

Calcium-41 ^{41}Ca is formed in the upper surface of the soil through the action of cosmic-neutrons on non-radioactive ^{40}Ca. It is subsequently taken up by plant tissues and becomes incorporated into animal bone (Aitken, 1990). Decay begins upon subsequent burial and hence, in theory, there is a means for establishing time of burial of bone material (e.g. Henning *et al*., 1987). The method is still experimental, however, and numerous technical problems remain to be resolved before ^{41}Ca/^{40}Ca concentrations in bone can be accepted as the basis of a practical dating technique (Middleton *et al*., 1989).

5.3.8.2 Short-lived radioactive isotopes

Radioisotopes with much shorter half-lives that have been used in dating Late Quaternary events include ^{210}Pb (half-life 22.26 years), ^{137}Cs (30 years) and ^{32}Si (*c*. 300 years) (Olsson, 1986).

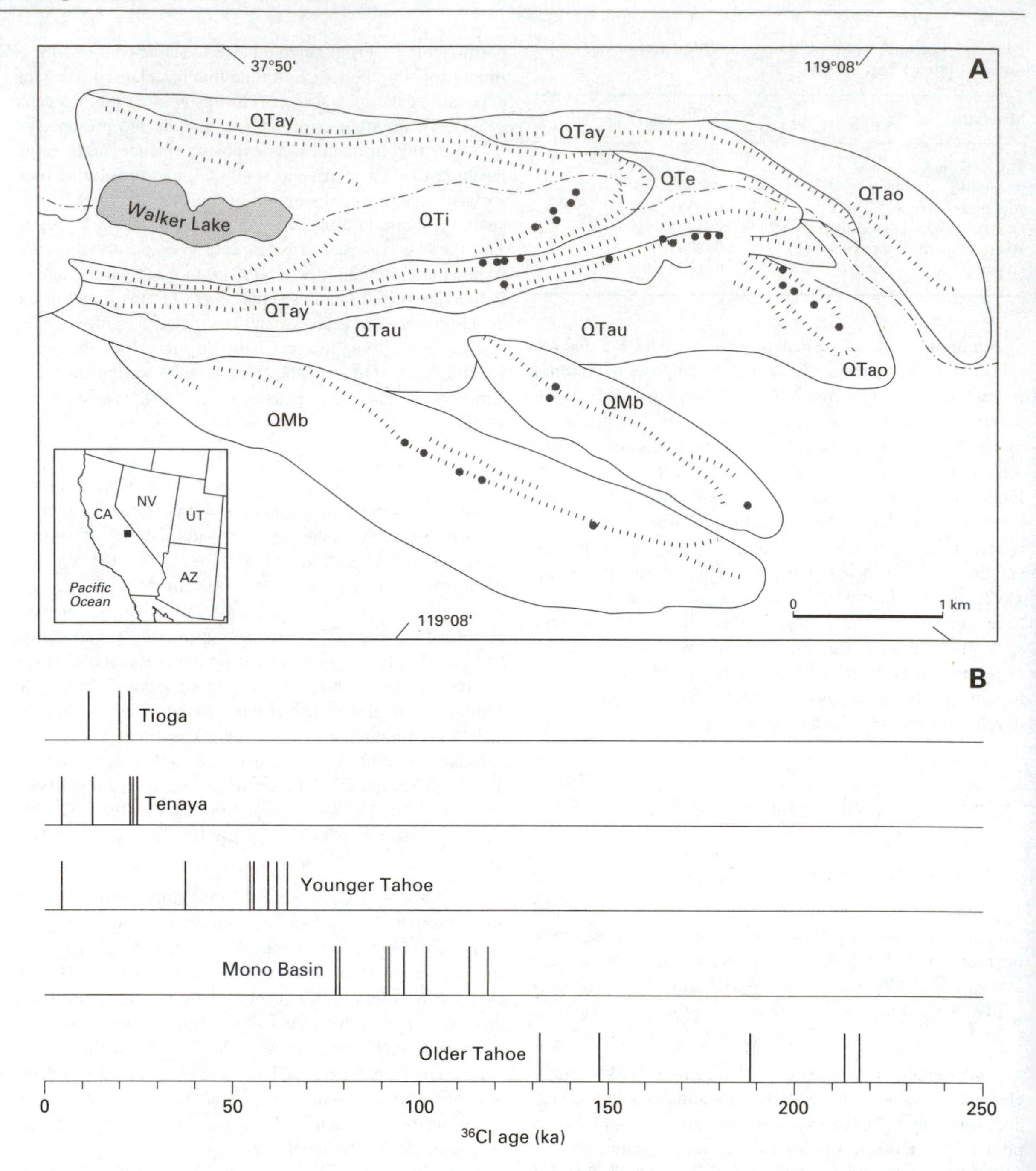

Figure 5.9 A. Map of the glacial deposits at Bloody Canyon, eastern Sierra Nevada, USA. QTi, Tioga deposits; QTe, Tenaya deposits; QTay, younger Tahoe deposits; QTao, older Tahoe deposits; QTau, undifferentiated Tahoe deposits; QMb, Mono Basin deposits. The hachures indicate the moraine crests, while the dots show sample locations of individual boulders. B. Distribution of cosmogenic ^{36}Cl boulder ages among the moraines at Bloody Canyon which suggest episodes of glaciation at around 21, 24, 65, 115, 145 and 200 ka BP (after Phillips *et al.*, 1990). A subsequent revision of the production constants for cosmogenic ^{36}Cl, however, provides the following revised ages for the five Bloody Canyon morraines. These are: Tioga 15.5–17 ka; Tenaya 18–19.5 ka; younger Tahoe 45–50 ka; Mono Basin 75–90 ka; older Tahoe 140–175 ka (F. M. Phillips, personal communication, 1995).

Lead-210 Radioactive decay of radon gas (^{222}Rn), which is part of the U-series decay chain (see above), produces a series of daughter nuclides, one of which is ^{210}Pb. This unstable isotope is removed from the atmosphere to accumulate in lacustrine and marine sediments, and in soils, peats and glacier ice, where it subsequently decays to stable ^{206}Pb over an interval of *c.* 150 years. By measuring the ratio of ^{210}Pb to ^{206}Pb in a column of sediment in relation to depth, and assuming that the atmospheric flux of ^{210}Pb has remained constant, the time that has elapsed since the lead was deposited can be determined (Olsson, 1986). In this way, the rate of sediment accumulation can be established. The main problem encountered with this method is that most sediments contain small amounts of ^{210}Pb derived from the decay of uranium or its daughters, and this 'supported' ^{210}Pb must be determined and subtracted from the 'unsupported' ^{210}Pb produced in the atmosphere. The technique has been applied mainly in limnological studies of, for example, laminated lake sediments (Appleby *et al.*, 1979), lake sedimentation rates (Appleby & Oldfield, 1983) and recent human impact on lake ecosystems (Varvas & Punning, 1993). ^{210}Pb dating has also been employed in the dating of peat (El-Daoushy, 1986) and cores from alpine glaciers and polar ice sheets (Stauffer, 1989).

Caesium-137 ^{37}Cs is an artificially generated radioactive nuclide that has only been produced in significant quantities as a result of thermonuclear weapons testing which began in 1945. ^{137}Cs has been used in the dating of lake sediments and peats (Pennington *et al.*, 1973; Longmore *et al.*, 1983) and in the estimation of rates of ground retreat on mining spoil heaps (Higgitt *et al.*, 1994), and it may also be used as an environmental tracer to indicate source areas of material reaching lakes from around the catchments (Wise, 1980).

Silicon-32 Cosmic-ray produced ^{32}Si has the *potential* for dating a range of recent materials, including marine sediments, groundwater and glacier ice (Grootes, 1984), over a time range of *c.* 1.5 ka. As such, it may eventually bridge the gap between ^{210}Pb and ^{14}C (Olsson, 1986), although a number of technical problems remain to be resolved before this can be considered a reliable dating method (Stauffer, 1989).

5.4 Incremental dating methods

Incremental dating methods are those based on regular additions of material to organic tissue or to sedimentary sequences. Those which have been most widely used are **dendrochronology** (tree-ring dating), **varve chronology** and **lichenometry**. In addition, **annual layers in glacier ice** form a basis for dating. These techniques are restricted in application largely to the Holocene, although both varves and ice-layer evidence have been used in the dating of pre-Holocene sequences.

5.4.1 Dendrochronology

5.4.1.1 General principles

In most softwood (coniferous) trees, new water and food-conducting cells (tracheids) are added to the outer perimeter of the trunk each growth season, following an inactive period in winter. The new cells that begin to grow in the spring tend to be larger and more thin-walled than those produced in late summer, as a result of heavier demands on water supply early in the growth season. Later in the year the cells become gradually smaller and develop thicker walls. There is normally, therefore, a distinct 'line' between successive annual increments of wood growth (Figure 5.10), and counting of these lines (**tree-rings**) allows the age of the tree to be established.

A greater variation in cellular structure is exhibited by hardwood (deciduous) trees. Hardwoods can be divided into **ring-porous** types where the spring vessels are normally

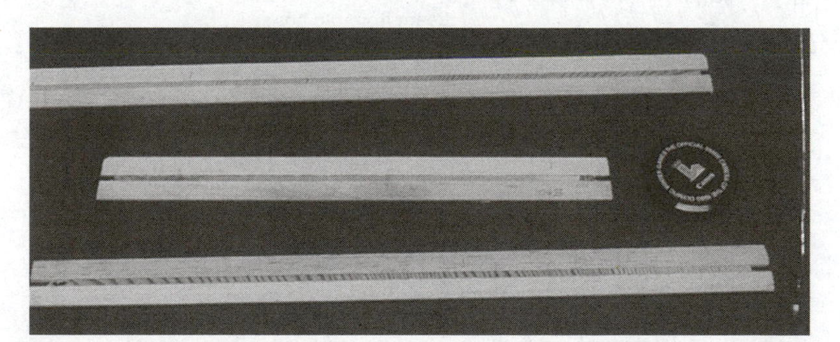

Figure 5.10 Radial bore samples, obtained from pine trees, which have been mounted, cross-cut and polished in preparation for dendrochronological analysis (width measurements, standardisation and cross-matching – photo: John Lowe).

distinctly larger than those of the summer wood (e.g. oak, ash, elm) and **diffuse-porous** types in which the pores are more uniform in size (e.g. beech, birch, alder, lime). There is, therefore, a considerable variation in the appearance of tree-rings between species, and not all trees show clearly defined annual bands. The most useful genera, or at least those most widely employed in dendrochronological work, are oak and pine (Schweingruber, 1988).

Tree-ring width is seldom uniform, for tree growth is influenced by a range of environmental factors, variations in which will produce different physiological responses within the trees (Fritts, 1976). The most important determining factor for many trees is climate. Under conditions of stress, growth is retarded, and a narrow tree-ring will result; conversely, under more favourable conditions, growth rates are increased, and wider annual rings are produced. As a consequence, climatic variations over very short and precisely dated timescales can often be inferred, an area of study known as **dendroclimatology** (Hughes *et al*., 1982).

5.4.1.2 Dendrochronological procedures

(a) Measurement Sub-fossil and dead trees can be cut so that a complete cross-section can be examined, allowing comparisons of ring width to be made in several radial directions. Living trees can be sampled with a metal increment corer which extracts small-diameter cylinders of wood from the tree trunk. In the laboratory samples are dried, polished and mounted prior to examination (Figure 5.10), or else cut with a sharp blade while still damp. Counting and measuring can be carried out by visual inspection of the rings under normal magnification, or on a moving stage under a binocular microscope. However, many tree-ring laboratories now employ more sophisticated electronic measuring equipment linked to both micro- and mainframe computers (Cook & Kairiukistis, 1990). An alternative technique is that of **X-ray densitometry** in which sections of wood are X-rayed and the negatives are then scanned by a beam of light and a photo-cell. The amount of light that is transmitted through the negative is determined by the character of the negative image which, in turn, reflects the density of the wood (Schweingrüber, 1988). Density variations are often more reliable than ring widths as climatic indicators (see below).

(b) Crossdating Ring patterns within trees from a limited geographical area can be matched through the technique of **crossdating**. Climatic variations within a particular area will be reflected in the trees from that area in characteristic ring-width patterns. Distinctive rings, or groups of rings, form markers and these can be used as a basis for cross-matching between trees of overlapping age range. In Figure 5.11, for example, a sequence of tree-rings from a living tree (A) contains a group of much narrower rings, perhaps reflecting drought conditions around 1930.

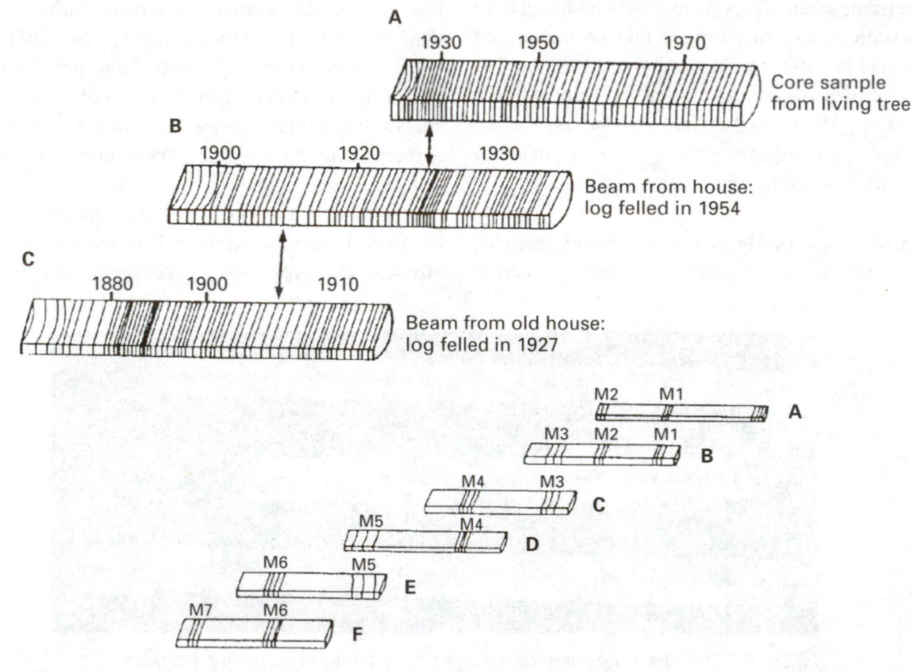

Figure 5.11 Core matching of tree-rings (crossdating). M = marker groups of tree-rings. For explanation see text.

This group can also be identified in a record from a tree which was felled in 1954 to provide beams for a building (B), and a match has also been established between B and a ring series in a beam taken from an older building (C). In this way, the tree-ring record can be extended back in time. Where sub-fossil logs are available (preserved, for example, in peat bogs), crossdating allows a link to be established between these and the historical or living record, and tree-ring chronologies spanning several thousands of years can therefore be constructed (see below). A number of computer-aided techniques have been developed to aid in the process of crossdating older material (Holmes, 1983).

(c) Standardisation Trees grow more vigorously in youth than in old age, and as a result there is usually a reduction in ring-width with age. Crossdating can therefore be difficult where ring-width variations resulting from the effect of limiting environmental factors are complicated or even masked by ring-width variations due to age. A further problem is that the width of tree-rings often varies with the height of the trunk. Each tree-ring series is therefore standardised, by transforming the measured ring-width values to **ring-width indices**. A range of statistical techniques is now employed in the production of ring-width indices (Schweingruber, 1988; Cook *et al.*, 1990). One approach is to fit a regression line to the measured ring-width values, and this provides an indication of the general decline in tree ring-width with age (Figure 5.12A). The measured ring-width value for each year is then divided by the value for that year obtained from the regression curve. The resulting ring-width indices (Figure 5.12B) have been corrected for the ageing effect and therefore fluctuations in the tree-ring curves reflect the influence of environmental factors only. Indices are required from a number of trees in a locality if accurate crossdating is to be achieved.

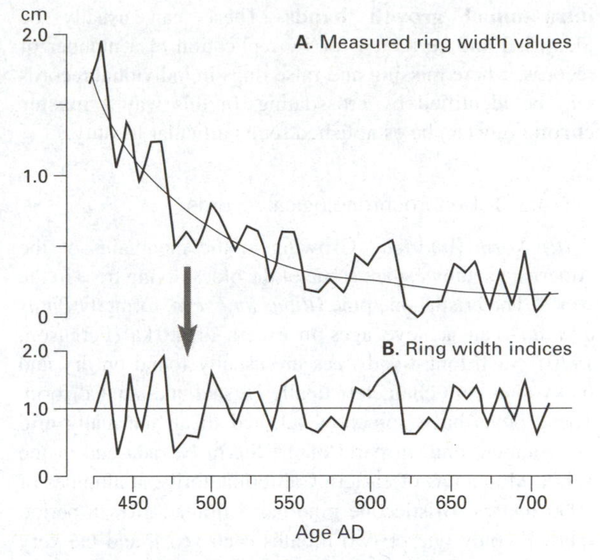

Figure 5.12 Standardisation of ring-width measurements by using regression curves to remove growth trends and hence to retain only the annual signal (after Baillie, 1982).

(d) Complacent and sensitive rings Local site conditions exert a major influence on tree growth and therefore also on tree-ring width. Some trees will experience more stress than others, depending on such factors as slope of the ground surface, water-retentive capacity of soils and sub-soil stratigraphy, relative amount of shade and exposure, genetic characteristics and so on. If a tree exists in a situation where there is a constant and adequate supply of water, and where it is protected by shading from the extremes of temperature, then tree-ring widths in such a specimen may show little variation, reflecting the lack of interruption in physiological activities. A tree-ring series of this type is termed a **complacent series**, and is of little value in crossdating as distinctive markers cannot be identified. Dendrochronologists therefore prefer to select trees from stressed situations, where some climatic factor has been critical to growth. Wood production will be reduced at times of stress, and this is shown by a **sensitive series** of rings, reflecting a clear and immediate response to some limiting factor. Tree-ring chronologies have therefore been derived largely, from semi-arid or arid sites where low moisture or high temperatures produce stressed conditions, or from sites at higher altitudes and latitudes where low temperatures in particular have restricted growth. It is also necessary to compare records from single species, since different species have different responses to environmental factors. In living specimens it is relatively easy to check for comparable site conditions affecting individuals, but this is obviously more difficult to monitor where sub-fossil specimens are used, especially where these are not in the position of growth.

(e) Missing and false rings Because trees are deliberately selected from stressed situations, there is always the possibility that during years of extreme climatic conditions a tree may fail to manufacture new cells, or may only produce new material on restricted parts of the trunk. These are referred to as **missing** and **partial rings** respectively. Partial rings may be absent from a record where only part of the trunk can be examined, for example in a wood core, a beam or a partly destroyed sub-fossil trunk. On the other hand, more than one growth layer can develop in a particular growth season where more severe conditions prevail for a short while during late spring or early summer, after the spring cells have commenced growth. This may produce a change in structure that resembles an annual boundary, and such layers are referred to as **false rings** or

intra-annual growth bands. These can usually be identified and corrected for by replication of a number of records, where missing and false rings in individual records will be identified by crossdating. In this way a **master chronology** can be established for a particular locality.

5.4.1.3 Dendrochronological records

(a) North America Growing in the mountains of the American southwest are some of the oldest living trees in the world. The bristlecone pine (*Pinus longaeva*, formerly *Pinus aristata*) can achieve ages in excess of 4.0 ka (Ferguson, 1970). Such long-lived trees are usually found on dry and rocky sites in a characteristically twisted and stunted form. These pines have somehow adapted to an unusually arid environment and, in parts of the Sierra Nevada and in the White Mountains of eastern California, thrive at altitudes of 4000 metres. Bristlecone pine has a limited growth period (perhaps only one or two months each year), and the very narrow rings produce a pattern that is highly sensitive to climatic variations. By crossdating between living and dead wood and then between sub-fossil samples, a continuous master chronology has been developed which now extends back to 8681 years BP (Ferguson & Graybill, 1983). Other relatively long tree-ring chronologies have been established in North America on species of *Sequoia*, Douglas fir (*Pseudotsuga*) and pine. Although the longest of these, based on foxtail pine (*Pinus balfouriana*), presently extends back only to 3031 BP (Scuderi, 1987), work currently in progress suggests that the record may eventually be extended beyond 6 ka (Scuderi, 1990). Much shorter chronologies, typically spanning the last 1.5 ka, have been obtained from a range of tree types elsewhere in the United States, in Canada and in Alaska (Brubaker & Cook, 1984; Parker *et al*., 1984)

(b) Western Europe In western Europe, there are no trees with an age range comparable to the bristlecone pine, and hence long chronologies have only been established by the laborious crossdating of large numbers of sub-fossil wood samples whose ring patterns often span no more than 100–200 years. The most commonly used species are the oaks (*Quercus robur* and *Quercus petraea*) and the Scots pine (*Pinus sylvestris*), the dendrochronological records being based on wood samples obtained from fen and raised bog peats, or from river gravels. The longest chronologies so far developed have been on oak, with a record from Ireland extending back to 7272 BP (Pilcher *et al*., 1984), and two continuous German oak chronologies, one stretching back to 6255 BC/8205 BP (Leuschner & Delorme, 1988), and the other to 8021 BC/9971 BP (Becker, 1993). A 7000-year oak chronology has also been established for lowland England (Baillie & Brown, 1988). In addition, a number of '**floating chronologies**' have been developed, i.e. tree-ring sequences not tied *directly* to historically dated or living wood, but which can be related to the master chronologies by other means, such as ^{14}C dating. Floating chronologies have been used, for example, to provide a timescale for events in the British Neolithic (Hillam *et al*., 1990), and to establish the age composition of former tree stands (Bridge *et al*., 1990). In Europe, the oak chronology referred to above has been linked into a floating chronology based on fossil pine which extends the European master dendrochronology back to 11 370 calendar years BP (Becker, 1993; Becker & Kromer, 1993). As in North America, there are many examples from Europe of much shorter chronologies (mainly on species of pine and spruce) extending back over the last 1.5 ka (Bartholin, 1984; Schweingruber *et al*., 1988; Briffa *et al*., 1990)

5.4.1.4 Dendroclimatology

Dendroclimatology is the science of reconstructing past climatic conditions and histories from tree-rings, and is generally considered to be part of the wider discipline of dendrochronology (Hughes *et al*., 1982). The importance of dendroclimatological records is that palaeoclimatic information can be precisely dated, and inferences can even be made about seasonal variations. Further, as a proxy record of climate it enables climatic data to be extended beyond the limits of records from instrumental measurements (Figure 5.13).

Palaeoclimatic reconstructions based on tree rings rest on assumed or demonstrable relationships between ring-width, or some other ring characteristic, and climatic parameters. These relationships are often complicated by lag effects between climatic inputs and tree response, for trees have the ability to store food reserves and water for several years and this stored material may then be used in adverse years. Because the vigour of a tree in any single year is determined by environmental influences both during that year and during previous years, a sequence of annual ring widths is **autocorrelated**, although the degree of autocorrelation can often be discerned by statistical tests (Fritts, 1982). Indeed, sophisticated multivariate techniques are essential for unravelling the complex climate–tree growth relationships (Cook, 1982). Once the linkages between climatic and tree-growth parameters have been established, however, inferences can be made about past climate for the timespan of the dendrochronological record. Dendroclimatology has become an integral tool in the study of environmental change, particularly over the course of the last millennium, allowing inferences to be made about, *inter alia*, temporal variations in summer temperature (Briffa *et al*., 1988), precipitation régimes (D'Arrigo & Jacoby, 1991), linkages

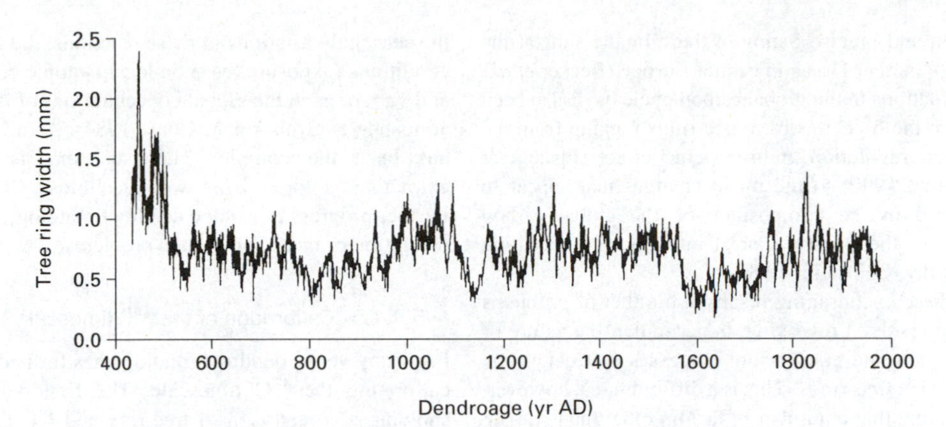

Figure 5.13 Dendrochronological series spanning almost 1.6 ka for the Torneträsk region of northern Sweden. Note the significant decline in ring widths at around AD 1100 and AD 1570, the latter reflecting the climatic deterioration of the Little Ice Age (after Bartholin, 1984).

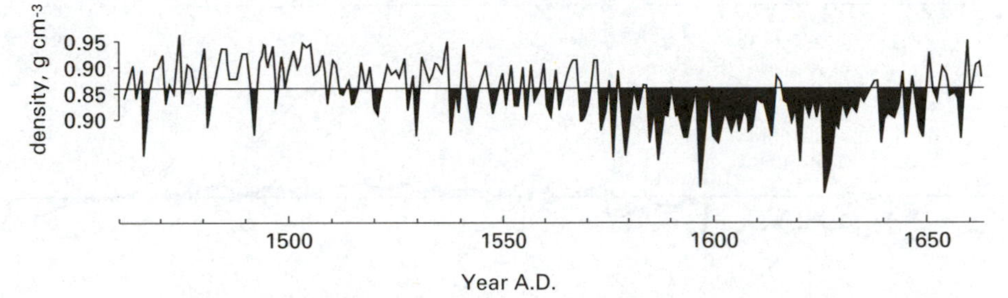

Figure 5.14 Curve of raw maximum latewood density values expressed as deviations from the mean for a tree (*Picea abies*) growing during the period AD 1460–1660 in the Lauenen area of the Swiss Bernese Oberland. The period of below-average growth during the Little Ice Age (*c.* 1570 onwards) is clearly visible (after Schweingruber *et al.*, 1988).

between glacier behaviour and climate (Scuderi, 1987), and relationships between climatic change and late Holocene volcanism (Baillie & Munro, 1988; Scuderi, 1990).

More recently, it has been found that variations in wood density may provide clearer palaeoclimatic information than simple ring-width measurements (Figure 5.14). As wood densities respond primarily to conditions during restricted periods (e.g. late summer), they offer the prospect for finer resolution of climatic relationships than ring widths which integrate the effects of conditions over several seasons (Brubaker & Cook, 1984). In Europe, relationships have been noted between maximum latewood density and late summer temperatures (Hughes *et al.*, 1984; Schweingrüber *et al.*, 1988), although in eastern North America a closer correlation has been detected between maximum wood density and spring temperatures (Conkey, 1986). There may also be a relationship between latewood density values and late-summer drought conditions (Conkey, 1979), while periodicities (60, 125 and 200 years), perhaps related to variations in solar influence, have also been detected in tree-ring density records (Stuiver *et al.*, 1991) Although much remains to be learned about the precise nature of the climatic signal in wood density variations, density measurements are becoming increasingly widely employed alongside conventional ring-width values in dendroclimatic research (e.g. Schweingrüber *et al.*, 1988; Briffa *et al.*, 1990).

A further climatic index derived from tree-ring data is based on stable isotope ratios of oxygen, carbon and hydrogen in wood cellulose, a field of research known as **isotope dendroclimatology**. The hydrogen and oxygen isotopic composition in plants is determined by the isotopic composition of the water used by the plants; moreover, there is a direct relationship between temperature and the D:H and ^{18}O:^{16}O ratios in precipitation. Hence δ^2H, $\delta^{18}O$ and also $\delta^{13}C$ values in tree-rings can, in theory, be employed as proxy climatic indicators (Bradley, 1985). Recent examples of this approach include the identification of Holocene 'pluvial' episodes in the Cairngorm Mountains, Scotland (Dubois & Ferguson, 1985), the measurement of global warming over the past 100 years (Epstein & Krishnamurthy, 1990), and the identification and precise dating of the climatic shift at the close of the Younger Dryas in central Europe (Becker *et al.*, 1991). Fluctuations in the global carbon cycle have also been inferred from the $\delta^{13}C$ record in tree-rings ranging from the post-industrial revolution anthropogenic effect (Epstein & Krishnamurthy, 1990) to the major changes that appear to have occurred in the composition of the global carbon reservoir at the last glacial–interglacial transition (Krishnamurthy & Epstein, 1990).

Despite these encouraging results, a number of problems remain to be resolved involving, in particular, the nature of the fractionation and equilibrium processes that determine isotopic ratios in tree-rings. This is a difficult area, however, for it is apparent that a number of factors other than climate can influence isotopic concentrations in tree-rings. These include the rates of cellular reactions such as photosynthesis, the seasonal variations in these reactions, the effects of site conditions (exposure, etc.) on local isotopic concentrations, and variations in the global concentrations of isotopes in the atmosphere (Brubaker & Cook, 1984). A further difficulty has been the complex chemistry that is involved in extracting isotopes from wood cellulose (Baillie, 1982). Further progress in isotope dendroclimatology depends on a resolution of these and related problems.

5.4.1.5 Calibration of the ^{14}C timescale

For many years, dendrochronology has formed the basis for calibrating the ^{14}C timescale. The first detailed dataset showing a comparison of tree-ring and ^{14}C dates over the last 7.0 ka was published in 1970 (Suess, 1970). This calibration curve (Figure 5.15) was important in that it

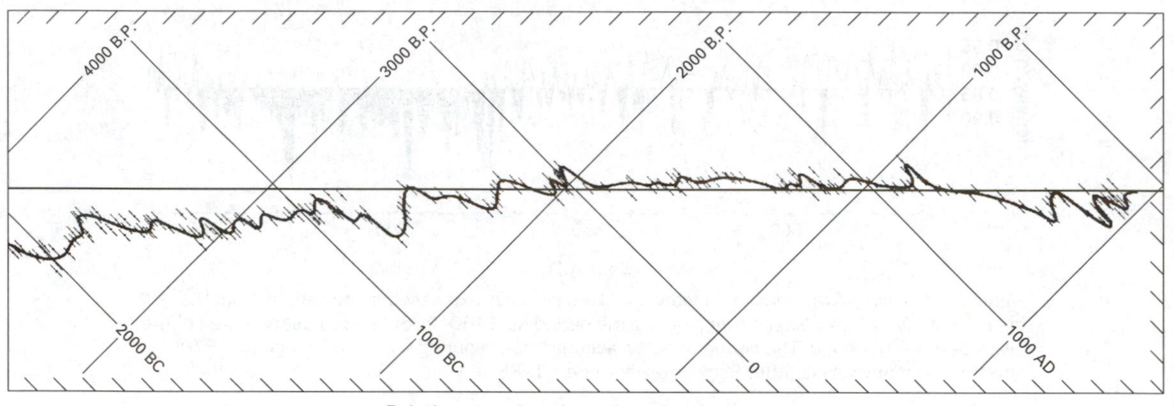

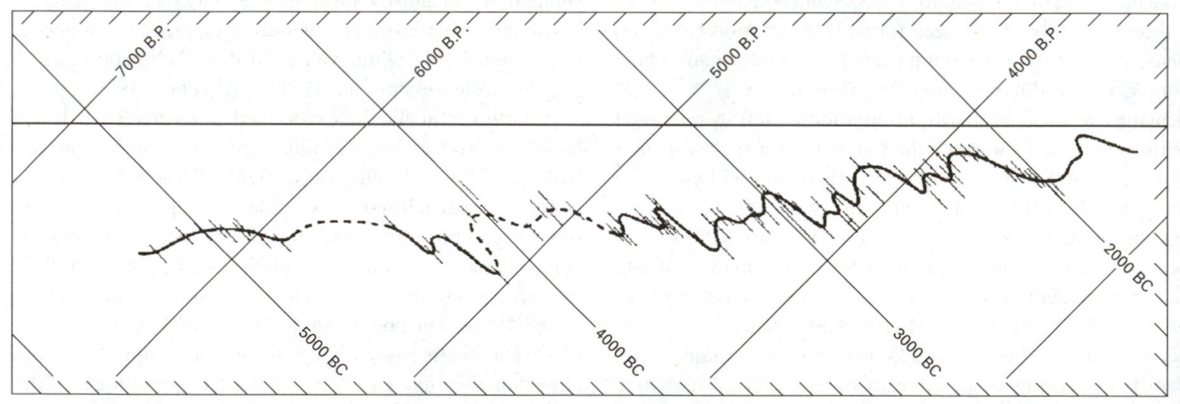

Figure 5.15 Tree-ring calibration of conventional ^{14}C ages, using different wood samples (including *Sequoiadendron giganteum*, *Pinus longaeva* and *Pinus ponderosa*) dated in ^{14}C laboratories in Arizona, Pennsylvania, Cambridge, Groningen, Copenhagen and Heidelberg (after Suess, 1970).

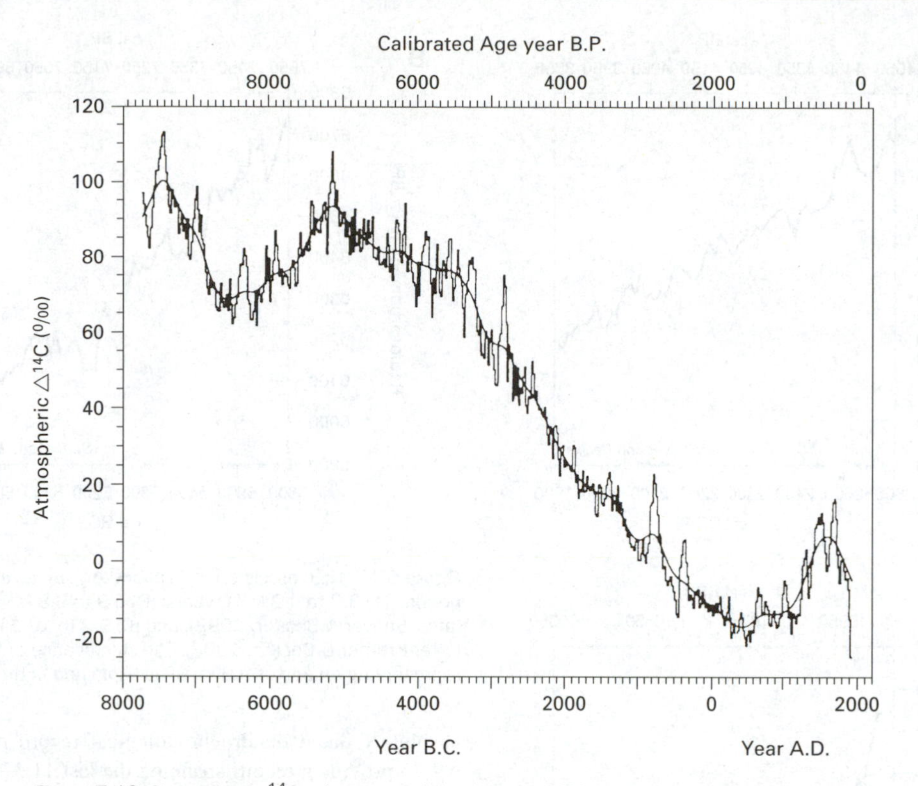

Figure 5.16 Atmospheric ^{14}C values per mil for the past 9.7 ka based on dendrochronologically dated wood, each data point (bar) covering a 20-year interval (after Stuiver *et al.*, 1991).

provided the first unequivocal evidence for temporal fluctuations in atmospheric ^{14}C activity, for dendrochronological and ^{14}C ages increasingly diverge so that wood with a ^{14}C date of 5.0 ka BP corresponds approximately to 5.8 ka on the dendrochronological timescale. If the tree-rings are accepted as reflecting calendar years, then the implication is that between 5 and 6 k calendar years BP ^{14}C levels must have been significantly higher by comparison with present-day values (Figure 5.16).

Since the publication of the Suess curve, numerous attempts have been made to develop a more secure basis for calibration based on additional data points and improved dating techniques (e.g. Damon *et al.*, 1972; Stuiver, 1982; Pearson *et al.*, 1986). The construction of high-resolution dendrochronological records, combined with developments in ^{14}C dating that allowed the production of dates with a statistically limited precision (standard errors of less than 20 years), particularly in the laboratories at Belfast and Seattle, eventually led to the publication of a series of high-precision calibration curves (Figure 5.17) extending back to around 8.0 ka BP (Stuiver & Pearson, 1986, 1993; Pearson & Stuiver, 1986, 1993). These are based on dated wood samples of oak (*Quercus petraea* and *Quercus robur*) from Ireland, Scotland and England, Douglas fir (*Pseudotsuga menziesii*) from the US Pacific northwest, and sequoia (*Sequoiadendron giganteum*) from California. Although these curves confirm the general trend that was apparent in the original Suess curve, they reveal a much more complicated pattern of ^{14}C variation. In practice, this can make calibration difficult to apply for the pertubations may result in ^{14}C dates having more than one calendar age. In Figure 5.17C, for example, the calibration curve crosses the line representing the mean value of a radiocarbon date of 9.6 ± 100 ka BP at no fewer than *seven* points. This means that when the 1σ confidence limit on the radiocarbon date is taken into acount, the calibrated age could lie anywhere between 10 500 and 10 980 ka BP. When the 2σ confidence limit is taken into account, this age range will more than double.

In some instances, however, the short-term fluctuations in the radiocarbon calibration curve can form a basis for very accurate dating (Pearson, 1986). The technique of **wiggle matching** involves the dating of a series of samples spanning a very short time interval (perhaps 100–150 years) whose dendrochronological ages are known. High-precision ^{14}C dates are obtained from the samples and these form a short section of the calibration curve which can then be

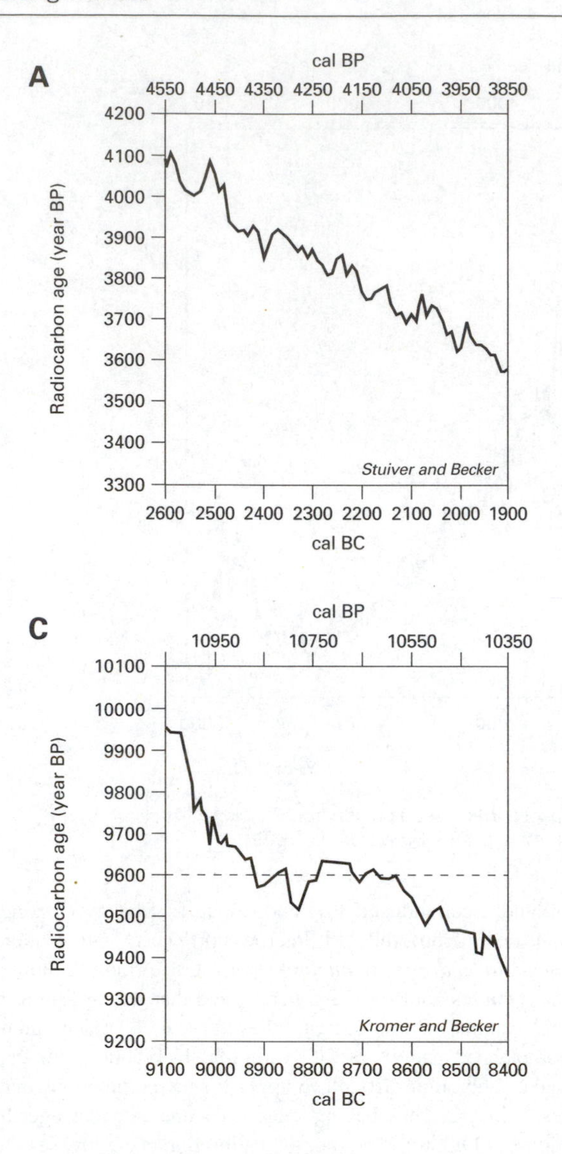

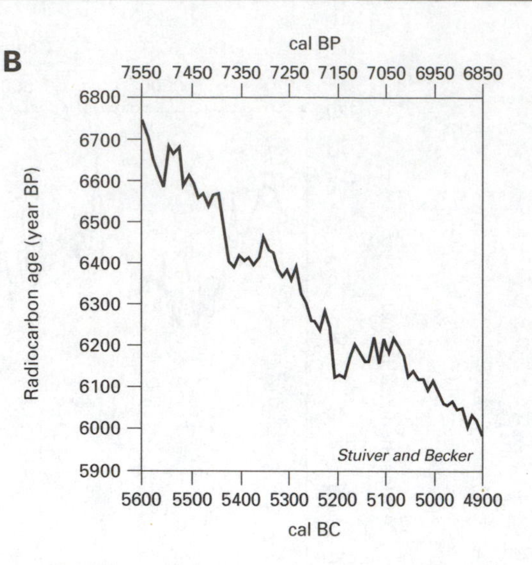

Figure 5.17 High-precision ^{14}C calibration curves covering the periods (A) 3.3 to 4.2 k ^{14}C years; (B) 5.9 to 6.8 k ^{14}C years (after Stuiver & Becker, 1993); and (C) 9.2 to 10.1 k ^{14}C years (after Kromer & Becker, 1993). The significance of the line representing an age of 9.6 ka BP is explained in the text.

matched against the full calibration curve. In some cases, it may prove possible to match the dated section to within a few years on the calendar timescale (Pilcher, 1991a).

Beyond the limits of present calibration curves (around 9.0 ka), some estimation of the extent of divergence between ^{14}C and calendar ages has been obtained by linking 'floating chronologies' with the older portions of continuous dendrochronogical series. This approach has enabled tentative age calibration to be established back to 13.3 ka BP (Stuiver *et al*., 1986), although these data tend to lack the precision of Holocene records. In Germany, however, a 1605-year floating Lateglacial and early Holocene pine (*Pinus sylvestris*) chronology has been crossdated with an absolutely dated dendrochronological record of Holocene oak to provide a record spanning the last 11.37 ka (Becker, 1993; Becker & Kromer, 1993). On this chronology, the age for the Lateglacial–Holocene boundary is 11.0–11.1 ka BP, compared with the conventional ^{14}C age of around 10.0 ka BP. This suggests a further divergence between calendar and ^{14}C, a trend which has been confirmed in the period back to 30 ka BP by comparisons between AMS ^{14}C and U-series dates on fossil corals (Bard *et al*., 1990b, 1993).

The large dendrochronological databases that have now been established have enabled the development of readily available computer software for calibrating ^{14}C dates. Examples include the CALIB 3.0 Age Calibration Program from the University of Washington (Stuiver & Reimer, 1993), and the CAL15 program from the University of Groningen (van der Plicht, 1993).

5.4.2 Varve chronology

5.4.2.1 The nature of varved sediments

Rhythmic accumulations of sediments, forming bands of laminae of fine sands, silts or clay, are common in the geological record. Often the laminae are arranged in couplets, with relatively coarse-grained layers alternating regularly with finer-grained bands (Figure 5.18). Such sediments are usually referred to as **rhythmites**, except where the laminations arise because of annual variations in the supply of sediment, in which case they are termed **varves** (after the Swedish '*varv*' meaning a lap, turn or revolution). Because they are deposited annually, varves can be used as a means of dating, for time intervals can be calculated and a floating chronology established. Moreover, if a varve sequence can be tied to the calendar timescale, then it may be possible to assign calendar dates to the varve record.

(a) Glaciolacustrine varves Large supplies of sediment are deposited in proglacial lakes and shallow seas, as a result of rapid spring and summer ice melt. The coarsest particles are deposited first on the lake or sea bed, leaving the finer clay particles in suspension. During winter, when the lake or sea adjacent to the shore is frozen, the suspended clay particles gradually settle out to produce a clay lamina that contrasts markedly with the coarser summer layer. Glaciolacustrine varves have usually formed the basis for varve chronologies.

The potential of such sediments as a means of dating was first recognised by the Swedish geologist Gerard De Geer who, as early as 1884, was investigating the exposed varve sequences in the Stockholm area (Lundqvist, 1985). He discovered that as the last ice sheet wasted northwards across southern Sweden, it left behind a complex of moraines and proglacial lakes. In many places, overlaps of varve 'histories' had developed, the uppermost varves in one lake sequence being of the same age as the lower varves in a close neighbour. In a now classic publication, De Geer (1912) first presented measurements of individual varves and curves of relative thickness for each site investigated. Comparisons between sites enabled correlations to be established using similar principles to those now employed in the matching of ring-width series in dendrochronology. By methodically extending his master chronology northwards and southwards across Sweden, De Geer compiled a long varve sequence which was believed to extend back continuously to the beginnings of deglaciation beyond 13 ka BP (De Geer, 1940). Although De Geer's original varve chronology has since undergone extensive revision, this seminal work laid the foundations for all subsequent work on ice recession in the Baltic region (see below).

Figure 5.18 A. Early Holocene clastic varves (with drop-stone) at Lavaksenharju, Siikainen, western Finland (photo: Raimo Kujansuu).

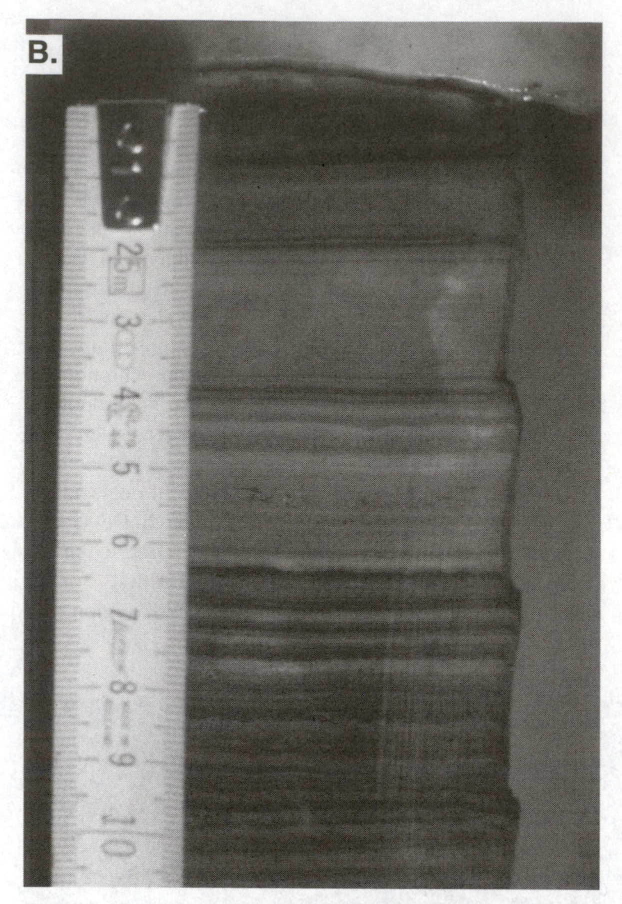

Figure 5.18 B. Varves from Lake Heinälampi, Finland, consisting of thin 'summer' layers rich in planktonic diatoms (light-coloured layers) and 'autumn–winter–spring' diatom-poor layers (photo: H. Simola). Several seasonal diatom 'blooms' are discernible in the 'summer' layers, which are also rich in chrysophyte 'cysts'. The three uppermost, very thick varves contain several intra-annual layers of silt brought into the lake by spring snowmelt floods. The sudden change in varve type and thickness indicates an abrupt change in sediment flux within the lake catchment.

By contrast with De Geer's pioneering work, most contemporary varve counting is carried out in the laboratory, either on core samples or on monoliths taken from open sections. Measurement and counting of varved sediments can sometimes be performed directly using conventional light microscopy, but other techniques involving X-radiography, scanning electron microscopy, photography and thin-section analysis, typically on frozen sections of core or monolith material, are now widely employed (O'Sullivan, 1983; Saarnisto, 1986).

(b) Other varved sediments Studies of recent sedimentation in lakes and in the sea have shown that annual rhythmites are common in many temperate areas far removed from nearby glacier activity (O'Sullivan, 1983). As a general rule, both sedimentation and biomass production are affected by seasonal variations in lakes, and in certain temperate lakes chemical precipitation can also vary seasonally. In many lakes, reworking of sediments by currents or by bottom-dwelling organisms prevents the formation of laminations (**holomictic lakes**). In deep lakes with oxygen-deficient basal water, however, the numbers of bottom-dwelling fauna are restricted, vertical water circulation does not extend to the bottom of the water column (**meromictic lakes**), and fine laminations may therefore be preserved. Where seasonal variations in the accumulation of organic detritus occur, couplets formed as a result are termed **organic varves** (Figure 5.18B). In some lakes, seasonal variation in precipitation of $CaCO_3$ leads to the development of light summer layers which alternate with dark winter layers rich in organic humus (Peglar *et al.*, 1984; Lotter, 1989). Diatom blooms during spring and early summer can also produce annual laminations in lake sediments, and these have been observed in both interglacial (Turner, 1975a) and more recent (Simola *et al.*, 1981) sequences. Seasonal variation in iron oxide precipitation may also lead to the formation of laminations (Renberg, 1981a). In some lakes, variations in the accumulation of organic matter or of chemical precipitates are superimposed on particle-size variations, thereby emphasising contrasts between winter and summer layers (Saarnisto, 1986).

5.4.2.2 Sources of error in varve counting

Problems in the development of varve chronologies are similar in many respects to those encountered in dendrochronology. Local site factors can lead to incorrect estimates of age, and also cause difficulties in the correlation (crossdating) between individual varve sequences. For example, adverse weather conditions in particular years can lead to a reduced input of sediment into a proglacial lake, or to reduced biomass or chemical precipitation in summer, so that an individual varve fails to develop or is too thin to be recognised. An annual increment of sediment may also be absent because of the intermittent erosion of bottom layers at some sites. Alternatively, more than one set of laminae (**sub-laminations**) can develop within an annual increment of sediment reflecting, for example, episodic sedimentation from intermittent local wind-driven currents (Catto, 1987). Normally, sub-laminations are found in the coarse member reflecting, for example, diurnal variations, with coarser materials being deposited by day and finer sediments deposited at night (Ringberg, 1984). However, sand layers can also form in the winter silt and clay layers of glaciolacustrine varves through the operation of turbidity currents (Shaw & Archer, 1978). Interpretation of the varve record in lacustrine environments may be further complicated by the products of intermittent slumping or large flood events (Leonard, 1986). Finally, the strict seasonality of couplet deposition has been questioned by studies from Switzerland (Lambert & Hsu, 1979) and Arctic Canada (Gilbert & Church, 1983), both of which provide evidence of the essentially non-annual nature of certain lacustrine sedimentary sequences.

As in the development of a tree-ring chronology, crossdating of varve sequences involves the analysis of a number of records from each locality, in order to exclude or at least minimise the problems arising from missing or false varves. Matching of relative varve thicknesses is only possible over limited geographical areas as climatic variations may produce quite different varve sequences even in adjacent regions. Because sections are not always conveniently located, long-distance correlations (**teleconnections**) have been attempted, and these often led to errors in the varve chronologies. Subsequently, varves have been obtained from terrestrial, lake and shallow-water marine localities using special corers which enable sediments to be frozen *in situ*, thereby ensuring recovery of undisturbed, finely laminated sediments (Renberg, 1981b). Coring of unexposed varve sequences has introduced a much greater flexibility into sampling designs for work on varve chronologies. In addition, correlation between varve sequences is increasingly being based on other parameters, particularly the palaeomagnetic properties of the sediments (Björck *et al.*, 1987; Sandgren *et al.*, 1988: see below).

5.4.2.3 Applications of varve chronologies

(a) Patterns of regional deglaciation The development of a regional varve chronology spanning a long period of deglaciation is so far unique to Scandinavia. In the British Isles and throughout most of western Europe, relatively few varved sequences have been recorded, and consequently there has been little interest in the use of varves as a basis for dating ice recession. In North America, by contrast, the complex system of proglacial lakes that developed around the southern margins of the Laurentide ice sheet left extensive suites of lacustrine sediment stretching intermittently from Ontario to the Canadian Prairies (Teller, 1995). Yet, despite the pioneering work of Antevs (1931, 1953) who developed tentative varve chronologies for Lake Agassiz and for the Great Lakes region, no regionally applicable chronology has been established. This reflects the general abandonment of varve counting as a dating method in North America following the advent of radiocarbon dating, the lack of good exposures and the discontinuity of the sedimentary record in a number of key areas (Klassen, 1984).

In Scandinavia, however, following on from the early work of De Geer (1912) and Sauramo (1918, 1923), glaciolacustrine varves have continued to attract interest as a means of dating the wastage of the last ice sheet. The starting point for De Geer's timescale was a key reference varve in a section in the Indalsälven Valley of north–central Sweden, which he considered to reflect the sudden input of large quantities of meltwater following the bipartition of the remaining Scandinavian ice sheet (Fromm, 1985). By reference to this 'zero' varve year, which was taken to mark the change from glacial to postglacial conditions, older varves were given negative numbers and younger varves positive numbers. Subsequently, De Geer's master chronology has been repeatedly revised as more data have become available (e.g. Tauber, 1970; Lundqvist, 1980). The recent connection of the upper part of the varve sequence to the present (1978), however, which has enabled a calendrical age of 9238 BP to be assigned to the 'zero year' (Cato, 1985) has placed the varve chronology on a much more secure foundation (Strömberg, 1985a). In some areas, the master chronology now extends back to *c.* 11.5 ka BP (Strömberg, 1985b). Use of the varve sequence in association with other chronological techniques, such as ^{14}C dating, palaeomagnetism and pollen stratigraphy (e.g. Björck *et al.*, 1987; Wohlfarth *et al.*, 1993), has enabled the deglacial history in parts of southern Sweden to be reconstructed in considerable detail (Figure 5.19). Moreover the temporal framework for ice recession throughout the Baltic region as a whole (Figure 5.20) is still, in large measure, provided by the clay-varve chronology.

(b) Duration of Quaternary time periods Varve chronology provides a means of estimating the duration of Quaternary time intervals. For example, analysis of organic varves in the Hoxnian Interglacial deposits at Marks Tey in southeast England enabled an estimate of 20–25 ka to be obtained for the duration of the interglacial episode (Turner, 1975a). Counting of annually laminated Lateglacial and early Holocene sediments in Swiss lakes suggests varve ages of 700–800, *c.* 400 and *c.* 900 calendar years respectively for the duration of the Bølling, Allerød and Younger Dryas chronozones (Lotter, 1991; Lotter *et al.*, 1992). In Sweden, the Younger Dryas is represented by 900–1000 varves (Björck *et al.*, 1987), while in a varved lake sequence from Poland, the duration of the Younger Dryas has been estimated at 1140 ± 20 calendar years (Goslar *et al.*, 1993). The analysis of laminated sediments also enables the duration of particular Quaternary events to be established. For example, at Diss Mere in southeast England, the *Ulmus* decline, a feature common in all northwest European pollen records and ^{14}C dated to *c.* 5 ka BP, has been traced through

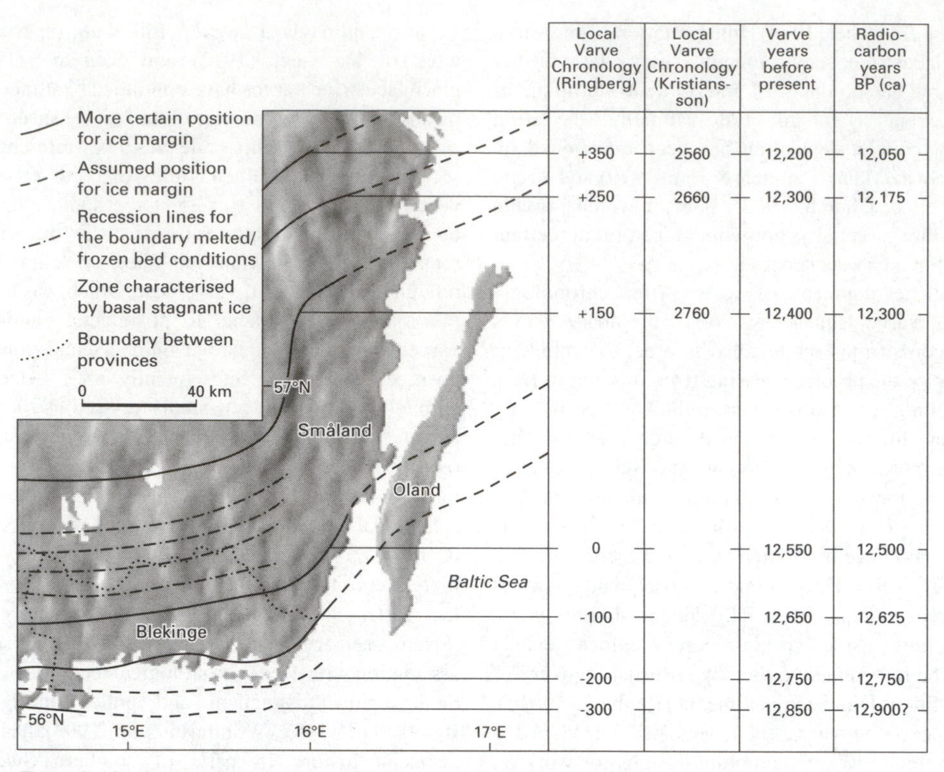

Figure 5.19. Deglacial chronology of Blekinge, south Småland, Sweden, showing the relationship between varve and ^{14}C timescales (after Björck *et al.*, 1988).

a series of annually laminated sediments. *Ulmus* pollen values fall by 73 per cent in six years, such a rapid rate being consistent with the hypothesis that the elm decline resulted from pathogenic attack (Peglar, 1993a; Peglar & Birks, 1993).

(c) Calibration of the ^{14}C timescale Where sufficient organic material has accumulated in varved sediments to allow radiocarbon measurement, sequences of varves can be used as an independent means of calibrating the radiocarbon timescale. In the organic varve sequence from Lake of the Clouds, Minnesota, for example, ^{14}C and varve ages display a broad measure of comparability over the course of the past 2.5 ka (Figure 5.21), but thereafter the two timescales diverge so that at around 5.5 ka BP, one ^{14}C year is the equivalent of *c.* 1.26 calendar years (Stuiver, 1971). This trend closely parallels that found in the dendrochronological records (see above) and confirms the hypothesis of increased atmospheric ^{14}C activity during the mid-Holocene (Figure 5.16).

For the early Holocene and Lateglacial periods, Stuiver *et al.* (1991) have shown how the Lake of the Clouds varve records can be combined with ^{14}C-dated pollen-stratigraphic and plant macrofossil evidence from Switzerland (Zbinden *et al.*, 1989), dendrochronological data from Germany (Becker & Kromer 1986), and Greenland ice-core years (Hammer *et al.*, 1986), to reconstruct the trend of atmospheric ^{14}C variation back to *c.* 14 k calibrated yr BP (Figure 5.22). This indicates significantly higher levels of ^{14}C activity throughout the Lateglacial and early Holocene, with a maximum during the Younger Dryas–Holocene transition (A), and a minimum during the Lateglacial Interstadial (B). In southern Sweden, where AMS ^{14}C dating has been carried out on plant macrofossils from varved sediments, thereby allowing a direct comparison between varve (calendar) and ^{14}C years, the evidence suggests that 'clay varve years' exceed terrestrial ^{14}C years by *c.* 900 years at the end of, and by 1100–1200 years at the beginning of, the Younger Dryas chronozone (Wohlfarth *et al.*, 1993). Prior to that, the differences are less pronounced, and the two timescales may converge around 12.7–12.8 ka BP (Björck *et al.*, 1987; Wohlfarth *et al.*, 1993), reflecting the abrupt decline in atmospheric ^{14}C activity between 12.2 and 12.8 ka BP shown in Figure 5.22. On the Swedish varve chronology (Strömberg, 1985a), the Lateglacial–Holocene boundary is placed at *c.* 10.7 ka BP (as opposed to *c.*

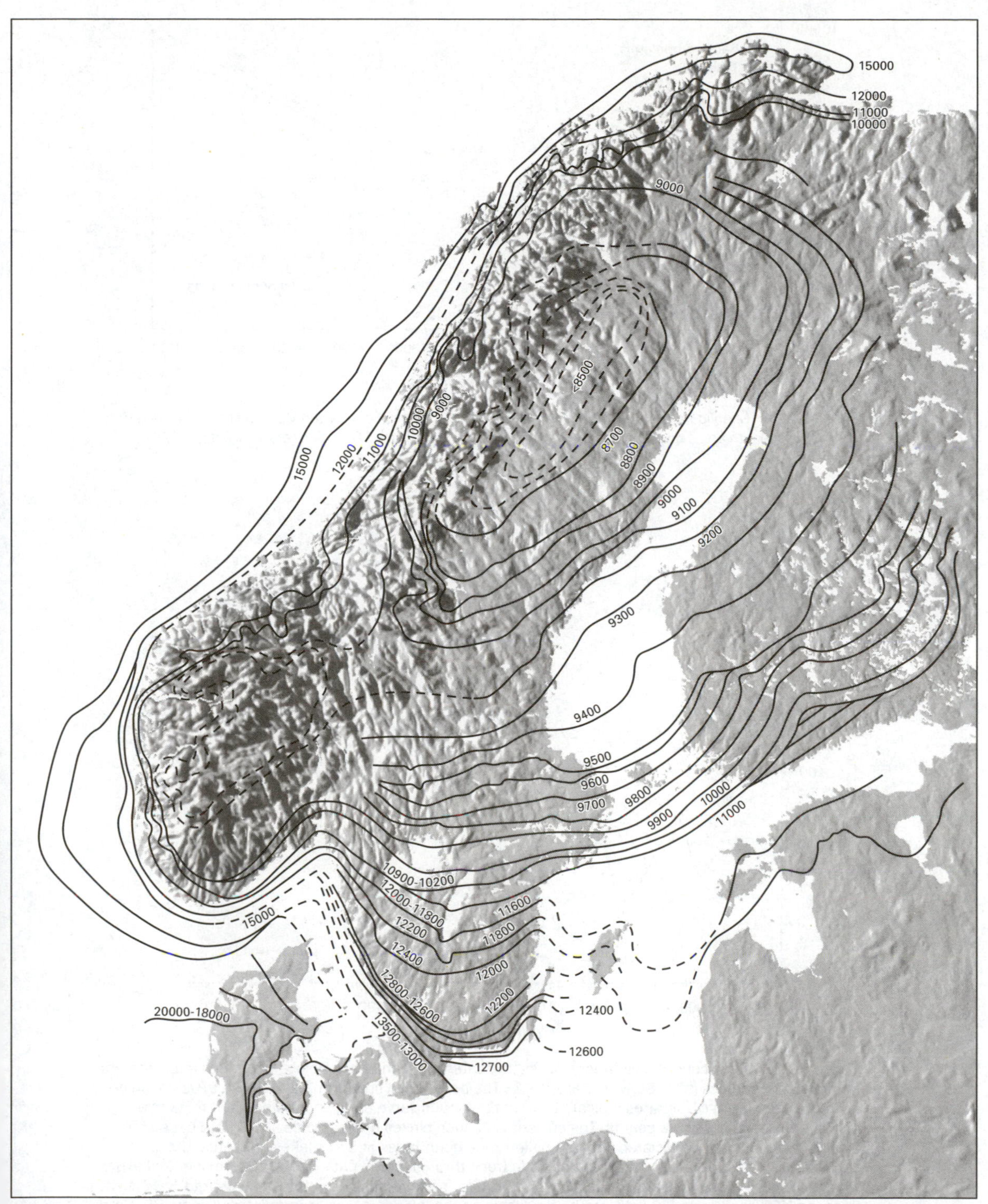

Figure 5.20 The pattern of ice retreat across Scandinavia. The dates from northern Sweden and southwards along the Baltic coast are based on clay-varve chronology and should, following Cato (1985), be corrected by +365 years. Other dates are based on ^{14}C determinations (after Lundqvist, 1986a).

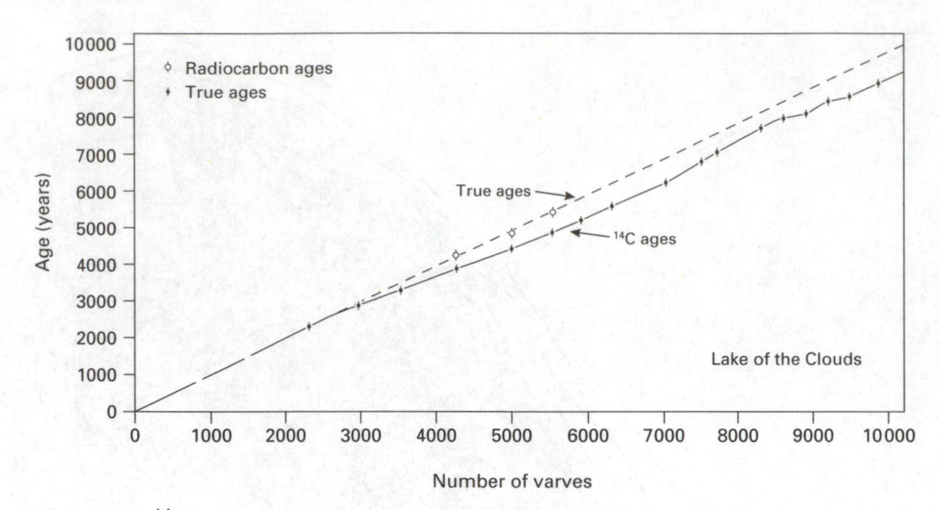

Figure 5.21 ^{14}C (solid points) and tree-ring ages (open circles) of Lake of the Clouds sediment as a function of the number of varves counted. Each ^{14}C sample contains about 50 varves (after Stuiver, 1970).

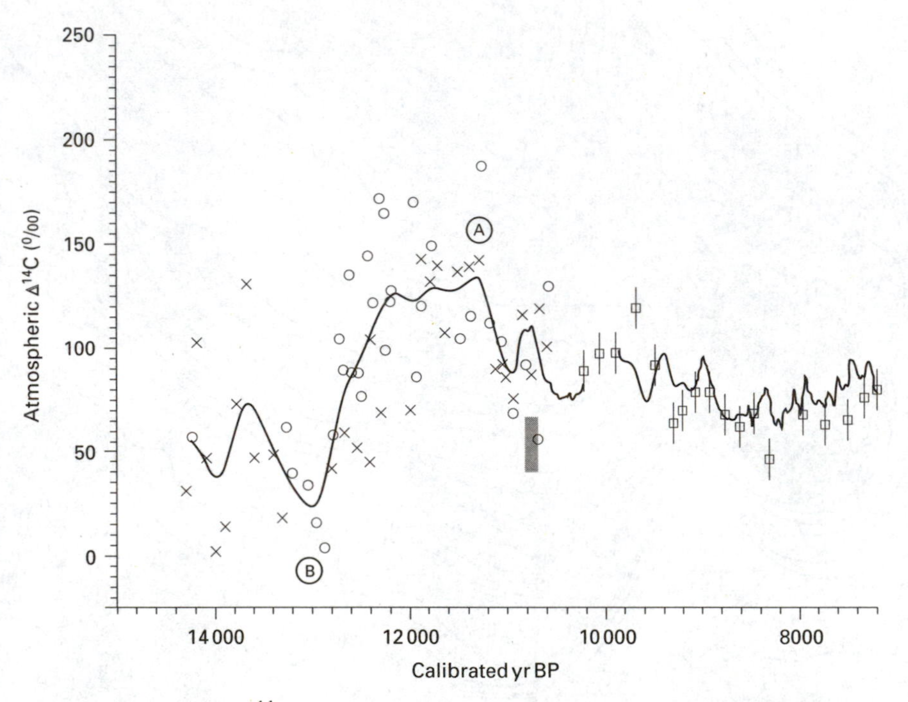

Figure 5.22 Atmospheric ^{14}C variations during the Lateglacial and early Holocene periods inferred from four different sources (after Stuiver *et al.*, 1991). The heavy solid line from 7.2–10.0 ka BP is based on absolute tree-ring ages: squares indicate Lake of the Clouds varve data; and the hatched rectangle denotes Greenland ice-core counts. The crosses and open circles are macrofossil ^{14}C data based on Swedish varve counts associated with the pollen zone boundaries in Swiss lake sediments; the continuous line from 10.2 to 14.2 ka shows the trend through these data. The $\Delta^{14}C$ term (vertical axis), which indirectly represents atmospheric $^{14}C/^{12}C$ ratios, is the relative deviation of measured ^{14}C activity from the NBS oxalic acid ^{14}C activity. The timescale on the horizontal axis is in calibrated years BP. Note how the largest deviation from the standard (*c.* 125–135‰) occurs at the Younger Dryas–Holocene transition (A), while the minimum deviation (*c.* 25‰) is recorded during the Lateglacial Interstadial (B).

10 ka BP on the conventional ^{14}C timescale). Other independent age determinations, however, place the boundary at an earlier calendrical date. For example, the Younger Dryas boundary has been placed at 11.0–11.1 ka BP in German dendrochronological records (Becker & Kromer, 1993) and at *c.* 11.2 varve years in a Polish lake sequence (Rozanski et al., 1992), while in the Greenland ice cores the boundary is dated at *c.* 11.5 k ice-accumulation yrs BP (Alley *et al.*, 1993; Dansgaard *et al.*, 1993).

5.4.3 Lichenometry

Lichens are complex organisms consisting of algae and fungi living together symbiotically. The algae provide carbohydrates via photosynthesis, while the fungi provide the protective environment in which the algal cells can function (Bradley, 1985). The use of lichens in dating was pioneered by Beschel (1973) and rests on the principle that there is a direct relationship between lichen size and age. Where a surface has been recently exposed to lichen colonisation, providing (a) that the growth patterns of the lichens are known and (b) that no major time lapse has occurred between surface exposure and lichen colonisation, an estimate of the age of the substrate can be made. Some lichen species (e.g. *Rhizocarpon geographicum*) will continue to grow for several thousand years and therefore, in theory, lichenometry is a technique that may be applicable to most of the Holocene. In practice, however, the dating limit is around 4500 years in extremely cold and dry continental regions, such as west Greenland, whereas in the majority of cases the age range for lichenometry as a dating technique is 500 years or less (Innes, 1985; Matthews, 1992).

Lichenometry has been most widely employed in the dating of glacier recession. Lichen size (usually maximum diameter of the largest lichen) is first established for morainic surfaces of known age (dated, for example, by radiometric methods or by historical evidence such as old photographs) and a lichen growth curve can then be constructed based on these '**fixed points**' (Figure 5.23). Surfaces of unknown age can then be dated by relating lichen diameters on those surfaces to the growth-rate curve and deriving a calendar age. In this way, a detailed deglacial chronology can be established for an area. The method has been used to establish glacial chronologies, particularly for the Little Ice Age period, in many parts of the world including Scandinavia (Erikstad & Sollid, 1986; Ballantyne, 1990), the North American Cordillera (Osborn & Luckman, 1988), Alaska (Calkin, 1988), South America (Rodbell, 1992b) and New Zealand (Gellatly, 1982).

In addition to the establishment of deglacial chronologies, lichenometry has a range of other applications. These include the provision of a timescale for plant colonisation of newly exposed substrates in proglacial areas (Matthews, 1992); the development of a chronology for raised beach deposits (Birkenmajer, 1981) and associated archaeological features (Broadbent & Bergqvist, 1986); the relative dating of proglacial river terraces (Thompson & Jones, 1986), slope

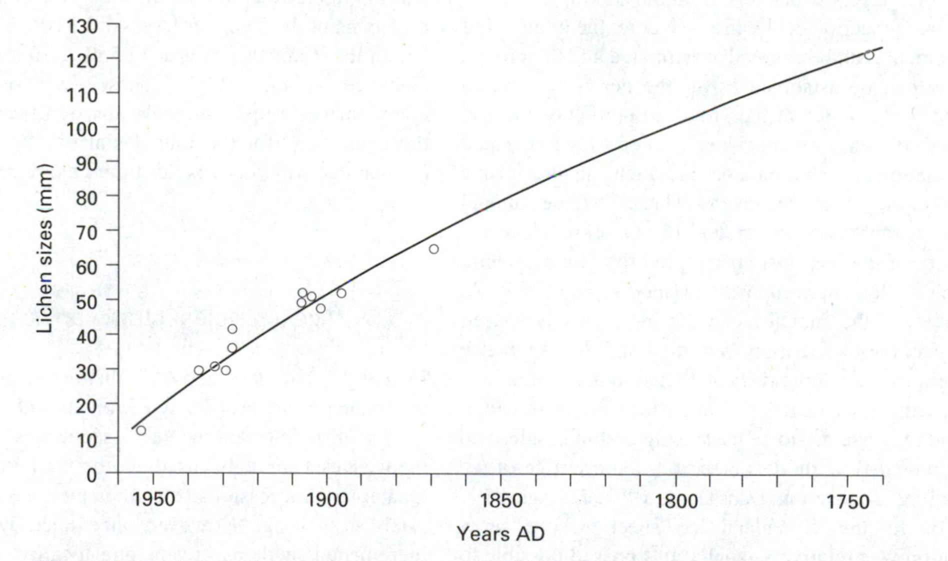

Figure 5.23 Lichenometric growth curve for western Norway, based on lichen-size measurements on newly deglaciated surfaces of known age (fixed points). The age of unknown surfaces (moraines, etc.) can then be obtained by measuring average lichen size on those surfaces (*y* axis) and reading off the appropriate age on the *x* axis (after Erikstad & Sollid, 1986).

deposits (André, 1986) and debris-flow activity (Innes, 1983); and the dating of earthquake-induced surface features in areas of recent seismic activity (Smirnova & Nikonov, 1990).

Not all lichens are suitable for lichenometrical purposes, however; only those that show a gradual and progressive rate of growth can be employed. Moreover, lichen growth is affected by local environmental conditions that vary with both latitude and altitude, for example air temperature, day length and snow cover (Benedict, 1990). A lichen growth curve must, therefore, be constructed for specific lichens and will only be applicable to a limited geographical area (Innes, 1985). Further, it is assumed that in the study of glacier recession, for example, there is no significant delay in lichen colonisation of exposed surfaces following ice retreat. This, however, can never be proved and must always remain a source of uncertainty. Conversely, lichens have been reported growing on actively forming medial moraines (Matthews, 1973), and therefore in some areas lichen growth must have *preceded* ice wastage. Finally, problems have been encountered in sampling and measurement of lichens in the field and very thorough preliminary investigations are required to establish reproducibility of results within any one region (Innes, 1985).

5.4.4 Annual layers in glacier ice

Analysis of the upper levels of polar ice sheets often reveals clearly defined layers which reflect annual additions of snow to the ice mass (section 3.11). In an ice core, the numbers of these increments can be counted or estimated and therefore a chronology can be established over the depth of core. In addition to the changes in the visual properties of the ice, closely spaced measurements along ice cores have revealed regular variations in other parameters which can also form a basis for dating the ice layers. These include annual variations in stable isotope ratios ($\delta^{18}O$, δD), electrical conductivity of the ice, dust content, microparticle content, and chemical element composition (Budd *et al.*, 1989). At depth, however, the annual layers are more closely spaced and they become increasingly deformed and diffuse, hence annual variations are more difficult to distinguish. Older ice, therefore, cannot be dated by straightforward incremental means, and recourse has to be made to age estimates derived from theoretical flow models based on a knowledge of ice dynamics (e.g. Johnsen & Dansgaard, 1992; Dansgaard *et al.*, 1993). In the Greenland ice sheet, where snow accumulation was relatively rapid, it has proved possible to develop a timescale in the GRIP (Greenland Ice Core Project) core back to *c.* 14.5 ka BP by counting annual layers down from the surface (Johnsen *et al.*, 1992). In the GISP2 (Greenland Ice-Sheet Project 2) core, every annual layer has been counted for the past 17.4 ka, and a chronology back to 40.5 ka BP established partly by counting recognisable ice layers and partly by interpolation between these horizons (Taylor *et al.*, 1993a). By contrast, in some areas of the Antarctic ice sheet where annual accumulation rates are low, removal processes at the surface and diffusion of isotopic components in firn tend to smooth out the seasonal variations of the isotopic signal (Whillans & Grootes, 1985). In these circumstances, ice flow models may have to be employed for estimating the ages of both Holocene and pre-Holocene sections of ice cores (Jouzel *et al.*, 1992).

The most significant potential contribution to Quaternary chronology offered by incremental dating of ice cores is in the establishment of a timescale for climatic changes during the Lateglacial period (14–9 k ^{14}C years BP). In the GISP2 core, for example, proxy records of climatic change are provided by variations in dust concentration (indicating increased aeolian activity during colder episodes) and changes in thickness of annual ice layers, reflecting variations in accumulation rate (i.e. snowfall). Within these data (Figure 7.21) the Lateglacial climatic episodes are clearly recognisable and can be dated in ice-accumulation years: the transition from the Oldest Dryas to the Bølling Interstadial occurs at 14 680 ± 400 ice-core years BP; the end of the Younger Dryas is dated at 11 640 ± 250 BP; and the duration of the Younger Dryas Stadial is estimated at 1300 ± 70 BP (Alley *et al.*, 1993; Taylor *et al.*, 1993a). In the GRIP core, annual layer counting dates the onset of the Bølling Interstadial at 14 450 ± 200 ice-core years BP, the beginning of the Younger Dryas at 12 700 ± 100 BP and the end of the Younger Dryas at 11 550 ± 70 ice-core years BP (Johnsen *et al.*, 1992). These two sets of broadly complementary age estimates are significantly older than those provided for the Lateglacial by ^{14}C dating, but are comparable with the U-series timescale referred to above.

5.5 Age-equivalent stratigraphic markers

In many Quaternary deposits, distinctive marker horizon: are found that are broadly synchronous and form time plane across different sedimentary sequences. The horizon themselves cannot be used in the first instance, to dat Quaternary successions, for other methods are required t establish their age. However, once dated by radiometric c incremental methods at any one locality, they allow ag estimates to be extended to other sequences where th marker horizon is present. As such, they form an indire means of dating. Moreover, in view of their ofte widespread distribution, they also form a basis for stratigraphic subdivision and time–stratigraphic correlation (see Chapter 6).

Three methods of dating using age-equivalent stratigraphic markers are considered here: **palaeomagnetism**, which is based on the changes in the earth's magnetic field preserved in rocks and sediments; **tephrochronology**, the use of volcanic ash layers as a means of dating; and **oxygen isotope chronostratigraphy** which employs globally synchronous changes in the oxygen isotope signal in deep-ocean sediments. Other marker horizons which are more widely employed in stratigraphic subdivision and correlation than in dating, such as palaeosols and shorelines, are discussed in section 6.3.2.

Table 5.6 Time constants of the geomagnetic field (after Thompson, 1978).

Geomagnetic changes	Duration (years)
Change in average frequency of polarity inversions	5×10^7
Time between successive polarity inversions	10^7, 10^6, 10^5
Intensity and direction fluctuations of dipole and non-dipole fields (secular variation)	10^4, 10^3, 10^2
11-year sunspot cycle	10^1
Annual variation	10^0, 10^{-1}
Diurnal variation	10^{-2}
Magnetic storms	10^{-3}
Micropulsations	10^{-4}

5.5.1 Palaeomagnetism

5.5.1.1 Geomagnetic field and remanent magnetism

The earth's magnetic field varies constantly both in field strength and in polarity direction. Variations range in periodicity from milliseconds to tens of millions of years (Table 5.6), the shorter-lived phenomena resulting from external influences such as variations in solar radiation, the influence of magnetic storms, etc., and the longer-lived changes from internal geophysical factors. Rocks and unconsolidated sediments containing magnetic minerals are magnetised during formation, and individual crystals or particles can often reveal this **natural remanent magnetism (NRM)** which is a reflection of the geomagnetic field at the time of rock or sediment formation (Thompson, 1991).

Volcanic rocks contain minerals with various ferromagnetic properties. Before cooling, high temperatures lead to thermal fluctuations of ions, and in the absence of an external magnetic field a random orientation of ions would result. The temperature below which the ambient geomagnetic field would be retained is known as the **Curie Point**. Below this thermal agitation is insufficient to destroy the ferromagnetic alignment induced by the geomagnetic field. As long as the crystals are not reheated above the Curie Point, then this **thermoremanent magnetisation (TRM)** will be retained by the crystal lattice.

Sedimentary rocks and unconsolidated sediments accumulating on the sea floor or in lakes also contain evidence of former geomagnetic fields, for a record is preserved in the alignment of ferromagnetic sedimentary particles as they settle in water or in water-saturated sediments. The resulting **detrital remanent magnetism (DRM)**, though weak, can easily be measured. In sediments, NRM can also be acquired by chemical action, where the crystallisation of ferromagnetic oxides results in a **chemical remanent magnetism (CRM)**. This process may occur later than, and under a different magnetic field from, that of DRM in the same sediment unit. In this way, a **secondary magnetisation** is introduced into both volcanic and sedimentary rocks which serves to complicate the study of palaeomagnetic variations (Tarling, 1983; Thompson & Oldfield, 1986).

At any one point and at any one moment in time the earth's magnetic field can be resolved into three components:

1. **Declination** This is the angle between magnetic north and geographic (true) north.
2. **Inclination** A freely suspended needle at the surface of the earth will align with the prevailing magnetic field (declination). The inclination of the needle is the amount of dip exhibited by the needle relative to the horizontal. The inclination value varies from 0° at the magnetic equator to 90° at the magnetic poles.
3. **Intensity** This refers to the strength of the geomagnetic field. At the present day the field strength at the geomagnetic poles is twice that at the geomagnetic equator. The strength of the field can be estimated in the following way. Suppose a magnetic needle is fixed to a horizontal axle, so that the axle passes through the centre of gravity of the needle and is orientated along magnetic east–west. If the needle is allowed to swing freely it would eventually stabilise at the angle of magnetic dip. Magnetic intensity can be measured by the amount of torque required to prevent the needle returning to the angle of magnetic dip after it has been rotated through 90°.

5.5.1.2 Magnetostratigraphy

The study of variations in magnetic properties through a sequence of rocks or sediments is termed **magnetostratigraphy**. Geomagnetic field variation can be detected in rocks or sediments that contain even small amounts of magnetic minerals, and the identification of synchronous changes in, for example, declination or inclination in different sequences provides a means of relative dating and correlation. Where sediments have accumulated rapidly, short-term **secular variations** in the earth's field can be employed, but for sediments that have accumulated more slowly (e.g. deep-sea sediments) or for sequences of volcanic rocks, longer-term **field reversals** are used. In addition, however, **mineral magnetic 'potential'**, that is the concentration and magnetic susceptibility of magnetic minerals, also varies within sediments, and under certain circumstances this too can provide a basis for correlation.

(a) Secular variations Direct measurements of values of field declination, inclination and intensity have been collected over the past 400 years for London (Malin & Bullard, 1981) and for other major centres including Rome and Boston (Figure 5.24). In London, for example, declination has varied from 11°E in 1570 AD to 24°W in 1820 AD. Since then declination has decreased to the present value of 5°W, and it continues to decrease at a rate of 9 minutes each year (Thompson & Oldfield, 1986). Magnetic inclination has varied from a maximum of over 74° in 1700 AD and is now near to 66°. Figure 5.24 shows that secular magnetic variations in London and Rome have been relatively similar whereas at Boston, some 5000 km from Europe, the pattern has been very different. Beyond the historical period, proxy records of geomagnetic variations can be obtained from an analysis of archaeological materials such as bricks, tiles and pottery (**archaeomagnetic measurements**). Geomagnetic fluctuations appear to have continued throughout earth history, and records for a number of locations allow the plotting of the position of the geomagnetic poles for different times in the past (Tarling, 1983). Variations have also been detected in field intensity. Over the past 11 ka, for

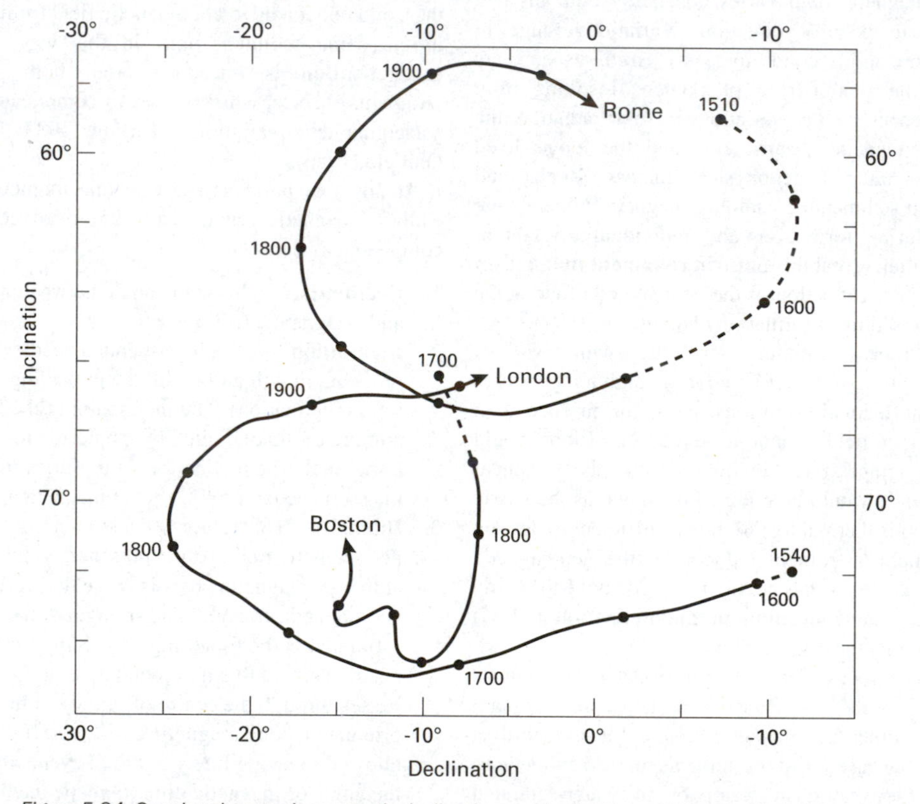

Figure 5.24 Secular changes in magnetic declination and inclination as observed in London, Rome and Boston. The solid curves begin at the time of the first declination measurements at each locality. The earlier inclination changes (dashed lines) are based on archaeomagnetic data (after Thompson & Oldfield, 1986).

example, pronounced changes in field intensity have occurred with durations of the order of 10^4 years. The dipole field intensity has varied by a factor of two between a minimum at 6.5 ka BP and a maximum at 2.8 ka BP (Thompson & Oldfield, 1986).

Lake sediments that have accumulated during the last 10 ka or so often contain a very detailed record of secular geomagnetic variations, and these can be used as an indirect means of dating. High-resolution magnetostratigraphic sequences, which can be dated by independent methods such as ^{14}C and pollen stratigraphy (Thompson & Edwards, 1982; Snowball & Thompson, 1990), constitute type profiles against which other secular magnetic records can be matched. Detailed analysis of lake sediments from many parts of the world has enabled master geomagnetic curves to be established (Figure 5.25), which are likely to be applicable to sediments found up to 1000–2000 km from the type sites (Thompson, 1986). Core matching can be achieved using the

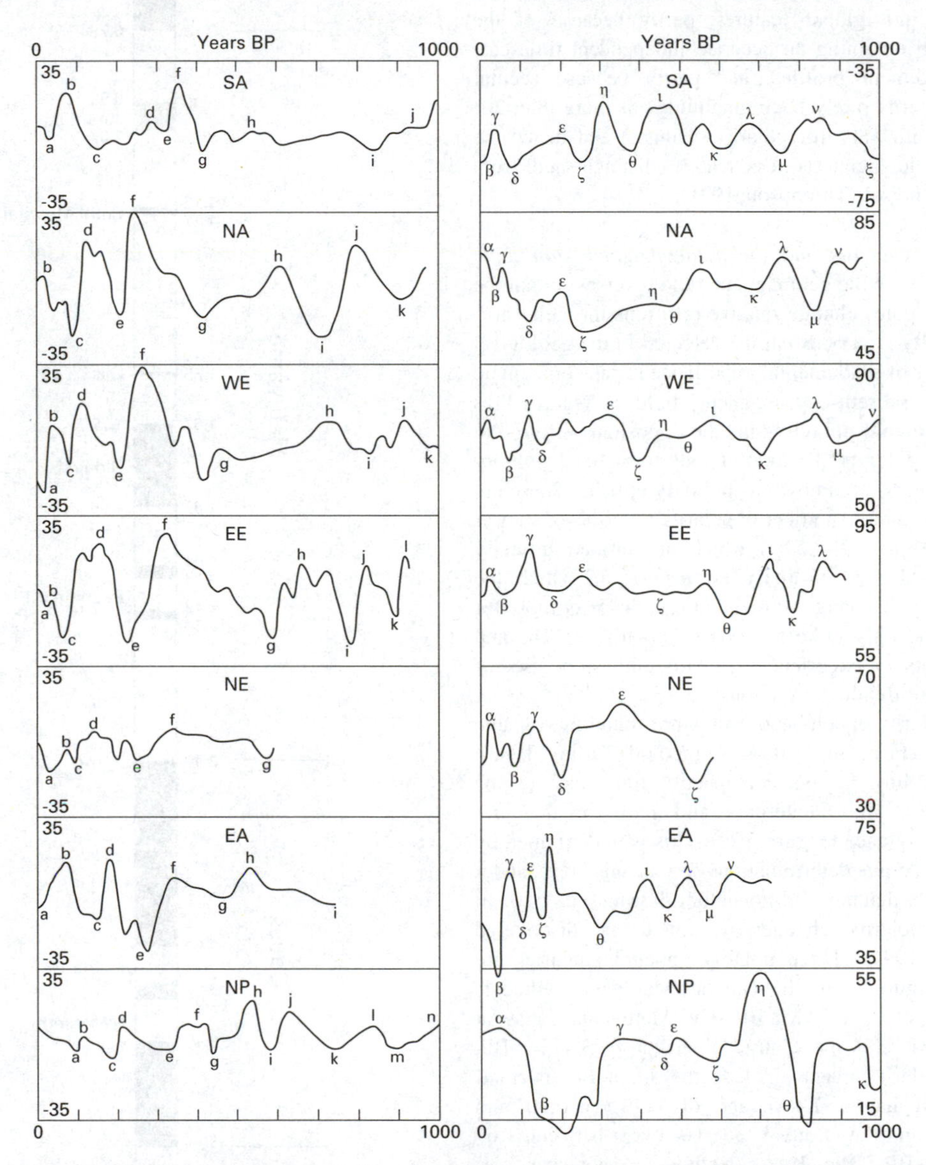

Figure 5.25 Regional declination (left) and inclination (right) master curves. Tree-ring calibrated timescale in years BP. SA: South Australia; NA: North America; WE: Western Europe; EE: Eastern Europe; NE: Near East; EA: East Asia; NP: North Pacific (after Thompson, 1986).

distinctive **turning points** in the magnetic profiles. Secular magnetic variations have also been employed in the matching of Lateglacial clay-varve sequences in southern Sweden (Björck *et al.*, 1987; Sandgren *et al.*, 1988), while on a much longer timescale a palaeomagnetic record of secular variation spanning some 500 ka has been obtained from a loess sequence in North Island, New Zealand (Pillans & Wright, 1990). However, these secular variation approaches to magnetostratigraphy are more difficult to apply than polarity reversals (see below), partly because secular geomagnetic changes are not global features, partly because of the difficulties in obtaining an accurate independent timescale for lake sediment profiles, and partly because secular variation patterns rarely have amplitudes of more than 20° (compared with 180° for polarity changes) and hence the palaeomagnetic signal is less readily distinguished from background 'noise' (Thompson, 1991).

(b) Field reversals and the palaeomagnetic timescale From time to time the geomagnetic field reverses so that the geomagnetic poles change relative positions through 180°. These **polarity reversals** can be detected in the geological record and are of fundamental importance in palaeomagnetic studies. The present-day magnetic field is regarded as possessing **normal polarity**, and the opposite is referred to as **reversed polarity**. Periods of long-term fixed polarity (10^5 to 10^7 years) are known as **polarity epochs**. These are interrupted by a large number of polarity reversals of shorter duration (10^4 to 10^5 years) which are termed **polarity events**, and also by **polarity excursions**, in which the geomagnetic pole changes direction through 45° or more for a short period only (1 ka to 100 ka). Polarity epochs and polarity events are experienced globally and can be used as a basis for worldwide correlations.

Where polarity epochs and events are found in volcanic rocks they can be dated by the K–Ar method (Tarling, 1983), thereby enabling a palaeomagnetic timescale to be established for the Quaternary and parts of the pre-Quaternary sequence (Figure 5.26). This is underpinned by some 170 K–Ar age determinations on volcanic lavas, and it seems unlikely that any additional long-lasting (i.e. >20 ka) Pleistocene polarity changes remain to be discovered (Thompson, 1991). Three polarity epoch boundaries are shown on Figure 5.26: the Brunhes/Matuyama which is K–Ar dated at *c.* 0.73 Ma BP, the Matuyama/Gauss at *c.* 2.47 Ma BP and the Gauss/Gilbert at *c.* 3.41 Ma BP. Important polarity events include the Jaramillo 'normal' event, which has a K–Ar age of between 0.90 and 0.97 Ma BP, and the Olduvai 'normal' event between 1.67 and 1.87 Ma BP. The K–Ar method is, however, not

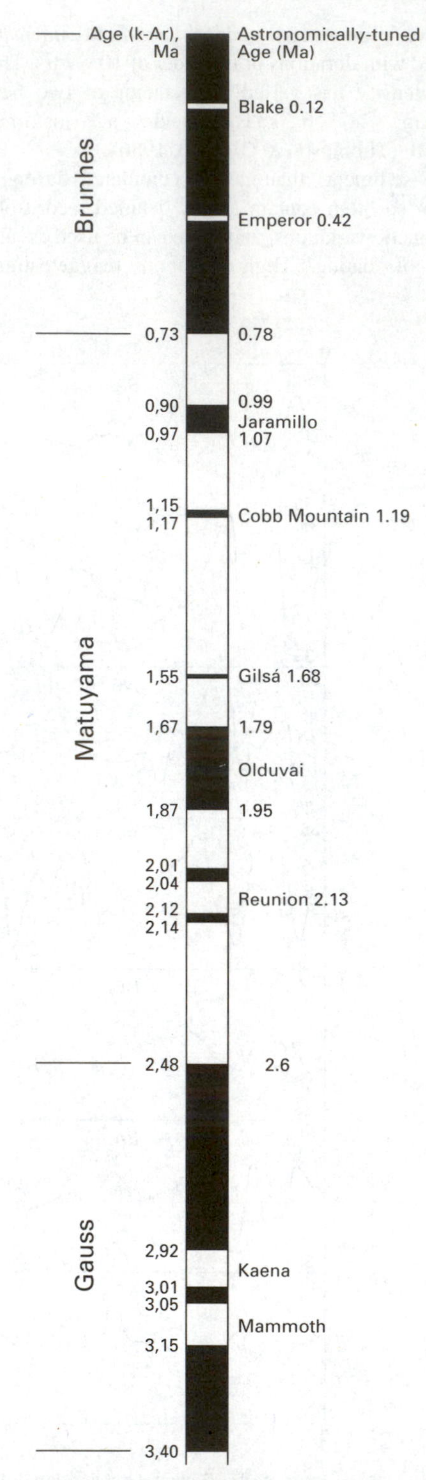

Figure 5.26 The palaeomagnetic timescale of the last 3.5 Ma. Shaded areas indicate periods of normal polarity; unshaded areas show episodes of reversed polarity. K–Ar ages are shown on the left; astronomically-tuned absolute ages are on the right (after Mankinen & Dalrymple, 1979; Funnell, 1995). For explanation, see text.

sufficiently precise to date some of the relatively short-lived polarity events, and their positions on the palaeomagnetic timescale have therefore been established by extrapolation based on the ages of epoch boundaries. As a consequence, the dating of polarity events tends to be less secure (Berggren *et al.*, 1980).

More recently, however, a different approach has been adopted to dating the palaeomagnetic timescale, involving the ocean sediment record. As sediments accumulate on the deep-ocean floors, individual particles adopt the direction of the earth's magnetic field, and hence a continuous record of geomagnetic changes is preserved within the sediment sequence. Astronomical tuning of the oxygen isotope signal obtained from the microfossil record within these sediments (sections 5.5.3 and 6.3.2.7) provides the basis for a timescale for the geomagnetic changes that is, of course, independent of that based on K–Ar dating of volcanic rocks (e.g. Valet & Meynadier, 1993; Funnell, 1995). Ages of the polarity epochs and principal polarity events that have been obtained using this method are shown on the right of Figure 5.26. Overall, the dates tend to be older than those based on K–Ar. Hence, the Brunhes–Matuyama boundary is dated at *c.* 0.78 Ma BP and the Matuyama–Gauss boundary at *c.* 2.6 ka BP, while the ages of the Jaramillo and Olduvai events are 0.99 to 1.07 Ma and 1.79 to 1.95 Ma BP respectively. The resolution and length of the polarity timescale in the ocean sediment record is, however, dependent on rates of sedimentation. For example, in parts of the oceans experiencing comparatively slow sedimentation rates (e.g. 0.1 to 1.2 m per 1000 years), a core may contain a complete record of Pleistocene polarity changes. Often, however, the timescale will be compressed, and it may prove difficult to identify the shorter-lived polarity events. Conversely, in some areas of the oceans, such as the North Atlantic, where sedimentation rates are rapid (e.g. 0.25 to 0.5 m per 1000 years), coring may fail to reach the first major geomagnetic boundary, the Brunhes/Matuyama transition.

The use of palaeomagnetic stratigraphy as a means of correlating between individual deep-sea cores, and also between the marine and terrestrial records, is discussed further in Chapter 6.

(c) Magnetic susceptibility, isothermal remanent magnetism and coercivity A number of magnetic characteristics of sediments do not depend on variations in the earth's magnetic field, but rather reflect the nature and origins of magnetic minerals in the sediments. While these properties cannot in themselves be used as a basis for a relative chronology, they may provide valuable palaeoenvironmental information relating to contexts of deposition. Moreover, they offer considerable potential as a means of correlation in a wide variety of depositional contexts and across a range of timescales, where variations in mineral magnetic properties reflect broadly synchronous environmental changes.

1. **Mineral magnetic susceptibility** is the ratio of the magnetisation produced in a substance to the intensity of the magnetic field to which it has been subjected. In effect, susceptibility is the degree to which a substance can be magnetised (Thompson & Oldfield, 1986). Variations in natural magnetic assemblages, which will be reflected in susceptibility measurements, have been used to make inferences about a number of environmental processes including sediment transport in rivers (Oldfield, 1983), sediment flux and erosion in lake catchments (Dearing *et al.*, 1981), anthropogenically induced erosion (Sandgren & Fredskild, 1991) and sources of atmospheric dust (Oldfield, 1991). However, magnetic susceptibility has also been widely used in association with other magnetic properties as a basis for correlating between cores from Holocene lake sediment sequences (Dearing, 1986). On a longer timescale, magnetic susceptibility measurements have been obtained from TL-dated loess/palaeosol sequences from China extending back over 130 ka (Kukla *et al.*, 1988; An *et al.*, 1991). The magnetic susceptibility signal, which is usually higher in soils than in loessic sediments reflecting the production or concentration of magnetic minerals with pedogenesis (Forman, 1991), therefore forms a basis for correlation between widely separated profiles (Figure 5.27).
2. **Isothermal remanent magnetisation (IRM)** is the magnetic moment activated in and retained by a sample placed in a magnetic field at room temperature. With a gradual increase in the strength of the field, IRM will increase non-linearly until **saturation isothermal remanent magnetisation (SIRM)** is reached. This is the level at which, for any particular sample, a further increase in the magnetic field will not result in any increase in IRM, and is dependent upon the sizes and types of magnetic minerals present in a sample (Thompson, 1986). SIRM measurements have been employed in a range of studies, including sediment yield and erosion rates around lake catchments (Bloemendal *et al.*, 1979), heavy metal pollution (Oldfield & Maher, 1989), and magnetic characterisation of glacial diamicts (Walden *et al.*, 1987) and deep-sea sediments (Bloemendal & de Menocal, 1989). SIRM values have also been used in the correlation of Holocene lake sediment cores (Snowball & Thompson, 1990).
3. **Coercivity of IRM** is the reversed field strength required to reduce the remanent magnetism to zero

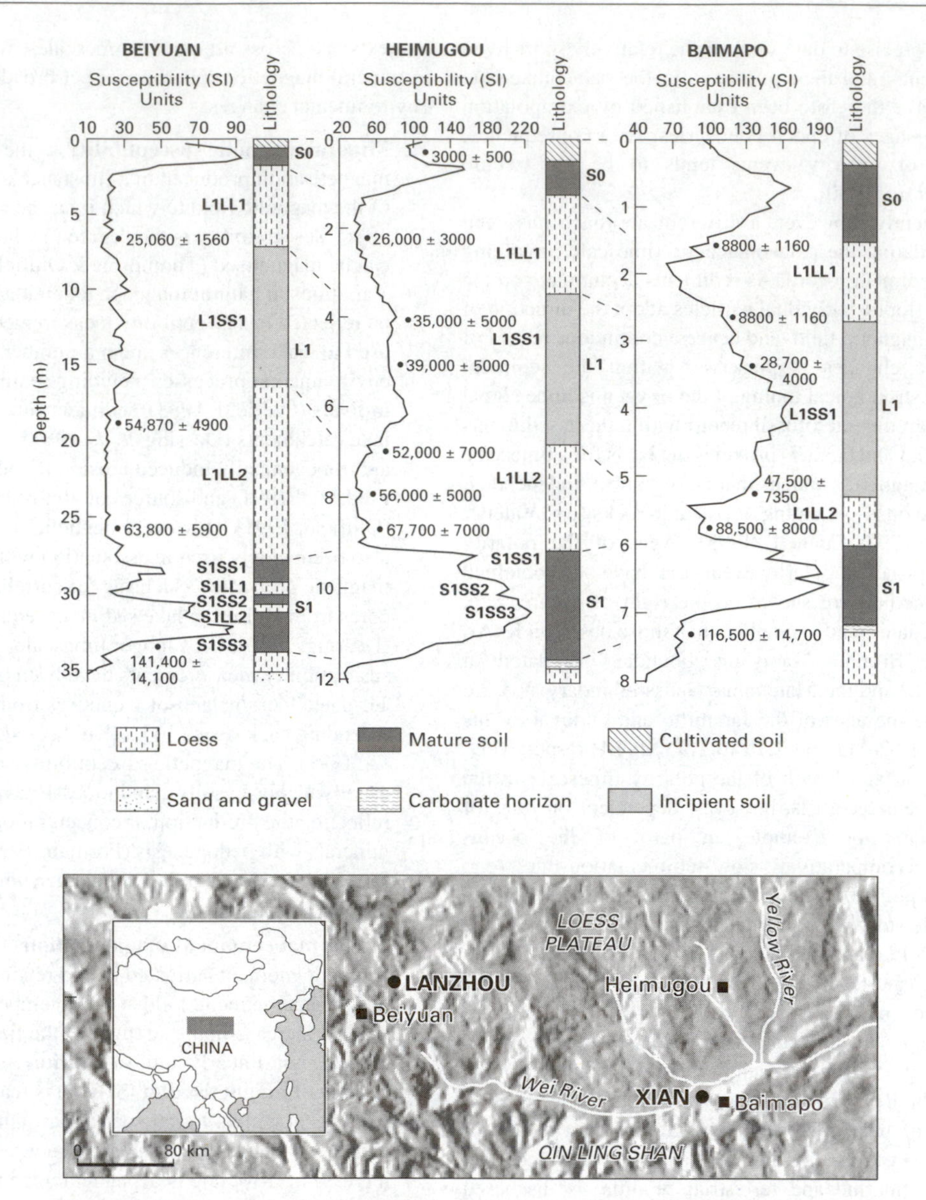

Figure 5.27 Magnetic susceptibility profiles through three loess sequences in China covering the last 130 ka. The chronology is based on TL dating. Location of profiles is shown on inset map (after An *et al.*, 1991).

after saturation. Low coercivity values appear to be characteristic of large-grained magnetite, while high coercivities tend to be associated with fine-grained haematite. In practice, coercivity curves seem to differentiate between assemblages of soil and sediment types very sensitively (Thompson, 1986), and may offer a further means whereby correlation between sediment sequences can be effected.

5.5.2 Tephrochronology

Following a volcanic eruption ash or tephra is often spread rapidly over a relatively wide area and forms a thin cover over contemporaneous peat surfaces, lake floor sediments, estuarine sediments, river terraces, etc. Thin ash layers have also been found in deep-sea sediments. Ash beds often stan

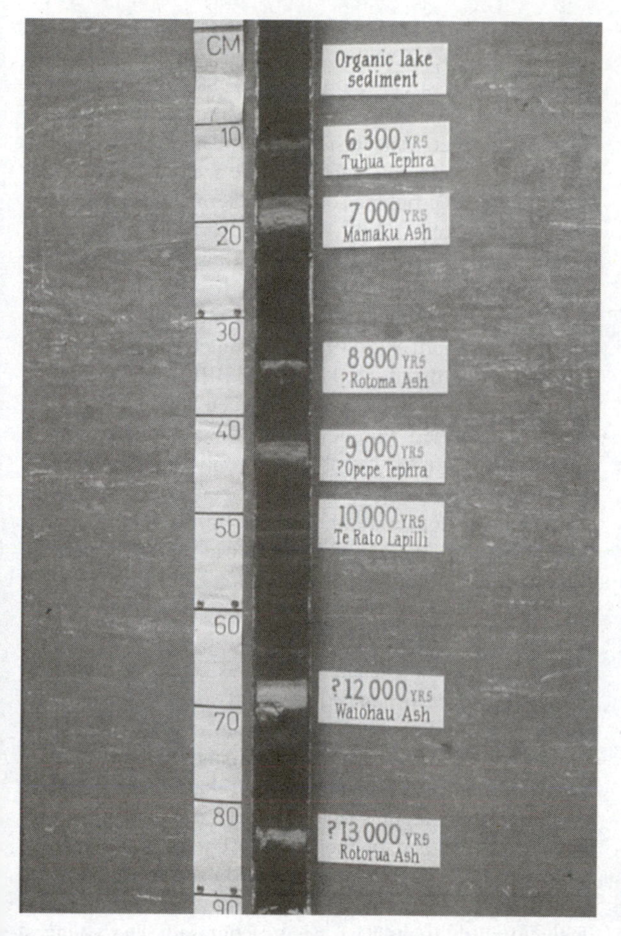

Figure 5.28 Part of a sediment core from Lake Rotomanuka, Waikato Region, North Island, New Zealand, showing a number of radiocarbon-dated tephra layers (photo: David J. Lowe).

out as distinctive light-coloured horizons in sedimentary sequences and can be identified in cores (Figure 5.28) by a variety of methods, including granulometric characteristics, petrographical and mineralogical properties, and geochemical signatures (Einarsson, 1986). Not only do these serve to distinguish between ashes, they may also enable source area to be established. For example, the 10.6 ka Vedde Ash (Figure 5.29) found in western Norway and North Atlantic ocean cores has been attributed to the Katla–Eldgjá volcanic system in southern Iceland, as both the ash and the local volcanics have been shown to possess an unusually high TiO_2 content (Mangerud *et al.*, 1984). The age of an ash bed can be established by radiocarbon dating of associated organic material such as wood, peat or lake sediments (Mangerud *et al.*, 1984, 1986), or for older deposits by K–Ar/^{40}Ar–^{39}Ar (Bogaard *et al.*, 1989), fission track (Meyer *et al.*, 1991), TL (Berger, 1988) or ESR dating (Imai *et al.*, 1985) of some of the primary mineral constituents. Where tephras are found in ice cores, the date of the ash fall can be obtained by counting annual layers in the ice (Hammer *et al.*, 1980). Other means whereby tephra can be dated include stratigraphical position in relation to dated tephra layers, palaeomagnetic correlations, annually laminated sediments, biostratigraphical methods (e.g. pollen analysis) and relationships to oxygen isotope stage boundaries in deep ocean sediments (Einarsson, 1986).

Tephrochronology has now been employed in many areas of the world. In North America, successive phases of volcanism have resulted in the accumulation of ash layers of widely differing ages in the sedimentary records of the Western Cordillera and adjacent areas of the high plains. These range from the Pearlette Ashes, which date from around 2 million years to 600 ka years BP (Izett, 1981), to the more recent ash falls which include the Glacier Peak (*c.* 11.25 ka BP), Mazama (*c.* 6.8 ka BP) and St Helens 'Y' (*c.* 3.4 ka BP) Ashes (Beget, 1984: Figure 5.30). These tephras have been particularly valuable in reconstructing a glacial chronology for the western United States, the older ashes, for example, providing chronometric control on interbedded till sequences (Richmond & Fullerton, 1986b), while the more recent ashes have been used to establish

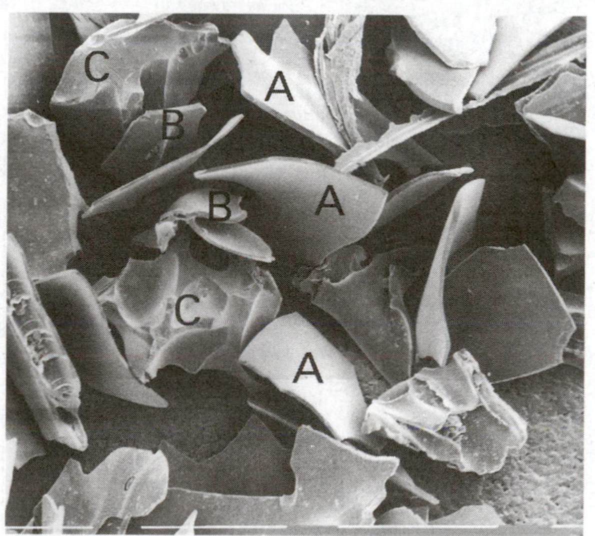

Figure 5.29 Scanning electron microscope photograph of some of the types of glass shards that occur in the Vedde Ash Bed in western Norway (white bars at bottom = 100 μm). A. Thin, slightly curved platy shards, representing fragments of broken walls of gas bubbles. B. Oblong fragments characterised by three- to four-winged cross-sections which are fragments of the seams between bubbles. (Types A and B are the most common in the Vedde Ash Bed of Norway.) C. Dark-brown, blocky vesicular particles (photo: Jan Mangerud; after Mangerud *et al.*, 1984).

terminal Wisconsinan and Holocene glacial chronologies for the mountain regions of the northwest USA and western Canada (Davis, 1988; Osborn & Luckman, 1988). Elsewhere, the Old Crow tephra of Alaska forms an important chronostratigraphic marker predating the last interglacial (Westgate *et al.*, 1985), while Late Holocene tephras in both Alaska and Antarctica provide key marker horizons for studies of climatic and environmental change in

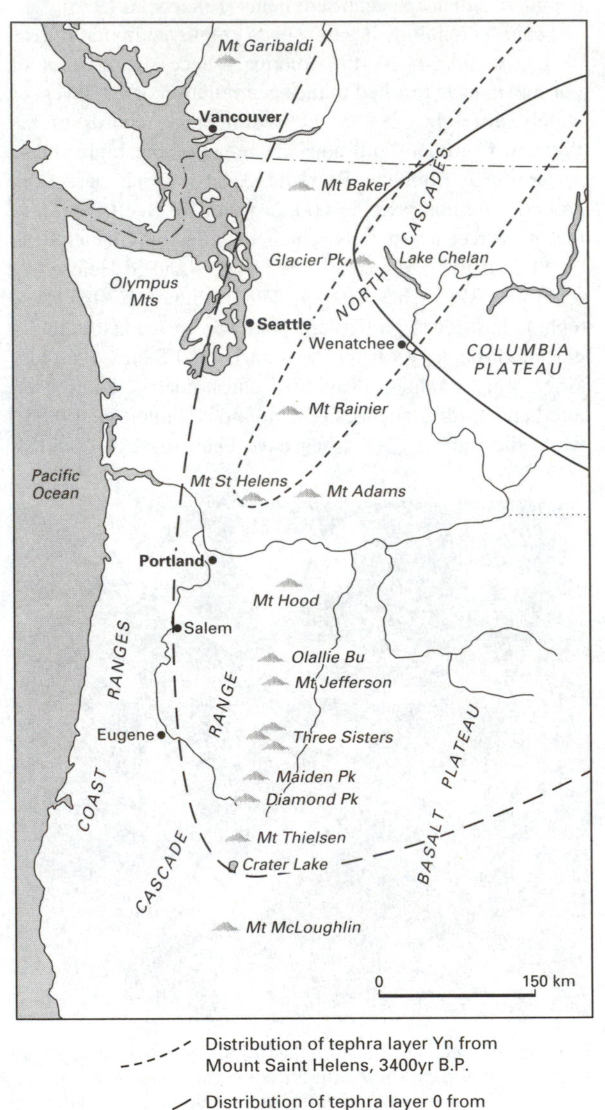

Figure 5.30 The distribution of Glacier Peak, Mazama and St Helens 'Y' tephras in the mountains of Washington and Oregon, northwest USA (after Beget, 1984).

these high latitude regions (Riehle *et al.*, 1990; Björck *et al.*, 1991). In New Zealand a tephrochronological record spanning the last 17 ka (Lowe, 1988) offers a time-stratigraphic framework for the reconstruction of Late Quaternary vegetational history (Newnham & Lowe, 1991).

In Europe, numerous tephras have been described from the Eifel volcanic field in southwest Germany (Brunnacker, 1975). K–Ar and $^{40}Ar/^{39}Ar$ dating show that these are predominantly of Middle Pleistocene age, and include the widely distributed Hüttenberg tephra (*c.* 215 ka BP) which appears to have been deposited during a glacial–interglacial transition (Bogaard *et al.*, 1989). The much younger Laacher See ash (^{14}C dated to 11 ka ± 50 BP) from the same volcanic centre (Figure 5.31) has been found throughout central and northern Europe (Bogaard & Schmincke, 1985), and has proved to be a valuable isochronous marker horizon, particularly in studies of Lateglacial lake sediments (Ammann & Lotter, 1989; Lotter, 1991). In the Massif Central region of France, widespread tephra layers of Allerød (*c.* 11.4 ka BP) and early Holocene (*c.* 8.5 ka BP) age have also been identified in lacustrine and fluvial sequences (Juvigné *et al.*, 1992).

The majority of Quaternary tephras found in northwest Europe, however, originate in Iceland, and are found as far afield as Scandinavia (Thorarinsson, 1981), Greenland (Hammer *et al.*, 1980), mainland Scotland (Dugmore, 1989; Blackford *et al.*, 1992) and Northern Ireland (Pilcher & Hall, 1992). Icelandic tephras have also been discovered in cores taken from the bed of the North Atlantic (Ruddiman & McIntyre, 1977; Duplessy *et al.*, 1981) and from the North Sea (Long & Morton, 1987). Within the ocean cores, distinctive ash zones have been recognised. The youngest (North Atlantic Ash Zone 1) dates from the last glacial–interglacial transition and contains four distinct tephras dated at *c.* 10.8 ka, 10.6 ka (the widespread Vedde Ash: Figure 5.29), 9.3 ka and 8.9 ka BP (Björck *et al.*, 1992b), while earlier ash zones have been recognised in oxygen isotope stages 5, 7, 9 and 11, i.e. in interglacial sections of ocean cores (Sejrup *et al.*, 1989; Sjøholm *et al.*, 1991). These ash zones appear to represent periods of increased explosive volcanism on Iceland, which has been linked to pressure release in the crust following regional deglaciation (Anderson, 1987). The tephras constitute key chronostratigraphic marker horizons and offer a basis for correlation between individual ocean cores and also between the marine and terrestrial records.

These results show that tephrochronology is a technique of considerable potential, both as a correlative tool and in the development of local and regional chronologies. It must be emphasised, however, that at best it can only form a basis for regionally applicable schemes of relative dating, for individual ash layers are restricted spatially by such factor

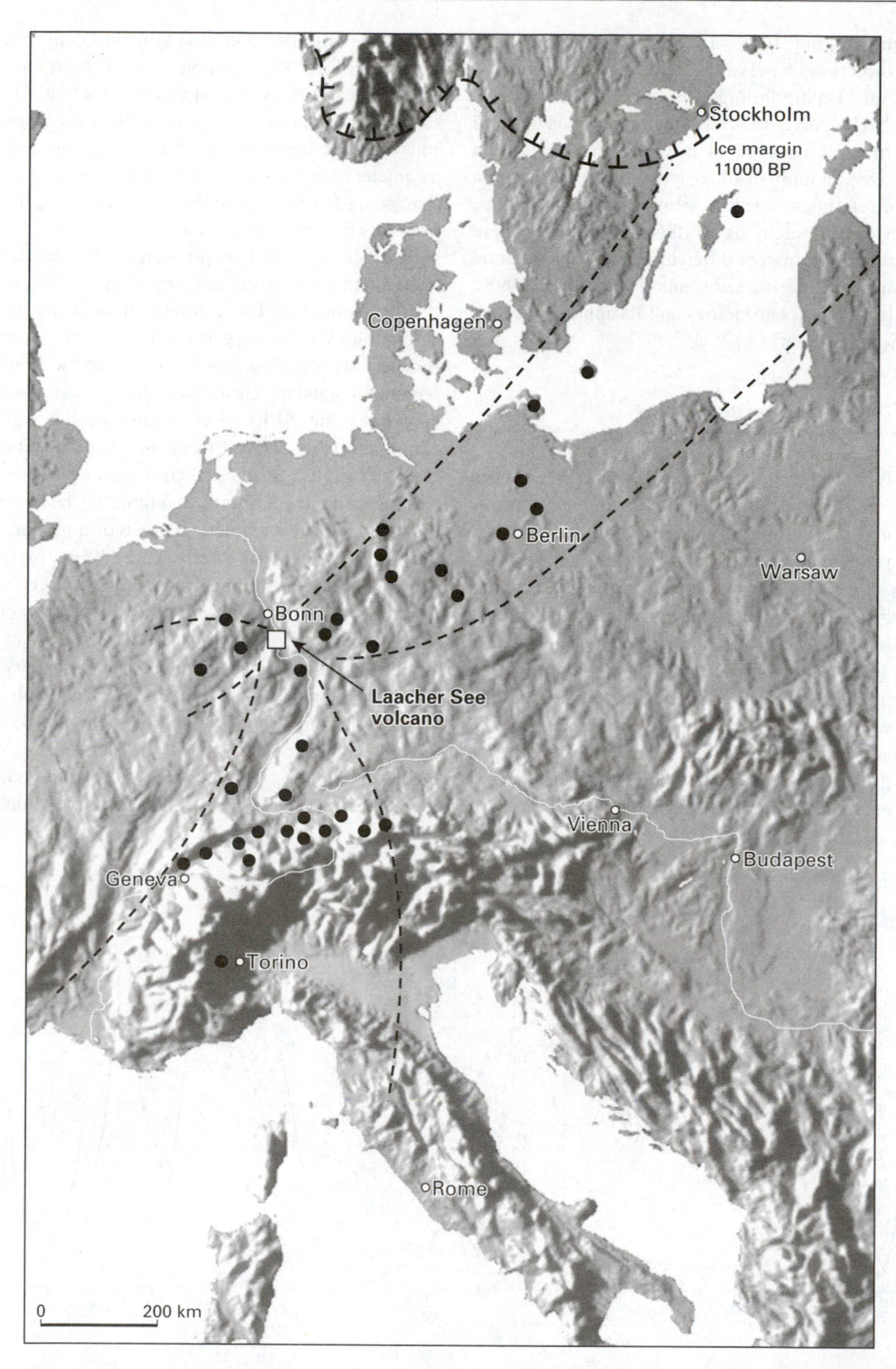

Figure 5.31 The areal distribution of the Laacher See tephra (11 ka BP). Dots mark sites where ash has been recorded; the heavy broken line indicates the outer margins of volcanic airfall (after Bogaard & Schmincke, 1985).

as the magnitude and type of volcanic explosion, the strength and direction of prevailing winds, the particle size of the tephra, etc. Tephra falling over northern and southern oceans may also be prevented from reaching the sea bed and hence the marine sediment record, by ice cover, at the ocean surface. Moreover, a number of analytical problems remain to be resolved relating to specific identifications of tephras, especially in areas such as northwest Europe where numerous Icelandic tephras of different age have been found in both terrestrial and marine sediments (Hunt & Hill, 1993). Further details on tephrochronology and its applications can be found in Self and Sparks (1981).

5.5.3 Oxygen isotope chronology

It has already been seen (section 3.10) that the oxygen isotope trace in deep-ocean sediments represents a proxy record of long-term climatic change. Moreover, as the isotopic signal is geographically consistent and can be replicated in cores taken from widely separated localities (Figure 1.4), inflections in the isotopic profiles are essentially time-parallel events, and constitute age-equivalent marker horizons. As will be shown in section 6.2.3.5, these form the basis for a globally applicable scheme of **oxygen isotope stages**. However, it is also possible to derive ages for the major isotopically defined horizons. Analysis of oxygen isotope records using time-series techniques, such as Spectral Analysis, has revealed periodicities that can be related to the operation of the astronomical (Milankovitch) variables (Hay *et al.*, 1976). The isotopic signal therefore represent astronomical forcing at frequencies of 23 ka, 41 ka and 10 ka, reflecting the separate influences of precession, axial til and orbital eccentricity (section 1.6). Since these orbita parameters are constant and their frequency is known, the provide a basis for timing the cycles reflected in the isotopi records (Berger, 1978). Hence, the age of each cycl represented by the isotopic stages can be calculated b extrapolating back from the present day.

This method of using orbital frequencies to develop chronology for the oxygen isotope record is termed **orbita tuning**, and was first employed by Imbrie *et al.* (1984) t produce a standard chronology for oxygen isotope record known as the **SPECMAP timescale**. A high-resolutio chronology was derived from the amalgamation of severa isotopic records ('stacked' records), and the composite curv was then smoothed, filtered and tuned to the known cycles o the astronomical variables. The use of a number of isotopi profiles was designed to eliminate 'noise' from errors th could have been contained within a single isotopic recor (Prell *et al.*, 1986). Six radiometrically dated horizons (fiv based on ^{14}C; one based on K–Ar, i.e. th Brunhes–Matuyama geomagnetic boundary) forme reference points for the initial timescale. Subsequently, high-resolution chronostratigraphic isotopic record has bee developed for the last 300 ka of the Pleistocene (Martinson *e al.*, 1987: Figure 5.32 and Table 5.7), while other orbitall tuned timescales have been obtained for the Middle and Earl

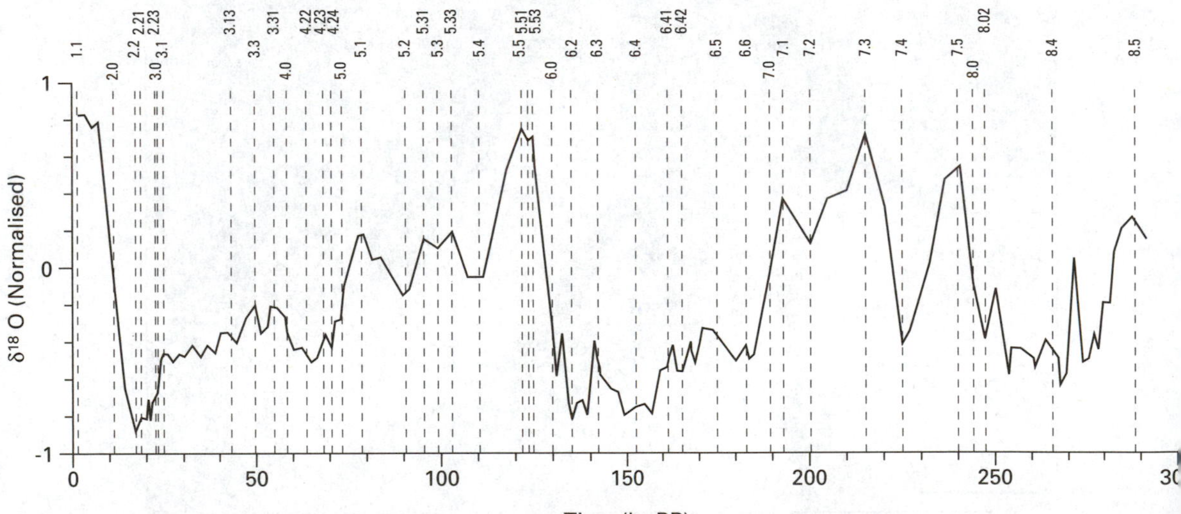

Figure 5.32 Orbitally driven chronostratigraphy for a stacked deep-ocean isotopic record spanning the last 300 000 years (after Martinson *et al.*, 1987). The numbered vertical lines indicate identifiable features of the record, the dates of which are shown in Table 5.7. The basis of the numbering scheme is explained in section 6.2.3.5.

Table 5.7 Age estimates for the events in the stacked isotopic record shown in Figure 5.32 (from Martinson *et el*., 1987).

Event	Depth (cm)	Age (years)	Error (years)	$\delta^{18}O$
1.1	13.0	2 320	2 110	0.86
2.0	80.0	12 050	3 140	−0.06
2.2	129.0	17 850	1 370	−0.84
2.21	146.0	19 220	1 390	−0.76
2.23	219.0	23 170	4 190	−0.65
3.0	238.0	24 110	4 930	−0.60
3.1	264.0	25 420	5 900	−0.42
3.13	375.0	43 880	4 710	−0.34
3.31	469.0	55 450	5 030	−0.17
4.0	495.0	58 960	5 560	−0.32
4.22	520.0	64 090	6 350	−0.43
4.24	570.0	70 820	3 950	−0.40
5.0	600.0	73 910	2 590	−0.18
5.1	651.0	79 250	3 580	0.22
5.2	699.0	90 950	6 830	−0.11
5.3	733.0	99 380	3 410	0.14
5.33	743.0	103 290	3 410	0.21
5.4	780.0	110 790	6 280	−0.02
5.5	818.0	122 560	2 410	0.74
5.51	825.0	123 820	2 620	0.69
5.53	836.0	125 910	2 920	0.66
6.0	865.0	129 840	3 050	−0.28
6.2	901.0	135 100	4 240	−0.83
6.3	959.0	142 280	5 280	−0.50
6.4	1 005.0	152 580	9 910	−0.73
6.41	1 041.0	161 340	8 860	−0.51
6.42	1 069.0	165 350	8 390	−0.53
6.5	1 117.0	175 050	9 840	−0.36
6.6	1 170.0	183 300	5 740	−0.48
7.0	1 205.0	189 610	2 310	−0.02
7.1	1 220.0	193 070	2 020	−0.38
7.2	1 240.0	200 570	4 960	0.13
7.3	1 270.0	215 540	1 420	0.73
7.4	1 299.0	224 890	1 210	−0.40
7.5	1 359.0	240 190	6 340	0.53
8.0	1 370.0	244 180	7 110	−0.11
8.02	1 380.0	247 600	6 250	−0.43
8.4	1 460.0	265 670	7 720	−0.44
8.5	1 585.0	288 540	3 520	0.27

Pleistocene periods (Ruddiman *et al*., 1986, 1989; Raymo *et al*., 1989; Shackleton *et al*., 1990). More recently, an orbitally tuned timescale has been developed for Chinese loess deposits spanning the whole of the last 2.5 Ma (Ding *et al*., 1994).

Although orbital tuning provides a global reference standard and offers an innovative method for the dating of deep-ocean and other long sediment records, the approach is not without its problems (Patience & Kroon, 1991). These include uncertainties over the factors that contribute to the overall isotopic signal in the carbonate tests of Foraminifera; questions relating to the reliability of reference radiometric determinations; statistical difficulties surrounding the application of time-series analysis; problems arising from poor stratigraphic resolution, bioturbation and reworking of sediments; and the difficulties of recognition and precise definition of stage boundaries in some isotopic profiles. Hence, Patience and Kroon conclude (p. 222) that while 'Oxygen isotope ratio curves are potentially a powerful tool for dating and correlating ocean sediments through time, … they must be used in conjunction with other corroborative evidence such as carbon isotope ratios, magnetic susceptibility, geochemistry and power spectral analysis to enable the production of a reliable age model.'

5.6 Relative chronology based on processes of chemical alteration

Fossils, sediments and rocks are affected by a number of chemical reactions that are partly time-dependent. Upon the death of an organism, tissues are broken down by a variety of chemical processes to produce compounds of a more simple chemical structure; the surfaces of fossils or minerals may be altered by the effects of hydration or the accumulation of precipitates of certain chemicals in groundwaters, while weathering and pedogenic processes will gradually effect visible changes on rock and sediment surfaces. In all of these cases, the degree of alteration brought about by different chemical reactions increases with time, and this therefore offers a basis for relative dating. Some of the more widely used techniques are considered in the following sections.

5.6.1 Amino-acid geochronology

Living bone consists of approximately 23 per cent collagen (protein-bearing) fibrils bound within a phosphatic–calcitic matrix. Proteins can survive in bones and shells for extremely long periods[3] but undergo a number of molecular changes. The discovery of protein residues in fossil bones and shells was first reported by Abelson (1956), and since then the study of protein transformation in the geological record has developed rapidy. Some of the chemical changes in proteins that occur after the death of organisms are time-dependent, and thus the characteristics of certain protein residues from the Quaternary record provide the basis for a relative chronology.

5.6.1.1 Chemistry of proteins

Proteins are large and complex molecules and are basic ingredients of all living organisms. They are composed of **amino acids**, which have the generalised chemical formula shown in Figure 5.33A. The 'R' linkage differs for each amino acid, ranging from a simple hydrogen atom in glycine, to a methyl group (CH_3) in alanine, and to highly complex chemical structures in other amino acids. About twenty amino acids are commonly found in proteins. Proteins form through the combination of several amino acids into **peptide chains,** each amino acid linked to the next by a **peptide bond** following the loss of a water molecule (Figure 5.33B and C). A chain of amino acids is referred to as the **primary structure** of a protein, which is defined by the number of amino acids and the order in which they are arranged in the protein molecule. Proteins typically consist of complex chains (**polypeptide chains**) of amino acids, and if several arrange themselves in parallel, a degree of molecular stability is achieved, and the resulting peptide arrangement is referred to as **secondary structure**. Complex folding of polypeptide chains produces three-dimensional **tertiary structures**, and polymeric aggregation of several protein sub-units leads to complex coiled, folded and branched **quaternary structures**. Some of the larger proteins may contain up to 3000 amino-acid residues. It is clear, therefore, that the number of ways in which the twenty common amino acids can be joined together is almost infinite, and it is this that leads to the large number of distinctive proteins found in living organisms (Hare *et al.*, 1980).

With the exception of glycine, all amino acids commonly found in proteins can exist in two molecular forms (**isomeric forms**). The chemical and biochemical properties of the two forms of amino acid are similar, but they rotate plane-polarised light in opposite directions. Effectively, they constitute two non-superimposable mirror images (Figure 5.33D), rather like left and right hands, and these optical isomers are referred to as **L-amino acids** and **D-amino acids**. The carbon atom at the centre of the isomers (the **chiral carbon atom**) forms the point of asymmetry and allows the development of the two optical isomers (Sykes, 1991). The biological significance of this distinction between the L- and D-configuration in amino acids is that only L-isomers occur in living (active) proteins. D-isomers can occur in a free state, as components of non-protein structures, and in fossil organic materials as a result of the breakdown of proteins (see below).

5.6.1.2 Amino-acid diagenesis

Chemical alteration in proteinaceous residues following the death of an organism results in the disruption of peptide chains to release free amino acids. Where proteins are exposed to the atmosphere or to biological processes, very rapid degradation will take place, but if they are protected by skeletal hard parts so that a 'closed' system prevails, then much slower chemical alterations occur. Some reaction times are in the range of 50 ka to a few million years whereas other amino-acid reactions operate over a timescale of only a few thousand years.

Two diagenetic processes in particular reflect relative age of fossils. First, peptide bonds are broken by **hydrolysis**, and this eventually releases free amino acids, so that the ratio of free amino acids to peptide-bound acids increases with time. Although this index should, in theory, provide a basis for estimating the age of fossil material, it is difficult to apply in practice. The second mechanism is called **racemisation** (or **epimerisation**) in which the L-isomers are converted (interconverted) to the corresponding D-isomer, and the reaction continues until equilibrium is achieved. If the rate of interconversion can be determined, the time that has elapsed since the death of an organism can be calculated from the ratio of D/L amino acids in the fossil material. It is this aspect of amino-acid diagenesis that forms the basis for most geochronological applications, with the most widely used reaction being the epimerisation of the protein amino acid L-isoleucine (L-Ile) to its non-protein diastereoisomer D-alloisoleucine (D-aIle). In modern shell material, the ratio is effectively zero and progressive inversion proceeds to an equilibrium value of *c.* 1.30 (Bowen *et al.*, 1985).

Although in theory amino-acid determinations can be made on any protein-bearing materials, such as hair, teeth, Foraminifera, shells, bones and tusks, in practice suitable samples are restricted to those with 'tight' skeletal carbonate matrices. The most widely used media have been molluscan shells which appear to form a closed system that is affected by a minimum of external factors. Research has tended to concentrate on marine Mollusca (e.g. Bowen *et al.*, 1985; Hearty *et al.*, 1986), although increasingly amino-acid ratios are being derived from terrestrial molluscs (e.g. McCoy, 1987; Bowen *et al.*, 1989). Bone proteins, by contrast, are complex and highly soluble, and are poorly suited to the amino-acid method, although tooth enamel may provide a more reliable medium than bone (Bada, 1985). Amino-acid ratios have been obtained from cave speleothem, where microscopic organic residues derived originally from surface soils have been co-precipitated with calcite from percolating groundwaters (Lauritzen *et al.*, 1986). By calibrating epimerisation rates against U-series dates obtained from the speleothems, amino-acid ratios can be used to date speleothem material over timespans beyond the range of conventional dating methods (Lauritzen *et al.* 1994). Amino-acid ratios have also been reported from wood samples (Pillans, 1983), although care needs to be

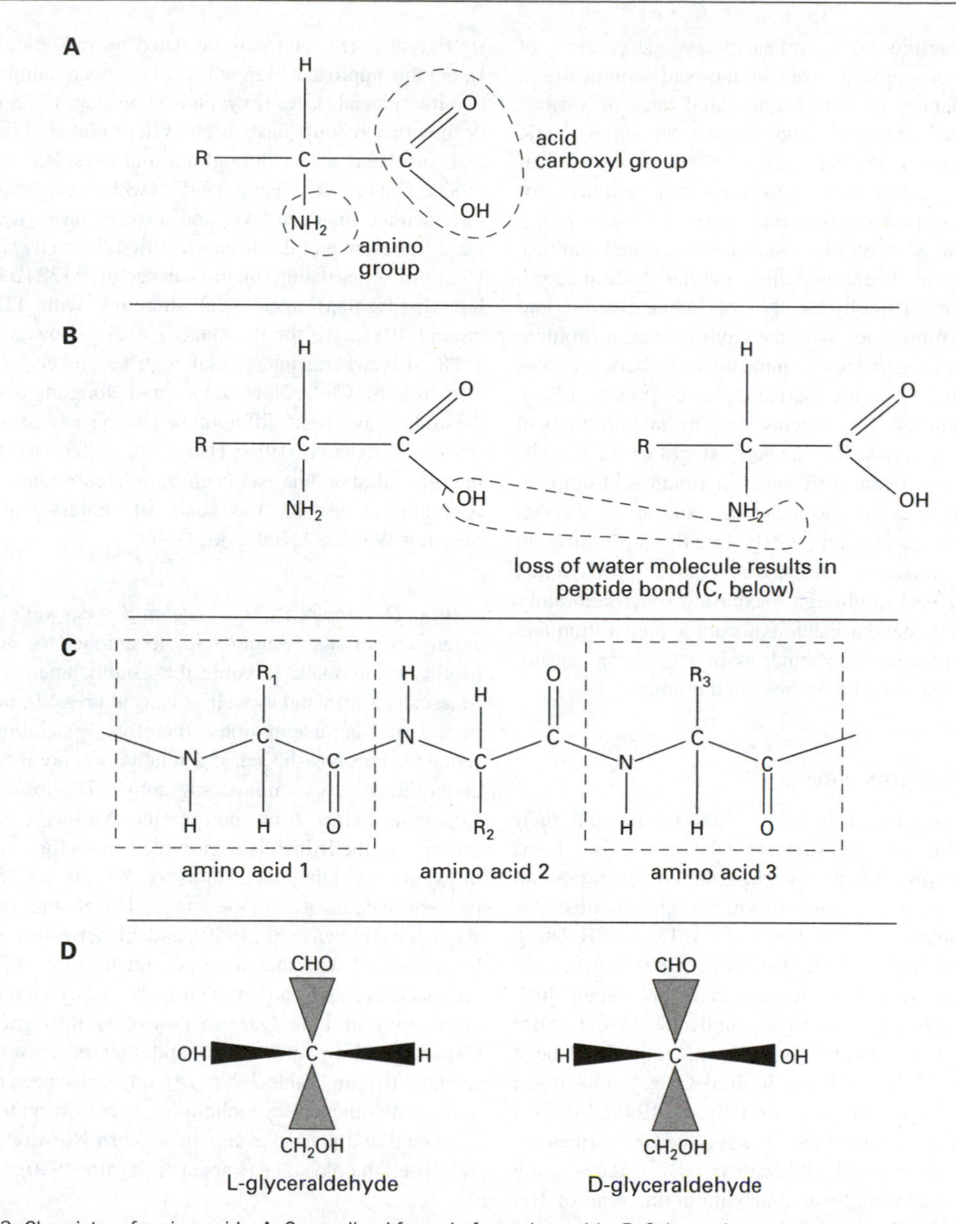

Figure 5.33 Chemistry of amino acids. A. Generalised formula for amino acids. B. Schematic representation of chemical combination of amino acids through loss of water molecule to form a peptide bond. C. Combination of carboxyl and amino groups between amino acids to form peptide chains. D. Isomers of glyceraldehyde; stippled wedges represent bonds extending behind the plane of the page; bold wedges represent bonds extending in front of the page.

taken in evaluating the results from such non-carbonate materials (Sykes, 1991).

Amino-acid measurements are made using either ion-exchange chromatography or gas–liquid chromatography, and kinetic models are used to construct the trend of protein diagenesis. The relationship between D/L ratio and time is complicated by a number of factors, however. First, all reactions are temperature-dependent, with epimerisation proceeding more rapidly at higher temperatures. In practice, this means that meaningful comparisons of amino-acid ratios can only be made between sites that have experienced a similar climatic, and particularly thermal, history. Second, reaction rates depend on the permeability of the shell or bone matrix and on the original amino-acid composition and type of protein structures (e.g. primary, secondary, etc.) of samples. As a consequence, amino-acid composition, relative abundances of amino acids, and rates of various amino-acid reactions are genus – and, in some cases, species – dependent (Miller & Hare, 1980). In the case of marine molluscs, however, corrections can be made for interspecies variation in epimerisation rates (Bowen *et al.*, 1985), since rates of epimerisation for individual mollusc species have now been quantified (Miller & Mangerud, 1985). A third difficulty is that of contamination and exchange of amino acids with the environment, a problem that appears to be particularly intractable for bone samples although much less so for marine molluscs (Sykes, 1991). Fourth, degradation of proteins due to microbiological activity, particularly during the early stages of diagenesis, may result in significant differences in amino-acid ratios in molluscs from sites of the same age and same thermal histories (Sejrup & Haugen, 1994). Finally, differences in results from different laboratories have been noted (Wehmiller, 1984), although increasing inter-laboratory comparisons and standardisation of sample preparation and measurement procedures should, as in ^{14}C dating, ensure that this becomes less of a problem in the future.

5.6.1.3 Aminostratigraphy

The use of amino-acid ratios to rank fossils and their associated sediments according to relative age is termed **aminostratigraphy.** Moreover, where amino-acid ratios can be calibrated against samples of known age obtained, for example, by means of ^{14}C, uranium series, TL or ESR dating (e.g. Bowen & Sykes, 1988; Hsu *et al.*, 1989), a timescale based on amino-acid evidence can be established. Aminostratigraphy appears to be applicable to the entire timespan of the Quaternary, for isoleucine ratios have been obtained from Mollusca from the Red Crag, some of the earliest British Quaternary deposits (Bowen, 1991). Indeed a range of several million years is suggested by isoleucine ratios from tooth enamel, although published dates using aspartic acid in bone indicate a maximum timespan of 100 ka (Aitken, 1990). Some of the principal applications of aminostratigraphy are as follows.

(a) Correlation and dating of interglacial shorelines One of the most successful applications of this technique has been in the resolution and correlation of high sea-level stands. This involves the development of regionally applicable **aminozones**, each of which is characterised by a distinctive range of D/L ratios for marine Mollusca, and each reflects a particular high sea-level (i.e. interglacial) event. These aminozones are constrained by geomorphology, stratigraphy, etc., and may be dated by radiometric means. Using this approach, high sea-level events relating to the last two interglacial stages (oxygen isotope stages 5e and 7 of the deep-ocean record) have been differentiated (Figure 5.34) and correlated at sites throughout southwest Britain (Davies, 1983; Davies & Keen, 1985), while evidence for an intermediate high sea level and also for an earlier (isotope stage 9?) event has also been identified (Bowen *et al.*, 1985). Uranium-series dating suggests an age of *c.* 122 ka BP for the last (Ipswichian) interglacial shoreline with TL ages of around 191 ka BP for the stage 7 event (Bowen & Sykes, 1988). Elsewhere, interglacial high sea levels of different ages in Peru, Chile, New Guinea and along the east coast of Australia have been differentiated using aminostratigraphy (Hearty & Aharon, 1988; Hsu *et al.*, 1989), while the last interglacial shoreline has been correlated around the entire Australian coast on the basis of amino-acid evidence (Murray-Wallace & Belperio, 1991).

(b) Development of regional chronologies The extension of the amino-acid technique to non-marine Mollusca has made possible the establishment of relative timescales for inland as well as coastal areas. In many areas of the mid-latitude regions, therefore, glacial/interglacial sequence can now be set in a relative temporal framework underpinned by aminostratigraphy. D-alloisoleucine/L-isoleucine ratios from non-marine Mollusca from sites throughout the British Isles provide a basis for a chronology of events extending back some 600 ka (Figure 5.35), i.e. to the equivalent of isotope stage 15 of the deep-ocean sequence (Bowen *et al.*, 1989), and this terrestrial record can be correlated with that from coastal localities (Table 5.8). Aminostratigraphy has proved to be a key element in the correlation of Late Quaternary events throughout Arctic Canada (Andrews *et al.*, 1986) and river terrace sequences in lowland Britain (Table 2.2, p. 75). It has also been employed, along with other dating techniques, to establish the sequence of glacial and marine events in western Norway during the Weichselian cold stage (Larsen & Sejrup, 1990).

(c) Land–sea correlations Aminostratigraphic evidence offers a basis for correlation between marine and terrestrial records at the continental scale. For example, Miller and Mangerud (1985) used D/L ratios in marine Mollusca from sites throughout northwest Europe and in nearby Arctic regions to develop a chronology of interglacial episodes which could be correlated with the conventional Cromerian, Holsteinian and Eemian stages of the European terrestrial sequence. By implication, they also assigned ages to the glacial episodes between those warm stages. Subsequently eight high sea-level events have been recognised on the bas

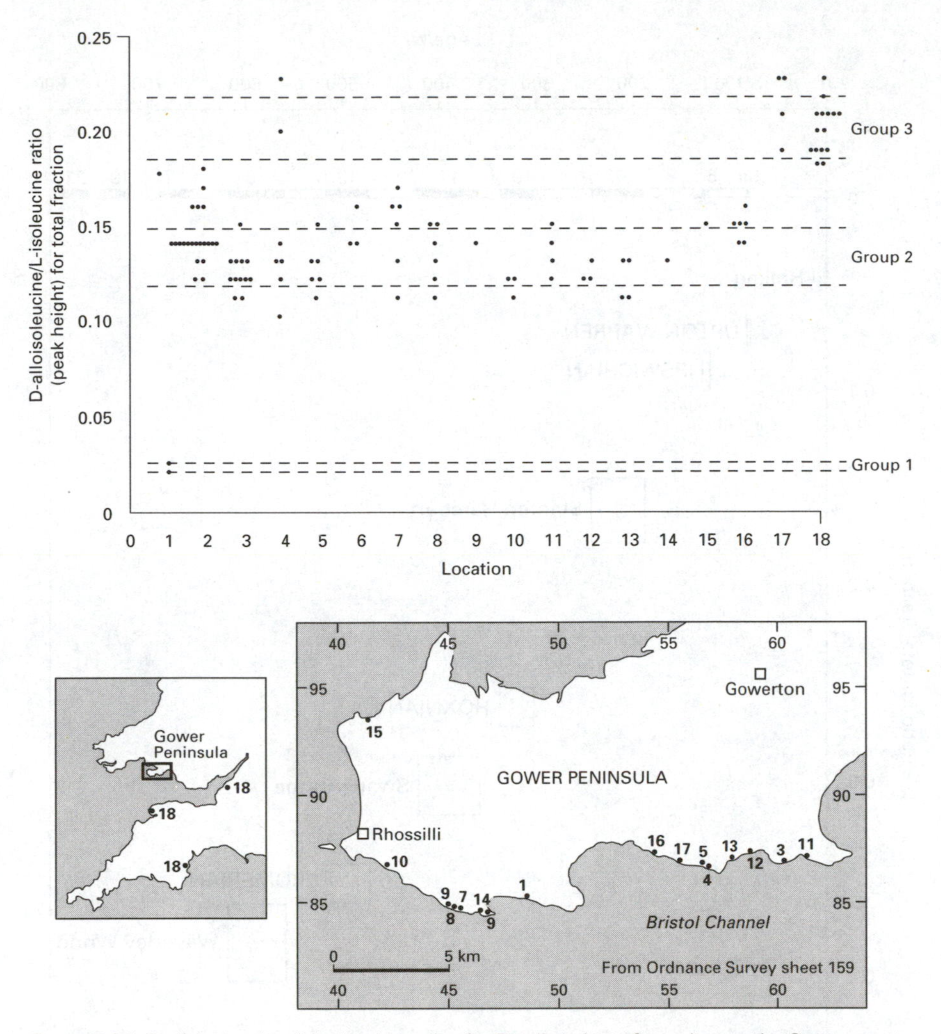

Figure 5.34 D-alloisoleucine/L-isoleucine ratios for *Patella vulgata* from sites on the Gower coast, South Wales, and north and south Devon in southwest England. Sample numbers (*y* axis) relate to sample points on the inset map. Three groups (aminozones) are apparent in the data: Group 1 has a mean ratio of 0.0255 ± 0.0015 and is a control sample from the modern Gower beach; Group 2 has a mean ratio of 0.134 ± 0.016 and dates from the last interglacial (OI stage 5e); Group 3 has a mean value of 0.203 ± 0.016 and dates from the previous interglacial (OI stage 7) (after Davies, 1983).

of amino-acid data at sites throughout northwest Europe, the oldest of which appears to equate with isotope stage 13 of the deep-sea sequence (Figure 5.36). These data can be used to constrain the timing of Middle and Late Pleistocene events in the British Isles and Scandinavia during the Middle and Late Pleistocene (Bowen & Sykes, 1988).

(d) Resolution of populations of mixed ages Assemblages of marine molluscs often comprise elements of different ages, and this may not be apparent from visual inspection of the fossil materials. Such mixed or reworked populations have been detected in beach deposits, for example, on the basis of amino-acid evidence, where materials of greater antiquity are reflected in higher than average D/L ratios (Bowen *et al.*, 1985).

(e) Screening of fossils for ^{14}C dating Radiocarbon dates in excess of about 40 ka BP are generally considered to be unreliable and the dating process is expensive and time-consuming. By comparison, amino-acid analysis is less costly and relatively rapid, and may therefore be used to 'screen' fossils prior to radiocarbon dating. This could be of

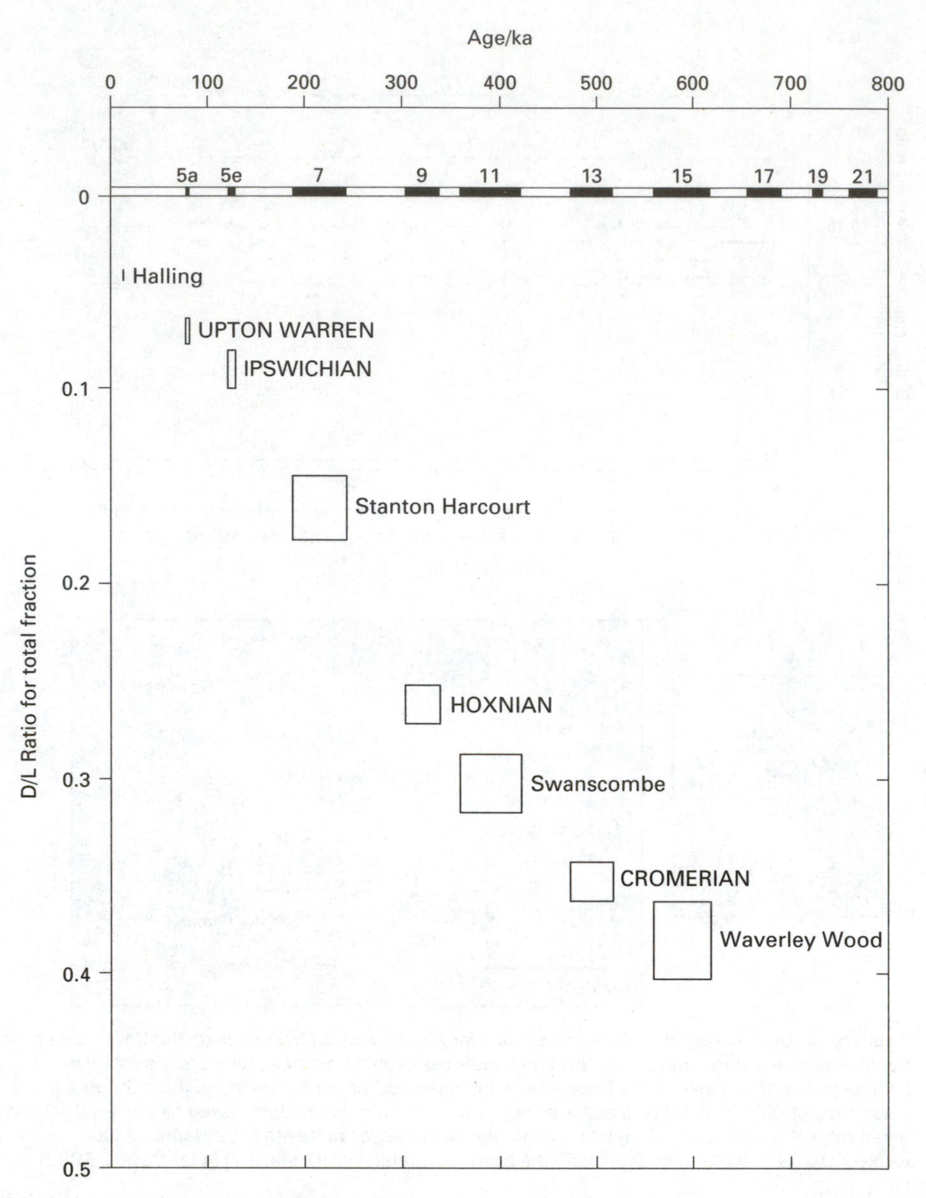

Figure 5.35 Correlation of D-alloisoleucine/L-isoleucine molluscan non-marine events in lowland Britain with marine oxygen isotope stages. Each rectangle is one standard deviation (after Bowen *et al.*, 1989).

particular value in the dating of mollusc-bearing tills, in regions such as Arctic Canada where mollusc valves are the most widely used materials in radiocarbon dating, but where reworking of molluscs and mixing of molluscan populations has often occurred (Blake, 1980).

(f) Palaeothermometry Since racemisation rates are temperature-dependent, where the exact age of fossil protein can be established amino-acid ratios can be used to infer temperature conditions that have affected the proteins since the commencement of diagenesis (McCoy, 1987). Although this has limited application where fossils have experienced a number of major temperature oscillations, aspartic acid D/L ratios in bristlecone pines, for example, have formed the basis for specific palaeotemperature inferences (Zumberge *et al.*, 1980).

Table 5.8 Amino-acid geochronology (calibrated epimerisation of L-isoleucine) of marine (sea-level changes), non-marine (interglacials) and glacial events in the British Isles. Odd-numbered oxygen isotope stages reflect low global ice volume (interglacials); even numbers indicate high continental ice volume (glacial stages) – see section 6.2.3.5. Marine Mollusca epimerise at either slow (*Littorina*) or moderate (*Arctica* or *Macoma*) rates (after Bowen, 1991).

Standard stages and events to be defined	$\delta^{18}O$ stages	Amino acid epimerisation			Glaciation
		Marine		Non-marine	
		Moderate	Slow		
					Loch
		0.05			Lomond
DEVENSIAN Late	2			0.04	
		0.07			Late
					Devensian
DEVENSIAN Middle	3	0.09			
	4				Early
DEVENSIAN Early					Devensian
	5				
IPSWICHIAN	5e	0.16	0.1	0.1	
	6				
Stanton	7	0.2	0.16	0.16	
Harcourt	8				Paviland
HOXNIAN	9	0.28	0.22	0.24	
	10				
Swanscombe	11	0.38	0.28	0.3	
ANGLIAN	12				Anglian
CROMERIAN	13			0.34	KESGRAVE FORMATION
Waverley Wood	15			0.38	Up to 4 upland glaciations
	Brunhes (approximate position)				
0.73 Ma	Matuyama				
BAVENTIAN					? Upland ice
Norwich Crag		0.8			
Red Crag		1.09			
	2.4 Ma				Ice rafting

5.6.2 Fluorine, uranium and nitrogen content of fossil bones

Hydroxyapatite, the principal mineral constituent of bones and teeth (section 4.12), absorbs fluorine from groundwater progressively over time. The rate at which the fluorine content increases varies from locality to locality, but bones that have been buried for the same length of time in a particular deposit will have approximately the same fluorine content. As the fluorine fixed in the bone is not readily removed, a specimen that is washed out of an older deposit will show a much higher fluorine content than bones that are contemporaneous with the bed, while bones from a younger stratum will have accumulated substantially less fluorine. Hence the technique is useful in the relative dating of bones in a deposit, and it also enables intrusive (i.e. younger or older) material in a fossil assemblage to be identified (Haddy & Hanson, 1982; Demetsopoulos *et al.*, 1983). The fluorine to phosphate ratio is established by X-ray diffraction techniques. The analysis of uranium incorporated into fossil bones from groundwater has been employed in a similar way to the fluorine method, the former having the advantage that the counting of uranium emissions does not involve the destruction of the bone material. The relative ages of fossil bones in assemblages can also be established by analysing the nitrogen content of fossil bone, for as proteinaceous materials disappear from bone collagen, so

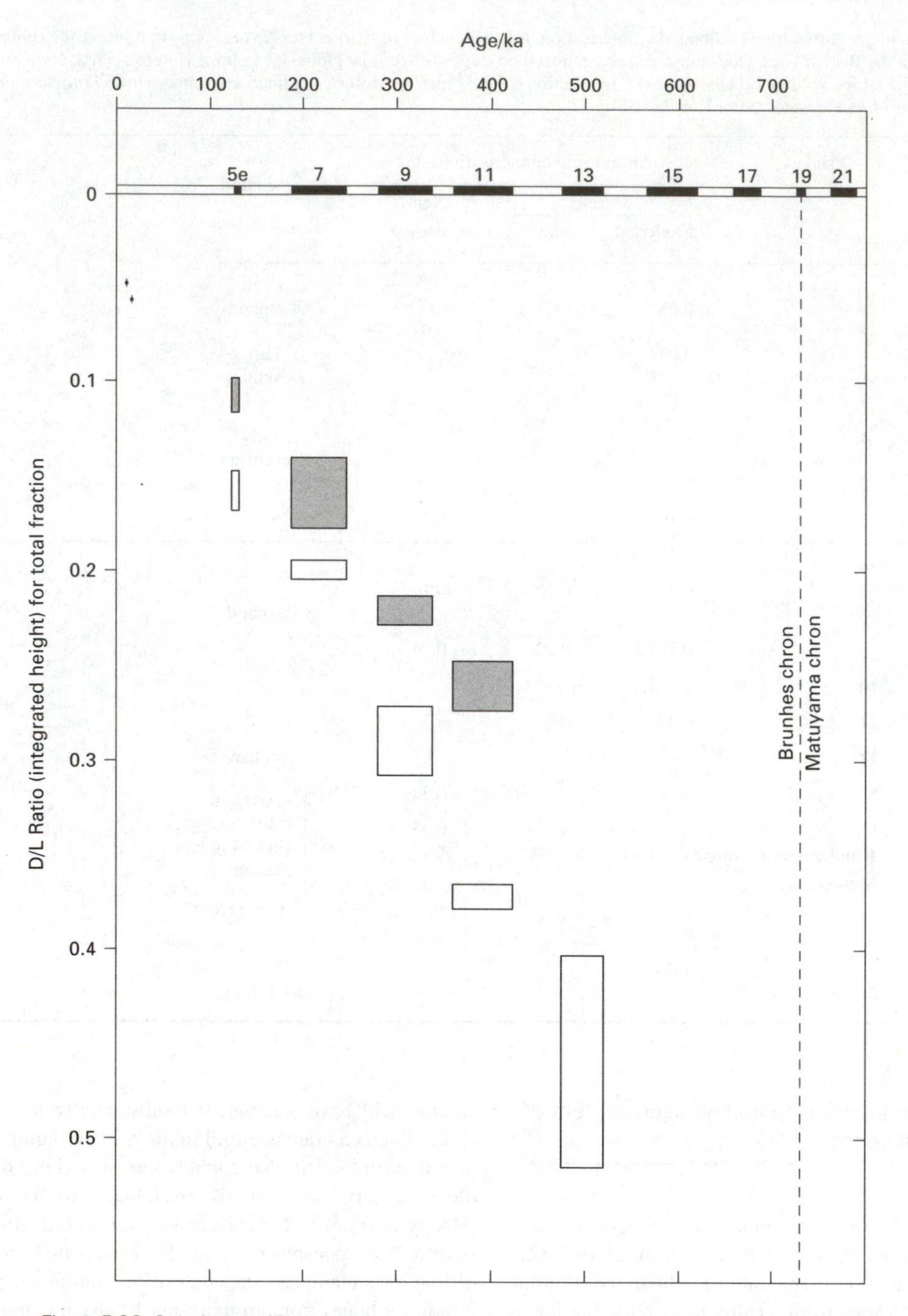

Figure 5.36 Correlation of D-alloisoleucine/L-isoleucine high sea-level events at sites in northwest Europe with oxygen isotope stages on the SPECMAP timescale of Imbrie *et al.* (1984). The rectangles show one D/L standard deviation correlated with odd-numbered oxygen isotope stages. Shaded boxes show the *Littorina* D/L data; unshaded boxes the *Macoma–Arctica* D/L data (after Bowen & Sykes, 1988).

too will nitrogen. Hence decreasing nitrogen content will reflect increasing age (Oakley, 1980).

A further means of dating using fluorine content of bone involves the establishment of **fluorine profiles** through sections of bone using a nuclear microprobe. In older bones, the profile is relatively flat, whereas in younger samples, the concentration of fluorine falls off steeply with increasing distance into the bone. The shape of the profile is therefore age-dependent and offers the basis for dating. The technique can be used to establish relative age in mixed bone assemblages, as a screening process in the selection of samples for ^{14}C dating, and as a means of approximate dating of very recent materials, i.e. over the timespan of the last few hundred years (Aitken, 1990). However, these methods suffer from the same problems of non-uniform alteration and low levels of accuracy and reliability as other chemical methods considered in this section.

5.6.3 Obsidian hydration

Freshly exposed surfaces of obsidian (a form of volcanic glass) absorb water from their surroundings to form a hydration layer known as **perlite** (Trembour & Friedman, 1984). This external rind should not be confused with the patina that develops on many materials as a result of chemical weathering (see below). Hydration layers require detailed examination by high resolution optical microscopy, for the thickness of such layers is typically less than 20 μm. The precise thickness of the hydration layer reflects the length of time since exposure of an obsidian surface and the rate at which water has diffused into that surface. This, in turn, depends on temperature during the time of burial and on the chemical composition of the obsidian (primarily the percentage silica content).

Initially, variations in thickness of obsidian hydration layers were used to establish relative ages of, for example, rhyolite flows (Friedman, 1968), archaeological artefacts (Michels & Bebrich, 1971) and glacial events (Pierce *et al.*, 1976), with more sophisticated chronologies depending on calibration using samples of known age. More recently, however, experimental work has shown that ages can be derived directly from obsidian hydration measurements, providing that the effective burial temperature and the rate of hydration at that temperature can be established (Stevenson *et al.*, 1989b). Using this approach, ages ranging from 200 years to 100 ka BP have been obtained (Aitken, 1990), a number of which have been found to be compatible with those derived from other chronologies (Lynch & Stevenson, 1992; Ambrose, 1994). However, experimental imprecision in measuring the hydration layer constrains recent ages, while uncertainty over the quantification of effective hydration temperature (EHT) is a major factor currently limiting both the accuracy and precision of older dates (Aitken, 1990).

5.6.4 Weathering characteristics of rock surfaces

A range of rock-weathering parameters has been employed to assign ages to rock surfaces (Brookes, 1982). Degree of weathering of stones or boulders, which is a reflection of time, is indicated by in a range of characteristics including the relative decomposition or exfoliation of boulders (Figure 5.37), lack of soluble materials (e.g. limestones) on older surfaces, the crumbly nature of sandstone or volcanic clasts on older drifts, and the the relative concentration of more durable materials (e.g. quartz, chert, siliceous rocks) at the surface. The extent of weathering may also be established by recording the sound produced by boulders being struck with a hammer. Fresh boulders produce a sharp ring and a strong hammer rebound whereas weathered stones emit a dull sound and a weaker recoil (Mahaney *et al.*, 1984). More quantitative estimates may be obtained from instruments such as a **Schmidt hammer**, where rock hardness is measured by the distance of rebound of a spring-loaded mass (McCarroll, 1989), or by using a **microseismic timer** to determine the compressional wave (P-wave) velocity through stones or boulders, the velocity of the P-wave being determined by the soundness of the rock material (Crook, 1986). The thickness of the **weathering rind**, the outer layers of boulders or stones which have been oxidised and discoloured by iron-bearing minerals, can also be used to establish relative order of antiquity of rock surfaces (Mahaney *et al.*, 1984). An additional approach, known as **cation-ratio dating**, involves establishing the relative stability of chemical constituents of **rock varnish**, the coating of clay minerals, manganese, iron oxides, etc., that forms on rock surfaces (Dorn & Oberlander, 1982). Certain bases, notably potassium and calcium, are easily mobilised whereas others, such as titanium, are more stable. Hence the ratio of K + Ca to Ti in rock varnish decreases with time and provides an indication of age (Dorn, 1983, 1989). Moreover, rock varnish also contains minute quantities of organic material, and where sufficient material can be obtained for AMS ^{14}C dating, the relative chronologies obtained from cation-ratio dating can be calibrated to a radiometric timescale (Dorn *et al.*, 1987).

Variations in rock weathering characteristics have been most widely used in the establishment of Late Quaternary glacial chronologies (Figure 5.38) where moraines and other glacial features have been placed in relative order of age on the basis of, *inter alia*, degree of weathering of contained

Figure 5.37 Deeply weathered boulder exposed in Late Wisconsinan moraine, Allens Park, Colorado, USA (photo: John Lowe).

rocks and boulders (Colman & Pierce, 1986; Shiraiwa & Watanabe, 1991). Rock surface weathering variations have also been used to delimit the upper level of glacial erosion on mountain summits (the trimline: section 2.3.1), from which the former vertical extent of ice sheets can be reconstructed (McCarroll & Nesje, 1993; Nesje *et al.*, 1994). However, this type of relative dating is not without its problems. Qualitative estimates of weathering tend to be subjective and reproducibility of results between workers has often been poor (Burke & Birkeland, 1979). Instrumental and experimental difficulties also continue to pose problems in the derivation of quantitative weathering measurements (McCarroll, 1987). Rates of weathering are not linear, but tend to decrease with age (Colman, 1982), while degree of weathering varies according to local conditions (altitude, climate, aspect, etc.), bedrock composition in particular being a decisive factor (Karlén, 1988). The incorporation of material from older weathering régimes into younger glacial contexts is a further problem in the interpretation of rock weathering data (McCarroll, 1990). Particular criticism has been directed at cation-ratio dating, where questions have been raised about the precise mechanisms involved in cation-ratio variations (Reneau & Raymond, 1991), about the nature of the chemical analyses involved in the measurement of cation-ratios (Bierman & Gillespie, 1991), and about the long-term stability of micrometre-thick varnished rock surfaces (Bierman & Gillespie, 1994). These and other uncertainties suggest that rock weathering characteristics are perhaps best employed in conjunction with other dating techniques, such as lichenometry, ^{14}C dating and tephrochronology.

5.6.5 Pedogenesis

Degree of pedogenic development has been used as a basis for relative chronologies of glacial events and also as a means of correlation (section 6.3.2.3). This approach rests on the concept of a **soil chronosequence**, i.e. a series of related soils developed when all factors of soil formation except time (climate, organisms, parent material and topography) are held more or less constant. Hence, contrasts between soil profiles, such as grain size variations, physical and chemical properties, micromorphology and depth of soil development, can be interpreted as a function of time, and therefore provide a basis for relative dating (Birkeland, 1984). Although qualitative assessments of soil development have proved useful in the establishment of local glacial chronologies (e.g. Mahaney *et al.*, 1981), a more quantitative methodology has been adopted in more recent research. Examples include **Soil Development Indices** which characterise soil properties (structure, rubification, clay content, moisture consistency, etc.) and enable statistical comparison to be made between different soil profiles (Harden & Taylor, 1983). Such indices can then be employed in **soil chronofunctions** in which changes in soil properties are modelled as functions of time (Birkeland, 1984; Rodbell, 1990). These provide a quantitative framework for the construction of timescales of soil development.

Pedogenic contrasts as a basis for relative chronologies have been most widely employed in North America. In many formerly glaciated areas of the Western Cordillera, for example, systematic changes in morphology and soil development reflect increasing age of parent material. These include the sequence and thickness of genetic horizons; increase in clay content of the B horizon; depth of oxidation of the B and C horizons; and clay mineral alteration (Shroba & Birkeland, 1983). A clear trend is often apparent in pedogenesis with older soils showing evidence of, for example, deeper profiles, thicker horizons and increased clay accumulation (Figure 5.39). Using this type of evidence, it has been possible to assign relative ages to moraine sequences extending back 140 000 years (Birkeland, 1985; Colman & Pierce, 1986). A recent development has been the analysis of **soil catena chronosequences**, i.e. the evolution of soils at different positions on slopes of moraines of different ages, in order to provide a more secure basis for pedologically derived timescales of glacier activity (Birkeland & Burke, 1988). Elsewhere, contrasts in soil development on morainic substrates have been used to establish Late Pleistocene and Holocene glacial chronologies in Scandinavia (Mellor, 1985), New Zealand (Rodbell, 1990) and the Himalayas (Shiraiwa & Watanabe, 1991).

The major difficulty in using degree of pedogenesis as a basis for chronology lies in the requirement that all soil-forming factors, apart from time, must be held constant, for numerous studies have shown that factors such as topography and vegetation are frequently as important in controlling soil genesis (Reider, 1983). Rates of pedogenesis also appear to be strongly influenced by regional climatic parameters (Birkeland *et al.*, 1989) and this imposes further constraints on models of soil evolution where time is considered to be the major independent variable. An additional problem, which particularly affects older (pre-Holocene) glacial substrates, is that episodic erosion and redeposition can result in soils that may be an order of magnitude younger than underlying tills (Rodbell, 1990). As with weathering characteristics of rock surfaces, therefore, it

Figure 5.38 The Mueller Glacier, Mt Cook, New Zealand, showing ages of moraine suites determined by relative dating methods, including weathering characteristics of bedrock surfaces (photo: Chris Bradley; after Gellatly *et al.*, 1988).

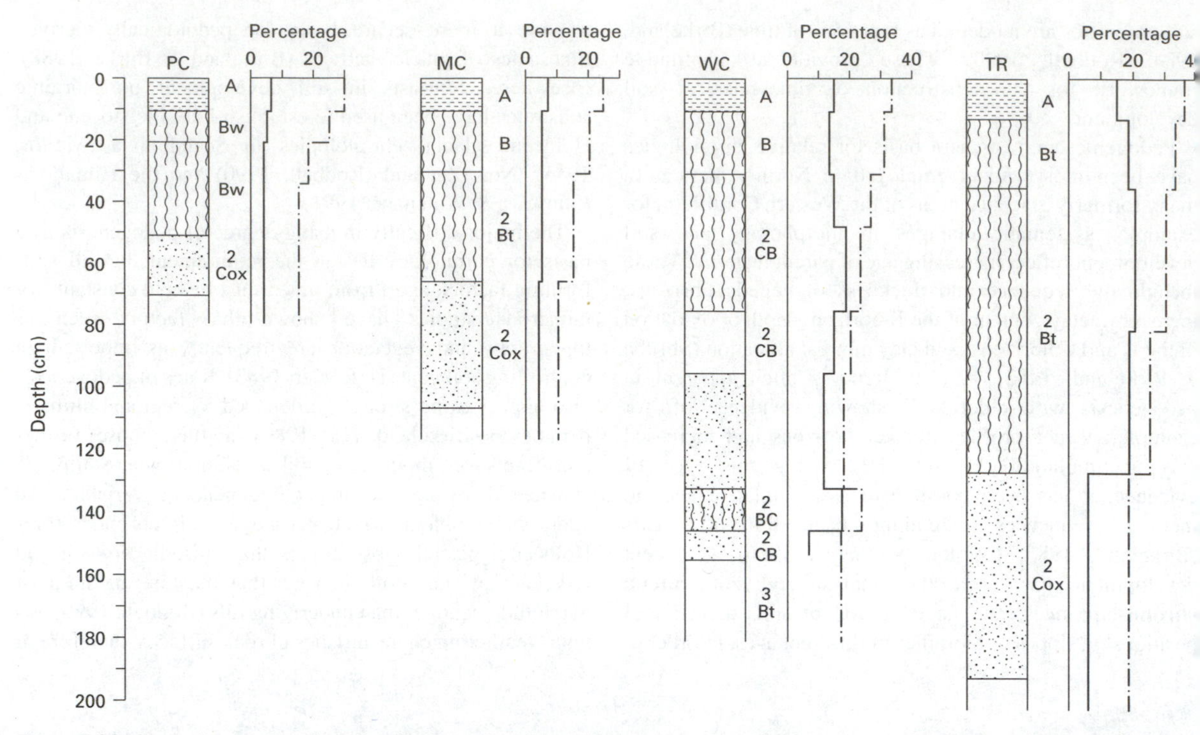

Figure 5.39 Generalised soil profiles developed in till of four different ages near McCall, Idaho, northwest USA. Labels to the right of each profile are soil horizon designations (Bt: horizon of pedogenic clay accumulation; Bw: rubified horizon; Cox: oxidised horizon). Grain size profiles are for clay (solid line) and silt (dashed line). Density of the symbol for B horizons indicates the degree of soil structure development. PC. Pilgrim Cove soil (*c.* 14 ka BP); MC, McCall soil (*c.* 20 ka BP); WC, Williams Creek soil (*c.* 60 ka BP); TR, Timber Ridge soil (150–140 ka BP) (after Colman & Pierce, 1986).

seems advisable to employ pedogenic development in an auxiliary, as opposed to a primary, rôle in the establishment of relative chronologies.

5.7 Conclusions

Although techniques that establish age equivalence on the basis of stratigraphic markers, and methods that determinate the relative order of antiquity of rocks or fossils, are being increasingly used in Quaternary research, the most important methods for establishing the age of Quaternary events are likely to remain the radiometric and incremental techniques. Because they allow events to be dated in years, their application has often been referred to as 'absolute dating'. However, this term has not been used here as it implies a level of exactitude that is rarely attainable at present. Nor have we used 'geochronometric', which has frequently been employed to describe dating techniques from all of the categories described above, including methods which, at best, only provide relative age relationships at a very broad scale. Again, a degree of precision is implied by the use of this term which is often not reflected in the results produced by many of the dating methods.

The major problem now in Quaternary dating would appear to be how to achieve a greater degree of reliability in age estimates based on methods that have, in a number of cases, somewhat uncertain foundations. Progress in this respect depends on replication of results from a given method, but also, perhaps more importantly, on the application of more than one technique to the dating of a specific fossil assemblage, body of material, stratigraphic horizon or event. Although the methods reviewed in this chapter have gone some way towards providing a realistic dating framework, particularly for the Late Quaternary, a major objective in future research must be the development of even more sophisticated dating techniques, thereby providing greater scope for the calibration of existing methods. Bearing these points in mind, we now turn to the wider question of stratigraphic subdivision and correlation of the Quaternary record.

Notes

1. The **principle of superposition** states that, in the absence of evidence for disturbance or reworking, the overlying sediments in a sequence are younger than those lying beneath them. This applies both to lithological units, such as tills, solifluction deposits, etc., and to biostratigraphic units, such as pollen or molluscan assemblage zones. In this way, pollen analysis, for example, can be employed as a relative dating technique.
2. Since this chapter was completed, a Third International Intercomparison (TIRI) has been completed, the preliminary results of which were reported to the 15th International Radiocarbon Conference in Glasgow, UK, in August 1994 (Gulliksen & Scott, 1995). The full results of TIRI will be available shortly.
3. For example, protein residues have been found in Ordovician and Devonian shells, and collagen-like proteins have been recovered from Cretaceous dinosaur bones (Wyckoff, 1980).

CHAPTER 6 Approaches to Quaternary stratigraphy and correlation

6.1 Introduction

In previous chapters the various methods employed in palaeoenvironmental reconstruction have been examined and the techniques used in the dating of Quaternary events have been discussed. Of particular interest to the Quaternary scientist, however, is the way in which environments have changed through time and how such information can be gained from the stratigraphic record. There are two aspects to the interpretation of that record, namely the ordering of the evidence at any one locality into a time sequence (*temporal dimension*) and the relation of the evidence at one locality to that at another (*spatial dimension*). The temporal dimension involves principles of **stratigraphy**, while the spatial dimension involves principles of **correlation**. A proper understanding of the procedures involved in these two aspects of geological investigation is fundamental to a correct interpretation of Quaternary environmental change.

If the stratigraphic record were complete at all places on the earth's surface, and if every important horizon could be dated accurately, then units of the record could be arranged into stratigraphic order based on increments of time. In this way a **time-stratigraphic** framework for environmental change could be developed, and correlation between even the most widely separated localities would present few problems. In practice, however, this is not possible. The terrestrial record is far from complete, and dating control cannot always be established, due either to the absence of suitable dating materials, or to limitations of the actual techniques employed (Chapter 5). Only in the deep oceans, and in certain exceptional terrestrial situations, such as deep lakes or regions with thick loess deposits, are long stratigraphic records preserved. Even these, however, may allow only a generalised dating framework to be created. For the most part, therefore, the Quaternary stratigrapher is confronted by a highly fragmented and partial stratigraphic record which can be dated securely only in certain circumstances and over limited time ranges. As a consequence, very careful palaeoenvironmental interpretation and stratigraphic evaluation are required before sequences can be ordered at one place and correctly related to those at another.

In this chapter the bases of, and the procedures involved in, Quaternary stratigraphy and correlation are examined, and the means whereby time-stratigraphic correlation can be achieved are assessed. This discussion relates essentially to the terrestrial record, for it is the terrestrial evidence that provides the greatest obstacles to the application of conventional stratigraphic and correlation procedures. Deep-ocean sediments are discussed in a separate section. The chapter concludes with an examination of the bases for correlation between the marine and terrestrial records.

6.2 Stratigraphic subdivision

6.2.1 Principles of Quaternary stratigraphy

Stratigraphy is the study of the chronological order of rocks and sediments, and also of the sequence of events reflected within them (Doyle *et al.*, 1994). The fundamental building blocks are units of geological strata that can be recognised on the basis of visible attributes, such as colour, grain-size variations or structural elements (bedding, deformation structures, etc.). These units constitute the basic elements of **lithostratigraphy**, which involves the ranking or ordering of local rock or sediment successions according to observable variations in lithology. Increasingly, however, *instrumental* techniques are being employed in the classification and subsequent analysis of geological units (Whittaker *et al.*, 1991). These enable sequences of rock or sediment to be described in terms of **magnetostratigraphy** (stratigraphic subdivision based on magnetic properties), **seismic stratigraphy** (using seismic properties such as resistivity measurements), **chemical stratigraphy** (variations in chemical properties of rocks and sediments), **geophysical data logs** and so on. In addition, there are methods that have particular applications in Quaternary research, such as **oxygen isotope stratigraphy** (a form of chemical stratigraphy) and **electrical resistivity measurements** on ice cores (a geophysical signal). Where clearly defined individual 'events' are recorded in the stratigraphic record, such as magnetic reversals or volcanic eruptions (evident in tephra layers), these can form the basis for an **event stratigraphy**. As the event horizons are essentially time-parallel (**isochronous**), they offer a means of correlation at a range of spatial scales (see below).

Modern stratigraphical investigations normally involve a range of such descriptive methods, and geological successions are usually subdivided after integration of all of the available evidence. In most situations, however, it is still the visible features that are of primary importance in the formal subdivision of stratigraphic sequences: the parameters derived from instrumentation, such as geophysical or chemical data, provide supplementary information only. In some important cases, however, visible lithological variations may be absent or imperceptible, while important stratigraphic changes, which may be crucial in subsequent palaeoenvironmental reconstructions, can be detected only by instrumental techniques. The isotopic and trace gas signals in both deep-ocean sediments and ice cores (sections 3.10 and 3.11) are important examples of such instrumentally derived stratigraphic records.

There are a number of other important ways in which stratigraphic units can be subdivided and classified, however. **Biostratigraphy** involves the classification of sediment units according to observable variations in fossil content, and enables a sediment sequence to be divided into **biostratigraphic units** or **biozones**, each characterised by a distinctive fossil assemblage. In Quaternary stratigraphy, landforms can be classified according to their relative order of age (**morphostratigraphy**), with each landform or landform suite constituting a distinct **morphostratigraphic unit**. **Climatostratigraphy** is concerned with the division of a stratigraphic sequence into **geological–climatic units** on the basis of inferred changes in climate, while **chronostratigraphy** involves the classification of stratigraphic units according to inferred age, the interval of time during which a geological unit has developed being referred to as a **geochronological unit**. Climatostratigraphy and chronostratigraphy are essentially *inferential* methods of stratigraphic subdivision, whereas both lithostratigraphy and biostratigraphy are more direct, being based on either observable or measurable properties of the sediment record.

Codes of practice have been produced to aid the geologist in the task of stratigraphic subdivision and classification (e.g. Hedberg, 1976; Holland *et al.*, 1978; North American Commission on Stratigraphic Nomenclature, 1983; Whittaker *et al.*, 1991; Salvador, 1994). These aim to provide a clear and unambiguous terminological framework, as well as guiding principles on stratigraphic procedures. All of these codes, however, have been developed for use in pre-Quaternary geology, and have major limitations when applied to Quaternary stratigraphy. This is because Quaternary sediment sequences can be examined in far greater detail and at a much higher level of temporal resolution than their pre-Quaternary counterparts. The result is that the Quaternary scientist has to deal with sedimentary sequences of unusual geological complexity, and this gives rise to problems of classification, interpretation and correlation that are not encountered in investigations of the earlier geological record. For example, boundaries between lithological units are placed at positions of lithological change (see below), but in many Quaternary sequences these cut across the limits of fossil ranges (reflected in biozones) and the boundaries of other types of stratigraphic unit (Figure 6.1). Moreover, they frequently cut across time-horizons (**time-transgression**) which poses particular problems in the correlation of stratigraphic sequences at the scale of resolution required in Quaternary research (see below).

6.2.2 Stratotypes

According to conventional geological procedures, a locality where a particular stratigraphic unit is clearly and fully recorded, or where its lower boundary is securely defined, is termed a **type site** or **stratotype**. This can then be used as a reference standard against which other sections, where the equivalent unit or its boundaries are only partially or poorly

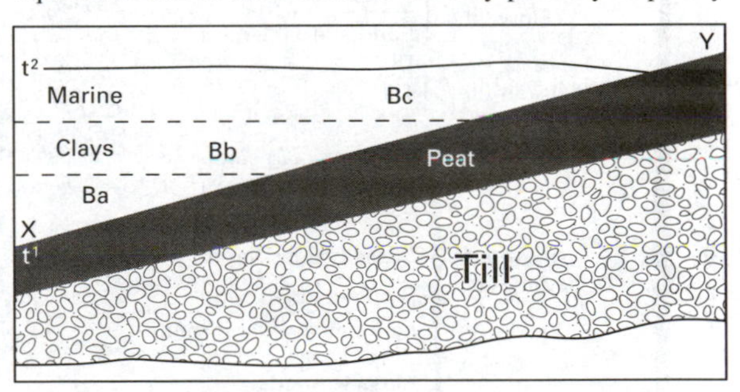

Figure 6.1 Time-transgression in lithostratigraphic and biostratigraphic boundaries. A till unit is overlain by a peat layer which has, in turn, been buried by marine clays. The marine deposits accumulated during a gradual marine transgression (t^1 to t^2) and therefore the lithostratigraphic boundary between peat and the overlying marine sediments (X–Y) will be time-transgressive. During the deposition of the clays, changes in fossil content have occurred represented by biozones Ba, Bb and Bc. The lithostratigraphic boundary X–Y therefore cuts across the biozone boundaries.

represented, can be compared. The type section should be accessible and durable, so that it may be available for further study, and perhaps re-assessment where necessary.

Ideally, all stratigraphic units should be defined with reference to a stratotype. However, this is often difficult in Quaternary stratigraphy because of the highly fragmented nature of the terrestrial stratigraphic record, the very considerable spatial variation in type and thickness of Quaternary sediments, the limited lateral continuity of many Quaternary deposits, the spatial and temporal contrasts in Quaternary environments as reflected in the biostratigraphic record and, above all, the markedly time-transgressive nature of terrestrial stratigraphic boundaries (Figure 6.1). In many situations, therefore, a stratotype will have little more than local application and, as a consequence, the concept of the stratotype has been less widely adopted by Quaternary workers than by geologists dealing with older segments of the stratigraphic record. However, in the interests of clarity and precision, a strong case can be made for the more widespread adoption of Quaternary stratotypes than has been the practice in the past (West, 1989). Certainly, if effective communication between scientists is to be achieved with respect to the subdivision of the Quaternary stratigraphic record, the establishment of regional stratotypes where the lithostratigraphic and biostratigraphic units constitute clear reference standards is an integral component of that process (Rose, 1989).

6.2.3 Elements of Quaternary stratigraphy

6.2.3.1 Lithostratigraphy

In theory, lithostratigraphic units should be recognised and defined on the basis of sediment properties alone, such as colour (using Munsell and other colour charts), particle shape and size, and grain-size variations. In practice, however, Quaternary scientists have also tended to classify sedimentary units on the basis of inferred mode of origin (till, aeolian sand, glaciofluvial gravel, etc.). The problem with such genetically based classifications is that if the mode of origin is incorrectly inferred, then the subsequent stratigraphic reconstruction may be undermined. Hence, many Quaternary geologists now prefer the use of sedimentological terms which are free from such genetic connotations. One such term is **diamicton**, which simply refers to any sedimentary unit with a heterogeneous mix of particle sizes from clay through to boulders. The mode of genesis of a diamicton is inferred at a later stage in the analysis on the basis of such diagnostic criteria as fabric data, details of clast size and shape and micromorphological properties. When confronted by a complex stratigraphic sequence (e.g. Figure 6.2), therefore, the approach should be first to identify **beds**, which are individual sediment units or bodies that are considered to have originated during the same depositional event. For example, careful examination

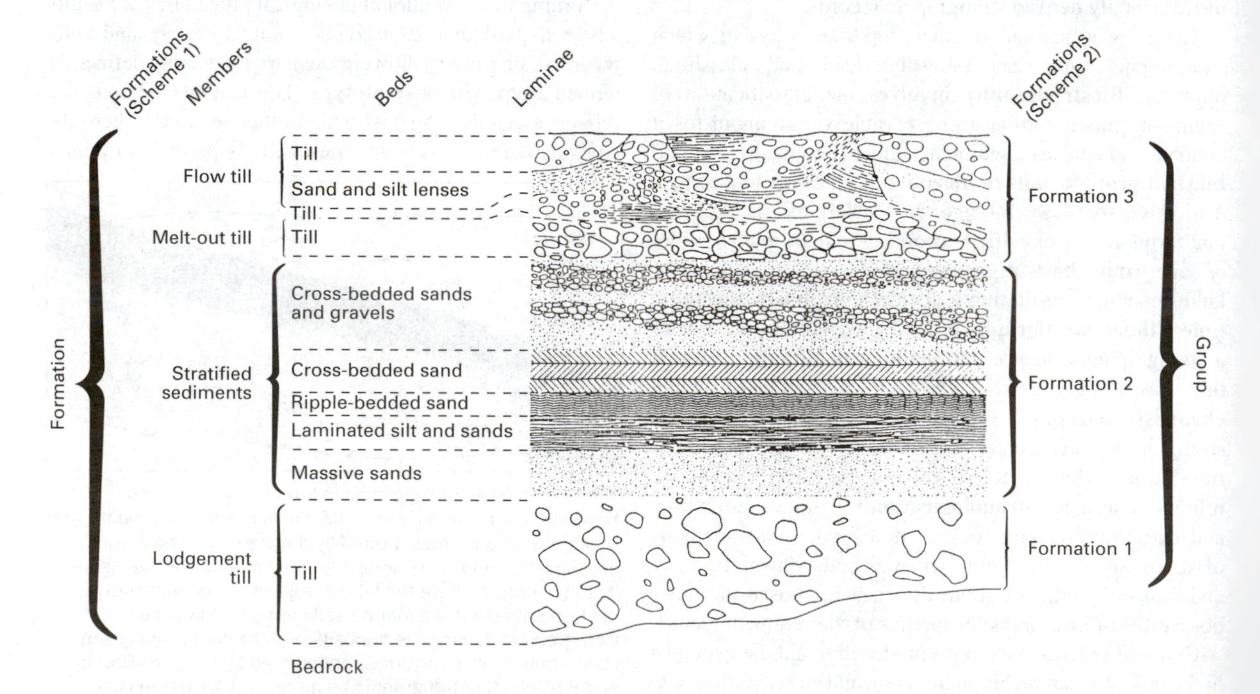

Figure 6.2 Lithostratigraphic subdivision of a glacigenic sequence.

of a sand unit may reveal a series of large- or small-scale cross-bedded layers, each reflecting changes in the environment of deposition. Similarly, alternations between sand and gravel beds, layers of clay-dominated beds, and diamicton units (which frequently form individual beds), may indicate important variations in mode of deposition. In some instances, even finer subdivisions of the stratigraphy may be possible where, for example, very thin layers of perhaps 1 cm or less (**laminae**) occur throughout a bed, each reflecting a very short-lived sedimentary episode. Once such individual units have been identified, the second stage is to establish mode of genesis. In other words *description* and *classification* should always precede *interpretation* of the stratigraphic record.

Where several beds may be shown to be related, in that they have accumulated through similar depositional processes, they constitute larger lithostratigraphic units known as **members**. Hence, the stratified sediments in Figure 6.2 are grouped together to form a member which can be differentiated lithologically from the overlying and underlying unstratified diamicton (till) members. Members that accumulated sequentially during a major depositional event (e.g. a glacial episode, or a marine transgression) form lithostratigraphic units of higher rank termed **formations**. Two or more contiguous formations which possess common lithological properties (particle size characteristics, clast litholologies, etc.) may be aggregated into **groups**. Distinctive lithostratigraphic units that appear regularly in sediment records, and provide a basis for correlation between sequences (such as tephra layers), are referred to as **markers**. Units of formation or group rank, displaying common properties or associations of members and beds, may also constitute markers. In such cases, these higher ranking lithostratigraphic units are given formal status and may be accorded proper names. In southeast England, for example, extensive spreads of glaciofluvial gravels dating from the Anglian Glacial Stage have been termed the 'Barham *Formation*'. A more extensive suite of river sediments that formed over 1.2 Ma prior to the Anglian Glaciation, the 'Kesgrave Sands and Gravels', accumulated over many warm and cold stages. This lithostratigraphic unit, which contains two major formations (Colchester and Sudbury) and which can be traced over large areas of the River Thames basin and East Anglia, is referred to as the 'Kesgrave *Group*' (Bridgland, 1988; Whiteman & Rose, 1992). Further examples of the ways in which these stratigraphic procedures have been applied in Quaternary research can be found in Gibbard (1985, 1994), Rose & Schlüchter (1989), Rose *et al.* (1985), McCabe & Dardis (1989) and Walker *et al.* (1992).

A recent approach to stratigraphic subdivision stems from detailed studies that have been undertaken on contemporary depositional environments. These have shown that different sedimentary contexts (estuarine, glacial, fluvial, aeolian, etc.) give rise to distinctive associations of sedimentary units or **facies** (Miall, 1984; Walker, 1984; section 3.3). Key diagnostic properties (colour, grain-size variations, geometry, bedding, external contacts, etc.) can be established for each facies, and these data can then be used to define **lithofacies** in rock or sediment sequences, i.e. ancient equivalents of modern facies types. Facies may be subdivided into **subfacies** or grouped together into **facies associations**, while at the regional scale they can be considered three-dimensionally in terms of **facies architecture** (Miall, 1984). The facies concept has been widely used in pre-Quaternary geology, but its application in Quaternary stratigraphy is a relatively recent innovation and, to date, it has been most frequently employed in the analysis of glacigenic sequences (e.g. Eyles *et al.*, 1983; Eyles & Miall, 1984; McCabe & Dardis, 1989; section 3.3). As in conventional lithostratigraphic procedures, however, it is important to begin with a descriptive facies analysis, using the sort of criteria shown in Table 6.1. Non-genetic terms such as diamict should be applied, and a shorthand notation (e.g. Gm for massive gravels; Sh for horizontally laminated sands) should be used to describe each lithofacies. Interpretation of the sediment record in terms of mode of genesis then constitutes the second stage of the analysis (Hambrey, 1994).

The level of lithostratigraphic subdivision depends on the local complexity of the record, and on what are frequently subjective decisions as to the nature or rank of a particular stratigraphic unit. Also, the interpretation of the Quaternary depositional record is often complicated by the fact that apparently similar sediments and sedimentary sequences can result from different geological processes. Fluvial, glaciofluvial and aeolian sediments, for example, can appear lithologically similar despite the fact that they have accumulated in markedly different depositional environments. Many Quaternary sedimentary sequences (such as fluvial deposits) show considerable lateral and vertical variation, and these local facies changes frequently pose problems of interpretation and classification using traditional lithostratigraphic criteria (e.g. Maddy & Green, 1989; Whiteman & Kemp, 1990). Even where the origin of the sediments can be established, the subdivision of very complex sequences into lithostratigraphic units is not always straightforward. In Figure 6.2, for example, the beds of stratified sediments can be grouped together to form a member which can be differentiated lithologically from the unstratified diamictons above and below. Analysis shows the sediments to have been derived from episodes of glacial and glaciofluvial activity during a single, glacial event, and hence the whole sequence can be classified as a formation

Table 6.1 Principal descriptive criteria used in defining lithofacies in glacigenic sequences (after Hambrey, 1994).

Lithology	Bedding characteristics	Bedding geometry	Sedimentary structures	Boundary relations
Diamict(on/ite)	Massive	Sheet	Grading: normal	Sharp
Gravel	Weakly stratified	Discontinuous	reverse	Gradational
Sand(stone)	Well stratified	Lensoid	coarse-tail	Disconformable
Mud(stone)	Laminated	Draped	Cross-bedding: tabular	Unconformable
	Rhythmic lamination	Prograding	trough	
	Wispy stratification		Lonestones (dropstones)	
	Inclined stratification		Clast supported	
			Matrix supported	
			Clast concentrations:	
			layers	
			pockets	
			Ripples	
			Scours	
			Load structures	
			Mottling (= bioturbation)	

(scheme 1). Alternatively, the upper and lower diamictons and the intervening stratified unit might be regarded as representing discrete depositional events and under this interpretation each could be classified as units of formation rank (scheme 2). Consistency of interpretation between individual workers may, therefore, be difficult to achieve. Nevertheless, insofar as it emphasises the need for careful analysis and interpretation, this approach still offers the only adequate means of subdividing the Quaternary rock-stratigraphic record.

6.2.3.2 Biostratigraphy

Biostratigraphic classification organises rock strata into units based on the variety and abundance of fossils. Biostratigraphic units are usually termed **biozones** (Whittaker *et al.*, 1991) and the following types are commonly employed in stratigraphic classification (Figure 6.3):

1. **Total range biozone** A group of strata containing the full stratigraphic and geographical range of a particular fossil or group of fossils.
2. **Acme biozone** A group of strata based on the acme or maximum development of a particular taxon.
3. **Partial range biozone** The stratigraphic interval described by that part of the range of a particular taxon which lies above that of a second taxon and below that of a third. Hence in Figure 6.3, the PBR of taxon b is that part of the range of taxon b which lies above the range of taxon a and below that of taxon c.
4. **Concurrent range biozone** Defined on the basis of the overlapping ranges of several taxa.
5. **Consecutive range biozone** Where speciation changes can be clearly established (**phylogenic lineage**), biozones can be defined on the basis of the consecutive ranges of fossils in an evolving lineage.
6. **Assemblage biozone** Biozones defined on the basis of a characteristic mix and relative abundance of fossil types.

In pre-Quaternary geology, the biostratigraphic record is subdivided largely on the basis of evolutionary changes in organisms, and thus acme biozones and range biozones reflect those episodes of geological time where a species appeared, thrived for a while, and then died away. Acme and range biozones constitute valuable stratigraphic markers, and often form the basis for correlation. Although there is some evidence of evolutionary development in the Quaternary record, for example within the rodents (Gamble, 1994) and beetles (Elias, 1994), evolutionary changes are not common over much of the Quaternary biological record, and thus acme and range biozones are less widely used than in pre-Quaternary geology. Quaternary biostratigraphy is, therefore, based almost entirely on **assemblage biozones** and these largely reflect the ecological response of organisms to environmental change rather than evolutionary changes in flora and fauna. Unlike evolutionary trends, however, ecological changes are both reversible and repeatable. Hence Quaternary assemblage biozones of markedly different ages may contain essentially the same mix of fossils, in which case they are potentially flawed as stratigraphic tools. This situation differs from that in pre-Quaternary strata where evolutionary changes over much longer time periods have resulted in assemblage biozones that can be unique.

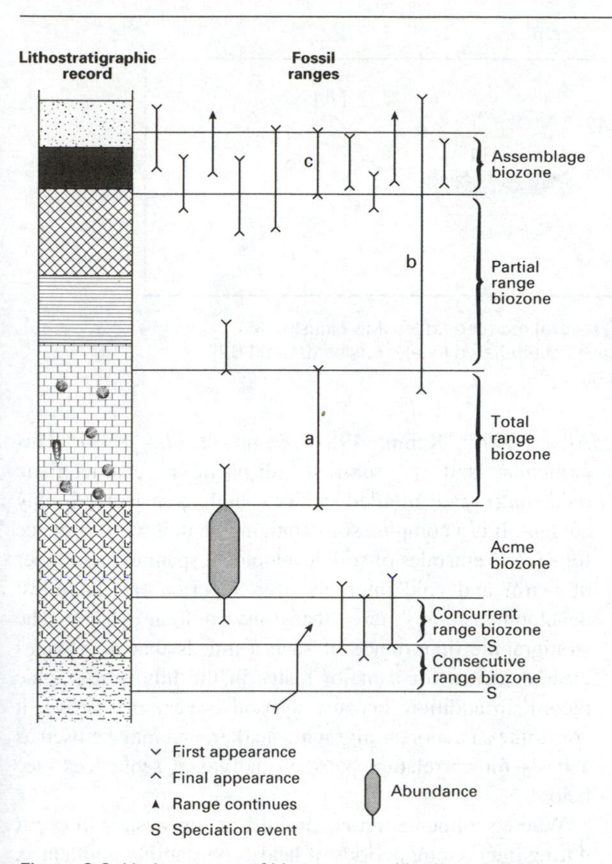

Figure 6.3 Various types of biozones used in the subdivision and correlation of strata (based on Holland *et al.*, 1978, p. 14) The **consecutive** and **concurrent ranges** are defined on the basis of the appearance and disappearance of taxa, normally in the context of evolutionary lineages. The **acme biozone** is defined on the basis of maximum representation of one or more taxa. The limits of the **total range biozone** are defined by the first appearance and disappearance of one or more taxa. The lower boundary of the **partial range biozone** is defined by the first appearance of one taxon, while the upper boundary is defined by the first appearance of other taxa. The **assemblage biozone** is defined on the basis of the characteristic mix and relative abundance of particular taxa.

Because of the time-transgressive nature of climatic and environmental change, the boundaries of biozones (e.g. pollen assemblage zones, molluscan assemblage zones, diatom assemblage zones) cut across time horizons and frequently transgress the boundaries of other stratigraphic units. In the deep oceans, however, where sedimentation rates are slow, assemblage biozone boundaries may appear to be time-parallel. In reality, however, all assemblage biozone boundaries examined at the resolution required in Quaternary science must be time-transgressive, since it takes time for organisms to adapt to changes in environmental conditions. Whether the degree of time-transgression can be ascertained depends on the nature of the changes and, as in all biostratigraphical investigations, on the temporal resolution of the stratigraphic record. Difficulties may also be encountered in the establishment of Quaternary biozones as a result of the reworking and selective destruction of fossils, and the problems of identification and ecological interpretation of fossil assemblages (see Chapter 4).

6.2.3.3 Morphostratigraphy

Morphostratigraphy is rarely discussed in stratigraphic codes, yet is an essential stratigraphic method in Quaternary science. A morphostratigraphic unit has been defined as 'a body of rock that is identified primarily from the surface form it displays' (Willman and Frye, 1970). This definition was coined for use by geological surveyors unable to unravel the complexities of glacial stratigraphy on the basis of lithology alone. In Quaternary research, a range of morphological features can be recognised that possess distinctive forms and which are often diagnostic of specific geological processes. These include moraine ridges, displaced shoreline landforms or beaches, and fluvial and glaciofluvial terraces. A morphostratigraphic unit may therefore be defined as that part of the lithostratigraphic record represented by a specific morphological feature. Often a series of landforms reflect a temporal succession of phases of formation, such as river terrace sequences, flights of raised beaches, or sequences of moraine ridges. As a distinct category of stratigraphy, morphostratigraphic units are integral components of the Quaternary stratigraphic record.

An example of the role of morphostratigraphy and its relation to the lithostratigraphic record is shown in Figure 6.4. Here the landforms have been categorised on the basis of morphology and geomorphic context. Between the moraine ridges are lake basins (B1, B2) that contain limnic sediments and peats. In order to provide a full account of the stratigraphy sequence in this area, the morphological evidence must be integrated with the lithostratigraphic and biostratigraphic record. One interpretation is that the moraines formed sequentially during a period of glacier retreat, with M1 being the oldest and M4 the youngest. If this was so, the base of the lake sediments in B1 would post-date M1, but pre-date the formation of M2. Similarly, the base of the limnic deposits in B2 would post-date both M2 and M3. The complete stratigraphic record would therefore consist of morphostratigraphic unit (m.u.) M1 > (=older than) B1 > m.u. M2 > m.u. M3 > B2 > m.u. M4. Alternatively, all of the ridges might have formed in a very short period of time and hence the accumulation of lake sediments in both B1 and B2 post-dates the entire morphostratigraphic sequence. Only the application of other

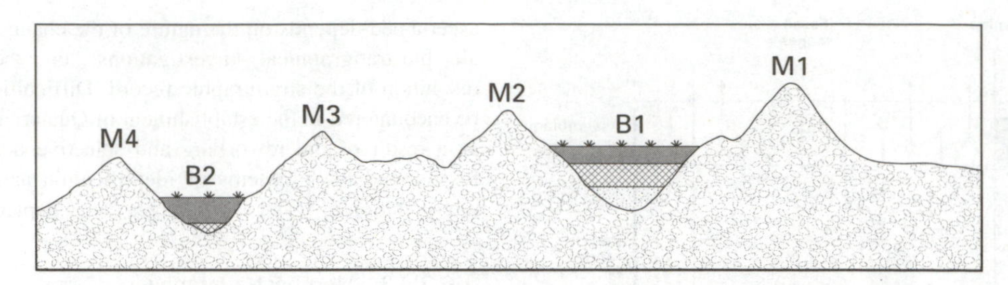

Figure 6.4 Schematic diagram showing, in profile, several moraine ridges (M1 oldest to M4 youngest) between which sediment successions have accumulated in lake basins (B1 and B2) following deglaciation. For further explanation see text.

methods (e.g. biostratigraphic evidence, radiometric dating) can resolve such an issue.

6.2.3.4 Soil stratigraphy

Palaeosols (section 3.5) occur in many Quaternary sediment sequences. Well-developed palaeosols that evolved during a specific soil-forming interval and which possess sufficiently distinctive characteristics to enable them to be traced over a wide area can be considered as **soil stratigraphic units** (Morrison, 1978). A soil stratigraphic unit was initially defined by the American Commission on Stratigraphic Nomenclature (1961) as 'a soil with physical features and stratigraphic relationships that permit its consistent recognition and mapping'. A better term than 'physical' might be 'pedological', since soils are now identified and categorised on the basis of a range of properties, including chemical, magnetic and micromorphological characteristics. The North American Commission on Stratigraphic Nomenclature (1983) subsequently introduced the term **pedostratigraphic unit** to describe 'a buried, three-dimensional body of rock that consists of one or more differentiated pedologic horizons' (p. 864), and has recommended that the term **geosol** replace the term soil in stratigraphic usage. Here, however, we retain the terms 'palaeosol' and 'soil-stratigraphic unit' to refer, respectively, to fossil soils and to the stratigraphic units that they represent, since these terms are still most widely employed in the Quaternary literature.

Soil-forming processes will commence in most areas immediately after surfaces become exposed to the atmosphere. The subsequent degree of pedogenesis will depend, however, on a range of site and climatic factors, as well as on regional history which includes length of exposure of the land surface. In many areas, the typical soil features of horizonation may not be present, while in other situations very distinct and clearly demarcated horizons will develop. Many soils are polygenetic, such as the Valley Farm Soil of East Anglia and southeast England (Rose & Allen, 1977; Kemp, 1987; Kemp *et al.*, 1993). This particular soil is rubified (displays a characteristic reddening) and mottled with a high translocated clay content. It is a complex soil-stratigraphic unit with evidence for several episodes of soil development spanning a number of warm and cold intervals over a period (at its fullest development) of more than one million years. The stratigraphic importance of such a unit is that it provides clear evidence for a major hiatus in the lithostratigraphic record. In addition, because the soil is very distinctive, it constitutes a major stratigraphic marker, and may be used as a basis for correlation between individual sequences (see below).

Where sedimentation is episodic, pedogenesis will occur during intervening periods of landscape stability and hence, over time, sequences of soils may form. This is most clearly demonstrated in the loess regions of the world where numerous soil-stratigraphic units are frequently found interbedded with suites of aeolian sediment (Kukla, 1987a; Kukla & An, 1989). In the Baoji loess section of the Chinese Loess Plateau, for example, up to 37 separate palaeosols (Figure 3.28) have been identified, the lowest of which has been dated to around 2.5 Ma BP (Rutter & Ding, 1993). The palaeosols have formed through a combination of carbonate eluviation and illuviation, clay translocation, pseudogleiziation and rubification. Most of them display Bt horizons which are considered to have developed under forest and/or steppe–forest environments. This sequence, like those elsewhere in China and in other loess regions, reflects a long history of dust (loess) deposition with intermittent episodes of stability during which vegetation became established and the soils formed. The soil-stratigraphic units within these long, quasi-continuous sequences offer the prospect of correlation between the different loess regions on the basis of magnetostratigraphy, marker layers and pedostratigraphic units (Rutter *et al.*, 1990; Ding *et al.* 1993), and also between loess accumulations and the deep-ocean sediment record (Rutter, 1992). These aspects will be examined in more detail later in the chapter.

6.2.3.5 Oxygen isotope stratigraphy

It has already been shown that, in the deep oceans of the world, long sequences of relatively undisturbed sediments are preserved that may extend back to the beginning of the Quaternary and, indeed, into the Tertiary. Within these sediments the microfauna and flora contain a record of changing oxygen isotope ratios that not only provide evidence for former glacial and interglacial oscillations (section 3.10) but also form the basis for stratigraphic subdivision and long- distance correlation.

Following the pioneering work of Emiliani (1955), isotopic profiles from deep-ocean cores have been divided into **isotopic stages** (Figure 6.5), with the parts of the curve interpreted as representing warm stages (lighter $\delta^{18}O$ values) being given odd numbers, and the cold stages (heavier $\delta^{18}O$ values) even numbers (section 3.10). Oxygen Isotope (OI) stage 1 represents the Holocene period, and higher numbers indicate successively older cold and warm stages. The allocation of stage numbers is therefore a 'count from the top' division of a sinusoidal curve, the underlying assumptions being (i) that inflections in the curve reflect ice sheet growth and decay with a minor variation caused by temperature variations (Mix & Ruddiman, 1985), and (ii) that no major interruption in sedimentation has occurred. In the majority of cases, the stage numbers refer to conventional glacial and interglacial episodes inferred from the terrestrial Quaternary stratigraphic record, although it should be noted that OI stage 3 is anomalous in that, although recognised as a 'warm' stage, it is regarded only as a period of interstadial status. In addition, the later 'interglacial' stages have been subdivided into separate warmer and colder episodes. OI stage 5 contains five substages, with 5a, 5c and 5e interpreted as warmer phases while 5b and 5d are believed to indicate colder intervals. The last (Eemian, Ipswichian, Sangamon) interglacial (as defined on land) is considered to be represented in the isotope record by substage 5e (Shackleton, 1977). Similarly, OI stage 7 is now generally subdivided into two warm intervals (7a and 7c), with substage 7b reflecting an intervening cooler episode (Ruddiman & McIntyre, 1982).

A subsequent development in oxygen isotope stratigraphy has been to give events within isotopic stages a decimal notation, so that negative (interglacial) and positive (glacial) excursions have odd and even numbers, respectively (Imbrie *et al.*, 1984; Martinson *et al.*, 1987). Hence, the negative events in OI stage 5 which were previously referred to as 5a, 5c and 5e are designated 5.1, 5.3 and 5.5; the positive events previously named 5b and 5d are designated 5.2 and 5.4. The boundary between OI stages 5 and 6 is now designated 6.0. The rationale underlying this scheme is that the exact stratigraphic level of a peak or trough in the isotope curve can be defined unambiguously in one curve and correlated precisely with the same level in another. The way in which this notation scheme has been applied to the last 300 ka of the isotope record is shown in Figure 5.32.

Emiliani (1955) initially recognised 16 OI stages in cores from the Atlantic and Pacific Oceans, and this was subsequently extended to 23 following the analysis of the now famous core V28-238[1] (Figure 6.5) from the Solomon Plateau in the western Pacific (Shackleton & Opdyke, 1973). Twenty-two of these stages can be widely recognised in the ocean sediment record spanning approximately the last 800 ka (Imbrie *et al.*, 1984) which implies that something of the order of ten interglacials (or near-interglacials) and ten glacial episodes have occurred during that time interval. The total number of isotopic stages formally identified in Quaternary deep-ocean cores now extends down to OI stage 116, dated to approximately 2.73 Ma (Patience & Kroon, 1991). These stages can be related to the palaeomagnetic timescale (Figure 5.26) with the Brunhes–Matuyama boundary located in OI stage 20, the reversal marking the onset of the Olduvai chron at the base of OI stage 63 (Figure 6.10 below) and the Gauss–Matuyama boundary in OI stage

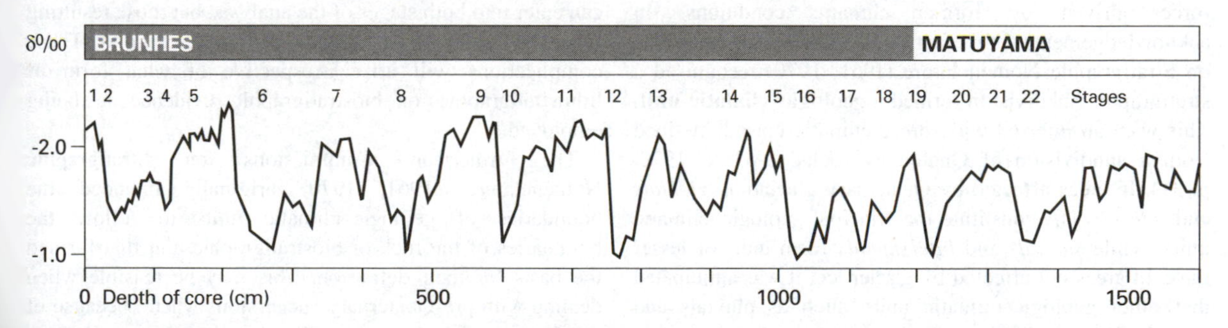

Figure 6.5 Oxygen isotope trace from deep-sea sediment core V28-238. The Brunhes–Matuyama geomagnetic boundary has been dated to *c.* 735 ka by K–Ar and *c.* 780 ka by astronomical tuning (Figure 5.26). Odd- and even-numbered isotopic stages represent interglacial and glacial episodes respectively (after Shackleton & Opdyke, 1973).

104 (Ruddiman *et al.*, 1986; Williams *et al.*, 1988; Shackleton *et al.*, 1990). What is particularly remarkable about the deep-sea isotopic signal is that the pattern is geographically consistent, and can be replicated in cores from different oceanic areas (Figure 1.4), thus emphasising the fact that the oxygen isotope signal provides a proxy climatic record of global significance (see section 6.3).

As noted above (section 3.10.3), a distinctive feature of most deep-ocean oxygen isotope profiles is the 'saw-tooth' appearance of the curves (Figures 1.4 and 6.5), with the most rapid isotopic changes occurring at the end of glacial episodes (Figure 3.45). These seemingly rapid changes from inferred glacial to interglacial conditions have been referred to as **terminations** (van Donk, 1976; Broecker, 1984) and constitute major **events** (see above) in the isotope stratigraphy. As with the isotopic stages, these terminations have been numbered from the top down. Hence, Termination I refers to the sudden relative decrease in $\delta^{18}O$ of the world's oceans at the end of the last glacial stage (OI stage 2). As Isotopic stage 3 is not considered to be of full interglacial rank, however, Termination II is the transition between OI stages 6 and 5 and not between 2 and 3. With the development of high-resolution composite (stacked) isotopic profiles, the most recent termination has been subdivided into Termination Ia (dated mid-point 13 ^{14}C ka BP) and Termination Ib (dated mid-point 9.5 ^{14}C ka BP), and these now constitute key marker horizons for correlating the oceanic isotopic record at the end of the last cold stage (Mix & Ruddiman, 1985; Mix, 1987). The use of oxygen isotope stratigraphy for time-stratigraphic correlation is considered later in the chapter.

6.2.3.6 Climatostratigraphy

A universally employed basis for the subdivision of the Quaternary is climatic change, for the characteristics of the stratigraphic record that are distinctive, and therefore used in the identification of stratigraphic units, frequently reflect forces driven by former climatic conditions. In acknowledgement of this fact, the American Commissions on Stratigraphic Nomenclature (1961, 1970) recognised a stratigraphic subdivision termed a **geologic–climatic unit**. This was 'an inferred widespread climatic episode defined from a subdivision of Quaternary rocks' (ACSN, 1970, p. 31). In areas affected by Quaternary glaciation, *glacials* and *interglacials* constitute the principal geologic–climatic units, while *stadials* and *interstadials* form units of lesser rank. In areas not affected by glacier ice, it was anticipated that other geologic–climatic units, such as pluvials and interpluvials, would be established.

Geologic–climatic units are undoubtedly useful concepts and, insofar as the Quaternary sequence in mid- and high latitudes tends to be subdivided into glacials and interglacials, they form the basis for stratigraphic subdivision at the regional and continental scales. However, because climatic change is time-transgressive, the boundaries of geologic–climatic units are diachronous. It is not, therefore, appropriate to use geologic–climatic terms (glacial, interglacial, etc.) and chronostratigraphic terms (stage, substage) interchangeably, although this is often done even at the formal level. For example, in a formerly glaciated region, the sedimentary record of the presence of ice may comprise only materials that were deposited during deglaciation (see Figure 6.8 below). In this case, therefore, the actual time interval represented in the stratigraphic record by a suite of glacigenic sediments (the geologic–climatic unit) reflects only a small part of the period of time of the glacial/cold stage, the duration of which has been inferred from other lines of evidence. In a lake sediment sequence, by contrast, the entire cold stage may be reflected in the biostratigraphic record. In this case, the time interval of the cold stage (inferred from the pollen stratigraphy) and represented by the geologic–climatic unit (a suite of minerogenic sediments) will be virtually the same. Reconciling diachronous climatic signals which may have contrasting manifestations in different geologic units is therefore a major difficulty in climatostratigraphy.

The fundamental problem, of course, is that it is not climate that is directly recorded in the stratigraphic record, but *manifestations* of climate, namely the results of climatic influences on, for example, biota, soils, sediments and glaciers. Climatic reconstructions are, therefore, two steps removed from the observable data, and at each stage in the analysis, interpretation is required. If, for example, pollen assemblage zones are being used as the basis for geologic–climatic units, the first step is to infer vegetational communities and patterns of vegetational change from the pollen record, and the second is to use these reconstructions to infer climatic changes. Interpretations are therefore being made which themselves are based on interpretations. Errors can enter into both stages of the analysis, but those resulting from the first will be compounded in the second. These complications will arise irrespective of what form of lithostratigraphic or biostratigraphic evidence is being employed.

The American Commissions on Stratigraphic Nomenclature (1961, 1970) originally intended the boundaries of geologic–climatic units to follow the boundaries of the rock or biostratigraphic unit that formed the basis for their definition. This may be feasible when dealing with pre-Quaternary successions where, because of the long time intervals and low stratigraphic resolutions involved, climatic change appears in the geological record to have been virtually instantaneous. In Quaternary sequences,

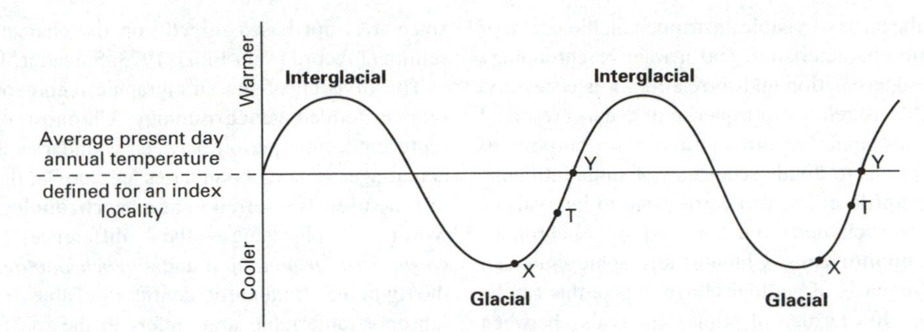

Figure 6.6 Different ways of defining the onset of an interglacial. The curve is schematic and represents temperature oscillations between glacials and interglacials. The three hypothetical points of onset (X, T, Y) are explained in the text.

however, where temporal resolution is usually so much greater, stratigraphic boundaries based on evidence of climatic change are much more difficult to locate, and certainly more difficult to correlate widely. Consider, for example, Figure 6.6 which depicts temperature changes through a glacial–interglacial cycle and represents one of the climatic cycles reflected in the oxygen isotope signal (Figure 6.5). At what point on the curve does the geologic–climatic unit of the interglacial begin? It could be argued that the boundary should be placed at that point on the curve where temperature increases following a thermal minimum (point X). Alternatively, the boundary could be located where a temperature similar to that of today is achieved (point Y). A third view would be to place the boundary at a point where a particular temperature threshold is crossed (point T) as indicated, for example, by the first occurrence of a certain indicator species in the fossil record. The problem, of course, is that we are attempting to place an essentially fixed boundary on what is a continuum of climatic change. Any of the above could be used as a basis for defining the onset of an interglacial but, because of the diachronous nature of climatic change, it is unlikely that an interglacial established in one region will be an equivalent geologic–climatic unit to that in another. Similar problems are encountered in determining the end of an interglacial. Moreover, the extent to which climatic episodes will be represented in the stratigraphic record will depend both on the amplitude and duration of climatic shifts, and also on the sensitivity of the proxy evidence from which climatic inferences are drawn.

Further complications arise when different types of evidence are being employed in environmental reconstruction. In Figure 6.7 a transgressive sequence of marine sediments has been divided on the basis of a number of lithostratigraphic and biostratigraphic criteria. Each particular line of evidence may indicate a different geologic–climatic unit, however. On the basis of the molluscan records, for example, the boundary between warm and cold climatic conditions may be placed between assemblage zones 1 and 2; the pollen data may indicate a climatic change between pollen zones 4 and 5, while the boundary between cold and temperate Coleoptera may occur between assemblage zones 1 and 2. These contrasts reflect different response rates of biota to climatic change and also, insofar as the taxa are derived from both marine and terrestrial environments, time-delay between atmospheric and oceanic temperature changes. Within a single sequence, therefore, the boundary of a geologic–climatic unit can be placed at any one of several levels. Once again, the problems that this can pose for inter-regional comparisons are clear.

The designation of geologic–climatic units is therefore frequently intuitive, often arbitrary, and much less precise than other forms of stratigraphic subdivision. Indeed, this has been recognised in the latest code of practice produced by the North American Commission on Stratigraphic Nomenclature (1983), where the concept of geologic–climatic units has been abandoned because '...inferences regarding climate are subjective and too tenuous a basis for the definition of formal geologic units' (p. 849). That stance is unhelpful to the Quaternary researcher, however, since climatic change is undoubtedly the dominant characteristic of the Quaternary, and it is difficult to envisage a stratigraphic scheme that does not explicitly acknowledge this fact. Oxygen isotope stages, for example, are **formal** definitions based on climatostratigraphic units, and they are pivotal to Quaternary stratigraphic subdivision and correlations (see below). Hence, despite the problems of definition, geologic–climatic units seem destined to remain an integral tool in Quaternary stratigraphy.

6.2.3.7 Chronostratigraphy

Chronostratigraphy is the classification of the stratigraphic record in terms of time. Once stratigraphic units have been established on the basis of visible, instrumental, biological or inferred climatic characteristics, and a relative chronology determined by superposition and correlation, it is necessary, wherever possible, to relate the sequence of events preserved in the rock-stratigraphic record to time. The purpose of chronostratigraphy is to divide sequences of strata into units (**chronostratigraphic units**) that correspond to intervals of geological time. Such units are bounded by isochronous surfaces or **chronohorizons**. Chronostratigraphic units can be defined on the basis of geological age, where this can be established, or in terms of time intervals between isochronous horizons. In other words, where chronostratigraphic comparisons are being effected between sites, reference can be made to the time period encompassed by two designated stratigraphic horizons, even where the actual age of each horizon is unknown and many of the rocks are missing. For example, reference can be made to the time period encompassed by two biozone or lithological boundaries. If these boundaries prove to be time-parallel, they can be employed as chronostratigraphic boundaries, and the biozone or lithofacies or, indeed, any other stratigraphic unit can be considered as a chronostratigraphic unit. Chronostratigraphy, like climatostratigraphy, is inferential, since it is not based directly on the characteristics of the sediment record (Vita-Finzi, 1973; Salvador, 1994).

The division of the stratigraphic record on the basis of time is termed **geochronology**. Chronostratigraphic units represent a time period between boundaries and, where the actual ages of such boundaries are known, then the interval of time itself is referred to as a **geochronological unit**. It is worth emphasising the difference between a *chronostratigraphic unit* and a *geochronological unit* since the two are frequently confused (Table 6.2). The term 'chronostratigraphic unit' refers to the rock sequence laid down during a particular time interval. Hence the term 'Quaternary system' is used to describe the rocks and sediments that have accumulated over a time range that is called the Quaternary. The geochronological equivalent (**geochronological unit**), that is the time interval itself, is referred to as the 'Quaternary period' (North American Commission on Stratigraphic Nomenclature, 1983).

The basic working units in Quaternary stratigraphy are **stages**, for they are the smallest subdivisions of the standard stratigraphic hierarchy that can be recognised on both a regional and inter-regional scale. Conventional stages in pre-Quaternary time typically range from 3 to 10 million

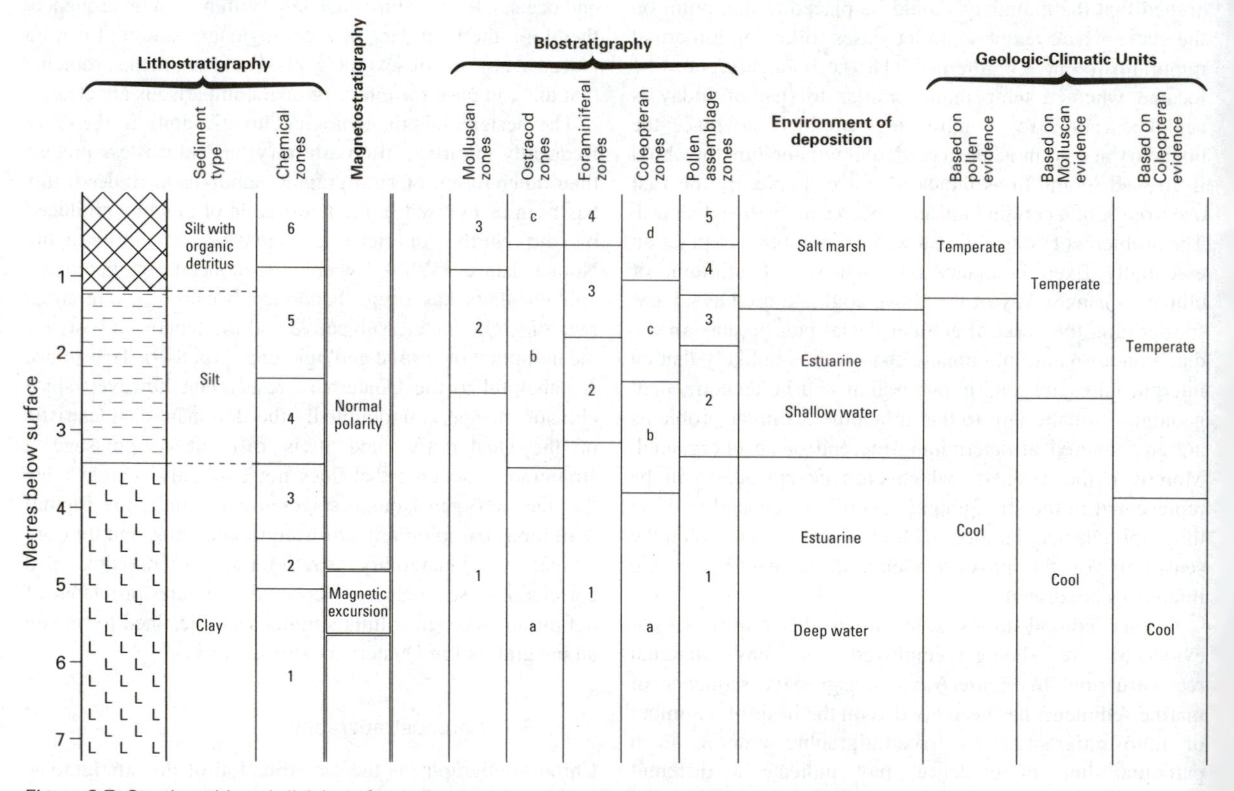

Figure 6.7 Stratigraphic subdivision of a sedimentary sequence based on different criteria.

Table 6.2 Conventional hierarchy of chronostratigraphic and geochronological units (after Hedberg, 1976).

Chronostratigraphic	Geochronological	Examples
Eonothem	Eon	Phanerozoic
Erathem	Era	Cenozoic
System	Period	Quaternary
Series	Epoch	Pleistocene
Stage	Age	Devensian
Chronozone	Chron	Younger Dryas

years, but in the Quaternary stages are measured in tens of thousands of years, a level of precision that is required in order to understand Quaternary stratigraphic processes. Stages can be divided into smaller units or substages. The basic chronostratigraphic unit is the **chronozone**, the timespan of which is usually defined in terms of the timespan of a previously designated stratigraphic unit such as a formation, or a member, or a biozone (e.g. Rose, 1985). Chronozones have been most widely employed in Quaternary research where biozones have been dated by radiometric methods, for example radiocarbon-dated pollen assemblage zones (Moore *et al.*, 1991).

The division of Quaternary strata into chronostratigraphic units is seldom straightforward, however, and problems arise that are not encountered in the earlier geological record. In the main, these result from the relatively short timespan of the Quaternary, and from the fact that the much finer divisions of the Quaternary stratigraphic sequence are of equal importance to the coarser subdivisions of the earlier geological record. Only a small number of geochronologic dating methods cover the whole of Quaternary time, and of those that do, few produce consistently reliable results. Also, dating aberrations are likely to pose particular problems in Quaternary research because of the more limited time range involved. Moreover, if the statistical uncertainty associated with the dates (i.e. the quoted ± value) is as great or greater than the stratigraphic intervals that are under examination, then clearly these age determinations are insufficiently precise either to define the boundaries of such units or to distinguish subdivisions within those units (Lowe & Gray, 1980). A further difficulty arises over the recognition of isochronous horizons in the stratigraphic record. In pre-Quaternary strata, lithostratigraphic and biostratigraphic boundaries, although inherently time-transgressive, *appear* to be synchronous when set against the vast span of geological time. As such, they are frequently used as time-stratigraphic markers. In the Quaternary stratigraphic record, however, most boundaries are time-transgressive. Nevertheless, some Quaternary stratigraphic units or boundaries do provide a basis for time-stratigraphic correlation, and some of these will now be discussed.

6.3 Time-stratigraphic correlation

6.3.1 Principles of Quaternary correlation

The stratigraphic methods outlined above all provide a basis for **correlation**, that is the relationship of stratigraphic sequences or events at one locality to those at another. Throughout most of the geological column, lithostratigraphic and biostratigraphic boundaries are, at the level of resolution applied, time-parallel, and are therefore regarded as being almost of equivalent status to chronostratigraphic units in time-stratigraphic subdivision and subsequent correlation. This assumption cannot, however, be made in the correlation of Quaternary successions, except perhaps at the local scale. Not only are lithostratigraphic and biostratigraphic boundaries time-transgressive (see above), but the repetitive nature of Quaternary climatic change has meant that at any given locality, similar depositional sequences may be preserved that are of markedly different age. In view of the fragmented and highly diverse nature of the Quaternary stratigraphic record, the difficulties in effecting meaningful correlations between often widely separated sites are considerable.

Some examples of the complications that can arise in the correlation of Quaternary successions are illustrated in Figure 6.8. This shows, in a diagrammatic way, the extent of glaciation over time, the shaded area representing the time period during which ice covered the ground at increasing distance from the ice dispersal centre. Site 1 was affected by glacier ice for almost the whole of the 'glacial' time interval; sites 2 and 3 were occupied by ice for a shorter period, while site 4, and especially site 5, were ice-covered for only a small proportion of the glacial episode. Sites 2, 3 and 4 also experienced interstadial conditions between successive glacial advances. At each site, the environmental history is recorded in sequences of glacigenic sediments, interbedded at sites 2, 3 and 4, with organic deposits (peats or soils).

At site 1, which for much of the glacial interval lay some distance up-glacier, erosion probably dominated and the only record of glacier activity is a thin layer of till deposited during glacier wastage. By contrast, a complex sequence of glaciogenic sediments accumulated at site 5 near the ice margin. There are two points to note here: first, the thickest and most complex sequence of sediments is preserved at the site that was glaciated for the shortest length of time and, secondly, although site 1 was covered by glacier ice before

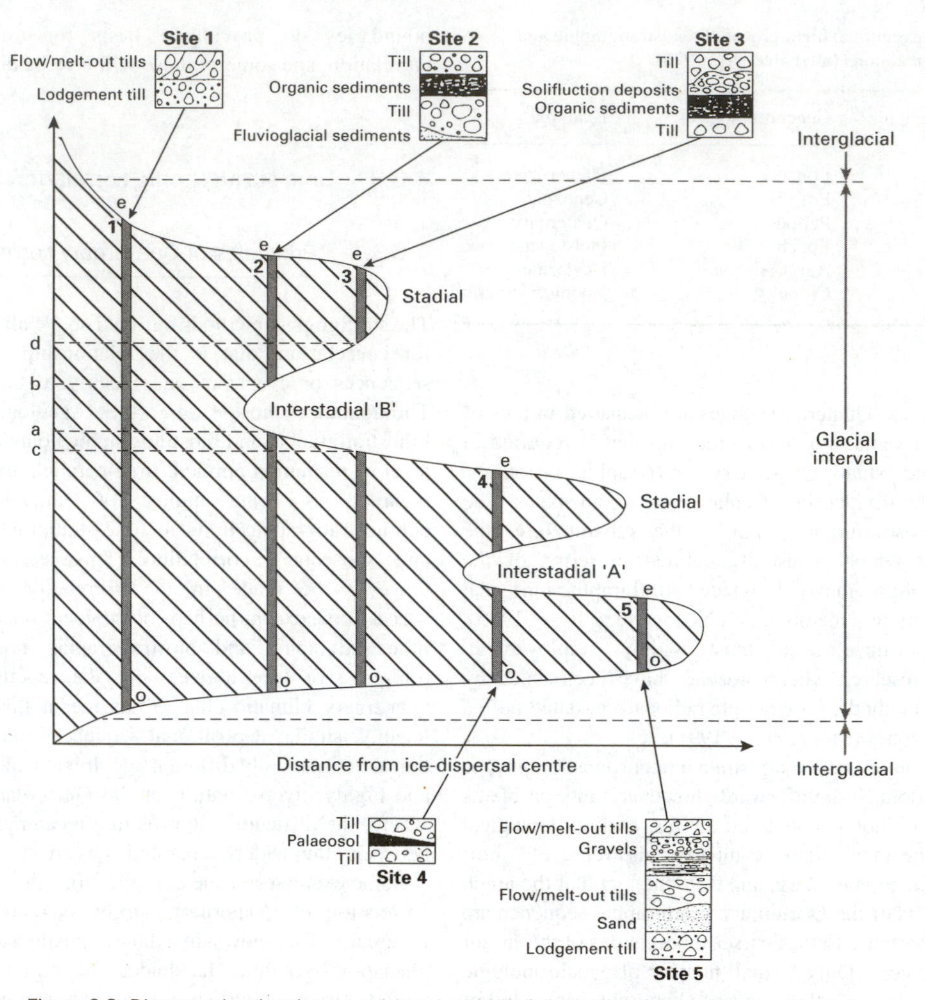

Figure 6.8 Diagram showing the onset (o) and end (e) of glaciation at sites at increasing distance from the ice-dispersal centre (modified from Andrews, 1979). Possible glacigenic sequences and relative sediment thicknesses at each site are also indicated. For further explanation see text.

site 5, on a time-stratigraphic basis, the deposits *preserved* at site 1 may, in fact, be *younger* than those at site 5. At sites 2, 3 and 4, two till units are preserved, but from the stratigraphic evidence alone it is far from clear which till unit is the correlative of the glaciogenic sequence recorded at site 5. Moreover, because sites 2, 3 and 4 have experienced two periods of glaciation separated by an interstadial interval, almost identical stratigraphic sequences (till/organic sediments/till) have developed. In the absence of other evidence, correlation might therefore be unwittingly effected between these sequences. Yet, as the diagram shows, the sequence of deposits at site 4 relates to an earlier period of time and to a stadial/interstadial oscillation which is different from that at sites 2 and 3. Where units of similar composition and mode of origin but of very different ages are incorrectly correlated, such errors of correspondence are referred to as **homotaxial errors**.

This type of error can also arise in biostratigraphy. It is entirely possible, indeed likely, that a very similar environment existed during both interstadials A and B, and this would be reflected in the fossil evidence preserved in the organic horizons at sites 2, 3 and 4. Again, the inference of a single interstadial based, for example, on the pollen evidence at the three sites would be homotaxially incorrect. Finally, the diagram illustrates the time-transgressive nature not only of the lithostratigraphic and biostratigraphic units, but also of the geologic–climatic units. At site 2, the geologic–climatic unit (interstadial) inferred from the fossil content of the organic deposit spans the time interval a–b, while at site 3 the same interstadial covers the time interval c–d.

These complications arise partly because the deposits of varying ages are not arranged vertically in order of superposition, and partly because lateral correlation often has to be made between numerous short-lived depositional records. The above example relates to a single glacial interval, yet when it is recalled that at least twenty glacial/interglacial cycles affected the mid- and high latitude regions of the world during the Middle and Late Quaternary, the scale of the problem begins to emerge. Moreover, it is not only in formerly glaciated areas that difficulties are encountered. Repeated climatic changes had a profound effect on those regions that lay beyond the margins of the ice sheets, and there too the complicated erosional and depositional history of the Quaternary presents the stratigrapher with major problems of correlation. Clearly, therefore, at the regional and continental scales, lithostratigraphy, biostratigraphy and climatostratigraphy are not sufficiently sensitive tools with which to effect meaningful time-stratigraphic correlations. A geochronological basis is required, although as yet no single radiometric dating method has been developed that is applicable to the whole of the Quaternary period at the required level of temporal resolution. There are, however, a number of ways in which sequences may be correlated on a time-stratigraphic basis, and some of these are examined in the following section.

6.3.2 Elements of time-stratigraphic correlation

6.3.2.1 Palaeomagnetic correlation

Magnetostratigraphy utilizes stratigraphical variations in the magnetic properties of rocks as a basis for geological correlation (Hailwood, 1989). The palaeomagnetic record can be divided into a series of **magnetozones** (units of rock with a specific magnetic character), whose boundaries are clearly defined due to the relatively abrupt changes in the earth's magnetic field (see section 5.5.1). Because these changes are experienced globally palaeomagnetic stratigraphy offers one of the most promising methods for establishing worldwide correlation of Quaternary events. It is, moreover, the most frequently used means of correlating the marine and terrestrial records (see below). On the other hand, the method is restricted to certain rock types (e.g. volcanic rocks) and specific depositional environments (e.g. lake sites and aeolian contexts), and the precise nature and timing of some palaeomagnetic events remain to be established. Further, it is a relatively 'coarse' record at present, although a much finer resolution (e.g. the determination of global or regional paleomagnetic changes within the Holocene) may ultimately be achieved.

6.3.2.2 Correlation using tephra layers

Tephra layers (section 5.5.2) constitute marker horizons in the stratigraphic record that are essentially isochronous, and hence have the potential to serve as the basis for time-stratigraphic correlation between sediments formed on land, in lakes and on the sea-bottom (Einarsson, 1986). In recent years, considerable advances have been made in the recognition, extraction and identification to source of tephra materials from a range of sedimentary contexts (e.g. Westgate & Briggs, 1985; Dugmore, 1989: Dugmore & Newton, 1992). However, correlation on the basis of tephra deposits is not as universally applicable as that based on palaeomagnetism (such as the Brunhes–Matuyama boundary, for example), for individual ash beds have relatively limited geographical ranges, and no tephra has yet been discovered that is sufficiently extensive to be useful for intercontinental correlation. Overall, therefore, although tephra deposits are potentially excellent time-datum surfaces, in practice tephrochronology can only form a basis for correlation at the local or regional scale.

6.3.2.3 Correlation using palaeosols

Palaeosols (section 3.5) have been most widely employed as a correlative tool in North America where it has traditionally been assumed that episodes of rapid soil profile development during interglacials and certain interstadials alternated with periods of negligible profile development or arctic soil formation in the intervening cold phases. Hence, well-developed buried soils (e.g. the Sangamon Soil) have been considered indicative of interglacial episodes and used as key marker horizons for time-stratigraphic correlation (e.g. Birkeland *et al*., 1971). In recent years, however, a more cautious approach has been adopted in the use of buried palaeosols in stratigraphy and correlation, particularly at the inter-regional and continental scales (Richmond & Fullerton, 1986b), for it is now recognised that degree of soil development as a direct function of time can only be assumed where other soil-forming factors (parent material, climate, slope and biological factors) can be shown to have been constant (Boardman, 1985). Moreover, insofar as most soils are polygenetic, any buried profile may be the product of more than one phase of pedogenesis. Indeed, the Sangamon Soil of North America, hitherto widely considered to be of last interglacial age, is now acknowledged to be the product of pedogenesis not only during the last interglacial, but perhaps also during preceding and succeeding cold stages (Follmer, 1983).

Despite these limitations, however, palaeosols have been used successfully in the development of local histories and, in certain circumstances, in wider correlative schemes. For example, the polygenetic Valley Farm Soil (see above) and the Barham Soil (periglacial) form important marker horizons in the Quaternary stratigraphy of large areas of southeast England (Rose *et al.*, 1985), while interglacial fossil soils of Middle and Late Pleistocene age constitute key elements in the stratigraphic record of western Europe and have been employed as a basis for inter-site correlation (e.g. Lautridou *et al.*, 1986; Šibrava, 1986). In the American Midwest, the Farmdale and Sidney Palaeosols, which have similar pedological properties and a ^{14}C age of 28–25 ka BP (Follmer, 1983), also form important stratigraphic markers that can be used to correlate Late Wisconsinan events at the local and regional scales.

6.3.2.4 Shoreline correlation

Marine shorelines and deposits may also, in certain contexts, provide a basis for time-stratigraphic correlation. If world sea levels are stable for long enough to allow the development of shoreline features, then essentially isochronous reference surfaces will form which, in theory, should provide a basis for inter-regional schemes of correlation. At the local scale, for example, isochronous shorelines can often be traced around a coastline, even in regions affected by glacial isostasy or tectonic uplift (Pillans, 1987). If the shoreline cuts certain sedimentary units, but is overlain by others, then the shoreline can be used as a time-stratigraphic reference plane for separating sedimentary units of different age. At the regional scale, sedimentary units formed as a result of catastrophic marine events, such as tsunamis, constitute marker horizons for correlating between marine and terrestrial sequences (Dawson *et al.*, 1988). At the broader scale, evidence of former sea levels (raised beaches and corals, submerged corals, submerged clifflines, etc.) may provide a basis both for inter-regional correlation and for linking terrestrial and deep-sea records. Problems can arise in correlation, however, from complexities in the coastal records induced by variations in amounts and rates of local tectonic activity (e.g. Pirazzoli *et al.*, 1994), and also from the phenomenon of geoidal eustasy (Devoy, 1987a; section 2.5).

6.3.2.5 Correlation on the basis of radiometric dating

These methods, which were considered in section 5.3, are an extremely important independent means of long-distance correlation, for they form time planes across the stratigraphic record against which the time-transgressive litho-, bio- and morphostratigraphic boundaries can be measured. There is, however, a substantial body of opinion in support of the view that the subdivision of the Quaternary should rest on the stratigraphic record, with radiometric dates being merely a means whereby that record can be underpinned (Morrison, 1968). Certainly, any radiometric age determination applies only to the locality and to the horizon from which it was taken and it can only be related to other successions on the basis of the observed stratigraphic sequence at the different localities. As has already been shown, no radiometric date is free from analytical errors – in some cases the error (±) associated with the date may be so great that the date cannot be used in time-stratigraphic correlation. Equally, all dated samples are prone to errors of, for example, isotopic exchange and contamination. Every date should not only be carefully checked against other age determinations, but also thoroughly evaluated in the light of its stratigraphic context before it is used as an aid in correlation. Overall, radiometric dating is perhaps best regarded as a means of corroborating and validating other stratigraphic and correlative procedures, rather than as the primary basis for time-stratigraphic correlation.

6.3.2.6 Event stratigraphy and correlation

Events may be broadly defined as comparatively rare and geologically short-lived occurrences which have left some trace in the rock records, and which may therefore be employed as a means of correlation (Whittaker *et al.*, 1991). The products of some events (tephra from volcanic eruptions; tsunami deposits in marine sequences) have already been discussed, but others might include the geological manifestations of storms, floods, earthquakes, mass movement and turbidity flows, although clearly not all of these will be sufficiently widespread to constitute a basis for inter-regional correlation. On the other hand, the layers of carbonate-rich debris in North Atlantic ocean sediments reflecting major ice-rafting events ('Heinrich Events') during the course of the last cold stage (Bond *et al.*, 1992b) offer a means both for inter-core correlation, and also for linking oceanic and terrestrial sequences (Chapter 7). Other characteristics of the geological record may also offer a basis for broader time-stratigraphic correlation. These include palaeomagnetic changes preserved in Quaternary sediments (see above and section 5.5.1), and the use of a range of geochemical evidence, including stable isotopes of oxygen, carbon, sulphur and strontium as stratigraphic markers (Odin *et al.*, 1982). An unusual application involving unstable isotopes would be to employ the apparent 'plateaux' in atmospheric ^{14}C production during the Lateglacial and early Holocene (Ammann & Lotter, 1989; section 5.3.2.3) as a basis for time-stratigraphic correlation, working on the assumption that individual plateaux reflect essentially isochronous and universal marker horizons (Lowe, 1993b). Certain biological events may also be used for correlation purposes. For example, the sudden decline in

Ulmus in Holocene pollen records from western Europe, which has been dated to around 5 ka BP (Huntley & Birks, 1983) and appears to reflect a widespread biotic catastrophe (Peglar & Birks, 1993), could be employed in time-stratigraphic correlation. In earlier Quaternary sequences, mammalian fossils are often important biostratigraphic indicators (Lister, 1992), and mammalian extinction events might also offer a basis for regional and inter-regional correlation.

6.3.2.7 Correlation using the marine oxygen isotope record

Oxygen isotope profiles from deep-ocean cores (section 3.10) represent a proxy record of long-term climatic change in which, because the mixing time of the oceans is relatively short, the isotopic stage boundaries, especially the terminations, are essentially synchronous. They therefore constitute key marker horizons within the ocean sediment sequences and offer the potential for the development of a high-resolution correlative chronology that is globally applicable (Jansen, 1989). Moreover, the stage boundaries can be dated by the technique of **orbital tuning** (section 5.5.3), and hence the SPECMAP and related timescales constitute a type sequence for the Quaternary against which other isotopic profiles can be compared (Shackleton *et al.*, 1990).

Deep-ocean records undoubtedly hold a number of advantages over terrestrial sequences from the point of view of stratigraphic subdivision and correlation. First, the sediments from which the isotopic data have been obtained *appear* to be relatively undisturbed. Second, a common technique (oxygen isotope analysis) can be used to compare profiles from widely scattered localities on the deep-ocean floors. Third, the terminations can be used as universal reference points in inter-core correlation. Fourth, although the isotopic changes are a consequence of climatic changes, and are therefore time-transgressive, this to a very large extent is masked by the slow rate of sediment accumulation. As a consequence, isotopic stage boundaries and terminations can be interpreted as *essentially* time-parallel horizons. Fifth, orbital tuning enables a timeframe to be established for the isotope curves, with key levels in the cores being dated on the basis of periodicities obtained from astronomical calculations. Moreover, key horizons (e.g. the Brunhes–Matuyama boundary) can also be dated by the independent method of palaeomagnetic stratigraphy.

Interpretation of the isotopic evidence is not always straightforward, however. Continuity of sedimentation can never be proved, and it is questionable whether, in practice, gaps in the sedimentary record can ever be reliably detected. Problems also arise over poor stratigraphic resolution, over bioturbation and reworking of sediments, and over the recognition and precise definition of stage boundaries in some isotopic profiles (Patience & Kroon, 1991; section 3.10). Moreover, although the profiles can be dated by orbital tuning and correlated by reference to the palaeomagnetic timescale, correlation between individual isotopic profiles is based largely on a 'count from the top' principle, and hence the possibilities of homotaxial error are always present. A further potential source of error arises from the differentiation between interglacials and interstadials. It has already been shown that OI stage 3 is regarded as being of interstadial rather than interglacial status. The possibility cannot, therefore, be excluded that some previous interglacials and interstadials have been confused, particularly in the earlier part of the Quaternary record. If this has occurred, then it clearly has implications for the status attached to isotope stages, and could lead to problems in correlating between continental records and the marine oxygen isotope sequence (section 6.3.3). By the same token, the complexities of OI stages 5 and 7 (see above) are surely not unique to the upper parts of the isotopic sequence. Consistency of interpretation between different isotopic profiles may not, therefore, be easy to achieve, particularly in those cores where stratigraphic resolution is poor.

Although the above difficulties have yet to be satisfactorily resolved, there is no doubt that the isotopic trace in the ocean sediments provides a remarkable record of Quaternary climatic change, and it is now widely accepted that this, and not the terrestrial sequence, provides the basic framework for a global scheme of Quaternary correlation. Because the isotopic stages are a reflection of climatic change, they are essentially geologic–climatic units and, as such, should have correlatives in the terrestrial record. Clearly, if meaningful global correlations are to be effected, a basis for correlation between the deep-ocean and terrestrial successions is required, and it is to the ways whereby this may be achieved that attention is finally directed.

6.3.3 Correlation between continental, marine and ice-core records

If correlations are to be established with the oceanic successions, terrestrial sequences must possess certain characteristic features. First, a lengthy stratigraphic record must be available. Second, evidence of climatic change in the stratigraphic record must be clear and unequivocal so that the climatic signal can be compared directly with that in the oxygen isotope profiles. Third, contexts should be sought where there are firm grounds for believing that the record of sedimentation is continuous, or more or less continuous. Fourth, correlations will be more secure if a time-frame can be established based, for example, on radiometric methods such as radiocarbon, uranium series, fission-track or potassium–argon dating, or on age equivalence techniques such as palaeomagnetism or tephrochronology.

Depositional sequences that possess these particular characteristics are least likely to be found in areas formerly occupied by glacier ice, yet ironically most of the schemes of stratigraphic subdivision and global correlation have been based upon evidence from such regions. More recently, however, the focus of attention has shifted to those parts of the world that lay beyond the margins of the Quaternary ice sheets and, in the main, it is the stratigraphic records from these areas that now offer the greatest potential for correlating the marine and terrestrial sequences. The most promising depositional contexts appear to be tectonic basins, deep lakes, caves, loess accumulations and marine sequences. Ice core records provide a further basis for correlation, although these are applicable only to the last 160 ka BP or so.

6.3.3.1 Tectonic basins

In some of the great tectonic basins of the world, sediments have been accumulating in a variety of depositional contexts for several millions of years. In the Carpathian Basin beneath the present Hungarian Plain, for example, are more than 600 m of Quaternary and Tertiary sediments, the earliest of which exceed 2.5 Ma in age. Pollen, molluscan and ostracod analyses underpinned by a high-resolution palaeomagnetic record suggest that warmer conditions prevailed until approximately 1.5 million years ago, and the onset of consistently cooler conditions began around 900 ka, a pattern that is similar in outline to faunal and isotopic changes in the oceans (Cooke, 1981). In the North Sea basin, the Central Graben of the southern Netherlands contains a sequence of marine and terrestrial sediments extending back to the Praetiglian Stage, the base of which has been dated to *c.* 2.3 Ma BP and marks the lower limit of the Pleistocene in this area (Gibbard *et al.*, 1991). Although a detailed correlation remains to be established, there are a number of similarities between the climatic curve for the Netherlands derived mainly from pollen-stratigraphic evidence (Figure 6.9) and the isotope trace from the deep ocean record (de Jong, 1988). In the Southern Hemisphere, the Wanganui Basin in New Zealand contains up to 4000 m of Plio–Pleistocene shallow water marine strata. Correlations with the marine oxygen isotope record (Figure 6.10) are based on a combination of biostratigraphy, magnetostratigraphy and fission-track dating of interbedded rhyolite tephra (Pillans, 1991).

6.3.3.2 Lake sediment records

Within the last twenty years or so, cores have been obtained from a number of deep lake sequences that extend back to the beginning of the Quaternary and, in some cases, into the Tertiary. These include Lake Biwa in Japan (Fuji, 1988), Lake George in southeast Australia (Kershaw, 1991), Hula Lake in the Jordan Rift Valley, Israel (Horowitz, 1989), and the Tenaghi Phillipon peat bog in Macedonia, Northern Greece (Van der Weil & Wijmstra, 1987a, 1987b). A range of proxy data, including pollen, plant macrofossils, charcoal fragments, Mollusca, diatoms, sediment chemistry and palaeomagnetic properties, have been obtained from these profiles, and correlations have been proposed between some of these sequences and the deep-ocean oxygen isotope trace (Figure 6.11). One of the most impressive records, however, is that from Funza on the high plain of Bogotá, Colombian Andes (Hooghiemstra, 1984). This vast subsiding intermontane basin, which covers an area of some 1100 km^2, formerly contained a lake in which over 580 m of sediment have accumulated. Palynological evidence from the Funza cores provides a high-resolution continuous record of changes in the Andean vegetation belts and hence climatic fluctuations over the course of the last 3.5 Ma (Hooghiemstra, 1989). Comparison between the Funza arboreal pollen record and the high-resolution oxygen-isotope profile from the Ocean Drilling Programme Site 677 for the past 1.2 Ma (Figure 6.12) shows a remarkable correspondence, and offers the prospect for a detailed land–sea correlation which may eventually span the whole Quaternary period (Hooghiemstra & Sarmiento, 1991).

In addition to the exceptionally long records like Funza, a number of present and former lakes contain Late Quaternary sedimentary sequences spanning at least the last interglacial–glacial cycle. In eastern France, for example, the mires of La Grande Pile in the Vosges and Les Echets near Lyon contain a continuous record of peat and lacustrine sediment accumulation spanning the last 140 ka (Woillard, 1978; Beaulieu & Reille, 1984, 1992a). A link between the Grande Pile sequence and the marine isotope record was first attempted on the basis of a radiocarbon-dated pollen record (Woillard & Mook, 1982). Subsequently climatic proxies based on the palynological data from both La Grande Pile and Les Echets have been correlated with the deep-sea oxygen isotope trace and tuned to the SPECMAP timescale (Guiot *et al.*, 1989; Seret *et al.*, 1992). These data (Figure 6.13) show clear evidence of interglacial/interstadial conditions (Eemian; St Germain I; St Germain II) corresponding with OI substages 5e, 5c and 5a respectively and significantly reduced temperatures during the terrestrial equivalents of OI stages 2 and 4.

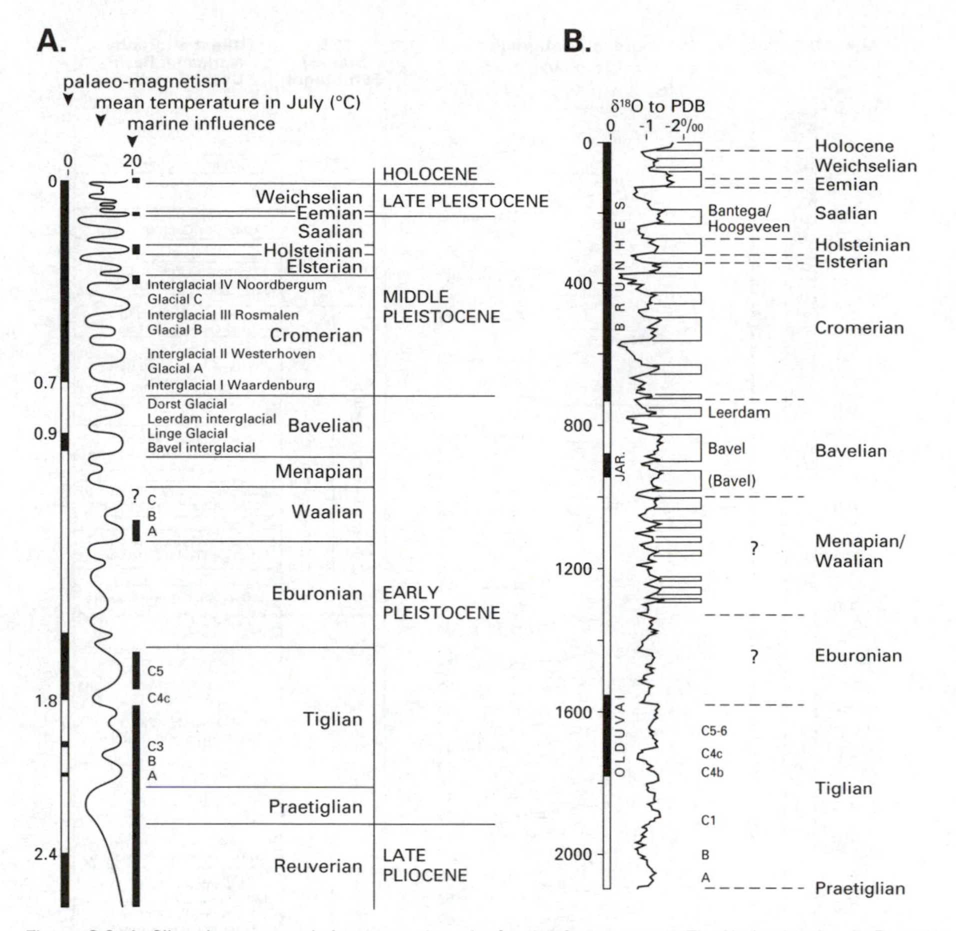

Figure 6.9 A. Climatic curve and chronostratigraphy for the Quaternary in The Netherlands. B. Tentative correlation of that climatic curve with the isotopic stages in deep-sea core V28-239 (from de Jong, 1988).

In other parts of Europe, notably in the Eifel region of Germany, in southern Italy and in the Massif Central, biostratigraphic records spanning at least the past 100 ka have been obtained from **maars**, extinct volcanic craters which now contain either open or infilled lakes. One of the most intensively investigated of these sites, Lac du Bouchet near Le Puy in central France (Figure 6.14), has yielded a pollen and palaeomagnetic record that extends back to the Eemian Interglacial, and which can be correlated with marine OI stages 1–5e (Creer, 1991; Beaulieu *et al.*, 1991; Thouveny *et al.*, 1994). In Italy, maar sequences up to 250 ka in age have been investigated palynologically (Follieri *et al.*, 1988) and correlated with palaeoclimatic variations inferred from oxygen isotope stratigraphy (Narcisi *et al.*, 1992), while on the Devès plateaux of the Massif Central, an exceptional record of five climatic cycles extending back into the Middle Pleistocene is preserved in the Praclaux maar, the earliest interglacial episode in the sequence (the equivalent of the Holsteinian of Northern Europe) being correlated with OI stage 11 of the deep-sea sequence (Beaulieu & Reille, 1992b).

The pluvial lake sequences in low latitude regions (sections 2.7.1 and 3.6), which provide evidence for fluctuations in the water budget of low latitude lakes, can also be used as a basis for linking marine and terrestrial evidence during the later part of the Quaternary sequence. In the Arava Rift Valley of Israel, for example, the correlatives of OI stages 5, 6 and 7 of the deep-sea record can be recognised in a sequence of U-series dated travertines and spring deposits that provided evidence of two warmer and wetter phases and an intervening episode of aridity (Livnat & Kronfeld, 1985). In the western United States, fluctuations in the water level of the Lake Lahontan system over the course of the last 300 ka have been correlated with

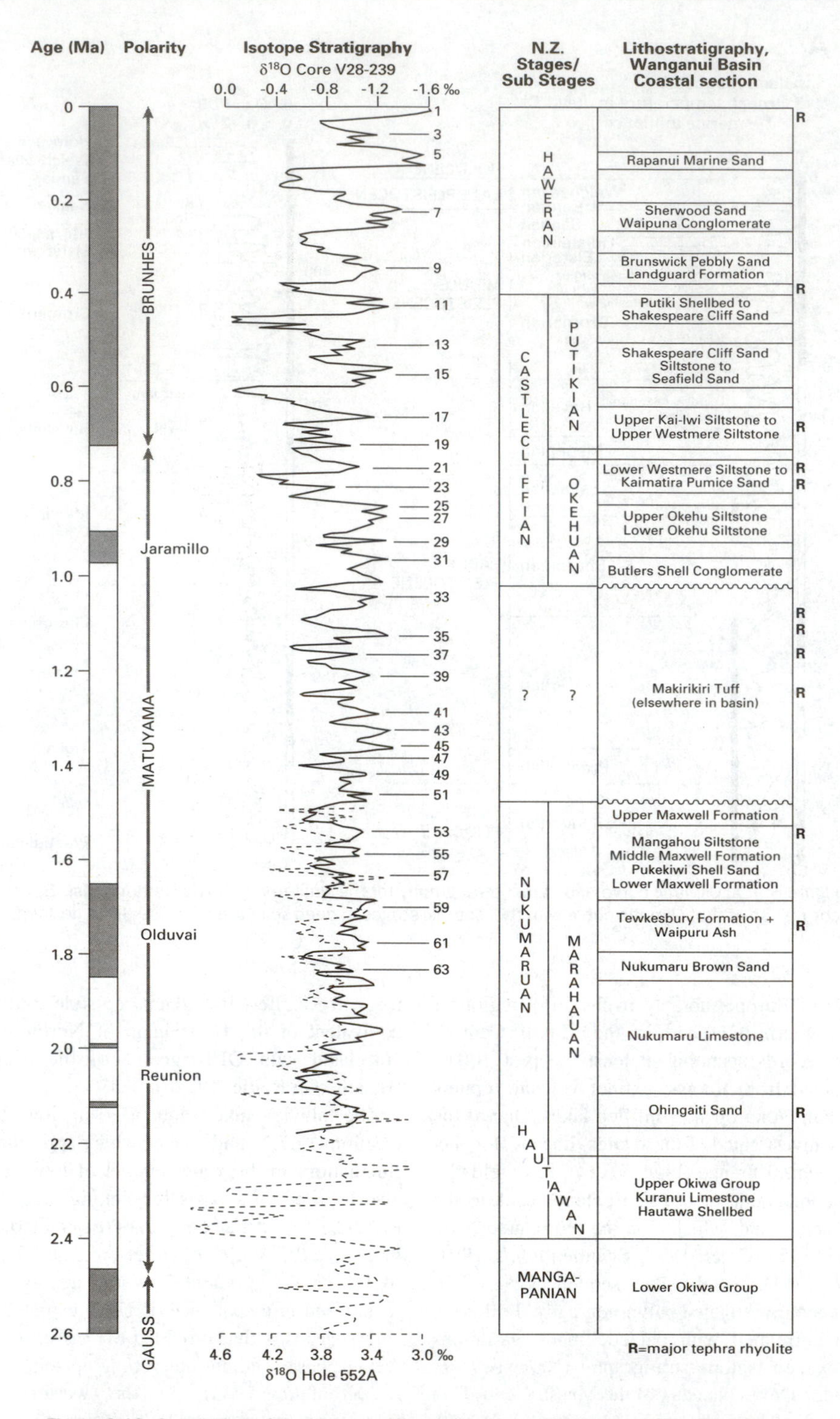

Figure 6.10 Correlation of the Wanganui Basin sequence, North Island, New Zealand, and the oxygen isotope stratigraphy of deep-sea cores V28-239 and 552A (from Pillans, 1991).

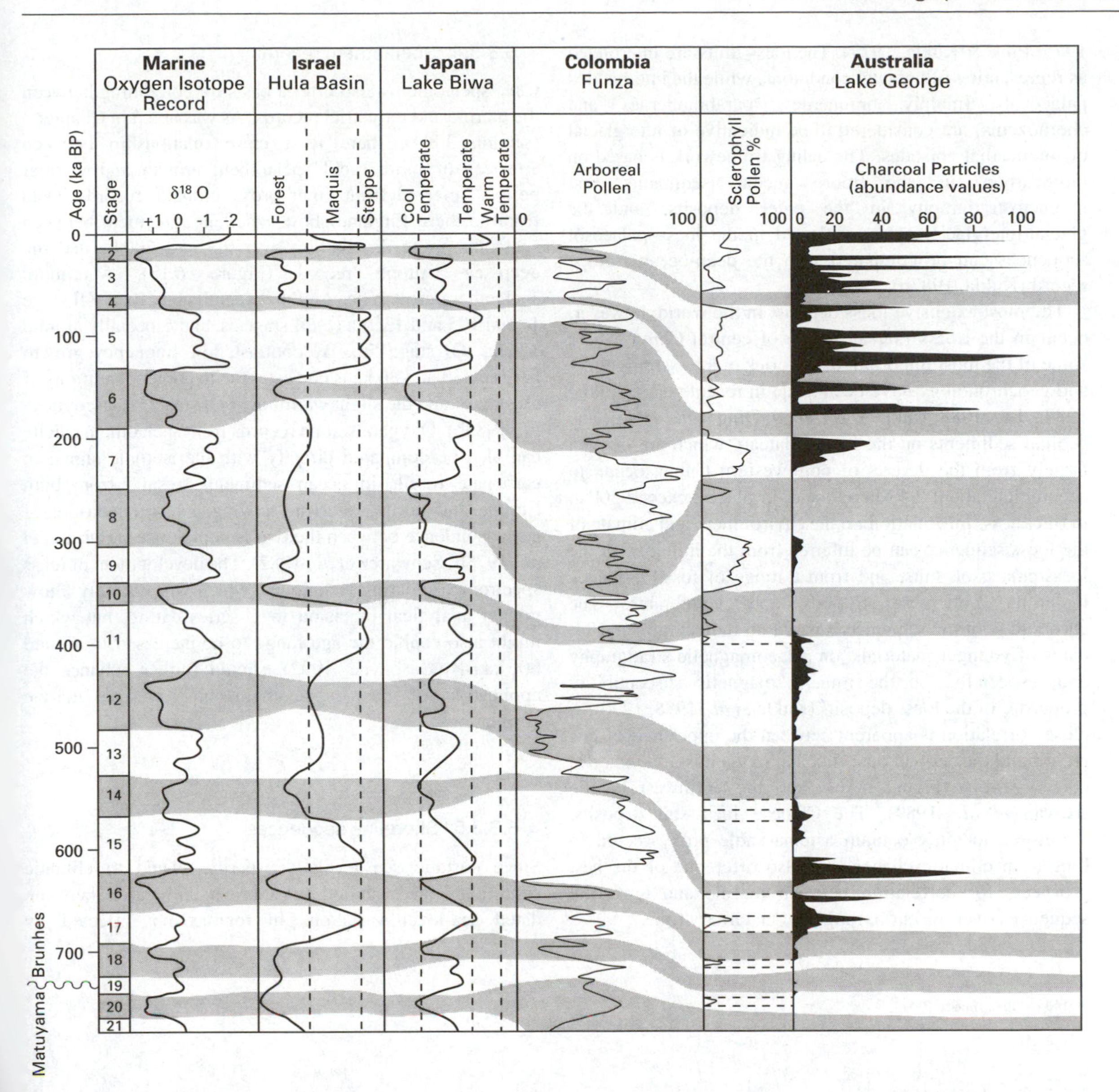

Figure 6.11 Comparison of four Quaternary records – Hula Basin, Israel; Lake Biwa, Japan; Funza, Colombian Andes; and Lake George, Australia – in relation to the marine oxygen isotope record over the last 800 ka (after Williams *et al.*, 1993a).

the stacked isotope record and the orbitally based marine chronostratigraphy (Lao & Benson, 1988). This comparison shows lake high-stands occurring at or shortly after times of maximum ice-sheet size as reflected in the isotopic trace (Figure 6.15). Other Late Quaternary lake cycles recorded in sediments from the Great Basin have been similarly correlated with the $\delta^{18}O$ signal in deep-sea cores (Benson & Thompson, 1987; Oviatt *et al.*, 1987).

6.3.3.3 Loess deposits

In some areas of the world, long aeolian sedimentary sequences provide a basis for Quaternary land–sea correlation. In the Czech and Slovak Republics and parts of Austria, sequences of interbedded soil and loessic units provide a record of glacial and interglacial conditions stretching back to the early part of the Quaternary (Kukla,

1975; Fink & Kukla, 1977). The loess units are interpreted as representing full glacial conditions, while the interbedded palaeosols (mainly braunerdes, parabraunerdes and chernozems) are considered to be indicative of interglacial or interstadial episodes. The dating framework is based on radiocarbon in the more recent sediments and magnetostratigraphy in the older deposits, and the glacial/interglacial cycles reflected in the loess–palaeosol sequences can be compared with the deep-ocean isotope record (Kukla, 1987b).

The most extensive loess deposits in the world, however, occur in the Loess Plateau region of central China, where some of the most important discoveries in loess stratigraphy and geochronology have been made in recent years (Kukla, 1987a; Liu, 1988; Kukla & An, 1989; Ding *et al.*, 1994). The aeolian sediments on the Loess Plateau, which are derived largely from the deserts of northwestern China, began to accumulate about 2.4 Ma ago and, in places, exceed 300 m in thickness. Information on the environment and climate of the loess sequence can be inferred from the lithology of the loess/palaeosol units, and from a range of fossil evidence including pollen, gastropods and vertebrates. The chronology of the sequence is based on radiocarbon and TL dates of younger materials, on palaeomagnetic stratigraphy and, especially, on the mineral magnetic susceptibility properties of the loess deposits (Kukla *et al.*, 1988, 1990). A close correlation is apparent between the upper part of this record and the aeolian dust flux and oxygen isotope trace in core V21-146 (Figure 6.16) from the northwest Pacific (Hovan *et al.*, 1989). The Chinese loess-soil deposits, therefore, not only contain a remarkable proxy record of long-term climatic change, but also offer one of the best prospects for correlation between marine and terrestrial sequences over the entire range of Quaternary time.

6.3.3.4 Speleothem records

Cave speleothems also offer a basis for correlating between the marine and terrestrial records. As was noted in Chapter 3 (section 3.8.4), there is a close relationship between episodes of more rapid speleothem growth and warmer periods inferred from other proxy climatic records. Data from northern England show excellent agreement between speleothem age distribution over the past 200 ka and the deep-sea isotope record (Figure 6.17). Maximum speleothem growth occurred between 0 and 10 ka (OI stage 1) and 105 and 135 ka (~OI stage 5), and especially around 120 ka (OI stage 5e). By contrast, low frequency growth between 35 and 80 ka is consistent with cooler conditions of OI stage 4 and the slight warming of OI stage 3 (Gascoyne *et al.*, 1983b). Oxygen isotope records from speleothem calcite can also be compared directly with the isotopic signal in carbonate fossils in ocean sediments, results from both Europe and north America showing a broad measure of correspondence between the two isotopic traces (Harmon *et al.*, 1978b; Schwarcz *et al.*, 1982). The development of mass spectrometric dating techniques, which will not only allow greater analytical precision in U-series dating, but which might also enable the age range to be increased to around 600 ka BP (Gascoyne, 1992), should further enhance the application of cave speleothems in terrestrial–marine correlations.

6.3.3.5 Shoreline sequences

Since eustatic sea levels are partly related to climatic conditions, there should be a broad correlation between dated sea-level variations in regions not affected by

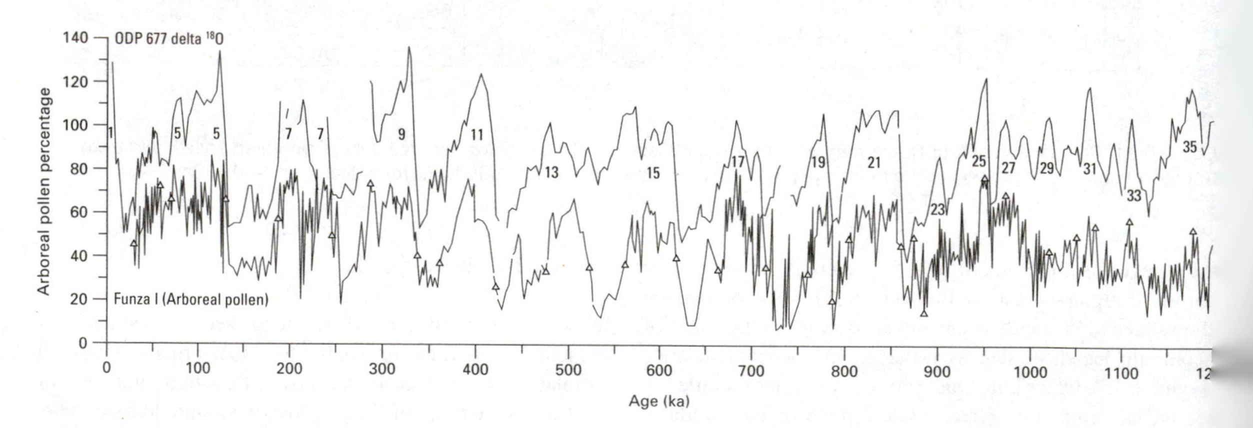

Figure 6.12 Tentative correlation of the pollen record of Funza 1 and the oxygen isotope record of ODP Site 677 (Shackleton *et al.*, 1990) over the last 1.2 Ma (after Hooghiemstra & Sarmiento, 1991).

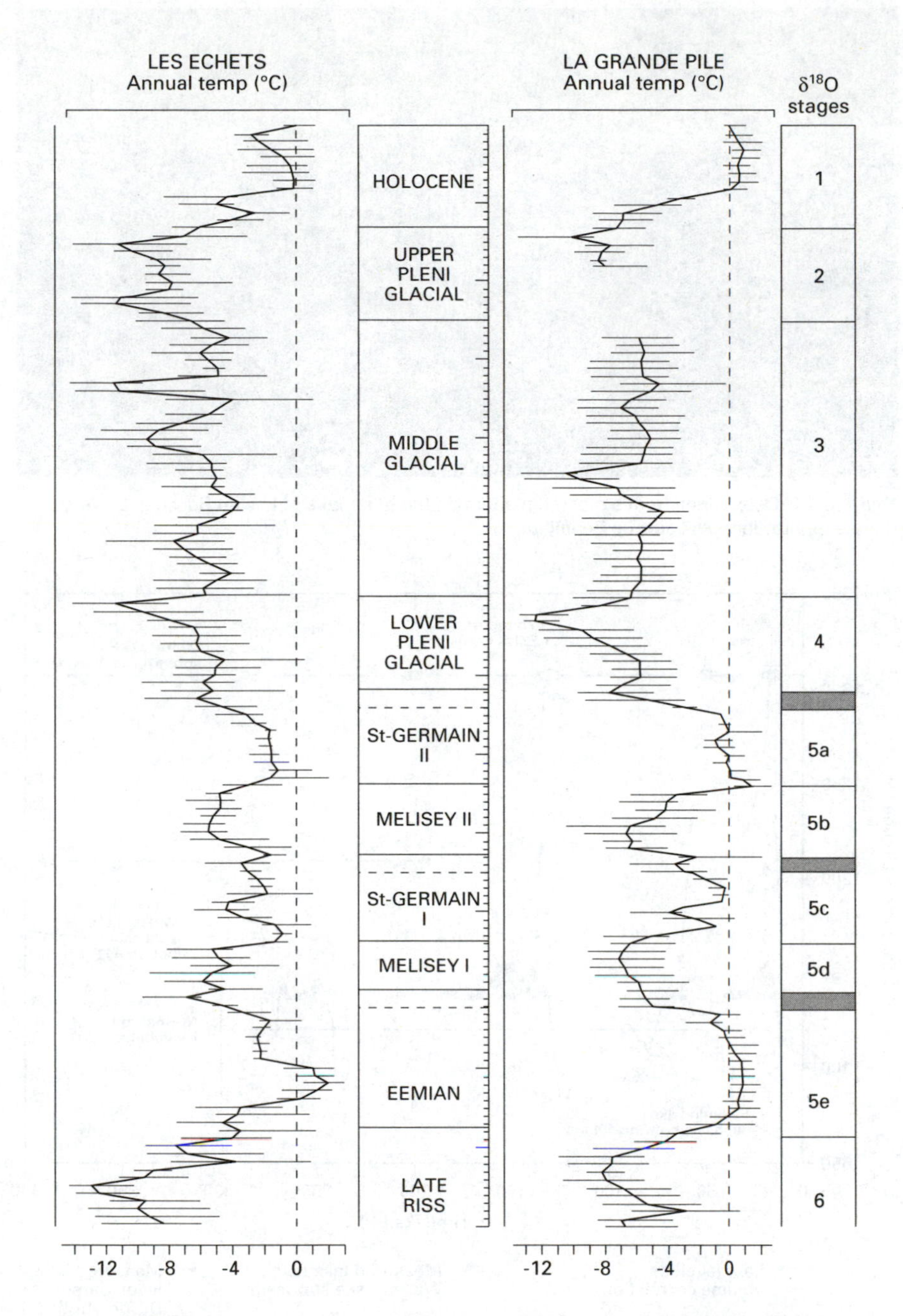

Figure 6.13 Reconstructed mean annual temperatures from Les Echets and La Grande Pile and the oxygen isotope trace from the deep-ocean record (from Guiot *et al.*, 1989).

Figure 6.14 Deep coring from a moored platform in the maar lake of Lac du Bouchet, Le Velay, France (photo: Jacques-Louis de Beaulieu).

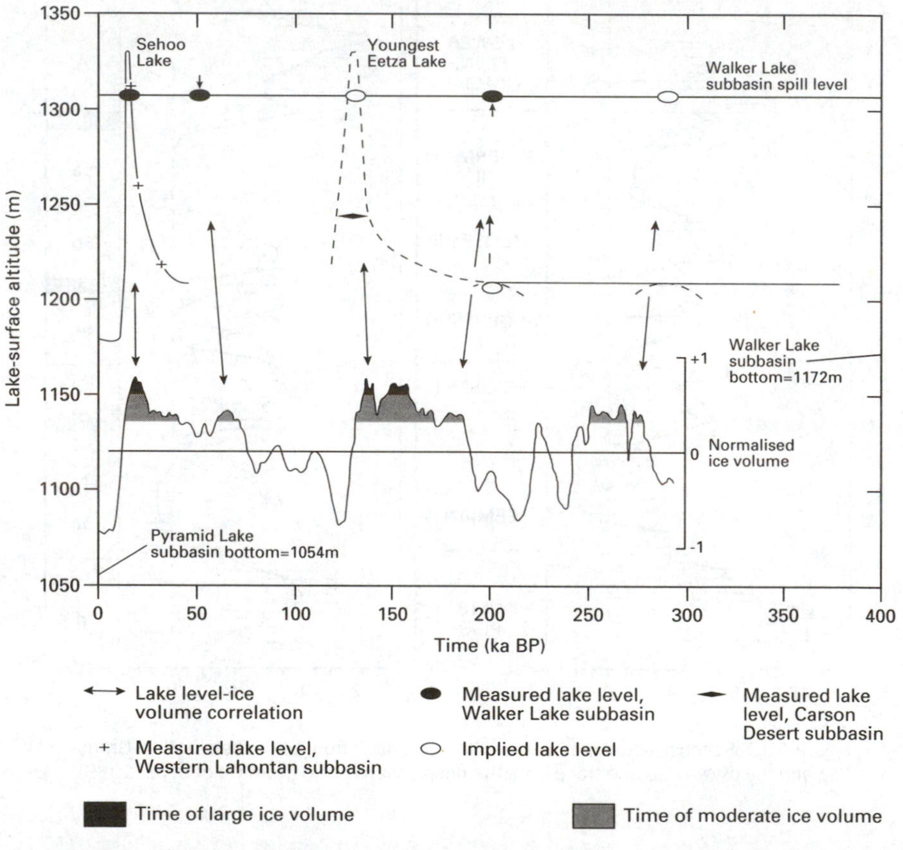

Figure 6.15 Chronology of lake-level changes in the Lahontan Basin, western United States, compared with fluctuations in global ice volumes inferred from the oxygen isotope trace in deep-ocean sediments. Note how high lake levels occurred around, or shortly after, times of maximum ice-sheet size (after Lao & Benson, 1988).

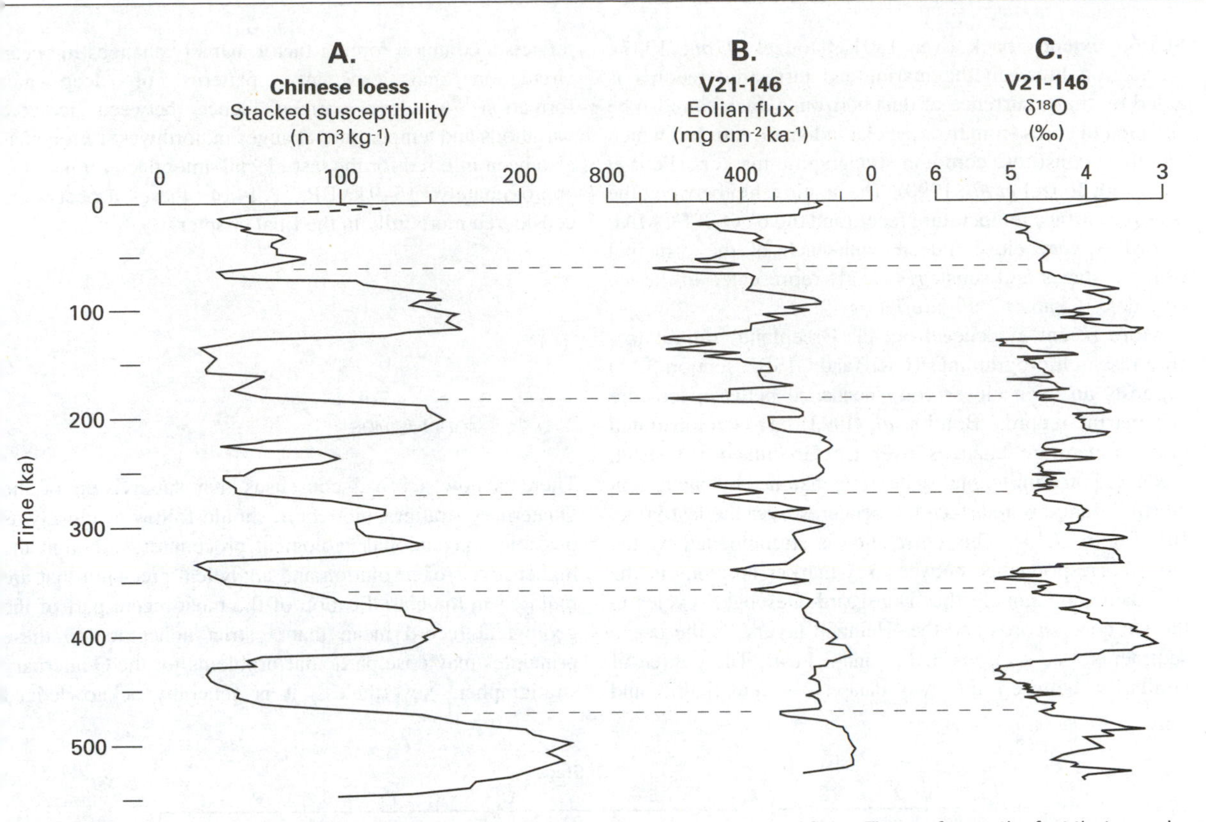

Figure 6.16 A. The stacked loess magnetic susceptibility record from the Loess Plateau of China. The n refers to the fact that samples used for magnetic susceptibility varied in volume; the typical sample size was 8 $m^3\ kg^{-1}$. B. The aeolian dust flux in northwest Pacific core V21-146. C. The $\delta^{18}O$ record from the same core plotted on the SPECMAP timescale (after Kukla *et al.*, 1990).

tectonic disturbance or glacio-isostasy and the main glacial/interglacial cycles recognised in the marine isotopic record. In particular, times of high sea level should equate with episodes of warmer climate and low stands with cooler intervals, although climatic change and sea-level variations may be slightly out of phase since, for example, time is required for the melting of the ice sheets following climatic improvement. In general, however, coastal localities that contain long histories of sea-level change offer a basis for correlation between oceanic and terrestrial records. Of particular importance are the coral reef sequences that are found in many mid/low-latitude regions of the world, which can be dated by radiometric means and which contain evidence of alternating high and low stands that can be related to the marine isotope signal. Raised reef tracts in Barbados, for example, have been assigned to high sea stands from successive interglacials on the basis of U-series and ESR ages and correlated with OI stages 5, 7, 9, 11, 13 and 15, a record extending back over 500 ka (Radtke *et al.*, 1988). In a number of areas, it has also been possible to correlate sea-level changes and OI substages. Data from Bermuda indicate high sea levels at *c.* 125 ka, 105 ka and 85 ka, equivalents of OI substages 5e, 5c and 5a respectively (Harmon *et al.*, 1981), while the Barbados reef sequence shows similar evidence of high sea levels at around these dates (Bard *et al.*, 1990a). In North Island, New Zealand, interbedded marine and terrestrial deposits preserve a record of relative sea-level changes during OI stage 7, with two separate high stands of the sea during OI substages 7a and 7c separated by an episode of low sea-level during substage 7b (Pillans *et al.*, 1988).

6.3.3.6 Ice-core records

Insofar as $\delta^{18}O$ values in ice cores are the reciprocal of those in the marine record (section 3.11), the two signals constitute an ideal basis for correlating between marine and terrestrial sequences. In the core from Camp Century, Greenland, Dansgaard *et al.* (1982) were able to identify the equivalents of OI stages 1 to 5e, with the isotopically warmest ^{18}O values recorded in levels equating to OI substage 5e. In Antarctica, where the record from Vostok

Station extends back over 160 ka (Jouzel *et al.*, 1987), correlation between the marine and terrestrial records is aided by the occurrence of dust horizons which can also be detected in cores from the subpolar Indian Ocean, and which therefore constitute common stratigraphic markers (Petit *et al.*, 1990; Jouzel *et al.*, 1990). The relationship between the isotopic surface temperature record and the $\delta^{18}O$ SPECMAP record is very close indeed, with each of the principal isotopic stages and substages clearly represented in the ice core data (Chapter 7, Figure 7.1).

More recent evidence from the Greenland summit ice-core research programmes (GRIP and GISP2: section 3.11) suggests an even closer correspondence between ice-core and marine records. Bond *et al.* (1993) have demonstrated that temperature changes over the Greenland ice sheet, measured on a millennial scale, were matched by changes in North Atlantic sea-surface temperatures over the last 90 ka BP (Figure 7.14). This correlation is strengthened by the close correspondence between key marker horizons in the two data sets, namely the 'Dansgaard–Oeschger' cycles in the ice core records and the 'Heinrich layers' in the ocean sediments (see sections 3.10.1 and 3.11.4). The degree of similarity between the two datasets is remarkable and reflects a common forcing factor, namely changes in ocean circulation and especially patterns of deep-water formation. A close correspondence between ice-core variations and temperature changes in northwest Europe has also been inferred for the last glacial–interglacial transition, approximately 15–9 ka BP. All of these aspects are considered more fully in the final chapter.

6.4 Conclusions

There is now a broad consensus that subdivision of the Quaternary stratigraphic record should follow as closely as possible conventional geological procedures, although the higher levels of resolution and analytical precision that are required in the classification of the most recent part of the geological record mean that a strict adherence to these principles may pose particular problems for the Quaternary stratigrapher. Nevertheless, it is generally acknowledged

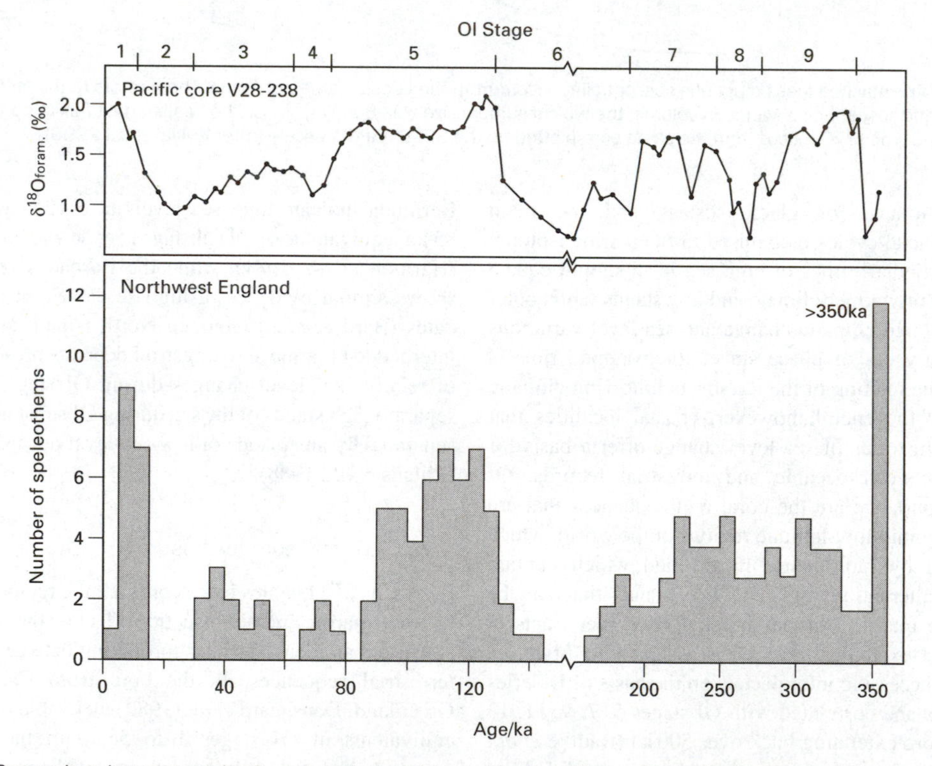

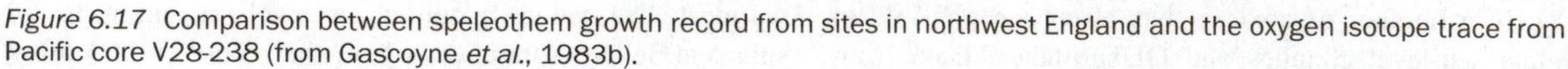
Figure 6.17 Comparison between speleothem growth record from sites in northwest England and the oxygen isotope trace from Pacific core V28-238 (from Gascoyne *et al.*, 1983b).

that, wherever possible, lithostratigraphy should comprise the basic building blocks of the Quaternary terrestrial stratigraphic record, augmented by empirical evidence from bio- and morphostratigraphy along with inferential data from chrono- and climatostratigraphy. Ultimately, geologic–climatic units will be established, and these form the basis for correlation at the regional and continental scales. Geologic–climatic units are also the basis for correlating marine and terrestrial sequences. The essential reference standard for global correlation, however, is the oxygen isotope stratigraphy of the deep-ocean sediments, especially the orbitally tuned SPECMAP timescale, and it is against this record that future stratigraphic schemes will be measured.

Note

1. In deep-sea core investigations, the core numbers code the research vessel, journey number and core number raised on a particular voyage. For example, V28-238 indicates that the core was the 238th obtained during the 28th cruise of the research vessel *Vema*.

CHAPTER 7

The last interglacial–glacial cycle: 130–10 ka BP

7.1 Introduction

Thus far, we have considered the different types of evidence that can be used to reconstruct Quaternary environments, the means by which a timescale for environmental change can be established, and the stratigraphic procedures that enable sedimentary records to be interpreted and meaningful correlations to be effected between often widely scattered localities. In this final chapter, the aim is to demonstrate how these different methods and approaches can be employed to produce an overview of the environmental conditions that prevailed during a particular segment of Quaternary time. Data syntheses at the continental or global scales are important aspects of contemporary Quaternary science for, in addition to providing snapshots of past environmental conditions, they offer a means of cross-checking evidence from different proxy sources, they focus attention on the linkages between processes and components of the global environmental system (e.g. between the glacial, oceanic and terrestrial realms), and they may provide new insights into the causes of environmental changes. All of these aspects are exemplified by the CLIMAP and COHMAP projects (section 4.11), while other similar programmes are considered further in this chapter.

In order to develop these themes, we have selected the interval from *c*. 130 to 10 ka BP, a period which encompasses the last interglacial–glacial cycle of the Quaternary record, and we have done so for the following reasons. First, a considerable body of data has been assembled for this time period, a great deal of which has been produced over the course of the last two decades. The evidence is extremely diverse, ranging from fossil evidence on land to isotopic profiles from polar ice sheets and deep-ocean floors. Nevertheless, a broad measure of agreement between the different proxy records is beginning to emerge and this lends considerable strength to subsequent palaeoenvironmental reconstructions. Secondly, continuous time-series data extending back to, and in some cases beyond, 130 ka BP are available from such diverse records as pollen diagrams from lake sediments, dust and gas profiles from polar ice sheets, and stable isotope traces in deep-ocean sequences. These provide an essential framework for the integration and subsequent interpretation of more fragmented geomorphological and stratigraphical evidence. Moreover, the quality of the data, particularly those from sediments that accumulated after the last cold stage, is such that modelling of a number of different environmental components (ice sheets, vegetation, atmospheric circulation) is now possible. Thirdly, the earth experienced a wide spectrum of environmental conditions during this time period, including episodes of warm and cold conditions ('interglacial' and 'glacial' in the conventional terminology) as well as a number of short-lived climatic oscillations ('stadials' and 'interstadials'). The detailed record of the last 130 ka, therefore, provides a useful template for the interpretation of environmental changes during earlier interglacial–glacial cycles. Fourthly, although a number of problems relating to the dating of events remain to be resolved, a range of geochronological methods has produced a relatively robust timescale for the last 130 ka.[1] Finally, the last 130 ka coincides with the evolution of anatomically and genetically modern humans (Gamble, 1994), and hence the analysis and reconstruction of Quaternary environments during this time period increasingly becomes the study of the environmental context of human evolution. As a consequence, environmental reconstructions for this period are likely to be of interest to a wide range of disciplines, and provide a framework not only for geological, geographical, botanical and zoological investigations, but also for research in the fields of prehistoric archaeology and anthropology.

A number of developments in Quaternary science over recent years make such a review timely. Data from long cores recovered from the Greenland and Antarctic ice sheets have provided startling new insights into the climatic history of the past 130 ka (section 3.11); recent evidence from the deep-ocean floors points to massive reorganisations of ocean circulation systems, often over timescales of less than 100 years, and these have exerted a profound influence on global climates; continuous pollen diagrams are now available from a number of deep lake sites which make up a unique record of terrestrial environmental change extending back to the last interglacial and beyond; and, finally, data from all of these sources form the basis for the construction of increasingly sophisticated **environmental simulation models**, which constitute an exciting new approach to the study of land–ice–atmosphere interactions over the course of the last 130 ka. The development of powerful computers has enabled simulations of the climate system in particular to be constructed at the global scale (Wright *et al.*, 1993). Such **General Circulation Models** (or **GCMs**) are able to simulate climatic states very different from those that exist at present, and can also be used to evaluate the sensitivity of the global climate system to changes in a range of environmental conditions (Street-Perrott, 1991).

These and related aspects of the recent Quaternary record are explored during the course of this chapter. However, what follows is not intended to be an exhaustive examination of the sequence of events during the last 130 ka, for the literature is too voluminous and the constraints of space too restrictive to attempt such an all-embracing overview. For the same reasons, a global review of the evidence lies beyond the scope of the present chapter, and hence the perspective that has been adopted is essentially a Northern Hemisphere one, with the greatest emphasis being placed on events around the North Atlantic basin.

7.2 The stratigraphic framework for the last 130 ka

As was pointed out in Chapter 6, the oxygen isotope signal from the deep-ocean floors is the standard for global correlation. The large number of isotopic records that are now available for the last 130 ka, and which show essentially similar isotopic changes, have been 'normalised' to produce a generalised or standard curve (Figure 7.1A), with a timescale derived by 'orbital tuning' (section 5.5.3). The main features of this curve are:

(a) a marked shift from isotopically heavy to lighter values at Termination II (*c.* 130 ka BP);

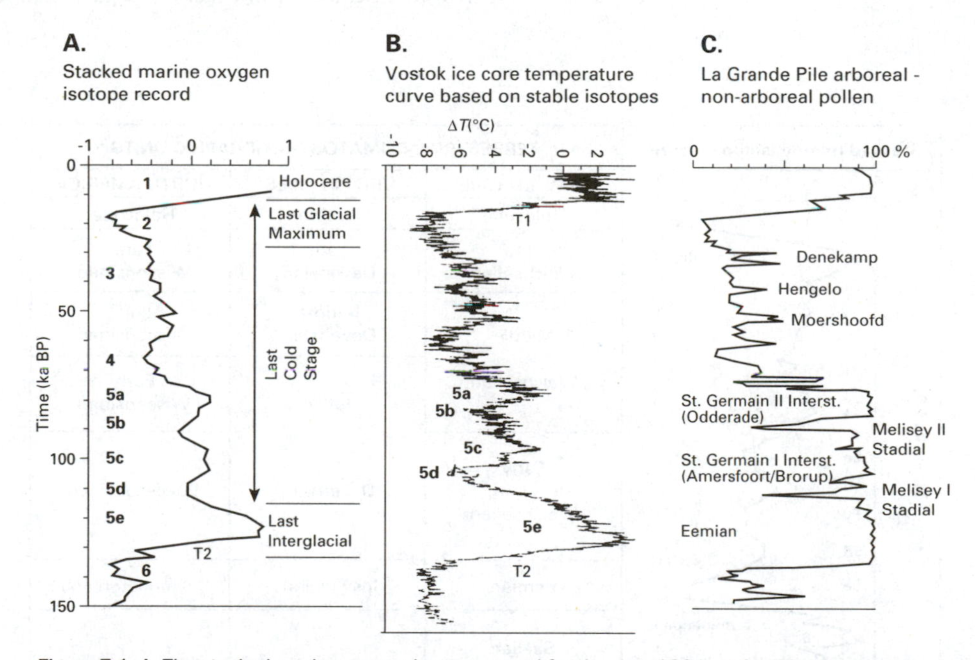

Figure 7.1 A. The stacked marine oxygen isotope record for the past 130 ka (after Martinson *et al.*, 1987). B. The Vostok ice-core temperature curve for the same period (after Jouzel *et al.*, 1987). C. The arboreal/non-arboreal pollen record from La Grande Pile, Vosges, France (after Beaulieu & Reille, 1992a).

(b) the peak of isotopically light values marking oxygen isotope (OI) stage 5e (5.5 of Martinson *et al.*, 1987) of the last interglacial;
(c) the gradual shift to isotopically heavy values throughout OI stages 5d (5.4) to 3, with isotopically lighter episodes in stages 5c (5.3), 5a (5.1) and 3;
(d) the very heavy isotopic values that characterise OI stage 2 (the last glacial maximum);
(e) the abrupt shift to isotopically light values at Termination I, the beginning of the Holocene

This isotopic record, which is essentially a reflection of changes in global ice volumes (section 3.10), is also a climatic proxy for the past 130 ka. Indeed the close correspondence between the marine isotope sequence and climatic records derived from terrestrial data sources (Figure 7.1B and C) indicates that the marine isotope signal in ocean sediments gives a reasonable indication of the main temperature trends in the Northern Hemisphere during the course of the last interglacial–glacial cycle. Moreover, as the boundaries of the isotopic stages and substages are broadly time-parallel, oxygen isotope profiles provide a stratigraphic framework for this time period. The following discussion is therefore structured largely in terms of the nomenclature of the marine oxygen isotope sequence. The climatostratigraphic equivalents of the isotopic stages and substages in the terrestrial records are shown in Figure 7.2.

7.3 The last interglacial (OI substage 5e)

7.3.1 Defining the last interglacial

The beginning of the last interglacial is reflected in the marine records by the abrupt shift to lighter isotopic values (Termination II) around, or shortly after, 130 ka BP (Figure 7.2). Although there is some evidence to suggest that the OI stage 6/5 transition may have been marked by short-lived climatic oscillations (Seidenkrantz, 1993), these were relatively minor compared with the magnitude and apparent speed of the change from full glacial to interglacial conditions. A similar rapid climatic amelioration is also apparent in other proxy records (Figure 7.1). Termination II therefore provides a clear marker horizon for defining the lower boundary of the last interglacial, and for correlating marine and terrestrial records. Establishing the upper boundary of the last interglacial has not proved to be quite so straightforward, however, because the return to full glacial conditions appears to have been more protracted, and was interrupted by pronounced climatic oscillations (section 7.4). Indeed, in the early stages of the development of the oxygen isotope stratigraphy, there was a tendency to regard the whole of OI stage 5 as the marine equivalent of the terrestrial Eemian, Ipswichian or Sangamonian Interglacial

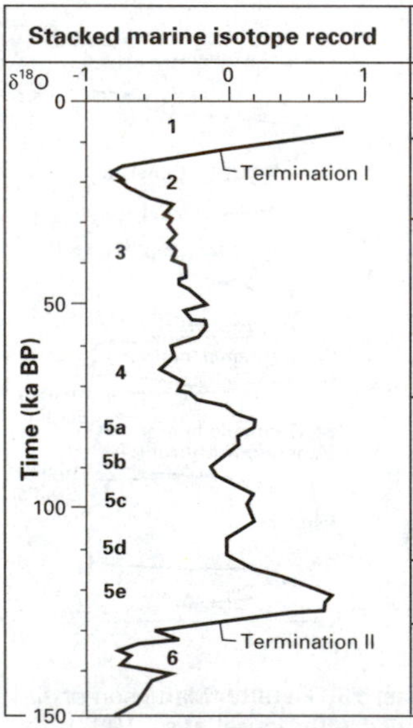

Stacked marine isotope record (stage)	N. W. EUROPE	BRITISH ISLES	NORTH AMERICA
1	Holocene	Flandrian	Holocene
2 (Termination I)	Late Weichselian	Late Devensian	Late Wisconsinan
3	Middle Weichselian	Middle Devensian	Middle Wisconsinan
4	Middle Weichselian	Early Devensian	Early Wisconsinan
5a, 5b, 5c, 5d	Early Weichselian	Early Devensian	Eowisconsinan
5e	Eemian	Ipswichian	Sangamonian
6 (Termination II)	Saalian		

Figure 7.2 The stacked marine oxygen isotope record for the last 130 ka (Martinson *et al.*, 1987) and terrestrial climatostratigraphic units from northwest Europe and North America.

(Figure 7.2). However, better resolution of marine isotope records has led to a five-fold subdivision of OI stage 5 (substages 5a to 5e) and it is now generally accepted that only the first part of the stage is of true interglacial status; hence the last interglacial *sensu stricto* is represented by OI substage 5e (section 6.2.3.5). The end of the last interglacial is therefore reflected in the marine isotope sequence by the substage 5e/5d boundary.

The transition from the last interglacial to the succeeding cold stage has attracted considerable research interest, for it is the most recent example in the geological record of a termination of a major warm episode. Moreover, the last interglacial–glacial transition provides the nearest natural analogue for the end of the present interglacial, and hence understanding the nature, timing and environmental consequences of climatic change at the end of the previous warm stage is an essential prerequisite for the development of models of future climatic scenarios in which human influence is likely to be an additional and possibly crucial factor (Houghton *et al.*, 1990, 1992).

7.3.2 Proxy records from the last interglacial

A range of proxy data suggests that, throughout the mid- and high latitude regions of the Northern Hemisphere, climate at the last interglacial maximum was considerably warmer than that of the present day, with palaeobotanical records pointing to extensive woodland over much of Europe and North America. Mean annual temperatures 1–2°C higher than at present are indicated by pollen evidence from France and northwest Germany (Guiot *et al.*, 1989; Guiot, 1990), while summer temperatures 2°C above those of the present day are implied by palaeobotanical data from western Norway, The Netherlands and Italy (Mangerud *et al.*, 1981; Zagwijn, 1989; Follieri *et al.*, 1989). Fossil coleopteran assemblages suggest summer temperatures only slightly above those of the present in southern Europe (Figure 7.3A), 2–3°C above present values in lowland Britain (Coope, 1974), but as much as 5°C above contemporary levels in northern Sweden (Lindroth & Coope, 1971). Evidence from North America is broadly in agreement with that from Europe, pollen data indicating temperatures ~1.5°C above Holocene levels in California (Figure 7.3B) and at least 4°C warmer than at present in Atlantic Canada (Mott, 1990).

Marine and ice-core data from the last interglacial generally support the climatic interpretations based on terrestrial evidence. Estimates of sea-surface temperatures (SSTs) for the 5e thermal maximum (Figure 7.4) range from 22°C (4°C warmer than present) at latitude 45°N, to *c.* 5°C (similar to present) at 76°N (Sejrup & Larsen, 1991; McManus *et al.*, 1994). Oxygen isotope data from the GRIP ice core suggest that temperatures in Greenland were ~2°C warmer than now during the warmest stages of the Eemian (GRIP Members, 1993), while a similar temperature difference between the last interglacial and the present is evident in isotopic profiles from Antarctica (Jouzel *et al.*, 1990).

The terrestrial, marine and ice-core data are consistent in indicating that the thermal maximum was achieved early in OI substage 5e, after which temperatures declined progressively. However, recent data from a range of sources suggest that the last interglacial was also characterised by marked temperature fluctuations. The GRIP ice-core record spanning OI substage 5e shows three distinct warm episodes (5e1, 5e3 and 5e5: Figure 7.5A) separated by two intervals of much lower temperatures (5e2 and 5e4). Similar evidence has been found in Eemian pollen and sediment profiles from sites in Europe (Field *et al.*, 1994; Thouveny *et al.*, 1994: Figure 7.5B), in marine benthic Foraminifera from shelf sediments in northwest Europe (Seidenkrantz *et al.*, 1995), and in marine $\delta^{18}O$ records and reconstructed sea-surface temperature data from the southern part of the North Atlantic (Eglinton *et al.*, 1992). However, an even greater degree of climatic instability during the last interglacial may be reflected in the GRIP ice-core record, for the isotopic trace suggests, in addition to the changes outlined above, a series of short-lived (70–750 years) and abrupt climatic oscillations during which temperatures may have fallen by up to 14°C within a matter of decades (Dansgaard *et al.*, 1993). Prior to this discovery, there had been a general assumption that interglacials were episodes of relative climatic stability, and that temperature changes of this magnitude were not possible over such short timescales. The GRIP data challenge these assumptions and raise the question as to whether such abrupt natural changes could occur under the present interglacial climatic régime. The economic, political, social and environmental consequences of such catastrophic temperature fluctuations would clearly be very far-reaching.

7.3.3 Dating the last interglacial

At present, there is a difference of opinion over the age and duration of the last interglacial. According to the SPECMAP timescale, OI substage 5e (event 5.5: Figure 5.32) began at 128 ± 3 ka BP and ended around 115 ka BP (Imbrie *et al.*, 1989). Dates that are in broad agreement with this chronology have been obtained from the Vostok ice-core record (Petit *et al.*, 1990), from coral reef sequences and marine speleothem records from both the Caribbean and the Pacific (Harmon *et al.*, 1983; Bard *et al.*, 1990a; Lundberg &

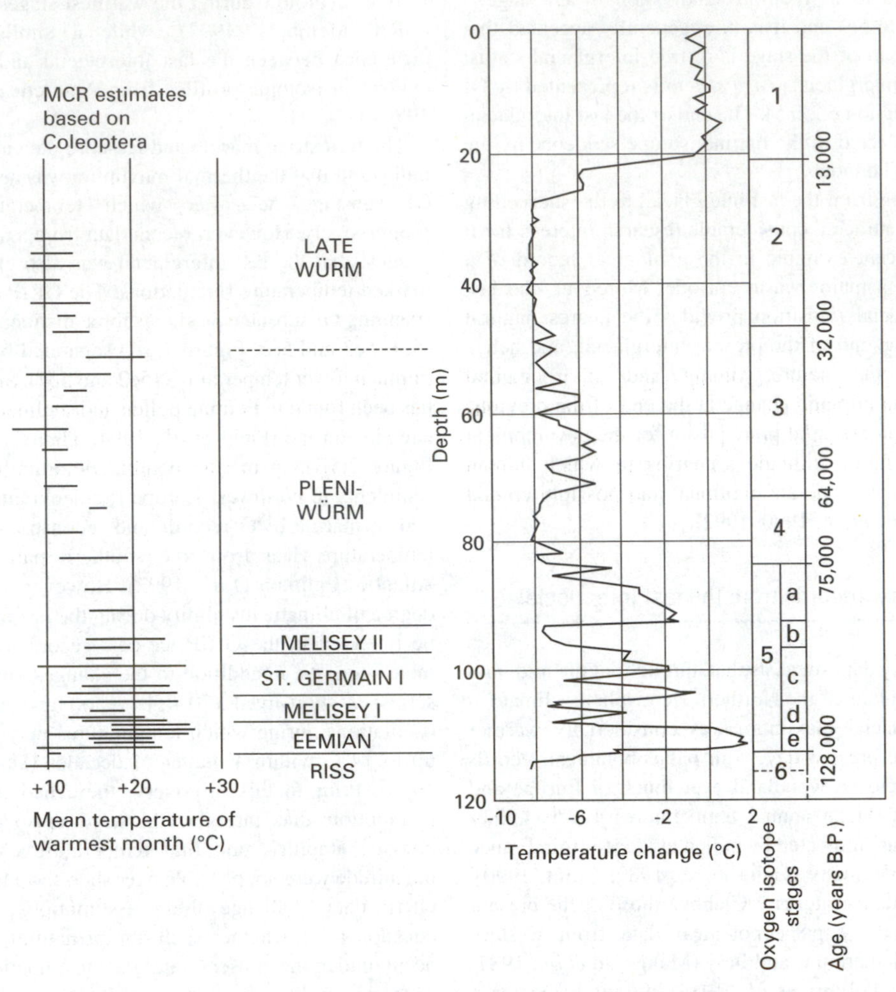

Figure 7.3 A. Mutual climatic range temperature reconstructions based on coleopteran data from La Grande Pile, Vosges, France (after Ponel, 1995). B. Reconstructed temperature changes (relative to modern values) based on pollen data from the Clear Lake site, California (after Adam & West, 1983).

Ford, 1994), and from the last interglacial soil in the Baoji loess–soil sequence on the Loess Plateau of China (Ding *et al.*, 1994). Annually laminated lake sediments from sites in western Europe suggest that an interglacial climate persisted for 8–10 ka (Turner, 1975a; Field *et al.*, 1994), an estimate that again matches that based on the SPECMAP chronology (Keigwin *et al.*, 1994). However, other evidence points to a longer duration for the last interglacial and, in particular, to an earlier onset. An alternative chronology for the Vostok ice core, for example, places the beginning of the last interglacial at around 140 ± 15 ka BP (Jouzel *et al.*, 1989), while in the GRIP ice core, the equivalent of OI substage 5e spans the time interval 133–114 ka BP (Dansgaard *et al.*, 1993). Other data that suggest an earlier beginning (*c.* 133–140 ka BP) for the last interglacial include climatic records in cave calcite/speleothem from Nevada (Winograd *et al.*, 1992; Coplen *et al.*, 1994) and Britain (Gascoyne *et al.*, 1983b), and models of eustatic sea-level variations prior to and during the last interglacial period (Lambeck & Nakada, 1992).

Age estimates for the duration of the last interglacial, therefore, appear to be polarised into a 'shorter chronology' (*c.* 10 ka or so) and a 'longer chronology' spanning almost twice that time range. One explanation for this temporal

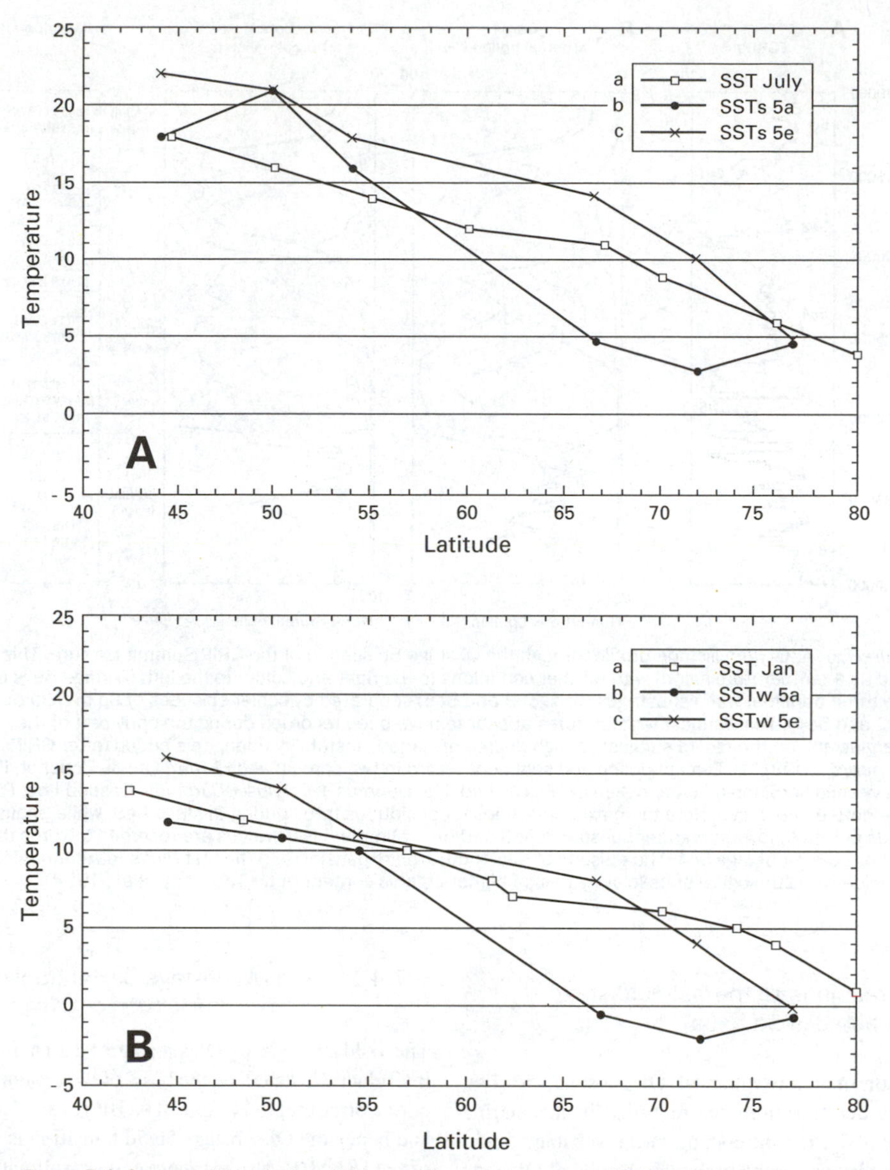

Figure 7.4 Summer (A) and winter (B) sea-surface temperatures (SSTs) in the North Atlantic along a transect from Svalbard (right) to the latitude of northern Spain (left) for (a) the present day; (b) OI stage 5e; (c) OI stage 5a (from Sejrup & Larsen, 1991).

discrepancy is that climatic instability in the early part of the last interglacial (Figure 7.5) may have delayed the melting of the OI stage 6 (Saalian) ice sheets in Eurasia and North America, causing both the deep-ocean isotope signal and the interglacial sea-level rise to lag behind the global climatic shift that is evident in other records (Dansgaard *et al.*, 1993). Alternatively, differences between the two timespans may reflect shortcomings in the datasets including, for example, variable levels of resolution in different stratigraphic contexts (e.g. the often poor stratigraphic resolution in deep marine sequences compared with the higher resolution of some lake sites); problems of a uniform definition of what constitutes the beginning or end of an 'interglacial' in different proxy records (section 6.2.3.6); and errors in dating techniques, particularly where conventional $^{230}Th/^{234}U$ has been employed (Gallup *et al.*, 1994).

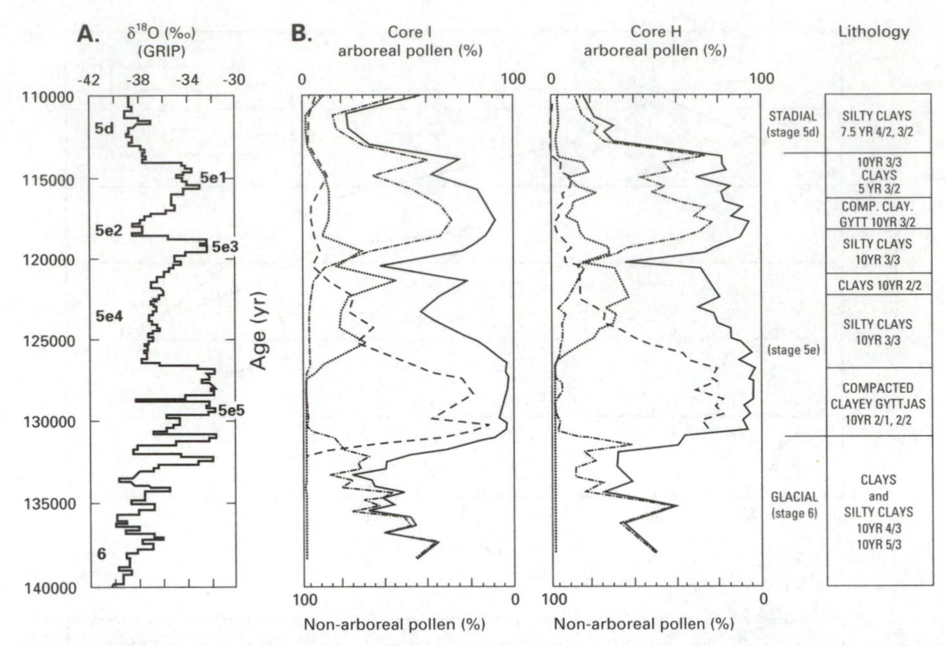

Figure 7.5 A. Oxygen isotope profile through the OI stage 5e section of the GRIP Summit ice core. This can be read as a temperature record with warmer conditions to the right and colder to the left. OI stage 5e is divided into three principal warm substages (5e1, 5e3 and 5e5) separated by cooler episodes of up to 5 ka duration (5e2 and 5e4). The warmest temperatures appear to have been recorded during the early part of the interglacial, and the record suggests a high degree of climatic instability during this period (after GRIP Members, 1993). B. Eemian pollen and sediment record in two cores (I and H) from Lac du Bouchet, France. Heavy solid line: non-arboreal pollen (NAP). Dashed line: *Quercus* + *Corylus* + *Carpinus*. Dashed line: *Pinus*. Fine dotted line: *Picea*. Note the maximum values for deciduous trees during OI stage 5e5, while peaks in the *Pinus* curves follow the warmer substages 5e3 and 5e1. Minimum NAP values are recorded towards the end of the cooler substage 5e4. The episode of deciduous tree expansion and the first *Pinus* maximum also coincide with lithological units (clay-gyttjas) of higher organic content (after Thouveny *et al.*, 1994).

7.4 The transition to the last cold stage (OI substages 5d to 5a)

Following the thermal maximum of OI substage 5e, two marked climatic deteriorations are recorded in the marine isotope records: a significant cooling at the substage 5e/5d boundary, and a further abrupt cooling at the end of OI stage 5. During the course of OI stage 5, however, there is also evidence for two well-defined climatic ameliorations after OI substage 5e, reflected in OI substages 5c and 5a. A similar sequence has subsequently been found in continental and ice-core records (Figure 7.1). In North America, the period covered by OI substages 5d to 5a is termed the 'Eowisconsin' (Richmond & Fullerton, 1986a), while in Europe it is generally referred to as the Early Weichselian (Behre, 1989). The entire transition from interglacial conditions at the peak of OI substage 5e to full glacial conditions in OI stage 4 spans a time interval of more than 50 ka.

7.4.1 The OI substage 5e/5d transition

The cold episode of OI substage 5d that followed the last interglacial is dated, on the basis of the orbitally tuned marine chronostratigraphy, to *c.* 110 ka BP (event 5.4: Figure 5.32), and hence the OI substage 5e/5d transition is assigned an age of ~115 ka BP. This estimate agrees well with the GRIP and Vostok ice-core timescales, and also with radiometric dates from terrestrial sites (Gascoyne *et al.*, 1983b; Winograd *et al.*, 1992). The scale of the climatic change is most clearly reflected in the oxygen isotope records from the deep oceans and from the ice sheets. In the stacked (normalised) $\delta^{18}O$ curve the isotopic shift between OI substages 5e and 5d (events 5.5 and 5.4) is of the order of 0.8‰, approximately 50 per cent of the observed oxygen isotopic difference between the last glacial (OI stage 2) and the present interglacial (Figure 5.32). In the GRIP ice core, the isotopic change is of the order of 6‰ (Dansgaard *et al.*, 1993), which equates to a decline in air temperature of ~9°C (Johnsen *e al.*, 1992), while a decline in air temperature of ~10°C is reflected in the Vostok ice-core record (Figure 7.1). Climatic reconstructions based on pollen data from sites in France and the southwest United States suggest a drop in mean annual temperature of *c.* 8–9°C (Guiot *et al.*, 1989; Figure 7.3B). In the subtropical North Atlantic, sea-surface temperatures (SSTs) fell by ~6°C (Eglinton *et al.*, 1992), while further north a sharp decline in SSTs was accompanied by a marked increase in ice-rafted debris (IRD) from the expanding Northern Hemisphere glaciers and ice sheets (Heinrich, 1988; McManus *et al.*, 1994).

The pronounced shift to heavier isotopic values at the 5e/5d boundary in marine records mainly reflects an increase in the storage of $H_2{}^{16}O$, which implies a significant increase in volumes of continental land ice (section 3.10). Precisely when the great ice masses began to develop over North America and Eurasia, and the rate of ice sheet expansion, cannot be established with certainty, but falling sea levels on the coast of western Norway during the later part of the Eemian suggest that global land ice volumes had begun to increase before the end of the last interglacial. The Norwegian evidence also indicates that the initial fall in sea level following the isotopic peak of 5e was due to ice accumulation outside Scandinavia, probably in North America and/or Antarctica (Mangerud, 1991a). The extent of subsequent glacier and ice sheet development is reflected in the overall drop in eustatic sea level of more than 60 m from the peak of OI substage 5e to the trough of 5d (Shackleton, 1987). The build-up of land ice was so rapid that by the end of substage 5d, global land-ice volume was already some 50 per cent of that of the last glacial maximum (at *c.* 18–22 ka BP: section 7.5).

In North America, the Laurentide ice sheet began to develop over Keewatin, Labrador and Baffin Island and then advanced into the eastern Canadian Arctic early in OI stage 5, possibly during substage 5d (Clark *et al.*, 1993). A major glacial advance in the mountains of western North America has also been dated to substage 5d (Richmond & Fullerton, 1986b). TL dating of glacial evidence from western Norway indicates a substantial ice advance (and therefore the existence of a large ice mass) during OI substage 5d (Mangerud, 1991b; Figure 7.6), while amino-acid ratios

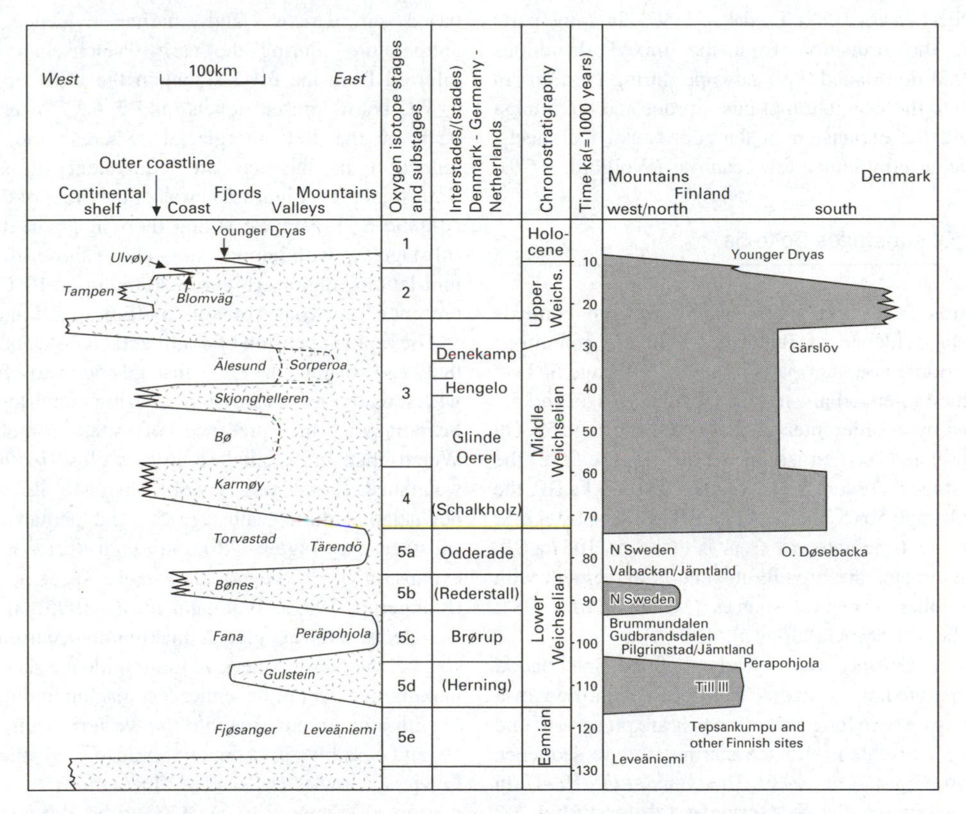

Figure 7.6 Schematic glaciation curve for the last interglacial–glacial cycle in Scandinavia. The left-hand curve is for the west side of the mountains, the right-hand curve for the eastern and southern side of the mountains (after Mangerud, 1991a; Baumann *et al.*, 1995).

suggest an ice advance at *c.* 98–108 ka BP (Larsen & Sejrup, 1990). Foraminiferal evidence from the North Atlantic suggests that SSTs remained relatively high for some time after the end of OI substage 5e, and so the rapid build-up of the Laurentide ice sheet has been attributed to the juxtaposition of a rapidly cooling ('glacial') landmass and a still warm ('interglacial') ocean, the latter constituting a ready moisture source for the expanding continental glaciers (Ruddiman *et al.*, 1980).

Detailed records of the transition from interglacial to full glacial conditions are found in deep lake sites in western and southern Europe. In these records, the terrestrial equivalent of marine OI substage 5d is termed the **Melisey I Stadial** in France (Woillard, 1978), the **Herning Stadial** in Germany (Behre, 1989) and the **Gulstein Stadial** in western Norway (Mangerud, 1989; Table 7.1 and Figure 7.6). Pollen data from high-resolution sequences, such as La Grande Pile in France (Figure 6.13), indicate that during this interval the deciduous forests of the Eemian were replaced first by boreal woodland and eventually by open tundra (Beaulieu & Reille, 1989, 1992a). In southern Europe, the interglacial deciduous woodland gave way to a landscape of tundra and steppe (Follieri *et al.*, 1989; Tzedakis, 1993). In some parts of Europe, the transition from the mixed deciduous woodland that dominated the landscape during the Eemian Interglacial, to the vegetation of pine, spruce and birch taiga that heralded the expansion of the continental ice sheets, may have occurred within a few centuries (Woillard, 1979).

7.4.2 OI substages 5c to 5a

Proxy records from both marine and terrestrial contexts provide clear evidence of fluctuating climatic conditions during the middle and later parts of marine OI stage 5. Two clearly defined interstadials, marine OI substages 5a and 5c, are separated by a colder interval, marine OI substage 5b. On the orbitally tuned oxygen isotope record (Figure 5.32) the peak of substage 5a (event 5.1) is centred on *c.* 79 ka BP, the trough of substage 5b (5.2) on *c.* 91 ka BP, while substage 5c (5.3) spans the time interval from *c.* 96 ka to 103 ka BP. These age estimates are broadly in general agreement with those from other terrestrial sources (Miller *et al.*, 1983; GRIP Members, 1993; Gallup *et al.*, 1994).

In western Europe, two clearly defined interstadial episodes separated by a stadial have been identified in a number of sites where long pollen records are preserved, and these can be correlated with the marine isotope sequence (Behre, 1989; Guiot *et al.*, 1989; Thouveny *et al.*, 1994). In France, for example, the **St Germain I Interstadial** has been equated with OI substage 5c; the **Melisey II Stadial** with 5b; and the **St Germain II Interstadial** with 5a. Possible correlatives of these episodes in other European sequences are shown in Table 7.1. The vegetation cover of Europe during the warmer intervals varied from mixed woodland in southern areas to boreal woodland further north, while in the intervening stadial phase, woodland was replaced by dwarf shrub tundra and steppe. Climatic reconstructions based on pollen and coleopteran data suggest that mean annual temperatures in France were only a little lower (by 1–2°C) than those of today during the St Germain I and II Interstadials, but were around 5–7°C lower during the Melisey II Stadial (Guiot *et al.*, 1989, 1992; Ponel, 1995). Very similar estimates have been obtained for the equivalent parts of the Clear Lake sequence in California (Figure 7.3B). Proxy records from other parts of Europe indicate that summer temperatures may have been 2–3°C below present in Scandinavia (Helle *et al.*, 1981; Lundqvist, 1986a), and perhaps as much as 4°C lower in Britain and The Netherlands (Coope, 1959; Andersen, 1961). In both interstadials, however, winter temperatures appear to have been well below present values, suggesting a greater degree of continentality than at the present day.

These reconstructions correspond very closely to those based on ice-core and marine evidence. Maximum temperatures during the early Weichselian interstadials inferred from the $\delta^{18}O$ signal in the GRIP ice core were 2–3°C below present levels, and 5–6°C lower than at the peak of the last interglacial, whereas the temperature reduction in the ice-core equivalent of substage 5c (interstadial 23: Figure 7.7) was of the order of 6–7°C. North Atlantic SSTs suggest strong thermal gradients during OI substage 5a, with temperatures above those of today in the mid-latitude regions (Figure 7.4) but up to 10°C cooler than present off the coasts of Norway (Sejrup & Larsen, 1991).

The rapid expansion of continental ice sheets and glaciers that had begun in OI substage 5d was followed by widespread ice recession during substage 5c. In Scandinavia, the presence of sites containing Early Weichselian interstadial sediments close to the mountain watershed (Lundqvist, 1986b) suggests almost complete deglaciation during substage 5c, and perhaps also during substage 5a (Figure 7.6), although there was a major expansion of the Fennoscandian ice sheet in substage 5b (Mangerud, 1991a; Baumann *et al.*, 1995). In contrast to Europe, where the glacial maximum occurred during OI stage 2, the North American Laurentide ice sheet reached its maximum extent of the entire last glaciation during OI stage 5, with a major advance into the western Arctic in substage 5b, and possibly also a contemporaneous advance into the St Lawrence lowlands. Marked fluctuations in the ice sheet margin also appear to have occurred during this period, however. Late in stage 5, for example, the Hudson Bay lowland may have been ice free, and there was a major

Table 7.1 Correlation of Late Quaternary stratigraphies of Europe with the deep-sea oxygen isotope stages (after Behre, 1989; Ehlers *et al.*, 1991; Baumann *et al.*, 1995).

Marine oxygen isotope stages	Stage	Britain	Stage	France	Netherlands Denmark	Germany	Western Norway
4			Middle Weichs.			Shalkholz	Karmøy
5a	Early Devensian	Brimpton	Early Weichselian	St Germain II	Odderade	Odderade	Torvastad
5b				Melisey II		Rederstall	Bønes
5c		Chelford		St Germain I	Brørup/Amersfoort	Brørup	Fana
5b				Melisey I		Herning	Gulstein
5e		Ipswichian		Eemian	Eemian	Eemian	Eemian (Fjøsanger)

retreat (during 5a?) on Baffin Island. Fossil records from these regions indicate that, following deglaciation, climatic conditions were very similar to those of the present (Clark *et al.*, 1993). The marine oxygen isotope signal implies global sea levels more than 50 m below those of the present during OI substage 5b (Figure 3.46), and 20–25 m below present-day levels in substages 5a and 5c (Shackleton, 1987). However, dated marine terraces from tectonically stable areas suggest that for stage 5a at least, a sea-level lowering of 13–18 m may be a more realistic estimate (Richards *et al.*, 1994). If, as seems likely, the ice cover over Scandinavia was significantly reduced during substages 5c and 5a, the sea-level data indicate that relatively large ice masses persisted elsewhere (presumably in North America), although the dimensions of these ice sheets cannot be determined on present evidence.

7.4.3 The OI stage 5/4 transition

The end of OI stage 5 is dated to around 70 ka BP on the SPECMAP timescale (Figure 5.32) and to about 75–80 ka BP in the GRIP ice core (Figure 7.7). In the marine $\delta^{18}O$ profile, the difference in isotopic composition of ocean waters between the peak of OI substage 5a and the trough of stage 4 was of the order of 0.6–0.7‰, while in the GRIP and Vostok ice-core records, the isotopic shift is of the order of 6–8‰. The oxygen isotopic signal reflects a further build-up of continental ice masses and a complementary fall in eustatic sea level of more than 60 m (Shackleton, 1987). The shift to heavier isotopic values in the marine record is equivalent to a cooling of ~10°C, an estimate which is consistent with climatic reconstructions from European

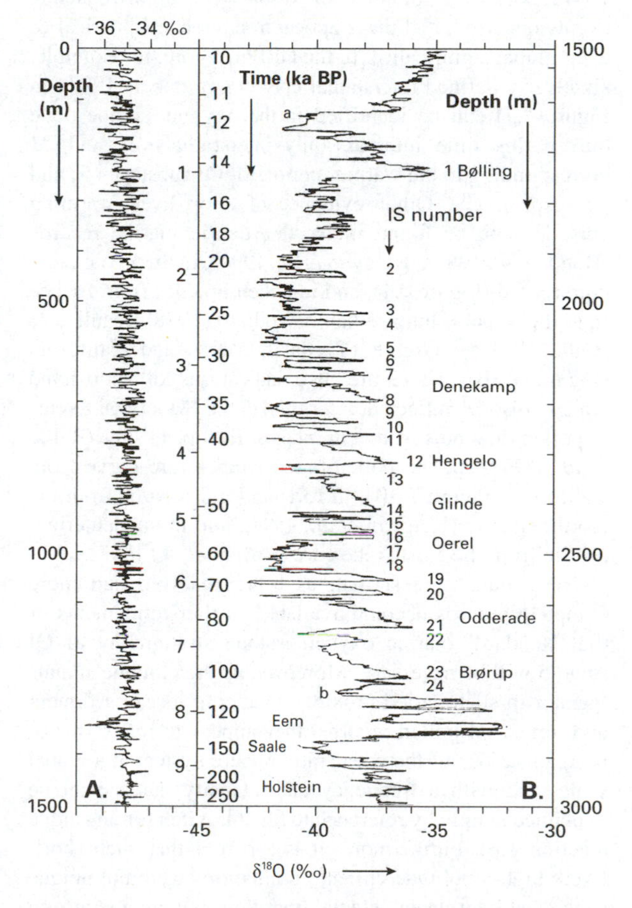

Figure 7.7 The continuous GRIP Summit $\delta^{18}O$ record, showing 24 separate interstadial events during the course of the last cold stage, and suggested correlations with European interstadial episodes (after Dansgaard *et al.*, 1993).

pollen profiles (Guiot *et al.*, 1989, 1992). In California, however, the drop in temperature was nearer 7°C (Figure 7.3B). SST estimates for the North Atlantic indicate a sharp decline at the stage 5/4 boundary (Bond *et al.*, 1993), perhaps by as much as 5–6°C in the lower latitudes (Eglinton *et al.*, 1992). Pollen records show that the mixed woodland of southern Europe was replaced by open steppe, while in the north and west, boreal woodland gave way to a landscape of taiga and barren tundra.

7.4.4 Short-lived 'events' during the last interglacial-glacial transition

Recent high-resolution records suggest that, within the standard isotope stratigraphy for the time period 70–115 ka BP, a number of distinctive, high-frequency events occurred, and these appear to have hemispherical or even global significance. In the GRIP ice core, for example, six clearly defined interstadial episodes (numbers 19–24 in Figure 7.7) can be identified in the oxygen isotope trace during this time interval, only interstadials 21 and 23 corresponding to the warmer periods of OI substages 5a and 5c respectively. Other evidence of short-lived warming episodes can be found in North Atlantic marine records (Bond *et al.*, 1992a; Keigwin *et al.*, 1994), in the Vostok ice-core record (Figure 7.1), and in pollen profiles from France (e.g. the 'Ognon Interstadials'; Woillard, 1978; Beaulieu & Reille, 1992a), Greece (Tzedakis, 1993) and California (Figure 7.3B). There are also indications of short-lived colder episodes reflected, *inter alia*, in the 'Montaigu Event' in pollen diagrams from a number of European sites (Reille *et al.*, 1992), in the pollen-based temperature curve from California (Figure 7.3B), in foraminiferal records from the North Atlantic (Keigwin *et al.*, 1994), and in the deuterium profile from the Vostok ice core (Sowers *et al.*, 1993).

These data suggest that, as better resolved and more complete records become available, further refinements of the 'standard' marine oxygen isotope stratigraphy of OI stage 5 will be necessary. Moreover, as the climatic signals register in such varied proxies as ice-core, ocean sediment and terrestrial pollen records, they appear to reflect major reorganisations of the ocean–atmosphere system at a global scale, and with a frequency and intensity that cannot be explained simply by reference to Milankovitch forcing alone (section 1.6). Furthermore, it is apparent that such short-lived, high-amplitude climatic oscillations were not unique to the last interglacial–glacial transition, but are a recurring feature of the record of the last cold stage (see below). The possible causal mechanisms behind these abrupt climatic events are considered below (section 8).

7.5 The last cold stage (OI stages 4 to 2)

The last cold stage, spanning the time interval from *c.* 75 ka BP to 10 ka BP and termed the **Weichselian** in northern Europe, the **Würmian** in central Europe, the **Devensian** in Britain and the **Wisconsinan** in North America, is represented by OI stages 4, 3 and 2 in the marine isotope record (Figure 7.1). Although essentially an episode of cold climatic conditions in the mid- and high latitude regions, it was also characterised by a number of short-lived warmer intervals of interstadial status. In this section, we examine the evidence for environmental change in the circum-Atlantic region during the last cold stage, focusing on data from marine, terrestrial, glacial and ice-core records.

7.5.1 Events during the last cold stage: the marine record

The oxygen isotope trace from deep-ocean cores contains two isotopically 'heavy' excursions (stages 4 and 2) and an intervening stage (3) in which lighter isotopic values are recorded. On the orbitally tuned timescale (Figure 5.32), the OI stage 4/3 and 3/2 boundaries are dated to *c.* 58 ka BP and 23 ka BP respectively. Termination I, which marks the end of the last cold stage, is dated at *c.* 11 ka BP (Martinson *et al.*, 1987). The isotopic signal suggests that a major increase in continental ice volumes occurred during OI stage 4, and especially during OI stage 2 (see below), but that OI stage 3 was an interval when global ice volumes were somewhat reduced. The marine isotope record (Figure 3.46) implies that sea levels rose from below −75 m to approximately −50 m between OI stages 4 and 3, followed by an overall sea-level fall to *c.* −120 m in OI stage 2 (Shackleton, 1987). Uranium-series dates from Barbados confirm a sea-level fall to at least −85 m at *c.* 30 ka BP (Bard *et al.*, 1990a).

Many of the more recent marine oxygen isotopic records, however, contain a greater degree of variability than the three-fold sequence of OI stages 4, 3 and 2 described above. In Figure 3.46, for example, there are at least six excursions from lighter to heavier isotopic values between *c.* 60 and 20 ka BP, suggesting episodic fluctuations in overall global ice volumes. Other marine proxies also point to significant environmental changes during this period. In the North Atlantic, for example, planktonic foraminiferal evidence indicates marked changes in SSTs, particularly between 40° and 55°N (Figure 7.8). Moreover, layers of sediments rich in ice-rafted debris (IRD) and usually poor in Foraminifera have been found in sediment cores from the mid- and high

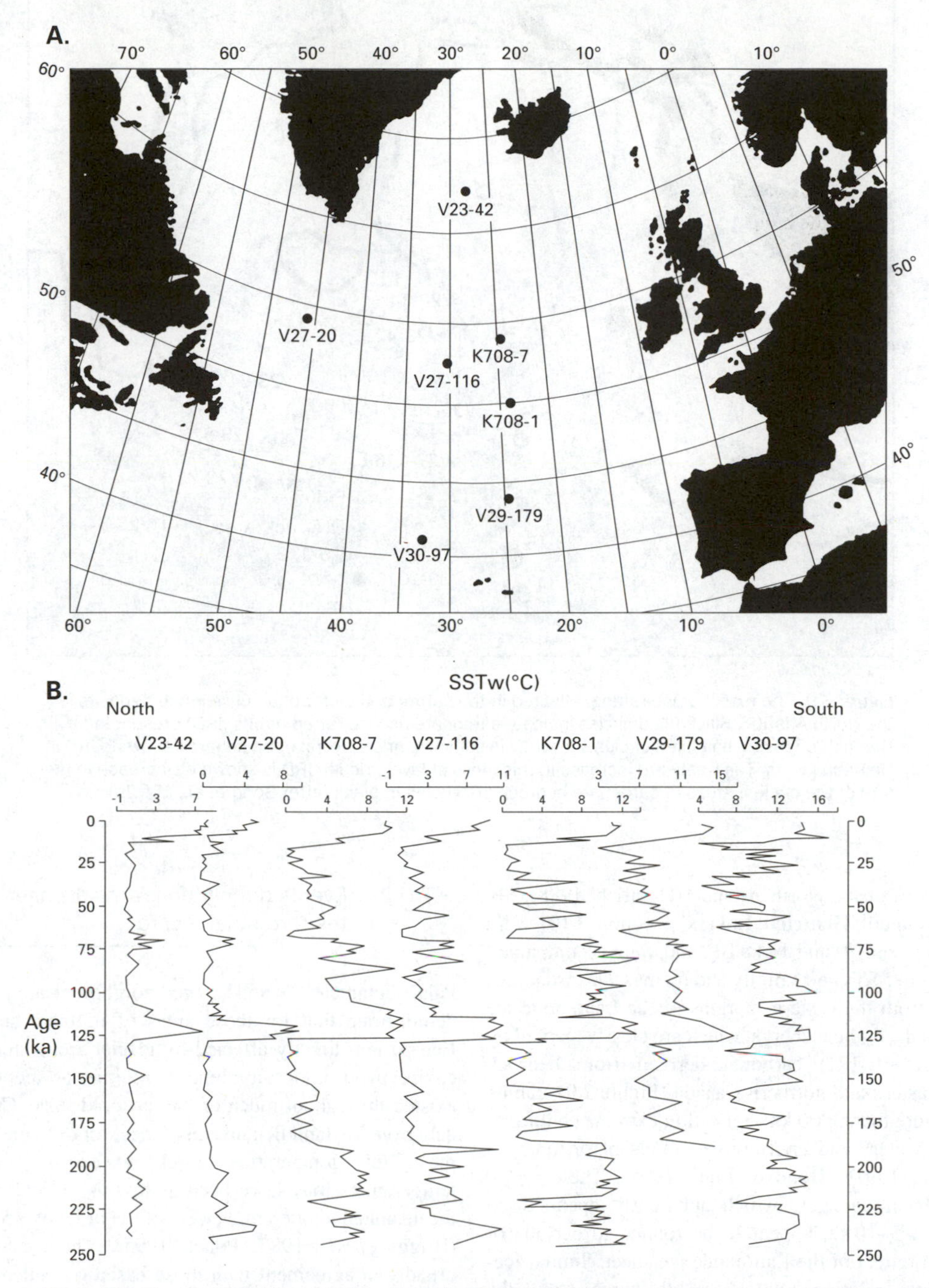

Figure 7.8 Changes in estimated winter SSTs (in °C) in North Atlantic cores over the last 130 ka: A location of cores; B north–south transect of SST trends (after Ruddiman, 1987).

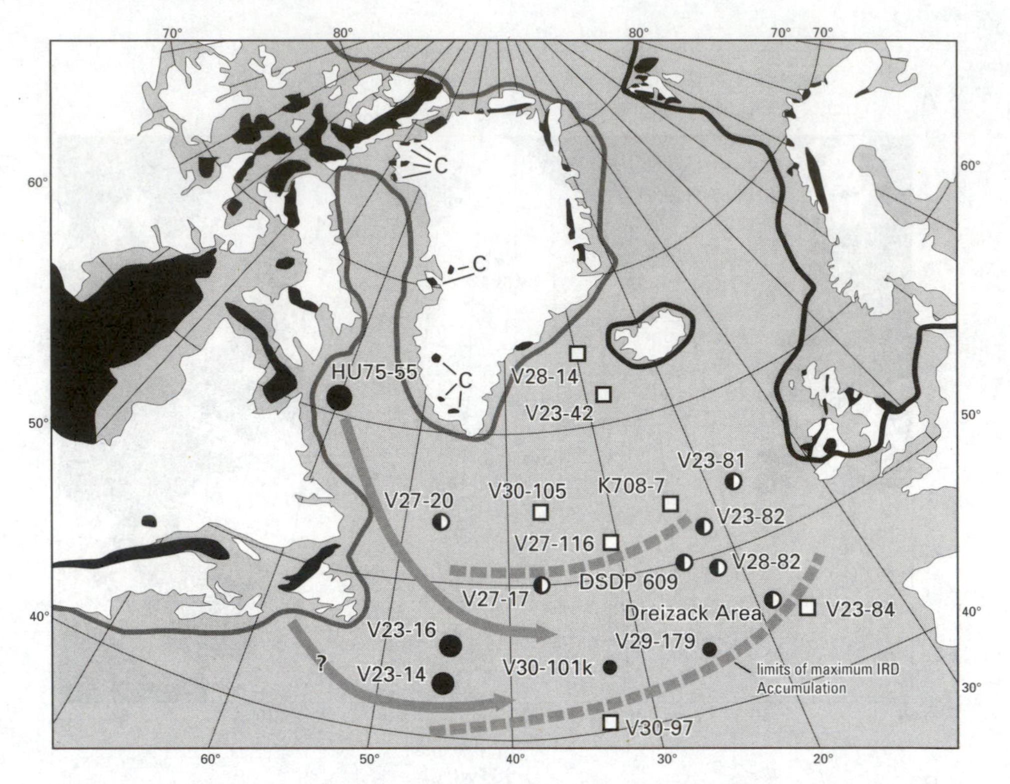

Figure 7.9 The extent of ice-rafting reflected in the nature and distribution of Heinrich deposits in the North Atlantic. Black filled circles indicate carbonate-rich ice-rafted debris (IRD) present in all Heinrich deposits; half-filled circles show IRD in some Heinrich layers; open squares show no IRD in Heinrich layers. The westward increase in thickness of layers rich in IRD is shown by increase in the size of the circles. Areas of carbonate bedrock are shown in black (after Bond *et al.*, 1992b).

latitude regions of the North Atlantic (Heinrich, 1988). Six of these so-called '**Heinrich Layers**' (section 3.10) were deposited between 70 and 14 ka BP, and were accompanied by decreases in SSTs and salinity and by massive discharges of icebergs from the eastern margins of the Laurentide ice sheet. The paths of the icebergs, which are clearly marked by the presence of IRD carbonate-derived from bedrock sources in eastern and northern Canada (Figure 7.9), can be traced for more than 3000 km, reflecting extreme cooling of the surface waters and enormous amounts of drifting ice (Bond *et al.*, 1992b; Bond & Lotti, 1995). These major oceanographical changes, which apparently operated on timescales of 5–10 ka, appear to be related to periods of advance and retreat of the Laurentide ice sheet. Similar ice-rafting events relating to iceberg discharge from the Fennoscandian Ice Sheet also occurred in the northeast Atlantic (Baumann *et al.*, 1995; Fronval *et al.*, 1995). The linkages between ice-sheet fluctuation, ice-rafting episodes and climate change are discussed more fully below.

7.5.2 Events during the last cold stage: the terrestrial record

Palaeobotanical and geomorphological evidence demonstrate that for those areas of northern and western Europe not directly affected by glacier ice, a steppe/tundra environment under a climatic régime of arctic severity existed throughout much of the last cold stage. Coleopteran data from lowland Britain and France, for example, show that mean July temperatures rarely exceeded 10°C, winter temperatures may have been as low as −25°C, and hence mean annual temperatures were of the order of −5 to −10°C (Briggs *et al.*, 1985; Ponel, 1995). These estimates are broadly in agreement with those based on pollen analytical data from eastern France (Guiot *et al.*, 1989). There are indications, however, that several short-lived warmer episodes of interstadial character punctuated this period of otherwise unremitting cold, when barren steppe and tundra was replaced by shrub tundra (Figure 7.10). A synthesis of data from northern and western Europe (principally from The Netherlands and northern Germany) reveals evidence for five such interstadials (Behre, 1989): **Oerel** (58–54 ka BP); **Glinde** (51–48 ka BP); **Moershoofd** (46–44 ka BP); **Hengelo** (39–36 ka BP); and **Denekamp** (32–28 ka BP) (Behre & van der Plicht, 1992). Although the biostratigraphy of the terrestrial record appears relatively secure, the reliability of some of the age determinations has been questioned (Evin, 1992) as the older dates in particular lie at, or beyond, five half-lives of ^{14}C, which is the normal limit of the radiocarbon dating method (section 5.3.2). In southern Europe, continuous pollen records from the last cold stage show fluctuations in the AP/NAP ratio reflecting cycles of vegetation change from transitional parkland through grassland steppe communities to desert steppe vegetation and subsequently back to open parkland (Tzedakis, 1993). Between five and seven of these vegetation cycles may be present in the pollen records. Similar clearly defined AP/NAP oscillations are apparent in long pollen records from France and Italy (Watts, 1985a; Guiot *et al.*, 1989; Follieri *et al.*, 1989).

In North America, the Clear Lake palaeotemperature record (Figure 7.3B) shows evidence of six or possibly seven climatic oscillations between *c.* 65 and 32 ka BP, the magnitude of the temperature shift ranging from 1–3°C. In a pollen profile from Florida (Figure 7.11), abrupt changes in *Pinus* percentages suggest five major climatic oscillations (mainly from wetter to drier episodes) between 50 and 20 ka BP, which are correlated with Heinrich events H5 to H1 (Grimm *et al.*, 1993). Elsewhere, however, the fragmented stratigraphic record has yielded evidence for only two clearly defined interstadial events during the Early and Middle Wisconsinan, namely the Port Talbot (*c.* 64–40 ka BP) and Plum Point (*c.* 35–23 ka BP) Interstades of the Great Lakes region (Karrow, 1984). In the Cascade Mountains of British Columbia, depression of glacier equilibrium line altitudes by 850–900 m during the last glacial maximum suggests a reduction in mean annual temperature of 4–5°C, while in the Rocky Mountains mean annual air temperature lowering may have been two to three times as great (Porter *et al.*, 1983).

7.5.3 Events during the last cold stage: the ice-sheet record

In Chapter 2 we examined the evidence that can be used to reconstruct both the extent of the last ice sheets in Europe and North America and the pattern of ice retreat. We also discussed the ways in which models of those ice sheets can be constructed. Here we consider briefly the history of these northern ice sheets during the course of the last cold stage.

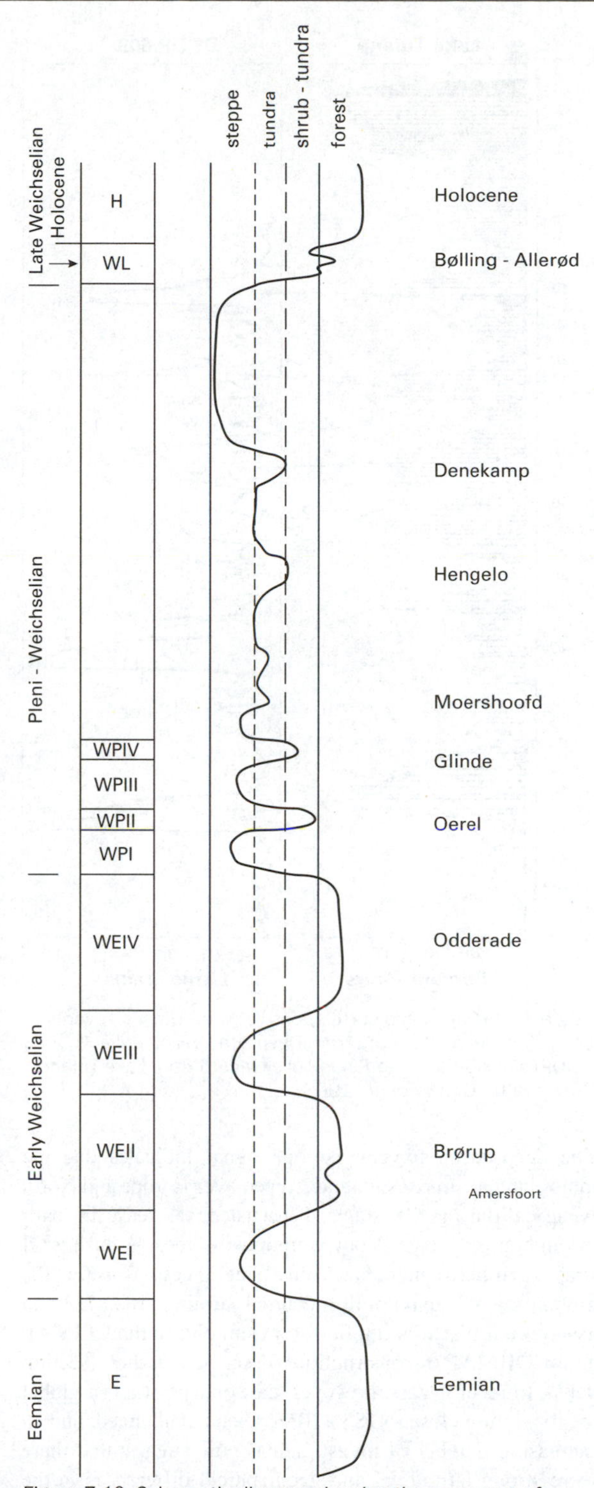

Figure 7.10 Schematic diagram showing the sequence of Weichselian interstadials and their vegetational characteristics in northern and western Europe (after Behre, 1989).

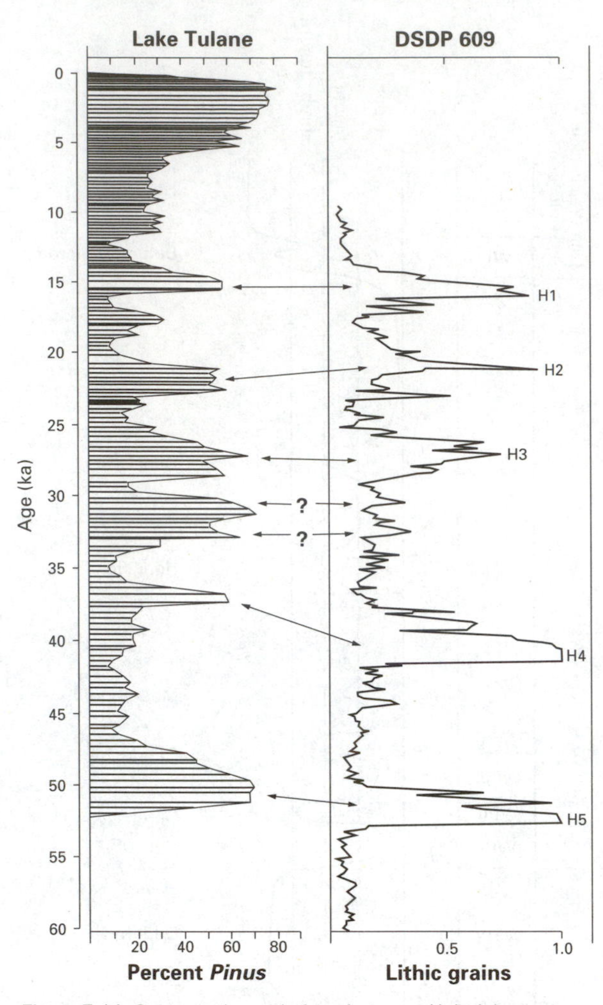

Figure 7.11 Suggested correlations between Heinrich events 1–5 reflected in lithic data from Deep Sea Drilling Project (DSDP) site 609 and the *Pinus* pollen record from Lake Tulane, Florida (after Grimm *et al.*, 1993).

The deep-ocean oxygen isotope signal indicates that the major land ice masses that developed over Europe and North America during OI stage 4 persisted, at least in part, throughout OI stage 3 but were most extensive during OI stage 2. Hitherto, there has been a general consensus that the global glacial maximum occurred around 18 ka BP, an assumption that was implicit, for example, in the CLIMAP and COHMAP reconstructions described earlier (section 4.11). In recent years, however, the concept of a last global ice maximum close to 18 ka BP has been challenged, and an increasing number of investigations now suggest that there were strong latitudinal and geographical differences in the timing of the culmination and retreat of the last mid-latitude ice sheets (Sejrup *et al.*, 1994).

7.5.3.1 The European ice sheets

Evidence from Scandinavia indicates extensive glaciation during OI stage 4 and early in OI stage 3 when the margins of the Fennoscandian ice sheet expanded westwards to the Continental Shelf (Figure 7.6), southwards into Denmark and southeastwards into Poland (Drozdowski & Fedorowicz, 1987; Houmark-Nielsen, 1989; Figure 2.5). The ice maximum occurred some time after 63 ka BP (**Karmøy Stadial**) and was terminated by a marked deglacial event (**Bø Interstadial**) at *c.* 54 ka BP. A readvance of the ice margin (**Skjongherren Advance**) between 47 and 43 ka BP was followed by widespread deglaciation during the **Ålesund Interstadial** which has been dated to *c.* 35–29 ka BP (Larsen *et al.*, 1987). This chronology is based on a combination of palaeomagnetic, TL, U-series and radiocarbon dates. Each of the major readvance episodes is reflected in significant increases in IRD content in Norwegian Sea sediments (Baumann *et al.*, 1995). Along the northwest fringes of the Fennoscandian ice sheet, the glacial maximum occurred at *c.* 22 ka BP (Møller *et al.*, 1992), while data from the northern North Sea point to a glacial maximum around the southwest margins of the ice sheet between 29 and 22 ka BP (Sejrup *et al.*, 1994). Subsequent wastage of the ice sheet was interrupted by a further readvance between 18 and 15 ka BP, while later halt/readvance phases occurred at *c.* 12.4–12.1 ka BP and during the Younger Dryas Stadial between 11 and 10.2 ka BP (Vorren *et al.*, 1988; Baumann *et al.*, 1995).

The chronology of glaciation during the last cold stage in Britain is less secure than in Scandinavia. Moreover, the extent and, indeed, the existence of glacier ice during the Early Devensian has yet to be established (Ehlers *et al.*, 1991). In view of the evidence for ice build-up elsewhere, and the fact that the marine isotope record indicates that 68 per cent of the global land ice volume that existed during the last glacial maximum (OI stage 2) formed during OI stage 4 (Shackleton, 1987), it seems highly likely that the uplands of Britain were glaciated during the Early Devensian (Boulton *et al.*, 1991). However, stratigraphic evidence in support of such a hypothesis remains frustratingly elusive (Worsley, 1991). A large part of Scotland (and, by implication, the entire British Isles) appears to have been ice-free towards the end of the Middle Devensian (~OI stage 3), and at sites where radiometrically dated organic materials underlie till the absence of age determinations younger than 25 ka implies expansion of the ice sheet only after that time (Boulton *et al.*, 1991). The glacial maximum appears to have occurred between 22 and 18 ka BP (Eyles & McCabe, 1989; Eyles *et al.*, 1994), although the precise timing of this event is still uncertain and, indeed, the extent to which it was synchronous around the ice-sheet margins remains to be established.

Whether the British and Scandinavian ice sheets coalesced during the last cold stage has been actively debated. In recent years, the consensus view has tended to be that this was not the case (e.g. Sutherland, 1984b; Sejrup *et al.*, 1987, 1991), although recent evidence from the northern North Sea raises the possibility that the British and Scandinavian ice sheets were confluent at the last glacial maximum (*c.* 29–22 ka BP) in the area between northeast Scotland and western Norway. The wastage of the Late Devensian ice sheet seems to have been interrupted by a number of readvance episodes. In the Irish Sea region, for example, the 'Drumlin Readvance' has been dated to around 17 ka BP (McCabe, 1985), while along the eastern margins of the ice sheet a major readvance or ice-sheet surge (Eyles *et al.*, 1994) occurred between 18 and 15 ka BP, an event that may coincide with renewed glacier activity elsewhere in the North Sea region (Sejrup *et al.*, 1994). Thereafter ice wastage seems to have been rapid, with almost all of the British Isles, including much of Scotland, deglaciated by *c.* 13 ka BP. During the Loch Lomond or Younger Dryas Stadial, small glaciers formed anew in many upland areas, although in the Scottish Highlands remnants of the last ice sheet may have survived the preceding Lateglacial Interstadial (13–11 ka BP) to become rejuvenated during the Stadial (Boulton *et al.*, 1991).

7.5.3.2 The North American ice sheets

As noted above, the Laurentide ice sheet first developed during OI stage 5 over Keewatin, northern Quebec/Labrador and Baffin Island, and reached its maximum extent along its northern margin at that time (Figure 7.12). Early in OI stage 4, the ice sheet advanced across southern Quebec and into the St Lawrence lowlands, and may also have extended southwards into the Ontario basin and northern New England (Clark *et al.*, 1993). Laurentide glaciation was

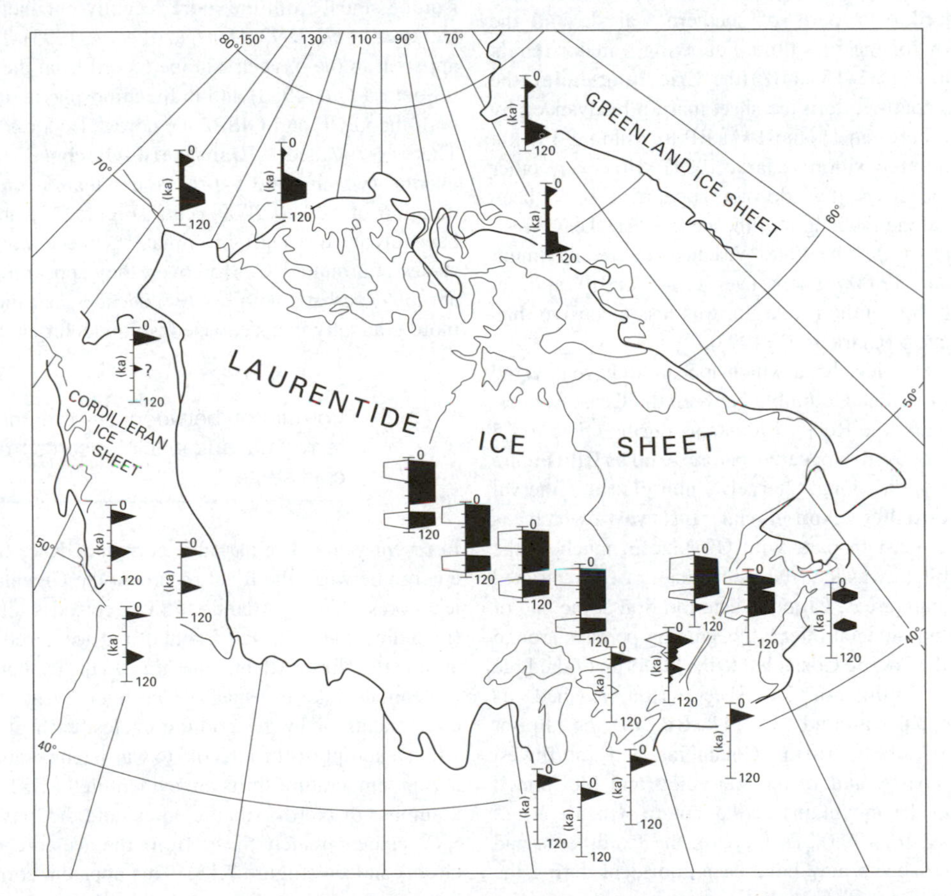

Figure 7.12 Schematic time–distance diagrams of ice-sheet advance and retreat from areas covered by the Laurentide and Cordilleran ice sheets (after Clark *et al.*, 1993).

extensive in the Great Lakes region, but more limited glaciation appears to have occurred further west (Richmond & Fullerton, 1986b). In the east, a satellite ice cap developed over Nova Scotia, while in the far west, the Cordilleran ice sheet advanced across British Columbia into the northern Puget Lowland (Clark *et al.*, 1993). The position of the northern margin of the Laurentide ice sheet at that time has yet to be established, although evidence from sediment cores in Baffin Bay suggests a substantial decrease in ice cover in that region during OI stage 4.

During the Middle Wisconsinan (broadly the equivalent of OI stage 3), the Laurentide ice sheet retreated from its maximum position in the western Arctic some time prior to 48 ka BP. The extent of the retreat is not known, but the Keewatin sector of the ice sheet to the west of Hudson Bay may have remained as a relatively stable ice body, and a major ice divide also persisted in the Foxe Basin area to the west of Baffin Island (Clark *et al.*, 1993). During the Late Wisconsinan (OI stage 2), the Laurentide ice sheet reached its southernmost extent between 21 and 18 ka BP, and the ice sheet advanced over parts of southern Canada and the northern USA for the first time. Following a major retreat that ended in *c.* 15.5–15 ka BP (the **Erie Interstade**), the southern and southwestern ice sheet margin readvanced by up to 800 km between 15 and 14 ka BP to within *c.* 150 km of its maximum position (Clark, 1994). However, other readvances occurred that do not appear to have been synchronous along the length of the ice front (Dyke & Prest, 1987). In the Arctic, the Late Wisconsinan ice maximum occurred some 12 000 years later (13–8 ka BP) and, in contrast with the southern margin, was less extensive than during OI stage 5 (Clark *et al.*, 1993).

The Cordilleran ice sheet, which appears to have covered a large area of British Columbia between the Coast Ranges, the Cascades and the Rocky Mountains during OI stages 5 and 4, had largely disappeared before *c.* 60 ka BP (Figure 7.12). During a long, largely non-glacial interval (**Olympia/Boutellier Non-glacial Interval**) which is correlative, at least in part, with OI stage 3, much of the western cordillera was ice free, and glaciers were restricted to high mountain areas. Climatic deterioration at the end of the Olympia non-glacial interval, beginning perhaps around 29 ka BP on the Pacific Coast, led to the build-up of the Late Wisconsinan Cordilleran ice sheet. This reached its maximum extent around 15–14 ka BP in the Fraser Glaciation of the British Columbia and northwest Washington coasts, and in the Macauley and McConnell Glaciations in the mountains of the Yukon (Clague *et al.*, 1992; Clark *et al.*, 1993). To the east, the Cordilleran and Laurentide ice sheets may have been confluent during the Late Wisconsinan, although the timing of such episodes remains to be established (Vincent & Klassen, 1989). Further north in Alaska, a local ice cap formed on the Brooks Range (Hamilton, 1986), while independent ice caps developed and subsequently coalesced over the Aleutian Islands (Thorson & Hamilton, 1986). To the south of the Cordilleran ice sheet, mountain glaciers developed during both the Early and Late Wisconsinan, the Late Wisconsinan maximum being dated in a number of areas at around 22 ka BP (Richmond, 1986b).

7.5.4 Events during the last cold stage: the ice-core record

Irregular, but well-defined episodes of relatively mild climatic conditions during the middle and later parts of the last cold stage have been identified over the past 20 years or so in oxygen isotope profiles from the Greenland ice sheet (Johnsen *et al.*, 1972; Dansgaard *et al.*, 1982), and these have been most emphatically demonstrated by the oxygen isotope signal from the more recently obtained GRIP core (Johnsen *et al.*, 1992; Dansgaard *et al.*, 1993). They are also apparent in the oxygen isotope record from the GISP2 core (Grootes *et al.*, 1993) and in the atmospheric dust record in both the GRIP and GISP2 ice cores (Taylor *et al.*, 1993b). These so-called **'Dansgaard–Oeschger interstadial events'** (which were 5–6°C colder than at present) had a duration of 500 to 2000 years (Figure 7.7) and appear to have involved a shift in climate between warm and cold stages of around 7°C. Moreover, they appear to have begun abruptly, perhaps within a few decades, but they terminate more gradually in a stepwise fashion (Johnsen *et al.*, 1992).

7.5.5 Correlation between ice-core, marine, terrestrial and glacial records from the last cold stage

In recent years, a remarkable correspondence has begun to emerge between the $\delta^{18}O$ record in the Greenland Summit ice cores, North Atlantic SSTs, episodes of ice-rafting (including the Heinrich Events) and ice-sheet fluctuations around the North Atlantic margins. In the ice-core signal, the millennial-scale Dansgaard–Oeschger events seem to be components of longer cooling cycles, each of which ends with an abrupt shift from cold to warm temperatures. Similar abrupt temperature shifts are evident in the SST record from a number of North Atlantic cores, and the longer cycles of SST change match those from the ice-core record very closely indeed (Figure 7.13). This apparent correspondence between the ocean and ice-core records suggests that, for the last 80 ka at least, the atmosphere and ocean surface were a

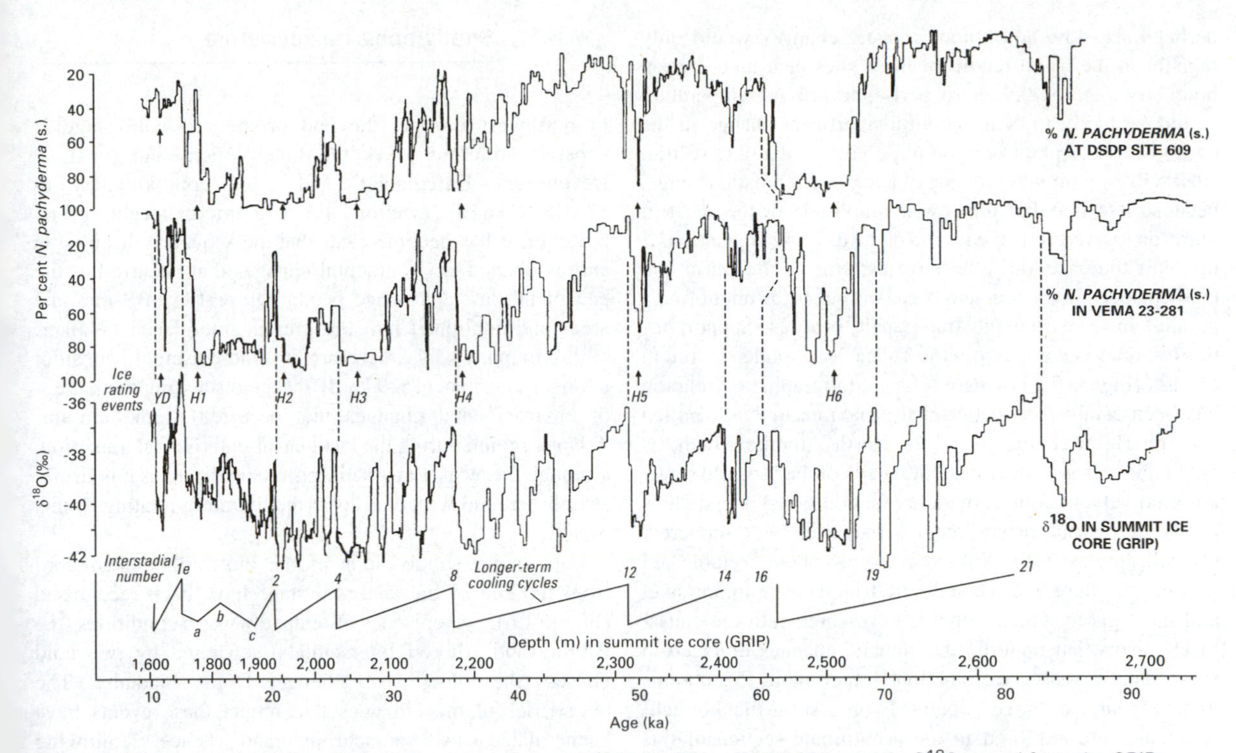

Figure 7.13 Correlation between foraminiferal records from DSDP site 609 and V23-281, and the $\delta^{18}O$ record from the GRIP Summit ice core. Also shown are the long-term cooling cycles, defined by grouping the Dansgaard–Oeschger cycles, and Heinrich events 1–6 (H1–H6) (after Bond *et al.*, 1993).

coupled system that was repeatedly undergoing massive reorganisations on timescales of centuries or even less (Bond *et al.*, 1993; Bond & Lotti, 1995).

There appears to be an equally close relationship between the Heinrich Events, ice-sheet fluctuations and the Greenland ice-core records. Heinrich Layers 2 and 1, for example, dated to 21–19 and 15–14 ka BP in North Atlantic sediment sequences, are broadly synchronous with the two most recent major advances along the southern margin of the Laurentide ice sheet (Clark, 1994), while the four most recent Heinrich Events can be correlated with episodes of maximum iceberg calving from the Scandinavian ice sheet (Baumann *et al.*, 1995). Moreover, data from the western United States indicate that advances and retreats of the Cordilleran glaciers were broadly in phase with episodes of expansion and collapse of the Laurentide ice sheet, and with the North Atlantic Heinrich events (Clark & Bartlein, 1995). This implies a link between the behaviour of the Cordilleran, Laurentide and Scandinavian ice sheets, and may even suggest a coupling mechanism for ice-sheet oscillations, where instability on one of the ice sheets can cause other ice sheets to behave in a coherent 'phase-locked' manner (Baumann *et al.*, 1995; Fronval *et al.*, 1995). The episodes of massive IRD influx into the North Atlantic can, in turn, be correlated with 'colder' intervals reflected in the $\delta^{18}O$ trace in the GRIP ice core (Bond & Lotti, 1995), providing further evidence of the linkages between the different components of the ocean, atmosphere and ice-sheet systems. Indeed, episodes of ice-rafting in the sub-arctic Pacific Ocean, which also show a temporal variability that is similar to the oxygen isotope record from the GRIP ice core, suggest that the rapid climatic changes in the circum-Atlantic region over the past 95 ka could, in fact, be a circumpolar phenomenon (Kotilainen & Shackleton, 1995).

Relating these major hemispherical (and perhaps global) climatic changes to the terrestrial (non-glacial) proxy records has proved to be more difficult, however. A time-stratigraphic correlation has been suggested between the major isotopically defined 'interstadials' of the ice-core record and European interstadial episodes (Figure 7.7), based on close similarities between ^{14}C dates on these events and the ice-core chronology (Dansgaard *et al.*, 1993). However, there appear to be other interstadials in the ice-core record for which there are no parallels in European pollen-stratigraphies. One explanation for the lack of evidence for such episodes is that relatively short-lived (and perhaps also low amplitude) climatic changes would only register in the fossil record of those sites near a vegetation boundary (*ecotone*), where a rise or fall in temperature would be likely to bring about a significant change in the composition of plant communities. Hence, many terrestrial sites will contain no evidence of a regional climatic change, because it was either too low in amplitude or too short in duration to overcome the inertia of the existing vegetation. If this was the case, only the most extreme of the warm and cold episodes in the ice- and ocean-core records might find a parallel in terrestrial biostratigraphic profiles. Support for this hypothesis comes from the Tulane Lake pollen record in Florida (Figure 7.11) where a time-stratigraphic correlation has been established between the maxima in *Pinus* pollen and the Heinrich Layers of the North Atlantic which, as noted above, represent the coldest parts of the last cold stage. Alternatively, of course, evidence for additional interstadials in the terrestrial proxy records may yet be discovered. Overall, however, the links that are now being established between ice-core, marine and terrestrial data are impressive, and have given added impetus to research into the causal mechanisms that underlie the climatic changes reflected in those data. Possible explanations for climatic fluctuations that appear to have operated on 'sub-Milankovitch' timescales are explored in the penultimate section of this chapter.

7.6 The last glacial–interglacial transition (OI stage 2/1)

The transition from the last cold stage to the present interglacial, reflected in the marine records by the OI stage 2/1 boundary (Termination I), is one of the most intensively studied episodes of the entire Quaternary period. The evidence can frequently be examined at a much higher temporal resolution than in earlier parts of the Quaternary sequence and, because the inferred environmental changes can be dated more precisely using, for example, radiocarbon, varve, ice-layer or tree-ring chronologies (e.g. Bard & Broecker, 1992), correlation between the different proxy records can be effected within a time-stratigraphic framework. As a consequence, former linkages between the various components of the earth–atmosphere system can be explored in much greater detail than has been possible hitherto. The stratigraphic record of this period is therefore of particular interest to the Quaternary scientist, for it constitutes our best archive of the way in which earth and atmospheric processes interact during the transition from a cold ('glacial mode') to a warm stage ('interglacial mode').

7.6.1 Stratigraphic nomenclature

In northwest Europe, the end of the last cold stage is generally referred to as the Late Weichselian (or Late Devensian) **'Lateglacial'** and dated approximately to 13–10 ^{14}C ka BP (section 1.5). In more recent years, however, it has become clear that the sequence of climatic and associated environmental changes that occurred at the end of the last cold stage began before 13 ka BP in some areas and continued into the present interglacial (Walker, 1995). In many ways, therefore, it is more useful to consider a longer timespan (15–9 ka BP) when discussing the pattern of environmental changes that occurred in the circum-Atlantic region during the last glacial–interglacial transition, although the term 'Lateglacial' is retained here as it is firmly established and widely used in the literature relating to this period.

Four major subdivisions of the biostratigraphic record from the end of the last cold stage have been recognised (Figure 1.6): two episodes of relatively warm conditions (the Bølling and Allerød Interstadials) separated by two cold phases (the Older and Younger Dryas Stadials). The boundaries of the biozones that reflect these events have been dated by radiocarbon and hence, following conventional geological procedures (section 6.2.3.7), the subdivisions have been formally defined as *chronozones* (Mangerud *et al.*, 1974) and dated as follows: Bølling (13–12 ka BP), Older Dryas (12–11.8 ka BP), Allerød (11.8–11 ka BP) and Younger Dryas (11–10 ka BP).

This stratigraphic scheme was intended to be used strictly in a chronostratigraphic sense (Bølling *chronozone*, Older Dryas *chronozone*, etc.), but problems have arisen in its application, partly because of the problems of ^{14}C dating (section 5.3.2) which do not permit the chronohorizons to be dated with adequate precision or consistency, and partly because of the ambiguity that has arisen in the use of chronozones themselves. In the biostratigraphic record from northwest Europe, the chronozones are all radiocarbon-dated *biozones*. As noted above (section 6.2.3.7) biozone boundaries tend to be time-transgressive (because they reflect biological response to climatic change which is spatially and temporally diachronous), whereas the boundaries of chronozones are time-parallel. Yet, in discussions of the European Lateglacial, biozones and chronozones have tended to be used interchangeably, and this has inevitably led to confusion. For example, the 'Younger Dryas' *chronozone* is that part of the stratigraphic record spanning the time interval between 11 and 10 ka BP. By contrast, the Younger Dryas *biozone* is that part of the stratigraphic record characterised by a fossil assemblage indicative of cold conditions. In some areas, the Younger

Dryas biozone begins before 11 ka BP; in other areas, the onset is later. Chronostratigraphically, therefore, the two types of zone boundary will seldom coincide (Walker, 1995). By the same token, the beginning and end of the *Younger Dryas Stadial*, which is a geologic–climatic unit (section 6.2.3.6), are also time-transgressive.

These difficulties arise because conventional geological procedures are being applied to stratigraphic records of unusually high temporal resolution, in which evidence of rapid and often short-lived climatic changes is preserved (Lowe & Gray, 1980). The last glacial–interglacial transition, therefore, provides an example of a part of the Quaternary record where the stratigraphic practices set out in Chapter 6 cannot easily be followed. Accordingly, many scientists studying this time period prefer to use an informal and more loosely dated climatostratigraphic subdivision in which four distinct climatic episodes can be recognised in the terrestrial record: an initial cold phase prior to 13 ka BP, when the first signs of warming are apparent (sometimes referred to as the **Oldest Dryas**); a period of significantly warmer conditions from *c.* 13–11 ka BP (**Lateglacial Interstadial**); a second cold episode from *c.* 11–10 ka BP (**Younger Dryas** or **Late Dryas Stadial**); and the climatic amelioration of the early Holocene. In Britain, the terms **Windermere Interstadial** (13–11 ka BP) and **Loch Lomond Stadial** (11–10 ka BP) are employed. In North America, a cold phase broadly correlative with the Younger Dryas of Europe has been recognised in biostratigraphic records from the eastern seaboards of Canada and the northeast United States, from parts of the Great Lakes region, and from sites along the western Pacific coast (Peteet, 1993, 1995). However, the preceding warm interval is less well constrained, in terms of both biostratigraphy and geochronology, than is the Lateglacial Interstadial in western Europe. Moreover, no formal stratigraphic designations have so far been proposed for the North American correlatives of the Lateglacial Interstadial or Younger Dryas Stadial of Europe.

7.6.2 Events during the last glacial–interglacial transition: the terrestrial record

Records of environmental and climatic change in the North Atlantic region during the last glacial–interglacial transition have been compiled as part of a collaborative research programme (*NASP: The North Atlantic Seaboard Programme*), a sub-project of IGCP-253 (International Geological Correlation Programme: '*Termination of the Pleistocene*'). In this programme a range of proxy data was employed, including pollen, plant macrofossils, fossil Coleoptera and Mollusca, isotopic evidence, and geological evidence including landforms (glacial, periglacial and aeolian features, for example) and sediments (glacial, periglacial, limnic, aeolian, etc.). Syntheses of the different proxy data sources formed the basis for detailed spatial and temporal reconstructions of environmental and climatic change around the North Atlantic margins during the last glacial–interglacial transition (Walker & Lowe, 1993; Peteet, 1993; Lowe, 1994; Peteet, 1995; Walker, 1995). The following account is based largely on these published sources.

7.6.2.1 The European Lateglacial record

Data from Europe indicate that the earliest warming occurred in southwest Europe around 15 ka BP, where an expansion in steppic plants (e.g. *Artemisia*) reflects initial climatic amelioration (Figure 7.14). There are indications in other areas of warming from around 13.5 ka BP, but the first clear evidence for sustained and widespread warming dates from around 13 ka BP. In lake sites, this is typically reflected in the transition from sterile, minerogenic sediments to deposits with increased organic content, accompanied by palaeobotanical evidence for the replacement of open-habitat herbaceous flora by dwarf shrub communities. The sudden appearance of the thermophilous blue-green algae *Gleotrichia* is also indicative of a rapid change to warmer conditions (Van Geel *et al.*, 1989). However, the most striking evidence for climatic improvement at this time is provided by fossil Coleoptera. At many sites, arctic assemblages are suddenly replaced by thermophilous faunas. Data from central and southern Britain, for example, suggest that temperatures at that time may have risen by as much as 1°C per decade with an overall rate of warming of 7°C per century (Coope & Brophy, 1972), while the temperature of the coldest month may have increased by up to 20°C (Atkinson *et al.*, 1987; Walker *et al.*, 1993).

The thermal maximum of the Lateglacial Interstadial (16–18°C July) occurred during 13–12.5 ka BP in the British Isles, The Netherlands, southwest Europe and lowland Switzerland, between 12.5 and 12 ka BP in southern Scandinavia and Germany, and between 11.5 and 11 ka BP in southwest and northern Norway (Figure 7.15). During the early part of the Interstadial in particular, marked climatic gradients developed over northwest Europe, with temperature differences of as much as 6–7°C within a few hundred kilometres (Coope & Lemdahl, 1995). These spatial and temporal contrasts reflect, above all, the cooling effects of the downwasting Scandinavian ice sheet and changing patterns of surface and deepwater (thermohaline circulation) in the North Atlantic (section 7.6.3). Comparison between the palaeobotanical and coleopteran records indicates that, for the first 1000 years of the

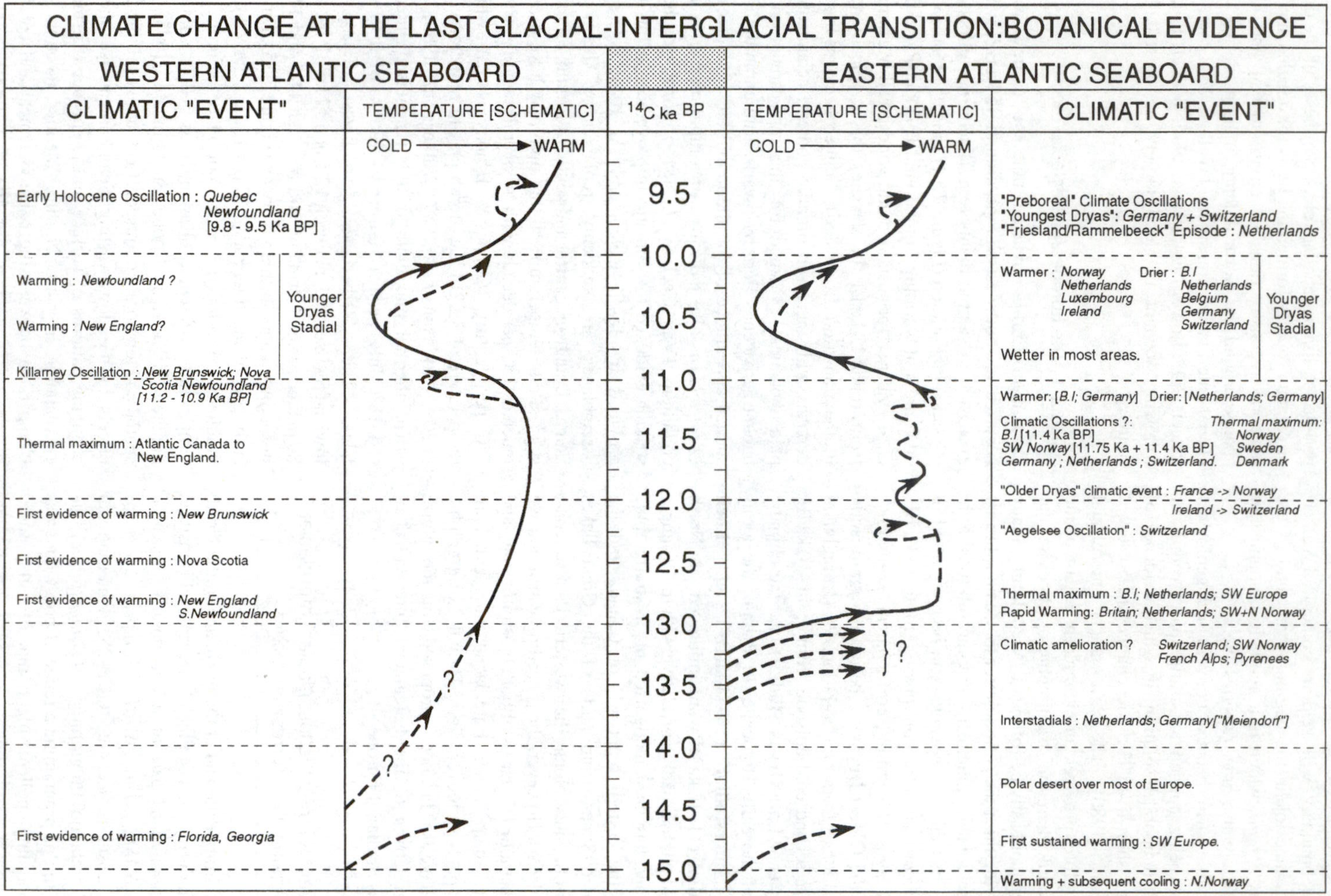

CLIMATE CHANGE AT THE LAST GLACIAL-INTERGLACIAL TRANSITION: BOTANICAL EVIDENCE				
WESTERN ATLANTIC SEABOARD			EASTERN ATLANTIC SEABOARD	
CLIMATIC "EVENT"	TEMPERATURE [SCHEMATIC]	^{14}C ka BP	TEMPERATURE [SCHEMATIC]	CLIMATIC "EVENT"
	COLD → WARM		COLD → WARM	
Early Holocene Oscillation : *Quebec Newfoundland* [9.8 - 9.5 Ka BP]		9.5		"Preboreal" Climate Oscillations "Youngest Dryas": *Germany + Switzerland* "Friesland/Rammelbeeck" Episode : *Netherlands*
Warming : *Newfoundland ?* Warming : *New England?* (Younger Dryas Stadial)		10.0 – 10.5		Warmer : *Norway Netherlands Luxembourg Ireland* Drier : *B.I Netherlands Belgium Germany Switzerland* Wetter in most areas. (Younger Dryas Stadial)
Killarney Oscillation : *New Brunswick; Nova Scotia Newfoundland* [11.2 - 10.9 Ka BP] Thermal maximum : Atlantic Canada to New England.		11.0 – 11.5		Warmer: [*B.I; Germany*] Drier: [*Netherlands; Germany*] Climatic Oscillations ?: *B.I* [11.4 Ka BP] *SW Norway* [11.75 Ka + 11.4 Ka BP] *Germany ; Netherlands ; Switzerland.* Thermal maximum: *Norway Sweden Denmark*
First evidence of warming : *New Brunswick* First evidence of warming : Nova Scotia First evidence of warming : *New England S.Newfoundland*		12.0 – 12.5		"Older Dryas" climatic event : *France -> Norway Ireland -> Switzerland* "Aegelsee Oscillation" : *Switzerland* Thermal maximum : *B.I; Netherlands; SW Europe* Rapid Warming: *Britain; Netherlands; SW+N Norway*
	?	13.0 – 13.5	?	Climatic amelioration ? *Switzerland; SW Norway French Alps; Pyrenees* Interstadials : *Netherlands; Germany["Meiendorf"]*
First evidence of warming : *Florida, Georgia*	?	14.0 – 14.5		Polar desert over most of Europe. First sustained warming : *SW Europe.*
		15.0		Warming + subsequent cooling : *N.Norway*

Figure 7.14 Climatic changes in Europe and eastern North America during the last glacial–interglacial transition (from various sources).

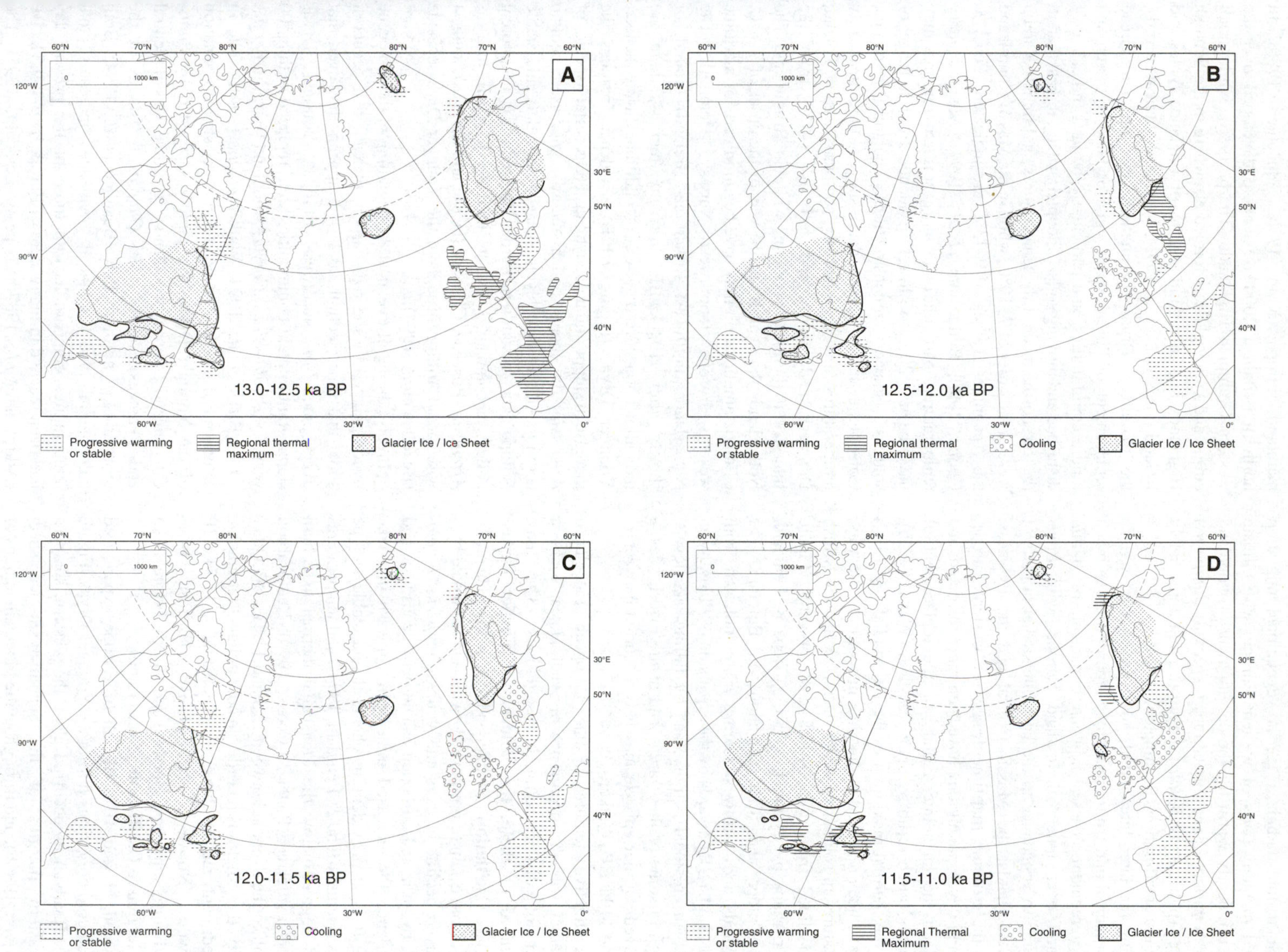

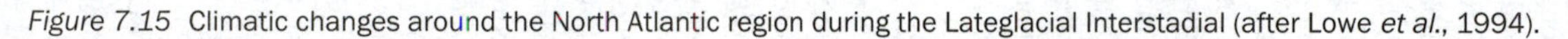

Figure 7.15 Climatic changes around the North Atlantic region during the Lateglacial Interstadial (after Lowe *et al.*, 1994).

Lateglacial Interstadial, a lag in vegetational migration occurred in many parts of western Europe, with climate being sufficiently warm for the development of woodland but, for a variety of reasons, shrubs (especially *Juniperus*) and trees (particularly *Betula*) did not respond immediately to climatic amelioration (Pennington, 1986).

During the later Interstadial, however, climatic signals from botanical and from other proxy records were in much closer agreement. These show an expansion of *Betula* woodland over much of lowland Europe, with *Pinus* increasing in abundance in the later Interstadial in parts of The Netherlands, the Massif Central and the Pyrenees. Along the extreme maritime fringes of western Europe, woodland development may have been more restricted, and in Ireland the later Interstadial saw the replacement of juniper and birch scrub by open grassland (Watts, 1985b). While the thermal maximum of the Interstadial (10–13°C) may have been achieved at this time in southwest Norway, temperatures in southern Sweden, Denmark and the British Isles declined from around 11.8 ka BP onwards (Figure 7.16), perhaps by as much as 5°C (Coope & Lemdahl, 1995). Colder and drier conditions in Belgium, The Netherlands and northwest Germany are indicated from around 11.3 ka BP onwards by the expansion of *Pinus*, by an increase in erosional activity, and by falling lake levels, although in some of these areas, and in the British Isles, a short-lived warmer episode may have occurred between *c.* 11.3 and 11 ka BP (Walker, 1995).

A distinctive feature of the European Lateglacial Interstadial was the occurrence of what appear to be a number of distinct climatic oscillations. The most pronounced of these, the '**Older**' or '**Earlier Dryas**' ('**Aegelsee Oscillation**': Switzerland) can be traced in proxy records from Spain to northern Norway, and is marked by a short-lived decline in trees and shrubs (*Juniperus, Betula, Pinus*), by an increase in open-habitat herbaceous taxa, and by evidence for increased minerogenic inwash into lake basins. It is also evident in chemical and stable isotope records (Hammarlund & Lemdahl, 1994), and temperatures several degrees below the Interstadial thermal maximum have been inferred (Ponel & Coope, 1990). Radiocarbon dates place the event at between 12.3 and 11.8 ka BP. Other short-lived oscillations recorded in both pollen-stratigraphic and sediment records have been dated at *c.* 11.7 ka BP (southwest Norway), *c.* 11.4 ka BP (British Isles, Luxembourg, southwest Norway) and shortly before 11 ka BP ('**Gerzensee Oscillation**': Switzerland).

Between 11 and 10 ka BP, all of Europe experienced renewed cold conditions during the Younger Dryas (Figure 7.17A). Readvances occurred along the western, southern and eastern margins of the Scandinavian ice sheet (Lundqvist, 1986b), and throughout the mountain regions of mainland Europe (Kohl, 1986; Schlüchter, 1986). In northern Britain a 2000 km^2 icefield developed over the western Grampian Highlands (Thorp, 1986), while small cirque and valley glaciers formed elsewhere in the Scottish Highlands and in upland regions of England, Wales and Ireland (Gray & Coxon, 1991). In palaeobotanical records, the Younger Dryas is marked by the widespread replacement of boreal shrub and woodland by scrub and tundra communities, and particularly by taxa indicative of disturbed and moving soils. Coleopteran data suggest that, in the far north, mean July temperatures in Finnish Karelia were of the order of 7–10°C, while in the Massif Central and the Apennines, values of 9–13°C are indicated (Ponel & Coope, 1990; Bondestam *et al.*, 1994). Mean January temperatures, particularly in more northern regions, may have been −20°C or lower. Evidence from a range of sources (glacier reconstructions, botanical records, Coleoptera, stable isotope data) suggest that mean July temperatures in western Norway, southern Sweden, Poland, Denmark and Switzerland fell by 5–6°C from the Interstadial thermal maximum, while in Britain and The Netherlands the overall decline was of the order of 7–8°C. Climatic amelioration may have begun in parts of maritime western Europe (southwest Norway, Denmark, The Netherlands, northwest Britain and Ireland) around, or shortly after, 10.5 ka BP. In many regions, there are indications of a shift to more arid conditions during the later Younger Dryas (Figure 7.17B). Evidence includes the increase in steppe and halophytic taxa (e.g. certain species of *Artemisia*) in many pollen records, increased aeolian activity, falls in lake levels, palaeoglacier data and stable isotope data. In southern Europe, the reduction in arboreal pollen, accompanied by an increase in such taxa as *Artemisia*, Poaceae, Chenopodiaceae and *Ephedra*, suggest widespread aridity throughout the Younger Dryas episode (Beaulieu *et al.*, 1994).

Precise dating of events at the beginning of the Holocene is complicated by the radiocarbon 'plateau' that occurs at about that time (section 5.3.2.3). It appears, however, that over much of northwest Europe, Younger Dryas steppe–tundra plant communities were replaced within 500 years by *Betula–Pinus–Corylus* woodland (Berglund *et al.*, 1994; Birks *et al.*, 1994; Walker *et al.*, 1994), while in southern Europe, the *Artemisia*-dominated steppe vegetation of the Younger Dryas was rapidly succeeded by *Pinus*, *Corylus* and *Quercus* woodland (e.g. in the Massif Central and Pyrenees), or by *Abies–Quercus* forests as in the mountains of northern Italy (Beaulieu *et al.*, 1994). Both coleopteran and botanical evidence indicates that temperatures comparable with those of the present day had been reached in many areas of Europe by 9.8–9.5 ka BP (Atkinson *et al.*, 1987; Coope & Lemdahl, 1995). The

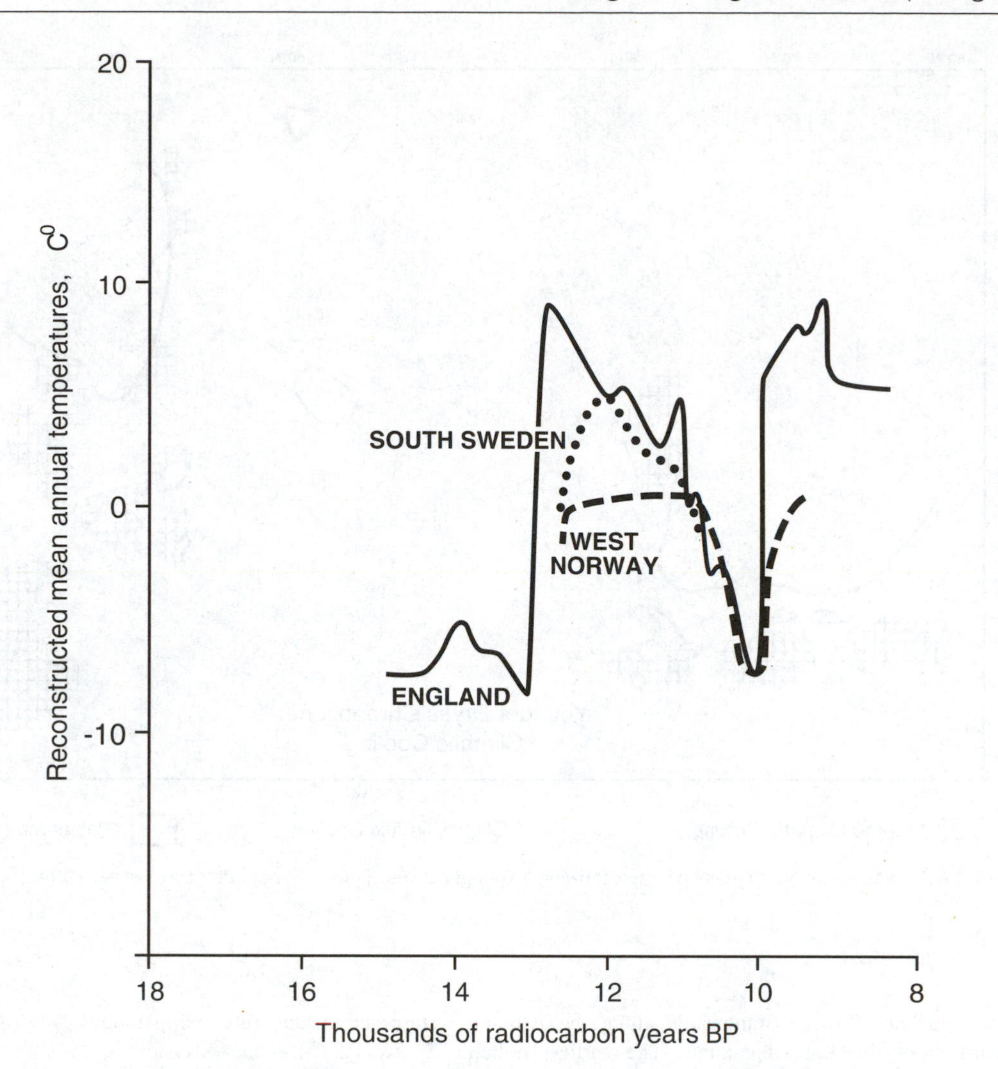

Figure 7.16 The course of climatic change in England, southern Sweden and western Norway during the last glacial-interglacial transition - the most probable values of palaeotemperature based on Mutual Climatic Range (MCR) (after Lowe *et al.*, 1995b).

emperature rise may not have been unidirectional, however, or short-lived climatic oscillations have been detected round 9.8 ka BP ('Friesland'/'Rammelbeek' phase: The Netherlands), 9.5 ka BP (Switzerland) and 9.5–9.1 ka BP western Norway). Other oscillations may also have ccurred ('Youngest Dryas' of northern Germany and the outhern Alps), but at present these remain undated Figure 7.14).

7.6.2.2 The North American Lateglacial record

Data from eastern North America suggest that climatic amelioration during the initial stages of the last glacial–interglacial transition lagged considerably behind equivalent northwest European latitudes (Figures 7.14 and 7.15), the continued cold being due partly to the chilling effects of the Laurentide ice sheet and partly to the influence

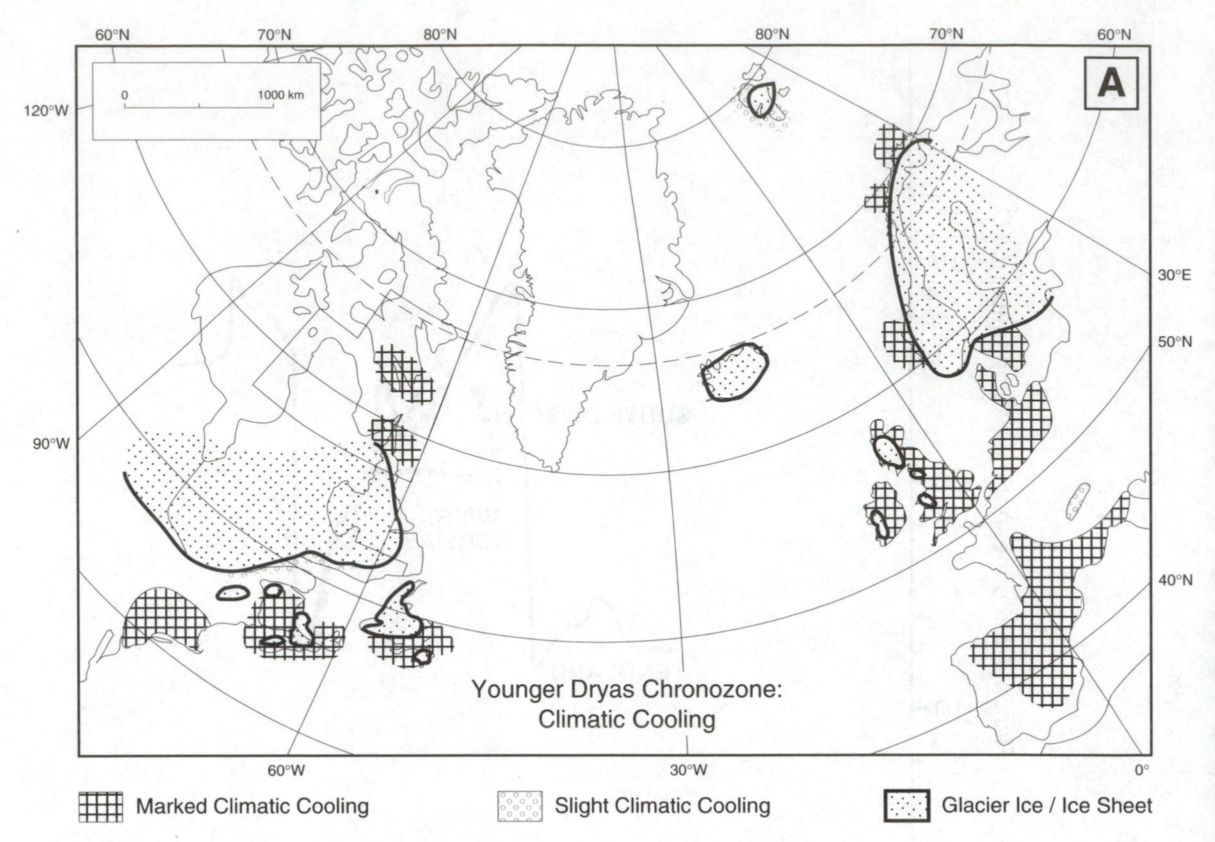

Figure 7.17A Areas of the North Atlantic region showing a Younger Dryas climatic signal (after Lowe *et al.*, 1994).

of cold currents flowing southwards along the Atlantic seaboard from the Labrador Strait. The earliest pollen evidence for climatic amelioration (increase in *Abies* and aquatic taxa; first occurrence of *Alnus*) dates from *c.* 17 ka BP in Kentucky and Virginia (Wilkins *et al.*, 1991; Kneller & Peteet, 1993). In Florida and Georgia, the replacement of *Pinus*-dominated forests by more mesophytic broadleaved trees shows that climatic amelioration was well under way by 14.5–13.5 ka BP (Watts & Stuiver, 1980), although a return to cold conditions along the southeast coast plain (reflected in increased *Picea* values) is apparent between 14 and 12 ka BP (Watts *et al.*, 1992). On the Allegheny Plateau and in the Till Plains region to the south of the Great Lakes, pollen-based transfer functions suggest mean July temperatures prior to 13 ka BP in the range 19–20°C and mean January temperatures of −14 to −17°C, reconstructions which are 4–5°C and 13–18°C respectively below current values (Shane & Anderson, 1993). Coleopteran evidence from the same area suggests mean July temperatures of *c.* 15°C around 14.5 ka BP (Morgan, 1987).

The record of vegetation change in eastern North American during the Lateglacial Interstadial is one of progressive afforestation from tundra to *Picea*-dominated boreal forest, and eventually to mixed deciduous woodland, reflecting gradual monotonic warming. In New England, palaeobotanical data point to a gradual rise in mean July temperatures from 12–13°C in the period 12.5–12 ka BP to 17°C+ at the thermal maximum between 11.5 and 11 ka BP (Peteet *et al.*, 1993, 1994). Climatic reconstructions from pollen response surfaces, however, imply somewhat higher temperatures at *c.* 12 ka BP, with mean values around, or above, 16°C throughout the northeast American seaboard (Webb *et al.*, 1993a). To the south of the Great Lakes, warming during the period 13–11 ka BP approached modern July conditions, but January temperatures remained below those of the present day (Shane & Anderson, 1993). Further north, in the Maritime Provinces of Canada, shrub tundra

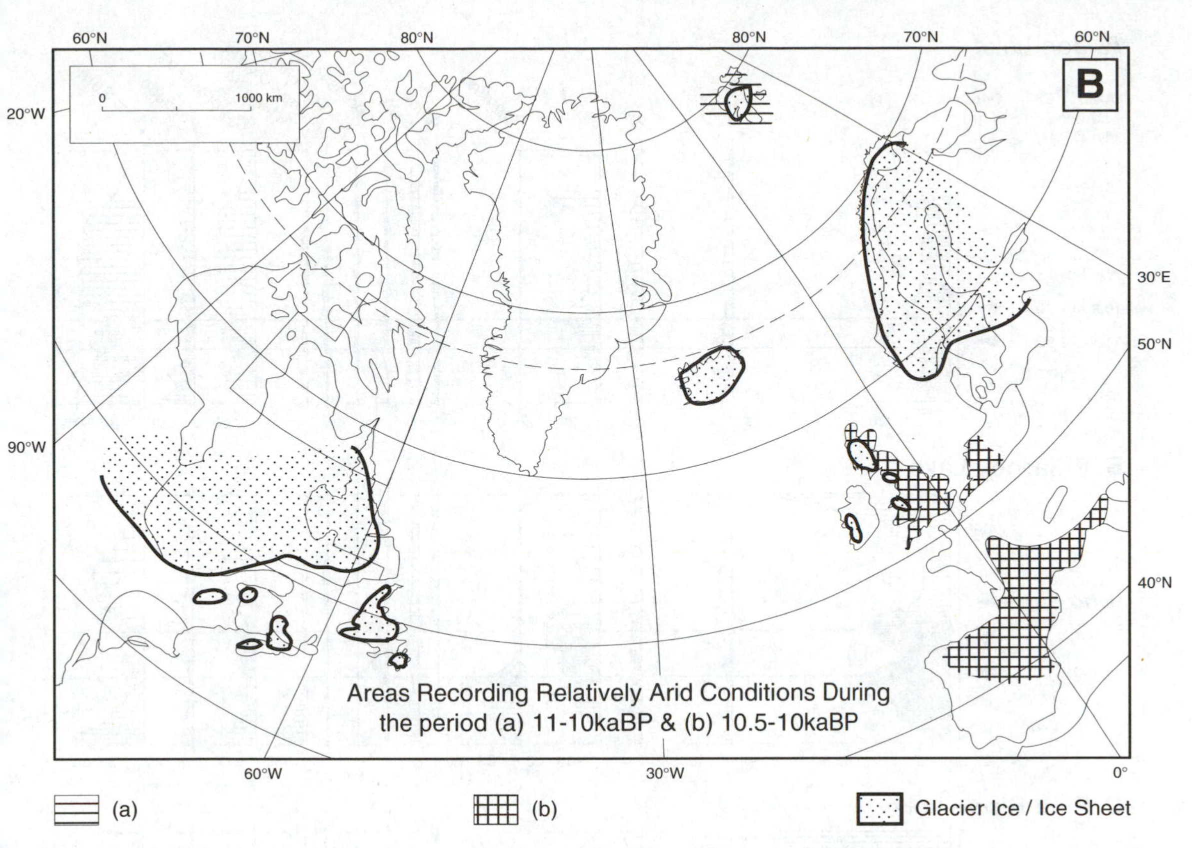

gure 7.17B Areas of the North Atlantic region showing evidence of aridity during the Younger Dryas (after Lowe *et al.*, 1994).

as succeeded by woodland (*Salix*, *Betula*, *Picea* and *opulus*) during the thermal maximum between 11.5 and 1.0 ka BP (Anderson & Macpherson, 1994; Mott, 1994). hironomid data from New Brunswick suggest that mean ıly lake-surface water temperatures at that time may have xceeded 17°C (Cwynar *et al.*, 1994).

To date, only one short-term climatic fluctuation prior to e Younger Dryas has been recorded in eastern North merica (Figure 7.14). This is the '**Killarney Oscillation**', hich has been detected throughout Atlantic Canada and ated to 11.2–10.9 ka BP (Levesque *et al.*, 1993). At ormerly wooded sites, this is reflected in abrupt declines in *opulus*, and an increase in *Betula glandulosa* and a variety f herbs (Figure 7.18). A short-lived climatic oscillation has lso been detected in Maine, where chironomid data show a ıll in summer pond-water temperatures of 2.1–2.6°C owards the end of the Lateglacial Interstadial (Cwynar & evesque, 1995).

The Younger Dryas cooling is most clearly registered in tes along the eastern seaboard between New Jersey and tlantic Canada (Figure 7.17). In southern New England, mixed forests gave way to boreal woodland at the start of the Younger Dryas cooling (Peteet *et al.*, 1990, 1993). Mean July temperatures of 13–14°C have been inferred for this phase, although the subsequent re-expansion of thermophilous trees in the period 10.5–10 ka BP suggests an increase to 15–16°C (Peteet *et al.*, 1994). In Maine, chironomid evidence indicates a fall in summer pond-water temperatures of 7–13°C during the Younger Dryas (Cwynar & Levesque, 1995). Further north in Maritime Canada, the Younger Dryas cooling is reflected in the replacement of *Betula* woodland by herbaceous communities between 10.8 and 10 ka BP (Mott *et al.*, 1986; Mayle *et al.*, 1993), although in Newfoundland there is evidence of warming from around 10.4 ka BP onwards (Anderson & Macpherson, 1994). Elsewhere in North America, an abrupt climatic downturn ending around 10.5 ka BP is recorded in pollen records from the Allegheny Plateau (Shane & Anderson, 1993), while in British Columbia the Younger Dryas Cooling may be reflected in a marked increase in mountain hemlock (*Tsuga mertensiana*) between 10.7 and 10 ka BP, and a decline in arboreal pollen and a complementary rise in

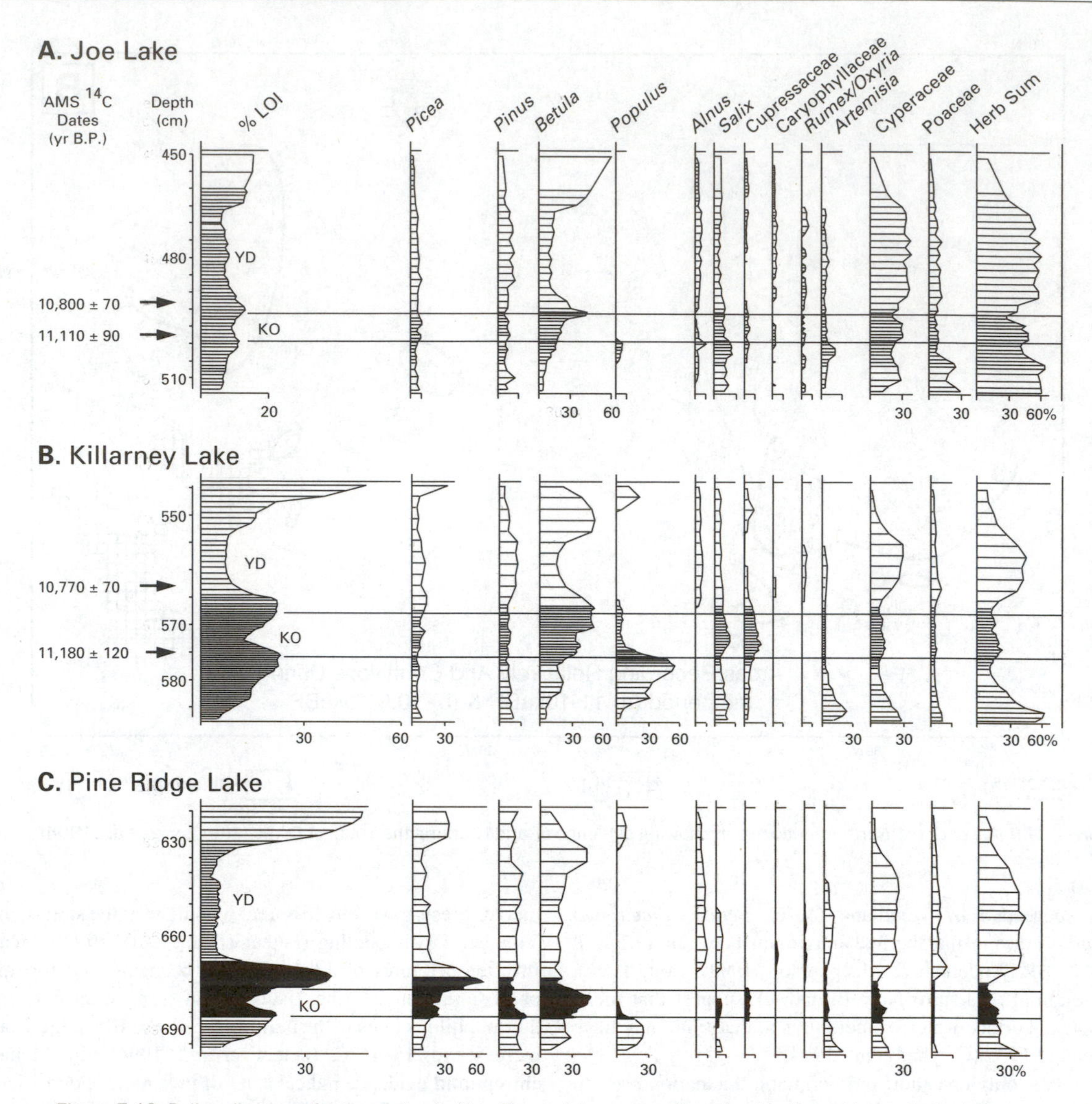

Figure 7.18 Pollen diagrams for three sites in New Brunswick, eastern Canada, showing evidence for both the 'Killarney Oscillation' KO) and the Younger Dryas event (YD) (after Levesque *et al.*, 1993).

herbaceous taxa (Mathewes, 1993).

The glacial history of the Younger Dryas in North America remains uncertain, for although readvances have been recorded around the southern and western margins of the Laurentide ice sheet, these may be attributable largely to surges (Dawson, 1992). However, the Marquette Advance in the Lake Superior and northern Michigan areas, which has been dated to *c.* 10 ka BP (Eschman & Mickelson, 1986), appears to have been an event of regional significance and can be traced eastwards towards the St Lawrence (Andrews, 1987). There, a readvance has been dated to between 11 and 10.4 ka BP (LaSalle & Shilts, 1993). To the east in Nova Scotia and Newfoundland, local glaciers and ice caps appear to have been reactivated during the Younger Dryas cold stage (Mott, 1994; Anderson & Macpherson, 1994).

Climatic improvement at the beginning of the Holocene appears to have begun before 10 ka BP along the eastern seaboard of North America, as palynological data show the rapid expansion of thermophilous woodland taxa around or prior to that date. In New England, temperatures in the period

10–9.5 ka BP rose to at least 18–19°C, only a degree or so below present levels (Peteet *et al.*, 1994), while further north in New Brunwick, chironomid data also indicate rapid warming during this time interval (Cwynar *et al.*, 1994). As in Europe, there are indications of a short-lived climatic oscillation (or oscillations) during the early Holocene in parts of eastern North America (Figure 7.15), notably in southern Quebec, in Newfoundland, and along the St Lawrence lowlands, where a reversal within the afforestation phase, dated at 9.5 ka BP, points to a strong and persistent cooling (Anderson & Macpherson, 1994; Richard, 1994).

7.6.3 Environmental changes during the last glacial–interglacial transition: the marine record

7.6.3.1 The North Atlantic Polar Front

Marine micropalaeontological evidence obtained during the 1970s showed how major changes had occurred in the position of the North Atlantic Polar Front during the course of the last glacial–interglacial transition (Ruddiman & McIntyre, 1981). The sequence of events seems to have been: (i) a rapid northward shift of the Polar Front, beginning around 13.5 ka BP, from the southerly limit (situated close to northern Portugal in the eastern North Atlantic sector) reached during the last cold stage; (ii) a subsequent southward migration of the front on its eastern flank to the latitude of southwest Ireland between about 11 and 10 ka BP; and (iii) a rapid northward movement at around 10 ka BP. The data suggested a much greater arc of N–S movement of the front in the eastern Atlantic by comparison with the western sector (Figure 7.19). The last deglacial warming of the mid-latitude North Atlantic (between *c.* 40° and 65°N) seems to have occurred in three distinct 'steps': the first at *c.* 13.5–13 ka BP in the east and northeast, the second at *c.* 10 ka BP in the central and northern areas, and the final warming between 9 and 6 ka BP in the western (Labrador Sea) sector. Rates of movement of the Polar Front and consequent changes in ocean temperatures during this period appear to have been dramatic. Initial reconstructions suggested rates of cooling or warming of 1–5°C ka^{-1} leading to sea-surface temperature changes of 7–11°C, while estimates of rates of movement of the front ranged from 200 to 1600 m yr^{-1} (Ruddiman *et al.*, 1977).

Subsequent evidence involving, in particular, the use of AMS radiocarbon dating of marine microfossils has not only confirmed the main elements of the Ruddiman and McIntyre model, but also supported the concept of extremely rapid rates of frontal movement. For example, data from the

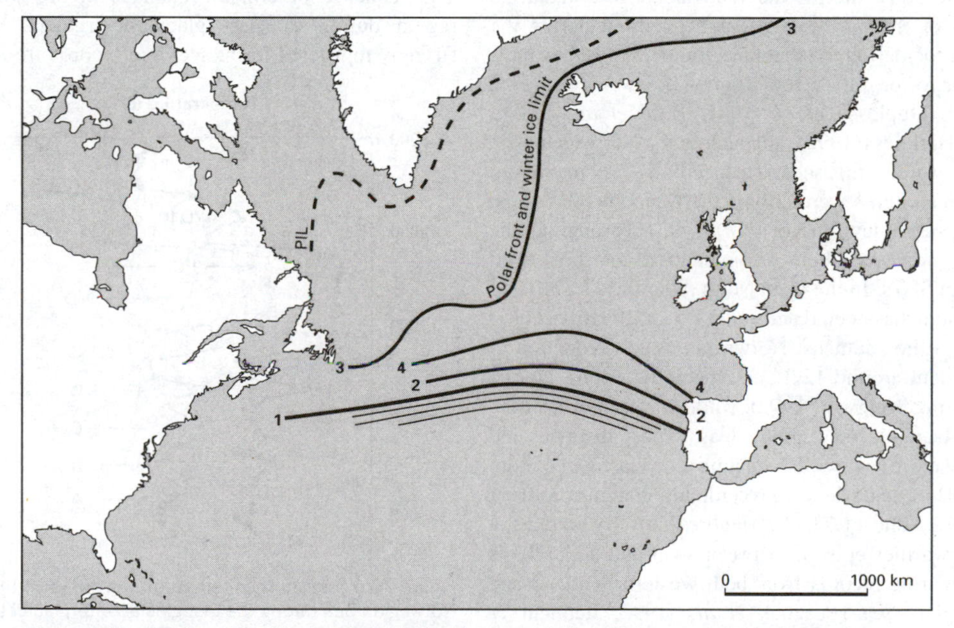

Figure 7.19 Position of the North Atlantic Polar Front and limits of winter sea ice during, and immediately prior to, the last glacial–interglacial transition (based on Ruddiman & McIntyre, 1981): 1, 20-16 ka BP; 2, 16-13 ka BP; 3, 13-11 ka BP; 4, 11-10 ka BP. Thin lines represent the pronounced thermal gradient to the south of the Polar Front; the dashed line marked PIL shows the approximate southern limit of pack ice at the present day.

eastern Atlantic suggest that, around 13 ka BP, the Polar Front retreated northwards between latitudes 35°N and 55°N (a distance of around 2000 km) at a rate of more than 2 km per year (Bard *et al.*, 1987b). Indeed, so rapid appears to have been the movement of the Front that warming seems to have been an almost synchronous event over large areas of the eastern North Atlantic (Koç Karpuz & Jansen, 1992). Even more rapid rates of movement (>5 km yr^{-1}) have been inferred for the Polar Front at the beginning and end of the Younger Dryas event (Bard *et al.*, 1987b).

7.6.3.2 Lateglacial sea-surface temperature changes

During the early part of the last glacial–interglacial transition (15–13.5 ka BP), surface temperatures of the northern North Atlantic appear to have been uniformly low as a result of meltwater influx from the northern sector of the Laurentide ice sheet and from decaying ice sheets in the Barents Sea area (Sarnthein *et al.*, 1992, 1994). The first signs of warming are found in the mid-North Atlantic (south of 55°N) around 14 ka BP (Veum *et al.*, 1992), and further north in the East Norwegian Sea around 13.4 ka BP (Lehman & Keigwin, 1992). In the latter area, the most rapid rise in SSTs occurred between 13.2 and 13 ka BP, diatom-based evidence indicating that the initial warming of surface waters was completed within 200 years.

Maximum SSTs during the Lateglacial Interstadial in many areas of the northeast Atlantic occurred during the early part of the period. At that time, temperatures may have been similar to, or only a few degrees below, those of the present day (Duplessy *et al.*, 1981; Bard *et al.*, 1987b; Peacock & Harkness, 1990), although in the Norwegian Sea, where meltwater influences reduced SSTs, maximum summer temperatures were only in the region 6–7°C, some 7°C below present levels (Koç *et al.*, 1993). Throughout the Interstadial, however, there are indications of short-lived reductions in SSTs. In the open North Atlantic, a significant cooling episode has been dated at *c.* 11.5 ka BP (Broecker *et al.*, 1988); in the southeast Norwegian Sea, a reduction in SSTs is evident around 12.3 and especially around 11.7 ka BP (Lehman & Keigwin, 1992); while further north in the Greenland–Iceland–Norwegian Seas, four distinct and increasingly more severe SST minima (each lasting no more than a couple of centuries) are recorded in diatom records at 12.9, 12.5, 12.3 and 11.7 ka BP (Figure 7.20). By contrast, a short-lived warmer episode between 11.3 and 11 ka BP is evident in marine records from both western Scotland and the Norwegian Sea (Peacock *et al.*, 1992; Lehman & Keigwin, 1992). The principal cause of these fluctuations in sea-surface temperatures appears to be short-lived pulses of meltwater discharge from the decaying northern ice sheets, although variability in heat flux caused by instabilities in thermohaline circulation may also have been a factor (section 7.7.2).

The Younger Dryas cooling is recorded in marine records from all parts of the Atlantic by a significant increase in ice-rafted debris and associated influx of cold meltwaters from both the North American and European ice sheets (Bond *et al.*, 1992b; Fronval *et al.*, 1995). The effect on sea-surface temperatures was dramatic. In the Norwegian Sea, a fall in summer SSTs of ⩾5°C in less than 40 years is reflected in microfossil records, with the expansion of sea-ice cover during the winter months. Further south, winter SSTs to the west of Ireland fell by 3–4°C below the Interstadial thermal maximum, while off Portugal the reduction may have been nearer 8°C (Bard *et al.*, 1987b). The temperature increase at the end of the Younger Dryas appears to have been equally abrupt. For example, in the Greenland–Iceland–Norwegian Seas, a rise in SSTs of 9°C within half a century has been recorded with summer values of 10–14°C and winter temperatures of 4–9°C (Koç *et al.*, 1993). A short-lived fall in SSTs of around 2°C during the early Holocene (~9.7 ka BP) has also been noted in this area (Lehman & Keigwin, 1992).

7.6.4 Ice-core records of environmental changes during the last glacial-interglacial transition

The sequence of climatic changes in the North Atlantic region during the last glacial–interglacial transition is strongly registered in the ice-core records from Greenland,

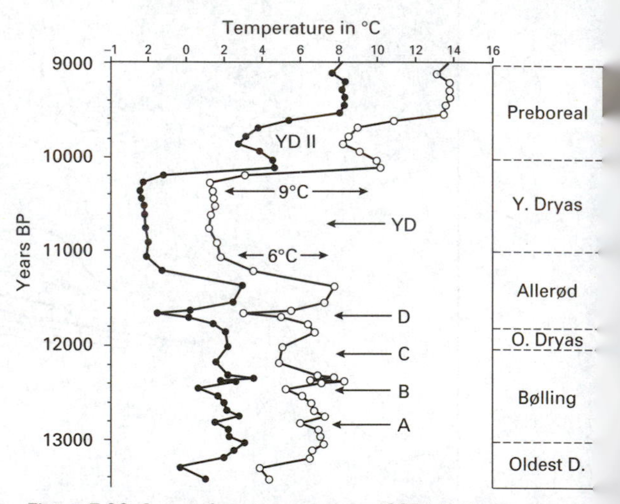

Figure 7.20 Sea-surface temperatures (SSTs) in the southeast Norwegian Sea during the last glacial-interglacial transition based on diatom data. Open circles – summer temperatures; closed circles – winter temperatures. Four short-lived colder episodes (A to D), each more severe than the previous one, occurred prior to the pronounced cooling of the Younger Dryas (after Koç Karpuz & Jansen, 1992).

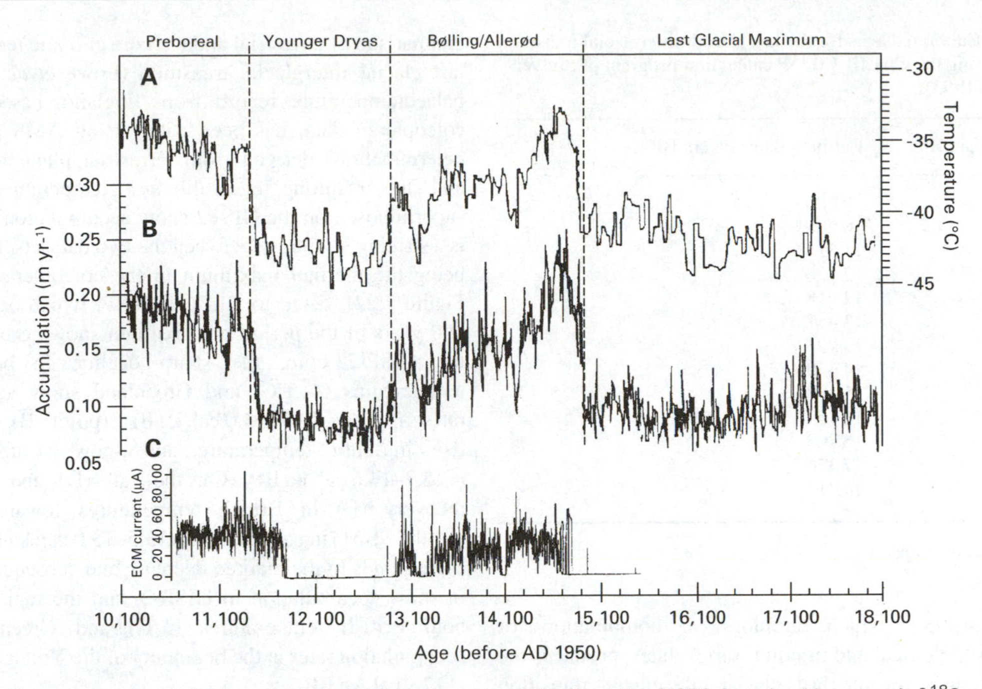

Figure 7.21 Three proxy records from the last glacial–interglacial transition from the GISP2 Greenland ice core. A. $\delta^{18}O$ values which have been converted into a temperature record. B. Snow accumulation (10-year averages). C. Electrical conductivity measurements (ECM) reflecting variations in atmospheric dust content (yearly averages) (after Kapsner *et al.*, 1995).

particularly the GISP2 and GRIP cores (section 3.11). Figure 7.21 shows three climatic proxies from the GISP2 core: (A) the $\delta^{18}O$ trace which has been calibrated to produce a temperature record (Kapsner *et al.*, 1995); (B) variations in snow accumulation rates, the higher rates of accumulation reflecting warmer intervals (Alley *et al.*, 1993); and (C) electrical conductivity measurements which reflect changes in the dust content of the atmosphere, episodes of higher dust content (lower ECM measurements) being correlated with colder phases (Taylor *et al.*, 1993a, 1993b). The similarities between the three curves are remarkable. All show the rapid climatic amelioration at the beginning of the Lateglacial; the thermal maximum during the early Interstadial, at which time air temperatures over Greenland were within a degree or so of the Holocene mean (Grootes *et al.*, 1993); the gradual deterioration during the mid- and late Interstadial, which was characterised by two, and possibly three, short-lived episodes of more severe conditions, each followed by a brief climatic recovery; the Younger Dryas cold phase when temperatures fell to levels 12–13°C below those of the Lateglacial thermal maximum, but during which rapid temperature variations of 5–7°C appear to have occurred; the abrupt amelioration at the beginning of the Holocene when the rise in temperatures may have been as rapid as 7°C in 50 years (Dansgaard *et al.*, 1989); and finally a marked climatic oscillation in the early Holocene (reflected in both $\delta^{18}O$ and snow accumulation records) when Greenland temperatures fluctuated by 3–4°C over a time interval of less than 200 years.

7.6.5 Correlation between terrestrial, marine and ice-core records of the last glacial–interglacial transition

Time-stratigraphic correlation between Lateglacial terrestrial, marine and ice-core records requires the development of a common timescale, for the Greenland ice core records are dated in 'ice-layer years' derived from the analysis of ice increments combined with glaciological parameters (section 5.4.4), whereas the chronology of terrestrial and marine sequences is based principally on radiocarbon (both conventional and AMS) dating. Calibration of the radiocarbon timescale, using uranium-series, dendrochronology and varve sequences (sections 5.3.4.4, 5.4.1.4 and 5.4.2.3) has shown that radiocarbon years significantly underestimate the true ages of Lateglacial events, although there are differences in the scale of the discrepancies indicated by the three methods. The most widely used calibration program (CALIB 3.0: Stuiver &

Table 7.2 Calibrated ages for the last glacial–interglacial transition obtained using the CALIB 3.0 ^{14}C calibration program of Stuiver & Reimer (1993).

^{14}C age (years BP)	Calibrated age (years BP)
9 000	9 980
9 500	10 588*
10 000	11 515*
10 500	12 420
11 000	12 918
11 500	13 417
12 000	13 993
12 500	14 649
13 000	15 438
13 500	16 163
14 000	16 793
14 500	17 370
15 000	17 917

*Mid point 2σ range.

Reimer, 1993), which employs a combination of dendrochronological and uranium-series dates, produces the calibrated ages for the last glacial–interglacial transition shown in Table 7.2. Although these may not match the 'ice-layer years' precisely, they offer the basis for a direct comparison between the ice-core, terrestrial and marine records.

This approach has been employed in Figure 7.22 to show a comparison of terrestrial and ice-core climatic records for the last glacial–interglacial transition (Lowe *et al.*, 1995a). A palaeotemperature record from England, based on fossil coleopteran data, has been dated using AMS radiocarbon determinations derived from terrestrial plant macrofossils, and the resulting age-calibrated temperature curve is superimposed on the GISP2 snow accumulation data. There is a striking similarity between the two datasets, key features being the thermal maximum in the coleopteran curve (A, Figure 7.22), dated to 14.7 cal. ka BP, which occurs within 100 years of the period of maximum snow accumulation in the GISP-2 core; the sharp decline in both British temperatures (4–5°C) and Greenland snow accumulation rates from *c.* 14.7–14.0 cal. ka BP (point B); the slight rise in both temperature and snow accumulation at *c.* 13.7–13.6 cal. ka BP (C); the fall (D) and subsequent recovery (E) in British temperatures towards the end of the Bø11ing-Allerød (*c.* 13.5–13.0 cal. ka BP) which corresponds to the marked decrease and subsequent increase in snow accumulation in GISP-2; and the rapid decline in both British temperatures (5°C) and Greenland snow accumulation rates at the beginning of the Younger Dryas (F: *c.* 12.9 cal. ka BP).

An alternative strategy is to use the data in Table 7.2 to convert the ice-layer years into ^{14}C years so that the ice-core record can be compared directly with a range of radiocarbon-dated proxies. Athough this is perhaps less satisfactory because of the difficulties that are involved in

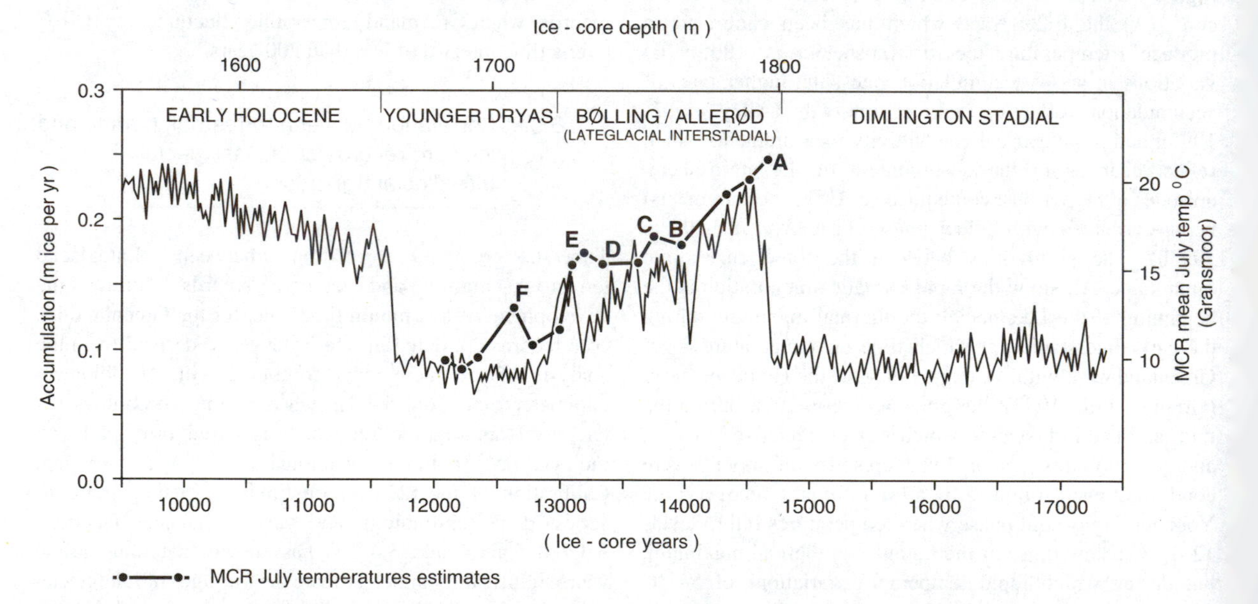

Figure 7.22 Comparison between Greenland snow accumulation rate data from the GISP2 ice core during the last glacial–interglacial transition, and palaeotemperature data based on fossil coleopteran records from England (MCR = Mutual Climatic Range) (after Lowe *et al.*, 1995a).

radiocarbon dating the last glacial–interglacial transition (section 5.3.2.3), the majority of proxy records from this period are underpinned by radiocarbon dates, and hence scientists working on this time period have tended to work with a radiocarbon timescale. Figure 7.23 shows a series of terrestrial, marine and ice-core records from the last glacial–interglacial transition set against such a timescale. The climatic trends revealed in these data are remarkably similar. Hence the initial warming prior to, or around, 13 ka BP is clearly recorded in the terrestrial records from southern and western Europe, in the marine evidence and, shortly thereafter, in the ice-core records. The gradual decline in temperature between *c.* 12.5 and 11 ka BP, the sequence of climatic oscillations (e.g. at *c.* 12.5, 12.0, 11.7 and 11.4 ka BP), each apparently of increasing severity, the cold episode of the Younger Dryas (*c.* 11–10.2 ka BP), and a short-lived oscillation during the early Holocene (9.8–9.5 ka BP), are also common elements in many of the proxy records.

Similarities are also beginning to emerge between the European and American Lateglacial sequences. For example, Kaiser (1994) has suggested that the **Two Creeks Interstadial** of Wisconsin, dated to 12.05–11.75 ^{14}C years BP, is the correlative of the 'Older Dryas' of Europe, and that the two events are contemporaneous with a major meltwater pulse from the Gulf of St Lawrence and with climatic or isotopic reversal in the Greenland ice cores. The 'Killarney Oscillation', dated at *c.* 11.2–10.9 ka BP in Maritime Canada, has been correlated with the immediate pre-Younger Dryas cooling events identified in European terrestrial records, in Atlantic marine microfossil records and in the Greenland ice cores, and has thus been termed the '**Amphi-Atlantic Oscillation**' (Levesque *et al.*, 1993). The Younger Dryas is found in pollen records not only from the northeastern seaboard of North America, but also in parts of the midwest, on the Pacific coast and, perhaps, even in Alaska (Peteet, 1995). These data suggest that the climatic changes that occurred in the North Atlantic region during the last glacial–interglacial transition were not just of regional but of hemispherical significance.

Examination of the shapes of the various proxy curves for the last glacial–interglacial transition, in particular the oxygen isotope and snow accumulation curves from the GISP2 core (Figure 7.21), shows that the signal is similar in almost every respect to those of climatic cycles (Dansgaard–Oeschger cycles) earlier in the last cold stage (section 7.4.1). The pattern of an abrupt increase in temperature, followed by a gradual temperature reduction, which in turn is terminated by a second abrupt rise in temperature, is common to the majority of these (Figure 7.7). From this perspective the '*Lateglacial Oscillation*' is the most recent of a series of long-term climatic fluctuations and, moreover, the Younger Dryas can be regarded simply as the end-member of a phase of temperature decline which commenced some 1500 ice-core years before the onset of the cold event itself. In this respect, therefore, the climatic oscillation during the last glacial–interglacial transition is not, as was once considered, a unique event (e.g. Mercer, 1969). On the other hand, the amplitude of the Lateglacial oscillation perhaps sets it apart from some of the earlier climatic fluctuations, while the Younger Dryas is also unusual in that it occurs during a Milankovitch radiation maximum (sections 1.6 and 7.8.1).

Perhaps the most significant point to emerge from a comparison of the various proxy records from the last glacial–interglacial transition in the North Atlantic region is that changes in such diverse parameters as ocean circulation, SSTs, Greenland snow accumulation rates and continental seasonal temperatures can occur almost simultaneously and extremely rapidly. Zahn (1994) has suggested that the mechanism which still best explains this 'orchestration' of events in the North Atlantic is that first proposed by Ruddiman and McIntyre (1973), i.e. the migrating Polar Front which is likened to a 'door' which swings open and shut. As the 'door' closes (the Polar Front migrates southwards), surface currents that normally flow along its southern flank (the 'Gulf Stream', North Atlantic atmospheric storm tracks) are deflected southwards, sea-ice cover expands and ice sheets grow. When the door 'opens' (the Polar Front migrates northwards), warmer waters can spread northeastwards and the ice sheets contract. The operation of this type of mechanism may also explain why changes in Greenland snow accumulation rates are broadly contemporaneous with changes in SSTs and with atmospheric circulation changes in Europe. It has been suggested that snow accumulation rates in Greenland are controlled not by changes in air temperature but by variations in atmospheric circulation (Kapsner *et al.*, 1995). These determine the position of the dominant storm tracks which, in turn, are governed by the position of the Polar Front. The idea is gaining credence, therefore, that changes in the position of the Polar Front are one surface expression of a range of processes that affect the entire ocean–atmosphere system of the North Atlantic virtually simultaneously. What is not clearly understood at present, however, is what triggers these sudden ocean circulation changes, a point that will be returned to in the final sections of this chapter.

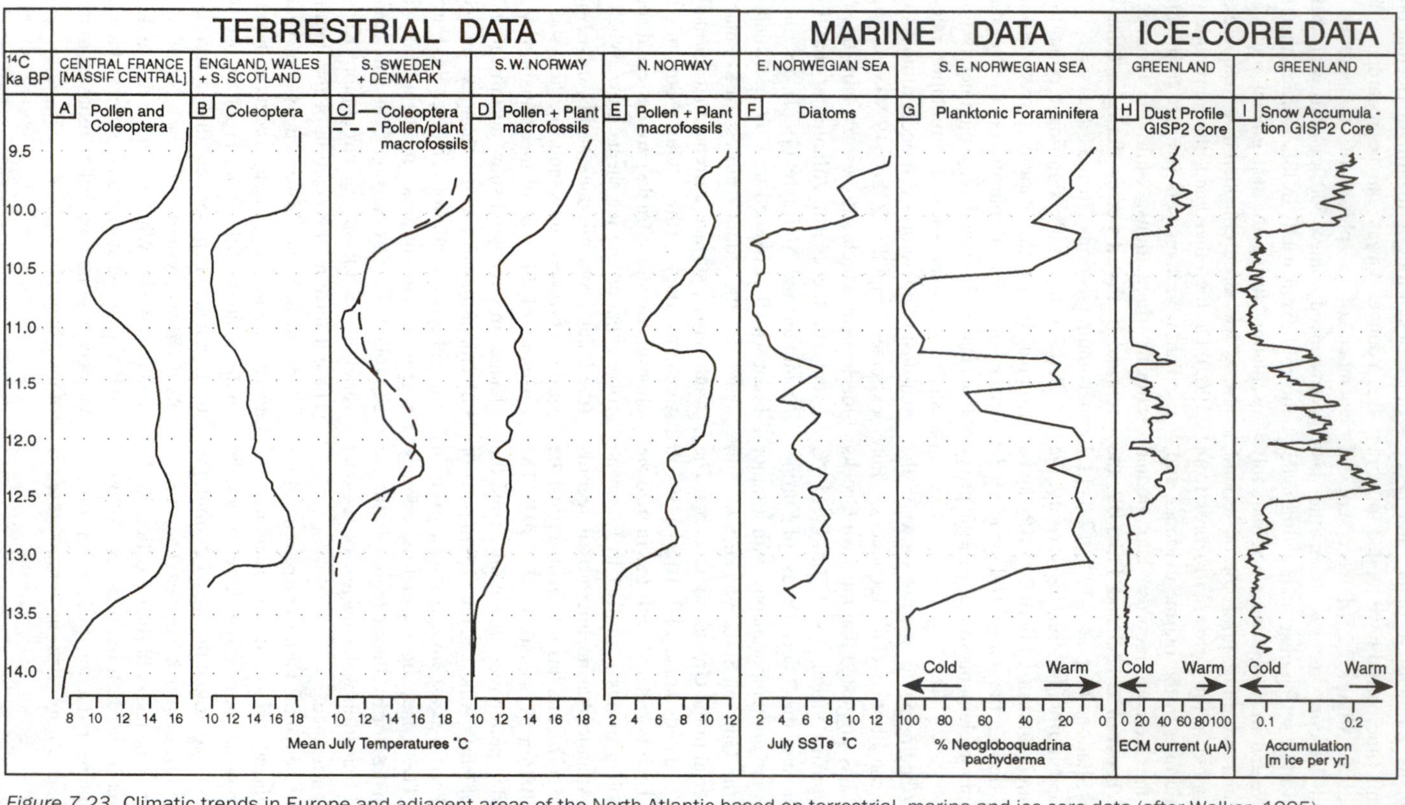

Figure 7.23 Climatic trends in Europe and adjacent areas of the North Atlantic based on terrestrial, marine and ice-core data (after Walker, 1995).

7.7 Atmospheric circulation: global circulation models

Numerous references have been made in preceding sections to ocean circulation and its effect on global climate, but there has been less discussion of atmospheric circulation. Yet atmospheric circulation is of equal, if not greater, importance in determining climatic patterns at both regional and global scales, while changes in atmospheric circulation are clearly of fundamental importance in determining the course of climatic change. Global atmospheric circulation is in a constant state of flux in response to both external (e.g. orbital parameters, variations in the solar constant) and internal (e.g. ocean circulation, changes in surface albedo) forcing mechanisms. Changes in atmospheric circulation can also arise through the operation of internal feedback mechanisms. This is because the atmospheric circulation system determines the scale and pattern of heat and moisture exchange, and hence the distribution and persistence of, for example, snow and sea-ice cover, desert conditions and monsoonal effects. These, in turn, affect biomass, atmospheric gas content and albedo, all of which exert an influence on climate.

One approach to gaining a better understanding of atmospheric circulation is to attempt to simulate the processes that are operating at the global scale by the construction of **general circulation models (GCMs)**. These are mathematical models (essentially computer simulations) of the mode of operation of the earth's atmosphere and the way it interacts with the hydrological cycle (Wright *et al.*, 1993). The results of these experiments have yielded new insights into the dynamic links between atmospheric circulation and climatic forcing mechanisms, such as mountain uplift (Ruddiman & Kutzbach, 1990), changes in vegetation cover (Foley *et al.*, 1994), volcanic activity (Bryson, 1989), desert dust (Joussaume, 1989), ocean circulation (Lautenschlager *et al.*, 1992) and glacier ice cover (Lautenschlager & Herterich, 1990). In the following section, we examine the development of GCMs in the context of climatic change at the end of the last cold stage and during the early part of the present Interglacial.

7.7.1 Global circulation models (GCMs)

GCMs attempt to simulate the complex three-dimensional structure and flows of the earth's atmosphere. Inevitably, however, the models are gross simplifications of the global atmospheric circulation system, since even the most powerful of computers cannot adequately represent its scale and complexity. The earth's surface is usually represented by a geometrical grid, the size of the grid cells normally varying between 4° × 5° and 11.5° × 11.25°, depending upon the complexity of the information to be computed and the capabilities of the computer employed. For each grid cell, a number of values can be plotted, each representing the average of a range of estimates for selected earth surface parameters. These fall into two categories. **Boundary conditions** are the prescribed surface values for such physical conditions as sea-surface temperatures (SSTs), surface albedo, radiation receipt, atmospheric transparency, sea-ice cover and topography (Kutzbach & Ruddiman, 1993). For simulations of modern climate, the input values are based on direct measurements of current physical parameters, while reconstructions of the boundary conditions that prevailed during earlier periods are based on estimates derived from proxy (geological) data. **Dynamic conditions** (flows) can then be added by parametrisation of surface processes including, for example, heat and moisture exchange gradients between surfaces, moisture convection in the atmosphere, Coriolis and shear constants, and atmospheric pressure equilibria (Street-Perrott, 1991; Wright *et al.*, 1993).

GCMs vary in both spatial and vertical resolution, and the ability to simulate the true complexity of atmospheric circulation patterns depends upon the sophistication of the available data, as well as on the capabilities of the computers employed. GCMs involve enormous computational loads and lengthy computer runs: most of the work is undertaken using supercomputers in only a few highly specialised centres, e.g. the National Center for Atmospheric Research (NCAR) in the USA; the Goddard Institute for Space Studies, USA, which has developed the GISS models; the UKMO – UK Meteorological Office; and the DKRZ (Deutsches Klima Rechen Zenter) at Hamburg, Germany.

7.7.2 Setting the boundary conditions for GCMs

Most GCM experiments have concentrated on modelling the climatic conditions of the last 18 ka BP, since this is the period for which the appropriate proxy data are most abundant. Atmospheric temperatures over land areas can be reconstructed using palaeoecological evidence (botanical records, coleopteran records, etc.), SSTs can be estimated from marine micropalaeontological data, glacier ice and sea-ice cover from glacial geomorphological and palaeoceanographic data, and atmospheric aerosol content from ice-core records. Orbitally induced solar radiation receipt can be calculated from radiation tables. Figure 7.24 shows in a schematic way how these boundary conditions

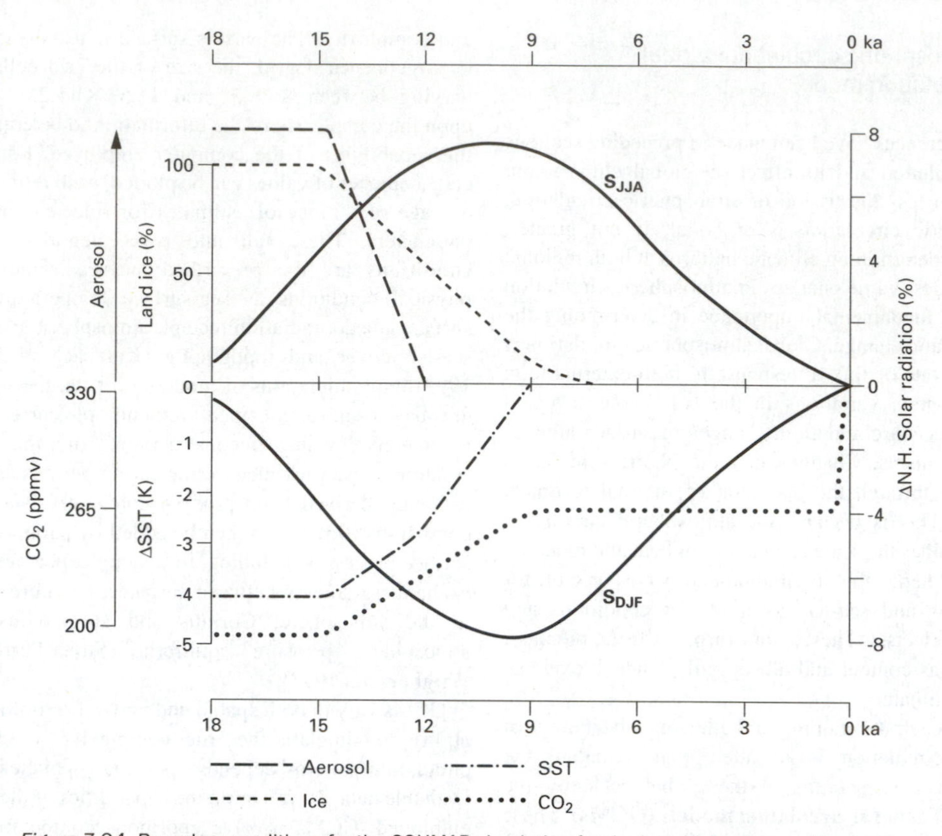

Figure 7.24 Boundary conditions for the COHMAP simulation for the last 18 ka. External forcing is shown for Northern Hemisphere solar radiation in June–August (S_{JJA}) and December–February (S_{DJF}) as the percentage difference from present-day radiation receipts. Internal boundary conditions include land ice as a percentage of 18 ka BP ice volume; global mean sea-surface temperatures (SST) expressed as a difference from present; excess glacial-age aerosol on an arbitrary scale; and atmospheric CO_2 concentration in parts per million by volume. The horizontal scale shows the times of the seven sets of simulation experiments (after Kutzbach & Webb, 1993).

may have varied at a global scale during the last 18 ka. The figure also illustrates how the earth's surface can be depicted in contrasting states or 'modes': (a) a 'glacial mode' at *c.* 18 ka BP (the last glacial maximum), when the earth's orbital configuration was similar to that of today, but ice cover was much greater, sea level lower, and SSTs well below those of the present; (b) a period of enhanced seasonality around 11–10 ka BP; (c) a period when climate was forced primarily by orbital effects (at *c.* 6 ka BP); and (d) the present-day situation. The proxy data for the period 18–0 ka BP therefore provide a basis for modelling the response of atmospheric circulation to different combinations of external and internal forcing mechanisms, and hence may also provide useful analogues for interpreting earlier climate cycles.

7.7.3 Running GCM experiments

Experiments using GCMs are of two basic types: **analogue experiments** and **sensitivity experiments** (Street-Perrott, 1991). **Analogue** (or **'realistic'**) **experiments** aim to simulate, as closely as possible, the conditions prevailing at the surface of the earth for any selected time period. The early GCMs of CLIMAP and COHMAP (section 4.11.6) were of this type (CLIMAP Project Members, 1976, 1981; Kutzbach & Guetter, 1986; COHMAP Members, 1988). These models used fixed SSTs, which is clearly unrealistic since the real oceans and atmosphere interact in a dynamic and mutually responsive way. More recent models are more sophisticated, and allow SSTs to vary within the model

(Manabe & Broccoli, 1985; Mitchell *et al.*, 1988), while some new GCMs also incorporate seasonal cycle variations, interactive soil moisture levels and changes in seasonal snow cover (Kutzbach *et al.*, 1993). Analogue experiments have demonstrated a number of important linkages in the global climate system including, for example, the close correspondence between enhanced summer insolation and the strength of the monsoon cells in the northern hemisphere (Kutzbach & Guetter, 1986), and the way in which the northern jet stream may have been deflected (or split) by the build-up of the Laurentide ice sheet (Figure 7.25), leading to marked changes in regional moisture distribution throughout the Northern Hemisphere (COHMAP Members, 1988; Kutzbach *et al.*, 1993). Other modelled effects include the downstream modification of climate in Europe caused by sea-surface temperature changes in the North Atlantic (Rind *et al.*, 1986), and the delay (by some 2–2.5 ka) of the Northern Hemisphere interglacial thermal maximum following the summer insolation peak at *c.* 10 ka BP (Figure 7.24) as a consequence of the delayed melting of the great northern ice sheets (Webb *et al.*, 1993b).

Sensitivity experiments, on the other hand, are used to assess the relative importance of different components of the system, such as insolation values, the level of CO_2 in the atmosphere and the prevailing surface boundary conditions. Inputs to the models in each of these categories may be deliberately exaggerated, one category at a time, in order to establish those factors to which the models are most responsive. By comparing the results of such model experiments with palaeoclimate reconstructions for specific 'time-slices' based upon proxy data, it may be possible to establish the primary forcing factors in past climatic changes. The results of such experiments have, however, frequently proved contradictory (Street-Perrott, 1991). For example, Hansen *et al.* (1984) found that the presence of land ice was the most important factor contributing to global cooling at 18 ka BP, whereas experiments by Broccoli and Manabe (1987) suggested that reduced atmospheric CO_2 levels were perhaps of greater importance. The differences between these results may reflect differences in the ways in which the models were constructed, but they may also indicate that in certain climate states (e.g. 'glacial climate mode' – see above), global climate is unusually sensitive to relatively minor changes in the system, and hence the course of climatic change may not be easy to predict. Indeed, Rind (1987) found that small variations in boundary conditions could force the climate system in a number of different directions. Moreover, some modelling experiments indicate that a major forcing factor, such as orbital radiation, can be temporarily overridden because of the sensitivity of the climate system to internal changes and feedback mechanisms during episodes of overall climatic instability.

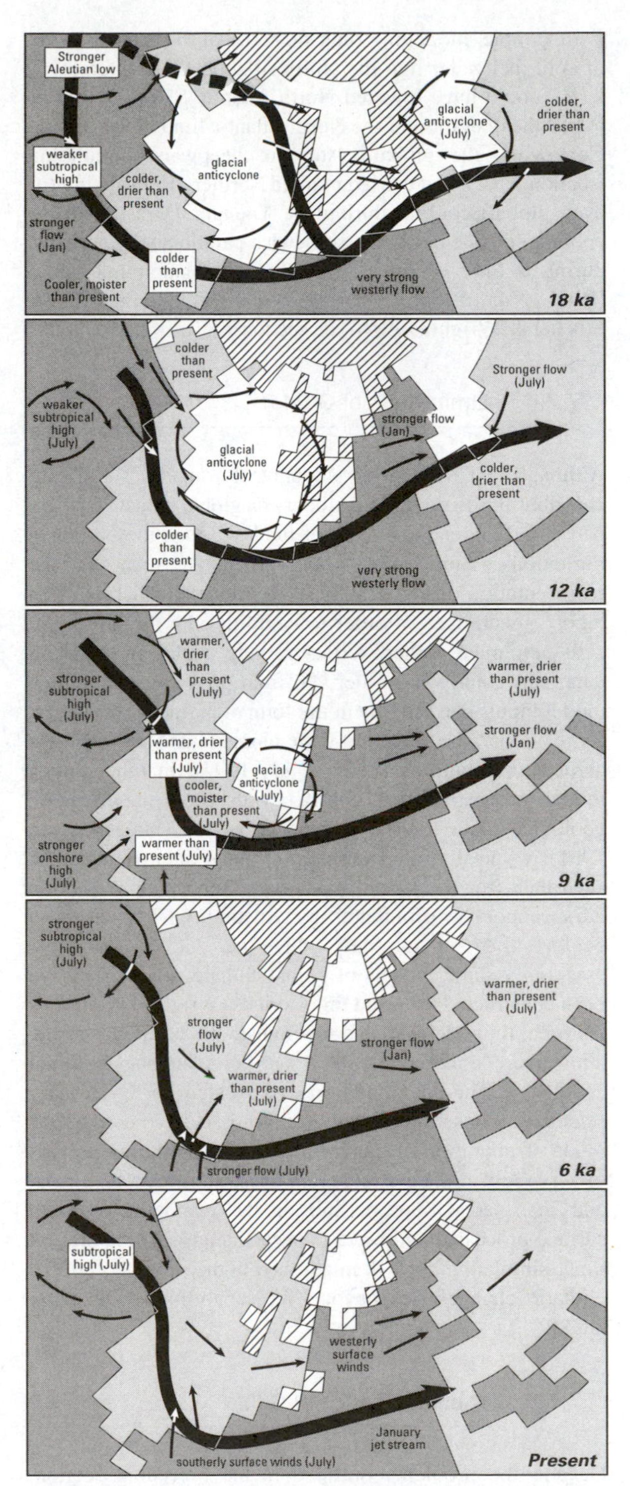

Figure 7.25 Palaeoclimatic model simulations for the last 18 ka (after Kutzbach *et al.*, 1993).

For example, modelling runs suggest that the sudden influx of cold meltwater into the North Atlantic shortly before 11 ka BP could have lowered North Atlantic SSTs, slowed or halted the operation of the North Atlantic limb of the 'Ocean Conveyor', the global system of deepwater circulation (section 7.8.2), and caused marked Northern Hemisphere (or even global) cooling during the Younger Dryas. Hence a cooling episode occurred, somewhat paradoxically perhaps, during a time of orbitally induced insolation maximum (Rind *et al.*, 1986; Overpeck *et al.*, 1989; Berger, 1990; Stocker & Wright, 1991 – see section 7.8.2)

7.7.4 Limitations of GCMs

Although GCMs have undoubtedly made significant contributions to developing theory on global climate change, particularly over the course of the last 18 ka, they do have limitations. Some are scale problems, such as the representation of topography in the models, which is often highly generalised in order to conform with the large grid cells employed. Others reflect varying quality in the input data, particularly those derived from proxy records. Some model inputs can only be in the form of a global average as, for example, is the case with biomass and atmospheric aerosol distributions (Peng *et al.*, 1994). In other cases, model reconstructions conflict with those based upon geological data for the same time period (Street-Perrott, 1991), while the rapid climatic changes often reflected in such data (for the Lateglacial period, for example: section 7.6) cannot be simulated by existing GCMs. From a geological point of view, therefore, GCMs constitute essentially static models of global climate, since they have been constructed for fixed time periods (Wright *et al.*, 1993). As such, they are designed to simulate the behaviour of the atmosphere under relatively fixed boundary and external conditions. This inevitably poses a problem for Quaternary scientists, whose interests are frequently centred both on the mode of operation of the earth's circulation during periods of very rapid climate change (e.g. transitions between glacial and interglacial stages), and on the responses of earth surface processes and biota to such changes. At the present time, simulating such dramatic and highly complex global climatic changes lies beyond the capabilities of existing GCMs.

7.7.5 The importance of GCMs

Despite the problems outlined in the preceding section, GCMs are playing an increasingly important role in climatic and palaeoclimatic research, and hence in Quaternary science. They provide a means of synthesising a wide range of environmental information at the global scale, of quantifying the major processes influencing global climate, and of conceptualising the cause and effect pathways involved. In terms of the last 18 ka, for example, the two most important cause and effect relationships to emerge so far from GCM experiments are (a) the more pronounced seasonality of Northern Hemisphere solar radiation between 12 and 6 ka BP which, in turn, increased the seasonality of Northern Hemisphere climates (resulting in greater climate extremes and enhanced monsoon circulation), and (b) the extent to which the growth of the last ice sheets and expanded sea-ice cover impacted on atmospheric circulation, and thereby influenced regional temperature gradients and moisture distribution in the middle to high latitudes (Webb *et al.*, 1993b).

GCMs have also focused attention on the important time lags that appear to occur between climate-forcing signals on the one hand and the response of climatic circulation on the other. For example, proxy records have demonstrated that the highest temperatures in both North America and Europe since the last glacial maximum occurred between 9 and 6 ka BP, significantly later than the maximum for summer insolation values (Figure 7.26). This appears to reflect the operation of major positive feedback mechanisms involving, *inter alia*, the delayed melting of the large ice sheets (Ruddiman *et al.*, 1989; Raymo *et al.*, 1990), meltwater fluxes in the North Atlantic and their effects on oceanic circulation (Broecker *et al.*, 1990; Crowley & Parkinson, 1988), and reduced atmospheric CO_2 content at, and shortly after, the Last Glacial Maximum (Stauffer *et al.*, 1988) perhaps related to reduced biomass in the oceans and on land (Huntley & Prentice, 1993). It could also (and almost certainly does) reflect a combination of these factors.

Despite the contributions that GCMs have undoubtedly made to our understanding of Late Quaternary palaeoclimates, it has to be borne in mind that GCMs are no more than mathematical experiments, scientific hypotheses that require some form of independent testing. The only way that this can be achieved is by 'hind-casting' climatic behaviour for selected times in the past and then comparing the results with geological reconstructions for the same period (Kutzbach, 1985; Saltzman, 1990; Wright *et al.*, 1993). Further insights into Late Quaternary climates, therefore, require a closer dialogue between climate modellers on the one hand and Quaternary scientists who collect and interpret proxy data on the other. The models are, of course, only as secure as the data upon which they are based, and against which they are ultimately to be tested. The onus therefore falls squarely on the Quaternary community to provide the necessary high quality, high resolution proxy records (such as those from the GRIP and

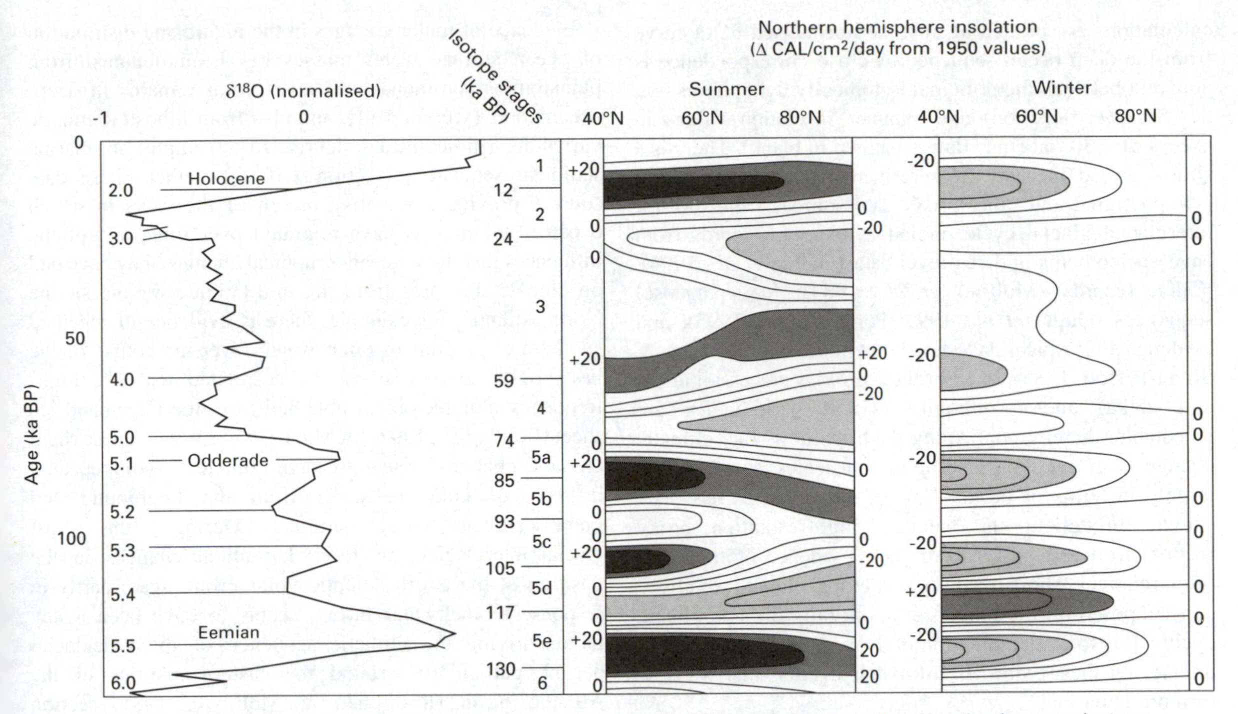

Figure 7.26 Northern Hemisphere insolation values (by comparison with 1950s values) plotted against the deep-sea isotope stratigraphy. Summer anomalies ΔCAL larger than $+30\ cm^{-2}\ day^{-1}$ are shown in black (after Sejrup & Larsen, 1991).

GISP2 Greenland ice core, for example) to enable climatic modelling to begin to provide answers to questions not only about the pattern of past climatic changes, but also about their causes. It is to this aspect of the last interglacial–glacial cycle that we finally turn our attention.

7.8 The search for causes

The causal factors behind the climatic changes that are manifest in the proxy records of the last 130 ka are examined in this section. A recurrent theme is the extent to which climatic changes are driven (or forced) by **external** factors, in particular by the operation of the astronomical variables (section 1.6), or by **internal** processes, i.e. by those that operate within the terrestrial–ocean–atmosphere system. A second thread that runs through the discussion is the influence of **feedback loops**, in other words the ways in which any one process may **amplify** or **modulate** the effects of another. We begin, however, with a short recapitulation of the fundamental component of Quaternary climatic change, the Astronomical Theory.

7.8.1 The Astronomical Theory of climatic change

The Astronomical Theory of climatic change was introduced in Chapter 1 (section 1.6). Often referred to as the **Milankovitch Hypothesis**, it is based on the premiss that variations in the geometry of the earth's orbit with respect to the sun, moon and other planets will govern the seasonal radiation cycle and thereby give rise to long-term fluctuations in climate. Although doubt has been expressed in some quarters about the role of **'orbital forcing'** as the major causal mechanism behind the repeated glacial–interglacial cycles of the Quaternary (e.g. Winograd *et al.*, 1988, 1992), it is now generally accepted that the astronomical variables of precession, obliquity and eccentricity, through their control of the seasonal and latitudinal distribution of solar radiation, drive the major climatic cycles (Imbrie *et al.*, 1992). Moreover, the influence of these orbital parameters on the earth's climatic system is readily detectable in a range of proxy records (Imbrie *et al.*, 1993). In Figure 7.26, for example, latitudinal variations in Northern Hemisphere insolation over the course of the last 130 ka derived from astronomical calculations are plotted against the normalised $\delta^{18}O$ curve from the deep ocean sequence. A close correspondence is apparent between the principal isotopically light phases (5e, 5c, 5a, 2/1 transition) and summer insolation values in excess of +30 Cal cm^{-2} day^{-1} (shown in black). There is a similar broad measure of agreement between the marine isotope signal and other proxy data sources from the last interglacial–glacial cycle, including isotopic records from cave speleothems and sea-level data (Gallup *et al*., 1994), pollen records (Molfino *et al*., 1984), loess/palaeosol sequences (Gupta *et al*., 1991; Porter & An, 1995), and evidence of tropical lake-level fluctuations (Kutzbach & Street-Perrott, 1985). In all of these proxies precessional (19 and 23 ka) and/or obliquity (41 ka) periodicities are detectable, thereby confirming the hypothesis that climatic change over medium and long timescales is explicable largely in terms of orbital forcing. However, as has been shown throughout this chapter, high-resolution proxy records from the past 130 ka show evidence of rapid but often relatively short-lived shifts in global climate which are superimposed on the longer-term, orbitally driven climatic cycles. It is to an examination of the possible causal factors underlying these '**sub-Milankovitch events**' that we now turn our attention.

7.8.2 Palaeoceanography and climatic change

Regional climate is closely linked to ocean circulation. At the present day, for example, the relatively mild, maritime climate of much of western Europe contrasts markedly with the more severe climatic régime of comparable latitudes in eastern Canada, the difference being attributable almost entirely to the northeastwards transfer of heat to western Europe along the Gulf Stream. Any change to the pattern of ocean circulation is likely, therefore, to have major consequences for regional and, indeed, hemispherical climates. Ocean water movements are driven by a combination of interconnected internal factors, such as density, temperature and salinity variations, as well as external factors, including frictional drag from the wind, freshwater influx from precipitation, rivers or glacial meltwater, and insolation variations. However, many of these variables will both influence, and be influenced by, others, and hence a complex system of feedback loops often makes it difficult to establish cause and effect in ocean–atmosphere–terrestrial relationships. Nevertheless it is apparent that over the past 130 ka, changes in ocean water movement have brought about significant regional climatic changes, and that these have operated at both Milankovitch and sub-Milankovitch frequencies.

Evidence of major changes in the nature and distribution of ocean surface water masses has been obtained from planktonic microfaunal and microfloral remains in deep-ocean cores (section 4.11), and also from lithostratigraphic variations (in ice-rafted debris, for example) in marine sediment sequences (section 3.10). Together, these data sources provide a sensitive record of the ways in which ocean water masses have migrated over time, and of the influences that these oceanographical changes may have had on climate. In cores from the mid-latitude regions of the North Atlantic. for example, there is evidence of repeated southern excursions of polar waters over the course of the last 110 ka, and these can be correlated with the high-frequency climatic signal obtained from the Greenland ice sheet (Bond *et al*., 1993; McManus *et al*., 1994). These rapid climatic changes appear to have coincided with repeated influxes of cold meltwater from the Laurentide and Fennoscandian ice sheets. During the last glacial–interglacial transition, latitudinal changes in the position of the North Atlantic Polar Front, again partly in response to meltwater influx, seems to have been a key factor driving the climatic sequence of the Lateglacial period, particularly around the eastern margins of the Atlantic basin (Ruddiman & McIntyre, 1981; section 7.6.3.1). Major meltwater discharge events occurred around 15.0–14.5, 13.5, 12.0 and 10.5 ka BP, and each can be associated with brief episodes of cooler climate in the North Atlantic region (Keigwin *et al*., 1991). The most pronounced of these was the Younger Dryas cooling when North Atlantic surface waters may have been further chilled by the diversion of Laurentide ice-sheet meltwater from the Mississippi to the St Lawrence drainage systems (Broecker *et al*., 1989; Overpeck *et al*., 1989). As noted above (section 7.6.3), these changes in surface water movements also seem to have occurred extremely rapidly. For instance, in the Norwegian Sea, sudden changes in the flow of warm Atlantic surface waters typically involved shifts in sea-surface temperatures of $\geqslant$5°C in less than 40 years, and these, in turn, appear to have led to equally large and rapid fluctuations in atmospheric temperatures (Lehman & Keigwin, 1992).

Although the movement of ocean water masses can be partly explained by such external forcing mechanisms as meltwater or freshwater influx, however, they are also a product of internal factors, notably circulation changes arising from salinity (density) and water temperature gradients, a process known as **thermohaline circulation**. Within the North Atlantic basin, this appears to operate by means of a **conveyor** system whereby water moves northward in the upper levels of the ocean, ultimately to sink around latitude 60°C to form a deep-water mass known as **North Atlantic Deep-Water (NADW)**. The return limb of the conveyor at depth transfers this deep water to the Southern Oceans. At the global scale, it has been suggested that differences in salt concentration between the Atlantic and Pacific Oceans drive a **Global Conveyor** (Figure 7.27) through which there is a mass transfer of dense, saline water from the Atlantic to the Pacific at depth, and a compensating countercurrent near the surface (Broecker *et al.*, 1985).

In the North Atlantic, the impact of NADW formation on climate is considerable. During the winter months 'ventilation' of the ocean by the sinking of surface waters releases heat to the atmosphere. As the northward-flowing waters have a temperature averaging about 10°C and the return flow at depth a temperature of around 2°C, about 8 calories are released for each cubic metre of water. This means that $\sim 5 \times 10^{21}$ cal of heat are released to the atmosphere each year, an amount equivalent to ~25 per cent of the solar heat reaching the surface of the Atlantic Ocean in the region north of 35°N (Broecker & Denton, 1990a). If the production of deep water in the northern North Atlantic could be varied, this would alter the rate of heat release and hence a mechanism would exist for generating regional climatic change. Indeed, a '*turn on and off*' *mechanism* has been used to describe the sudden changes in ocean circulation that appear to have occurred in the North Atlantic over the course of the last glacial–interglacial cycle (e.g. Broecker *et al.*, 1990). One way in which this process might operate is where transfer of salt by the conveyor and dilution by meltwater gradually drive down the salinity of the North Atlantic. The reduction in rate of ventilation (or sinking) of surface waters results in a gradual slowing down of the conveyor, lessening the amount of heat released to the atmosphere over the northern Atlantic. Eventually, a situation may be reached where the conveyor shuts down completely (Broecker, 1992).

Past changes in the production and circulation of deep-water can be inferred from variations in cadmium and in carbon isotope ratios in benthic Foraminifera, as these provide evidence of changes in nutrient budgets which are

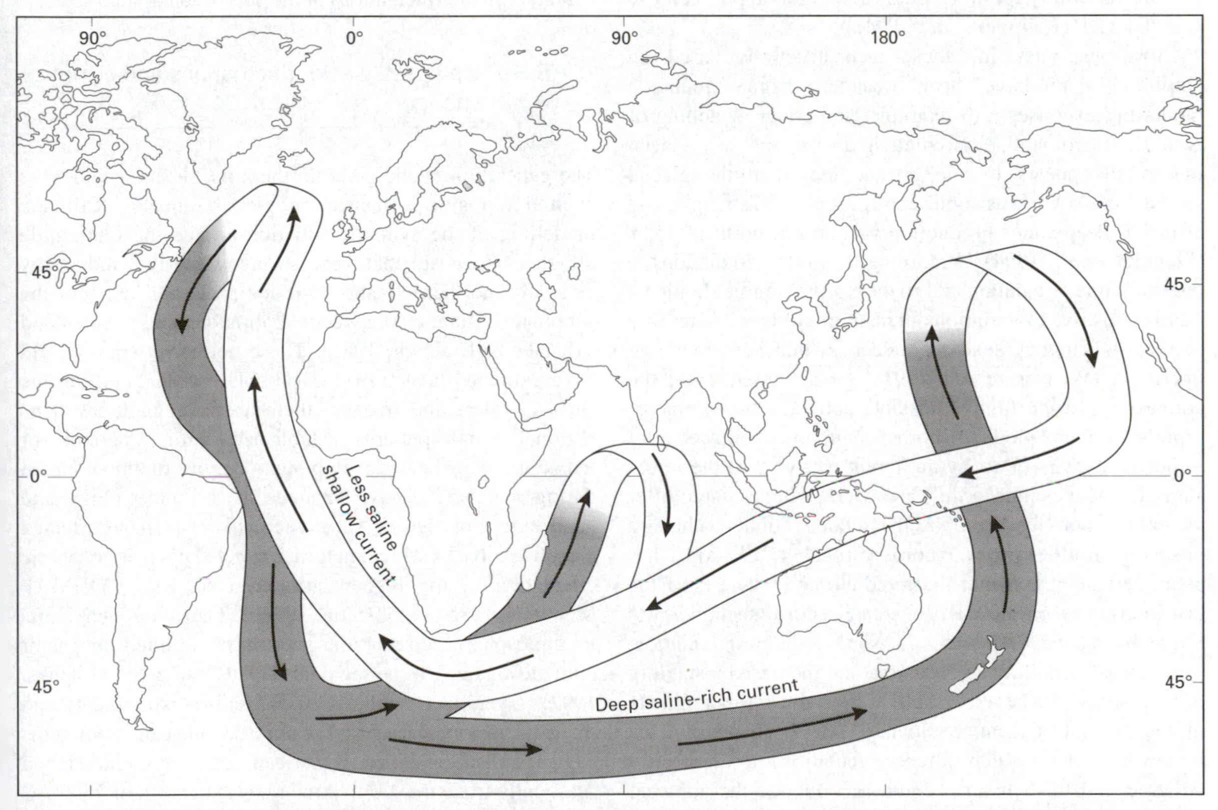

Figure 7.27 The large-scale salt transport system ('Ocean/Global Conveyor') operating in the present oceans. This compensates for the transport of water (as vapour) through the atmosphere from the Atlantic to the Pacific Ocean. Salt-laden deep water formed in the northern Atlantic flows down the length of the Atlantic and eventually northward into the deep Pacific. Some of this water upwells in the northern Pacific, bringing with it the salt left behind in the Atlantic due to vapour transport. Flow of the 'Atlantic Conveyor' may have been interrupted during cold episodes and was replaced by an alternative mode of operation (after Broecker & Denton, 1990a).

linked to deep-water changes (Boyle, 1988; Duplessy *et al.*, 1988). Near the OI stage 3/2 boundary, NADW production was similar to that of today, but outflow of NADW from the North Atlantic was significantly reduced at the last glacial maximum (Sarnthein *et al.*, 1994). During the last glacial–interglacial transition, NADW production was further decreased or eliminated, principally due to the episodes of meltwater discharge referred to above (Lehman *et al.*, 1991). Subsequently, however, the records show a sudden increase in NADW production which began in the northeast Atlantic around 12.8–12.5 ka BP (Sarnthein *et al.*, 1994). This sudden strengthening of the NADW thermohaline cell at a time when terrestrial proxy records also show signs of climatic amelioration (e.g. Lowe *et al.*, 1994) provides strong evidence for the importance of NADW in glacial–interglacial climate change (Charles & Fairbanks, 1992). The climatic influence of deep-water movement is also reflected in the close correlation that has been observed between NADW production and Greenland climate variability for much of the time interval between 130 and 70 ka BP (Keigwin *et al.*, 1994).

Close analogies for these reconstructions based on empirical data have been obtained from simulation modelling exercises. For example, numerical modelling of Atlantic thermohaline circulation during the last glacial maximum appears to support the idea that the glacial Atlantic conveyor was significantly reduced in strength, and also that deep-water production was absent north of 55°N (Fichefet *et al.*, 1994). According to model simulations, a thermohaline circulation cell in the North Atlantic should be highly sensitive to variations in input of cold meltwater (see above) which may lead to sudden switches in mode of operation (Weaver *et al.*, 1991). As a consequence, the conveyor may be highly unstable and be able to change rapidly from one mode to another, sometimes followed after a thousand years or so by an abrupt recovery to the initial state. If this was indeed the case, such switches could offer an explanation for some of the sudden climatic changes observed in other proxy records (Broecker, 1994a). It has also been suggested that the rapid climatic changes of the last interglacial in the GRIP ice-core record (section 7.3.1) might be related to changes in North Atlantic circulation. Simulation modelling suggests that the increased instability in the hydrological cycle, resulting from the warmer climate of the last interglacial, could have led to changes in the freshwater flux (which forces circulation) to the North Atlantic and this, in turn, could account for the apparent variability of the Eemian climate (Weaver & Hughes, 1994).

Variations in NADW formation, therefore, appear to have exercised a major influence on the climatic régime of the North Atlantic region over the course of the past 130 ka. However, not all of the climatic changes observed in proxy records are so readily explicable in terms of deep-water circulation changes. For example, although a temporary shut-down in the ocean conveyor has been advocated as the cause of the Younger Dryas climatic reversal between 11 and 10 ka BP (Broecker & Denton, 1990a; Broecker *et al.*, 1990), other workers have found no evidence of a reduction in NADW during this time interval (Jansen & Veum, 1990; Veum *et al.*, 1992). Indeed, Berger & Jansen (1995) specifically reject the conveyor-belt mechanism as a primary cause for the Younger Dryas event. Instead, they argue that the climatic warmings that preceded and succeeded the Younger Dryas were associated with meltwater pulses from wasting Northern Hemisphere ice sheets, each of which was accompanied by heat import into the northern areas of the North Atlantic through the operation of a fjord-like oceanic heat pump.[2] In this scenario, the Younger Dryas simply represents a pause in the process of deglaciation during which full glacial conditions were re-established. If correct, this suggests that ocean circulation changes are not the only forcing factor behind climatic fluctuations at the sub-Milankovitch scale.

7.8.3 Ice-sheet/glacier fluctuations and climatic change

The generation of the great northern ice sheets would have exerted a major influence on global climates. Climatic modelling of the synoptic situation around the Laurentide ice sheet suggests that very strong westerly winds blew along the northern flank of the ice sheet and out into the subpolar Atlantic between Labrador and Greenland (Manabe & Broccoli, 1985). These northerly winds would have extracted heat from the subpolar oceans, chilling the surface waters and freezing them at higher latitudes. This regional wind pattern, established during periods of maximum ice-sheet development, begins to break down during ice-sheet decay. Changes in the morphology and areal extent of the ice sheet seem likely to have affected albedo feedbacks which, in turn, impacted upon atmospheric circulation in the higher latitudes (see, e.g., COHMAP Members, 1988; Webb *et al.*, 1993b). The consequent shifts in atmospheric circulation may have resulted in major climatic changes at the regional scale (Charles & Fairbanks, 1992). Overall, a synthesis of modelling experiments and oxygen isotopic records from deep-ocean sediments suggests that the North American ice sheets chilled and seasonally froze the North Atlantic Ocean north of 45–50°C latitude at the 41 ka and 100 ka Milankovitch rhythms, but cooled the ocean southward into the middle latitudes mainly at the 100 ka periodicity (Ruddiman, 1987).

However, changes in the structure and configuration of the great ice sheets that are independent of the Milankovitch periodicities may also have been a forcing factor in Quaternary climatic change. For example, Wilson (1964) suggested that surging of the West Antarctic ice sheet towards the end of an interglacial would have led to an influx of up to 10^7 km^3 of ice into the world's oceans, thereby leading to global cooling and widespread glaciation. Such a dramatic surge would also have resulted in a global sea-level rise of 10–20 m, and evidence for late–last interglacial high sea-level stands has indeed been found at a number of sites in the Northern Hemisphere (Hollin, 1977; Aharon *et al.*, 1980). Surging of the East Antarctic ice sheet has also been proposed as a mechanism for initiating glacial episodes (Schubert & Yuen, 1982), and a surge event at ~95 ka BP has been associated with the dramatic cooling that occurred at the end of OI stage 5c (Hollin, 1980). Empirical evidence from Antarctica shows that marked fluctuations in the extent of grounded ice occurred throughout OI stage 5, and model simulations indicate sporadic, perhaps chaotic collapse (complete mobilisation) of the ice sheet over the course of the Middle and Late Quaternary (MacAyeal, 1992). This irregular behaviour appears to be unrelated to climatic forcing, but primarily reflects glaciological responses to basal till distribution which lubricates ice-sheet motion and therefore governs glacier behaviour.

More direct evidence of ice-sheet fluctuations during the course of the last 130 ka can be found in the **Heinrich Events**, the carbonate-rich layers of ice-rafted debris that occur in deep-ocean cores throughout the mid-latitude regions of the North Atlantic Ocean and which reflect episodes of massive iceberg discharge not only from the Laurentide ice sheet (Heinrich, 1988; Broecker, 1994b) but also from the Fennoscandian and British ice sheets (Baumann *et al.*, 1995; Rahman, 1995). The close correspondence between the Heinrich Events in the ocean cores, periods of reduced SSTs inferred from biostratigraphic evidence in marine cores, and the Dansgaard–Oeschger cycles in the GRIP Summit ice core has already been discussed (section 7.5.5). Between 20 and 80 ka BP, it appears that the shifts in ocean–atmosphere temperatures were grouped into a series of cooling cycles (**Bond Cycles**), each lasting 10–15 ka and containing a number of Dansgaard–Oeschger events, and each characterised by asymmetrical saw-tooth shapes (Figure 7.13). The ocean records suggest temperature changes of up to ~5°C in a matter of decades, a rate comparable with those reflected in the ice cores (Dansgaard *et al.*, 1993). Each cycle culminated in a cold stadial, which terminated with the enormous discharge of icebergs into the North Atlantic (a Heinrich Event), and this was followed by an abrupt shift to a warmer climatic régime (Bond *et al.*, 1993; Bond & Lotti, 1995). The relationship between the two most recent Heinrich Events (H1 and H2) and episodes of expansion of the circum-Atlantic ice sheets was noted above (section 7.5.5), although it is not clear at present whether all of the cycles reflect internal oscillations of the ice sheets (internal forcing) or are a result of external climatic forcing that caused ice sheets to grow, reach a position of instability and then collapse catastrophically (Broecker, 1994b). However, it does seem that the rapid warming at the end of each cycle *is* a direct result of ice-sheet behaviour. Subsequent landward retreat of the ice sheets would have reduced the flux of icebergs into the open ocean, and the resulting increase in surface salinity could have been sufficient to strengthen thermohaline circulation (see above) and raise the surface temperatures of North Atlantic waters (Bond *et al.*, 1993). The Dansgaard–Oeschger cycles that have been identified in ice-core records are believed to reflect marked fluctuations in air temperature, the warmer episodes correlating with the periods between Heinrich Events. It has been suggested that these reflect an interaction between oscillations of the ice sheets and variations in the strength of the ocean 'heat pump' which, in turn, is driven by variations in thermohaline circulation (Broecker *et al.*, 1990). Overall, there is now a considerable body of evidence to suggest that fluctuations in the margins of the great ice sheets were linked to fundamental changes in the ocean–atmosphere system, and these changes have been superimposed upon the climatic rhythms arising from the operation of the Milankovitch variables.

7.8.4 Variation in atmospheric gas content and climatic change

The most detailed records of former variations in atmospheric gas content have been obtained from polar ice cores. Data from Greenland (Neftel *et al.*, 1982; Stauffer *et al.*, 1984) and Antarctica (Barnola *et al.*, 1987, 1991) show that during the last cold stage, atmospheric CO_2 levels were significantly lower than during either the present or previous interglacials. A similar pattern is evident in ocean core records (Shackleton *et al.*, 1983, 1992). Markedly lower concentrations of other atmospheric gases, such as CH_4 (methane) and N_2O (nitrous oxide), have also been found in ice that accumulated during the last cold stage (Chappellaz *et al.*, 1990; Raynaud *et al.*, 1993). The amplitudes of the changes between glacial and interglacial episodes are considerable. During the last climatic transition (OI stage 2/1), for example, the atmospheric CO_2 level rose from ~200 ppmv (parts per million by volume) to an interglacial value of ~280 ppmv, the CH_4 concentration almost doubled from 350 to 650 ppbv (parts per billion by volume), while N_2O values increased from ~190 ppbv to ~270 ppbv (Figure 7.28).

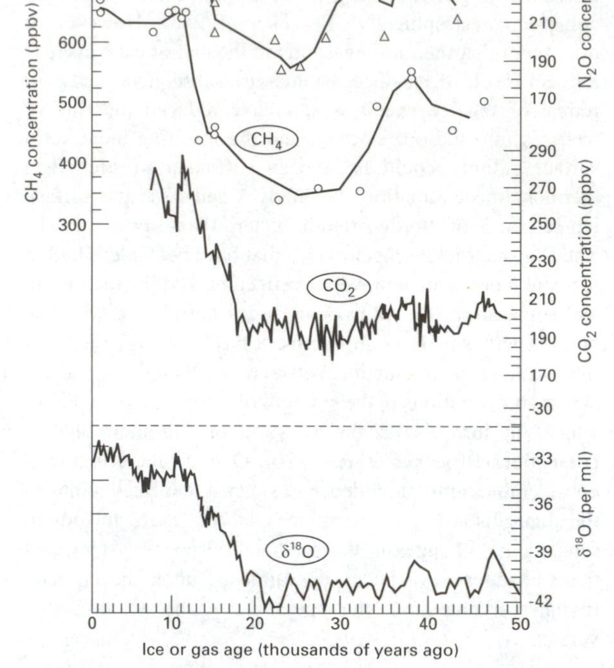

Figure 7.28 The Byrd ice-core record of CO_2, CH_4 and N_2O over the last 50 ka, with the $\delta^{18}O$ climatic record for comparison (after Raynaud *et al.*, 1993).

The mechanisms behind these atmospheric gas changes are not completely understood, but they appear to reflect complicated variations in rates of production and, in particular, in the operation of sources and sinks of the individual gases (Sundquist, 1993). In the case of CO_2, for example, the oceans would have constituted a major sink during glacial periods at times of reduced vegetation cover. Warming of the ocean at the end of the last glacial caused CO_2 to be degassed to the atmosphere, much of which eventually found its way into the sink of the terrestrial biosphere (vegetation and soils). However, a range of factors in the oceans, including, *inter alia*, changes in the rate of biological productivity, degree of alkalinity, pattern and vigour of ocean circulation, extent of sea-ice cover and rate of deposition of $CaCO_3$ in coral reefs, would have influenced the ocean–atmosphere flux of CO_2 at a glacial–interglacial transition. A key factor may be the abundance of nutrients in ocean waters, former concentrations of which can be inferred from variations in nitrate concentration in marine sediments. Recent data ($^{15}N/^{14}N$ ratios) from both the Arabian Sea and the North Pacific suggest that, during glacial periods, there was a significant increase in the nutrient inventory (Altabet *et al.*, 1995; Ganeshram *et al.*, 1995). This enhanced oceanic biological productivity and associated CO_2 storage could, in turn, have led to a reduction in CO_2 levels in the atmosphere.

As far as CH_4 is concerned, natural emissions occur from wetlands, animals, termites, biomass burning, oceans and lakes, while sinks include the oxidation by OH in the troposphere and bacterial consumption in aerated soils (Raynaud *et al.*, 1993). Of these, changes in wetland composition and extent (which are closely linked to continental moisture régimes) appear to have been the major driving force in long-term CH_4 changes (Chappellaz *et al.*, 1990; Blunier *et al.*, 1995). An additional source may be **gas hydrate**, the solid ice-like crystalline phase of methane which is stored in large quantities in sea-floor sediments (Haq, 1993). Sudden and massive expulsions of methane from gas hydrates that have become unstable have been suggested as a factor affecting climate at the last glacial–interglacial transition (Nisbet, 1992). Changes in the atmospheric N_2O budget reflect the loss of N_2O by photochemical decomposition in the atmosphere (the major sink), or fluctuations in nitrification/denitrification processes in soils and in ocean waters (the main natural sources) or both (Raynaud *et al.*, 1993).

What is especially striking about the record of gas concentration in polar ice cores is that the trends correspond very closely to those in the temperature curve based on the $\delta^{18}O$ record from the ice sheets (Figure 7.28). CO_2 and CH_4 are broadly in phase with the climatic signal during deglaciation periods; at the onset of a glacial, however, CH_4 remains in phase but CO_2 lags markedly behind (Raynaud *et al.*, 1992). This may reflect the abstraction of CO_2 from the atmosphere as the expansion of scrub and woodland into formerly glaciated regions led to a marked increase in global biomass and hence in the stored carbon inventory (Oeschger, 1992). The overall climate/gas relationship appears to be confirmed by spectral analysis of the CO_2 trace in the Vostok ice cores from Antarctica, which shows a clearly defined maximum at *c.* 21 ka and a secondary, less well-developed maximum at *c.* 41 ka BP, i.e. close to that predicted by the precessional and obliquity periodicities (Barnola *et al.*, 1987). A similar pattern is also evident in the CH_4 signal (Lorius *et al.*, 1990). These features strongly suggest that a coupling exists between atmospheric gas composition and major global climatic changes over the course of at least the last 130 ka, and also provide independent support for the Milankovitch Hypothesis. As CO_2, CH_4 and N_2O are all radiatively active gases (**'greenhouse gases'**), a causal link has been inferred between trace gases and climatic fluctuations, with changes in atmospheric gas composition amplifying orbitally driven long-term climatic variations.

Both empirical and model simulation data suggest that the influence of atmospheric gases can be considerable. It has been estimated, for example, that about half of the temperature change reflected in the Antarctic Vostok ice core record may be accounted for by the effects of variations in concentration of CO_2 and CH_4 (Lorius *et al.*, 1990), while model simulations suggest that 40–50 per cent of the last glacial–interglacial global-scale warming may be attributed to the greenhouse effect, i.e. direct radiative forcing and associated feedbacks (Jouzel *et al.*, 1990).

The atmospheric gas record for the past 130 ka also shows evidence of short-term climatic changes. In the Dye 3 core from Greenland, for example, marked fluctuations in CO_2 content during part of the the last cold stage (Stauffer *et al.*, 1984) parallel the Dansgaard–Oeschger events (Figure 7.7) reflected in the $\delta^{18}O$ signal, although similar features have yet to be detected in Antarctic ice core records (Dansgaard & Oeschger, 1989). By contrast, there is a marked oscillation in CH_4 content (from about 650 to 480 ppbv) in ice layers in Antarctic ice cores that broadly correspond to the Younger Dryas (Chapellaz *et al.*, 1990), and in the same horizons, a 'plateau' of relatively constant CO_2 values interrupts the overall deglacial CO_2 increase (Jouzel *et al.*, 1992; Figure 7.28). A lower value for CH_4 was also observed for the Younger Dryas period in the Dye 3 core (Dansgaard & Oeschger, 1989). Independent evidence of atmospheric gas changes during the last glacial–interglacial transition includes data from the carbon isotope record in peats (White *et al.*, 1994) and variations in stomatal density in fossil leaves (Beerling *et al.*, 1993). The former suggest sharp increases in atmospheric CO_2 at *c.* 12.8 and 10 ka BP, while the latter also suggest a marked decrease in atmospheric CO_2 at *c.* 11 ka BP.

Past fluctuations in the principal greenhouse gases therefore represent a complex response by the ocean–atmosphere–biosphere system to climatic change driven by orbital forcing. Through an intricate system of feedback loops these variations in atmospheric gas content appear to have amplified the orbitally induced climatic fluctuations, and may also have exerted a powerful influence on 'sub-Milankovitch' climatic events. They constitute one of the fundamental linkages through which external forcing factors are translated into climatic responses, and are therefore a vital element in any explanation of both long- and short-term climatic change.

7.8.5 Volcanic activity and climatic change

Short-term variations in climate have frequently been attributed to volcanic activity (Lamb, 1977; Grove, 1988). Large explosive volcanic events inject enormous quantities of fine ash and dust into the atmosphere, leading to localised and short-lived temperature reductions through the screening out of incoming radiation, a process which may be amplified by an increase in cloudiness as the dust particles act as foci for water droplets. More significant in terms of regional climatic change, however, is the release during an eruption of large volumes of sulphur volatiles. Once in the atmosphere, these are converted into sulphuric acid, and global dissemination of this aerosol results in cooling of the lower troposphere by the back-scattering of incoming long-wave radiation (Devine *et al.*, 1984). The mean residence time for sulphuric acid aerosols varies between one and five years (Schönwiese, 1988), but the effects can be pronounced. Following the eruption of Mt Pinatubo in the Philippines in 1991, for example, radiation receipt in the tropical regions declined by up to 10 per cent, while Northern Hemisphere surface temperatures fell by ~1.0°C (Dutton & Christy, 1992; Handler & Andsager, 1994). Within the historical period, some recent eruptions may have led to localised temperature reductions of up to 1.5°C (Porter, 1986). Records of past volcanism are reflected in acidity profiles from Greenland and Antarctic ice cores (section 3.11) and these data reveal a close relationship between climatic changes and episodes of volcanic activity, albeit over very short timescales (decades or less) and for the relatively recent past (Hammer *et al.*, 1980; Lyons *et al.*, 1990).

The links between earlier volcanic episodes and global climate change are more tenuous, however, although many large eruptions or clusters of eruptions appear to have coincided with climate transitions. For example, the great eruption of Mt Toba in Sumatra, which has been dated to *c.* 74 ka BP, occurred during the OI stage 5a–4 transition, at a time of rapid ice-sheet growth and falling sea level. It has been suggested that this was followed by a 'volcanic winter' during which surface temperatures fell by 3–5°C for several years. In high northern latitudes, summer temperatures may have declined by ≥10°C and, in areas adjacent to regions already covered by snow and ice, this temperature decline might have increased snow cover and sea-ice extent, thereby accelerating the global cooling already in progress (Rampino & Self, 1992, 1993). In the eastern Mediterranean, the widely disseminated Y-5 ash layer, which reflects a major eruptive event in southern Italy, has been dated to 25–30 ka BP (Cramp *et al.*, 1989) and therefore broadly coincides with the OI stage 3/2 transition. Numerous eruptive events seem to have characterised the stage 2/1 transition. In the North Atlantic, for example, Ash Zone 1, which constitutes a distinctive marker horizon in ocean sediments, contains at least four Icelandic tephras dating from *c.* 10.8 ka to 8.9 ka (Figure 5.29), while in Germany the Laacher See eruption occurred at *c.* 11 ka BP

(Figure 5.31). In the northwest USA, the Glacier Peak eruptions have been dated to between 12.7 and 11.2 ka BP (Busacca *et al.*, 1992), whereas in Alaska a major eruptive event occurred shortly after 11 ka BP (Riehle *et al.*, 1992).

Whether these correlations between volcanic activity and episodes of climatic change are simply coincidental is a matter of debate, but a number of scientists have suggested links between volcanic activity and Milankovitch parameters. For example, dating of major Indonesian eruptions during the Quaternary show a recurrence interval of *c.* 400 ka, close to the earth's 413 ka eccentricity cycle (section 1.6), while evidence from the Mediterranean, from Iceland and from Japan appears to show periodicities of ~23 ka and/or ~100 ka, i.e. similar to the earth's precessional and eccentricity cycles (Rampino & Self, 1993). There might, therefore, be a connection between volcanic activity and environmental variables which are themselves responding to orbital forcing. In glaciated regions that are volcanically active, such as Iceland, the increase in pressure caused by ice-loading may have prevented large-scale eruptions during a glacial episode (Anderson, 1987). During deglaciation, however, as the ice load was removed, the steeper thermal gradient within the crust and the corresponding higher pressures within the magma chambers could have led to an increase in explosive volcanism (Sejrup *et al.*, 1989), a process perhaps accelerated by stress fractures in the crust which allowed rapid ascent of magma to the surface (Hall, 1982). If this hypothesis is correct, episodes of volcanic activity can be related directly to the isostatic effects of ice-sheet fluctuations (Sigvaldason *et al.*, 1992). In other regions, stress changes in the crust produced by water loading and unloading on magma chambers as a result of Quaternary sea-level fluctuations have been suggested as a causal factor in explosive volcanism (Nakada & Yokose, 1992). Volcanic activity beneath ice sheets may also exert an influence on glacier behaviour. It has been suggested, for example, that heat flow from volcanically active centres may be affecting the present-day stability of the Antarctic ice sheet by acting as a controlling influence on ice-stream development (Blankenship *et al.*, 1993). As noted above, such instability may, in turn, have climatic consequences.

While volcanic activity alone clearly cannot explain the major climatic changes that are evident in the recent Quaternary record, there is now a growing body of opinion that episodes of explosive volcanism can amplify climatic changes that are already underway through the operation of a complex series of both positive and negative feedback loops. Hence, it has been suggested that if the climatic conditions of a volcanic winter occurred at a critical time in the orbitally driven insolation/greenhouse gas cycle, this could provide the kind of random forcing event required to flip the global climate system from an interglacial to a glacial mode (Rampino & Self, 1993). Certainly, the occurrence of the Toba eruption at the end of OI stage 5a, at a time when numerous proxy indicators point to a sharp fall in global temperatures (section 7.4), is curious. Equally intriguing is the coincidence between the upsurge in volcanic activity in the North Atlantic region at the end of the last cold stage and the short-lived but severe climatic reversal of the Younger Dryas event (section 7.6). It must be emphasised, however, that in both of these cases (and indeed in others also) the evidence remains circumstantial, and a clear cause and effect relationship between climatic change and volcanic activity remains to be established.

7.8.6 Variations in solar output and climatic change

Changes in radiative output from the sun have long been regarded as a major element in climatic change, and GCM simulations suggest that a 2 per cent variation in solar output would result in a 4°C change in surface temperatures (Hansen *et al.*, 1984). The principal indicators of variations in solar activity are **sunspots**, dark areas on the sun's surface (the photosphere) which are the central parts of active or disturbed regions reflecting convectional activity within the photosphere. Observations of sunspot occurrence over the last two centuries suggest an 11-year periodicity in solar activity (Figure 7.29), and this has subsequently been confirmed by satellite measurements and by astronomical observations (Schove, 1983). Other elements of solar activity, e.g. solar flares (eruptions on the photosphere) and solar UV flux, also appear to follow a similar cyclical pattern (Lean, 1984; Schmidt, 1986). The origins of these cycles are unknown, but they may relate to changes in the structure of the sun in response to magnetic field variations (Kuhn *et al.*, 1988), or they may in some way be linked to variations in solar radius (Ribes, 1990).

Records of past solar changes have been obtained from measurements of cosmogenic isotopes of ^{14}C and ^{10}Be. Higher levels of solar activity increase the strength of the **solar wind** (the stream of protons and electrons emitted by the sun), which deflects cosmic rays and results in a decrease in production of these cosmogenic isotopes. Long-term records of ^{10}Be production can be obtained from ice-core records (Beer *et al.*, 1992), whereas ^{14}C variations can be established by comparing ^{14}C dates from tree rings with the calendrical ages of the contemporaneous wood (Stuiver *et al.*, 1991; section 5.4.1.5). The close correspondence between the two records appears to confirm the hypothesis that short-term variations in ^{14}C and ^{10}Be reflect variations in their production rate due to solar modulation of cosmic

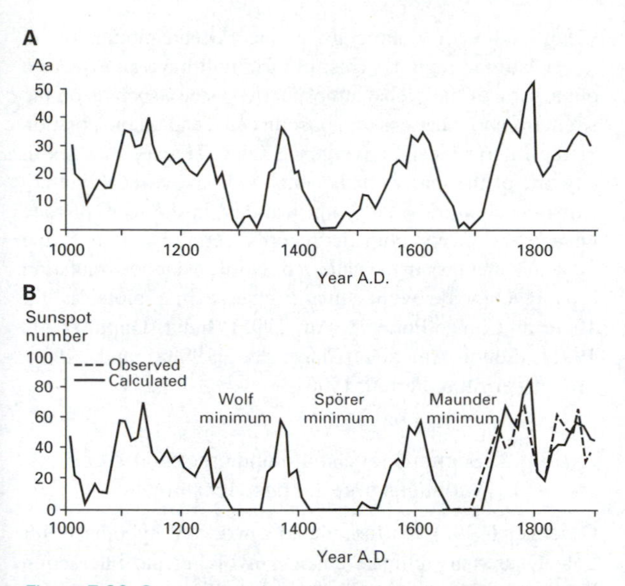

Figure 7.29 Sunspot numbers (b) and Aa indices (a) calculated from the ^{14}C record. The Aa indices reflect geomagnetic changes arising from variations in the solar wind. Changes in the geomagnetic parameters modulate the incoming cosmic ray flux which is responsible for ^{14}C production in the upper atmosphere. The dashed line shows the observed sunspot record. Both sunspot numbers and Aa indices are averaged over the 11-year solar cycle (after Stuiver & Quay, 1980).

radiation (Dansgaard & Oeschger, 1989). Spectral analysis of the ^{14}C record indicates apparent fluctuations in periodicity ranging from less than 100 to several thousand years. These include the 11- and 22-year (**Hale**) cycles, an 88-year (**Geisberg**) cycle, and ~200 and ~2500-year cycles (Rind & Overpeck, 1993). A 2300–2500-year cycle has been recognised in proxy climatic records from, *inter alia*, ice cores (Dansgaard *et al.*, 1984), ocean cores (Pestieux *et al.*, 1987), tree rings (Sonett & Finney, 1990), varved sediment sequences (Anderson, 1992) and lake level histories (Magny, 1993). A statistically significant correlation has also been established between global advance and retreat of alpine glaciers during the Holocene and variations in atmospheric ^{14}C concentration (Wigley & Kelly, 1990).

While there appears, therefore, to be some empirical support for a relationship between climate change and solar output variations, the precise linkages remain elusive. Moreover, solar forcing alone may not be capable of explaining all of the observed features of the climatic record. For example, during the most recent episode of reduced sunspot activity, the **Maunder Minimum**, which broadly coincides with the coldest part of the Little Ice Age during the seventeenth and eighteenth centuries (Grove, 1988), it has been estimated that the reduction in solar insolation was of the order of 0.25 per cent (Lean *et al.*, 1992). If GCM estimates of climate sensitivity are correct (see above), this is equivalent to a global temperature decline of ~0.5°C. Estimates for the Little Ice Age cooling based on empirical evidence, however, are of the order of 0.5 to 1.5°C (Rind & Overpeck, 1993), which suggests that other causal factors may need to be invoked to explain the full range of Little Ice Age cooling. One such factor may, once again, be ocean circulation, for in the context of the Maunder Minimum, it has been suggested that the minor solar-induced temperature decline was accompanied by higher precipitation levels around the North Atlantic which led eventually (after several decades) to reduced salinity levels in the upper layers of the ocean. This, in turn, triggered changes in the unstable thermohaline circulation which resulted in less deep-water formation, less upwelling and less northward heat transport along the Gulf Stream. The resulting colder conditions in the North Atlantic region, therefore, may have amplified the initial (solar-induced) climate signal (Stuiver & Brazunias, 1993). If this hypothesis is correct, it is a further demonstration of the ways in which external forcing mechanisms (in this case solar irradiance variations) may be amplified by internal feedback mechanisms.

7.8.7 Geodynamic factors

One other set of factors that may be influential in initiating or modulating global climatic fluctuations are those relating to the geodynamic balance of the earth. It has been suggested, for example, that changes in the distribution and volume of land ice and the associated rise and fall in eustatic sea level during glacial–interglacial cycles could have affected the angular momentum of the earth, resulting in deceleration of the earth's rotation during episodes of sea-level rise and accelerated rotation during low stands of sea level (Mörner, 1984, 1993a). This, in turn, could have led to changes in the rates and directions of flow of the major ocean surface currents, such as the Gulf Stream, the Labrador Current and the Humboldt Current. Such changes, it is argued, might explain some of the short-term global changes reflected in climatic records of the last millennium, as well as the pronounced climatic oscillations of the last glacial–interglacial transition (Mörner, 1993a, 1993b). However, it is difficult to test these hypotheses, since there is no proxy record of geodynamic changes. Moreover, these processes by themselves cannot explain the cyclical nature of the major Quaternary climatic changes. Nevertheless, variations in the earth's spin velocity or in other related geophysical effects may have acted as contributory factors which served to modulate or amplify climatic changes of the past 130 ka, and should not therefore be discounted.

7.8.8 Conceptual models of Late Quaternary climatic change

As more and more data become available, scientists are increasingly turning to simulation modelling in their attempts to seek an explanation for the causes of global climate change. Some of these modelling experiments have been outlined in the foregoing discussion. An essential precursor to such exercises, however, is the development of **conceptual models** which aim to highlight the linkages and feedback effects between different components of the system. Unlike GCMs (section 7.7), these are speculative process models which are a means whereby specific working hypotheses can be developed and subsequently tested both by the empirical (observational) evidence and by simulation modelling. Three examples of this type of model are described in this section.

7.8.8.1 Changes in Atlantic circulation during an interglacial–glacial cycle

Imbrie *et al.* (1992) have developed a conceptual model that examines the nature of the response of Atlantic Ocean circulation to solar radiation changes, and in which four dominant circulation modes develop:

1. During a warm (interglacial) stage, deep convection in the Atlantic is concentrated into three cells, each of moderate circulation strength: in the Nordic Seas (the **Nordic Heat Pump – NOR**); in the open ocean (the **Boreal Heat Pump – BOR**); and in the Antarctic Seas (**AA**).
2. During a pre-glacial episode that follows an interglacial stage, the Nordic Heat Pump is diminished in strength, or shut down, while the Antarctic circulation cell is enhanced.
3. During the full-glacial stage, the Boreal Heat Pump operates at a maximum, while the Antarctic cell remains stronger than during the interglacial stage.
4. During deglaciation, the Nordic cell is re-established, while the Boreal Heat Pump is enhanced by comparison with interglacial levels.

This model emphasises the way in which a change in one oceanic circulation component has an immediate knock-on effect throughout the entire system. If glacial ice cover and sea ice expand following a reduction in terrestrial radiation receipts (Milankovitch forcing), the Nordic heat pump is gradually eliminated and flow becomes concentrated in the Boreal cell. Loss of moisture from the free-ocean surface and further lowering of temperatures in the Nordic area will thus coincide with a shift in the position of the westerly winds and with a more invigorated ocean circulation in lower latitude regions. This, in turn, will have an effect on other parts of the global atmospheric system, such as on the strengths and locations of monsoon cells, and on the position of the Intertropical Convergence Zone. Hence, changes in any one of the major components will have effects that are experienced worldwide. Empirical data in the form of Late Quaternary proxy climatic records certainly support the view that changes in the pattern of North Atlantic circulation broadly coincide with climatic change in regions as far afield as China (Porter & An, 1995), India (Gupta *et al.*, 1991), South America (Clapperton, 1993a) and Africa (Street-Perrott & Perrott, 1990).

7.8.8.2 Scenario for climate and greenhouse gas variations over the past 150 ka

Oeschger (1992) has formulated a working hypothesis for Late Quaternary climate change involving the interaction between changes in ocean thermohaline circulation and fluctuations in atmospheric gases. It involves a shift between a glacial state (e.g. the Weichselian or Wisconsinan), when NADW formation is at a minimum, and warmer states (e.g. the Eemian or Holocene Interglacials), when NADW formation turned on. During the latter, sea level is higher, while atmospheric concentrations of CH_4 and CO_2 are similar to the present. According to this model, changing boundary conditions during the last interglacial–glacial transition, such as the distribution of solar irradiance, initiate changes in NADW formation and, perhaps, a complete shut-down for short periods. Changes in ocean chemistry and in the operation of the biological pump as sea levels fall cause a decline in the ocean–atmosphere CO_2 flux, but atmospheric CO_2 concentrations may be buffered by a net flux from terrestrial biomass as climatic deterioration reduces vegetation cover. This might explain why, at the end of the last interglacial, the overall decline in atmospheric CO_2 content appears to lag behind the general reduction in global temperature (Figure 3.51). By contrast, the decline in CH_4 closely paralleled climatic deterioration at that time, possibly because a general increase in aridity led to a global contraction in wetland and hence to a reduction in CH_4 flux to the atmosphere.

During a glacial period, NADW formation is mainly shut down, but changes in critical boundary conditions (solar irradiance distribution, sea-level change, variations in extent of ice cover, etc.) prompt the development of a chaotic–deterministic system in which NADW formation switches rapidly between 'on' and 'off' modes. The global climatic changes that are recorded in proxy records for the last cold stage (e.g. the Dansgaard–Oeschger cycles; section 7.5.4) may reflect a combination of solar irradiance changes and dampened and smoothed signals of the NADW switches. Dynamic changes in the ocean's biological pump (i.e. NADW fluctuations) are accompanied by switches in atmospheric CO_2 concentration between *c.* 200 (off) and 240 ppm (on), and a similar pattern is also apparent in the CH_4 records (Figure 7.28).

The onset of NADW formation at the beginning of the present interglacial resulted in the renewed transfer of heat into the high northern latitudes which reinforced the melting of the continental ice sheets. The CO_2 signal follows the global sea-level rise and reflects, in particular, the rearrangement of ocean chemistry and its influence on the CO_2 partial pressure in ocean surface waters. Superimposed on this general trend are CO_2 shifts similar to those observed during the period 80–30 ka BP related to changes in the global ocean's biological pump. Other factors may also come into play, however. For example, the fall in atmospheric CO_2 levels from the early Holocene maximum of around 280 ppm to values nearer 250 ppm (Figure 7.28) may reflect not only a reduction in the release of fossil CO_2 from the ocean surface following the onset of NADW formation (Sarnthein *et al.*, 1994), but also an increase in the amount of CO_2 extracted from the atmosphere by the rapidly expanding terrestrial biomass. Conceptual models like those of Oeschger challenge us to examine more closely the importance of gas exchanges between the atmosphere and other reservoirs, and the ways in which these, and their potential feedback mechanisms, can affect the global climate system.

7.8.8.3 Conceptual model of the onset of a glacial

Kukla and Gavin (1992) have developed a conceptual model of the ways in which the earth might respond to Milankovitch-induced insolation variations during the onset of a glacial episode. The model suggests that three insolation components are superimposed, and their combined effects drive the global climate inexorably towards a 'glacial mode'. The three components are: (i) a gradual decrease in insolation between July and November as a result of precessional changes; (ii) over the same period, radiation receipt in the high latitudes is further reduced as a result of obliquity changes; and (iii) seasonal insolation cycles in the low latitudes are reversed and, as a consequence, there is an *increase* in insolation during the spring. Kukla and Gavin identify nine important linkages in the global atmosphere–ocean system (Figure 7.30), and suggest the following potential chain of events as the cold climate intensifies. At first, the change from warm to cold conditions resulting from the Milankovitch-induced decrease in radiation is gradual, but the system is forced into a more rapid cooling by feedback processes. For example, the insolation decline in July and August in the high Arctic leads to more frequent outbreaks of polar air, and an increase in precipitation in the high northern latitudes. This, in turn, results in higher runoff into the Arctic Ocean, a decrease in salinity in the northern North Atlantic, and a reduction in NADW formation which further accelerates cooling in the higher latitudes. As meridional insolation gradients become more pronounced, the transport of warm water from the low latitudes to the middle latitudes, and the transfer of water vapour from the middle to the high northern latitudes increases sufficiently to support the build-up of glaciers. This combination of factors would enable the global climate system to shift rapidly from an interglacial to a glacial mode.

The model also emphasises the interplay between Milankovitch forcing and shorter-term climatic oscillations that operate on century to millennial timescales. The latter are considered to fluctuate around an unchanging long-term mean, and may reflect entirely random internal variations, the impacts of short-term changes in solar activity, volcanic eruptions, etc. In Kukla and Gavin's model, Milankovitch-induced variations magnify such short-term effects when they are in phase, and suppress those that are out of phase with orbital forcing. The model envisages little or no delay in the response of the global climate system to Milankovitch-induced insolation changes at the start of a glacial. It also suggests that insolation changes in the low latitudes are of equal importance to those in high latitudes, with ocean warming in the low and middle latitudes combining with atmospheric cooling in the high latitudes to increase the vigour of ocean–atmosphere and ocean–land moisture transfer. Finally, the model suggests that the insolation impacts on both hemispheres are broadly in phase in leading towards the glacial onset.

7.9 Concluding remarks

This chapter has reviewed the different lines of evidence for climatic change over the course of the past 130 ka, and has considered some of the more important factors that appear to have influenced global climates during this time period. Perhaps the most striking discovery in the palaeoclimatic research of the last decade is the apparent rapidity with which the global climate system can shift from one dominant mode to another. It now seems that the earth can switch from a 'glacial' to an 'interglacial' state within a matter of hundreds, as opposed to thousands, of years, a timescale that was considered almost inconceivable prior to the recent findings in the polar ice cores. Indeed, there are indications that wholesale atmospheric reorganisations, involving temperature changes of up to 10°C, can occur

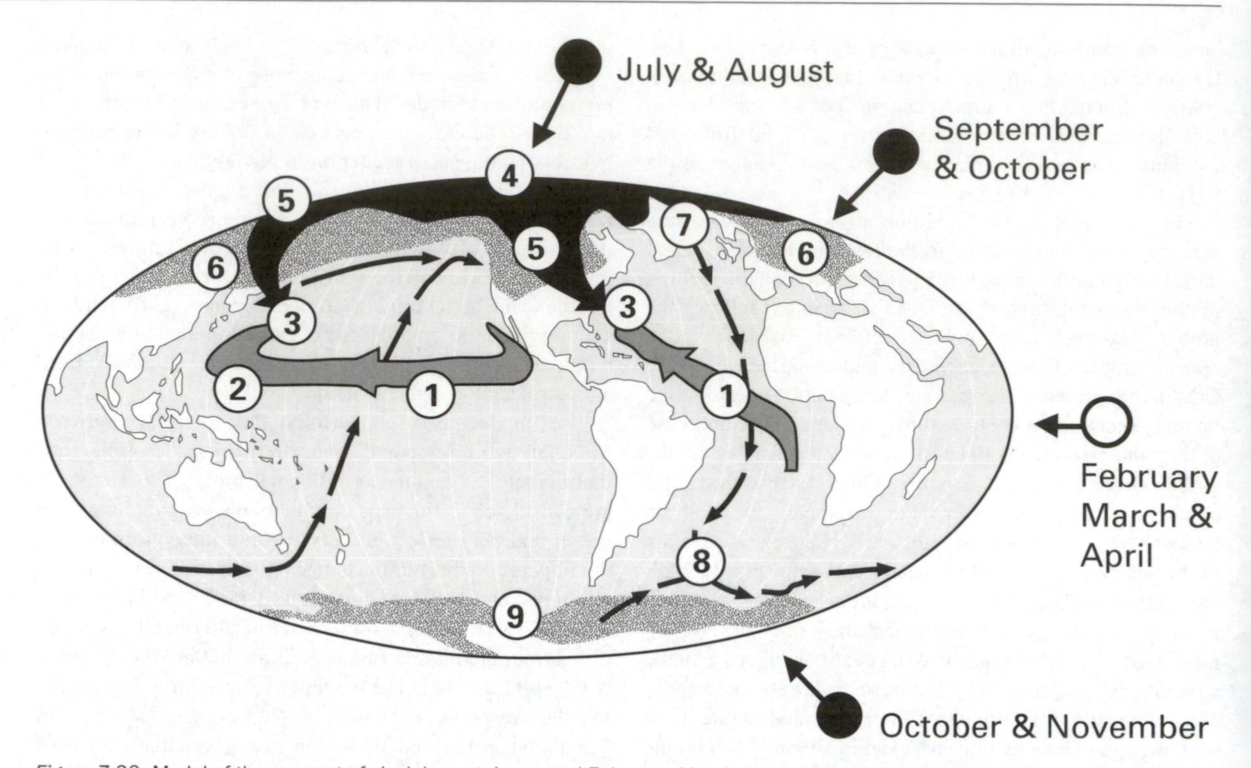

Figure 7.30 Model of the concept of glacial onset. Increased February, March and April insolation (open circle) warms the Pacific and Atlantic equatorial currents (1), leading to increased evaporation and high-latitude poleward moisture transport from the West Equatorial Pacific (2) and to increased transport of warm waters to the Kuroshio and Gulf Stream (3). Decreased insolation (filled circle) in July and August in the Arctic (4) and in September and October over the northern middle latitudes leads to the expansion of snow covers (6), to more frequent and deeper southward penetrations of polar air outbreaks (5) and more vigorous ocean mixing and moisture extraction along the northeastern coasts of Asia and North America (3). Increased precipitation and sea-ice export into the northern North Atlantic (7) reduces the deep-water circulation (8). Decreased insolation to subantarctic pack ice (9) in October and November delays the ice dissipation (after Kukla & Gavin, 1992).

within a few decades! Moreover, such shifts appear as more or less synchronous events in proxy records from ocean sediments, polar ice and terrestrial contexts, while the periodicity and general pattern of climatic changes are similarly manifest across these diverse data sources. Milankovitch-induced variations in insolation are generally acknowedged as the principal driving forces underlying glacial–interglacial cycles, but superimposed on these are a complex series of internal feedback mechanisms, including ocean heat transfer, albedo effects and gas exchanges between the lithosphere, biosphere and atmosphere. The question still remains, however, as to precisely *how* these various components interact to bring about climatic changes; in other words, how do the *linkages* between the different components of the ocean–atmosphere–terrestrial system function?

Two problems in particular still confront Quaternary scientists when trying to resolve this problem: the first is distinguishing between cause and effect, and the second is the establishment of the precise order of events. In attempting to explain the climatic sequence in the North Atlantic region over the past 130 ka, for example, separating the dependent from the independent variables in the context of atmospheric gas changes, ocean thermohaline changes, variations in meltwater flux, ice-sheet fluctuations, etc., is far from straightforward. The effect of both positive and negative feedback loops means that it is frequently difficult to avoid the 'chicken-and egg' dilemma of which component of the global system was primarily responsible for stimulating climate change. In addition it is often equally difficult to establish the precise temporal relationships in a sequence involving, *inter alia*, Heinrich events, shifts in the position of ice-sheet margins, changes in rates of snow accumulation in Greenland, ocean salinity and circulation changes, and fluctuations in atmospheric gas content. These problems arise partly because of inadequacies in the climatic proxies themselves (poor stratigraphic resolution, for example), but they also reflect a lack of precision in dating ınd correlation. Hence, although climatic changes often *ıppear* to be synchronous, the true temporal relationships of :ritical events may be blurred by deficiencies both in the ›asic datasets and in the various chronological methods that ıave been employed.

Refinements in both of these important areas will ındoubtedly come as the pace of research accelerates. Stratigraphic sequences are currently being investigated at a evel of resolution that was considered impractical only a ew decades ago, recent developments in geochronology ıold out the prospect of refinements to existing timescales, while new dating techniques will further underpin the leveloping temporal framework for environmental and :limatic change. Recent years have seen the establishment of a growing number of multidisciplinary research teams, as ı consequence of which the synthesis of proxy data from a ange of sources is now becoming routine. Above all, the ntroduction of both conceptual and simulation models epresents, perhaps, the most exciting new development in Quaternary science, for while the models may not yet be providing all of the answers, they are at least posing the questions and signalling the directions in which future research should proceed. The models, however, are only as good as the empirical data upon which they are based and against which, in turn, they must be tested. The task now facing the Quaternary research community, therefore, is the development of a palaeoenvironmental database of sufficient quality so that models can approximate even more closely the reality of the global climate machine. Only then, perhaps, will the final pieces of the jigsaw puzzle, i.e. the crucial linkages between the different components of the ocean–atmosphere–terrestrial system, begin to fall into place.

Notes

1. Radiocarbon is the most widely used technique for dating events during a substantial part of the last cold stage, and hence most published research employs a timescale based on ^{14}C. This practice has therefore been followed here. As noted in section 5.4, however, ^{14}C years do not always equate with calendar years. Hence there will be discrepancies between radiocarbon years and ages obtained using other dating methods (section 7.6.5).
2. Data from present-day Norwegian fjords indicate that, during the springtime, there is a net outflow of ice and surface meltwater, and that this is replaced by seawater which brings heat into the inner fjords. During peak deglaciation, it has been suggested that the Nordic Seas bounded by Norway, Greenland, Iceland and Svalbard behaved as one large fjord, in which the export of surface meltwaters and entrained salt-waters were compensated by a northward flow of subsurface waters that brought with them large quantities of heat. Reduction in meltwater flux from the Nordic Seas, as appears to have occurred during the Younger Dryas, would, therefore, have been accompanied by a reduction in northward heat transport, and a consequent fall in regional temperatures (Berger & Jansen, 1995).

Bibliography

Aaby, B. (1976): Cyclic climatic variations in climate over the past 5500 years reflected in raised bogs. *Nature*, **263**, 281–284.

Aaby, B. (1983): Forest development, soil genesis and human activity illustrated by pollen and hypha analysis of two neighbouring podzols in Draved Forest, Denmark. *Danmarks Geologiske Undersøgelse, ser. II*, **114**, 1–114.

Aaby, B. (1986): Palaeoecological studies of mires. In *Handbook of Holocene Palaeoecology and Palaeohydrology* (edited by B.E. Berglund), John Wiley & Sons, Chichester & New York, 145–164.

Aaby, B. & Berglund, B. (1986): Characterization of peat and lake deposits. In *Handbook of Holocene Palaeoecology and Palaeohydrology* (edited by B.E. Berglund), John Wiley & Sons, Chichester & New York, 231–246.

Aaby, B. & Digerfeldt, G. (1986): Sampling techniques for lakes and bogs. In *Handbook of Holocene Palaeoecology and Palaeohydrology* (edited by B.E. Berglund), John Wiley & Sons, Chichester & New York, 181–194.

Aarseth, I. & Mangerud, J. (1974): Younger Dryas end moraines between Hardangerfjorden and Sognefjorden, western Norway. *Boreas*, **3**, 3–22.

Abell, P.I. & Williams, M.A.J. (1989): Oxygen and carbon isotope ratios in gastropod shells as indicators of paleoenvironments in the Afar region of Ethiopia. *Palaeogeography, Palaeoclimatology, Palaeoecology*, **74**, 265–278.

Abelson, P.H. (1956): Palaeobiochemistry. *Scientific American*, **195**, 85–92.

Aber, J.S. (1991): The glaciation of northeastern Kansas. *Boreas*, **20**, 297–314.

Adam, D.P. & West, G.J. (1983): Temperature and precipitation estimates through the Last Glacial Cycle from Clear Lake, California, pollen data. *Science*, **219**, 168–170.

Adam, D.P., Sims, J.D. & Throckmorton, C.K. (1981): 130,000 year continuous pollen record from Clear Lake County, California. *Geology*, **9**, 373–377.

Adam, D.P., Sarna-Wojcicki, A.M., Rieck, H.J., Bradbury, J.P., Dean, W.E. & Forrester, R.M. (1989): Tulane Lake, California: the last 3 million years. *Palaeogeography, Palaeoclimatology, Palaeoecology*, **72**, 89–103.

Adamson, D.A., Gasse, F., Street, F.A. & Williams, M.A.J. (1980): Late Quaternary history of the Nile. *Nature*, **287**, 50–55.

Agarwal, R.P. & Bhoj, R. (1992): Evolution of the Kosi River fan, India: structural implications and geomorphic significance. *International Journal of Remote Sensing*, **13**, 1891–1901.

Agrawal, D.R., Dodia, R. & Seth, M. (1990): South Asia climate and environment at 18,000 BP. In *The World at 18,000 BP*, Vol. 2, *Low Latitudes* (edited by C.S. Gamble & O. Soffer), Unwin Hyman, London, 231–260.

Aguirre, E. & Pasini, G. (1985): The Pliocene–Pleistocene boundary. *Episodes*, **8**, 116–120.

Aharon, P. (1984): Implications of the coral-reef record from New Guinea concerning the astronomical theory of climatic change. In *Milankovitch and Climate* (edited by A. Berger, J. Imbrie, J. Hays, G. Kukla & B. Saltzman), Reidel, Dordrecht, 379–390.

Aharon, P., Chappell, J. & Compston, W. (1980): Stable isotope and sea-level data from New Guinea support Antarctic ice-surge theory of ice ages. *Nature*, **283**, 649–651.

Ahlman, H.W. son (1948): Glaciological research on the North Atlantic coasts. *Royal Geographical Society Research Series* **1**, 83 pp.

Aitken, M.J. (1985): *Thermoluminescence Dating*. Academic Press, London & New York.

Aitken, M.J. (1990): *Science-based Dating in Archaeology*. Longman, London.

Aitken, M.J. (1992): Optical dating. *Quaternary Science Reviews*, **11**, 127–132.

Aitken, M., Grün, R., Mejdahl, V., Miallier, D., Rendell, H., Wieser, A. & Wintle, A. (eds) (1994): Proceedings of the 7th International Specialist Seminar on Thermoluminescence and Electron Spin Resonance Dating. *Quaternary Science Reviews*, **13**, 403–684.

Alhonen, P., Kokkonen, J., Matiskainen, H. & Vuorinen, A. (1980): Applications of AAS and diatom analysis and stylistic studies of sub-neolithic pottery in Finland. *Bulletin of the Geological Society of Finland*, **52**, 193–206.

Alley, R.B. & MacAyeal, D.R. (1994): Ice-rafted debris associated with binge/purge oscillations of the Laurentide Ice Sheet. *Paleoceanography*, **9**, 503–511.

Alley, R.B., Blankenship, D.D., Bentley, C.R. & Rooney, S.T. (1986): Deformation of till beneath ice stream B, West Antarctica. *Nature*, **322**, 57–59.

Alley, R.B., Meese, D.A., Shuman, C.A., Gow, A.J., Taylor, K.C.,

Grootes, P.M., White, J.W.C., Ram, M., Waddington, E.D., Mayewski, P.A. & Zelinski, G.A. (1993): Abrupt increase in Greenland snow accumulation at the end of the Younger Dryas event. *Nature*, **362**, 527–529.

Allison, P.A. & Briggs, D.E.G. (1991): *Taphonomy: Releasing the Data Locked in the Fossil Record*. Plenum, New York.

Allison, T.D., Moeler, R.E. & Davis, M.B. (1986): Pollen in laminated sediments provides evidence for a mid-Holocene pathogen outbreak. *Ecology*, **67**, 1101–1105.

Altabet, M.A., Francois, R., Murra, D.W. & Prell, W.L. (1995): Climate-related variations in denitrification in the Arabian Sea from sediment $^{15}N/^{14}N$ ratios. *Nature*, **373**, 506–509.

Ambrose, W.R. (1994): Obsidian hydration dating of a Pleistocene age site from the Manus Islands, Papua New Guinea. *Quaternary Science Reviews*, **13**, 137–142.

American Commission on Stratigraphic Nomenclature (1961): Code of Stratigraphic Nomenclature. *American Association of Petroleum Geologists Bulletin*, **45**, 645–665.

American Commission on Stratigraphic Nomenclature (1970): Code of Stratigraphic Nomenclature (2nd edition). *American Association of Petroleum Geologists Bulletin*, **60**, 1–45.

Ammann, B. & Lotter, A.F. (1989): Late-Glacial radiocarbon- and palynostratigraphy on the Swiss Plateau. *Boreas*, **18**, 109–126.

An, Z., Kukla, G.J., Porter, S.C. & Xiao, J. (1991): Magnetic susceptibility evidence of monsoon variation on the Loess Plateau of central China during the last 130,000 years. *Quaternary Research*, **36**, 29–36.

Andel, T.H. van, Zangger, E. & Demitrack, A. (1990): Land use and soil erosion in prehistoric Greece. *Journal of Field Archaeology*, **17**, 379–396.

Andersen, B. (1992): Jens Esmark – a pioneer in glacial geology. *Boreas*, **21**, 97–102.

Andersen, S.Th. (1961): Vegetation and its environment in Denmark in the Early Weichselian glacial (last glacial). *Danmarks Geologiske Undersøgelse*, **R II 75**, 1–175.

Andersen, S.T. (1973): The differential pollen productivity of trees and its significance for the interpretation of a pollen diagram from a forested region. In *Quaternary Plant Ecology* (edited by H.J.B. Birks & R. G. West), Blackwell, Oxford, 109–115.

Anderson, I. (1987): Melting glaciers pull the plug of volcanoes? *New Scientist*, **113**, 30.

Anderson, N.J., Rippey, B. & Gibson, C.E. (1993): A comparison of sedimentary and diatom-inferred phophorous profiles: implications for defining pre-disturbance nutrient conditions. *Hydrobiologia*, **253**, 357–366.

Anderson, O.R. (1983): *Radiolaria*. Springer-Verlag, New York.

Anderson, R.Y. (1992): Possible connection between surface winds, solar activity and the Earth's magnetic field. *Nature*, **358**, 51–53.

Anderson, T. & Macpherson, J.B. (1994): Wisconsinan Late-glacial environmental change in Newfoundland: a regional review. *Journal of Quaternary Science*, **9**, 171–178.

André, M.F. (1986): Dating slope deposits and estimating rates of rock wall retreat in northwest Spitsbergen by lichenometry. *Geografiska Annaler*, **68A**, 65–75.

Andree, M., Moor, E. *et al.* (1984): ^{14}C dating of polar ice. *Nuclear Instruments and Methods, Physical Research*, **233**, 380–384.

Andree, M., Oeschger, H., Siegenthaler, U., Riesen, T., Moell, M., Ammann, B. & Tobolski, K. (1986): ^{14}C dating of plant macrofossils in lake sediment. *Radiocarbon*, **28**, 411–416.

Andrews, J.T. (1970): A geomorphological study of Postglacial uplift with particular reference to Arctic Canada. *Institute of British Geographers Special Publication*, **2**.

Andrews, J.T. (1975): *Glacial Systems*. Duxbury Press, Massachusetts.

Andrews, J.T. (1979): The present ice age: Cenozoic. In *The Winters of the World* (edited by B.S. John), David and Charles, London and North Pomfret (Vt), 173–218.

Andrews, J.T. (1982): On the reconstruction of Pleistocene ice sheets: a review. *Quaternary Science Reviews*, **1**, 1–30.

Andrews, J.T. (1987): The Late Wisconsin glaciation and deglaciation of the Laurentide Ice Sheet. In *The Geology of North America*. Volume K-3, *North America and Adjacent Oceans During the Last Deglaciation* (edited by W.F. Ruddiman & H.E. Wright, Jr), Geological Society of America, Boulder, Colorado, 13–37.

Andrews, J.T. (1990): Fjord to deep-sea sediment transfers along the northeastern Canadian continental margin: models and data. *Géographie Physique et Quaternaire*, **44**, 55–70.

Andrews, J.T. & Ives, J.D. (1978): 'Cockburn' nomenclature and the late Quaternary history of the eastern Canadian *Arctic. Arctic and Alpine Research*, **10**, 617–633.

Andrews, J.T. & Miller, G.H. (1985): Holocene sea-level variations within Frobisher Bay. In *Quaternary Environments: Eastern Canadian Arctic, Baffin Bay and Western Greenland* (edited by J.T. Andrews), Allen & Unwin, London, 585–607.

Andrews, J.T. & Tedesco, K. (1992): Detrital carbonate-rich sediments, northwestern Labrador Sea: implications for ice-sheet dynamics and iceberg rafting (Heinrich) events in the North Atlantic. *Geology*, **20**, 1087–1090.

Andrews, J.T., Stravers, J.A. & Miller, G.H. (1985): Patterns of glacial erosion and deposition around Cumberland Sound, Frobisher Bay and Hudson Strait, and the location of ice streams in the eastern Canadian Arctic. In *Models in Geomorphology* (edited by M. Waldenburg), Allen & Unwin, London, 93–117.

Andrews, J.T., Miller, G.H., Vincent, J.-S. & Shilts, W.W. (1986): Quaternary correlations in Arctic Canada. *Quaternary Science Reviews*, **5**, 243–249.

Angel, M.V. (1993): *Marine Planktonic Ostracods*. Field Studies Council, The Linnaean Society, London.

Antevs, E. (1931): Late-glacial correlations and ice recession in Manitoba. *Geological Society of Canada, Memoir* **168**, 1–76.

Antevs, E. (1953): Geochronology of the deglacial and neothermal ages. *Journal of Geology*, **61**, 195–230.

Appleby, P.G. & Oldfield, F. (1983): The assessment of ^{210}Pb data from sites with varying sedimentation rates. *Hydrobiologia*, **103**, 29–35.

Appleby, P.G., Oldfield, F., Thompson, R., Huttunen, P. & Tolonen, K. (1979): ^{210}Pb dating of annually laminated lake sediments from Finland. *Nature*, **280**, 53–55.

Atkinson, T.C., Harmon, R.S., Smart, P.J. & Waltham, A.C.

(1978): Paleoclimatic and geomorphic implications of $^{230}Th/^{234}U$ dates on speleothems from Britain. *Nature*, **272**, 24–28.

Atkinson, T.C., Briffa, K.R., Coope, G.R., Joachim, M. & Perry, D.W. (1986a): Climatic calibration of coleopteran data. In *Handbook of Holocene Palaeoecology and Palaeohydrology* (edited by B.E. Berglund), John Wiley, Chichester & New York, 851–858.

Atkinson, T.C., Lawson, T.J., Smart, P.L., Harmon, R.S. & Hess, J.W. (1986b): New data on speleothem deposition and palaeoclimate in Britain over the last forty thousand years. *Journal of Quaternary Science*, **1**, 67–72.

Atkinson, T.C., Briffa, K.R. & Coope, G.R. (1987): Seasonal temperatures in Britain during the last 22,000 years, reconstructed using beetle remains. *Nature*, **325**, 587–592.

Austin, W.E.N. & Kroon, D. (1996): Late glacial sedimentology, foraminifera and stable isotope stratigraphy of the Hebridean continental shelf, northwest Scotland. In *Late Quaternary Palaeoceanography of the North Atlantic Margins* (edited by J.T. Andrews, W.E.N. Austin, H. Bergsten & A.E. Jennings), Geological Society Publication No. 111, Oxford, 187–214.

Austin, W.E.N. & McCarroll, D. (1992): Foraminifera from the Irish Sea glacigenic deposits at Aberdaron, western Lleyn, North Wales: palaeoenvironmental implications. *Journal of Quaternary Science*, **7**, 311–318.

Bada, J.L. (1985): Amino acid racemisation dating of fossil bones. *Annual Review of Earth and Planetary Science*, **13**, 241–268.

Baillie, M. (1989): Do Irish bogs date the Shang dynasty? *Current Archaeology*, **117**, 310–313.

Baillie, M.G.L. (1982): *Tree-ring Dating and Archaeology*. Croom Helm, London & Canberra.

Baillie, M.G.L. & Brown, D.M. (1988): An overview of oak chronologies. *British Archaeological Reports (British Series)*, **196**, 543–548.

Baillie, M.G.L. & Munro, M.A.R. (1988): Irish tree-rings, Santorini and volcanic dust veils. *Nature*, **332**, 344–346.

Baker, A., Smart, P.L., Edwards, R.L. & Richards, D.A. (1993): Annual growth banding in a cave stalagmite. *Nature*, **364**, 518–520.

Baker, A., Smart, P.L. & Edwards, R.L. (1995): Paleoclimate implications of mass spectrometric dating of a British flowstone. *Geology*, **23**, 309–312.

Baker, R.G., Rhodes II, R.S., Schwert, D.P., Ashworth, A.C., Frest, T.J., Hallberg, G.R. & Janssens, J.A. (1986): A full-glacial biota from southeastern Iowa, USA. *Journal of Quaternary Science*, **1**, 91–108.

Ballantyne, C.K. (1984): The Late Devensian periglaciation of upland Scotland. *Quaternary Science Reviews*, **3**, 311–344.

Ballantyne C.K. (1989): The Loch Lomond Readvance on the Isle of Skye, Scotland: glacier reconstruction and palaeoclimatic implications. *Journal of Quaternary Science*, 4, 95–108.

Ballantyne, C.K. (1990): The Holocene glacial history of Lyngshalvöya, northern Norway: chronology and climatic implications. *Boreas*, **19**, 93–119.

Ballantyne, C.K. (1991): Periglacial features on the mountains of Skye. In *The Quaternary of the Isle of Skye* (edited by C.K. Ballantyne, D.I. Benn, J.J. Lowe & M.J.C. Walker), Quaternary Research Association, Cambridge, 68–81.

Ballantyne, C.K. (1994): The tors of the Cairngorms. *Scottish Geographical Magazine*, **110**, 54–59.

Ballantyne, C.K. & Benn, D.I. (1991): The glacial history of the Isle of Skye. In *The Quaternary of the Isle of Skye* (edited by C.K. Ballantyne, D.I. Benn, J.J. Lowe & M.J.C. Walker), Quaternary Research Association, Cambridge, 11–34.

Ballantyne, C.K. & Benn, D.I. (1994a): Paraglacial slope adjustment and resedimentation in response to recent glacier retreat, Fåbergstølsdalen, Norway. *Arctic and Alpine Research*, **26**, 255–269.

Ballantyne, C.K. & Benn, D.I. (1994b): Glaciological constraints on protalus rampart development. *Permafrost and Periglacial Landforms*, **5**, 145–153.

Ballantyne, C.K. & Gray, J.M. (1984): The Quaternary geomorphology of Scotland: the research contribution of J.B. Sissons. *Quaternary Science Reviews*, **3**, 259–289.

Ballantyne, C.K. & Harris, C. (1994): *The Periglaciation of Great Britain*. Cambridge University Press, Cambridge.

Ballantyne, C.K. & Kirkbride, M.P. (1986): The characteristics and significance of some Lateglacial protalus ramparts in upland Britain. *Earth Surface Processes and Landforms*, **11**, 659–671.

Bannister, A., Raymond, S. & Baker, R. (1992): *Surveying*, 6th edition. Longman, London.

Barber, K.E. (1981): *Peat Stratigraphy and Climate Change*. Balkema, Rotterdam.

Barber, K.E. (1993): Peatlands as scientific archives of past biodiversity. *Biodiversity and Conservation*, **2**, 474–489.

Barber, K.E., Chambers, F.M. & Maddy, D. (1993): Sensitive high-resolution records of Holocene palaeoclimate from ombrotrophic bogs. In *Palaeoclimate of the Last Glacial/Interglacial Cycle* (edited by B.M. Funnell & R.N.L. Kay), NERC., Swindon, 57–60.

Barber, K.E., Chambers, F.M., Maddy, D., Stoneman, R. & Brew, J.S. (1994): A sensitive high-resolution record of late Holocene climatic change from a raised bog in northern England. *The Holocene*, **4**, 198–205.

Bard, E. & Broecker, W.S. (ed.) (1992): *The Last Deglaciation: Absolute and Radiocarbon Chronologies*. Springer-Verlag, Berlin.

Bard, E., Arnold, M., Duprat, J. & Duplessy, J.-C. (1987a): Reconstruction of the last deglaciation: deconvolved records of $\delta^{18}O$ profiles, micropalaeontological variations and accelerator mass spectrometric ^{14}C dating. *Climate Dynamics*, **1**, 101–112.

Bard, E., Arnold, M., Duprat, J., Moyes, J. & Duplessy, J.-C. (1987b): Retreat velocity of the North Atlantic polar front during the last deglaciation determined by accelerator mass spectrometry. *Nature*, **328**, 791–794.

Bard, E., Hamelin, B. & Fairbanks, R.G. (1990a): U-Th ages obtained by mass spectrometry in corals from Barbados: sea level during the past 130,000 years. *Nature*, **346**, 456–458.

Bard, E., Hamelin, B., Fairbanks, R.G. & Zindler, A.(1990b): Calibration of the ^{14}C timescale over the past 30,000 years using mass spectrometric U-Th ages from Barbados. *Nature*, **345**, 405–410.

Bard, E., Arnold, M. & Duplessy, J.-C. (1991): Reconciling the sea level record of the last deglaciation with ^{18}O spectra from deep sea cores. In *Radiocarbon Dating: Recent Applications and Future Potential* (edited by J.J. Lowe), *Quaternary Proceedings*, **1**, Quaternary Research Association, Cambridge, 67–73.

Bard, E., Fairbanks, R.G., Arnold, M. & Hamelin, B. (1992): $^{230}Th/^{234}U$ and ^{14}C ages obtained by mass spectrometry on corals from Barbados (West Indies), Isabela (Galapagos) and Muroroa (French Polynesia). In The *Last Deglaciation: Absolute and Radiocarbon Chronologies* (edited by E. Bard & W.S. Broecker), *NATO ASI Series* **1**, **2**, Springer Verlag, Berlin, 103–110.

Bard, E., Arnold, M., Fairbanks, R.G. & Hamelin, B. (1993): ^{230}U, ^{234}U and ^{14}C ages obtained by mass spectrometry on corals. *Radiocarbon*, **35**, 191–199.

Bard, E., Arnold, M., Mangerud, J., Paterne, M., Labeyrie, L., Duprat, J., Mélières, M.-A., Sønstegaard, E. & Duplessy, J.-C. (1994): The North Atlantic atmosphere–sea-surface ^{14}C gradient during the Younger Dryas climatic event. *Earth and Planetary Science Letters*, **126**, 275–287.

Barnola, J.-M., Raynaud, D., Korotkevich, Y.S. & Lorius, C. (1987): Vostok ice core provides 160,000 year record of atmospheric CO_2. *Nature*, **329**, 408–414.

Barnola, J.-M., Pimienta, P., Raynaud, D. & Korotkevich, Y.S. (1991): CO_2–climate relationship as deduced from the Vostok ice core: a re-examination based on new measurements and on a re-evaluation of the air dating. *Tellus*, **43B**, 83–90.

Barsch, D. (1988): Rock glaciers. In *Advances in Periglacial Geomorphology* (edited by M.J. Clark), John Wiley, Chichester & New York, 69–90.

Bartholin, T.S. (1984): Dendrochronology in Sweden. In *Climatic Change on a Yearly to Millennial Basis* (edited by N.-A. Mörner & W. Karlén), Reidel, Dordrecht, 261–262.

Bartlein, P.J., Prentice, I.C. & Webb, T. III (1986): Climatic response surfaces from pollen data for some eastern North American taxa. *Journal of Biogeography*, **13**, 35–57.

Battarbee, R.W. (1978): Observations on the recent history of Lough Neagh and its drainage basin. *Philosophical Transactions of the Royal Society of London*, **B281**, 303–345.

Battarbee, R.W. (1984): Diatom analysis and the acidification of lakes. *Philosophical Transactions of the Royal Society of London*, **B305**, 451–477.

Battarbee, R.W. (1986): Diatom analysis. In *Handbook of Holocene Palaeoecology and Palaeohydrology* (edited by B.E. Berglund), John Wiley & Sons, Chichester & New York, 527–570.

Battarbee, R.W. (1988): The use of diatom analysis in archaeology: a review. *Journal of Archaeological Science*, **15**, 621–644.

Battarbee, R.W. (1991): Recent paleolimnology and diatom-based environmental reconstruction. In *Quaternary Landscapes* (edited by L.C.K. Shane & E.J. Cushing), University of Minnesota Press, Minneapolis, 129–174.

Battarbee, R.W. & Charles, D.F. (1987): The use of diatom assemblages in lake sediments as a means of assessing the timing, trends and causes of lake acidification. *Progress in Physical Geography*, **11**, 552–580.

Battarbee, R.W., Flower, R.J., Stevenson, A.C. & Rippey, B. (1985): Lake acidification in Galloway: a palaeoecological test of competing hypotheses. *Nature*, **314**, 350–352.

Battarbee, R.W., Smol, J.P. & Merilinen, J. (1986): Diatoms as indicators of pH: a historical review. In *Diatoms and Lake Acidity* (edited by J.P. Smol, R.W. Battarbee, R.B. Davis & J. Merilinen), W. Junk, The Hague, 141–168.

Battarbee, R.W., Mason, J., Renberg, I. & Talling, J.F. (1990): *Palaeolimnology and Lake Acidification*. Cambridge University Press, Cambridge.

Baumann, K.-H., Lackschewitz, K.S., Mangerud, J., Spielhagen, R.F., Wolf-Welling, T.C.W., Henrich, R. & Kassens, H. (1995): Reflection of Scandinavian Ice Sheet fluctuations in Norwegian Sea sediments during the last 150,000 years. *Quaternary Research*, 4**3**, 185–197.

Bé, A.W.H. & Tolderlund, D.S. (1971): Distribution and ecology of living planktonic foraminifera in surface waters of the Atlantic and Indian Oceans. In *Micropalaeontology of the Oceans* (edited by B.M. Funnell & W.R. Riedel), Cambridge University Press, Cambridge, 105–149.

Beasley, T., Cecil, L., Mann, L., Kubik, P.W., Sharma, P. & Gove, H.E. (1991): ^{36}Cl in the Snake River Plain aquifer: origin and implications. *Radiocarbon*, **33**, 174.

Beaulieu, J.-L. de & Reille, M. (1984): A long upper-Pleistocene pollen record from Les Echets near Lyon, France. *Boreas*, **13**, 111–132.

Beaulieu, J.-L. de & Reille, M. (1989): The transition from temperate phases to stadial in the long Upper Pleistocene sequence from Les Echets (France). *Palaeogeography, Palaeoclimatology, Palaeoecology*, **72**, 147–159.

Beaulieu, J.-L. de & Reille, M. (1992a): The last climatic cycle at La Grande Pile (Vosges, France): a new pollen profile. *Quaternary Science Reviews*, **11**, 431–438.

Beaulieu, J.L. de & Reille, M. (1992b): Pollen records of the last climatic cycles in the Devès volcano craters (Massif Central, France) II. Lac du Bouchet. In 8th International Palynological Congress, Aix-en-Provence, Program and Abstracts, 33.

Beaulieu, J.-L. de, Pons, A. & Reille, M. (1982): Recherches pollenanalytiques sur l'histoire de la végétation de la bordure nord du massif du Cantal (Massif Central, France). *Pollen et Spores*, **34**, 251–300.

Beaulieu, J.-L. de, Guiot, J. & Reille, M. (1991): Long European pollen records and quantitative reconstructions of the last climatic cycle. In *Proceedings of the International Workshop 'Future Climatic Change and Radioactive Waste Disposal'*, Climatic Research Unit, University of East Anglia, Norwich, 116–136.

Beaulieu, J.-L. de, Andrieu, V., Ponel, P., Reille, M. & Lowe, J.J. (1994): The Weichselian Late-glacial in southwestern Europe (Iberian Peninsula, Pyrenees, Massif Central, northern Italy). *Journal of Quaternary Science*, **9**, 101–107.

Beck, J.W., Edwards, R.L., Ito, E., Taylor, F.W., Recy, J., Rougerie, F., Joannot, P. & Henin, C. (1992): Sea-surface temperature from coral skeletal strontium/calcium ratios. *Science*, **257**, 644–647.

Becker, B. (1993): A 11,000-year German oak and pine dendrochronology for radiocarbon calibration. *Radiocarbon*, **35**, 201–213.

Becker, B. & Kromer, B. (1986): Extension of the Holocene dendrochronology by the preboreal pine series, 8800 to 10,100 BP. *Radiocarbon*, **28**, 961–968.

Becker, B. & Kromer, B (1993): The continental tree-ring record – absolute chronology, ^{14}C calibration and climatic change at 11 ka BP. *Palaeogeography, Palaeoclimatology, Palaeoecology*, **103**, 67–71.

Becker, B., Kromer, B.& Trimborn, P. (1991): A stable-isotope tree-ring timescale of the Late Glacial/Holocene boundary. *Nature*, **353**, 647–649.

Beer, J., Johnsen, S.J., Bonani, G., Finkel, R.C., Langway, C.C., Oeschger, H., Stauffer, B., Suter, M. & Woelfli, W. (1992): ^{10}Be peaks as time markers in polar ice cores. In *The Last Deglaciation: Absolute and Radiocarbon Chronologies* (edited by E. Bard & W.S. Broecker), *NATO ASI Series* **1, 2,** Springer Verlag, Berlin, 141–153..

Beerling, D.J. & Chaloner, W.G. (1993); Evolutionary responses of stomatal density to global carbon dioxide change. *Biological Journal of the Linnaean Society*, **48**, 343–353.

Beerling, D.J. & Woodward, F.I. (1993): Ecophysiological responses of plants to global environmental change since the Last Glacial Maximum. *New Phytologist*, **125**, 641–648.

Beerling, D.J., Chaloner, W.G., Huntley, B., Pearson, J.A., Tooley, M.J. & Woodward, F.I. (1992): Variations in the stomatal density of *Salix herbacea* L. under the changing atmospheric CO_2 concentrations of late- and post-glacial time. *Philosophical Transactions of the Royal Society of London*, **B336**, 215–224.

Beerling, D.J., Chaloner, W.G., Huntley, B., Pearson, J.A. & Tooley, M.J. (1993); Stomatal density responses to the glacial cycle of environmental change. *Philosophical Transactions of the Royal Society of London*, **B251**, 133–138.

Beget, J.E. (1984): Tephrochronology of Late Wisconsin deglaciation and Holocene glacier fluctuations near Glacier Peak, North Cascade Range, Washington. *Quaternary Research*, **21**, 304–316.

Behre, K.-E. (ed.) (1986): *Anthropogenic Indicators in Pollen Diagrams*. Balkema, Rotterdam.

Behre, K.-E. (1989): Biostratigraphy of the last glacial period in Europe. *Quaternary Science Reviews*, **8**, 25–44.

Behre, K.-E. & van der Plicht, J. (1992): Towards an absolute chronology for the last glacial period in Europe: radiocarbon dates from Oerel, northern Germany. *Vegetation History and Archaeobotany*, **1**, 111–117.

Behrensmeyer, A.K. (1984): Taphonomy and the fossil record. *American Scientist*, **72**, 558–566.

Behrensmeyer, A.K., Damuth, J.D., DiMichele, W.A., Potts, R., Sues, H.-D. & Wing, S.L. (1992): *Terrestrial Ecosystems Through Time*. University of Chicago Press, Chicago.

Bell, K. & Murton, J.B. (1995): A new indicator of glacial dispersal: lead isotopes. *Quaternary Science Reviews*, **14**, 275–288.

Bell, M. & Walker, M.J.C. (1992): *Late Quaternary Environmental Change: Physical and Human Perspectives*. Longman, London.

Benedict, J.B. (1990): Experiments on lichen growth. 1. Seasonal patterns and environmental controls. *Arctic and Alpine Research*, **22**, 244–253.

Bengtsson, L. & Enell, M. (1986): Chemical analysis. In *Handbook of Holocene Palaeoecology and Palaeohydrology* (edited by B.E. Berglund), John Wiley, Chichester & New York, 423–454.

Benn, D.I. (1992): The genesis and significance of 'hummocky moraine': evidence from the Isle of Skye, Scotland. *Quaternary Science Reviews*, 11, 781–800.

Benn, D.I. & Evans, D.J.A. (1996): The interpretation and classification of subglacially-deformed materials. *Quaternary Science Reviews*, **15**, 23–52.

Bennett, K.D. (1983): Postglacial expansion of forest trees in Norfolk, UK. *Nature*, **303**, 164–167.

Bennett, K.D. (1988): Post-glacial vegetation history: ecological considerations. In *Vegetation History* (edited by B. Huntley & T. Webb III), Kluwer, Dordrecht, 699–724.

Bennett, K.D. & Lamb, H.F. (1988): Holocene pollen sequences as a record of competitive interactions among tree populations. *Tree*, **3**, 141–145.

Bennett, K.D., Tzedakis, P.C. & Willis, K.J. (1991): Quaternary refugia of north European trees. *Journal of Biogeography*, **18**, 103–115.

Bennett, K.D., Boreham, S., Sharp, M.J. & Switsur, V.R. (1992): Holocene history of environment, vegetation and human settlement on Catta Ness, Lunnasting, Shetland. *Journal of Ecology*, **80**, 241–273.

Bennett, M.R. & Boulton, G.S. (1993): Deglaciation of the Younger Dryas or Loch Lomond Stadial ice-field in the northern Highlands, Scotland. *Journal of Quaternary Science*, **8**, 147–159.

Bennion, H. (1994): A diatom–phosphorous transfer function for shallow, eutrophic ponds in southeast England. *Hydrobiologia*, **275/276**, 391–410.

Benson, L. (1993): Factors affecting the ^{14}C ages of lacustrine carbonates: timing and duration of the last highstand lake in the Lahontan Basin. *Quaternary Research*, **30**, 177–189.

Benson, L. (1994): Carbonate deposition, Pyramid Lake sub-basin, Nevada. I. Sequence of formation and elevational distribution of carbonate deposits (tufas). *Palaeogeography, Palaeoclimatology, Palaeoecology*, **109**, 55–87.

Benson, L. & Thompson, R.S. (1987): The physical record of lakes in the Great Basin. In *The Geology of North America*, Volume K-3, *North America and Adjacent Oceans During the Last Deglaciation* (edited by W.F. Ruddiman & H.E. Wright, Jr), Geological Society of America, Boulder, Colorado, 241–260..

Benson, L.V. & Paillet, F.L. (1989): The use of total lake-surface area as an indicator of climatic change: examples from the Lahontan Basin. *Quaternary Research*, 32, 262–275.

Berger, A. (1978): Long-term variations of calorific insolation resulting from the earth's orbital elements. *Quaternary Research*, **9**, 139–167.

Berger, A. (1984): Accuracy and frequency stability of the earth's orbital elements during the Quaternary. In *Milankovitch and Climate* (edited by A. Berger, J. Imbrie, J. Hays, G. Kukla & B. Saltzman), Reidel, Dordrecht, 3–39.

Berger, G.W. (1987): Thermoluminescence dating of Pleistocene Old Crow tephra and adjacent loess, near Fairbanks, Alaska.

Canadian Journal of Earth Sciences, **24**, 1975–1984.

Berger, G.W. (1988): Dating Quaternary events by luminescence. *Geological Society of America Special Paper*, **227**, 13–49.

Berger, W.H. (1990): The Younger Dryas cold spell: a quest for causes. *Palaeogeography, Palaeoclimatology, Palaeoecology*, **89**, 219–237.

Berger, W.H. & Jansen, E. (1995): Younger Dryas episode: ice collapse and super-fjord heat pump. In *The Younger Dryas* (edited by S.R. Troelstra, J.E. van Hinte & G.M. Ganssen), Koninklijke Nederlandse Akademie van Wetenschappen, Amsterdam, 61–105.

Berggren, G. (1981): *Atlas of Seeds and Small Fruits of North-west European Plant Species with Morphological Descriptions. Part 3: Salicaceae–Cruciferae*. Swedish Natural Science Research Council, Stockholm, 258 pp.

Berggren, W.A. *et al.* (1980): Towards a Quaternary timescale. *Quaternary Research*, **13**, 277–302.

Berglund, B.E. (1979): The deglaciation of southern Sweden 13,500–10,000 BP. *Boreas*, **8**, 89–118.

Berglund, B.E. (ed.) (1986): *Handbook of Holocene Palaeoecology and Palaeohydrology*. John Wiley, Chichester & New York.

Berglund, B.E. (ed.) (1991): The cultural landscape during 6000 years in southern Sweden, *Ecological Bulletins* **41**, Munksgaard, Copenhagen.

Berglund, B.E. & Ralska-Jasiewiczowa, M. (1986): Pollen analysis and pollen diagrams. In *Handbook of Holocene Palaeoecology and Palaeohydrology* (edited by B.E. Berglund), John Wiley, Chichester & New York, 455–484.

Berglund, B.E., Lemdahl, G., Liedberg-Jönssen, B. & Perssen, T. (1984): Biotic responses to climatic changes during the time span 13,000 to 10,000 B.P. A case study from S.W. Sweden. In *Climatic Changes on a Yearly to Millennial Basis* (edited by N.-A. Mörner & W. Karlén), Reidel, Dordrecht, 25–36.

Berglund, B.E., Björck, S., Lemdahl, G. Bergsten, H., Nordberg, K. & Kolstrup, E. (1994): Late Weichselian environmental change in southern Sweden and Denmark. *Journal of Quaternary Science*, **9**, 127–132.

Bergsten, H. (1994): A high-resolution record of Lateglacial and early Holocene marine sediments from southwestern Sweden; with special emphasis on environmental changes close to the Pleistocene–Holocene transition and the influence of fresh water from the Baltic basin. *Journal of Quaternary Science*, **9**, 1–12.

Bergsten, H. & Dennegård, B. (1988): Late Weichselian–Holocene foraminiferal stratigraphy and palaeohydrographic changes in the Gothenburg area, southwestern Sweden. *Boreas*, **17**, 229–242.

Berry, M.E. (1990): Soil catena development on fault scarps of different ages, eastern escarpment of the Sierra Nevada, California. In *Soils and Landscape Evolution* (edited by P.L.K. Knuepfer & L.D. McFadden), Elsevier, Amsterdam, 333–350.

Berryman, K. (1987): Tectonic processes and their impact on the recording of relative sea-level data. In *Sea Surface Studies* (edited by R.J.N. Devoy), Croom Helm, London, 127–161.

Beschel, R.E. (1973): Lichens as a measure of the age of recent moraines. *Arctic and Alpine Research*, **5**, 303–309.

Betancourt, J.L., Van Devender, T.R. & Martin, P.S. (eds) (1990): *Packrat Middens – The Last 40,000 Years of Biotic Change*. University of Arizona Press, Tucson.

Bickerton, R.W. & Matthews, J.A. (1992): On the accuracy of lichenometric dates: an assessment based on the 'Little Ice Age' moraine sequence of Nigardsbreen, southern Norway. *The Holocene*, **2**, 227–237.

Bierman, P.R. & Gillespie, A.R. (1991): Accuracy of rock varnish chemical analyses; implications for cation-ratio dating. *Geology*, **19**, 196–199.

Bierman, P.R. & Gillespie, A.R. (1994): Evidence suggesting that methods of rock-varnish cation-ratio dating are neither comparable nor consistently reliable. *Quaternary Research*, **41**, 82–90.

Billard, A. (1987): *Analyse Critique de Stratotypes Quaternaire*. Edition of the Centre National de la Recherche Scientifique, Paris, 141 pp.

Billard, A., Derbyshire, E., Shaw, J. & Rolph, T. (1987): New data on the sedimentology and magnetostratigraphy of the loessic silts at Saint Vallier, Drôme, France. *Catena, Supplement* **9**, 117–128.

Birchfield, G.E. & Weertman, J. (1983): Topography, albedo–temperature feedback, and climatic sensitivity. *Science*, **219**, 284–285.

Birkeland, P.W. (1974): *Pedology, Weathering and Geomorphological Research*. Oxford University Press, London & New York.

Birkeland, P.W. (1984): *Soils and Geomorphology*. Oxford University Press, New York.

Birkeland, P.W. (1985): Quaternary soils in the western United States. In *Soils and Geomorphology* (edited by J. Boardman), John Wiley, Chichester & New York, 303–324.

Birkeland, P.W. (1992): Quaternary soil chronosequences in various environments – extremely arid to humid tropical. In *Weathering, Soils and Palaeosols* (edited by I.P. Martini & W. Chesworth), *Developments in Earth Surface Processes* **2**, Elsevier, Amsterdam & London, 261–282.

Birkeland, P.W. & Burke, R.M. (1988): Soil catena chronosequences on eastern Sierra Nevada moraines, California, U.S.A. *Arctic and Alpine Research*, **20**, 473–484.

Birkeland, P.W., Crandell, D.R. & Richmond, G.M. (1971): Status of correlation of Quaternary stratigraphic units in the western coterminus United States. *Quaternary Research*, **1**, 208–227.

Birkeland, P.W., Burke, R.M. & Benedict, J.B. (1989): Pedogenic gradients for iron and aluminium accumulation and phosphorus depletion in arctic and alpine soils as a function of time and climate. *Quaternary Research*, **32**, 193–204.

Birkeland, P.W., Berry, M.E. & Swanson, D.K. (1991): Use of soil catena field data for estimating relative ages of moraines. *Geology*, **19**, 281–283.

Birkenmajer, K. (1981): Lichenometric dating of raised beaches at Admiralty Bay, King George Island, South Shetland Islands, West Antarctica. *Bulletin of the Polish Academy of Sciences, Ser. Sci. Terre*, **29**, 119–128.

Birks, H.H. (1993): The importance of plant macrofossils in Late-glacial climatic reconstructions: an example from western Norway. *Quaternary Science Reviews*, **12**, 719–726.

Birks, H.H. & Mathewes, R.W. (1978): Studies in the vegetational history of Scotland. V. Late Devensian and early Flandrian pollen and macrofossil stratigraphy at Abernethy Forest, Inverness-shire. *New Phytologist*, **80**, 455–484.

Birks, H.H., Birks, H.J.B., Kaland, P. & Moe, D. (eds) (1988): *The Cultural Landscape – Past, Present, Future*. Cambridge University Press, Cambridge, 409–428.

Birks, H.H., Lemdahl, G., Svendsen, J.I. & Landvik, J.Y. (1993): Palaeoecology of a late Allerød peat bed at Godøy, western Norway. *Journal of Quaternary Science*, **8**, 95–183.

Birks, H.H., Paus, A., Svendsen, J.I., Alm, T., Mangerud, J. & Landvik, J.Y. (1994): Late Weichselian environmental change in Norway, including Svalbard. *Journal of Quaternary Science*, **9**, 133–146.

Birks, H.J.B. (1981): The use of pollen analysis in the reconstruction of past climates. In *Climate and History* (edited by T.M.L. Wigley, M.J. Ingram & G. Farmer), Cambridge University Press, Cambridge, 111–138.

Birks, H.J.B. (1986): Numerical zonation, comparison and correlation in Quaternary pollen-stratigraphical data. In *Handbook of Holocene Palaeoecology and Palaeohydrology* (edited by B.E. Berglund), John Wiley, Chichester & New York, 743–774.

Birks, H.J.B. & Birks, H.H. (1980): *Quaternary Palaeoecology*. Edward Arnold, London.

Birks, H.J.B. & Line, J.M. (1993): Glacial refugia of European trees – a matter of chance? *Dissertationes Botanicae*, **196**, 283–291.

Birks, H.J.B., Line, J.M., Juggins, S., Stevenson, A.C. & ter Braak, C.J.F. (1990a): Diatoms and pH reconstruction. *Philosophical Transactions of the Royal Society of London*, **B327**, 263–278.

Birks, H.J.B., Berge, F., Boyle, J.F. & Cumming, B.F. (1990b): A palaeoecological test of the land-use hypothesis for recent lake acidification in south-west Norway using hill-top lakes. *Journal of Palaeolimnology*, **4**, 69–85.

Bischoff, J.L. & Fitzpatrick, J.A. (1991): U-series dating of carbonates: an isochron technique using total sample dissolution. *Geochimica et Cosmochimica Acta*, **55**, 543–554.

Björck, S. & Digerfeldt, G. (1991): Allerød–Younger Dryas sea level changes in southwestern Sweden and their relation to the Baltic Ice Lake development. *Boreas*, **20**, 115–134.

Björck, S. & Håkansson, S. (1982): Radiocarbon dates from Late Weichselian lake sediments in South Sweden as a basis for chronostratigraphic subdivision. *Boreas*, **11**, 141–150.

Björck, S., Sandgren, P. & Holmquist, B. (1987): A magnetostratigraphic comparison between ^{14}C years and varve years during the Late Weichselian, indicating significant differences between time-scales. *Journal of Quaternary Science*, **2**, 133–140.

Björck, S. Berglund, B. & Digerfeldt, G. (1988): New aspects on the deglaciation chronology of South Sweden. *Geographia Polonica*, **55**, 37–49.

Björck, S., Sandgren, P. & Zale, R. (1991): Late Holocene tephrochronology of the Northern Antarctic Peninsula. *Quaternary Research*, **36**, 322–328.

Björck, S., Cato, I., Brunnberg, L. & Strömberg, B. (1992a): The clay-varve based Swedish time scale and its relation to the Late Weichselian radiocarbon chronology. In *The Last Deglaciation: Absolute and Radiocarbon Chronologies* (edited by E. Bard & W.S. Broecker), *NATO ASI Series*, **1, 2,** Springer-Verlag, Berlin, 25–43.

Björck, S., Ingólfsson, O., Haflidason, H., Hallsdóttir, M. & Anderson, N.J. (1992b): Lake Torfadalsvatn: a high resolution record of North Atlantic ash zone 1 and the last glacial–interglacial environmental changes in Iceland. *Boreas*, **21**, 15–22.

Blackford, J. (1993): Peat bogs as sources of proxy climatic data: past approaches and future research. In *Climate Change and Human Impact on the Landscape* (edited by F.M. Chambers), Chapman & Hall, London, 47–56.

Blackford, J. & Chambers, F.M. (1991): Proxy records of climate from blanket mires: evidence for a Dark Age (1400 BP) climatic deterioration in the British Isles. *The Holocene*, **1**, 63–67.

Blackford, J.J., Edwards, K.J., Dugmore, A.J., Cook, G.T. & Buckland, P. C. (1992): Icelandic volcanic ash and the mid-Holocene Scots pine (*Pinus sylvestris*) pollen decline in northern Scotland. *The Holocene*, **2**, 260–265.

Blackwell, B., Porat, N., Schwarcz, H.P. & Debénath, A. (1992): ESR dating of tooth enamel: comparison with $^{230}Th/^{234}U$ speleothem dates at La Chaise-de-Vouthon (Charente), France. *Quaternary Science Reviews*, **11**, 231–244.

Blake, W. Jr. (1980): Application of amino-acid ratios to studies of Quaternary geology in the High Arctic. In *Biogeochemistry of Amino Acids* (edited by P.E. Hare, T.C. Hoering & K. King, Jr), John Wiley, Chichester & New York, 453–461.

Blankenship, D.D. *et al.* (1993): Active volcanism beneath the West Antarctic ice sheet and implications for ice-sheet stability. *Nature*, **361**, 526–529.

Bloemendal, J. & de Menocal, P. (1989): Evidence for a change in the periodicity of tropical climate cycles at 2.4 Myr from whole-core magnetic susceptibility measurements. *Nature*, **432**, 897–900.

Bloemendal, J., Oldfield, F. & Thompson, R. (1979): Magnetic measurements used to assess sediment influx at Llyn Goddionduon. *Nature*, **280**, 50–53.

Bloemendal, J., King, J.W., Hunt, A., de Menocal, P.B. & Hayashida, A. (1993): Origin of the sedimentary magnetic record at Ocean Drilling Program sites on the Owen Ridge, western Arabian Sea. *Journal of Geophysical Research*, **98**, 4199–4219.

Blom, R. & Elachi, C. (1981): Spaceborne and airborne imaging radar observations of sand dunes. *Journal of Geophysical Research*, **86**, 3061–3073.

Blunier, T., Chappellaz, J., Schwander, J., Stauffer, B. & Raynaud, D. (1995): Variations in atmospheric methane concentration during the Holocene epoch. *Nature*, **374**, 46–49.

Boardman, J. (ed.) (1985): *Soils and Quaternary Landscape Evolution*. Wiley, Chichester.

Bockheim, J.G. (1995): Permafrost distribution in the Southern Circumpolar Region and its relation to the environment: a review and recommendations for further research. *Permafrost and Periglacial Processes*, **6**, 27–45.

Bogaard, P. van den & Schmincke, H.-U. (1985): Laacher See

Tephra: a widespread isochronous late Quaternary tephra layer in central and northern Europe. *Geological Society of America Bulletin*, **96**, 1554–1571.

Bogaard, P. van den, Hall, C.M., Schmincke, H.-U. & York, D. (1989): Precise single grain $^{40}Ar/^{39}Ar$ dating of a cold to warm climatic transition in Central Europe. *Nature*, **342**, 523–525.

Bohncke, S. & Wijmstra, L. (1988): Reconstruction of Late-glacial lake-level fluctuations in The Netherlands based on palaeobotanical analyses, geochemical results and pollen-density data. *Boreas*, **17**, 403–425.

Bohncke, S., Vandenberghe, J. & Wijmstra, T.A. (1988): Lake level changes and fluvial activity in the Late Glacial lowland valleys. In *Lake, Mire and River Environments* (edited by G. Lang & C. Schlüchter), Balkema, Rotterdam, 115–122.

Bohncke, S.J.P. & Vandenberghe, J. (1991): Palaeohydrological development in the Southern Netherlands during the last 15,000 years. In *Temperate Palaeohydrology* (edited by L. Starkel, K.J. Gregory & J.B. Thornes), John Wiley, Chichester & New York, 253–281.

Bolli, H.M., Saunders, J.B. & Perch-Nielsen, K. (eds) (1985): *Plankton Stratigraphy*. Cambridge University Press, Cambridge.

Bonani, G.B., Pollard, D. & Thompson, S.L. (1992): Effects of boreal forest vegetation on global climate. *Nature* **359**, 716–718.

Bond, G., Broecker, W. Lotti, R.S. & McManus, J. (1992a): Abrupt color changes in isotope stage 5 in North Atlantic deep sea cores: implications for rapid changes of climate-driven events. In *Start of a Glacial* (edited by G.J. Kukla & E. Went), *NATO ASI Series* **1, 3,** Springer-Verlag, Berlin and Heidelberg, 185–205.

Bond, G., Heinrich, H., Broecker, W., Labeyrie, L., McManus, J., Andrews, J., Huon, S., Jantschik, R., Clasen, S., Simet, C., Tedesco, K., Klas, M., Bonani, G. & Ivy, S. (1992b): Evidence for massive discharges of icebergs into the North Atlantic during the last glacial period. *Nature*, **360**, 245–249.

Bond, G., Broecker, W., Johnsen, S., McManus, J., Labeyrie, L., Jouzel, J. & Bonani, G. (1993): Correlations between climate records from North Atlantic sediments and Greenland ice. *Nature*, **365**, 143–147.

Bond, G.C. & Lotti, R. (1995): Iceberg discharge into the North Atlantic on millennnial time scales during the Last Glaciation. *Science*, **267**, 1005–1010.

Bondestam, K., Vasari, A., Vasari, Y., Lemdahl, G. & Eskonen, K. (1994): Younger Dryas and Preboreal in Salpausselkä Foreland, Finnish Karelia. *Dissertationes Botanicae*, **234**, 161–206.

Bonnefille, R., Roeland, J.C. & Guiot, J. (1990): Temperature and rainfall estimates for the past 40,000 years in equatorial Africa. *Nature*, **346**, 347–349.

Bonny, A.P. (1980): Seasonal and annual variations over 5 years in contemporary airborne pollen trapped at a Cumbrian lake. *Journal of Ecology*, **68**, 421–441.

Bonny, A.P. & Allen, P.V. (1984): Pollen recruitment to the sediments of an enclosed lake in Shropshire. In *Lake Sediments and Environmental History* (edited by E.Y. Howarth & J.W.G. Lund), Leicester University Press, Leicester, 231–259.

Böse, M. (1990): Reconstruction of ice flow directions south of the Baltic Sea during the Saalian and Weichselian glaciations. *Boreas*, **19**, 217–226.

Böse, M. (1991): A palaeoclimatic interpretation of frost-wedge casts and aeolian sand deposits in the lowlands between Rhine and Vistula in the Upper Pleniglacial and Late Glacial. *Zeitschrift für Geomorphologie*, Suppl. Bd. **90**, 15–28.

Bouchard, M.A. & Salonen, V.-P. (1990): Boulder transport in shield areas. In *Glacier Indicator Tracing* (edited by R. Kujansuu & M. Saarnisto), Balkema, Rotterdam, 87–108.

Boulton, G.S. (1972): Modern arctic glaciers as depositional models for former ice sheets. *Quarterly Journal of the Geological Society of London*, **128**, 361–393.

Boulton, G.S. (1974): Processes and patterns of glacial erosion. In *Glacial Geomorphology* (edited by D.R. Coates), State University of New York, Binghampton, 41–87.

Boulton, G.S. (1975): Processes and patterns of subglacial sedimentation: a theoretical approach. *In Ice Ages: Ancient and Modern* (edited by A.E. Wright & F. Mosely), Seel House Press, Liverpool, 7–42.

Boulton, G.S. (1977): A multiple till sequence formed by a Late Devensian Welsh ice cap: Glanllynnau, Gwynedd. *Cambria*, **4**, 10–31.

Boulton, G.S. (1980): Classification of till. *Quaternary Research Association (G.B.) Newsletter*, **31**, 1–12.

Boulton, G.S. (1987): A theory of drumlin formation by subglacial deformation. In *Drumlin Symposium* (edited by J. Menzies & J.Rose), Balkema, Rotterdam, 25–80.

Boulton, G.S. & Clark, C.D. (1990a): A highly mobile Laurentide ice sheet revealed by satellite images of glacial lineations. *Nature*, **346**, 813–817.

Boulton, G.S. & Clark, C.D. (1990b): The Laurentide ice sheet through the last glacial cycle: the topology of drift lineations as a key to the dynamic behaviour of former ice sheets. *Transactions of the Royal Society of Edinburgh: Earth Sciences*, **81**, 327–347.

Boulton, G.S. & Eyles, N. (1979): Sedimentation by valley glaciers: a model and genetic classification. In *Moraines and Varves* (edited by C. Schlüchter), Balkema, Rotterdam, 11–23.

Boulton, G.S., Jones, A.S., Clayton, K.M. & Kenning, M.J. (1977): A British ice sheet model and patterns of glacial erosion and deposition in Britain. In *British Quaternary Studies* (edited by F.W. Shotton), Oxford University Press, Oxford, 231–246.

Boulton, G.S., Smith, G.D. & Morland, L.W. (1984): The reconstruction of former ice sheets and their mass balance characteristics using a non-linearly viscous flow model. *Journal of Glaciology*, **30**, 140–152.

Boulton, G.S., Smith, G.D., Jones, A.S. & Newsome, J. (1985): Glacial geology and glaciology of the last mid-latitude ice sheets. *Journal of the Geological Society of London*, **142**, 447–474.

Boulton, G.S., Peacock, J.D. & Sutherland, D.G. (1991): Quaternary. In *The Geology of Scotland* (edited by G.Y. Craig), 3rd edition, The Geological Society, London, 503–542.

Bowen, D.Q. (1981): The 'South Wales End Moraine': fifty years after. In *The Quaternary in Britain* (edited by J. Neale & J. Flenley), Pergamon Press, Oxford, 60–67.

Bowen, D.Q. (1991): Time and space in the glacial sediment

systems of the British Isles. In *Glacial Deposits in Great Britain and Ireland* (edited by J. Ehlers, P.L. Gibbard & J. Rose), Balkema, Rotterdam, 3–11.

Bowen, D.Q. & Sykes, G.A. (1988): Correlations of marine events and glaciations on the northeast Atlantic margin. *Philosophical Transactions of the Royal Society, London*, **B318**, 619–635.

Bowen, D.Q., Sykes, G.A., Reeves, A. Miller, G.H., Andrews, J.T., Brew, J.S. & Hare, P.E. (1985): Amino acid geochronology of raised beaches in south west Britain. *Quaternary Science Reviews*, **4**, 279–318.

Bowen, D.Q., Rose, J., McCabe, A.M. & Sutherland, D.G. (1986): Correlation of Quaternary glaciations in England, Ireland, Scotland and Wales. *Quaternary Science Reviews*, **5**, 299–340.

Bowen, D.Q., Hughes, S. Sykes, G.A. & Miller, G.H. (1989): Land sea correlations in the Pleistocene based on isoleucine epimerisation in non marine molluscs. *Nature*, **350**, 49–51.

Bowen, D.Q., Sykes, G.A., Maddy, D., Bridgland, D.R. & Lewis, S.G. (1995); Aminostratigraphy and amino-acid geochronology of English lowland valleys: the lower Thames in context. In *The Quaternary of the Lower Reaches of the Thames: Field Guide* (edited by D.R. Bridgland, P.A. Allen & B.A. Haggart), Quaternary Research Association, London, 61–63.

Bowler, J.M.& Polach, H.A. (1971): Radiocarbon analysis of soil carbonates: an evaluation from palaeosols in south-eastern Australia. In *Paleopedology* (edited by D. A. Yaalon), International Soil Science and Israel University Press, Jerusalem, 97–108.

Bowler, J.M., Huang, Q., Chen, K., Head, M.J. & Yuan, B. (1986): Radiocarbon dating of playa-lake hydrologic changes: examples from northwestern China and central Australia. *Palaeogeography, Palaeoclimatology, Palaeoecology*, **54**, 241–260.

Bowman, S. (1990): *Radiocarbon Dating*. British Museum, London.

Boyle, E.A. (1988): Cadmium chemical tracer of deepwater. *Paleoceanography*, **3**, 471–489.

Boyle, E.A. & Keigwin, L. (1987): North Atlantic thermohaline circulation during the past 20,000 years linked to high-latitude surface temperature. *Nature*, **330**, 35–40.

Bradbury, J.P. (1971): Palaeolimnology of Lake Texcoco, Mexico. Evidence from diatoms. *Limnology and Oceanography*, **16**, 180–200.

Bradbury, J.P., Leyden, B. *et al.* (1981): Late-Quaternary environmental history of Lake Valencia, Venezuela. *Science*, **214**, 1299–1305.

Bradley, R.S. (1985): *Quaternary Paleoclimatology*. Allen & Unwin, London & Boston.

Bradley, W.H. *et al.* (1941): Geology and biology of North Atlantic deep-sea cores. *U.S. Geological Survey Professional Papers*, **196**, 163 pp.

Bradshaw, R.H.W. (1981): Modern pollen representation factors for woods in south-east England. *Journal of Ecology*, **69**, 45–70.

Brasier, M.D. (1980): *Microfossils*. George Allen & Unwin, London & Boston.

Brenningkmeier, C.A.M., van Geel, B. & Mook, W.G. (1982): Variations in the D/H and $^{18}O/^{16}O$ ratio in cellulose extracted from a peat bog core. *Earth and Planetary Science Letters*, **61**, 283–290.

Bridge, M.C., Haggart, B.A. & Lowe, J.J. (1990): The history and palaeoclimatic significance of subfossil remains of *Pinus sylvestris* in blanket peats from Scotland. *Journal of Ecology*, **78**, 77–99.

Bridgland, B., D'Ollier, B., Gibbard, P.L. & Roe, H.M. (1993): Correlation of Thames terrace deposits between the lower Thames, eastern Essex and the submerged offshore continuation of the Thames–Medway Valley. *Proceedings of the Geologists' Association*, **104**, 51–58.

Bridgland, D. (1986): *Clast Lithological Analysis*. Technical Guide No. 3, Quaternary Research Association, Cambridge.

Bridgland, D.R. (1988): The Pleistocene fluvial stratigraphy and paleogeography of Essex. *Proceedings of the Geologists' Association*, **99**, 291–314.

Bridgland, D.R. (ed.) (1994): *Quaternary of the Thames*. Chapman & Hall, London.

Bridgland, D.R. (1995): The Quaternary sequence in the eastern Thames basin: problems of correlation. In *The Quaternary of the Lower Reaches of the Thames: Field Guide* (edited by D.R. Bridgland, P.A. Allen & B.A. Haggart), Quaternary Research Association, London, 35–52.

Bridgland, D.R., Allen, P.A. & Haggart, B.A. (eds) (1995): *The Quaternary of the Lower Reaches of the Thames: Field Guide*. Quaternary Research Association, London.

Briffa, K.R., Jones, P.D., Pilcher, J.R. & Hughes, M.K. (1988): Reconstructing summer temperatures in northern Fennoscandia back to A.D. 1700 using tree-ring data from Scots Pine. *Arctic and Alpine Research*, **20**, 385–394.

Briffa, K.R., Bartholin, T.S., Eckstein, D., Jones, P.D., Karlén, W., Schweingrüber, F.H. & Zetterberg, P. (1990): A 1,400-year tree-ring record of summer temperatures in Fennoscandia. *Nature*, **346**, 434–439.

Briggs, D.J. (1977): *Sediments*. Butterworths, London & Boston.

Briggs, D.J. & Gilbertson, D.D. (1980): Quaternary processes and environments in the upper Thames Valley. *Transactions of the Institute of British Geographers, New Series*, **5**, 53–65.

Briggs, D.J., Coope, G.R. & Gilbertson, D.D. (1985): The chronology and environmental framework of early man in the Upper Thames Valley. *British Archaeological Reports, British Series*, **137**, BAR, Oxford.

Briggs, D.J., Gilbertson, D.D. & Harris, A.L. (1990): Molluscan taphonomy in a braided river environment and its implications for studies of Quaternary cold-stage river deposits. *Journal of Biogeography*, **17**, 623–637.

Broadbent, N.D. (1979): *Coastal Resources and Settlement Stability. A critical analysis of a Mesolithic site complex in northern Sweden. Aun 3*. Archaeological Studies Institute of Northern European Archaeology, University of Uppsala, Uppsala Borgstroms tryckeri, 268 pp.

Broadbent, N.D. & Bergqvist, K.I. (1986): Lichenometric chronology and archaeological features on raised beaches: preliminary results from the Swedish North Bothnian coastal region. *Arctic and Alpine Research*, **18**, 297–306.

Broccoli, A.J. & Manabe, S. (1987): The influence of continental ice, atmospheric CO_2 and land albedo on the climate of the last glacial maximum. *Climate Dynamics*, **1**, 87–89.

Brodzikowski, K. & Van Loon, A.J. (1987): A systematic classification of glacial and periglacial environments, facies and deposits. *Earth-Science Reviews*, **24**, 297–381.

Brodzikowski, K. & Van Loon, A.J. (1991): *Glacigenic Sediments*. Developments in Sedimentology, **49**. Elsevier, Amsterdam.

Broecker, W.S. (1971): Calcite accumulation rates and glacial to interglacial changes in ocean mixing. In *The Late Cenozoic Glacial Ages* (edited by K.K. Turekian), Yale University Press, New Haven, 239–265.

Broecker, W.S. (1984): Terminations. In *Milankovitch and Climate* (edited by A. Berger, J. Imbrie, J. Hays, G. Kukla & B. Saltzman), Reidel, Dordrecht, 687–698.

Broecker, W.S. (1990): Salinity history of the northern Atlantic during the last deglaciation. *Paleoceanography*, **5**, 459–467.

Broecker, W.S. (1992): The strength of the Nordic heat pump. In *The Last Deglaciation: Absolute and Relative Chronologies* (edited by E. Bard & W.S. Broecker), *NATO ASI Series* **1, 2,** Springer-Verlag, Berlin, 173–181.

Broecker, W.S. (1994a): An unstable superconveyor. *Nature*, **367**, 414–415.

Broecker, W.S. (1994b): Massive iceberg discharges as triggers for global climate change. *Nature*, **372**, 421–424.

Broecker, W. S. & Bender, M.L. (1972): Age determinations on marine strandlines. In *Calibration of Hominid Evolution* (edited by W.W. Bishop & J.A. Miller), Scottish Academic Press, Edinburgh: Wenner Grend Foundation, New York, 19–38.

Broecker, W.S. & Denton, G.H. (1990a): The role of ocean-atmosphere reorganisations in glacial cycles. *Quaternary Science Reviews*, **9**, 305–341.

Broecker, W.S. & Denton, G.H. (1990b): What drives glacial cycles? *Scientific American*, **262**, 42–50.

Broecker, W.S., Peteet, D.M., & Rind, D. (1985): Does the ocean–atmosphere system have more than one stable mode of operation? *Nature*, **315**, 21–25.

Broecker, W.S., Andrée, M., Wölfli, W., Oeschger, H., Bonani, G., Kennett, J. & Peteet, D. (1988): The chronology of the last deglaciation: implications to the cause of the Younger Dryas event. *Paleoceanography*, **3**, 1–19.

Broecker, W.S., Kennett, J.P., Flower, B.P., Teller, J.T., Trumboe, S., Bonani, G. & Wölfli, W. (1989): Routing of meltwater from the Laurentide Ice Sheet during the Younger Dryas cold episode. *Nature*, **341**, 318–321.

Broecker, W.S., Bond, G. & Klas, M. (1990): A salt oscillator in the glacial North Atlantic? 1. The concept. *Paleoceanography*, **5**, 469–477.

Broecker, W.S., Peng, T.H., Trumbore, S., Bonani, G. & Wölfli, W. (1993): The distribution of radiocarbon in the glacial oceans. *Global Biogeochemical Cycles*, **4**, 103–117.

Bromehead, C.E.N. (1925): *The Geology of North London*. Memoir of the Geological Survey of the British Isles.

Bronger, A. & Catt, J.A. (1989): *Palaeopedology, Catena Supplement* **16**, Catena Verlag, Cremlingen, Germany.

Bronger, A. & Heinkele, Th. (1989): Paleosol sequences as witnesses of Pleistocene climatic history. In *Palaeopedology*, (edited by A. Bronger & J.A. Catt), *Catena Supplement*, **16**, Catena Verlag, Cremlingen, Germany, 163–186.

Bronger, A., Winter, R., Derevjanko, O. & Aldag, S. (1995): Loess–palaeosol sequences in Tadjikistan as a palaeoclimatic record of the Quaternary in Central Asia. *Quaternary Proceedings*, John Wiley, Chichester, **4**, 69–82.

Brookes, I.A. (1982): Dating methods of Pleistocene deposits and their problems. VIII, Weathering. *Geoscience Canada*, **99**, 188–199.

Brookes, I.A. (1993): Geomorphology and Quaternary geology of the Dakhla Oasis region, Egypt. *Quaternary Science Reviews*, **12**, 529–552.

Brown, I.M. (1990): Quaternary glaciations of New Guinea. *Quaternary Science Reviews*, **9**, 273–280.

Brown, I.M. (1993): Pattern of deglaciation of the last (Late Devensian) Scottish ice sheet: evidence from ice-marginal deposits in the Dee valley, northeast Scotland. *Journal of Quaternary Science*, **8**, 235–250.

Brown, R.J.E., Johnston, G.H., Mackay, J.R., Morgenstern, N.R. & Smith, W.W. (1981): Permafrost distribution and terrain characteristics. In *Permafrost Engineering* (edited by G.H. Johnston), Wiley, Toronto, 31–72.

Brown, T.A., Nelson, D.E., Mathews, R.W., Vogel, J.S. & Southon, J.R. (1989): Radiocarbon dating of pollen by accelerator mass spectrometry. *Quaternary Research*, **32**, 205–212.

Brubaker, L.B. & Cook, E.R. (1984): Tree-ring studies of Holocene environments. In *Late Quaternary Environments of the United States*, Volume 2, *The Holocene* (edited by H.E. Wright, Jr), Longman, London, 222–235.

Brunnacker, K. (1975): The Mid-Pleistocene of the Rhine Basin. In *After the Australopithecines* (edited by K.W. Butzer & G.L. Isaac), Mouton Press, The Hague, 189–224.

Bryant, I.D. (1983): The utilisation of Arctic river analogue studies in the interpretation of periglacial river sediments from southern Britain. In *Background to Palaeohydrology* (edited by K.J. Gregory), John Wiley, Chichester & New York, 413–431.

Bryant, R.H. & Carpenter, C.P. (1987): Ramparted ground ice depressions in Britain and Ireland. In *Periglacial Processes and Landforms in Britain and Ireland* (edited by J. Boardman), Cambridge University Press, Cambridge, 183–190.

Bryson, R.A. (1989): Late Quaternary volcanic modulation of Milankovitch climate forcing. *Theoretical Applied Climatology*, **39**, 115–125.

Buckland, P.C. (1979): Thorne Moors: a palaeoecological study of a Bronze Age site (a contribution to the history of the British insect fauna). *University of Birmingham, Department of Geography Occasional Publication* **8**, 173 pp.

Buckland, P.C. (1981): The early dispersal of insect pests of stored products as indicated by archaeological records. *Journal of Stored Product Research*, **17**, 1–12.

Buckland, P.C. & Coope, G.R. (1991): *A Bibliography and Literature Review of Quaternary Entomology*, Collins, Sheffield.

Buckland, W. (1822): An account of an assemblage of fossil teeth and bones discovered in a cave at Kirkdale. *Philosophical Transactions of the Royal Society, London*, **122**, 171–236.

Buckland, W. (1840–41): On the evidences of glaciers in Scotland and Northern England. *Proceedings of the Geological Society of London*, **3**, 332–337, 345–348.

Budd, W.F. & Smith, I.N. (1981): The growth and retreat of ice sheets in response to orbital radiation changes. In *Sea Level, Ice and Climatic Change*, International Association of Hydrological Sciences, Publication Number 141, 369–410.

Budd, W.F., Andrews, J.T., Finkel, R.C., Fireman, E.L., Graf, W., Hammer, C.U., Jouzel, J., Raynaud, D.P., Reeh, N., Shoji, H., Stauffer, B.R. & Weertman, J. (1989): Group Report. How can an ice core chronology be established? In *The Environmental Record in Glaciers and Ice Sheets* (edited by H. Oeschger & C.C. Langway, Jr.), John Wiley, Chichester & New York, 177–192.

Bull, P.A. (1983): Chemical sedimentation in caves. In *Chemical Sediments and Geomorphology: Precipitates and Residua in the Near-Surface Environment* (edited by A.S. Goudie & K. Pye), Academic Press, London, 301–320.

Burbidge, G.H., French, H.M. & Rust, B.R. (1988): Water escape fissures resembling ice-wedge casts in Late Quaternary subaqueous outwash near St. Lazare, Québec, Canada. *Boreas*, **17**, 33–40.

Burgess, C. (1989): Volcanoes, catastrophe and the global crisis of the late second millenium BC. *Current Archaeology*, **117**, 325–329.

Burgis, M.J. & Morris, P. (1987): *The Natural History of Lakes*. Cambridge University Press, Cambridge.

Burke, R.M. & Birkeland, P.W. (1979): Re-evaluation of multiparameter relative dating techniques and their application to the glacial sequence along the eastern escarpment of the Sierra Nevada, California. *Quaternary Research*, **11**, 21–51.

Burney, D.A., Brook, G.A. & Cowart, J.B. (1994): A Holocene pollen record for the Kalahari Desert of Botswana from a U-series dated speleothem. *The Holocene*, **4**, 225–232.

Busacca, A.J., Nelstead, K.T., McDonald, E.V. & Purser, M.D. (1992): Correlation of distal tephra layers in loess in the channeled scabland and palouse of Washington state. *Quaternary Research*, **37**, 281–303.

Butterfield, B.G. & Meylan, B.A. (1980): *Three-dimensional Structure of Wood: An Ultrasonic Approach*. Chapman & Hall, London & New York.

Butzer, K.W. (1980): Holocene alluvial sequences: problems of dating and correlation. In *Timescales in Geomorphology* (edited by R.A. Cullingford, D.A. Davidson & J. Lewin), John Wiley, Chichester & New York, 131–142.

Calkin, P.E. (1988): Holocene glaciation of Alaska (and adjoining Yukon Territory, Canada). *Quaternary Science Reviews*, **7**, 159–184.

Campy, M. & Chaline, J. (1993): Missing records and depositional breaks in French Late Pleistocene cave sediments. *Quaternary Research*, **40**, 318–331.

Carew, J.L. & Mylroie, J.E. (1995): Quaternary tectonic stability of the Bahamian Archipelago: evidence from fossil coral reefs and flank margin caves. *Quaternary Science Reviews*, **14**, 145–154.

Carter, R.W.G. (1992): Sea-level changes: past, present and future. *Quaternary Proceedings*, **2**, 111–132.

Carter, R.W.G., Devoy, R.J.N. & Shaw, J. (1989): Late Holocene sea levels in Ireland. *Journal of Quaternary Science*, **4**, 7–24.

Carter, S.P. (1990): The stratification and taphonomy of shells in calcareous soils: implications for land snail analysis in archaeology. *Journal of Archaeological Science*, **17**, 495–507.

Caseldine, C.J. & Hatton, J. (1993): The development of high moorland on Dartmoor: fire and the influence of Mesolithic activity on vegetation change. In *Climatic Change and Human Impact on the Landscape* (edited by F.M. Chambers), Chapman & Hall, London, 109–118.

Caseldine, C.J. & Stötter, J. (1993): 'Little Ice Age' glaciation of Tröllaskagi peninsula, northern Iceland: climatic implications for reconstructed equilibrium line altitudes (ELAs). *The Holocene*, **3**, 357–366.

Casey, R.E. (1993): Radiolaria. In *Fossil Prokaryotes and Protists* (edited by J.H. Lipps), Basil Blackwell, London.

Cato, I. (1985): The definitive connection of the Swedish geochronological time scale with the present, and the new date of the zero year in Döviken, northern Sweden. *Boreas*, **14**, 117–122.

Catt, J.A. (1977): Loess and coversands. *In British Quaternary Studies – Recent Advances* (edited by F.M. Shotton), Oxford University Press, Oxford, 221–229.

Catt, J.A. (1986): *Soils and Quaternary Geology*. Clarendon Press, Oxford.

Catt, J.A. (1987): Effects of the Devensian cold stage on soil characteristics and distribution in eastern England. In *Periglacial Processes and Landforms in Britain and Ireland* (edited by J. Boardman), Cambridge University Press, Cambridge, 145–152.

Catt, J.A. (1988): Soils of the Plio-Pleistocene: do they distinguish types of interglacial? *Philosophical Transactions of the Royal Society*, **B318**, 539–557.

Catt, J.A. (1990): Palaeopedology Manual. *Quaternary International*, **6**, 1–95.

Catt, J.A. (1991): The Quaternary history and glacial deposits of East Yorkshire. In *Glacial Deposits in Great Britain and Ireland* (edited by J. Ehlers, P.L. Gibbard & J. Rose), Balkema, Rotterdam, 185–191.

Catt, J.A., Bateman, R.M., Wintle, A.G. & Murphy, C.P. (1987): The 'loess' section at Borden, Kent, SE England. *Journal of Quaternary Science*, **2**, 141–147.

Catto, N. (1987): Lacustrine sedimentation in a proglacial environment, Caribou River Valley, Yukon, Canada. *Boreas*, **16**, 197–206.

Catto, N.R. (1990): Clast fabric of diamictons associated with some roches moutonnées. *Boreas*, **19**, 289–296.

Chambers, F.M. (1993): *Climate Change and Human Impact on the Landscape*, Chapman & Hall, London.

Chappell, J. (1974): Late Quaternary glacio- and hydro-isostasy on a layered earth. *Quaternary Research*, **4**, 405–428.

Chappell, J. (1983): A revised sea-level record for the last 300,000 years from Papua New Guinea. *Search*, **14**, 99–101.

Chappell, J. (1987): Ocean volume change and the history of sea water. In *Sea Surface Studies* (edited by R.J.N. Devoy), Croom Helm, London, 33–56.

Chappell, J. (1991): Late Quaternary environmental changes in eastern and central Australia, and their climatic interpretation. *Quaternary Science Reviews*, **10**, 377–390.

Chappellaz, J., Barnola, J.M., Raynaud, D., Korotkevich, Y.S. & Lorius, C. (1990): Ice-core record of atmospheric methane over the past 160,000 years. *Nature*, **345**, 127–131.

Charles, C.D. & Fairbanks, R.G. (1992): Evidence from Southern Ocean sediments for the effect of North-Atlantic deep-water flux on climate. *Nature*, **355**, 416–419.

Charleson, R., Lovelock, J., Andreae, M. & Warren, S. (1987): Oceanic phytoplankton, atmospheric sulfur, cloud albedo and climate. *Nature*, **326**, 655–661.

Charlesworth, J.K. (1928): The glacial retreat from central and southern Ireland. *Journal of the Geological Society of London*, **84**, 293–344.

Charlesworth, J.K. (1929): The South Wales End-Moraine. *Journal of the Geological Society of London*, **85**, 335–358.

Charman, D. (1994): Patterned fen development in northern Scotland: developing a hypothesis from palaeoecological data. *Journal of Quaternary Science*, **9**, 285–297.

Charman, D.J. (1992): Blanket mire formation at the Cross Lochs, Sutherland, northern Scotland. *Boreas*, **21**, 53–72.

Chen, X.Y. (1995): Geomorphology, stratigraphy and thermoluminescence dating of the lunette dune at Lake Victoria, western New South Wales. *Palaeogeography, Palaeoclimatology, Palaeoecology*, **113**, 69–86.

Chepalyga, A.L. (1984): Inland sea basins. In *Late Quaternary Environments of the Soviet Union* (edited by A.A. Velitchko), Longman, London, 229–247.

Chivas, A.R., De Deckker, P. & Shelley, J.M.C. (1986): Magnesium content of non-marine ostracod shells: a new palaeosalinometer and palaeothermometer. *Palaeogeography, Palaeoclimatology, Palaeoecology*, **54**, 43–61.

Chorley, R.J., Dunn, A.J. & Beckinsale, R.P. (1964): *A History of the Study of Landforms*, Volume 1. Methuen, London.

Churcher, C.S. & Wilson, M.C. (1990): Vertebrates. In *Methods in Quaternary Ecology* (edited by B.G. Warner), *Geoscience Canada, Reprint Series* **5**, 127–148.

Cifelli, R. & Benier, C.-S. (1976):Distribution of planktonic Foraminifera in the North Atlantic. *Journal of Foraminiferal Research*, **6**, 158–273

Clague, J.J., Easterbrook, D.J., Hughes, O.L. & Matthews, J.V. Jr. (1992): The Sangamonian and Early Wisconsinan stages in western Canada and northwestern United States. In *The Last Interglacial–Glacial Transition in North America* (edited by P.U. Clark & P.D. Lea), *Geological Society of America Special Paper*, **270**, 1–11, Boulder, Colorado.

Clapperton, C.M. (1986): Glacial geomorphology, Quaternary glacial sequences and palaeoclimatic inferences in the Ecuadorian Andes. In *International Geomorphology* (edited by V. Gardiner), John Wiley, Chichester & New York, 843–870.

Clapperton, C.M. (1993a): Glacier advances in the Andes at 12,500–10,000 yr BP: implications for mechanisms of Late-glacial climatic change. *Journal of Quaternary Science*, **8**, 197–215.

Clapperton, C.M. (1993b): Nature of environmental changes in South America at the Last Glacial Maximum. *Palaeogeography, Palaeoclimatology, Palaeoecology*, **101**, 189–208.

Clark, C.D. (1993): Mega-scale glacial lineations and cross-cutting ice-flow landforms. *Earth Surface Processes and Landforms*, **18**, 1–29.

Clark, I.D. & Fontes, J.-C. (1990): Paleoclimatic reconstruction in Northern Oman based on carbonates from hyperalkaline groundwaters. *Quaternary Research*, **33**, 320–336.

Clark, J.A., Farrell, W.E. & Peltier, W.R. (1978): Global changes in post-glacial sea level: a numerical calculation. *Quaternary Research*, **9**, 265–287.

Clark, J.G.D. (1954): *Excavations at Star Carr, an Early Mesolithic Site at Seamer, near Scarborough, Yorkshire*. Cambridge University Press, Cambridge.

Clark, M.J. (ed.) (1988): *Advances in Periglacial Geomorphology*. John Wiley, Chichester & New York.

Clark, P.U. (1994): Unstable behaviour of the Laurentide Ice Sheet over deforming sediment and its implications for climate change. *Quaternary Research*, **41**, 19–25.

Clark, P.U. & Bartlein, P.J. (1995): Correlation of late Pleistocene glaciation in the western United States with North Atlantic Heinrich events. *Geology*, **23**, 483–486.

Clark, P.U., Clague, J.J., Curry, B.B., Dreimanis, A., Hicock, S.R., Miller, G.H., Berger, G.W., Eyles, N., Lamothe, M., Miller, B.B., Mott, R.J., Oldale, R.N., Stea, R.R., Szabo, J.P., Thorleifson, L.H. & Vincent, J.-S. (1993): Initiation and development of the Laurentide and Cordilleran Ice Sheets following the last interglaciation. *Quaternary Science Reviews*, **12**, 79–114.

Clarke, G.K.C. (1987) Subglacial till: a physical framework for its properties and processes. *Journal of Geophysical Research*, **92**, 8942–8984.

Clarke, M.L. (1994): Infra-red stimulated luminescence ages from aeolian sand and alluvial fan deposits from the Eastern Mojave Desert, California. *Quaternary Geochronology, Quaternary Science Reviews*, **13**, 553–538.

Clayton, K.M. (1977): River terraces. In *British Quaternary Studies: Recent Advances* (edited by F.W. Shotton), Oxford University Press, Oxford, 153–167.

Clayton, L. & Moran, S.R. (1982): Chronology of late Wisconsinan glaciation in middle North America. *Quaternary Science Reviews*, **1**, 55–82.

Clayton, L., Teller, J.T. & Attig, J.W. (1985): Surging of the south-western part of the Laurentide Ice Sheet. *Boreas*, **14**, 235–242.

CLIMAP Project Members (1976): The surface of ice-age earth. *Science*, **191**, 1131–1137.

CLIMAP Project Members (1981): Seasonal reconstructions of the earth's surface at the last glacial maximum. *Geological Society of America Map and Chart Series*, **MC-36**.

Cline, R.M. & Hays, J.D. (1976): Investigation of late Quaternary paleoceanography and paleoclimatology. *Geological Society of America Memoirs*, **145**.

Close, F., Marten, M. & Sutton, C. (1987): *The Particle Explosion*. Oxford University Press, London.

Clymo, R.S. (1984): The limits of peat bog growth. *Philosophical Transactions of the Royal Society of London*, **B303**, 605–654.

Clymo, R.S. (1991): Peat growth. In *Quaternary Landscapes* (edited by L.C.K. Shane & E.J. Cushing), University of Minnesota Press, Minneapolis, 76–112.

Cohen, A.S. & Thouin, C. (1987): Nearshore carbonate deposits in Lake Tanganyika. *Geology*, **15**, 414–418.

COHMAP Members (1988): Climatic changes of the last 18,000 years: observations and model simulations. *Science*, **241**, 1043–1052.

Colinvaux, P. (1987): Amazon diversity in light of the palaeoecological record. *Quaternary Science Reviews*, **6**, 93–114.

Colinvaux, P.A., Miller, M.C., Kam-Biu, L., Steinitz-Kannan, M. & Frost, I. (1985): Discovery of permanent Amazon lakes and hydraulic disturbance in the upper Amazon basin. *Nature*, **313**, 42–45.

Colman, S.M. (1982): Chemical weathering of basalts and andesites: evidence from weathering rinds. *United States Geological Survey Professional Paper*, **1246**.

Colman, S.M. & Dethier, D.P. (eds) (1986): *Rates of Chemical Weathering of Rocks and Minerals*. Academic Press, Orlando.

Colman, S.M. & Pierce, K.L. (1986): Glacial sequence near McCall, Idaho: weathering rinds, soil development, morphology and other relative age criteria. *Quaternary Research*, **25**, 25–42.

Conkey, L. (1979): Response of tree-ring density to climate in Maine, USA. *Tree-Ring Bulletin*, **39**, 29–38.

Conkey, L.E. (1986): Red spruce tree-ring widths and densities in eastern North America as indicators of past climate. *Quaternary Research*, **26**, 232–243.

Cook, E. (1982): A prospectus on the development of tree-ring network in eastern North America. In *Climate from Tree Rings* (edited by M.K. Hughes, P.M. Kelly, J. Pilcher & V.C. LaMarche, Jr.), Cambridge University Press, 126–134.

Cook, E.R. & Kairiukistis, L.A. (eds) (1990): *Methods of Dendrochronology*. Kluwer, Dordrecht.

Cook, E.R., Briffa, K.R., Shiyatov, S.G. & Mazepa, V.S. (1990): Tree-ring standardisation and growth-trend estimation. In *Methods of Dendrochronology* (edited by E.R. Cook & L.A. Kairiukstis), Kluwer, Dordrecht, 104–123.

Cooke, H.B.S. (1981): Age control of Quaternary sedimentary/climatic record from deep boreholes in the Great Hungarian Plain. In *Quaternary Palaeoclimatology* (edited by W.C. Mahaney), GeoAbstracts, Norwich, 1–12.

Cooke, H.J. (1984): The evidence from northern Botswana of late Quaternary climatic change. In *Late Quaternary Palaeoclimates of the Southern Hemisphere* (edited by J.G. Vogel), Balkema, Rotterdam, 265–278.

Cooke, R.U. & Doornkamp, J.C. (1990): *Geomorphology in Environmental Management. A New Introduction*, 2nd edition. Clarendon Press, Oxford.

Coope, G.R. (1959): A late Pleistocene insect fauna from Chelford, Cheshire. *Proceedings of the Royal Society, London*, **B151**, 70–86.

Coope, G.R. (1967): The value of Quaternary insect faunas in the interpetation of ancient ecology and climate. In *Quaternary Paleoecology* (edited by E.J. Cushing & H.E.Wright Jr.), Yale University Press, New Haven, 359–380.

Coope, G.R. (1974): Interglacial Coleoptera from Bobbitshole, Ipswich. *Quarterly Journal of the Geological Society of London*, **130**, 333–340.

Coope, G.R. (1977a): Fossil coleopteran assemblages as sensitive indicators of climatic changes during the Devensian (last) cold stage. *Philosophical Transactions of the Royal Society of London*, B280, 313–340.

Coope, G.R. (1977b): Quaternary Coleoptera as aids in the interpretation of environmental history. In *British Quaternary Studies – Recent Advances* (edited by F.W. Shotton), Oxford University Press, Oxford, 55–68.

Coope, G.R. (1986a): Coleoptera analysis. In *Handbook of Holocene Palaeoecology and Palaeohydrology* (edited by B.E. Berglund), John Wiley, Chichester & New York, 703–713.

Coope, G.R. (1986b): The invasion and colonization of the North Atlantic islands: a palaeoecological solution to a biogeographic problem. *Philosophical Transactions of the Royal Society of London*, **B314**, 619–635.

Coope, G.R. (1987): The response of Late Quaternary insect communities to sudden climatic changes. In *Organization of Communities, Past and Present* (edited by J.H.R. Gee & P.S. Giller), Blackwell, Oxford & Boston, 421–438.

Coope, G.R. (1990): The invasion of Northern Europe during the Pleistocene by Mediterranean species of Coleoptera. In *Biological Invasions in Europe and the Mediterranean Basin* (edited by F. di Castri, A.J. Hansen & M. Debussche), Kluwer, Dordrecht, 203–215.

Coope, G.R. (1994a): The response of insect faunas to glacial–interglacial climatic fluctuations. *Philosophical Transactions of the Royal Society of London*, **B344**, 19–26.

Coope, G.R. (1994b): Insect faunas in ice age environments: why so little extinction? In *Estimation of Extinction Rates* (edited by R. May & J. Lawton), Oxford University Press, Oxford, 55–74.

Coope, G.R. & Brophy, J.A. (1972): Late Glacial environmental changes indicated by a coleopteran succession from North Wales. *Boreas*, **1**, 97–142.

Coope, G.R. & Lemdahl, G. (1995): Regional differences in the Lateglacial climate of northern Europe based on coleopteran analysis. *Journal of Quaternary Science*, **10**, 391–395.

Coope, G.R. & Lister, A.M. (1987): Late-glacial mammoth skeletons from Condover, Shropshire. *Nature*, **330**, 472–474.

Coope, G.R. & Pennington, W. (1977): The Windermere Interstadial of the Late Devensian. *Philosophical Transactions of the Royal Society, London*, **B280**, 337–339.

Coplen, T.B. (1995): Discontinuance of SMOW and PDB. *Nature*, **375**, 285.

Coplen, T.B., Winograd, I.J., Landwehr, J.M. & Riggs, A.C. (1994): 500,000-year stable isotopic record from Devil's Hole, Nevada. *Science*, **263**, 361–364.

Cornwall, I.W. (1974): *Bones for the Archaeologist*. Dent, London.

Courty, M.A., Goldberg, P. & Macphail, R.I. (1989): *Soils and Micromorphology in Archaeology*. Manuals in Archaeology Series, Cambridge University Press, Cambridge.

Courty, M.A., Macphail, R.I. & Wattez, J. (1991): Soil micromorphological indicators of pastoralism, with special reference to Arene Candide, Finale Ligure, Italy. *Rivista di Studi Liguri*, **A. LVI**, 127–150.

Coxon, P. & O'Callaghan, P. (1987): The distribution and age of pingo remnants in Ireland. In *Periglacial Processes and Landforms in Britain and Ireland* (edited by J. Boardman),

Cambridge University Press, Cambridge, 195–202.

Cracknell, A.P. & Hayes, L.W.B. (1991): *Introduction to Remote Sensing*. Taylor & Francis, London.

Craig, H. (1961): Standard for reporting concentrations of D and ^{18}O in natural waters. *Science*, **133**, 1833–1834.

Cramp, A., Vitaliano C. & Collins, M.B. (1989): Identification and dispersion of the Campanian ash layer (Y-5) in the sediments of the Eastern Mediterranean. *Geomarine Letters*, **9**, 19–25.

Creer, K.M. (1991): Dating of a maar lake sediment sequence covering the Last Glacial Cycle. In *Radiocarbon Dating: Recent Applications and Future Potentia*l (edited by J.J. Lowe), *Quaternary Proceedings*, Quaternary Research Association, Cambridge, **1**, 75–87.

Cremaschi, M., Marchetti, M. & Ravazzi, C. (1994): Geomorphological evidence for land surfaces cleared from forest in the central Po plain (northern Italy) during the Roman period. In *Evaluation of Land Surfaces Cleared from Forests in the Mediterranean Region During the Time of the Roman Empire* (edited by B. Frenzel, L. Reisch & M.M. Weiß), Gustav Fischer Verlag, Stuttgart & New York, 119–132.

Cronberg, G. (1986): Blue-green algae, green algae and chrysophyceae in sediments. In *Handbook of Holocene Palaeoecology and Palaeohydrology* (edited by B.E. Berglund), John Wiley, Chichester & New York, 507–526..

Cronin, S.P., Lamb H.F. & Whittington, R.J. (1993a): Seismic reflection and sonar survey as an aid to the investigation of lake sediment stratigraphy: a case study from upland Wales. In *Geomorphology and Sedimentology of Lakes and Reservoirs* (edited by J. McManus & R.W. Duck), John Wiley, Chichester & New York, 181–203.

Cronin, T.M., Whatley, R., Wood, A., Tsukagoshi, A., Ikeya, N., Brouwers, E.M. & Briggs, W.M. Jr. (1993b): Microfaunal evidence for elevated Pliocene temperatures in the Arctic Ocean. *Paleoceanography*, **8**, 161–173.

Crook, R. (1986): Relative dating of Quaternary deposits based on P-wave velocities in weathered granitic clasts. *Quaternary Research*, **25**, 281–292.

Croot, D.G. (ed.) (1988): *Glaciotectonics*. Balkema, Rotterdam.

Cropper, J.P. (1979): Tree-ring skeleton plotting by computer. *Tree-Ring Bulletin*, **39**, 47–59.

Crosskey, R.W. & Taylor, B.J. (1986): Fossil blackflies from Pleistocene interglacial deposits in Norfolk, England. (Diptera: Simuliidae.) *Systematic Entomology*, **11**, 401–412.

Crowley, T.J. (1993): Use and misuse of the Geologic 'Analogs' Concept. In *Global Changes in the Perspective of the Past* (edited by J.A. Eddy & H. Oeschger), John Wiley, Chichester & New York, 17–27.

Crowley, T.J. & Parkinson, C.L. (1988): Late Pleistocene variations in Antarctic sea ice. II: Effect of interhemispheric deep-ocean heat exchange. *Climate Dynamics*, **3**, 93.

Cuffey, K.M., Alley, R.B., Grootes, P.M. & Anandakrishna, S. (1992): Towards using borehole temperatures to calibrate an isotopic paleothermometer in central Greenland. *Palaeogeography, Palaeoclimatology, Palaeoecology*, **98**, 265–268.

Cullingford, R.A. & Smith, D.E. (1966): Late-glacial shorelines in eastern Fife. *Transactions of the Institute of British Geographers*, **39**, 31–51.

Cullingford, R.A., Caseldine, C.J. & Gotts, P.E. (1980): Early Flandrian land and sea-level changes in Lower Strathearn. *Nature*, **284**, 159–161.

Cullingford, R.A., Caseldine, C.J. & Gotts, P.E. (1989): Evidence of early Flandrian tidal surges in Lower Strathearn, Scotland. *Journal of Quaternary Science*, **4**, 51–60.

Cumming, B.F., Smol, J.P. & Birks, H.J.B. (1991): The relationship between sedimentary chrysophyte scales (Chrysophyceae and Synurophyceae) and limnological characteristics in 25 Norwegian lakes. *Nordic Journal of Botany*, **2**, 231–242.

Cumming, B.F., Smol, J.P. & Birks, H.J.B. (1992): Scaled chrysophytes (Chrysophyceae and Synurophyceae) from Adirondack drainage lakes and their relationship to environmental variables. *Journal of Phycology*, **28**, 162–178.

Cummins, R.H. (1994): Taphonomic processes in modern freshwater molluscan death assemblages: implications for the freshwater fossil record. *Palaeogeography, Palaeoclimatology, Palaeoecology*, **108**, 55–73.

Curran, P.J. (1985): *Principles of Remote Sensing*. John Wiley, Chichester & New York.

Currant, A. (1989): The Quaternary origins of the modern British mammalian fauna. *Biological Journal of the Linnaean Society*, **38**, 23–30.

Cushing, E.J. (1967a): Late Wisconsin pollen stratigraphy and the glacial sequence in Minnesota. In *Quaternary Palaeoecology* (edited by E.J. Cushing & H.E. Wright), Yale University Press, New Haven, 59–88.

Cushing, E.J. (1967b): Evidence for differential pollen preservation in late Quaternary sediments in Minnesota. *Review Palaeobotany Palynology*, **4**, 87–101.

Cushing, E.J. & Wright, H.E. (eds) (1967): *Quaternary Paleoecology*. Yale University Press, New Haven.

Cwynar, L.C. & Levesque, A. (1995): Chironomid evidence for Late-Glacial climatic reversal in Maine. *Quaternary Research*, **43**, 405–413.

Cwynar, L.C., Levesque, A.J., Mayle, F.E. & Walker, I. (1994): Wisconsinan Late-glacial environmental change in New Brunswick: a regional synthesis. *Journal of Quaternary Science*, **9**, 161–164.

Dahl, S.O. & Nesje, A. (1992): Palaeoclimatic implications based on equilibrium-line altitude depressions of reconstructed Younger Dryas and Holocene cirque glaciers in inner Nordfjord, western Norway. *Palaeogeography, Palaeoclimatology, Palaeoecology*, **94**, 87–97.

Dahms, D.E. (1994): Mid-Holocene erosion of soil catenas on moraines near the type Pinedale Till, Wind River Range, Wyoming. *Quaternary Research*, **42**, 41–48.

Damon, P.E., Donahue, D.J. *et al.* (1989): Radiocarbon dating of the Shroud of Turin. *Nature*, **337**, 611–615.

Damon, P.E., Long, A. & Wallick, E.J. (1972): Dendrochronological calibration of the ^{14}C time-scale: causal factors and implications. *Proceedings of the 8th International Conference on Radiocarbon Dating*, Royal Society of New Zealand, Wellington, Volume 1, A28–A43.

Dansgaard, W. & Oeschger, H. (1989): Past environmental long-term records from the Arctic. In *The Environmental Record in Glaciers and Ice Sheets* (edited by H. Oeschger & C.C. Langway, Jr.), John Wiley, Chichester & New York, 287–317.

Dansgaard, W. & Tauber, H. (1969): Glacier oxygen-18 content and Pleistocene ocean temperatures. *Science*, **166**, 499–502.

Dansgaard, W., Johnsen, S.J., Moller, J. & Langway, C.C. Jr. (1969): One thousand centuries of climatic record from Camp Century on the Greenland ice sheet. *Science*, **166**, 377–381.

Dansgaard, W. *et al.* (1971): Climatic record revealed by the Camp Century ice core. In *The Late Cenozoic Glacial Ages* (edited by K.K. Turekian), Yale University Press, New Haven, 37–56.

Dansgaard, W., Johnsen, S.J., Reeh, N., Gundestrup, N., Clausen, H.B. & Hammer, C.U. (1975): Climatic change, Norsemen and modern man. *Nature*, **255**, 24–28.

Dansgaard, W., Clausen, H.B., Gundestrup, N., Hammer, C.U., Johnsen, S.J., Kristinsdottir, P.M. & Reeh, N. (1982): A new Greenland deep ice core. *Science*, **218**, 1273–1277.

Dansgaard, W., Johnsen, S.J., Clausen, H.B., Dahl-Jensen, D., Gundestrup, N., Hammer, C.U. & Oeschger, H. (1984): North Atlantic oscillations revealed by deep Greenland ice cores. *Geophysical Monographs*, **29**, 288–298.

Dansgaard, W., White, J.W.C. & Johnsen, S.J. (1989): The abrupt termination of the Younger Dryas climatic event. *Nature*, **33**, 532–534.

Dansgaard, W., Johnsen, S.J., Clausen, H.B., Dahl-Jensen, D., Gundestrup, N.S., Hammer, C.U., Hvidberg, C.S., Steffensen, J.P., Sveinbjörnsdottir, A.E., Jouzel, J. & Bond, G. (1993): Evidence for general instability of past climate from a 250–kyr ice-core record. *Nature*, **364**, 218–220.

D'Arrigo, R.D. & Jacoby, G.C. (1991): A 1000-year record of winter precipitation from northwestern New Mexico, USA: a reconstruction from tree-rings and its relation to El Niño and the Southern Oscillation. *The Holocene*, **1**, 95–101.

Davies, G.L. (1968): *The Earth in Decay*. MacDonald, London.

Davies, H.C., Dobson, M.R. & Whittington, R.J. (1984): A revised seismic stratigraphy for Quaternary deposits on the inner continental shelf west of Scotland between 55°30'N and 57°30' N. *Boreas*, **13**, 49–66.

Davies, K.H. (1983): Amino acid analysis of Pleistocene marine molluscs from the Gower Peninsula. *Nature*, **302**, 137–139.

Davies, K.H. & Keen, D.H. (1985): The age of Pleistocene marine deposits at Portland, Dorset. *Proceedings of the Geologists' Association*, **96**, 217–225.

Davis, M.B. (1976): Erosion rates and land-use history in southern Michigan. *Environmental Conservation*, **3**, 139–148.

Davis, M.B. (1981): Outbreaks of forest pathogens in Quaternary history. *Proceedings of VI International Conference on Palynology, Lucknow, India*, **3**, 216–227.

Davis, O.K. (ed.) (1990): Palaeoenvironments of arid lands. *Palaeogeography, Palaeoclimatology, Palaeoecology*, **74**, 187–388.

Davis, P.T. (1988): Holocene glacier fluctuations in the American Cordillera. *Quaternary Science Reviews*, **7**, 129–158.

Davis, P.T. & Osborn, G. (eds) (1988): Holocene glacier fluctuations. *Quaternary Science Reviews*, **7**, 113–242.

Davis, S.J.M. (1987): *The Archaeology of Animals*. Batsford, London.

Dawson, A.G. (1992): *Ice Age Earth: Late Quaternary Geology and Climate*. Routledge, London.

Dawson, A.G., Long, D. & Smith, D.E. (1988): The Storegga Slides: evidence from eastern Scotland for a possible tsunami. *Marine Geology*, **82**, 271–276.

Dawson, M.R. & Gardiner, V. (1987): River terraces: the general model and a palaeohydrological and sedimentological interpretation of the terraces of the lower Severn. In *Palaeohydrology in Practice* (edited by K.J. Gregory, J. Lewin & J.B. Thornes), John Wiley, Chichester & New York, 269–305.

De Angelis, M., Barkov, N.I. & Petroy, V.N. (1987): Aerosol concentrations over the last climatic cycle (160 kyr) from an Antarctic ice core. *Nature*, **325**, 318–321.

De Deckker, P. (1988a): Biological and sedimentary facies of Australian salt lakes. *Palaeogeography, Palaeoclimatology, Palaeoecology*, **62**, 237–270.

De Deckker, P. (1988b): An account of the techniques using ostracodes in palaeolimnology in Australia. *Palaeogeography, Palaeoclimatology, Palaeoecology*, **62**, 463–475.

De Deckker, P., Colin, J.P. & Peypouquet, J.P. (eds) (1988): *Ostracoda in the Earth Sciences*. Elsevier, Amsterdam.

De Gans, W. (1988): Pingo scars and their identifications. In *Advances in Periglacial Geomorphology* (edited by M.J. Clark), John Wiley, Chichester & New York, 299–322.

De Geer, G. (1912): A geochronology of the last 12,000 years. *XIth International Geological Congress, Stockholm*, **1**, 241–253.

De Geer, G. (1940): *Geochronologia Suecica Principles*. Kungliga Svenska Vetenskapliga Akademica, Handlingen, 3:18:6, 367 pp.

De Jong, J. (1988): Climatic variability during the past three million years, as indicated by vegetational evolution in northwest Europe and with emphasis on data from The Netherlands. *Philosophical Transactions of the Royal Society, London*, **B318**, 603–617.

De Rouffignac, C., Coope, G.R., Keen, D.H., Lister, A.M., Maddy, D., Robinson, J.E. & Walker, M.J.C. (1995): Pleistocene interglacial deposits at Upper Strensham, Worcestershire, England. *Journal of Quaternary Science*, **10**, 15–32.

De Vernal, A., Causse, C., Hillaire-Marcel, C., Mott, R.J. & Occhietti, S. (1986): Palynostratigraphy and Th/U ages of upper Pleistocene interglacial and interstadial deposits on Cape Breton Island, eastern Canada. *Geology*, **14**, 554–557.

De Vries, H. (1958): Variation in concentration of radiocarbon with time and location on earth. *Koninkijk Nederlandse Akademie von Wetenschappen, Amsterdam, Proc.*, **B61**, 94–102.

Deacon, M. (1973): The voyage of HMS Challenger. In *Oceanography: Contemporary Readings in Ocean Sciences* (edited by R.G. Pirie), Oxford University Press, London, 24–44.

Dearing, J. (1986): Core correlation and total sediment influx. In *Handbook of Holocene Palaeoecology and Palaeohydrology* (edited by B.E Berglund), John Wiley, Chichester & New York, 247–270.

Dearing, J.A. & Foster, I.D.L. (1986): Lake sediments and palaeohydrological studies. In *Handbook of Holocene*

Palaeoecology and Palaeohydrology (edited by B.E. Berglund), John Wiley, Chichester & New York, 67–90.

Dearing, J.A., Elner, J.K. & Happey-Wood, C.M. (1981): Recent sediment influx and erosional processes in a Welsh upland lake-catchment based on magnetic susceptibility measurements. *Quaternary Research*, **16**, 356–372.

Debenham, N.C. (1983): Reliability of thermoluminescence dating of stalagmite calcite. *Nature*, **304**, 154–156.

Delcourt, H.R. & Delcourt, P.A. (1991): *Quaternary Ecology: A Palaeoecological Perspective*, Chapman & Hall, London.

Delorme, L.D. (1990): Freshwater Ostracodes. In *Methods in Quaternary Ecology* (edited by B.G. Warner), *Geoscience Canada, Reprint Series* **5**, 93–100.

Demetsopoulos, J.C., Burleigh, R. & Oakley, K.P. (1983): Relative and absolute dating of the human skeleton from Galley Hill, Kent. *Journal of Archaeological Science*, **10**, 12–134.

Denton, G.H. & Hughes, T.J. (eds) (1981): *The Last Great Ice Sheets*. John Wiley, New York

Derbyshire, E. (ed.) (1995): Wind blown sediments in the Quaternary record. *Quaternary Proceedings*, **4**, John Wiley, Chichester & New York.

Derbyshire, E., Billard, A., Van-Vliet Lanoë, B., Lautridou, J.-P. & Cremaschi, M. (1988): Loess and palaeoenvironment: some results of a European joint programme of research. *Journal of Quaternary Science*, **3**, 147–169.

Derbyshire, E., Keen, D., Kemp, R.A., Rolph, T.A., Shaw, J. & Meng, X.M. (1995): Loess–palaeosol sequences as recorders of palaeoclimatic variations during the last glacial–interglacial cycle: some problems of correlation in North–Central China. *Quaternary Proceedings*, **4**, 7–18.

Devine, J.D., Sigurdsson, H., Davis, A.N. & Self, S. (1984): Estimates of sulfur and chlorine yield to the atmosphere from volcanic eruptions and potential climatic effects. *Journal of Geophysical Research*, **89**, 6309–6325.

Devoy, R.J.N. (1979): Flandrian sea-level changes and vegetational history of the lower Thames Estuary. *Philosophical Transactions of the Royal Society, London*, **B285**, 355–407.

Devoy, R.J. N. (1985): The problems of a Late Quaternary land bridge between Britain and Ireland. *Quaternary Science Reviews*, **4**, 43–58.

Devoy R.J.N. (ed.) (1987a): Introduction: first principles and the scope of sea-surface studies. In *Sea Surface Studies* (edited by R.J.N. Devoy), Croom Helm, London, 1–30.

Devoy, R.J.N. (ed.) (1987b): *Sea Surface Studies*. Croom Helm, London.

Devoy, R.J.N. (1987c): Hydrocarbon exploration and biostratigraphy: the application of sea-level studies. In *Sea Surface Studies* (edited by R.J.N. Devoy), Croom Helm, London, 531–568.

Devoy, R.J.N. (1987d): Sea-level changes during the Holocene: The North Atlantic. In *Sea Surface Studies* (edited by R.J.N. Devoy), Croom Helm, London, 294–347.

Dickinson, W. (1975): Recurrence surfaces in Rusland Moss, Cumbria (formerly north Lancashire). *Journal of Ecology*, **63**, 913–935.

Dickson, J.H. (1986): Bryophyte analysis. In *Handbook of Palaeoecology and Palaeohydrology* (edited by B.E. Berglund), John Wiley, Chichester & New York, 627–644.

Digerfeldt, G. (1975): Post-glacial water-level changes in Lake Växjösjön, central southern Sweden. *Geologiska Föreningens i Stockholm Förhandlinger*, **97**, 167–173.

Digerfeldt, G. (1986): Studies on past lake-level fluctuations. In *Handbook of Holocene Palaeoecology and Palaeohydrology* (edited by B.E. Berglund), John Wiley, Chichester & New York, 127–143.

Digerfeldt, G. (1988): Reconstruction and regional correlation of Holocene lake-level fluctuations in Lake Bysjön, South Sweden. *Boreas*, **17**, 237–263.

Dijkstra, T., Derbyshire, E. & Meng, X.M. (1993): Neotectonics and mass movements in the loess of north–central China. *Quaternary Proceedings*, **3**, 93–110.

Dimbleby, G.W. (1985): *The Palynology of Archaeological Sites*. Academic Press, London.

Ding, Z., Liu, T. & Rutter, N. (1993): Pedostratigraphy of Chinese loess deposits and climatic cycles in the last 2.5 Ma. *Catena*, **20**, 73–91.

Ding, Z., Yu, Z., Rutter, N.W. & Liu, T. (1994): Towards an orbital time scale for Chinese loess deposits. *Quaternary Science Reviews*, **13**, 39–70.

Dixit, A.S., Dixit, S.S. & Smol, J.P. (1991): Multivariable environmental inferences based on diatom assemblages from Sudbury (Canada) lakes. *Freshwater Biology*, **26**, 251–266.

Dixit, S.S., Dixit, A.S. & Smol, J.P. (1989): Relationship between chrysophyte assemblages and environmental variables in seventy-two Sudbury lakes as examined by canonical correspondence analysis (CCA). *Canadian Journal of Fisheries and Aquatic Science*, **46**, 1667–1676.

Dixit, S.S., Cumming, B.F., Birks, H.J.B., Smol, J.P., Kingston, J.C., Uatala, A.J., Charles, D.F. & Camburn, K.E. (1993): Diatom assemblages from Adirondack lakes (New York, USA) and the development of inference models for retrospective environmental assessment. *Journal of Palaeolimnology*, **8**, 27–47.

Dixon, J.C. & Abrahams, A.D. (eds) (1992); *Periglacial Geomorphology*. John Wiley, Chichester & New York.

Dodonov, A.E. (1991): Loess of Central Asia. *GeoJournal*, **24**, 185–194.

Dohrenwend, S.G., Wells, S.G. & McFadden, L.D. (1986): Geomorphic and stratigraphic indicators of Neogene–Quaternary climatic change in arid and semi-arid environments. *Geology*, **14**, 263–264.

Dorn, R.I. (1983): Cation-ratio dating: a new rock varnish age-determination technique. *Quaternary Research*, **20**, 49–73.

Dorn, R.I. (1989): Cation-ratio dating of rock varnish. A geographic assessment. *Progress in Physical Geography*, **13**, 559–596.

Dorn, R.I. & Oberlander, T.M. (1982): Rock varnish. *Progress in Physical Geography*, **6**, 317–367.

Dorn, R.J., Turrin, B.D., Jull, A.J.T., Linick, T.W. & Donahue, D.J. (1987): Radiocarbon and cation-ratio ages for rock varnish on Tioga and Tahoe morainal boulders of Pine Creek, eastern Sierra Nevada, California, and their paleoclimatic implications. *Quaternary Research*, **28**, 38–49.

Dowdeswell, J.A. & Scourse, J.D. (eds) (1990): *Glacimarine Environments: Processes and Sediments*. Special Publication of the Geological Society, No. 53, London.

Dowdeswell, J.A. & Sharp, M. (1986): Characterization of pebble fabrics in modern terrestrial glacigenic sediments. *Sedimentology*, **33**, 699–710.

Dowsett, H.J. & Cronin, T.M. (1990): High eustatic sea level during the Middle Pliocene: evidence from the southeastern Atlantic coastal plain. *Geology*, **18**, 435–438.

Doyle, P., Bennett, M.R. & Baxter, A.N. (1994): *The Key to Earth History: An Introduction to Stratigraphy*. John Wiley, Chichester & New York.

Dredge, L.A. (1988): Drift carbonate on the Canadian Shield. II. Carbonate dispersal and ice-flow patterns in northern Manitoba. *Canadian Journal of Earth Sciences*, **25**, 783–787.

Dredge, L.A. & Cowan, W.R. (1989): Quaternary geology of the southwestern Canadian Shield. In *Quaternary Geology of Canada and Greenland* (edited by R.J. Fulton), Geological Survey of Canada, Geology of Canada, No. 1, Ottawa, 214–235.

Dreimanis, A. (1993): Small to medium-sized glacitectonic structures in till and in its substratum and their comparison with mass movement structures. *Quaternary International*, **18**, 69–79.

Dreimanis, A., Livrand, E. & Raukas, A. (1989): Glacially redeposited pollen in tills of southern Ontario, Canada. *Canadian Journal of Earth Sciences*, **26**, 1667–1676.

Drewry, D.J. (1986): *Glacial Geologic Processes*. Edward Arnold, London.

Drozdowski, E. & Fedorowicz, S. (1987): Stratigraphy of Vistulian glaciogenic deposits and corresponding thermoluminescence dates in the lower Vistula region, northern Poland. *Boreas*, **16**, 139–153.

Dubois, A.D. & Ferguson, D.K. (1985): The climatic history of pine in the Cairngorms based on radiocarbon dates and stable isotope analysis, with an account of the events leading up to its colonisation. *Review of Palaeobotany & Palynology*, **46**, 55–80.

Dugmore, A.J. (1989): Icelandic volcanic ash in Scotland. *Scottish Geographical Magazine*, **105**, 168–172.

Dugmore, A.J. & Newton, A.J. (1992): Thin tephra layers in peat revealed by x-radiography. *Journal of Archaeological Science*, **19**, 151–161.

Dumas, B., Gueremy, P., Lhenaff, R. & Raffy, J. (1993): Land uplift, stepped marine terraces and raised shorelines on the Calabrian coast of Messina Strait, Italy. *Earth Surface Processes and Landforms*, **18**, 241–256.

Dumayne, L. & Barber, K.E. (1994): The impact of the Romans on the environment of northern England: pollen data from three sites close to Hadrian's Wall. *The Holocene*, **4**, 165–173.

Dunbar, R.B., Wellington, G.M., Colgan, M.W. & Glynn, P.W. (1994): Eastern Pacific sea surface temperature since 1600 AD: the $\delta^{18}O$ record of climate variability in Galapagos corals. *Paleoceanography*, **9**, 291–316.

Duplessy, J.-C. (1978): Isotope studies. In *Climatic Change* (edited by J.R. Gribbin), Cambridge University Press, Cambridge, 46–67.

Duplessy, J.-C., Delibrias, G., Turon, J.L., Pujol, C. & Duprat, J. (1981): Deglacial warming of the northeastern Atlantic Ocean: correlation with the palaeoclimatic evolution of the European continent. *Palaeogeography, Palaeoclimatology, Palaeoecology*, **35**, 121–144.

Duplessy, J.-C., Shackleton, N.J., Fairbanks, R.G., Labeyrie, L., Oppo, D. & Kallell, N. (1988): Deepwater source variations during the last climatic cycle and their impact on the global deep-water circulation. *Paleoceanography*, **3**, 343–360.

Duplessy, J.-C., Labeyrie, L., Juillet-Leclerc, A. & Duprat, J. (1992): A new method to reconstruct sea surface salinity: application to the North Atlantic Ocean during the Younger Dryas. In *The Last Deglaciation: Absolute and Radiocarbon Chronologies* (edited by E. Bard & W.S. Broecker), *NATO ASI Series* **1, 2,** Springer-Verlag, Berlin, 201–217.

Dupont, L.D. (1993): Vegetation zones in NW Africa during the Brunhes Chron reconstructed from marine palynological data. *Quaternary Science Reviews*, **12**, 189–202.

Dupont, L.M. (1986): Temperature and rainfall variation in the Holocene based on comparative palaeoecology and isotope geology of a hummock and hollow (Bourtangerveen, The Netherlands). *Review of Palaeobotany and Palynology*, **48**, 71–159.

Dupont, L.M. (1989): Palynology of the last 680,000 years of ODP Site 658 (off NW Africa): fluctuations in palaeowind systems. In *Palaeoclimatology and Palaeometeorology: Modern and Past Patterns of Global Atmospheric Transport* (edited by M. Leinen & M. Sarnthein), Kluwer, Dordrecht, 779–794.

Dutton, E.G. & Christy, J.R. (1992): Solar radiative forcing at selected locations and evidence for global lower tropospheric cooling following the eruptions of El Chichón and Pinatubo. *Geophysical Research Letters*, **19**, 2313–2316.

Dyke, A.S. & Prest, V.K. (1987): Late Wisconsin and Holocene history of the Laurentide Ice Sheet. *Géographie Physique et Quaternaire*, **41**, 237–263.

Dymond, J., Suess, E. & Lyle, M. (1992): Barium in deep-sea sediment: a geochemical proxy for palaeoproductivity. *Paleoceanography*, **7**, 163–181.

Eardley, A.J. *et al.* (1973): Lake cycles in the Lake Bonneville Basin, Utah. *Bulletin of the Geological Society of America*, **84**, 211–216.

Easterbrook, D.J. (1986): Stratigraphy and chronology of Quaternary deposits of the Puget Lowland and Olympic Mountains of Washington and the Cascade Mountains of Washington and Oregon. *Quaternary Science Reviews*, **5**, 135–159.

Ebert, J.I. & Hitchcock, R.K. (1978): Ancient Lake MakGadikgadi, Botswana: mapping, measurement and palaeoclimatic significance. In *Palaeoecology of Africa and the Surrounding Islands* (edited by E.M. Van Zinderen Bakker), Balkema, Rotterdam, 47–56.

Eden, D.N. & Furkert, R.J. (eds) (1988): *Loess: Its Distribution, Geology and Soils*. Balkema, Rotterdam.

Edwards, K.J. & MacDonald, G.M. (1991): Holocene palynology: II. Human influence and vegetation change. *Progress in Physical Geography*, **15**, 364–391.

Edwards, K.J. & Rowntree, K.M. (1980): Radiocarbon and palaeoenvironmental evidence for changing rates of erosion at a Flandrian stage site in Scotland. In *Timescales in Geomorphology* (edited by R.A. Cullingford, D.A. Davidson & J. Lewin), Wiley, Chichester, 207–223.

Edwards, K.J., Hirons, K.R. & Newell, P.J. (1991): The palaeoecological and prehistoric context of minerogenic layers in blanket peat: a study from Loch Dee, southwest Scotland. *The Holocene*, **1**, 29–39.

Edwards, L.R., Chen, J.H., Ku T.L. & Wasserburg, G.J. (1987): Precise timing of the last interglacial period from mass-spectrometric determinations of thorium-230 in corals. *Science*, **236**, 1647–1553.

Eglinton, G., Bradshaw, S.A., Rosell, A., Sarnthein, M., Pflaumann, U. & Tiedemann, R. (1992): Molecular record of secular sea surface temperature changes on 100-year timescales for glacial terminations I, II and IV. *Nature*, 356, 423–426.

Ehlers, J. & Wingfield, R. (1991): The extension of the Late Weichselian/Late Devensian ice sheets in the North Sea Basin. *Journal of Quaternary Science*, **6**, 313–326.

Ehlers, J., Gibbard, P.J. & Rose, J. (eds) (1991): *Glacial Deposits in Great Britain and Ireland*. Balkema, Rotterdam.

Ehrlich, P. & Ehrlich, A. (1981): *Extinction*. Ballantine Books, New York.

Einarsson, T. (1986): Tephrochronology. In *Handbook of Holocene Palaeoecology and Palaeohydrology* (edited by B.E. Berglund), John Wiley, Chichester & New York, 329–342.

El-Daoushy, F. (1986): The value of ^{210}Pb in dating Scandinavian aquatic and peat deposits. *Radiocarbon*, **28**, 1031–1040.

Elias, S.A. (1992): Late Quaternary zoogeography of the Chihuahuan Desert insect fauna, based on fossil records from packrat middens. *Journal of Biogeography*, **19**, 285–298.

Elias, S.A. (1994): *Quaternary Insects and Their Environments*. Smithsonian Institution Press, Washington & London.

Elias, S.A. & Toolin, L.J. (1990): Accelerator dating of a mixed assemblage of Late Pleistocene insect fossils from the Lamb Spring site, Colorado. *Quaternary Research*, **33**, 122–126.

Ellis, A. E. (1978): British freshwater bivalve Mollusca. *Linnaean Synopses of the British Fauna, New Series*, **11**, 1–109.

Embleton, C. & King, C.A.M. (1975): *Glacial Geomorphology*. Edward Arnold, London.

Emery, K.O. & Aubrey, D.G. (1991): *Sea Levels, Land Levels and Tide Gauges*. Springer-Verlag, Berlin.

Emiliani, C. (1955): Pleistocene temperatures. *Journal of Geology*, **63**, 538–575.

Engen, S. (1978): *Stochastic Abundance Models*. Chapman & Hall, London, & Wiley, New York.

Engstrom, D.R. & Nelson, S.R. (1991): Paleosalinity from trace metals in fossil ostracodes compared with observational records at Devil's Lake, North Dakota, USA. *Palaeogeography, Palaeoclimatology, Palaeoecology*, **83**, 295–312.

Epstein, S. & Krishnamurthy, R.V. (1990): Environmental information in the isotopic record in trees. *Philosophical Transactions of the Royal Society, London*. **A330**, 427–439.

Erdtman, G. (1969): *Handbook of Palynology*. Munksgaard, Copenhagen.

Ericson, D.B. & Wollin, G. (1968): Pleistocene climates and chronology in deep-sea sediments. *Science*, **162**, 1227–1234.

Erikstad, L. & Sollid, L. (1986): Neoglaciation in South Norway using lichenometric methods. *Norsk Geografisk Tidsskrift*, **40**, 85–105.

Eronen, M. (1983): Late Weichselian and Holocene shore displacement in Finland. In *Shorelines and Isostasy* (edited by D.E. Smith & A.G. Dawson), Academic Press, London, 183–208.

Eronen, M. & Vesajoki, H. (1988): Deglaciation pattern indicated by the ice-marginal formations in Northern Karelia, eastern Finland. *Boreas*, **17**, 317–327.

Eschman, D.F. & Mickelson D.M. (1986): Correlation of Quaternary deposits of the Huron, Lake Michigan and Green Bay lobes in Michigan and Wisconsin. *Quaternary Science Reviews*, **5**, 53–58.

Eugster, H.P. & Hardie, L.A. (1978): Saline lakes. In *Lakes, Chemistry, Geology, Physics* (edited by A. Lerman), Springer-Verlag, New York, 237–294.

Evans, D.J.A. (1990): The last glaciation and relative sea-level history of Northwest Ellesmere Island, Canadian high arctic. *Journal of Quaternary Science*, **5**, 67–82.

Evans, D.J.A., Owen, L.A. & Roberts, D. (1995): Stratigraphy and sedimentology of Devensian Dimlington Stadial glacial deposits, east Yorkshire, England. *Journal of Quaternary Science*, **10**, 241–265.

Evans, J.G. (1972): *Land Snails in Archaeology*. Seminar Press, London.

Evin, J. (1992): Validity of the radiocarbon dates beyond 35,000 years BP. *Palaeogeography, Palaeoclimatology, Palaeoecology*, **90**, 71–78.

Eyles, N. (1983): Modern Icelandic glaciers as depositional models for 'hummocky moraine' in the Scottish highlands. In *Tills and Related Deposits* (edited by E.B. Evenson, C. Schlüchter & J. Rabassa), Balkema, Rotterdam, 47–59.

Eyles, N. (1993): Earth's glacial record and its tectonic setting. *Earth-Science Reviews*, **35**, 1–248.

Eyles, N. & Eyles, C.H. (1992): Glacial depositional systems. In *Facies Models: Response to Sea Level Change* (edited by R.G. Walker & N.P. James), Geological Association of Canada, St John's, Newfoundland, 73–100.

Eyles, N. & McCabe, A.M. (1989): The Late Devensian (<22,000 BP) Irish Sea Basin: the sedimentary record of a collapsed ice sheet margin. *Quaternary Science Reviews*, **8**, 307–351.

Eyles, N. & Miall, A.D. (1984): Glacial facies models. In *Facies Models* (edited by R.G. Walker), Geological Association of Canada, Toronto, 15–38.

Eyles, N., Sladen, H.A. & Gilroy, S. (1982): A depositional model for stratigraphic complexes and facies superimposition in lodgement till. *Boreas*, **11**, 317–333.

Eyles, N., Eyles, C.H. & Miall, A.D. (1983): Lithofacies types and vertical profile models: an alternative approach to the description and environmental interpretation of glacial diamict and diamictite sequences. *Sedimentology*, **30**, 393–410.

Eyles, N., McCabe, A.M. & Bowen, D.Q. (1994): The stratigraphic and sedimentological significance of Late Devensian ice sheet surging in Holderness, Yorkshire, U.K. *Quaternary Science Reviews*, **13**, 727–759.

Faegri, K. & Iversen, J. (1989): *Textbook of Pollen Analysis* (4th edition, with K. Krzywinski). John Wiley, Chichester & New York.

Fagan, B.M. (1991): *Ancient North America*. Thames & Hudson, London.

Faïn J. *et al.* (1992): Proceedings of the 6th International Specialist Seminar on Thermoluminescence and Electron Spin Resonance Dating. *Quaternary Science Reviews*, **11**, 1–274.

Fairbanks, R.G. (1989): A 17,000-year glacio-eustatic sea level record: influence of glacial melting rates on the Younger Dryas event and deep-ocean circulation. *Nature*, **342**, 637–642.

Fairbridge, R.W. (1983): Isostasy and eustasy. In *Shorelines and Isostasy* (edited by D.E. Smith & A.G. Dawson), Academic Press, London, 3–28.

Faure, H. & Leroux, M. (1990): Are there solar signals in the African monsoon and rainfall? *Philosophical Transactions of the Royal Society of London*, **A330**, 177.

Feng, Z.-D., Johnson, W.C., Sprowl, D.R. & Lu, Y. (1994): Loess accumulation and soil formation in Central Kansas, United States, during the past 400,000 years. *Earth Surface Processes and Landforms*, **19**, 55–67.

Fenton, M.M., Moran, S.R., Teller, J.T. & Clayton, L. (1983): Quaternary stratigraphy and history in the southern part of the Lake Agassiz Basin. In *Glacial Lake Agassiz* (edited by J.T. Teller & L. Clayton), University of Toronto Press, Geological Association of Canada Special Paper No. 26, 49–74.

Ferguson, C.W. (1970): Dendrochronology of bristlecone pine, *Pinus aristata*: establishment of a 7484-year chronology in the White Mountains of eastern-central California. In *Radiocarbon Variations and Absolute Chronology* (edited by I.U. Olsson), John Wiley, Chichester & New York, 237–259.

Ferguson, C.W. & Graybill, D.A. (1983): Dendrochronology of bristlecone pine: a progress report. *Radiocarbon*, **25**, 287–288.

Fernlund, J.M.R. (1993): The long-singular ridges of the Halland Coastal Moraines, south-western Sweden. *Journal of Quaternary Science*, **8**, 67–78.

Feyling-Hanssen, R.W., Jorgensen, J.A., Knudsen, K.L. & Anderson, A.-L.L. (1971): Late Quaternary foraminifera from Vendsyssel, Denmark and Sandnes, Norway. *Bulletin of the Geology Society of Denmark*, **21**, 67–317.

Fichefet, T., Hovine, S. & Duplessy, J.-C. (1994): A model study of the Atlantic thermohaline circulation during the last glacial maximum. *Nature*, **372**, 252–254.

Field, M.H., Huntley, B. & Müller, H. (1994): Eemian climatic fluctuations observed in a European pollen record. *Nature*, **371**, 779–783.

Fink, J. & Kukla, G.J. (1977): Pleistocene climates in central Europe: at least 17 interglacials after the Olduvai event. *Quaternary Research*, **7**, 363–371.

Firth, C.R., Smith, D.E. & Cullingford, R.A. (1993): Late Devensian and Holocene glacio-isostatic uplift patterns in Scotland. In *Neotectonics: Recent Advances* (edited by L.A. Owen, I. Stewart & C. Vita-Finzi), *Quaternary Proceedings* **3**, Quaternary Research Association, Cambridge, 1–13.

Fisher, D.A., Reeh, N. & Langley, K. (1985): Objective reconstructions of the late-Wisconsin Laurentide Ice Sheet and the significance of deformable beds. *Géographie Physique et Quaternaire*, **39**, 229–238.

FitzPatrick, E.A. (1993): *Soil Microscopy and Micromorphology*. Wiley, Chichester.

Flint, R.F. (1965): Introduction: historical perspectives. In *The Quaternary of the United States* (edited by H.E. Wright & D.G. Frey), Princeton University Press, Princeton, New Jersey, 3–11.

Flint, R.F. (1971): *Glacial and Quaternary Geology*. John Wiley, New York.

Flohn, H. (1974): Background of a geophysical model of the initiation of the next glaciation, *Quaternary Research*, **4**, 385–404.

Florin, M.-B. (1946): Clypeusfloran i postglaciala fornsjlagerfljder i stra Mellansverige. *Geologiska Föreningens i Stockholm Förhandlingar*, **68**, 429–458.

Foley, J.A., Kutzbach, J.E., Coe, M.T. & Levis, S. (1994): Feedbacks between climate and boreal forest during the Holocene epoch. *Nature*, **371**, 52–54.

Follieri, M., Magri, D. & Sadori, L. (1988): 250,000-year record from Valle di Castiglione (Roma). *Pollen et Spores*, **30**, 329–356.

Follieri, M., Magri, D. & Sadori, L. (1989): Pollen-stratigraphical synthesis from Valle di Castiglione (Roma). *Quaternary International*, **3/4**, 81–84.

Follmer, L.R. (1983): Sangamon and Wisconsinan pedogenesis in the Midwestern United States. In *Late Quaternary Environments of the United States*, Volume 1, *The Late Pleistocene* (edited by H.E. Wright, Jr.), Longman, London, 138–144.

Forbes, E. (1846): On the connexion between the distribution of the existing fauna and flora of the British Isles, and the geological changes which have affected their area, especially during the epoch of the Northern Drift. *Memoir of the Geological Survey of Great Britain*, **1**, 336–432.

Ford, D.C. & Williams, P.W. (1989): *Karst Geomorphology and Hydrology*. Unwin Hyman, London.

Forman, S. L. (1989): Applications and limitations of thermoluminescence to date Quaternary sediments. *Quaternary International*, **1**, 47–59.

Forman, S.L. (1991): Late Pleistocene chronology of loess deposition near Luochuan, China. *Quaternary Research*, **36**, 19–28.

Forman, S.L., Bettis III, E.A., Kemmis, T.J. & Miller, B.B. (1992): Chronologic evidence for multiple periods of loess deposition during the late Pleistocene in the Missouri and Mississippi River Valley, United States. *Palaeogeography, Palaeoclimatology, Palaeoecology*, **93**, 71–83.

Forman, S.L., Smith, R.P., Hackett, W.R., Tullis, J.A. & McDaniel, P.A. (1993): Timing of late Quaternary glaciations in the western United States based on the age of loess on the eastern Snake River Plain, Idaho. *Quaternary Research*, **40**, 30–37.

Foss, P. (1987):The distribution and formation of Irish peatlands. In *The IPCC Guide to Irish Peatlands* (edited by C, O'Connell), Irish Peatland Conservation Council, Ireland, 5–12.

Fowler, A.J., Gillespie. R. & Hedges. R.E.M. (1986): Radiocarbon dating of sediments. *Radiocarbon*, **28**, 441–450.

Francois, R., Bacon, M.P. & Suman, D.O. (1990): Thorium-230 profiling in deep-sea sediments: high resolution records of flux and dissolution of carbonate in the equatorial Atlantic during the last 24,000 years. *Paleoceanography*, **5**, 761–787.

Freden, S.C. & Gordon, F. (1983): Landsat satellites. In *Manual of Remote Sensing*, 2nd edition (edited by R.N. Colwell), American Society of Photogrammetry, Falls Church, Virginia, 517–570.

Fredskild, B. (1973): Studies in the vegetational history of Greenland. *Meddelelser om Grønland*, **198**, 1–245.

Fredskild, B. & Wagner, P. (1974): Pollen and fragments of plant tissue in core samples from the Greenland Ice Cap. *Boreas*, **3**, 105–108.

French, H. & Harry, D.G. (1988): Nature and origin of ground ice, Sandgills Moraine, southwest Banks Island, western Canadian Arctic. *Journal of Quaternary Science*, **3**, 19–30.

French, H.M. (1996): *The Periglacial Environment* 2nd ed. Longman, Harlow.

French, H.M. (1987): Periglacial processes and landforms in the western Canadian Arctic. In *Periglacial Processes and Landforms in Britain and Ireland* (edited by J. Boardman), Cambridge University Press, Cambridge, 27–43.

French, H.M. & Koster, A.E. (eds) (1988): Periglacial phenomena: ancient and modern. *Journal of Quaternary Science*, **3**, 1–110.

Frenzel, B., Matthews, J.A. & Gläser, B. (eds) (1993): *Solifluction and Climatic Variation in the Holocene*. Gustav Fischer Verlag, Stuttgart.

Frey, D.G. (1986): Cladocera analysis. In *Handbook of Holocene Palaeoecology and Palaeohydrology* (edited by B.E. Berglund), John Wiley, Chichester & New York, 667–692.

Friedman, I. (1968): Hydration rate dates rhyolite flows. *Science*, **159**, 878–880.

Fritts, H.C. (1976): *Tree Rings and Climate*. Academic Press, New York.

Fritts, H.C. (1982): An overview of dendroclimatic techniques, procedures and prospects. In *Climate and Tree Rings* (edited by M.K. Hughes, P.M. Kelly, J.R. Pilcher & V.C. LaMarche, Jr.), Cambridge University Press, Cambridge, 191–198.

Fritz, S.C. (1989): Lake development and limnological response to prehistoric and historic land use in Diss, Norfolk, UK. *Journal of Ecology*, **77**, 182–202.

Fritz, S.C. (1990): Twentieth century salinity and water-level fluctuations in Devil's Lake, North Dakota: a test of a diatom-based transfer function. *Limnology and Oceanography*, **35**, 1771–1781.

Fritz, S.C. & Carlson, R.E. (1982): Stratigraphic diatom and chemical evidence for acid strip-mine recovery. *Water, Air and Soil Pollution*, **17**, 151–163.

Fritz, S.C., Juggins, S., Battarbee, R.W. & Engstrom, D.R. (1991): Reconstruction of past changes in salinity and climate using a diatom-based transfer function. *Nature*, **352**, 702–704.

Fromm, E. (1985): Chronological calculation of the varve zero in Sweden. *Boreas*, **14**, 123–126.

Fronval, T., Jansen, E., Bloemendal, J. & Johnsen, S. (1995): Oceanic evidence for coherent fluctuations in Fennoscandian and Laurentide ice sheets on millenium timescales. *Nature*, **374**, 443–446.

Fryberger, S.G. (1980): Dune forms and wind regime, Mauretania, West Africa: implications for past climate. In *Palaeoecology of Africa* (edited by E.M. Van Zinderen Bakker & J.A. Coetzee), Balkema, Rotterdam, 79–96.

Fuji, N. (1988): Palaeovegetation and palaeoclimatic changes around Lake Biwa, Japan, during the last *ca.* 3 million years. *Quaternary Science Reviews*, **7**, 21–28.

Fullerton, D.S. (1986): Chronology and correlation of glacial deposits in the Sierra Nevada, California. *Quaternary Science Reviews*, **5**, 161–169.

Fullerton, D.S. & Colton, R.B. (1986): Stratigraphy and correlation of the glacial deposits on the Montana Plains. *Quaternary Science Reviews*, **5**, 69–82.

Fulton, R.J., Fenton, M.M. & Rutter, N.W. (1986): Summary of Quaternary stratigraphy and history, Western Canada. *Quaternary Science Reviews*, **5**, 229–241.

Funnell, B.M. (1995): Global sea-level and the (pen-) insularity of late Cenozoic Britain. In *Island Britain; a Quaternary Perspective* (edited by R.C. Preece), The Geological Society, Bath, 3–14.

Funnell, B.M. & West, R.G. (1977): Preglacial Pleistocene deposits of East Anglia. In *British Quaternary Studies – Recent Advances* (edited by F.W. Shotton), Oxford University Press, Oxford, 247–265.

Gaillard, M.-J. & Lemdahl, G. (1994): Lateglacial insect assemblages from Grand-Marais, south-western Switzerland – climatic implications and comparison with pollen and plant macrofossil data. *Dissertationes Botanicae*, **234**, 287–308.

Gaillard, M.-J., Birks, H.J.B., Emanuelsson, U. & Berglund, B.E. (1992): Modern pollen/land-use relationships as an aid in the reconstruction of past land-uses and cultural landscapes: an example from south Sweden. *Vegetation History and Archaeobotany*, **1**, 3–17.

Gale, S.J. & Hoare, P.J. (1991): *Quaternary Sediments*. Belhaven Press, London.

Gallup, C.D., Edwards, R.L. & Johnson, R.G. (1994): The timing of high sea levels over the past 200,000 years. *Science*, **263**, 796–800.

Gamble, C. (1993): *Timewalkers. The Prehistory of Global Colonisation.* Alan Sutton, Stroud.

Gamble, C. (1994): Time for Boxgrove Man. *Nature*, **369**, 275–276.

Ganeshram, R.S., Pedersen, T.F., Calvert, S.E. & Murray, J.W. (1995): Large changes in oceanic nutrient inventories from glacial to interglacial periods. *Nature*, **376**, 755–758.

Gardiner, V. & Dackombe, R.V. (1983): *Geomorphological Field Manual*. Allen & Unwin, London.

Gascoyne, M. (1992): Palaeoclimate determination from cave calcite deposits. *Quaternary Science Reviews*, **11**, 609–632.

Gascoyne, M., Ford, D.C. & Schwarcz, H.P. (1981): Late Pleistocene chronology and paleoclimate of Vancouver Island determined from cave deposits. *Canadian Journal of Earth Sciences*, **18**, 1643–1652.

Gascoyne, M., Ford, D.C. & Schwarcz, H.P. (1983a): Rates of cave and landform development in the Yorkshire Dales from speleothem age data. *Earth Surface Processes and Landforms*, **8**, 557–568.

Gascoyne, M., Schwarcz, H.P. & Ford, D.C. (1983b): Uranium-series ages of speleothem from northwest England correlation with Quaternary climate. *Philosophical Transactions of the Royal Society, London*, **B301**, 143–164.

Gasse, F. (1987): Diatoms for reconstructing palaeoenvironments and palaeohydrology in tropical semi-arid zones. *Hydrobiologia*, **154**, 127–163.

Gasse, F. & Tekaia, F. (1983): Transfer functions for estimating palaeoecological conditions (pH) from East African diatoms. *Hydrobiologia*, **103**, 85–90.

Gasse, F. *et al.* (1987): Biological remains, geochemistry and stable isotopes for the reconstruction of environmental and hydrological changes in Holocene lakes from North Sahara. *Palaeogeography, Palaeoclimatology, Palaeoecology*, **60**, 1–46.

Gasse, F., Tehet, R., Durand, A., Gilbert, E. & Fontes, J.-Ch. (1990): The arid–humid transition in the Sahara and the Sahel during the last deglaciation. *Nature*, **346**, 141–146.

Gaylord, D.R. (1990): Holocene palaeoclimatic fluctuations revealed from dune and interdune strata in Wyoming. *Journal of Arid Environments*, **18**, 123–138.

Gee, H. (1993): The distinction between postcranial bones of *Bos primigenius* Bojanus, 1827 and *Bison priscus* Bojanus, 1827 from the British Pleistocene and the taxonomic status of *Bos* and *Bison*. *Journal of Quaternary Science*, **8**, 79–92.

Gellatly, A.F. (1982): The use of lichenometry as a relative-age dating method with specific reference to Mount Cook National Park, New Zealand. *New Zealand Journal of Botany*, **20**, 343–353.

Gellatly, A.F., Chinn, T.J.H. & Röthlisberger, F. (1988): Holocene glacier variations in New Zealand: a review. *Quaternary Science Reviews*, **7**, 227–242.

Gemmell, J.C., Smart, D. & Sugden, D.E. (1986): Striae and former ice-flow directions in Snowdonia, North Wales. *Geographical Journal*, **152**, 19–29.

Genthon, C., Barnola, J.M., Raynaud, D., Lorius, C., Jouzel, J., Barkov, N.I., Korotkevich, Y.S. & Kotlyakov, V.M. (1987): Vostok ice core: climatic response to CO_2 and orbital forcing changes over the last climatic cycle. *Nature*, **329**, 414–418.

Gentner, W., Glass, B.P., Storzer, D. & Wagner, G.A. (1970): Fission track ages and ages of deep-sea microtektites. *Science*, **168**, 359–361

Gerrard, A.J. (1990): Soil variations on hillslopes in humid temperate climates. In *Soils and Landscape Evolution* (edited by P.L.K. Knuepfer & L.D. McFadden), Elsevier, Amsterdam, 225–244.

Geyh, M.A. (1990): ^{14}C dating of loess. *Quaternary International*, **7/8**, 115–118.

Geyh, M.A., Benzier, J.H. & Roeschmann, G. (1971): Problems of dating Pleistocene and Holocene soils by radiometric methods. In *Palaeopedology* (edited by D.A. Yaalon), International Society of Soil Science and Israel University Press, Jerusalem, 63–75.

Geyh, M.A., Roeschmann, G., Wijmastra, T.A. & Middeldorp, A.A. (1983): The unreliability of ^{14}C dates obtained from buried sandy podzols. *Radiocarbon*, **25**, 409–416.

Ghenea, C. & Mihailescu, N. (1991): Palaeogeography of the lower Danube Valley and the Danube Delta during the last 15,000 years. In *Temperate Palaeohydrology* (edited by L. Starkel, K.J. Gregory & J.B. Thornes), John Wiley, Chichester & New York, 343–364.

Giardino, J.R. & Vitek, J.D. (1988): The significance of rock glaciers in the glacial–periglacial landscape continuum. *Journal of Quaternary Science*, **3**, 97–103.

Giardino, J.R., Shroder, J.F. & Vitek, J.D. (eds) (1987): *Rock Glaciers*. Allen & Unwin, London.

Gibbard, P.L. (1985): *The Pleistocene History of the Middle Thames Valley*. Cambridge University Press, Cambridge.

Gibbard, P.L. (1989): The geomorphology of a part of the Middle Thames forty years on: a reappraisal of the work of F. Kenneth Hare. *Proceedings of the Geologists' Association*, **100**, 481–504.

Gibbard, P.L. (1994): *Pleistocene History of the Lower Thames Valley*. Cambridge University Press, Cambridge.

Gibbard, P.L. (1995): Palaeogeographical evolution of the Lower Thames. In *The Quaternary of the Lower Reaches of the Thames: Field Guide* (edited by D.R. Bridgland, P.A. Allen & B.A. Haggart), Quaternary Research Association, London, 5–34.

Gibbard, P.L. & Allen, L.G. (1994): Drainange evolution in south and east England during the Pleistocene. *Terra Nova*, **6**, 444–452.

Gibbard, P.L., Coope, G.R., Hall, A.R., Preece, R.C. & Robinson, J.E. (1982): Middle Devensian deposits beneath the 'Upper Floodplain' terrace of the River Thames at Kempton Park, Sunbury, Surrey, England. *Proceedings of the Geologists' Association*, **93**, 275–289.

Gibbard, P.L., Wintle, A.G. & Catt, J.A. (1987): Age and origin of clayey silt 'brickearth' in west London, England. *Journal of Quaternary Science*, **2**, 3–9.

Gibbard, P.L., West, R.G., Zagwijn, W.H., Balson, P.S., Burger, A.W., Funnell, B.M., Jeffery, D.H., de Jong, J., van Kolfschoten, T., Lister, A.M., Meier, T., Norton, P.E.P., Preece, R.C., Rose, J., Stuart, A.J., Whiteman, C.A. & Zalasiewicz, J. (1991): Early and Middle Pleistocene correlations in the Southern North Sea Basin. *Quaternary Science Reviews*, **10**, 23–52.

Gilbert, R. & Church, M. (1983): Contemporary sedimentary environments on Baffin Island, N.W.T., Canada: reconnaissance of lakes on the Cumberland Peninsula. *Arctic and Alpine Research*, **15**, 321–332.

Gillot, P.Y., Chiesa, S., Pasquare, G. & Vezzoli, L. (1982): $<$30,000–yr K/Ar dating of the volcano-tectonic horst of the Isle of Ischia, Gulf of Naples. *Nature*, **299**, 242–244.

Gleadow, J.W. (1980): Fission track age of the KBS Tuff and associated hominid remains in northern Kenya. *Nature*, **284**, 225–230.

Godwin, H. (1962): Half-life of radiocarbon. *Nature*, **195**, 944.

Godwin, H. (1975): *History of the British Flora*, 2nd edition. Cambridge University Press, Cambridge.

Godwin, H. (1981): *The Archives of the Peat Bogs*. Cambridge University Press, Cambridge.

Goede, A. & Harmon, R.S. (1983): Radiometric dating of Tasmanian speleothems – evidence of cave sediments and climatic change. *Journal of the Geological Society of Australia*, **30**, 89–100.

Goede, A., Harmon, R.S., Atkinson, T.C. & Rowe, P.J. (1990): Pleistocene climatic change in Southern Australia and its effect on speleothem deposition in some Nullarbor caves. *Journal of Quaternary Science*, **5**, 29–38.

Goh, K.M. & Molloy, B.J.P. (1979): Contaminants in charcoals used for radiocarbon dating. *New Zealand Journal of Soil Science*, **27**, 89–100.

Goldthwait, R.P. (1971): Introduction to till, today. In *Till: a Symposium* (edited by R.P. Goldthwait), Ohio State University Press, 3–26.

Goodfriend, G.A. (1992): The use of land snail shells in palaeoenvironmental reconstruction. *Quaternary Science Reviews*, **11**, 665–685.

Goodfriend, G.A. & Mitterer, R.M. (1993): A 45,000 year record of a tropical lowland biota: the land snail fauna from cave sediments at Coco Ree, Jamaica. *Geological Society of America Bulletin*, **105**, 18–29.

Gordon, A.D. & Birks, H.J.B. (1972): Numerical methods in Quaternary palaeoecology. I. Zonation of pollen diagrams. *New Phytologist*, **71**, 961–979.

Gordon, A.D. & Birks, H.J.B. (1985): *Numerical Methods in Quaternary Pollen Analysis*. Academic Press, London & New York.

Gordon, D., Smart, P.L., Ford, D.C., Andrews, J.N., Atkinson, T.C., Rowe, P.J. & Christopher, N.S.T. (1989): Dating Late Pleistocene interglacial and interstadial periods in the United Kingdom from speleothem growth frequency. *Quaternary Research*, **31**, 14–26.

Gordon, J.E. (1981): Ice-scoured topography and its relationships to bedrock structure and ice movement in parts of Northern Scotland and West Greenland. *Geografiska Annaler*, **63A**, 55–65.

Gordon, J.E. (1993): Agassiz Rock. In *Quaternary of Scotland* (edited by J.E. Gordon & D.G. Sutherland), Geological Conservation Review Series No. 3, Chapman & Hall, London, 565–568.

Gordon, J.E. & Sutherland, D.G. (1993): The Quaternary in Scotland. In *Quaternary of Scotland* (edited by J.E. Gordon & D.G. Sutherland), Geological Conservation Review Series No. 3, Chapman & Hall, London, 11–47.

Gordon, J.E., Whalley, W.B., Gellatly, A.F. & Vere, D.M. (1992): The formation of glacial flutes: assessment of models with evidence from Lyngsdalen, North Norway. *Quaternary Science Reviews*, **11**, 709–731.

Goslar, T., Kuc, T., Ralska-Jasiewiczowa, M., Różánski, K., Arnold, M., Bard, E., van Geel, B., Pazdur, M.F., Szeroczyńska K., Wicik, B., Więckowski, K. & Walanus, A. (1993): High-resolution lacustrine record of the Late Glacial/Holocene transition in central Europe. *Quaternary Science Reviews*, **12**, 287–294.

Goudie, A.S. (1973): *Duricrusts in Tropical and Subtropical Landscapes*. Clarendon Press, Oxford.

Goudie, A.S. (1983): Calcrete. In *Chemical Sediments and Geomorphology* (edited by A.S. Goudie & K. Pye), Academic Press, London & New York, 93–131.

Goudie, A.S. (ed.) (1990): *Geomorphological Techniques*, 2nd edition. Unwin Hyman, London & Boston.

Goudie, A.S. (1992): *Environmental Change*, 3rd edition. Oxford University Press, Oxford.

Goudie, A.S. (1994): Dryland degradation. In *The Changing Global Environment* (edited by N. Roberts), Blackwell, Oxford, 351–368.

Gove, H.E. (1987): Tandem-accelerator mass-spectrometry measurements of ^{36}Cl, ^{129}I and osmium isotopes in diverse natural samples. *Philosophical Transactions of the Royal Society, London*, **A323**, 103–119.

Gray, J. (ed.) (1988): Aspects of freshwater palaeoecology and biogeography. *Palaeogeography, Palaeoclimatology, Palaeoecology*, special issue, **62**, 1–623.

Gray, J.M. (1981): Large-scale geomorphological field mapping: teaching the first stage. *Journal of Geography in Higher Education*, **5**, 37–44.

Gray, J.M. (1982): The last glaciers (Loch Lomond Advance) in Snowdonia, N. Wales. *Geological Journal*, **17**, 111–133.

Gray, J.M. (1991): Glaciofluvial landforms. In *Glacial Deposits in Great Britain and Ireland* (edited by J. Ehlers, P.L. Gibbard & J. Rose), Balkema, Rotterdam, 443–454.

Gray, J.M. (1992): Loch Etive kame and outwash terrace system. In *The South-West Scottish Highlands. Field Guide* (edited by M.J.C. Walker, J.M. Gray & J.J. Lowe), Quaternary Research Association, Cambridge, 35–38.

Gray, J.M. (1995): Influence of Southern Upland ice on glacio-isostatic rebound in Scotland: the Main Rock Platform in the Firth of Clyde. *Boreas*, **24**, 30–36.

Gray, J.M. & Coxon, P. (1991): The Loch Lomond Stadial glaciation in Britain and Ireland. In *Glacial Deposits in Great Britain and Ireland* (edited by J. Ehlers, P.L. Gibbard & J. Rose), Balkema, Rotterdam, 89–105.

Gray, J.M. & Lowe, J.J. (1982): Problems in the interpretation of small-scale erosional forms on glaciated bedrock surfaces: examples from Snowdonia, North Wales. *Proceedings of the Geologists' Association*, **93**, 403–414.

Green, H.S. (1984): *Pontnewydd Cave*. National Museum of Wales, Cardiff.

Grichuk, V.P. (1969): An attempt to reconstruct certain elements of the climate of the Northern Hemisphere in the Atlantic Period of the Holocene. In *Proceedings of VIII INQUA Congress, Paris* (edited by K. Golotsen), Izd-vo Nauka, Moscow, 41–57.

Grimm, E.C., Jacobson, G.L., Watts, W.A., Hansen, B.C.S. & Maasch, K.A. (1993): A 50,000-year record of climatic oscillations from Florida and its temporal correlation with the Heinrich Events. *Science*, **261**, 198–200.

GRIP Greenland Ice-core Project Members (1993): Climate instability during the last interglacial period recorded in the GRIP ice core. *Nature*, **364**, 203–207.

Grootes, P.M. (1978): Carbon-14 timescale extended: comparison of chronologies. *Science*, **200**, 11–15.

Grootes, P.M. (1984): Radioactive isotopes in the Holocene. In *Late-Quaternary Environments of the United States*, Volume 2, *The Holocene* (edited by H.E.Wright, Jr.), Longman, London, 86–105.

Grootes, P.M., Stuiver, M., White, J.W.C., Johnsen, S.J. & Jouzel, J. (1993): Comparison of oxygen isotope records from the GISP2 and GRIP Greenland ice cores. *Nature*, **366**, 552–554.

Grosjean, M. (1994): Paleohydrology of the Laguna Lejia (north

Chilean Altiplano) and climate implications for late-glacial times. *Palaeogeography, Palaeoclimatology, Palaeoecology*, **109**, 89–100.

Grossman, E.L. (1984): Stable isotope fractionation in live benthic foraminifera from the southern California borderland. *Palaeogeography, Palaeoclimatology, Palaeoecology*, **47**, 301–327.

Grousset, F.E., Rognon, P., Coude-Gaussen, G. & Pedemay, P. (1992): Origins of peri-Saharan dust deposits traced by their Nd and Sr isotopic composition. *Palaeogeography, Palaeoclimatology, Palaeoecology*, **93**, 203–212.

Grousset, F.E., Labeyrie, J.A., Sinko, M., Cremer, M., Bond, G., Duprat, E., Cortijo, E. & Huon, S. (1993): Patterns of ice-rafted detritus in the glacial North Atlantic (40–55°N). *Paleoceanography*, **8**, 175–192.

Grove, A.T. & Warren, A. (1968): Quaternary landforms and climate on the south side of the Sahara. *Geographical Journal*, **134**, 194–208.

Grove, J.M. (1988): *The Little Ice Age*. Methuen, London.

Grube, F., Christensen, S., Vollmer, T., Duphorn, K., Klostermann, J. & Menke, B. (1986): Glaciations in North-West Germany. *Quaternary Science Reviews*, **5**, 347–358.

Grün, R. (1989): Electron spin resonance (ESR) dating. *Quaternary International*, **1**, 65–109.

Grün, R., Radtke, U. & Omura, A. (1992): ESR and U-series analyses on corals from Huon Peninsula, New Guinea. *Quaternary Science Reviews*, **11**, 197–202.

Guiot, J. (1990): Methodology of the last climatic cycle reconstruction in France from pollen data. *Palaeogeography, Palaeoclimatology, Palaeoecology*, **80**, 46–69.

Guiot, J., Pons, A., Beaulieu, J.-L. de & Reille, M. (1989): A 140,000-year continental climatic reconstruction from two European pollen records. *Nature*, **338**, 309–313.

Guiot, J., Beaulieu, J.-L. de, Reille, M. & Pons, A. (1992): Calibration of the climatic signal in a new pollen sequence from Grande Pile. *Climate Dynamics*, **6**, 259–264.

Guiot, J., Beaulieu, J.-L. de, Cheddadi, R., David, F., Ponel, P. & Reille, M. (1993): The climate in western Europe during the last glacial/interglacial cycle derived from pollen and insect remains. *Palaeogeography, Palaeoclimatology, Palaeoecology*, **103**, 73–93.

Gulliksen, S. & Scott, M. (1995): Report of the TIRI Workshop, Saturday 13 August, 1994. *Radiocarbon*, **37**, 820–821.

Gupta, S.K., Sharma, P., Juyal, N. & Agrawal, D.P. (1991): Loess–palaeosol sequence in Kashmir: correlation of mineral magnetic stratigraphy with the marine palaeoclimatic record. *Journal of Quaternary Science*, **6**, 3–12.

Haake, F.-W. & Pflaumann, U. (1989): Late Pleistocene foraminiferal stratigraphy on the Vring Plateau, Norwegian Sea. *Boreas*, **18**, 343–356.

Haddy, A. & Hanson, A. (1982): Nitrogen and fluorine dating of Moundville skeletal samples. *Archaeometry*, **24**, 37–44.

Haeberli, W. (1985): Creep of mountain permafrost: internal structure and flow of alpine rock glaciers. *Mitteilungen der Versuchsanstalt für Wasserbau, Hydrologie und Glaziologie*, **77**, 142 pp.

Hailwood, E.A. (1989): *Magnetostratigraphy*. Geological Society of London Special Report No. 19.

Haines-Young, R. (1994): Remote sensing of environmental change. In *The Changing Global Environment* (edited by N. Roberts), Blackwell, Oxford, 22–43.

Håkansson, L. & Jansson, M. (1983): *Principles of Lake Sedimentology*, Springer-Verlag, Berlin.

Håkansson, S. (1983): A reservoir age for the coastal waters of Iceland. *Geologiska Föreningens Stockholm Förhandlingar*, **105**, 64–67.

Haldorsen, S., Jørgensen, P., Rappol, M. & Riezebos, P.A. (1989): Composition and source of the clay-sized fraction of Saalian till in The Netherlands. *Boreas*, **18**, 89–98.

Hall, C.M. & York, D. (1984): The applicability of dating young volcanics. In *Quaternary Dating Methods* (edited by W.C. Mahaney), Elsevier, Amsterdam, 67–74.

Hall, K. (1982): Rapid deglaciation as an initiator of volcanic activity: an hypothesis. *Earth Surface Processes and Landforms*, **7**, 45–51.

Hallberg, G.R. (1986): Pre-Wisconsin glacial stratigraphy of the Central Plains region in Iowa, Nebraska, Kansas and Missouri. *Quaternary Science Reviews*, **5**, 11–16.

Hallberg, G.R. & Kemmis, T.J. (1986): Stratigraphy and correlation of the glacial deposits of the Des Moines and James Lobes and adjacent areas in North Dakota, South Dakota, Minnesota and Iowa. *Quaternary Science Reviews*, **5**, 65–68.

Hambrey, M. (1994): *Glacial Environments*. UCL Press, London.

Hambrey, M. & Alean, J. (1992): *Glaciers*. Cambridge University Press, Cambridge.

Hamilton, T.D. (1986): Correlation of Quaternary glacial deposits in Alaska. *Quaternary Science Reviews*, **5**, 171–180.

Hammarlund, D. & Lemdahl, G. (1994): A Late Weichselian stable isotope stratigraphy compared with biostratigraphical data: a case study from southern Sweden. *Journal of Quaternary Science*, **9**, 13–31.

Hammer, C.U. (1989): Dating by physical and chemical seasonal variations and reference horizons. In *The Environmental Record in Glaciers and Ice Sheets* (edited by F. Oeschger & C.C. Langway, Jr.), Wiley, Chichester, 99–122.

Hammer, C.U., Clausen, H.B. & Dansgaard, W. (1980): Greenland ice-sheet evidence of post-glacial volcanism and its climatic impact. *Nature*, **288**, 230–235.

Hammer, C.U., Clausen, H.B. & Tauber, H. (1986): Ice-core dating of the Pleistocene–Holocene boundary applied to a calibration of the ^{14}C timescale. *Radiocarbon*, **28**, 284–291.

Handler, P. & Andsager, K. (1994): El Niño, volcanism and global climate. *Human Ecology*, **22**, 37–57.

Hann, B.J. (1990): Cladocera. In *Methods in Quaternary Ecology* (edited by B.G. Warner), Geoscience Canada, Reprint Series No. 5, 81–91.

Hann, B.J. & Warner, B.G. (1987): Late Quaternary Cladocera from coastal British Columbia, Canada: a record of climatic or limnologic change? *Archiv für Hydrobiologie*, **110**, 161–177.

Hansen, J., Lacis, A., Rind, D., Russell, G., Stone, P., Fung, I., Ruedy, R. & Lerner, J. (1984): Climate sensitivity: analysis of feedback mechanisms. In *Climate Processes and Climatic Sensitivity* (edited by J.E. Hansen & T. Takahashi), *Geophysical Monogram Series*, **29**, 130–163.

Haq, B.U. (1993): Deep-sea response to eustatic change and significance of gas hydrates for continental margin stratigraphy. *Special Publications of the International Association of Sedimentologists*, **18**, 93–106.

Haq, B.U., Hardenbol, J. & Vail, P.R. (1987): Chronology of fluctuating sea levels since the Triassic. *Science*, **235**, 1156–1167.

Harden, J.W. & Taylor, E.M. (1983): A quantitative comparison of soil development in four climatic regimes. *Quaternary Research*, **20**, 342–359.

Hare, F.K. (1947): The geomorphology of parts of the Middle Thames. *Proceedings of the Geologists' Association*, **58**, 294–339.

Hare, P.E., Hoering, T.C. & King, K. Jr. (eds) (1980): *Biogeochemistry of Amino Acids*. John Wiley, Chichester & New York.

Harkness, D.D. (1975): The role of the archaeologist in C-14 measurement. In *Radiocarbon: Calibration and Prehistory* (edited by T. Watkins), Edinburgh University Press, Edinburgh, 128–135

Harkness, D.D. (1979): Radiocarbon dates from Antarctica. *British Antarctic Survey Bulletin*, **47**, 43–59.

Harkness, D.D. (1983): The extent of natural ^{14}C deficiency in the coastal environment of the United Kingdom. *PACT*, **8**, 351–364.

Harland, W.B., Armstrong, R.L., Cox, A.V., Craig, L.E., Smith, A.G. & Smith, D.G. (1990): *A Geological Time Scale 1989*. Cambridge University Press, Cambridge.

Harmon, R.S., Schwarcz, H.P. & Ford, D.C. (1978a): Late Pleistocene sea level history of Bermuda. *Quaternary Research*, **9**, 205–218.

Harmon, R.S., Thompson, P., Schwarcz, H.P. & Ford, D.C. (1978b): Late Pleistocene palaeoclimates of North America as inferred from stable isotope studies of cave speleothems. *Quaternary Research*, **9**, 54–70.

Harmon, R.S., Land, L.S., Mitterer, R.M., Garrett, P., Schwarcz, H.P. & Larson, G.J. (1981): Bermuda sea level during the last interglacial. *Nature*, **289**, 481–483.

Harmon, R.S., Mitterer, R.M., Kriausakal, N., Land, L.S., Schwarcz, H.P., Garrett, P., Larson, G.J., Vacher, H.L. & Rowe, M. (1983): U-series and amino-acid racemisation geochronology of Bermuda: implications for eustatic sea-level fluctuations over the past 250,000 years. *Palaeogeography, Palaeoclimatology, Palaeoecology*, **44**, 41–70.

Harris, C. & Donnelly, R. (1991): The glacial deposits of South Wales. In *Glacial Deposits in Great Britain and Ireland* (edited by J. Ehlers, P.L. Gibbard & J. Rose), Balkema, Rotterdam, 279–290.

Harris, R. (1987): *Satellite Remote Sensing*. Routledge, Kegan & Paul, London.

Harris, S.A. (1986): *The Permafrost Environment*. Croom Helm, London.

Harris, S.A. (1994): Chronostratigraphy of glaciations and permafrost episodes in the Cordillera of western North America. *Progress in Physical Geography*, **18**, 366–395.

Harris, S.A., French, H.M., Heginbottom, J.A., Johnston, G.H., Ladanyi, B., Sego, D.C. & van Everdingen, R.O. (1988): *Glossary of Permafrost and Related Ground-Ice Terms*. National Research Council of Canada Technical Memorandum 142, Ottawa.

Harrison, J.B.J., McFadden, L.D. & Weldon III, R.J. (1990): Spatial soil variability in the Cajon Pass chronosequence: implications for the use of soils as a geochronological tool. In *Soils and Landscape Evolution* (edited by P.L.K. Knuepfer & L.D. McFadden), Elsevier, Amsterdam, 399–416.

Harrison, S.P. & Digerfeldt, G. (1993): European lakes as palaeohydrological and palaeoclimatic indicators. *Quaternary Science Reviews*, **12**, 233–248.

Harrison, S.P. & Dodson, J. (1993): Climates of Australia and New Guinea since 18,000 yr BP. In *Global Climates since the Last Glacial Maximum* (edited by H.E. Wright, Jr. *et al.*), University of Minnesota Press, Minneapolis, 265–292.

Harry, D.G. (1988): Ground ice and permafrost. In *Advances in Periglacial Geomorphology* (edited by M.J. Clark), John Wiley, Chichester & New York, 113–149.

Harry, D.G. & Gozdzik, J.S. (1988): Ice wedges: growth, thaw transformation and palaeoenvironmental significance. *Journal of Quaternary Science*, **3**, 39–55.

Hart, J.K. (1990): Proglacial glaciotectonic deformation and the origin of the Cromer Ridge push moraine complex, North Norfolk, England. *Boreas*, **19**, 165–180.

Hart, J.K. (1995a): Drumlin formation in southern Anglesey and Avon, northwest Wales. *Journal of Quaternary Science*, **10**, 3–14.

Hart, J.K. (1995b): Subglacial erosion, deposition and deformation associated with deformable beds. *Progress in Physical Geography*, **19**, 173–191.

Hart, J.K. & Boulton, G.S. (1991): The interrelation of glaciotectonic and glaciodepositional processes within the glacial environment. *Quaternary Science Reviews*, **10**, 335–350.

Harvey, A.M. & Renwick, W.H. (1987): Holocene alluvial fan and terrace formation in the Bowland Fells, northwest England. *Earth Surface Processes and Landforms*, **12**, 249–257.

Harvey, L.D.D. (1980): Solar variability as a contributing factor to Holocene climatic change. *Progress in Physical Geography*, **4**, 487–530.

Hastenrath, S. & Kutzbach, J.E. (1985):Late Pleistocene climate and water budget of the South American altiplano. *Quaternary Research*, **24**, 249–256.

Havinga, A.J. (1964): Investigation into the differential corrosion susceptibility of pollen and spores. *Pollen et Spores*, **4**, 621–635.

Havinga, A.J. (1985): A 20-year experimental investigation into the differential corrosion susceptibility of pollen and spores in various soil types. *Pollen et Spores*, **26**, 541–558.

Haynes, C.V. Jr., Eyles, C.H., Pavlish, L.A., Ritchie, J.C. & Rybak, M. (1989): Holocene paleoecology of the eastern Sahara: Selima Oasis. *Quaternary Science Reviews*, **8**, 109–136.

Haynes, G. (1993): *Mammoths, Mastodonts and Elephants*. Cambridge University Press, Cambridge.

Haynes, J., Kiteley, R.J., Whatley, R.C. & Wilks, P.J. (1977): Microfaunas, microfloras and environmental stratigraphy of the Late Glacial and Holocene in Cardigan Bay. *Geological Journal*, **12**, 129–158.

Hays, J.D., Imbrie, J. & Shackleton, N.J. (1976): Variations in the earth's orbit: pacemaker of the Ice Ages. *Science*, **194**, 1121–1132.

Healy, T. (1981): Submarine terraces and morphology in the Kieler Bucht, western Baltic, and their relation to Quaternary events. *Boreas*, **10**, 209–217.

Hearty, P.J. (1987): New data on the Pleistocene of Mallorca. *Quaternary Science Reviews*, **6**, 245–258.

Hearty, P.J. & Aharon, P. (1988): Amino acid chronostratigraphy of late Quaternary coral reefs: Huon Peninsula, New Guinea and the Great Barrier Reef, Australia. *Geology*, **16**, 579–583.

Hearty, P.J. & Vacher, H.L. (1994): Quaternary stratigraphy of Bermuda: a high-resolution pre-Sangamonian rock record. *Quaternary Science Reviews*, **13**, 685–698.

Hearty, P.J., Miller, G.H., Stearnes, C.E. & Szabo, B.S. (1986): Amino-stratigraphy and Quaternary shorelines in the Mediterranean Basin. *Geological Society of America Bulletin*, **97**, 850–858.

Hedberg, H.D. (ed.) (1976): *International Stratigraphic Guide*. John Wiley, London & New York.

Hedges, R.E.M. (1991): AMS dating: present status and potential applications. In *Radiocarbon Dating: Recent Applications and Future Potential* (edited by J.J. Lowe), *Quaternary Proceedings*, **1**, Quaternary Research Association, Cambridge, 5–10.

Heer, O. (1865): *Die Urwelt der Schweiz*. F. Schulthess, Zürich.

Heide, K.M. & Bradshaw, R.H.W. (1982): The pollen–tree relationship within forests of Wisconsin and Upper Michigan, USA. *Review of Palaeobotany and Palynology*, **36**, 1–23.

Heijnis, H. & van der Plicht, J. (1992): Uranium/thorium dating of Late Pleistocene peat deposits in NW Europe, uranium/thorium isotope systematics and open-system behaviour of peat layers. *Chemical Geology (Isotope Geoscience Section)*, **94**, 161–171.

Heijnis, H., Ruddock, J. & Coxon, P. (1993): A uranium–thorium dated Late Eemian or Early Midlandian organic deposit from near Kilfenora between Spa and Fenit, Co. Kerry, Ireland. *Journal of Quaternary Science*, **8**, 31–44.

Heinrich, H. (1988): Origin and consequences of cyclic ice-rafting in the Northeast Atlantic Ocean during the past 130,000 years. *Quaternary Research*, **29**, 142–152.

Helle, M., Sønstergaard, E., Coope, G.R. & Rye, N. (1981): Early Weichselian peat at Brumunddal, southeastern Norway. *Boreas*, **10**, 369–379.

Henderson-Sellers, A., Dickinson, R.E., Durbidge, T.B., Kennedy, P.-J., McGuffie, K. & Pitman, A.J. (1993): Tropical deforestation: modelling local-to regional-scale climate change. *Journal of Geophysical Research* D – *Atmospheres*, **98**, 7289–7315.

Hengeveld, R. (1985): Dynamics of Dutch beetle species during the twentieth century (Coleoptera, Carabiadae). *Journal of Biogeography*, **12**, 389–411.

Hennig, G.J. & Grün, R. (1983): ESR dating in Quaternary geology. *Quaternary Science Reviews*, **2**, 157–238.

Hennig, G.J., Grün, R. & Brunnacker, K. (1983): Speleothems, travertines and palaeoclimates. *Quaternary Research*, **20**, 1–29.

Henning, W., Bell, W.A., Billquist, P.J., Glagola, B.G., Kutschera, W., Liu, A., Lucas, H.F., Paul, M., Rehm, K.E. & L'Yntema, J. (1987): Calcium-41 concentrations in terrestrial materials: prospects for dating Pleistocene samples. *Science*, **236**, 725–727.

Hey, R.W. (1986): A re-examination of the Northern Drift of Oxfordshire. *Proceedings of the Geologists' Association*, **97**, 291–301.

Hicock, S.R. (1992): Lobal interactions and rheologic superposition in subglacial till near Bradtville, Ontario, Canada. *Boreas*, **21**, 73–88.

Hicock, S.R. & Dreimanis, A. (1992): Deformation till in the Great Lakes region: implications for rapid flow along the south-central margin of the Laurentide Ice Sheet. *Canadian Journal of Earth Sciences*, **29**, 1565–1579.

Higgit, S.R., Oldfield, F. & Appleby, P.G. (1991): The record of land use change and soil erosion in the late Holocene sediments of the Petit Lac d'Annecy, eastern France. *The Holocene*, **1**, 14–28.

Higgitt, D.L., Walling, D.E. & Haigh, M.J. (1994): Estimating rates of ground retreat on mining spoils using caesium-137. *Applied Geography*, **14**, 294–307.

Hilgen, F.J. (1991): Astronomical calibration of Gauss to Matuyama sapropels in the Mediterranean and implications for the Geomagnetic Polarity Time Scale. *Earth and Planetary Science Letters*, **104**, 226–244.

Hill, A., Ward, S., Deino, A., Curtis, G. & Drake, R. (1992): Earliest Homo. *Nature*, **355**, 719–722.

Hillaire-Marcel, C. (1980): Multiple component postglacial emergence, eastern Hudson Bay, Canada. In *Earth Rheology, Isostasy and Eustasy* (edited by N.-A. Mörner), John Wiley, Chichester & New York, 215–230.

Hillaire-Marcel, C. & Causse, C. (1989): The late Pleistocene Laurentide glacier: Th/U dating of its major fluctuations and $\delta^{18}O$ range of the ice. *Quaternary Research*, **32**, 125–138.

Hillaire-Marcel, C.& Occhietti, S. (1980): Chronology, palaeogeography and palaeoclimatic significance of the late and post-glacial events in eastern Canada. *Zeitschrift für Geomorphologie*, **24**, 373–392.

Hillaire-Marcel, C., Carro, O. & Casanova, J. (1986a): ^{14}C and Th/U dating of Pleistocene and Holocene stromatolites from East African paleolakes. *Quaternary Research*, **25**, 312–329.

Hillaire-Marcel, C., Carro, O., Causse, C., Goy, J.-L. & Zazo, C. (1986b): Th/U dating of *Strombus bubonius*-bearing marine terraces in southeastern Spain. *Geology*, **14**, 613–616.

Hillaire-Marcel, C., Aucour, A.M., Bonnefille, R., Riollet, G., Vincens, A. & Williamson, D. (1989): ^{13}C/palynological evidence of differential residence times of organic carbon prior to its sedimentation in East African Rift lakes and peat bogs. *Quaternary Science Reviews*, **8**, 207–212.

Hillam, J., Groves, C.M., Brown, D.M., Baillie, M.G.L., Coles, J.M. & Coles, B.J. (1990): Dendrochronology of the English Neolithic. *Antiquity*, **64**, 210–220.

Hindmarsh, R.H. (1993): Modelling the dynamics of ice sheets. *Progress in Physical Geography*, **17**, 391–412.

Hirvas, H. & Nenonen, R.K. (1990): The till stratigraphy in Finland. *Geological Survey of Finland*, Special Paper **3**, 49–64.

Hirvas, H., Lagerbäck, R., Mäkinen, K., Nenonen, K., Olsen, L., Rodhe, L. & Thoresen, M. (1988): The Nordkalott Project – studies of Quaternary geology in northern Fennoscandia.

Boreas, **17**, 431–437.
Hjort, C. & Funder, S. (1974): The subfossil occurrence of *Mytilus edulis* L. in central East Greenland. *Boreas*, **3**, 23–33.
Hoang, C.T. & Taviani, M. (1991): Stratigraphic and tectonic implications of uranium-series-dated coral reefs from uplifted Red Sea islands. *Quaternary Research*, **35**, 264–273.
Hofmann, W. (1986): Chironomid analysis. In *Handbook of Holocene Palaeoecology and Palaeohydrology* (edited by B.E. Berglund), John Wiley, Chichester & New York, 715–727.
Hogg, A.H.A. (1980): *Surveying for Archaeologists and other Field Workers*. Croom Helm, London.
Holland, C.H. *et al.* (1978): *A Guide to Stratigraphical Procedure*. Geological Society of London, Special Report No. 10, 1–18.
Hollin, J.T. (1977): Thames interglacial sites, Ipswichian sea levels and Antarctic ice surges. *Boreas*, **6**, 33–52.
Hollin, J.T. (1980): Climate and sea level in isotope stage 5: an East Antarctic ice surge at ~95,000 BP? *Nature*, **283**, 629–633.
Holman, J.A., Stuart, A.J. & Clayden, J.D. (1990): A Middle Pleistocene herpetofauna from Cudmore Grove, Essex, England, and its palaeogeographic and palaeoclimatic implications. *Journal of Vertebrate Palaeontology*, **10**, 86–94.
Holmes, J.A. (1992): Nonmarine ostracods as Quaternary palaeoenvironmental indicators. *Progress in Physical Geography*, **16**, 405–431.
Holmes, J.A., Hales, P.E. & Street-Perrott, F.A. (1992): Trace-element chemistry of non-marine ostracods as a means of palaeolimnological reconstruction: an example from the Quaternary of Kashmir, northern India. *Chemical Geology*, **95**, 177–186.
Holmes, R.L. (1983): Computer-assisted quality control in tree-ring dating and measurement. *Tree-Ring Bulletin*, **43**, 69–75.
Holmgren, K., Lauritzen, S.-E. & Possnert, G. (1994): $^{230}Th/^{234}U$ and ^{14}C dating of a Late Pleistocene stalagmite in Lobatse II Cave, Botswana. *Quaternary Science Reviews*, **13**, 111–119.
Holyoak, D.T. (1984): Taphonomy of prospective plant macrofossils in a river catchment on Spitsbergen. *New Phytologist*, **98**, 405–423.
Holyoak, D.T. & Preece, R.C. (1985): Late Pleistocene interglacial deposits at Tattershall, Lincolnshire. *Philosophical Transactions of the Royal Society of London*, **B331**, 193–236.
Hooghiemstra, H. (1984): *Vegetational and Climatic History of the High Plain of Bogotá, Colombia: A Continuous Record of the Last 3.5 Million Years*. Cramer, Vaduz.
Hooghiemstra, H. (1988): The orbital-tuned marine oxygen isotope record applied to the Middle and Late Pleistocene pollen record of Funza (Colombian Andes). *Palaeogeography, Palaeoclimatology, Palaeoecology*, **66**, 9–17.
Hooghiemstra, H. (1989): Quaternary and Upper-Pliocene glaciations and forest development in the Tropical Andes: evidence from a long high-resolution pollen record from the sedimentary basin of Bogotá, Colombia. *Palaeogeography, Palaeoclimatology, Palaeoecology*, **72**, 11–26.
Hooghiemstra, H. & Sarmiento, G. (1991): Long continental pollen record from a tropical intermontane basin: Late Pliocene and Pleistocene history from a 540–meter core. *Episodes*, **14**, 107–115.
Hooghiemstra, H., Melice, J.L., Berger, A. & Shackleton, N.J. (1993): Frequency spectra and palaeoclimatic variability of the high-precision 30–1450 ka Funza 1 pollen record (Eastern Cordillera, Colombia). *Quaternary Science Reviews*, **12**, 141–156.
Hope, G. & Tulip, J. (1994): A long vegetation history from lowland Irian Jaya, Indonesia. *Palaeogeography, Palaeoclimatology, Palaeoecology*, **109**, 385–398.
Hopley, D. (1983): Deformation of the North Queensland continental shelf in the Late Quaternary. In *Shorelines and Isostasy* (edited by D.E. Smith & A.G. Dawson), Academic Press, London & New York, 347–368.
Hopley, D. (1987): Holocene sea-level changes in Australasia and the southern Pacific. In *Sea Surface Studies* (edited by R.J.N. Devoy), Croom Helm, London, 375–407.
Horowitz, A. (1989): Continuous pollen diagrams for the last 3.5 M.Y. from Israel: vegetation, climate and correlation with the oxygen isotope record. *Palaeogeography, Palaeoclimatology, Palaeoecology*, **72**, 63–78.
Horton, A. *et al.* (1992): The Hoxnian Interglacial deposits at Woodston, Peterborough. *Philosophical Transactions of the Royal Society of London*, **B338**, 131–164.
Hostetler, S.W., Giorgi, F., Bates, G.T. & Bartlein, P.J. (1994): Lake–atmosphere feedbacks associated with palaeolakes Bonneville and Lahontan. *Science*, **263**, 665–668.
Houghton, J.T., Jenkins, G.J. & Ephraums, J.J. (eds) (1990): *Climate Change. The IPCC Scientific Assessment*. Cambridge University Press, Cambridge.
Houghton, J.T., Callander, B.A. & Varney, S.K. (eds) (1992): *Climate Change 1992. The Supplementary Report to the IPCC Scientific Assessment.* Cambridge University Press, Cambridge.
Houmark-Nielsen, M. (1989): The last interglacial–glacial cycle in Denmark. *Quaternary International*, **3**, 31–39.
Housley, R.A. (1991): AMS dates from the Late Glacial and early Postglacial in north-west Europe: a review. In *The Late Glacial in North-West Europe* (edited by N. Barton, A.J. Roberts & D.A. Rowe), *CBA Research Report* **77**, 25–39.
Hovan, S.A., Rea, D.K., Pisias, N.G. & Shackleton, N.J. (1989): A direct link between the Chinese loess and marine O^{18} records: aeolian flux to the north Pacific. *Nature*, **340**, 296–298.
Hsu, J.T., Leonard, E.M. & Wehmiller, J.F. (1989): Aminostratigraphy of Peruvian and Chilean Quaternary marine terraces. *Quaternary Science Reviews*, **8**, 255–262.
Huddart, D. (1991): The glacial history and glacial deposits of the North and West Cumbrian lowlands. In *Glacial Deposits in Great Britain and Ireland* (edited by J. Ehlers, P.L. Gibbard & J. Rose), Balkema, Rotterdam, 151–168.
Huddart, D., Tooley, M.J. & Carter, P.A. (1977): The coasts of north-west England. In *The Quaternary History of the Irish Sea* (edited by C. Kidson & M.J. Tooley), Seel House Press, Liverpool, 119–154.
Hughes, M.K., Kelly, P.M., Pilcher, J.R. & LaMarche, V.C. Jr. (1982): *Climate from Tree-Rings*. Cambridge University Press, Cambridge.
Hughes, M.K., Schweingrüber, F.H., Cartwright, D. & Kelly, P.M. (1984): July–August temperature at Edinburgh between 1721 and 1975 from tree-ring density and width data. *Nature*, **308**, 341–344.

Hughes, T. (1987): Ice dynamics and deglaciation models when ice sheets collapsed. In *The Geology of North America*. Volume K-3, *North America and Adjacent Oceans During the Last Deglaciation* (edited by W.F. Ruddiman & H.E. Wright, Jr.), Geological Society of America, Boulder, Colorado, 183–220.

Humphrey, N.F., Kamb, B., Fahnestock, M. & Engelhardt, H. (1993): Characteristics of the bed of the lower Columbia Glacier, Alaska. *Journal of Geophysical Research*, **98**, 837–846.

Hunt, C.O. (1989): Taphonomy and the palynology of cave deposits. *Cave Science*, **16**, 83–89.

Hunt, C.O. (1993): Mollusc taphonomy in caves: a conceptual model. *Cave Science*, **20**, 45–49.

Hunt, J.B. & Hill, P.G. (1993): Tephra geochemistry: a discussion of some persistent analytical problems. *The Holocene*, **3**, 271–278.

Huntley, B. (1990): European vegetation history: palaeovegetation maps from pollen data – 13,000 yr BP to present. *Journal of Quaternary Science*, **5**, 103–122.

Huntley, B. (1991): How plants respond to climatic change: migration rates, individualism and the consequences for plant communities. *Annals of Botany*, **67**, 15–22.

Huntley, B. (1992): Pollen–climate response surfaces and the study of climate change. In *Applications of Quaternary Research* (edited by J.M. Gray), *Quaternary Proceedings*, **2**, Quaternary Research Association, Cambridge, 91–99.

Huntley, B. (1993a): Species-richness in north-temperate zone forests. *Journal of Biogeography*, **20**, 163–180.

Huntley, B. (1993b): The use of climate response surfaces to reconstruct palaeoclimate from Quaternary pollen and plant macrofossil data. *Philosophical Transactions of the Royal Society of London*, **B341**, 215–223.

Huntley, B. (1994): Late Devensian and Holocene palaeoecology and palaeo-environments of the Morrone Birkwoods, Aberdeenshire, Scotland. *Journal of Quaternary Science*, **9**, 311–336.

Huntley, B. & Birks, H.J.B. (1983): *An Atlas of Past and Present Pollen Maps for Europe: 0–13,000 Years Ago*. Cambridge University Press, Cambridge.

Huntley, B. & Prentice, I. C. (1993): Holocene vegetation and climates of Europe. In *Global Climates Since the Last Glacial Maximum* (edited by H.E. Wright, Jr., J.E. Kutzbach, T. Webb III, W.F. Ruddiman, F.A. Street-Perrott & P.J. Bartlein), University of Minnesota Press, Minneapolis & London, 136–168.

Huntley, B. & Webb III, T. (eds) (1988): *Vegetation History*. Kluwer, Dordrecht.

Huntley, D.J., Godfer-Smith, D.I. & Thewalt, M.L.W. (1985): Optical dating of sediments. *Nature*, **313**, 105–107.

Hurford, A.J. (1991): Fission track dating. In *Quaternary Dating Methods – a User's Guide* (edited by P.L. Smart and P.D. Frances), Technical Guide 4, Quaternary Research Association, Cambridge, UK, 84–107.

Hustedt, F. (1957): Die Diatomeenflora des Fluss-systems der Weser im Gebiet der Hansenstadt Bremen. *Abhandlungen herausgegeben vom Naturwissen schaflichen Verein zu Bremen*, **34**, 181–440.

Hütt, G., Jaek, I. & Tchonka, J. (1988): Optical dating: K-feldspars optical response simulation spectra. *Quaternary Science Reviews*, **7**, 381–385.

Hyvärinen, H. (1978): Use and definition of the term Flandrian. *Boreas*, **7**, 182.

Hyvärinen, H. & Alhonen, P. (1994): Holocene lake level changes in the Fennoscandian tree-line region, western Finnish Lapland: diatom and cladoceran evidence. *The Holocene*, **4**, 251–258.

Ignatius, H., Korpela, K. & Kajansuu, R. (1980): The deglaciation of Finland after 10,000 BP. *Boreas*, **9**, 217–228.

Imai, N., Shimokawa, K. & Hirota, M. (1985): ESR dating of volcanic ash. *Nature*, **314**, 81–83.

Imbrie, J. & Imbrie, K.P. (1979): *Ice Ages: Solving the Mystery*. MacMillan, London.

Imbrie, J. & Kipp, N. G. (1971): A new micropalaeontological method of quantitative palaeoclimatology: application to a Late Pleistocene Caribbean core. In *The Late Cenozoic Glacial Ages* (edited by K.K. Turekian), Yale University Press, New Haven, 71–181.

Imbrie, J., Van Donk, J. & Kipp, N.G. (1973): Palaeoclimatic investigations of a late Pleistocene Caribbean deep-sea core: comparisons of isotopic and faunal methods. *Quaternary Research*, **3**, 10–38.

Imbrie, J., Hays, J.D., Martinson, D.G., McIntyre, A., Mix, A.C., Morley, J.J., Pisias, N.G., Prell, W.L. & Shackleton, N.J. (1984): The orbital theory of Pleistocene climate: support from a revised chronology of the marine $\delta^{18}O$ record. In *Milankovitch and Climate* (edited by A. Berger, J. Imbrie, J. Hays, G. Kukla & B. Saltzman), Reidel, Dordrecht, 269–306.

Imbrie, J., McIntyre, A. & Mix, A. (1989): Oceanic response to orbital forcing in the late Quaternary: observational and experimental strategies. In *Climate and Geo-sciences* (edited by A. Berger, S. Schneider & J.C. Duplessy), Kluwer, Boston, 121–164.

Imbrie, J., Boyle, E.A., Clemens, S.C., Duffy, A., Howard, W.R., Kukla, G., Kutzbach, J., Martinson, D.G., McIntyre, A., Mix, A.C., Molfino, B., Morley, J.J., Peterson, L.C., Pisias, N.G., Prell, W.L., Raymo, M.E., Shackleton, N.J. & Toggweiler, J.R. (1992): On the structure and origin of major glaciation cycles 1. Linear responses to Milankovitch forcing. *Paleoceanography*, **7**, 701–738.

Imbrie, J., Berger, A. & Shackleton, N.J. (1993): Role of orbital forcing: a two-million-year perspective. In *Global Changes in the Perspective of the Past* (edited by J.A. Eddy & H. Oeschger), John Wiley, Chichester & New York, 263–277.

Innes, J. L. (1983): Lichenometric dating of debris flow activity in the Scottish Highlands. *Earth Surface Processes and Landforms*, **8**, 579–588.

Innes, J.L. (1985): Lichenometry. *Progress in Physical Geography*, **9**, 187–254.

Iriondo, M.H. (1995): The sediment–soil succession in the Pampean sand sea, Argentina. *XIV INQUA Abstracts, Terra Nostra,* Schriften 2/95, 121.

Iriondo, M.H. & Garcia, N.O. (1993): Climatic variations in the Argentine plains during the last 18,000 years. *Palaeogeography, Palaeoclimatology, Palaeoecology*, **101**, 209–220.

Ivanovich, M. & Harmon, R.S. (eds) (1995): *Uranium-series Disequilibrium: Applications to Earth, Marine and Environmental Sciences*, 2nd edition. Clarendon Press, Oxford.

Iversen, J. (1944): *Viscum, Hedera* and *Ilex* as climate indicators. *Geologiska Föreningens Stockholm Förhandlingar*, **66**, 463–483.

Ives, J.D. (1976): The Saglek moraines of northern Labrador: a commentary. *Arctic and Alpine Research*, **8**, 403–408.

Izett, G.A. (1981): Volcanic ash beds: recorders of upper Cenozoic silicic pyroclastic volcanism in the western United States. *Journal of Geophysical Research*, **86**, 10200–10222.

Izett, G.A. & Naeser, C.W. (1976): Age of the Bishop Tuff of eastern California as determined by the fission-track method. *Geology*, **4**, 587–590.

Jacobsen, G.L. & Bradshaw, R.H.W. (1981): The selection of sites for palaeo-vegetational studies. *Quaternary Research*, **16**, 80–96.

Jacobsen, G.L., Webb, T. & Grimm, E.C. (1987): Patterns and rates of change during the deglaciation of eastern North America. In *The Geology of North America* (edited by W.F. Ruddiman & H.E. Wright, Jr.), Geological Society of America, New York, 277–288.

Jäger, E. & Hunziker, J.C. (1979): *Lectures in Isotope Geology*. Springer-Verlag, Berlin & New York.

Jannik, N.O., Phillips, F.M., Smith, G.L. & Elmore, D.A. (1991): ^{36}Cl chronology of lacustrine sedimentation in the Pleistocene Owens Rivers system. *Bulletin of the Geological Society of America*, **103**, 1146–1159.

Jansen, E. (1989): The use of stable oxygen and carbon isotope stratigraphy as a dating tool. *Quaternary International*, **1**, 151–166.

Jansen, E. & Veum, T. (1990): Evidence for a two-step deglaciation and its impact on North Atlantic deep-water circulation. *Nature*, **343**, 612–616.

Jauhiainen, E. (1975): Morphometric analysis of drumlin fields in northern central Europe. *Boreas*, **4**, 219–230.

Jensen, K.A. & Knudsen, K.L. (1988): Quaternary foraminiferal stratigraphy in boring 81/29 from the central North Sea. *Boreas*, **17**, 273–288.

Jethick, E. & Allard, M. (1990): Soil wedge polygons in northern Québec: description and palaeoclimatic significance. *Boreas*, **19**, 289–368.

Jezek, K.C., Drinkwater, M.R., Crawford, J.P., Bindschandler, R. & Kwok, R. (1993): Analysis of synthetic aperture radar data collected over the southwestern Greenland ice sheet. *Journal of Glaciology*, **39**, 119–132.

Johnsen, S.J. & Dansgaard, W. (1992): On flow model dating of stable isotope records from Greenland ice cores. In *The Last Deglaciation: Absolute and Radiocarbon Chronologies* (edited by E. Bard & W.S. Broecker), *NATO ASI Series*, **1**, **2**, Springer-Verlag, Berlin, 13–23.

Johnsen, S.J., Dansgaard, W. Clausen, H.B. & Langway, C.C. Jr. (1972): Oxygen isotope profiles through the Antarctic and Greenland ice sheets. *Nature*, **235**, 429–343.

Johnsen, S.J., Clausen, H.B., Dansgaard, W., Fuhrer, K., Gundestrup, N., Hammer, C.U., Iversen, P., Jouzel, J., Stauffer, B. & Steffensen, J.P. (1992): Irregular glacial interstadials recorded in a new Greenland ice core. *Nature*, **359**, 311–313.

Johnsen, S.J., Clausen, H.B., Dansgaard, W., Gundestrup, N.S., Hammer, C.U. & Tauber, H. (1995): The Eem stable isotope record along the GRIP ice core and its interpretation. *Quaternary Research*, **43**, 117–124.

Johnson, M.D., Mickelson, D.M., Clayton, L. & Attig, J.W. (1995): Composition and genesis of glacial hummocks, western Wisconsin, USA. *Boreas*, **24**, 97–116.

Johnson, W.H. (1990): Ice-wedge casts and relict patterned ground in central Illinois and their palaeoenvironmental significance. *Quaternary Research*, **33**, 51–57.

Jolly, D., Bonnefille, R. & Roux, M. (1994): Numerical interpretation of a high resolution Holocene pollen record from Burundi. *Palaeogeography, Palaeoclimatology, Palaeoecology*, **109**, 357–370.

Jones, D.K.C. (1981): *South-east and Southern England*. Methuen, London & New York.

Jones, R.L. & Keen, D.H. (1993): *Pleistocene Environments in the British Isles*. Chapman & Hall, London.

Jones, V.J., Stevenson, A.C. & Battarbee, R.W. (1989): Acidification of lakes in Galloway, south-west Scotland: a diatom and pollen study of the Post-glacial history of the Round Loch of Glenhead. *Journal of Ecology*, 77, 1–23.

Jørgensen, P. (1977): Some properties of Norwegian tills. *Boreas*, **6**, 149–157.

Joussaume, S. (1989): Desert dust and climate: an investigation using a general circulation model. In *Palaeoclimatology and Palaeometeorology: Modern and Past Patterns of Global Atmospheric Transport* (edited by M. Leinen & M. Sarnthein), Kluwer, Dordrecht, 253–263.

Jouzel, J., Lorius, C., Petit, J.R., Genthon, C., Barkov, N.I., Kotlyakov, V.M. & Petrov, V.M. (1987): Vostok ice core: a continuous isotope temperature record over the last climatic cycle (160,000 years). *Nature*, **329**, 403–408.

Jouzel, J., Barkov, N.I., Barnola, J.M., Genthon, C., Korotkevitch, Y.S., Kotlyakov, V.M., Legrand, M., Lorius, C., Petit, J.P., Petrov, V.N., Raisbeck, G., Raynaud, D., Ritz, C. & Yiou, F. (1989): Global change over the last climatic cycle from the Vostok ice core record. *Quaternary International*, **2**, 15–24.

Jouzel, J., Petit, J.R. & Raynaud, D. (1990): Palaeoclimatic information from ice cores: the Vostok record. *Transactions of the Royal Society of Edinburgh: Earth Sciences*, **81**, 349–355.

Jouzel, J., Petit, J.R., Barkov, N.I., Barnola, J.M., Chappellaz, J., Ciais, P., Kotlyakov, V.M., Lorius, C., Petrov, V.N., Raynaud, D. & Ritz, C. (1992): The last deglaciation in Antarctica: further evidence of a 'Younger Dryas' type event. In *The Last Deglaciation: Absolute and Radiocarbon Chronologies* (edited by E. Bard & W.S. Broecker), *NATO ASI Series*, **2**, Springer-Verlag, Berlin, 229–266.

Jouzel, J., Barkov, N.I., Barnola, J.M., Bender, M., Chappellaz, J., Genthon, C., Kotlyakov, V.M., Lipenkov, V., Lorius, C., Petit, J.R., Raynaud, D., Raisbeck, G., Ritz, C., Sowers, T., Stievenard, M., Yiou, F. & Yiou, P. (1993): Extending the Vostok ice-core record of palaeoclimate to the penultimate glacial period. *Nature*, **364**, 407–412.

Juggins, S. (1988): A diatom/salinity transfer function for the Thames estuary and its application to waterfront archaeology. Unpublished PhD Thesis, University of London.

Juvigné, E.H., Kroonenberg, S.B., Veldkamp, A., El Arabi, A. & Vernet, G. (1992): Widespread Alleröd and Boreal trachyandesitic to trachytic tephra layers as stratigraphic markers in the Massif Central, France. *Quaternaire*, **3**, 137–146.

Kaiser, K.F. (1993): *Beiträge zur Klimageschichte vom späten Hochglazial bis ins frühe Holozän*. Eidgenössische Forschungsanstalt für Wald, Schnee und Landschaft, Birmensdorf, Switzerland.

Kaiser, K.F. (1994): Two Creeks Interstade dated through dendrochronology and AMS. *Quaternary Research*, **42**, 288–298.

Kaiser, K.F. & Eicher, U. (1987): Fossil pollen, molluscs and stable isotopes in the Dättnau Valley, Switzerland. *Boreas*, **16**, 293–304.

Kaland, P.E. (1986): The origin and management of Norwegian coastal heaths as affected by pollen analysis. In *Anthropogenic Indicators in Pollen Diagrams* (edited by K.-E. Behre), Balkema, Rotterdam, 19–38.

Kapsner, W.R., Alley, R.B., Shuman, C.A., Anandakrishnan, S. & Grootes, P.M. (1995): Dominant influence of atmospheric circulation on snow accumulation in Greenland over the past 18,000 years. *Nature*, **373**, 52–54.

Kapur, S., Veysel, S., Çavusgil, V.S., Senol, M., Gürel, N. & Fitzpatrick, E.A. (1990): Geomorphology and pedogenic evolution of Quaternary calcretes in the northern Adana Basin of Southern Turkey. *Zeitschrift für Geomorphologie*, **34**, 49–59.

Karlén, W. (1988): Scandinavian glacial and climatic fluctuations during the Holocene. *Quaternary Science Reviews*, **7**, 199–210.

Karrow, P.F. (1984): Quaternary stratigraphy and history, Great Lakes–St Lawrence region. In *Quaternary Stratigraphy of Canada – A Canadian Contribution to IGCP Project 24* (edited by R.J. Fulton), Geological Survey of Canada, Paper 84–10, 138–153.

Karte, J. (1983): Periglacial phenomena and their significance as climatic and edaphic indicators. *GeoJournal*, **7**, 329–340.

Kashiwaya, K., Atkinson, T.C. & Smart, P.L. (1991): Periodic variations in Late Pleistocene speleothem abundance in Britain. *Quaternary Research*, **35**, 190–196.

Kaufman, A. (1971): U-series dating of Dead Sea Basin carbonates. *Geochimica et Cosmochimica Acta*, **35**, 1269–1281.

Kaufman, A., Broecker, W.S., Ku, T.-L. & Thurker, D.L. (1971): The status of U-series methods of mollusk dating. *Geochimica et Cosmochimica Acta,* **35**, 1155–1183

Kayanne, H., Ishii, T., Matsumoto, E. & Yonekura, N. (1993): Late Holocene sea-level change on Rota and Guam, Mariana Islands, and its constraint on geophysical predictions. *Quaternary Research*, **40**, 189–200.

Keatinge, T.H. (1983): Development of pollen assemblage zones in soil profiles in southeastern England. *Boreas*, **12**, 1–12.

Keen, D.H. (1990): Significance of the record provided by Pleistocene fluvial deposits and their included faunas for palaeoenvironmental reconstruction and stratigraphy: case studies from the English Midlands. *Palaeogeography, Palaeoclimatology, Palaeoecology*, **80**, 25–34.

Keen, D.H. & Bridgland, D.R. (1986): An interglacial fauna from Avon No. 3 Terrace at Eckington, Worcestershire. *Proceedings of the Geologists' Association*, **97**, 303–307.

Keen, D.H., Jones, R.L., Evans, R.A. & Robinson, J.E. (1988): Faunal and floral assemblages from Bingley Bog, West Yorkshire, and their significance for Late Devensian and early Flandrian environmental changes. *Proceedings of the Yorkshire Geological Society*, **47**, 125–138.

Keigwin, L.D. (1978): Pliocene closing of the Isthmus of Panama, based on biostratigraphic evidence from nearby Pacific Ocean and Caribbean Sea cores. *Geology*, **6**, 630–634.

Keigwin, L.D., Jones, G.A. & Lehman, S.J. (1991): Deglacial meltwater discharge, North Atlantic deep circulation, and abrupt climatic change. *Journal of Geophysical Research*, **96**, 16811–16826.

Keigwin, L.D., Curry, W.B., Lehman, S.J. & Johnsen, S. (1994): The role of deep ocean in North Atlantic climate change between 70 and 130 kyr ago. *Nature*, **371**, 323–326.

Kellogg, T.B. (1976): Late Quaternary climatic changes: evidence from deep-sea cores of Norwegian and Greenland seas. In *Investigation of Late Quaternary Paleoceanography and Paleoclimatology* (edited by R.M. Cline & J.D. Hays), *Geological Society of America Memoir*, **145**, 77–110.

Kemp, A.E.S. & Baldauf, J. (1993): Vast Neogene laminated diatom mat deposits from the eastern equatorial Pacific Ocean. *Nature*, **362**, 141–144.

Kemp, A.S., Baldauf, J. & Pearce, R.B. (1994a): Origins and palaeoceanographic significance of laminated diatom ooze from the eastern equatorial Pacific Ocean (ODP Leg 138). In *Proceedings of the Ocean Drilling Program Scientific Results*, College Station, TX, 138.

Kemp, R.A. (1985a): *Soil Micromorphology and the Quaternary*. Technical Guide No. 2, Quaternary Research Association, Cambridge.

Kemp, R.A. (1985b): The cause of redness in some buried and non-buried soils in Eastern England. *Journal of Soil Science*, **36**, 329–334.

Kemp, R.A. (1987): The interpretation and environmental significance of a buried Middle Pleistocene soil near Ipswich Airport, Suffolk, England. *Philosophical Transactions of the Royal Society, London*, **B317**, 365–391.

Kemp, R.A. (1996): Role of micromorphology in paleopedological research. *Quaternary International* (in press).

Kemp, R.A., Whiteman, C.A. & Rose, J. (1993): Palaeoenvironmental and stratigraphic significance of the Valley Farm and Barham Soils in eastern England. *Quaternary Science Reviews*, **12**, 833–848.

Kemp, R.A., Jerz, H., Grottenhaler, W. & Preece, R.C. (1994b): Pedosedimentary fabrics of soils within loess and colluvium in southern England and southern Germany. In *Soil Micromorphology: Studies in Management and Genesis* (edited by A.J. Ringrose-Voase & G.S. Humphreys), Elsevier, Amsterdam, 207–219.

Kemp, R.A., Derbyshire, E., Meng, X.M., Chen, F. & Pan, B. (1995): Pedosedimentary reconstruction of a thick loess–paleosol sequence near Lanzhou in North-central China. *Quaternary Research*, **43**, 30–45.

Kennard, A.S. (1897): The post-Pliocene non-marine Mollusca of Essex. *Essex Naturalist*, **10**, 87–109.

Kennard, A.S. & Woodward, B.B. (1901): The post-Pliocene non-marine Mollusca of the south of England. *Proceedings of the Geologists' Association*, **17**, 213–260.

Kenward, H.K. (1975): Pitfalls in the environmental interpretation of insect death assemblages. *Journal of Archaeological Science*, **2**, 85–94.

Kenward, H.K. (1976): Reconstructing ancient ecological conditions from insect remains: some problems and an experimental approach. *Ecological Entomology*, **1**, 7–17.

Kenward, H.K. (1978): The value of insect remains as evidence of ecological conditions in archaeological sites. In *Research Problems in Zooarchaeology* (edited by D.R. Brothwell, K.D. Thomas & J. Clutton-Brock), Institute of Archaeology, University of London, Occasional Publication No. 3, 25–38.

Kenward, H.K. (1982): Insect communities and death assemblages past and present. In *Environmental Archaeology in the Urban Context* (edited by A.R. Hall and H.K. Kenward), Council for British Archaeology Research Report 43, 71–78.

Kenward, H.K. (1985): Outdoors-indoors? The outdoor component of archaeological insect assemblages. In *Palaeobiological Investigations: Research Design, Methods and Data* (edited by N.R.J. Fieller, D.D. Gilbertson & N.G.A. Ralph), Association for Environmental Archaeology Symposium 5B, 97–104.

Kerney, M.P. (1968): Britain's fauna of land mollusca and its relation to the post-glacial thermal optimum. *Symposia of the Zoological Society, London*, **22**, 273–291.

Kerney, M.P. (1977a): British Quaternary non-marine mollusca: a brief review. In *British Quaternary Studies – Recent Advances* (edited by F.W. Shotton), Oxford University Press, Oxford, 31–42.

Kerney, M.P. (1977b): A proposed zonation scheme for late-glacial and post-glacial deposits using land mollusca. *Journal of the Archaeological Society*, **4**, 387–390.

Kerney, M.P. & Cameron, R.A.D. (1979): *A Field Guide to the Land Snails of Britain and North-west Europe*. Collins, London.

Kerney, M.P., Preece, R.C. & Turner, C. (1980): Molluscan and plant biostratigraphy of some Late Devensian and Flandrian deposits, Kent. *Philosophical Transactions of the Royal Society of London*, **B291**, 1–43.

Kerschner, H. (1978): Palaeoclimatic inferences from late Würm rock glaciers, western Tyrol, Austria. *Arctic and Alpine Research*, **10**, 635–644.

Kerschner, H. (1985): Quantitative palaeoclimatic inferences from Late-glacial snowline, timberline and rock glacier data, Tyrolean Alps, Austria. *Zeitschrift für Gletscherkunde und Glazialgeologie*, **21**, 363–369.

Kershaw, A.P. (1991): Palynological evidence for Quaternary vegetation and environments of mainland Southeastern Australia. *Quaternary Science Reviews*, **10**, 391–404.

Kershaw, A.P. (1994): Pleistocene vegetation of the humid tropics of northeastern Queensland, Australia. *Palaeogeography, Palaeoclimatology, Palaeoecology*, **109**, 399–412.

Kidson, C., Gilbertson, D.D. *et al.* (1978): Interglacial marine deposits of the Somerset levels, south-west England. *Boreas*, **7**, 215–228.

Kidwell, S.M. & Bosence, W.J. (1991): Taphonomy and time-averaging of marine shelly faunas. In *Taphonomy: Releasing the Data Locked in the Fossil Record* (edited by P.A. Allison & D.E.G. Brigg), Plenum, New York, 115–209.

Killingly, J.S. (1983): Effects of diagenetic recrystallisation on $^{18}O/^{16}O$ values of deep sea sediments. *Nature*, **310**, 504–507.

King, L. (1983): High mountain permafrost in Scandinavia. *Proceedings of the 4th International Conference on Permafrost, Fairbanks, Alaska*, National Academy Press, Washington, 612–617.

King, L.H. (1993): Till in the marine environment. *Journal of Quaternary Science*, **8**, 347–358.

King, W.B.R. & Oakley, K.P. (1936): The Pleistocene succession in the lower part of the Thames valley. *Proceedings of the Prehistoric Society*, **2**, 52–76.

Kipp, N.G. (1976): New transfer function for estimating past sea-surface conditions from sea bed distribution of planktonic foraminiferal assemblages in the North Atlantic. *Geological Society of America Memoirs*, **145**, 3–41.

Klassen, R.W. (1984): Dating methods applicable to Late Glacial deposits of the Lake Agassiz basin, Manitoba. In *Quaternary Dating Methods* (edited by W.C. Mahaney), Elsevier, Amsterdam, 375–388.

Klatkowa, H. (1990): Synsedimentary frost cracks of the Warta cold sub-stage and their paleogeographical significance. *Quaternary Studies in Poland*, **9**, 33–50.

Klein, R., Loya, Y., Gvirtzman, G., Isdale, P.J. & Susic, M. (1990): Seasonal rainfall in the Sinai Desert during the late Quaternary from fluorescent bands in fossil corals. *Nature*, **345**, 145–147.

Klein, R.G. & Cruz-Uribe, K. (1984): *The Analysis of Animal Bones from Archaeological Sites*. University of Chicago Press, Chicago.

Kleman, J. (1994): Preservation of landforms under ice sheets and ice caps. *Geomorphology*, **9**, 19–32.

Kluiving, S.J. (1994): Glaciotectonics of the Itterbeck–Uelsen push moraines, Germany. *Journal of Quaternary Science*, **9**, 235–244.

Kluiving, S.J., Rappol, R. & Wateren, D. van der (1991): Till stratigraphy and ice movements in eastern Overijssel, The Netherlands. *Boreas*, **20**, 193–205.

Kneller, M. & Peteet, D. (1993): Late-Quaternary climate in the Ridge and Valley of Virginia, U.S.A.: changes in vegetation and depositional environment. *Quaternary Science Reviews*, **12**, 613–628.

Knox, C. (1984): Responses of river systems to Holocene climates. In *Late Quaternary Environments of the United States. 2. The Holocene* (edited by H.E. Wright, Jr.), Longman, London, 26–41.

Knudsen, K.L. (1985): Foraminiferal stratigraphy of Quaternary deposits in the Roar, Skjold and Dan Fields, central North Sea. *Boreas*, **14**, 325–332.

Knudsen, K.L. & Sejrup, H.P. (1993): Pleistocene stratigraphy in the Devils Hole area, central North Sea: foraminiferal and amino-acid evidence. *Journal of Quaternary Science*, **8**, 1–14.

Koç, N., Jansen, E. & Haflidason, H. (1993): Paleoceanographic reconstructions of surface ocean conditions in the Greenland,

Iceland and Norwegian Seas through the last 14 ka based on diatoms. *Quaternary Science Reviews*, **12**, 115–140.

Koç Karpuz, N. & Jansen, E. (1992): A high-resolution diatom record of the last deglaciation from the SE Norwegian Sea: documentation of rapid climatic changes. *Paleoceanography*, **7**, 499–520.

Koç Karpuz, N. & Schrader, H. (1990): Surface sediment diatom distribution and Holocene palaeotemperature variations in the Greenland, Iceland and Norwegian Sea. *Paleoceanography*, **5**, 557–580.

Kohl, H. (1986): Pleistocene glaciations in Austria. *Quaternary Science Reviews*, **5**, 421–427.

Kolstrup, E. (1987): Frost-wedge casts in western Jutland and their possible implications for European periglacial research. *Zeitschrift für Geomorphologie*, N.F., **31**, 449–461.

Kolstrup, E., Grün, R., Mejdahl, V., Packman, S.C. & Wintle, A.G. (1990): Stratigraphy and thermoluminescence dating of Late-glacial cover sands in Denmark. *Journal of Quaternary Science*, **5**, 207–224.

Koster, E.A. (1988): Ancient and modern cold-climate aeolian sand deposition: a review. *Journal of Quaternary Science*, **3**, 69–84.

Koster, E.A. & Dijkmans, J.W.A. (1988): Niveo-aeolian deposits and denivation forms, with special reference to the Great Kobuk sand dunes, northwestern Alaska. *Earth Surface Processes and Landforms*, **13**, 153–170.

Koster, E.A. & French, H.M. (eds) (1988): Periglacial processes and landforms. *Zeitschrift für Geomorphologie, Supplement Banud*, **71**, 1–155.

Koteff, C., Robinson, G.R., Goldsmith, R. & Thompson, W.B. (1993): Delayed postglacial uplift and synglacial sea levels in coastal central New England. *Quaternary Research*, **40**, 46–54.

Kotilainen, A.T. & Shackleton, N.J. (1995): Rapid climatic variability in the North Pacific Ocean during the past 95,000 years. *Nature*, **377**, 323–326.

Kotlyakov, V.M. & Krenke, A.N. (1979): The régime of the present-day glaciation of the Caucasus. *Zeitschrift für Gletscherkunde und Glaziolgedogie*, **15**, 7–21.

Koutaniemi, L. (1991): Glacio-isostatically adjusted palaeohydrology, the Rivers Ivalojoki and Oulankajolki, Northern Finland. In *Temperate Palaeohydrology* (edited by L. Starkel, K.J. Gregory & J.B. Thornes), John Wiley, Chichester & New York, 63–78.

Kramer, H.J. (1994): *Observation of the Earth and its Environment: Survey of Missions and Sensors*, 2nd edition. Springer-Verlag, Berlin.

Krishnamurthy, R.V. & Epstein, S. (1990): Glacial–interglacial excursion in the concentration of atmospheric CO_2: effect in the $^{13}C/^{12}C$ ratio of wood cellulose.*Tellus*, **42**, 423–434.

Krishnamurthy, R.V., Battacharya, S.K. & Kusumgar, S. (1986): Palaeoclimatic changes deduced from 13C/12C and C/N ratios of Karewa lake sediments. *Nature*, **323**, 150–153.

Kromer, B. & Becker, B. (1992): Tree-ring calibration at 10,000 BP. In *The Last Deglaciation: Absolute and Radiocarbon Chronologies* (edited by E. Bard & W.S. Broecker), *NATO ASI Series*, **1, 2**, Springer-Verlag, Berlin, 3–11.

Kromer, B. & Becker, B. (1993): German oak and pine ^{14}C calibration 7200–9400 BC. *Radiocarbon*, **35**, 125–137.

Kronfeld, J., Vogel, J.C., Rosenthal, E. & Weinstein-Evron, M. (1988): Age and palaeoclimatic implications of the Bet Shean travertines. *Quaternary Research*, **30**, 298–303.

Krzyszkowski, D. (1990): Middle and Late Weichselian stratigraphy and palaeoenvironments in central Poland. *Boreas*, **19**, 333–350.

Kuhn, J., Libbrecht, K.G. & Dicke, R. (1988): The surface temperature of the sun and changes in the solar constant. *Science*, **242**, 908–911.

Kujansuu, R. & Saarnisto, M. (eds) (1990): *Glacier Indicator Tracing*. Balkema, Rotterdam.

Kukla, G. (1975): Loess stratigraphy of central Europe. In *After the Australopithecines* (edited by K.W. Butzer & G.L. Isaac), Mouton Publishers, The Hague, 99–188.

Kukla, G. (1987a): Pleistocene climates in China and Europe compared to oxygen isotope record. *Palaeoecology of Africa*, **18**, 37–45.

Kukla, G. (1987b): Loess stratigraphy in central China. *Quaternary Science Reviews*, **6**, 191–219.

Kukla, G. (1989): Long continental records of climate – an introduction. *Palaeogeography, Palaeoclimatology, Palaeoecology*, **72**, 1–9.

Kukla, G. & An, Z.S. (1989): Loess stratigraphy in Central China. *Palaeogeography, Palaeoclimatology, Palaeoecology*, **72**, 203–225.

Kukla, G. & Gavin, J. (1992): Insolation regime of the warm to cold transitions. In *Start of a Glacial* (edited by G.J. Kukla & E. Went), Springer-Verlag, Berlin, 307–339.

Kukla, G., An, Z.S., Melice, J.L., Gavin, J. & Xiao, J.L. (1990): Magnetic susceptibility record of Chinese loess. *Transactions of the Royal Society of Edinburgh: Earth Sciences*, **81**, 263–288.

Kukla, G.J., Heller, F., Liu, X.M., Xu, T.C., Liu, T.S. & An, Z.S. (1988): Pleistocene climates in China dated by magnetic susceptibility. *Geology*, **16**, 811–814.

Kullenberg, B. (1955): Deep-sea coring. *Report of the Swedish Deep Sea Expeditions*, **4**, 35–96.

Kumpulainen, R.A. (1994): Fissure-fill and tunnel fill sediments: expressions of permafrost and increased hydrostatic pressure. *Journal of Quaternary Science*, **9**, 59–72.

Kurtén, B. (1968): *Pleistocene Mammals of Europe*. Aldine Publishing Company, Chicago.

Kutzbach, J.E. (1980): Estimates of past climate at palaeolake Chad, North Africa, based on a hydrological and energy balance model. *Quaternary Research*, **14**, 210–223.

Kutzbach, J. (1985): Modelling of paleoclimates. *Advances in Geophysics*, **28**, 159–196.

Kutzbach, J.E. & Guetter, P.J. (1986): The influence of changing orbital parameters and surface boundary conditions on climate simulations for the past 18,000 years. *Journal of Atmospheric Sciences*, **43**, 1726–1759.

Kutzbach, J.E. & Ruddiman, W.F. (1993): Model description, external forcing and surface boundary conditions. In *Global Climates Since the Last Glacial Maximum* (edited by H.E. Wright, Jr., J.E. Kutzbach, T. Webb III, W.F. Ruddiman, F.A. Street-Perrott & P.J. Bartlein), University of Minnesota Press, Minneapolis & London, 12–23.

Kutzbach, J.E. & Street-Perrott, F.A. (1985): Milankovitch forcing of fluctuations in the level of tropical lakes from 18–0 k yr BP. *Nature*, **317**, 1301–34.

Kutzbach, J.E. & Webb III, T. (1993): Conceptual basis for understanding Late Quaternary climates. In *Global Climates Since the Last Glacial Maximum* (edited by H.E. Wright, Jr., J.E. Kutzbach, T. Webb III, W.F. Ruddiman, F.A. Street-Perrott & P.J. Bartlein), University of Minnesota Press, Minneapolis & London, 5–11.

Kutzbach, J.E., Guetter, P.J., Behling, P.J. & Selin, R. (1993): Simulated climatic changes: results of the COHMAP climate-model experiments. In *Global Climates Since the Last Glacial Maximum* (edited by H.E. Wright, Jr., J.E. Kutzbach, T. Webb III, W.F. Ruddiman, F.A. Street-Perrott & P.J. Bartlein), University of Minnesota Press, Minneapolis & London, 24–93.

Lägerback, R. (1988): The Veiki moraines in northern Sweden – widespread evidence of an early Weichselian deglaciation. *Boreas*, **17**, 469–486.

Lagerlund, E., Knutsson, G., Åmark, M., Hebrand, M., Jönsson, L.-O., Karlgren, B., Kristiansson, J., Möller, P., Robison, J.M., Sandgren, P., Terne, Th. & Waldermarsson, D. (1983): The deglaciation pattern and dynamics in South Sweden, a preliminary report. *University of Lund, Department of Quaternary Geology, Report* **24**, 1–7.

Lajtha, K. & Michener, R.H. (eds) (1994): *Stable Isotopes in Ecology and Environmental Science*. Blackwell Scientific Publications, Oxford.

Lamb, H.H. (1977): *Climate: Present, Past and Future*. Methuen, London.

Lambeck, K.(1988): *Geophysical Geodesy*. Clarendon Press, Oxford.

Lambeck, K. (1991a): A model for Devensian and Flandrian glacial rebound and relative sea-level change in Scotland. In *Glacial Isostasy, Sea Level and Mantle Rheology* (edited by R. Sabadini, K. Lambeck & E. Boschi), Kluwer, Dordrecht, 33–62.

Lambeck, K. (1991b): Glacial rebound and sea-level change in the British Isles. *Terra Nova*, **3**, 379–389.

Lambeck, K. (1993a): Glacial rebound of the British Isles – I. Preliminary model results. *Geophysical Journal International*, **115**, 941–959.

Lambeck, K. (1993b): Glacial rebound of the British Isles – II. A high-resolution, high-precision model. *Geophysical Journal International*, **115**, 960–990.

Lambeck, K. (1995): Late Devensian and Holocene shorelines of the British Isles and North Sea from models of glacio-hydro-isostatic rebound. *Journal of the Geological Society, London*, **152**, 437–448.

Lambeck, K. & Nakada, M. (1992): Constraints on the age and duration of the last interglacial period and on sea-level variations. *Nature*, **357**, 125–128.

Lambert, A. & Hsu, K.J. (1979): Non-annual cycles of varve-like sediments in Walensee, Switzerland. *Sedimentology*, **26**, 453–461.

Lancaster, N. (1990): Palaeoclimatic evidence from sand seas. *Palaeogeography, Palaeoclimatology, Palaeoecology*, **74**, 279–290.

Landmesser, C.W., Johnson, T.C. & Wold, R.J. (1982): Seismic reflection study of recessional moraines beneath Lake Superior and their relationship to regional deglaciation. *Quaternary Research*, **17**, 173–190.

Langway, C.C. Jr. (1970): Stratigraphic analysis of a Deep Ice Core. *Geological Society of America, Special Paper* **125**.

Langway, C.C. Jr., Clausen, H.B. & Hammer, C.U. (1988): An inter-hemispheric volcanic time marker in ice cores from Greenland and Antarctica. *Annals of Glaciology*, **10**, 1–7.

Lao, Y. & Benson, L. (1988): Uranium-series age estimates and paleoclimatic significance of Pleistocene tufas from the Lahontan Basin, California and Nevada. *Quaternary Research*, **30**, 165–176.

Larsen, E. & Sejrup, H.P. (1990): Weichselian land–sea interactions: western Norwegian Sea. *Quaternary Science Reviews*, **9**, 85–98.

Larsen, E., Gulliksen, S., Lauritzen, S.-E., Lie, R., Løvlie, R. & Mangerud, J. (1987): Cave stratigraphy in western Norway: multiple Weichselian glaciations and interstadial vertebrate faunas. *Boreas*, **16**, 267–292.

Larsen, H.C., Saunders, A.D., Clift, P.D., Beget, J., Wei, W. & Spezzaferri, S., ODP Leg 152 Scientific Party (1994): Seven million years of glaciations in Greenland. *Science*, **264**, 952–955.

LaSalle, J. & Gauld, I.D. (eds) (1993): *Hymenoptera and Biodiversity*. CAB International, Wallingford.

LaSalle, P. & Shilts, W.W. (1993): Younger-Dryas age readvance of Laurentide ice into the Champlain Sea. *Boreas*, **22**, 25–37.

Lauritzen, S.-E. (1990): Uranium series dating of speleothems: a glacial chronology for Norway for the last 600 ka. *Striae*, **34**, 127–133.

Lauritzen, S.-E. (1993): Natural environmental change in karst: the Quaternary record. In *Karst Terrains: Environmental Changes and Human Impact* (edited by P.W. Williams), *Catena Supplement* **25**, Catena Verlag, Cremlingen-Destedt, Germany, 21–40.

Lauritzen, S.-E. (1995): High-resolution paleotemperature proxy record for the last interglaciation based on Norwegian speleothems. *Quaternary Research*, **43**, 133–146.

Lauritzen, S.-E., Ford, D.C. & Schwarcz, H.P. (1986): Humic substances in speleothem matrix – palaeoclimatic significance. *Proceedings of the 9th International Speleological Congress, Barcelona*, **1**, 77–79.

Lauritzen, S.-E., Løvlie, R., Moe, D. & Østbye, E. (1990): Paleoclimate deduced from a multidisciplinary study of a half-million year old stalagmite from Rana, Northern Norway. *Quaternary Research*, **34**, 306–316.

Lauritzen, S.-E., Haugen, J.E., Løvlie, R. & Gilje-Nielsen, H. (1994): Geochronological potential of isoleucine epimerisation in calcite speleothems. *Quaternary Research*, **41**, 52–58.

Lautenschlager, M. & Herterich, H. (1990): Atmospheric response to ice age conditions: climatology near earth's surface. *Journal of Geophysical Research*, **95**, 2547–2557.

Lautenschlager, M., Mikolajewicz, U., Maier-Reimer, E. & Heinze, C. (1992): Application of ocean models for the interpretation of Atmospheric Global Circulation Model experiments on the climate of the Last Glacial Maximum. *Paleoceanography*, **7**, 769–782.

Lautridou, J.-P., Monnier, J., Morzadec, M.T., Somme, J. & Tuffreau, A. (1986): The Pleistocene of northern France. *Quaternary Science Reviews*, **5**, 387–394.

Lea, P.D. & Waythomas, F.C. (1990): Late-Pleistocene eolian sand sheets in Alaska. *Quaternary Research*, **34**, 269–281.

Lean, J., Skumanich, A. & White, O. (1992): Estimating the sun's radiative output during the Maunder Minimum. *Geophysical Research Letters*, **19**, 1591–1594.

Lean, J.L. (1984): Solar ultraviolet irradiance variations and the earth's atmosphere. In *Climatic Change on a Yearly to Millennial Basis* (edited by N.-A. Morner & W. Karlén), Reidel, Dordrecht, 449–471.

Lees, B.G., Lu, Y. & Head, J. (1990): Reconnaissance thermoluminescence dating of northern Australian coastal dune systems. *Quaternary Research*, **34**, 169–185.

Leggett, J. (ed.) (1990): *Global Warming: The Greenpeace Report*. Oxford University Press, Oxford & New York.

Lehman, S.J. & Keigwin, L.D. (1992): Sudden changes in North Atlantic circulation during the last deglaciation. *Nature*, **356**, 757–762.

Lehman, S.J., Jones, G.A., Keigwin, L.D., Andersen, E.S., Butenko, G. & Østrmo, S.-J. (1991): Initiation of Fennoscandian ice-sheet retreat during the last deglaciation. *Nature*, **349**, 513–516.

Leiggi, P., May, P.J. & Horner, J.R. (eds) (1994): V*ertebrate Palaeontological Techniques*. Cambridge University Press, Cambridge.

Leigh, D.S. & Knox, J.C. (1994): Loess of the Upper Mississippi Valley Driftless Area. *Quaternary Research*, **42**, 30–40.

Lemdahl, G. (1991): A rapid climatic change at the end of the Younger Dryas in south Sweden – palaeoclimatic and palaeoenvironmental reconstructions based on fossil insect assemblages. *Palaeogeography, Palaeoclimatology, Palaeoecology*, **83**, 313–331.

Leonard, E. (1986): Varve studies at Hector Lake, Alberta, Canada, and the relationship between glacial activity and sedimentation. *Quaternary Research*, **25**, 199–214.

Leonard, E.M. (1989): Climatic change in the Colorado Rocky Mountains: estimates based on modern climate at Late Pleistocene equilibrium lines. *Arctic and Alpine Research*, **21**, 245–255.

Leroy, S. & Dupont, L. (1994): Development of vegetation and continental aridity in northwestern Africa during the late Pliocene: the pollen record of ODP Site 658. *Palaeogeography, Palaeoclimatology, Palaeoecology*, **109**, 295–316.

Leuschner, H.H. & Delorme, A. (1988): Tree-ring work in Göttingen. Absolute oak chronologies back to 6255 BC. *PACT*, **22**, 123–132.

Levesque, A., Mayle, F.E., Walker, I. & Cwynar, L.C. (1993): The Amphi-Atlantic Oscillation: a proposed Late-glacial climatic event. *Quaternary Science Reviews*, **12**, 629–644.

Levesque, A., Cwynar, L.C. & Walker, I. (1994): A multiproxy investigation of late-glacial climate and vegetation change at Pine Ridge Pond, SW New Brunswick, Canada. *Quaternary Research*, **42**, 316–327.

Lewkowicz, A.G. (1988): Slope processes. In *Advances in Periglacial Geomorphology* (edited by M.J. Clark), John Wiley, Chichester & New York, 325–368

Lézine, A.M. (1991): West African palaeoclimates during the last climatic cycle inferred from an Atlantic deep-sea pollen record. *Quaternary Research*, **35**, 456–463.

Lézine, A.M. & Casanova, J. (1991): Correlated oceanic and continental records demonstrate the past climate and hydrology of North Africa (0–140 kyr). *Geology*, **19**, 307–310.

Lézine, A.M. & Hooghiemstra, H. (1990): Land–sea comparisons during the last glacial–interglacial transition: pollen records from west tropical Africa. *Palaeogeography, Palaeoclimatology, Palaeoecology*, **79**, 313–331.

Lézine, A.M. & Vergnaud-Grazzini, C. (1993): Evidence of forest extension in West Africa since 22,000 BP: a pollen record from the eastern tropical Atlantic. *Quaternary Science Reviews*, **12**, 203–210.

Li, W.-X., Lundberg, J., Dickin, A., Ford, D.C., Schwarcz, H.P., McNutt, R. & Williams, D. (1989): High-precision mass spectrometric uranium-series dating of cave deposits and implications for palaeoclimate studies. *Nature*, **339**, 534–536.

Libby, W.F. (1955): *Radiocarbon Dating*, 2nd edition. University of Chicago Press, Chicago.

Liestøl, O. (1967): Storbreen glacier in Jotunheimen, Norway. *Norsk Polarinstitutt. Skrifter*, **141**, 63 pp.

Lillesand, T.M. & Kiefer, R.W. (1987): *Remote Sensing and Image Interpretation*, 2nd edition. John Wiley, Chichester & New York.

Lindroth, C.H. & Coope, G.R. (1971): The insects from the interglacial deposits at Leveäniemi. *Sveriges Geologiska Undersökning*, **C658**, 44–55.

Linick, W., Damon, P.E., Donahue, D.J. & Jull, A.J.T. (1989): Accelerator Mass Spectrometry: the new revolution in radiocarbon dating. *Quaternary International*, **1**, 1–6.

Linton, D.L. (1957): Radiating valleys in glaciated lands. In *Glaciers and Glacial Erosion* (edited by C. Embleton), Macmillan, London, 130–148.

Linton, D.L. (1963): The forms of glacial erosion. *Transactions of the Institute of British Geographers*, **33**, 1–28.

Lipps, J.H. (ed.) (1993): *Fossil Prokaryotes and Protists*. Basil Blackwell, London.

Lister, A.M. (1992): Mammalian fossils and Quaternary biostratigraphy. *Quaternary Science Reviews*, **11**, 329–344.

Lister, G.S., Kelts, K., Chen, K.Z., Jun-Quing, Y. & Niessen, F. (1991): Lake Quinghai, China: closed basin lake levels and the oxygen isotope record for Ostracoda since the latest Pleistocene. *Palaeogeography, Palaeoclimatology, Palaeoecology*, **84**, 141–162.

Liu, T.S. (1988): *Loess in China*. Springer-Verlag, Berlin.

Liu Tungsheng, Zhang Shouxin & Han Jiamao (1986): Stratigraphy and palaeoenvironmental changes in the loess of central China. *Quaternary Science Reviews*, **5**, 489–495.

Liu, X., Shaw, J., Liu, F., Heller, F. & Yuan, B. (1992): Magnetic mineralogy of Chinese loess and its significance. *Geophysical Journal International*, **108**, 301–308.

Liu, X., Shaw, J., Liu, T. & Heller, F. (1993): Magnetic susceptibility of the Chinese loess–palaeosol sequence: environmental change and pedogenesis. *Journal of the Geological Society*, **150**, 583–588.

Liu, X.M., Rolph, T.C., Bloemendal, J., Shaw, J. & Liu, T.S. (1994): Remanence characteristics of different magnetic grain size categories at Xifeng, central Chinese Loess Plateau. *Quaternary Research*, **42**, 162–165.

Liu, X.M., Rolph, T.C., Bloemendal, J., Shaw, J. & Liu, T.S. (1995): Quantitative estimates of palaeoprecipitation at Xifeng in the Loess Plateau of China. *Palaeogeography, Palaeoclimatology, Palaeoecology*, **113**, 243–248.

Lively, R.S. (1983): Late Quaternary U-series speleothem growth record from southeastern Minnesota. *Geology*, **11**, 259–262.

Livnat, A. & Kronfeld, J. (1985): Palaeoclimatic implications of U-series dates for lake sediments and travertines in the Arava Rift Valley, Israel. *Quaternary Research*, **24**, 164–172.

Loewe, F. (1971): Considerations on the origin of the Quaternary ice sheets of North America. *Arctic and Alpine Research*, **3**, 331–334.

Löffler, H. (1986): Ostracod analysis. In *Handbook of Holocene Palaeoecology and Palaeohydrology* (edited by B. Berglund), John Wiley, Chichester & New York, 693–702.

Long, A.J. & Shennan, I. (1993): Holocene relative sea-level and crustal movements in southeast and northeast England, UK. *Quaternary Proceedings*, **3**, 15–20.

Long, D. & Morton, A.C. (1987): An ash fall within the Loch Lomond Stadial. *Journal of Quaternary Science*, **2**, 87–96.

Long, D., Laban, C., Streif, H., Cameron, T.J.D. & Schüttenhelm, R.T.E. (1988): The sedimentary record of climatic variation in the southern North Sea. *Philosophical Transactions of the Royal Society, London*, **B318**, 523–537.

Longmore, M.E., O'Leary, B.M. & Rose, C.W. (1983): Caesium-137 profiles in the sediment of a partial-meromictic lake on Great Sandy Island (Fraser Island), Queensland, Australia. *Hydrobiologia*, **103**, 21–27.

Lord, A. (1980): Interpretation of Lateglacial marine environment of NW Europe by means of foraminifera. In *Studies in the Lateglacial of North-west Europe* (edited by J.J. Lowe, J.M. Gray and J.E. Robinson), Pergamon, Oxford, 103–114.

Lorius, C., Jouzel, J., Ritz, C., Merlivat, L., Barkov, N.I., Korotkevich, Y.S. & Kotlyakov, V.M. (1985): A 150,000 year climatic record from Antarctic ice. *Nature*, **316**, 591–596.

Lorius, C., Raisbeck, G., Jouzel, J. & Raynaud, D. (1989): Long-term environmental records from Antarctic ice cores. In *The Environmental Record in Glaciers and Ice Sheets* (edited by F. Oeschger & C.C. Langway, Jr.), John Wiley, Chichester & New York, 343–362.

Lorius, C., Jouzel, J., Raynaud, D., Hansen, J. & Le Treut, H. (1990): The ice-core record: climate sensitivity and future greenhouse warming. *Nature*, **347**, 139–145.

Lotter, A. (1989): Evidence of annual layering in Holocene sediments of Soppensee, Switzerland. *Aquatic Sciences*, **51**, 19–30.

Lotter, A. (1991): Absolute dating of the Late-Glacial period in Switzerland using annually laminated lake sediments. *Quaternary Research*, **35**, 321–330.

Lotter, A.F., Ammann, B., Beer, J., Hajdas, I. & Sturm, M. (1992): A step towards an absolute time-scale for the Late-glacial: annually laminated sediments from Soppensee (Switzerland). In *The Last Deglaciation: Absolute and Radiocarbon Chronologies* (edited by E. Bard & W.S. Broecker), *NATO ASI Series*, **1**, **2**, Springer-Verlag, Berlin, 45–67.

Lough, J.M. & Fritts, H.C. (1987): An assessment of the possible effects of volcanic eruptions on North American climate using tree-ring data, 1602 to 1900 A.D. *Climatic Change*, **10**, 219–239.

Lowe, D.J. (1988): Late Quaternary volcanism in New Zealand: towards an integrated record using distal airfall tephras in lakes and bogs. *Journal of Quaternary Science*, **3**, 111–120.

Lowe, J.J. (1982): Three Flandrian pollen profiles from the Teith Valley, Perthshire, Scotland. II. Analysis of deteriorated pollen. *New Phytologist*, **90**, 371–385.

Lowe, J.J. (1991): Stratigraphic resolution and radiocarbon dating of Devensian Lateglacial sediments. In *Radiocarbon Dating: Recent Applications and Future Potential* (edited by J.J. Lowe), *Quaternary Proceedings*, **1**, Quaternary Research Association, Cambridge, 19–26.

Lowe, J.J. (1993a): Isolating the climatic factors in early- and mid-Holocene palaeobotanical records from Scotland. In *Climate Change and Human Impact on the Landscape* (edited by F.M. Chambers), Chapman & Hall, London, 67–82.

Lowe, J.J. (1993b): Improvement in the dating and correlation of events during the Last Glacial/Interglacial Transition. *IGCP Project 253, Termination of the Pleistocene, Newsletter* **7**, 17.

Lowe, J.J. (ed.) (1994): North Atlantic Seaboard Programme IGCP-253: Climatic changes in areas adjacent to the North Atlantic during the last glacial–interglacial transition. *Journal of Quaternary Science*, **9**, 95–198.

Lowe, J.J. & Gray, J.M. (1980): The stratigraphic subdivision of the Lateglacial of north-west Europe. In *Studies in the Lateglacial of North-west Europe* (edited by J.J. Lowe, J.M. Gray & J.E. Robinson), Pergamon, Oxford, 157–175.

Lowe, J.J. & NASP Members (1995): Palaeoclimate of the North Atlantic seaboards during the last glacial/interglacial transition. *Quaternary International*, **28**, 51–62.

Lowe, J.J. & Walker, M.J.C. (1986a): Flandrian environmental history of the Isle of Mull, Scotland. II. Pollen analytical data from sites in western and northern Mull. *New Phytologist*, **103**, 417–436.

Lowe, J.J. & Walker, M.J.C. (1986b): Lateglacial and early Flandrian environmental history of the Isle of Mull, Inner Hebrides, Scotland. *Transactions of the Royal Society of Edinburgh: Earth Sciences*, **77**, 1–20.

Lowe, J.J. & Watson, C. (1993): Lateglacial and early Holocene pollen stratigraphy of the northern Apennines, Italy. *Quaternary Science Reviews*, **12**, 727–738.

Lowe, J.J., Lowe, S., Fowler, A.J., Hedges, R.E.M. & Austin, T.J.F. (1988): Comparison of accelerator and radiometric age measurements obtained from Late Devensian Lateglacial lake sediments from Llyn Gwernan, North Wales. *Boreas*, **17**, 355–369.

Lowe, J.J., Ammann, B., Birks, H.H., Björck, S., Coope, G.R., Cwynar, L.C., Beaulieu, J.-L. de, Mott, R.J., Peteet, D.M. & Walker, M.J.C. (1994): Climatic changes in areas adjacent to the North Atlantic during the last glacial–interglacial transition (14–9 ka BP). *Journal of Quaternary Science*, **9**, 185–198.

Lowe, J.J., Coope, G.R., Sheldrick, C., Harkness, D.D. & Walker, M.J.C. (1995a): Direct comparison of UK temperatures and Greenland snow accumulation rates, 15,000–12,000 yr ago. *Journal of Quaternary Science*, **10**, 175–180.

Lowe, J.J., Coope, G.R., Lerndahl, G. & Walker, M.J.C. (1995b): The Younger Dryas climate signal in land records from NW Europe. In *The Younger Dryas* (edited by S.R. Troelstra, J.E. Van Hinte & G.M. Ganssen), Koninklijke Nederlandse Akademie van Wetenschappen, Amsterdam, 3–26.

Lozek, V. (1986): Mollusca analysis. In *Handbook of Holocene Palaeoecology and Palaeohydrology* (edited by B.E. Berglund), John Wiley, Chichester & New York, 729–741.

Luckman, B.H. (1970): The Hereford Basin. In *The Glaciations of Wales and Adjoining Regions* (edited by C.A. Lewis), Longman, London, 175–196.

Luckman, B.H., Holdsworth, G. & Osborn, G.D. (1993): Neoglacial glacier fluctuations in the Canadian Rockies. *Quaternary Research*, **39**, 144–153.

Luly, J.G., Bowler, J.M. & Head, M.J. (1986): A radiocarbon chronology from the playa Lake Tyrrell, northwestern Victoria. *Palaeogeography, Palaeoclimatology, Palaeoecology*, **54**, 171–180.

Lumley, H. de (1976): *La Préhistoire Française*, CNRS, Paris.

Lundberg, J. & Ford, D.C. (1994): Late Pleistocene sea level change in the Bahamas from mass spectrometric U-series dating of submerged speleothem. *Quaternary Science Reviews*, **13**, 1–14.

Lundelius, E.L. (1976): Vertebrate palaeontology of the Pleistocene: an overview. *Geoscience and Man*, **13**, 45–59.

Lundqvist, J. (1980): The deglaciation of Sweden after 10,000 BP. *Boreas*, **9**, 229–238.

Lundqvist, J. (1985): The 1984 symposium on clay-varve chronology in Stockholm. *Boreas*, **14**, 97–100.

Lundqvist J. (1986a): Late Weichselian glaciation and deglaciation in Scandinavia. *Quaternary Science Reviews*, **5**, 269–292.

Lundqvist J. (1986b): Stratigraphy of the central area of the Scandinavian glaciation. *Quaternary Science Reviews*, **5**, 251–268.

Lunkka, J.P. (1994): Sedimentation and lithostratigraphy of the North Sea Drift and Lowestoft Till Formations in the coastal cliffs of NE Norfolk. *Journal of Quaternary Science*, **9**, 209–234.

Lynch, T.F. & Stevenson, C.M. (1992): Obsidian hydration dating and temperature controls in the Punta Negra region of northern Chile. *Quaternary Research*, **37**, 117–124.

Lyons, W.B., Mayewski, P.A., Spencer, M.J. & Twickler, M.S. (1990): A northern hemisphere volcanic chemistry record (1869–1984) and climatic implications using a South Greenland ice core. *Annals of Glaciology*, **14**, 176–182.

Maarleveld, G.C. (1976): Periglacial phenomena and mean annual temperatures during the last glacial time in The Netherlands. *Biuletyn Peryglacjalny*, **26**, 57–78.

Mabbutt, J.A. (1977): *Desert Landforms*. MIT Press, Cambridge, Massachusetts.

MacAyeal, D.R. (1992): Irregular oscillations of the West Antarctic ice sheet. *Nature*, **359**, 29–32.

MacDonald, G.M. (1987): Postglacial development of the subalpine–boreal transition forest in western Canada. *Journal of Ecology*, **75**, 303–320.

MacDonald, G.M. (1990): Palynology. In *Methods in Quaternary Ecology* (edited by B.G. Warner), Geoscience Canada, Reprint Series No. 5, 37–52.

MacDonald, G.M. & Ritchie, J.C. (1986): Modern pollen spectra from the western interior of Canada and the interpretation of Late Quaternary vegetation development. *New Phytologist*, **103**, 245–268.

MacFadden, B.J. (1994): *Fossil Horses: Systematics, Palaeobiology and Evolution of the Family Equidae*. Cambridge University Press, Cambridge.

Mackereth, F.J.H. (1965): Chemical investigation of lake sediments and their interpretation. *Proceedings of the Royal Society of London*, **B161**, 295–309.

Macklin, M.G., Pasmore, D.G., Stevenson, A.C., Cowley, D.C., Edwards, D.N. & O'Brien, C.F. (1991): Holocene alluviation and land-use change on Callaly Moor, Northumberland, England. *Journal of Quaternary Science*, **6**, 225–232.

Maddy, D. & Green, C.P. (1989): A unified approach to the stratigraphy of Pleistocene river basin sediments. *Quaternary Newsletter*, **59**, 8–13.

Maddy, D., Keen, D.H., Bridgland, D.R. & Green, C.P. (1991a): A revised model for the Pleistocene development of the River Avon, Warwickshire. *Journal of the Geological Society of London*, **148**, 473–484.

Maddy, D.M., Lewis, S.G. & Green, C.P. (1991b): A review of the stratigraphic significance of the Wolvercote Terrace of the upper Thames Valley. *Proceedings of the Geologists' Association*, **102**, 217–226.

Madole, R.F. (1995): Spatial and temporal patterns of Late Quaternary eolian deposition, eastern Colorado, USA. *Quaternary Science Reviews*, **14**, 155–177.

Magee, J.W., Bowler, J.M., Miller, G.H. & Williams, D.L.G. (1995): Stratigraphy, sedimentology, chronology and palaeohydrology of Quaternary lacustrine deposits at Madigan Gulf, Lake Eyre, South Australia. *Palaeogeography, Palaeoclimatology, Palaeoecology*, **113**, 3–42.

Maggi, R., Nisbet, R. & Barker, G. (eds) (1991): Archeologia della Pastorizia nell'Europa Meridionale, special edition of *Rivista di Studi Liguri* (2 vols), **A. LVI**.

Magny, M. (1993): Solar influences on Holocene climate changes, illustrated by correlation between past lake level fluctuations and the atmospheric ^{14}C record. *Quaternary Research*, **40**, 1–9.

Mahaney, W.C., Fahey, B.D. & Lloyd, D.T. (1981): Late Quaternary glacial deposits, soils and chronology, Hell Roaring Valley, Mount Adams, Cascade Range, Washington. *Arctic and Alpine Research*, **13**, 339–356.

Mahaney, W.C., Halvorson, D.L., Piegat, J. & Sanmugadas, K. (1984): Evaluation of dating methods used to assign ages in the Wind River and Teton Ranges, western Wyoming. In *Quaternary Dating Methods* (edited by W.C. Mahaney), Elsevier, Amsterdam, 355–374.

Maher, B.A. & Thompson, R. (1992): Paleoclimatic significance of the mineral magnetic record of the Chinese loess and palaeosols. *Quaternary Research*, **37**, 155–170.

Maher, B.A., Thompson, R. & Zhou, L.P. (1994): Spatial and temporal reconstruction of changes in the Asian palaeomonsoon: a new mineral magnetic approach. *Earth and Planetary Science Letters*, **125**, 461–471.

Maizels, J. (1987): Modelling of palaeohydrologic change during deglaciation. *Géographie Physique et Quaternaire*, **40**, 263–277.

Maizels, J. & Aitken, J. (1991): Palaeohydrological change during deglaciation in upland Britain: a case study from northeast Scotland. In *Temperate Palaeohydrology* (edited by L. Starkel, K.J. Gregory & J.B. Thornes), John Wiley, Chichester & New York, 105–145.

Malin, S.R.C. & Bullard, E.C. (1981): The direction of the earth's magnetic field at London 1570–1975. *Philosophical Transactions of the Royal Society London*, **A299**, 357–423.

Manabe, S. & Broccoli, A.J. (1985): The influence of continental ice sheets on the climate of an ice age. *Journal of Geophysical Research*, **90**, 2167–2190.

Mangerud, J. (1972): Radiocarbon dating of marine shells, including a discussion of apparent age of recent shells from Norway. *Boreas*, **1**, 143–172.

Mangerud, J. (1977): Late Weichselian marine sediments containing shells, foraminifera and pollen at Ågotnes, western Norway. *Norsk Geologiske Tidsskrift*, **57**, 23–54.

Mangerud, J. (1980): Ice-front variations of different parts of the Scandinavian ice sheet, 13,000 to 10,000 years BP. In *Studies in the Lateglacial of North-West Europe* (edited by J.J. Lowe, J.M. Gray & J.E. Robinson), Pergamon Press, Oxford, 23–30.

Mangerud, J. (1989): Correlation of the Eemian and the Weichselian with the Deep Sea Oxygen Isotope Stratigraphy. *Quaternary International* **3/4**, 1–4.

Mangerud, J. (1991a): The Scandinavian Ice Sheet through the last interglacial/glacial cycle. In *Klimageschichtliche Probleme der Letzen 130,000 Jahre* (edited by B. Frenzel), G. Fischer, Stuttgart & New York, 307–330.

Mangerud, J. (1991b): The last interglacial/glacial cycle in northern Europe. In *Quaternary Landscapes* (edited by L.C.K. Shane & E.J. Cushing), University of Minnesota Press, Minneapolis, 38–75.

Mangerud, J., Andersen, S.T., Berglund, B.E. & Donner, J.J. (1974): Quaternary stratigraphy of Norden: a proposal for terminology and classification. *Boreas*, **3**, 109–127.

Mangerud, J., Larsen, E., Longva, O. & Sønstegaard, E. (1979): Glacial history of western Norway 15,000–10,000 BP. *Boreas*, **8**, 179–187.

Mangerud, J., Sønstegaard, E., Sejrup, H.-P. & Haldorsen, S. (1981): A continuous Eemian–Early Weichselian sequence containing pollen and marine fossils at Fjøsanger, western Norway. *Boreas*, **10**, 137–208.

Mangerud, J., Lie, S.E., Furnes, H., Kristiansen, I.L. & Loemo, L. (1984): A Younger Dryas ash bed in Western Norway and its possible correlations with tephra cores from the Norwegian Sea and the North Atlantic. *Quaternary Research*, **21**, 85–104.

Mangerud, J., Furnes, H. & Johansen, H. (1986): A 9000-years old ash bed on the Faroe Islands. *Quaternary Research*, **26**, 262–265.

Mankinen, E.A. & Dalrymple, G.B. (1979): Revised geomagnetic polarity time scale for the interval 0–5 M.y. B.P. *Journal of Geophysical Research*, **84**, 615–626.

Mansikkaniemi, H. (1991): Regional case studies in southern Finland with reference to glacial rebound and Baltic regression. In *Temperate Palaeohydrology* (edited by L. Starkel, K.J. Gregory & J.B. Thornes), John Wiley, Chichester & New York, 79–104.

Marlowe, I.T., Brassell, S.C., Eglinton, G. & Green, J.C. (1990): Long-chain alkenones and alkyl alkenoates and the fossil coccolith record of marine sediments. *Chemical Geology*, **88**, 349–375.

Martin, C.W. (1993): Radiocarbon ages on late Pleistocene loess stratigraphy of Nebraska and Kansas, Central Great Plains, USA. *Quaternary Science Reviews*, **12**, 179–188.

Martin, J.H. & Fitzwater, S.E. (1990): Iron in Antarctic waters. *Nature*, **345**, 156–158.

Martin, P.S. (1984): Prehistoric overkill: a global model. In *Quaternary Extinctions: a Prehistoric Revolution* (edited by P.S. Martin & R.G.Klein), University of Arizona Press, Tucson.

Martin, P.S. (1990): 40,000 years of extinctions on the 'planet of doom'. *Palaeogeography, Palaeoclimatology, Palaeoecology*, **82**, 187–201.

Martin, P.S. & Klein, R.G. (eds) (1984): *Quaternary Extinctions: a Prehistoric Revolution*. University of Arizona Press, Tucson.

Martin, R.A. & Barnosky, A.D. (1993): *Morphological Change in Quaternary Mammals of North America*. Cambridge University Press, Cambridge.

Martini, I.P. & Chesworth, W. (eds) (1992): *Weathering, Soils and Palaeosols: Developments in Earth Surface Processes 2*, Elsevier, Amsterdam & London.

Martinson, D.G., Pisias, N.G., Hays, J.D., Imbrie, J., Moore, T.C. & Shackleton, N.J. (1987): Age dating and the orbital theory of the ice ages: development of a high resolution 0–300,000 year chronostratigraphy. *Quaternary Research*, **27**, 1–29.

Mason, J.A., Nater, E.A. & Hobbs, H.C. (1994): Transport direction of Wisconsinan Loess in Southeastern Minnesota. *Quaternary Research*, **41**, 44–51.

Mathewes, R.W. (1993): Evidence for Younger Dryas-age cooling on the north Pacific coast of America. *Quaternary Science Reviews*, **12**, 321–332.

Matsch, C.L. & Schneider, A.F. (1986): Stratigraphy and correlation of the glacial deposits of the glacial lobe complex in Minnesota and Northwestern Wisconsin. *Quaternary Science Reviews*, **5**, 59–64.

Matthews, J.A. (1973): Lichen growth on an active medial moraine, Jotunheimen, Norway. *Journal of Glaciology*, **65**, 305–313.

Matthews, J.A. (1985): Radiocarbon dating of surface and buried soils: principles, problems and prospects. In *Geomorphology and Soils* (edited by K.S. Richards, R.R. Arnett & S. Ellis), Allen & Unwin, London, 271–288.

Matthews, J.A. (1991): The late Neoglacial ('Little Ice Age') glacier maximum in southern Norway: new ^{14}C-dating evidence and climatic implications. *The Holocene*, **1**, 219–233.

Matthews, J.A. (1992): *The Ecology of Recently Deglaciated Terrain*. Cambridge Studies in Ecology, Cambridge University Press, Cambridge.

Matthews, J.A. & Dresser, P.Q. (1983): Intensive ^{14}C-dating of a buried palaeosol horizon. *Geologiska Föreningens Stockholm Förhandlingar*, **105**, 59–63.

Matthews, J.V. Jr. (1976): Insect fossils from the Beaufort Formation: geological and biological significance. *Geological Survey of Canada Papers*, **76**, 1B, 217–227.

Matthews, J.V. Jr. (1980): Tertiary land bridges and their climate: backdrop for development of the present Canadian insect fauna. *Canadian Entomologist*, **112**, 1089–1103.

Mayewski, P.A., Lyons, W.B., Spencer, M.J., Twickler, M.S., Buck, C.F. & Whitlow, S. (1990): An ice-core record of atmospheric response to anthropogenic sulphate and nitrate. *Nature*, **346**, 554–556.

Mayewski, P.A., Meeker, L.D. *et al.* (1993): Greenland ice core 'signal' characteristics offer expanded view of climate change. *Journal of Geophysical Research*, **98**, 12839–12847.

Mayhew, D.F. (1977): Avian predators as accumulators of fossil mammal material. *Boreas*, **6**, 25–31.

Mayle, F.E. & Cwynar, L.C. (1995): Impact of the Younger Dryas cooling event upon lowland vegetation of Maritime Canada. *Ecological Monographs*, **65**, 129–154,

Mayle, F.E., Levesque, A.J. & Cwynar, L.C. (1993): *Alnus* as an indicator taxon for the Younger Dryas in eastern North America. *Quaternary Science Reviews*, **12**, 295–306.

Mazaud, A., Laj, C., Bard, E., Arnold M. & Tric, E. (1992): A geomagnetic calibration of the radiocarbon timescale. In *The Last Deglaciation: Absolute and Relative Chronologies* (edited by E. Bard & W.S. Broecker), *NATO ASI Series* **1**, **2**, Springer-Verlag, Berlin, 163–169.

McAndrews, J.H. (1988): Human disturbance of North American forests and grasslands: the fossil record. In *Vegetation History* (edited by B. Huntley & T. Webb III), Kluwer, Dordrecht, 673–697.

McCabe, A.M. (1985): Glacial geomorphology. In *The Quaternary History of Ireland* (edited by K.J. Edwards & W.P. Warren), Academic Press, London, 67–93.

McCabe, A.M. (1991): The distribution and stratigraphy of drumlins in Ireland. In *Glacial Deposits in Great Britain and Ireland* (edited by J. Ehlers, P.L. Gibbard & J. Rose), Balkema, Rotterdam, 421–435.

McCabe, A.M. (1993): Drumlin bedforms and related ice-marginal depositional systems in Ireland. *Irish Geography*, **26**, 22–44.

McCabe, A.M. & Dardis, G.F. (1989): Sedimentology and depositional setting of Late Pleistocene drumlins, Galway Bay, Western Ireland. *Journal of Sedimentary Petrology*, **59**, 944–959.

McCarroll, D. (1987): The Schmidt hammer in geomorphology: five sources of instrument error. *British Geomorphological Research Group, Technical Bulletin*, **36**, 16–27.

McCarroll, D. (1989): Potential and limitations of the Schmidt hammer for relative-age dating: field tests on Neoglacial moraines, Jotunheimen, southern Norway. *Arctic and Alpine Research*, **21**, 268–275.

McCarroll, D. (1990): The age and origin of Neoglacial moraines in Jotunheimen, southern Norway: new evidence from weathering-based data. *Boreas*, **20**, 283–295.

McCarroll, D. (1991): Ice directions in western Lleyn and the status of the Gwynedd re-advance of the Last Irish Sea glacier. *Geological Journal*, **26**, 137–143.

McCarroll, D. & Harris, C. (1992): The glacigenic deposits of western Lleyn, north Wales: terrestrial or marine? *Journal of Quaternary Science*, **7**, 19–30.

McCarroll, D. & Nesje, A. (1993): The vertical extent of ice sheets in Nordfjord, western Norway measuring degree of rock surface weathering. *Boreas*, **22**, 255–265.

McCarroll, D., Ballantyne, C.K., Nesje, A. & Dahl, S.-O. (1995): Nunataks of the last ice sheet in north-west Scotland. *Boreas*, **24**, 305–323.

McCave, I.N., Manighetti, B. & Beveridge, N.A.S. (1995): Circulation in the glacial North Atlantic inferred from grain-size measurements. *Nature*, **374**, 149–152.

McCoy, W.D. (1987): The precision of amino acid geochronology and palaeothermometry. *Quaternary Science Reviews*, **6**, 43–54.

McDougal I. (1981): ^{40}Ar/^{39}Ar age spectra from the KBS Tuff, Koobi Fora Formation. *Nature*, **294**, 120–124.

McFadden, L.D. & Weldon, R.J. (1987): Rates and processes of soil development on Quaternary terraces in Cajon Pass, California. *Geological Society of America Bulletin*, **98**, 280–293.

McFarlane, M.J. (1983): Laterites. In *Chemical Sediments and Geomorphology* (edited by A.S. Goudie & K. Pye), Academic Press, London & New York, 7–58.

McGregor, G.R. (1992): Temporal and spatial characteristics of coastal rainfall anomalies in Papua New Guinea and their relationship to the Southern Oscillation. *International Journal of Climatology*, **12**, 449–468.

McIntyre, A. & Ruddiman, W.F. (1972): North-east Atlantic post-Eemian palaeoceanography: a predictive analog for the future. *Quaternary Research*, **2**, 350–354.

McIntyre, A., Ruddiman, W.F. & Jantzen, R. (1972): Southward penetration of the North Atlantic Polar Front and faunal and floral evidence of large scale surface water mass movements over the past 225,000 years. *Deep Sea Research*, **19**, 61–77.

McKee, P.M. *et al.* (1987): Sedimentation rates and sediment core profiles of ^{238}U and ^{232}Th decay chain radionuclides in a lake affected by uranium mining and milling. *Canadian Journal of Fisheries and Aquatic Sciences*, **44**, 390–398.

McKenzie, J.A. (1993): Pluvial conditions in the eastern Sahara following the penultimate deglaciation: implications for changes in atmospheric circulation patterns with global warming. *Palaeogeography, Palaeoclimatology, Palaeoecology*, **103**, 95–105.

McKenzie, K.G. & Jones, P.J. (eds) (1993): *Ostracoda in the Earth and Life Sciences*. Balkema, Rotterdam.

McManus, J.F., Bond, G.C., Broecker, W.S., Johnsen, S., Labeyrie, L. & Higgins, S. (1994): High-resolution climatic records from the North Atlantic during the last interglacial. *Nature*, **371**, 326–329.

Mead, J.I. & Meltzer, D.J. (eds) (1985): *Environments and Extinctions: Man in Late-glacial North America*. Centre for the Study of Early Man, Orono.

Meer, J.J.M. van der (1987): Micromorphology of glacial sediments as a tool in distinguishing genetic varieties of till. *Geological Survey of Finland, Special Paper*, **3**, 77–89.

Meer, J.J.M. van der (1993): Microscopic evidence of subglacial deformation. *Quaternary Science Reviews*, **12**, 553–587.

Meer, J.J.M. van der & Laban, C. (1990): Micromorphology of

some North Sea till samples, a pilot study. *Journal of Quaternary Science*, **5**, 95–101.

Meer, J.J.M. van der & Wicander, R. (1992): A Silurian–Devonian acritarch flora from Saalian till in The Netherlands. *Boreas*, **21**, 153–157.

Meierding, T.C. (1982): Late Pleistocene glacial equilibrium-line altitudes in the Colorado Front Range: a comparison of methods. *Quaternary Research*, **18**, 289–310.

Mejdahl, V. (1988): Long-term stability of the TL signal in alkali feldspars. *Quaternary Science Reviews*, **7**, 357–360.

Mellor, A. (1985): Soil chronosequences on Neoglacial moraine ridges, Jostedalsbreen and Jotunheimen, southern Norway: a quantitative pedogenic approach. In *Geomorphology and Soils* (edited by K.S. Richards, R.R. Arnett & S.Ellis), George Allen & Unwin, London, 289–308.

Menzies, J. (1978): A review of the literature on the formation and location of drumlins. *Earth Science Reviews*, **14**, 315–350.

Menzies, J. (1987): Towards a general hypothesis on the formation of drumlins. In *Drumlin Symposium* (edited by J. Menzies & J. Rose), Balkema, Rotterdam, 9–24.

Menzies, J. (1989): Drumlins – products of controlled or uncontrolled glaciodynamic response? *Quaternary Science Reviews*, **8**, 151–158.

Menzies, J. & Rose, J. (eds) (1987): *Drumlin Symposium*. Balkema, Rotterdam.

Mercer, J. (1969): The Allerød Oscillation: a European climatic anomaly? *Arctic and Alpine Research*, **1**, 227–234.

Merilinen, J. (1967): The diatom flora and the hydrogen-ion concentration of water. *Annales Botanica Fennica*, **4**, 51–58.

Merritt, J.W. (1992): The high-level marine shell-bearing deposits of Clava, Inverness-shire and their origin as glacial rafts. *Quaternary Science Reviews*, **11**, 759–780.

Meyer, E., Sarna-Wojcicki, A.M., Hillhouse, J.W., Woodward, M.J., Slate, J.L. & Sorg, D.H. (1991): Fission-track age (400,000 yr) of the Rockland tephra, based on inclusions of zircon grains lacking fossil fission tracks. *Quaternary Research*, **35**, 367–382.

Meyers, P.A., Takemura, K. & Horie, S. (1993): Reinterpretation of Late Quaternary sediment chronology of Lake Biwa, Japan, from correlation with marine glacial–interglacial cycles. *Quaternary Research*, **39**, 154–162.

Miall, A.D. (1984): *The Principles of Sedimentary Basin Analysis*. Springer-Verlag, Berlin.

Michels, J.W. & Bebrich C.A., (1971): Obsidian hydration dating. In *Dating Techniques for the Archaeologist* (edited by H.N. Michael & E.K. Ralph), MIT Press, Cambridge, Massachusetts, 164–221.

Mickelson, D.M., Clayton, L., Fullerton, D.S. & Borns, H. Jr. (1983): The Late Wisconsin glacial record of the Laurentide Ice Sheet in the United States. In *Late Quaternary Environments of the United States*, Volume 1, *The Late Pleistocene* (edited by S. C. Porter), Longman, London, 3–32.

Middleton, R. & Klein, J. (1987): ^{26}Al measurement and applications. *Philosophical Transactions of the Royal Society, London*, **A323**, 121–143.

Middleton, R., Fink, D., Klein, J. & Sharma, P. (1989): ^{41}Ca concentrations in modern bone and their implications for dating. *Radiocarbon*, **31**, 305–310.

Mifflin, M.D. & Wheat, M.M. (1979): Pluvial lakes and estimated pluvial climates of Nevada. *Nevada Bureau of Mines and Geology Bulletin*, **94**.

Mighall, T. & Chambers, F.M. (1993): The environmental impact of prehistoric mining at Copa Hill, Cwmystwyth, Wales. *The Holocene*, **4**, 260–264.

Miller, B.B. & Bajc, A.F. (1990): Non-marine molluscs. In *Methods in Quaternary Ecology* (edited by B.G. Warner), *Geoscience Canada Reprint Series* **5**, Geological Association of Canada, 101–112.

Miller, D.S. & Wagner, G.A. (1981): Fission-track ages applied to obsidian artifacts from South America using the plateau-annealing and the track-size age-correction techniques. *Nuclear Tracks*, **5**, 147–155.

Miller, G.H. & Hare, P.E. (1980): Amino acid geochemistry: integrity of the carbonate matrix and potential of molluscan fossils. In *Biogeochemistry of Amino Acids* (edited by P.E. Hare, T.C. Hoering & K. King, Jr.), John Wiley, Chichester & New York, 415–443.

Miller, G.H. & Mangerud, J. (1985): Aminostratigraphy of European marine interglacial deposits. *Quaternary Science Reviews*, **4**, 215–278.

Miller, G.H., Bradley, R.S. & Andrews, J.T. (1975): The glaciation level and lowest equilibrium line altitude in the high Canadian Arctic: maps and climatic interpretation. *Arctic and Alpine Research*, **7**, 155–168.

Miller, G.H., Sejrup, H.P., Mangerud, J. & Andersen, B. (1983): Amino acid ratios in Quaternary molluscs and foraminifera from western Norway: correlation, geochronology and palaeotemperature estimates. *Boreas*, **12**, 107–124.

Miller, U. (1982): Shore displacement and coastal dwelling in the Stockholm region during the past 5000 years. *Annales Academie Scientiarum Fennicae*, **A134**, 185–211.

Milne, G., Battarbee, R.W., Straker, V. & Yule, B. (1983): The London Thames in the mid-first century. *London and Middlesex Archaeological Society*, **34**, 19–30.

Milner, A.M. & Petts, G.E. (1994): Glacial rivers: physical habitat and ecology. *Freshwater Biology*, **32**, 295–307.

Mitchell, J.F.B., Grahame, N.S. & Needham, K.J. (1988): Climate simulations for 9000 BP: seasonal variations and effect of the Laurentide ice sheet. *Journal of Geophysical Research*, **93**, D7, 8283–8303.

Mitlehner, A.G. (1992): Palaeoenvironments of the Hoxnian Nar Valley Clay, Norfolk, England: evidence from an integrated study of diatoms and ostracods. *Journal of Quaternary Science*, **7**, 335–342.

Mix, A.C. (1987): The oxygen isotope record of glaciation. In *The Geology of North America*, Volume K-3, *North America and Adjacent Oceans During the Last Deglaciation* (edited by W.F. Ruddiman & H.E. Wright, Jr.), Geological Society of America, Boulder, Colorado, 111–135.

Mix, A.C. & Ruddiman, W.F. (1985): Structure and timing of the last deglaciation; oxygen isotope evidence. *Quaternary Science Reviews*, **4**, 59–108.

Mix, A.C., Ruddiman, W.F. & McIntyre, A. (1986): Late Quaternary paleoceanography of the tropical Atlantic: I. Spatial variability of annual mean sea-surface temperatures 0–20,000 years B.P. *Paleoceanography*, **1**, 43–66.

Miyata, T., Maeda, Y. *et al.* (1990): Evidence for a Holocene high sea-level stand, Vanau Levu, Fiji. *Quaternary Research*, **33**, 352–359.

Moers, M.E.C., Baas, M., Boon, J.J. & Leeuw, J.W. de (1990): Molecular characterisation of total organic matter and carbohydrates in peat samples from a Cypress swamp by pyrolysis-mass-spectrometry and wet chemical methods. *Biogeochemistry*, **11**, 251–277.

Mol, J., Vandenberghe, J., Kasse, K. & Stel, H. (1993): Periglacial microjointing and faulting in Weichselian fluvio-aeolian deposits. *Journal of Quaternary Science*, **8**, 15–30.

Molfino, B. (1990): Ocean model realizations and paleoceanographic data define tropical Atlantic Ocean response to orbitally forced changes in zonal and meridional winds. *EOS Transactions*, American Geophysical Union, **71** (43), 1367.

Molfino, B., Heusser, L.H. & Woillard, G.M. (1984): Frequency components of a Grande Pile pollen record: evidence of precessional orbital forcing. In *Milankovitch and Climate* (edited by A. Berger, J. Hays, J. Imbrie, G. Kukla & B. Saltzman), Reidel, Dordrecht, 391–404.

Møller, J.J., Danielson, T.K. & Fjalstad, S. (1992): Late Weichselian glacial maximum on Andøya, North Norway. *Boreas*, **21**, 1–14.

Molnar, P., Brown, E.T *et al.* (1994): Quaternary climatic change and the formation of river terraces across growing anticlines on the north flank of the Tien Shan, China. *Journal of Geology*, **102**, 583–602.

Monger, H.C., Daugherty, L.A. & Lindemann, W.C. (1991): Microbial precipitation of pedogenic calcite. *Geology*, **19**, 997–1000.

Mooers, H.D. (1989): Drumlin formation: a time-transgressive model. *Boreas*, **18**, 99–107.

Mook, W.G. (1986): Recommendations/resolutions adopted by the Twelfth International Radiocarbon Conference. *Radiocarbon*, **28**, 2A, 799.

Moore, P.D. (1986): Hydrological changes in mires. In *Handbook of Holocene Palaeoecology and Palaeohydrology* (edited by B.E. Berglund), John Wiley, Chichester & New York, 91–110.

Moore, P.D. (1993): Holocene paludification and hydrological changes as climate proxy data in Europe. In *Evaluation of Climate Proxy Data in Relation to the European Holocene* (edited by B. Frenzel, A. Pons & B. Gläser), Gustav Fischer Verlag, Stuttgart, 255–270.

Moore, P.D., Webb, J.A. & Collinson, M.D. (1991): *Pollen Analysis*. Blackwell, Oxford.

Morgan, A.V. (1987): Late Wisconsin and early Holocene environments of east–central North America based on assemblages of fossil Coleoptera. In *North America and Adjacent Oceans during the Last Deglaciation* (edited by W.F. Ruddiman & H.E. Wright, Jr.), *The Geology of North America*, Volume K-3, Geological Society of America, Boulder, Colorado, 353–370.

Morley, J.J. & Dworetzky, B.A. (1991): Evolving Pliocene–Pleistocene climate: a North Pacific perspective. *Quaternary Science Reviews*, **10**, 225–238.

Mörner, N.-A. (1976): Eustasy and geoid changes. *Journal of Geology*, **84**, 123–151.

Mörner, N.-A. (1980): Eustasy and geoid changes as a function of core/mantle changes. In *Earth Rheology, Isostasy and Eustasy* (edited by N.-A. Mörner), John Wiley, Chichester & New York, 535–553.

Mörner, N.-A. (1984): Planetary, solar, atmospheric, hydrospheric and endogene processes as origin of climatic changes on the Earth. In *Climatic Changes on a Yearly to Millennial Basis* (edited by N.-A. Mörner & W. Karlén), Reidel, Dordrecht, 483–507.

Mörner, N.-A. (1987a): Pre-Quaternary long-term changes in sea level. In *Sea Surface Studies* (edited by R.J.N. Devoy), Croom Helm, London, 233–241.

Mörner, N.-A. (1987b): Models of global sea level changes. In *Sea Level Changes* (edited by M.J. Tooley & I. Shennan), Blackwell, Oxford, 332–355.

Mörner, N.-A. (1987c): Quaternary sea-level changes: Northern hemisphere data. In *Sea Surface Studies* (edited by R.J.N Devoy), Croom Helm, London, 242–263.

Mörner, N.-A. (1993a): Global changes: the last millennia. *Global and Planetary Change*, **7**, 211–217.

Mörner, N.-A. (1993b): Global change: the high-amplitude changes 13–10 ka ago – novel aspects. *Global and Planetary Change*, **7**, 243–250.

Morowitz, H.J. (1991): Balancing species preservation and economic considerations. *Science*, **253**, 752–754.

Morrison, R.B. (1968): Means of time-stratigraphic subdivisions and long-distance correlation of Quaternary successions. In *Means of Correlating Quaternary Successions* (edited by R.B. Morrison & H.E. Wright, Jr.), University of Utah Press, Salt Lake City, 1–113.

Morrison, R.B. (1978): Quaternary soil stratigraphy – concepts, methods and problems. In *Quaternary Soils* (edited by W.C. Mahaney), GeoAbstracts, Norwich, 77–108.

Mott, R.J. (1990): Sangamonian forest history and climate in Atlantic Canada. *Géographie Physique et Quaternaire*, **44**, 257–270.

Mott, R.J. (1994): Wisconsinan Late-glacial environmental change in Nova Scotia: a regional synthesis. *Journal of Quaternary Science*, **9**, 155–160.

Mott, R.J., Grant, D.R., Stea, R. & Ochietti, S. (1986): Late-glacial climatic oscillation in Atlantic Canada equivalent to the Allerød/Younger Dryas event. *Nature*, **323**, 247–250.

Mourguiart, P. & Carbonel, P. (1994): A quantitative method of palaeo-lake level reconstruction using ostracod assemblages: an example from the Bolivian Altiplano. *Hydrobiologia*, **288**, 183–193.

Muhs, D.R. (1985): Age and palaeoclimatic significance of Holocene dune sands in northeastern Colorado. *Annals of the Association of American Geographers*, **75**, 566–582.

Muhs, D.R., Rosholt, J.N. & Bush, C.A. (1989): The uranium-trend dating method: principles and applications for southern California marine terrace deposits. *Quaternary International*, **1**, 19–34.

Muller, R.A. & MacDonald, G.J. (1995): Glacial cycles and orbital inclination. *Nature*, **377**, 107–108.

Muller, S.W. (1947): *Permafrost of Permanently Frozen Ground and Related Engineering Problems*. Ann Arbor, Michigan.

Munaut, A.-V. (1986): Dendrochronology applied to mire environments. In *Handbook of Holocene Palaeoecology and Palaeohydrology* (edited by B.E. Berglund), John Wiley, Chichester & New York, 317–385.

Munro, M.A.R. *et al.* (1990): Diatom quality control and data handling. *Philosophical Transactions of the Royal Society of London*, **B327**, 257–261.

Murray, J.W. (1991): *Ecology and Palaeoecology of Benthic Foraminifera*. Longman, London.

Murray-Wallace, C.V. & Belperio, A.P. (1991): The last interglacial shoreline in Australia – a review. *Quaternary Science Reviews*, **10**, 441–462.

Murton, J.B. & French, H.M. (1993): Thaw modification of frost-fissure wedges, Richards Island, Pleistocene Mackenzie Delta, western Arctic Canada. *Journal of Quaternary Science*, **8**, 185–196.

Mylroie, J.E. & Carew, J.L. (1988): Solution conduits as indicators of Late Quaternary sea level position. *Quaternary Science Reviews*, **7**, 55–64.

Naeser, C.W. & Naeser, N.D. (1988): Fission-track dating of Quaternary events. *Geological Society of America, Special Paper* **227**, 1–11.

Naeser, N.D. & Naeser, C.W. (1984): Fission-track dating. In *Quaternary Dating Methods* (edited by W.C. Mahaney), Elsevier, Amsterdam, 87–100.

Nagy, J. & Ofstad, K. (1980): Quaternary foraminifera and sediments in the Norwegian Channel. *Boreas*, **9**, 39–52.

Nakada, M. & Yokose, H. (1992): Ice age as a trigger of active Quaternary volcanism and tectonism. *Tectonophysics*, **212**, 321–329.

Nanson, G.C., Chen, X.Y. & Price, D.M. (1995): Aeolian and fluvial evidence of changing climate and wind patterns during the past 100 ka in the western Simpson Desert, Australia. *Palaeogeography, Palaeoclimatology, Palaeoecology*, **113**, 87–102.

Narcisi, B., Anselmi, B., Catalano, F., Dai Pra, G. & Magri, G. (1992): Lithostratigraphy of the 250,000 year record of lacustrine sediments from the Valle di Castiglione crater, Roma. *Quaternary Science Reviews*, **11**, 353–362.

Neale, J.W. (1988): Ostracods and palaeosalinity reconstruction. In *Ostrocoda in the Earth Sciences* (edited by P. De Deckker, J.P. Colin & J.P. Peypouquet), Elsevier, Amsterdam

Neftel, A., Oeschger, H., Schwander, J., Stauffer, B. & Zumbrunn, R. (1982): Ice core sample measurements give atmospheric CO_2 content during the past 40,000 years. *Nature*, **295**, 220–223.

Neftel, A., Moor, E., Oeschger, H. & Stauffer, B. (1985): Evidence from polar ice cores for the increase in atmospheric CO_2 in the past two centuries. *Nature*, **315**, 45–47.

Nesje, A. & Sejrup, H.-P. (1988): Late Weichselian/Devensian ice sheets in the North Sea and adjacent land areas. *Boreas*, **17**, 371–384.

Nesje, A., Kvamme, M., Ry, N. & Løvlie, R. (1991): Holocene glacial and climate history of the Jostedalsbreen region, western Norway: evidence from lake sediments and terrestrial deposits. *Quaternary Science Reviews*, **10**, 87–114.

Nesje, A., McCarroll, D. & Dahl, S.O. (1994): Degree of rock surface weathering as an indicator of ice-sheet thickness along an east–west transect across southern Norway. *Journal of Quaternary Science*, **9**, 337–37.

Newnham, R.M. & Lowe, D.J. (1991): Holocene vegetation and volcanic activity, Auckland Isthmus, New Zealand. *Journal of Quaternary Science*, **6**, 177–194.

Niessen, A.C., Koster, E.A. & Galloway, J.P. (1984): Periglacial sand dunes and eolian sand sheets: an annotated bibliography. *U.S. Geological Survey Open-file Report*, 84–167, 61 pp.

Nilssen, J.P. & Sandøy, S. (1990): Recent lake acidification and cladoceran dynamics: surface sediment and core analysis from lakes in Norway, Scotland and Sweden. In *Palaeolimnology and Lake Acidification* (edited by R.W. Battarbee, J. Mason, I. Renberg & J.P. Talling), Royal Society, London, 73–83.

Nilsson, T. (1983): *The Pleistocene*. Reidel, Dordrecht.

Ning, S., Aldahan, A.A., Haiping, Y., Possnert, G. & Königsson, L.-K. (1994): ^{10}Be in continental sediments from North China: probing into the last 5.4 Ma. *Quaternary Geochronology (Quaternary Science Reviews)*, **13**, 127–136.

Nisbet, E.G. (1990): The end of the ice age. *Canadian Journal of Earth Sciences*, **27**, 148–157.

Nisbet, E.G. (1992): Sources of atmospheric CH_4 in early postglacial time. *Journal of Geophysical Research*, **97**, 859–867.

Nordberg, K. (1989): *Sea-floor deposits, paleoecology and paleohydrography in the Kattegat during the later part of the Holocene*. Department of Geology, Chalmers University of Technology and University of Göteborg, A65, 205 pp.

North American Commission on Stratigraphic Nomenclature (1983): North American Stratigraphic Code. *American Association of Petroleum Geologists Bulletin*, **67**, 841–875.

Nyamweru, C.K. & Bowman, D. (1989): Climatic changes in the Chalbi Desert, North Kenya. *Journal of Quaternary Science*, **4**, 131–139.

Oakley, K.P. (1980): Relative dating of the fossil hominids of Europe. *Bulletin of the British Museum (Natural History), Geological Series*, **34**, 1–69.

Odin, G.S., Renard, M. & Grazzini, C.V. (1982): Geochemical events as a means of correlation. In *Numerical Dating in Stratigraphy* (edited by G.S. Odin), John Wiley, Chichester & New York, 37–71.

Oeschger, F. & Langway, C.C. Jr. (eds) (1989): *The Environmental Record in Glaciers and Ice Sheets*. John Wiley, Chichester & New York.

Oeschger, H. (1992): Working hypothesis for glaciation/deglaciation mechanisms. In *The Last Deglaciation: Absolute and Radiocarbon Chronologies* (edited by E. Bard & W.S. Broecker), *NATO ASI Series*, **1**, **2**, Springer-Verlag, Berlin, 273–289.

Ohmura, A., Kasser, P., & Funk, M. (1992): Climate at the equilibrium line of glaciers. *Journal of Glaciology*, **38**, 397–410.

Oldfield, F. (1978): Lakes and their drainage basins as units of sediment-based ecological study. *Progress in Physical Geography*, **1**, 460–504.

Oldfield, F. (1983): The role of palaeomagnetic studies in palaeohydrology. In *Background to Palaeohydrology* (edited by K.J. Gregory), John Wiley, Chichester & New York, 141–165.

Oldfield, F. (1991): Environmental magnetism – a personal perspective. *Quaternary Science Reviews*, **10**, 53–85.

Oldfield, F. (1993): Forward to the past: changing approaches to Quaternary palaeoecology. In *Climate Change and Human Impact on the Landscape* (edited by F.M. Chambers), Chapman & Hall, London, 13–22.

Oldfield, F. & Maher, B.A. (1989): Sediment source variations and lead-210 inventories in recent Potomac Estuary sediment cores. *Journal of Quaternary Science*, **4**, 189–200.

Ollier, C.D. (1991a): Laterite profiles, ferricrete and landscape evolution. *Zeitschrift für Geomorphologie*, **35**, 165–173.

Ollier, C.D. (1991b): Aspects of silcrete formation in Australia. *Zeitschrift für Geomorphologie*, **35**, 151–163.

Ollier, C.D. & Galloway, R.W. (1990): The laterite profile, ferricrete and unconformity. *Catena*, **17**, 99–109.

Olsson, I. (1986): Radiometric dating. In *Handbook of Holocene Palaeoecology and Palaeohydrology* (edited by B.E. Berglund), John Wiley, Chichester & New York, 273–312.

Oppo, D.W. & Fairbanks, R.G. (1990): Atlantic Ocean thermohaline circulation of the last 150,000 years: relationship to climate and atmospheric CO_2. *Paleoceanography*, **5**, 43–54.

Osborn, G. & Luckman, B.H. (1988): Holocene glacier fluctuations in the Canadian Cordillera (Alberta and British Columbia). *Quaternary Science Reviews*, **7**, 115–128.

Osborne, P.J. (1972): Insect faunas of Late Devensian and Flandrian age from Church Stretton, Shropshire. *Philosophical Transactions of the Royal Society of London*, **B263**, 327–367.

Ostlund, H.G. & Stuiver, M. (1980): GEOSECS Pacific Radiocarbon. *Radiocarbon*, **22**, 25–53.

O'Sullivan, P.E. (1983): Annually-laminated lake sediments and the study of Quaternary environmental changes – a review. *Quaternary Science Reviews*, **1**, 245–313.

Ota, Y. (1987): Sea-level changes during the Holocene: the northwest Pacific. In *Sea Surface Studies* (edited by R.J.N. Devoy), Croom Helm, London, 348–374.

Ota, Y., Hull, A.G. & Berryman, K.R. (1991): Coseismic uplift of Holocene marine terraces in the Pakare River area, eastern North Island, New Zealand. *Quaternary Research*, **35**, 331–346.

Otlet, R.L., Huxtable, G.& Sanderson, D.C.W. (1986): The development of practical systems for ^{14}C measurements of small samples using miniature counters. *Radiocarbon*, **28**, 603–614.

Overpeck, J.T., Peterson, L.C., Kipp, N., Imbrie, J. & Rind, D. (1989): Climate change in the circum-North Atlantic region during the last deglaciation. *Nature*, **338**, 553–557.

Oviatt, C.G. (1990): Age and palaeoclimatic significance of the Stansbury Shoreline of Lake Bonneville, Northeast Great Basin. *Quaternary Research*, **33**, 291–305.

Oviatt, C.G., McCoy, W.D. & Reider, R.G. (1987): Evidence for a shallow early or middle Wisconsin-age lake in the Bonneville Basin, Utah. *Quaternary Research*, **27**, 248–262.

Oviatt, C.G., Currey, D.R. & Sack, D. (1992): Radiocarbon chronology of Lake Bonneville, Eastern Great Basin, USA. *Palaeogeography, Palaeoclimatology, Palaeoecology*, **99**, 225–241.

Owen, L.A. (1994): Glacial and non-glacial diamictons in the Karakoram Mountains and western Himalayas. In *Formation and Deformation of Glacial Deposits* (edited by W.P. Warren & D.G. Croot), Balkema, Rotterdam & Brookfield, 9–28.

Owen, L.A. & Derbyshire, E. (1988): Glacially deformed diamictons in the Karakoram Mountains, northern Pakistan. In *Glaciotectonics: Forms and Processes* (edited by D. Croot), Balkema, Rotterdam, 149–176.

Owen, L.A., Stewart, I. & Vita-Finzi, C. (ed.) (1993): Neotectonics: recent advances. *Quaternary Proceedings*, **3**, Quaternary Research Association, Cambridge.

Owen, R. (1846): *A History of British Fossil Mammals, and Birds*. Van Voorst, London.

Palais, J.M., Germani, S. & Zielinski, G.A. (1992): Interhemispheric transport of volcanic ash from a 1259 AD volcanic eruption to the Greenland and Antarctic ice sheets. *Geophysical Research Letters*, **19**, 801–804.

Panagiotakopulu, E. & Buckland, P.C. (1991): Insect pests of stored products from Late Bronze Age, Santorini, Greece. *Journal of Stored Products Research*, **27**, 179–184.

Parker, M.L., Jozsa, L.A., Johnson, S.G. & Bramhall, P.A. (1984): Tree-ring dating in Canada and the Northwestern U.S. In *Quaternary Dating Methods* (edited by W.C. Mahaney), Elsevier, Amsterdam, 211–225.

Parks, D.A. & Rendell, H.M. (1992): Thermoluminescence dating and geochemistry of loessic deposits in southeast England. *Journal of Quaternary Science*, **7**, 99–108.

Paterson, W.S. B. (1994): *The Physics of Glaciers*, 3rd edition. Pergamon, Oxford.

Patience, A.J. & Kroon, D. (1991): Oxygen isotope chronostratigraphy. In *Quaternary Dating Methods – a User's Guide* (edited by P.L. Smart & P.D. Frances), Technical Guide 4, Quaternary Research Association, Cambridge, 199–228.

Patzold, J. (1984): Growth rhythms recorded in stable isotopes and density bands in the reef coral Porites lobata (Cebu, Philippines). *Coral Reefs*, **3**, 87–90.

Patzold, J. (1986): Temperature and CO_2 changes in tropical surface waters of the Philippines during the past 120 years: record in the stable isotopes of hermatypic corals. *Berichte-Reports, Geol.-Palaeontol. Institute, University of Kiel, Number 12*, 1–82.

Payne, A.J., Sugden, D.E. & Clapperton, C.M. (1989): Modelling the growth and decay of the Antarctic Peninsula ice sheet. *Quaternary Research*, **31**, 119–134.

Pazdur, A., Pazdur, M.F., Starkel, L. & Szulc, J. (1988): Stable isotopes of Holocene calcareous tufa in Southern Poland as palaeoclimatic indicators. *Quaternary Research*, **30**, 177–189.

Peacock J.D. (1987): A reassessment of the probable Loch Lomond Stade marine molluscan fauna at Garvel Park, Greenock. *Scottish Journal of Geology*, **23**, 93–103.

Peacock, J.D. (1989): Marine molluscs and Late Quaternary environmental studies with particular reference to the Late-glacial period in north-west Europe: a review. *Quaternary Science Reviews*, **8**, 179–192.

Peacock, J.D. (1993): Late Quaternary marine molluscs as palaeoenvironmental proxies: a compilation and assessment of basic numerical data for NE Atlantic species found in shallow water. *Quaternary Science Reviews*, **12**, 263–275.

Peacock, J.D. & Harkness, D.D. (1990): Radiocarbon ages and the full-glacial to Holocene transition in seas adjacent to Scotland and southern Scandinavia: a review. *Transactions of the Royal Society of Edinburgh: Earth Science*, **81**, 385–396.

Peacock, J.D., Graham, D.K., Robinson, J.E. & Wilkinson, I.P. (1977): Evolution and chronology of Late-glacial marine environments at Lochgilphead, Scotland. In *Studies in the Scottish Lateglacial Environment* (edited by J.M. Gray & J.J. Lowe), Pergamon, Oxford, 89–100.

Peacock, J.D., Graham, D.K. & Gregory, D.M. (1978): Late-glacial and post-glacial marine environments at Ardyne, Scotland, and their significance in the interpretation of the Clyde sea area. *Report of the Institute of Geological Sciences*, **78/17**, 1–25.

Peacock, J.D., Austin, W.E.N., Selby, I., Graham, D.K., Harland, R. & Wilkinson, I.P. (1992): Late Devensian and Flandrian palaeoenvironmental changes on the Scottish continental shelf west of the Outer Hebrides. *Journal of Quaternary Science*, **7**, 145–162.

Peake, D.S. (1961): Glacial changes in the Alyn river system and their significance in the glaciology of the north Welsh border. *Journal of the Geological Society, London*, **117**, 335–366.

Pearson, G.W. (1986): Precise calendrical dating of known growth period samples using a 'curve fitting' technique. *Radiocarbon*, **28**, 292–299.

Pearson, G.W & Stuiver, M. (1986): High-precision calibration of the radiocarbon timescale, 500–2500 BC. *Radiocarbon*, **28**, 839–862.

Pearson, G.W. & Stuiver, M. (1993): High-precision bidecadal calibration of the radiocarbon timescale, 500–2500 BC. *Radiocarbon*, **35**, 25–34.

Pearson, G.W., Pilcher, J.R., Baillie, M.G.L., Corbett, D.M. & Qua, F. (1986): High-precision ^{14}C measurement of Irish oaks to show the natural ^{14}C variations from AD 1840 to 5210 BC. *Radiocarbon*, **28**, 839–862.

Peck, R.M. (1974): A comparison of four absolute pollen preparation techniques. *New Phytologist*, **73**, 567–587.

Pécsi, M. (ed.) (1987): Loess and environment. *Catena Supplement* **9**, Catena Verlag, Cremlingen-Destedt.

Pécsi, M. (1990): Loess is not just the accumulation of dust. *Quaternary International*, **7/8**, 1–21.

Peel, D.A. (1994): The Greenland Ice-Core Project (GRIP): reducing uncertainties in climatic change? *NERC News*, April, 26–30.

Peglar, S.M. (1993a): The mid-Holocene *Ulmus* decline at Diss Mere, Norfolk, UK: a year-by-year pollen stratigraphy from annual laminations. *The Holocene*, **3**, 1–13.

Peglar, S.M. (1993b): The development of the cultural landscape around Diss Mere, Norfolk, U.K., during the past 7000 years. *Review of Palaeobotany and Palynology*, **76**, 1–47.

Peglar, S.M. & Birks, H.J.B. (1993): The mid-Holocene *Ulmus* fall at Diss Mere, south-east England – disease and human impact? *Vegetational History and Archaeobotany*, **2**, 61–68.

Peglar, S.M., Fritz, S.C., Alapieti, T., Saarnisto, M. & Birks, H.J.B. (1984): The composition and formation of laminated lake sediments in Diss Mere, Norfolk, England. *Boreas*, **13**, 13–28.

Peglar, S.M., Fritz, S.C. & Birks, H.J.B. (1989): Vegetation and land-use history at Diss, Norfolk. *Journal of Ecology*, **77**, 203–222.

Peltier, W.R. (1981): Ice age geodynamics. *Annual Review of Earth and Planetary Science Letters*, **9**, 199–216.

Peltier, W.R. (1987): Mechanisms of relative sea-level change and the geophysical responses to ice–water loading. In *Sea Surface Studies* (edited by R.J.N. Devoy), Croom Helm, London, 57–94.

Penck, A. & Brückner E. (1909): *Die Alpen im Eiszeitalter*. Tachnitz, Leipzig.

Pendall, E.G., Harden, J.W., Trumbore, S.E. & Chadwick, O.A. (1994): Isotopic approach to soil carbonate dynamics and implications for paleoclimatic interpretations. *Quaternary Research*, **42**, 60–71.

Peng, C.H., Guiot, J., Van Campo, E. & Cheddadi, R. (1994): The vegetation carbon storage in Europe since 6000 BP: reconstructions from pollen. *Journal of Biogeography*, **21**, 19–31.

Penney, D.N. (1993): Northern North Sea benthic Ostracoda: modern distributions and palaeoenvironmental significance. *The Holocene*, **3**, 241–264.

Pennington, W. (1980): Modern pollen samples from west Greenland and the interpretation of pollen data from the British Late-Glacial (Late Devensian). *New Phytologist*, **84**, 171–201.

Pennington, W. (1986): Lags in adjustment of vegetation to climate caused by the pace of soil developments: evidence from Britain. *Vegetatio*, **67**, 105–118.

Pennington, W. & Lishman, J.P. (1971): Iodine in lake sediments in northern England and Scotland. *Biological Reviews*, **46**, 279–313.

Pennington, W., Haworth, E.Y., Bonny, A.P. & Lishman, J.P. (1972): Lake sediments in northern Scotland. *Philosophical Transactions of the Royal Society of London*, **B264**, 191–294.

Pennington, W., Cambray, R.S. & Fisher, E.M. (1973): Observations on lake sediments using fallout ^{137}Cs as a tracer. *Nature*, **242**, 324–326.

Pestieux, P., Duplessy, J.-C. & Berger, A. (1987): Palaeoclimatic variability at frequencies ranging from 10^{-4} cycle per year to 10^{-3} cycle per year – evidence for nonlinear behaviour of the climate system. In *Climate, History, Periodicity and Predictability* (edited by M.R. Rampino, J.E. Sanders, W.S. Newman & L.K. Konigsson), Van Nostrand Reinhold, New York, 285–299.

Peteet, D.M. (ed.) (1993): Global Younger Dryas? *Quaternary Science Reviews*, **12**, 277–355.

Peteet, D.M. (1995): Global Younger Dryas? *Quaternary International*, **28**, 93–104.

Peteet, D.M., Vogel,, J.S., Nelson, D.E., Southon, J.R., Nickmann, R.J. & Heusser, L.E. (1990): Younger Dryas climatic reversal in northeastern USA? AMS ages for an old problem. *Quaternary Research*, **33**, 219–230.

Peteet, D.M., Daniels, R.A., Heusser, L.E., Vogel, J.S., Southon, J.R. & Nelson, D.E. (1993): Late-glacial pollen, macrofossils and fish remains in northeastern USA – the Younger Dryas Oscillation. *Quaternary Science Reviews*, **12**, 597–612.

Peteet, D.M., Daniels, R.A, Heusser, L.E., Vogel, J.S., Southon, J.R. & Nelson, D.E. (1994): Wisconsinan Late-glacial environmental change in southern New England: a regional synthesis. *Journal of Quaternary Science*, **9**, 151–154.

Petit, J.R., Mournier, L., Jouzel, J., Korotkevitch, Y.S., Kotlyakov, V.I. & Lorius, C. (1990): Palaeoclimatological and chronological implications of the Vostok core dust record. *Nature*, **343**, 56–58.

Petit-Maire, N. & Riser, J. (1983): *Sahara ou Sahel? Quaternaire récent du Bassin de Taoudenni (Mali)*. Laboratoire Géologie Quaternaire du C.N.R.S., Paris.

Petts, G.E., Moller, H. & Roux, A.L. (eds) (1989): *Historical Change of Large Alluvial Rivers: Western Europe*. John Wiley, Chichester & New York.

Peuraniemi, V. (1989): Till stratigraphy and ice movement directions in the Kitillä area, Finnish Lapland. *Boreas*, **18**, 145–158.

Peuranienui, V. (1990): Heavy minerals in glacial material. In *Glacier Indicator Tracing* (edited by R. Kujansuu & M. Saarnisto), Balkema, Rotterdam, 165–186.

Péwé, T.L. (1983): Alpine permafrost in the contiguous United States: a review. *Arctic and Alpine Research*, **15**, 145–156.

Phillips, F.M., Leavey, B.D., Jannik, N.O. & Kubik, P.W. (1986): The accumulation of cosmogenic chlorine-36 in rocks: a method for exposure dating. *Science*, **231**, 41–43.

Phillips, F.M., Zreda, M.G. Smith, S.S., Elmore, D., Kubik, P.W. & Sharma, P. (1990): Cosmogenic chlorine-36 chronology for glacial deposits at Bloody Canyon, eastern Sierra Nevada. *Science*, **248**, 1529–1532.

Phillips, F.M. *et al.* (1994): Surface exposure dating of glacial features in Great Britain using cosmogenic chlorine-36: preliminary results. *Mineralogical Magazine*, **58A**, 722–723.

Pickett, J.W., Ku, T.L., Thompson, C.H., Roman, D., Kelley, R.A. & Huang, Y.P. (1989): A review of age determinations on Pleistocene corals in eastern Australia. *Quaternary Research*, **31**, 392–395.

Pierce, K.L., Obradovich, J.D. & Friedman, I. (1976): Obsidian hydration dating and correlation of Bull Lake and Pinedale Glaciations near west Yellowstone, Montana. *Geological Society of America Bulletin*, **87**, 703–710.

Pilcher, J. (1991a): Radiocarbon dating for the Quaternary scientist. In *Radiocarbon Dating: Recent Applications and Future Potential* (edited by J.J. Lowe), *Quaternary Proceedings*, **1**, Quaternary Research Association, Cambridge, 27–34.

Pilcher, J. (1991b): Radiocarbon Dating. In *Quaternary Dating Methods – a User's Guide* (edited by P.L. Smart & P.D. Francis), Technical Guide 4, *Quaternary Research* Association, Cambridge, 16–36.

Pilcher, J.R. & Hall, V.A. (1992): Towards a tephrochronology for the Holocene of the north of Ireland. *The Holocene*, **2**, 255–259.

Pilcher, J.R., Baillie, M.G.L., Schmidt, B. & Becker, B. (1984): A 7272-year tree-ring chronology for western Europe. *Nature*, **312**, 150–152.

Pillans, B. (1983): Upper Quaternary marine terrace chronology and deformation, South Taranaki, New Zealand. *Geology*, **11**, 292–297.

Pillans, B. (1987): Quaternary sea-level changes: Southern Hemisphere data. In *Sea Surface Studies: A Global View* (edited by R.J.N. Devoy), Croom Helm, London, 264–293.

Pillans, B. (1991): New Zealand Quaternary stratigraphy: an overview. *Quaternary Science Reviews*, **10**, 405–418.

Pillans, B. & Wright, I. (1990): 500,000-year palaeomagnetic record from New Zealand loess. *Quaternary Research*, **33**, 178–187.

Pillans, B., Holgate, G. & McGlone, M. (1988): Climate and sea level during oxygen isotope stage 7b: on-land evidence from New Zealand. *Quaternary Research*, **29**, 176–185.

Piotrowski, J.A. (1987): Genesis of the Woodstock drumlin field, southern Ontario, Canada. *Boreas*, **16**, 249–266.

Pirazzoli, P.A., Stiros, S.C., Laborel, J., Laborel-Deguen, F., Arnold, M., Papageorgiou, S. & Morhange, C. (1994): Late-Holocene shoreline changes related to palaeoseismic events in the Ionian Islands, Greece. *The Holocene*, **4**, 397–405.

Pirozynski, K.A. (1990): Fungi. In *Methods in Quaternary Ecology* (edited by B.G. Warner), *Geoscience Canada, Reprint Series* **5**, 15–22.

Pisias, N.G. & Shackleton, N.J. (1984): Modelling the global climatic response to orbital forcing and amospheric carbon dioxide changes: a frequency domain approach. *Nature*, **310**, 757–759.

Pisias, N.G., Martinson, D.G., Moore, T.C., Shackleton, N.J., Prell, W., Hays, J.D. & Boden, G. (1984); High resolution stratigraphic correlation of benthic oxygen isotope records spanning the last 300,000 years. *Marine Geology*, **56**, 119–136.

Pissart, A. (1988): Pingos: an overview of the present state of knowledge. In *Advances in Periglacial Geomorphology* (edited by M.J. Clark), John Wiley, Chichester & New York, 279–297.

Pitman, W.C. III (1978): Relationship between eustasy and stratigraphic sequences. *Bulletin of the Geological Society of America*, **89**, 1389–1403.

Plassche, O. van de (ed.) (1986): *Sea-level Research: a Manual for the Collection and Evaluation of Data*. Geo Books, Norwich.

Polach, H.A. & Costin, A.B. (1971): Validity of soil organic matter radiocarbon dating: buried soils in the Snowy Mountains, southeastern Australia as example. In *Paleopedology: Origin, Nature and Dating of Paleosols* (edited by D.H. Yaalon), International Society of Soil Scientists and Israel University Press, Jerusalem, 89–96.

Ponel, P. (1995): Rissian, Eemian and Wurmian Coleoptera assemblages from La Grande Pile (Vosges, France). *Palaeogeography, Palaeoclimatology Palaeoecology*, **114**, 1–41.

Ponel, P. & Coope, G.R. (1990): Lateglacial and early Flandrian Coleoptera from La Taphanel, Massif Central, France: climatic and ecological implications. *Journal of Quaternary Science*, **5**, 235–250.

Ponel, P. & Lowe, J.J. (1992): Coleopteran, pollen and radiocarbon evidence from the Prato Spilla 'D' succession, N. Italy. *Comptes Rendues de l'Academie des Sciences, Paris, ser. II*, **315**, 1425–1431.

Pons, A., Guiot, J., Beaulieu, J.-L. de & Reille, M. (1992): Recent contributions to the climatology of the Last Glacial–Interglacial Cycle based on French pollen sequences. *Quaternary Science Reviews*, **11**, 439–448.

Porter, S.C. (1975): Equilibrium-line altitudes of late Quaternary glaciers in the Southern Alps, New Zealand. *Quaternary Research*, **5**, 27–47.

Porter, S.C. (ed.) (1983): *Late Quaternary Environments of the United States*, Volume 1, *The Late Pleistocene*. Longman, London.

Porter, S.C. (1986): Pattern and forcing of northern hemisphere glacier variations during the last millennium. *Quaternary Research*, **26**, 27–48.

Porter, S.C. & An, Z. (1995): Correlation between climatic events in the North Atlantic and China during the last glaciation. *Nature*, **375**, 305–307.

Porter, S.C., Pierce, K.L. & Hamilton, T.D. (1983): Late Wisconsin mountain glaciation in the western United States. In *Late Quaternary Environments of the United States*, Volume 1, *The Late Pleistocene* (edited by S.C. Porter), Longman, London, 71–111.

Post, L. von (1946): The prospects for pollen analysis in the study of the earth's climatic history. *New Phytologist*, **45**, 193–217.

Preece, R.C. (1980a): The biostratigraphy and dating of the tufa deposit at the Mesolithic site at Blashenwell, Dorset, England. *Journal of Archaeological Science*, **7**, 345–362.

Preece, R.C. (1980b): The biostratigraphy and dating of a Postglacial slope deposit at Gore Cliff, near Blackgang, Isle of Wight. *Journal of Archaeological Science*, **7**, 255–265.

Preece, R.C. (1981): The value of shell microsculpture as a guide to the identification of land mollusca from Quaternary deposits. *Journal of Conchology*, **30**, 331–337.

Preece, R.C. (1990): The molluscan fauna of the Middle Pleistocene interglacial deposits at Little Oakley, Essex, and its environmental and stratigraphical implications. *Philosophical Transactions of the Royal Society of London*, **B328**, 387–407.

Preece, R.C. (1993): Late Glacial and Post-glacial molluscan successions from the site of the Channel Tunnel in SE England. *Scripta Geologica, Special Issue* **2**, 387–395.

Preece, R.C. & Day, S.P. (1994): Comparison of post-glacial molluscan and vegetational successions from a radiocarbon-dated tufa sequence in Oxfordshire. *Journal of Biogeography*, **21**, 463–478.

Preece, R.C. & Robinson, J.E. (1984): Late Devensian and Flandrian environmental history of the Ancholme Valley, Lincolnshire: molluscan and ostracod evidence. *Journal of Biogeography*, **11**, 319–352.

Preece, R.C., Bennett, K.D. & Robinson, J.E. (1984): The biostratigraphy of an early Flandrian tufa at Inchrory, Glen Avon, Banffshire. *Scottish Journal of Geology*, **20**, 143–159.

Preece, R.C. (ed.) (1997): *Holywell Coombe, Folkestone: The Geology, Prehistory and Ecology of an English Chalk Valley* (in preparation).

Prell, W.L., Imbrie, J., Martinson, D.G., Morley, J., Pisias, N.G., Shackleton, N.J. & Streeter, H.F. (1986): Graphic correlation of oxygen isotope stratigraphy application to the late Quaternary. *Paleoceanography*, **1**, 137–162

Prell, W.L., Marvil, R.E. & Luther, M.E. (1990): Variability in up-welling fields in the northwestern Indian Ocean. 2. Data-model comparisons at 9000 years B.P. *Paleoceanography*, **5**, 447–457.

Prentice, I.C. & Solomon, A.M. (1991): Vegetation models and global change. In *Global Changes of the Past* (edited by R.S. Bradley), University Corporation for Atmospheric Research, Office of Interdisciplinary Earth Science, Boulder, Colorado, 365–383.

Prentice, I.C. & Webb, T. III (1986): Pollen percentages, tree abundances and the Fagerlind effect. *Journal of Quaternary Science*, **1**, 35–44.

Prentice, I.C., Bartlein, P.J. & Webb, T. III (1991): Vegetation and climate change in eastern North America since the last glacial maximum. *Ecology*, **72**, 2038–2056.

Prentice, I.C., Cramer, W., Harrison, S.P., Leemans, R., Monserud, R.A. & Soloman, A.M. (1992): A global biome model based on plant physiology and dominance, soil properties and climate. *Journal of Biogeography*, **19**, 117–134.

Prestwich, J. (1890): On the the relation of the Westleton Beds or pebbly sands of Suffolk, to those of Norfolk and their extension inland, etc. *Quarterly Journal of the Geological Society of London*, **46**, 120–154.

Priesnitz, K. (1988): Cryoplanation. In *Advances in Periglacial Geomorphology* (edited by M.J. Clark), John Wiley, Chichester & New York, 49–67.

Proctor, C.J. & Smart, P.L. (1991): A dated cave sediment record of Pleistocene transgressions on Berry Head, southwest England. *Journal of Quaternary Science*, **6**, 233–244.

Prospero, J.M., Glaccum, R.A. & Nees, R.T. (1981): Atmospheric transport of soil dust from S. Africa to S. America. *Nature*, **289**, 570–572.

Pullar, W.A., Pain, C.F. & Johns, R.J. (1967): Chronology of terraces, floodplains, fans and dunes in the Whakatone Valley. *Proceedings of the 5th New Zealand Geographical Conference*, 175–180.

Punning, J.-M., Martma, T., Kessel, H. & Vaikmäe, R. (1988): The isotopic composition of oxygen and carbon in the subfossil shells of the Baltic Sea as an indicator of palaeosalinity. *Boreas*, **17**, 27–32.

Punt, W. & Clarke, G.C.S. (eds) (1976): *Northwest European Pollen Flora* (series). Elsevier.

Pye, K. (ed.) (1993): The dynamics and environmental context of aeolian sedimentary systems. *Geological Society of London, Special Publication* **72**, Geological Society Publishing House, Bath.

Pye, K. & Tsoar, H. (1987): The mechanics and geological implications of dust transport and deposition in deserts with particular reference to loess formation and dune sand diagenesis in the northern Negev, Israel. In *Desert Sediments: Ancient and Modern* (edited by L. Frostick & I. Reid), Geological Society Special Publication **35**, Oxford, 139–156.

Qui, G. & Cheng, G. (1995): Permafrost in China: past and present. *Permafrost and Periglacial Processes*, **6**, 3–14.

Quinlan, G. (1985): A numerical model of postglacial relative sea level change near Baffin Island. In *Quaternary Environments: Eastern Canadian Arctic, Baffin Bay and Western Greenland* (edited by J.T. Andrews), Allen & Unwin, London, 560–584.

Quinn, T.M., Taylor, F.W. & Crowley, T.J. (1993): A 173 year stable isotope record from a tropical South Pacific coral. *Quaternary Science Reviews*, **12**, 407–418.

Radle, N., Keister, C.M. & Battarbee, R.W. (1989): Diatom, pollen and geochemical evidence for the palaeosalinity of Medicine Lake, South Dakota during the late Wisconsin and early Holocene. *Journal of Palaeolimnology*, **2**, 159–172.

Radtke, U., Grün, R. & Schwarcz, H.P. (1988): Electron spin resonance dating of the Pleistocene reef tracts of Barbados. *Quaternary Research*, **29**, 197–215.

Rae, A.M. & Ivanovich, M.L. (1986): Successful application of uranium series dating of fossil bone. *Applied Geochemistry*, **1**, 419–426.

Rae, A.M., Ivanovich, M. & Schwarcz, H.P. (1987): Absolute dating by uranium series disequilibrium of bones from the cave of La Chaise-de-Vouthon (Charente). *Earth Surface Processes and Landforms*, **12**, 543–550.

Rahman, A. (1995): Reworked nannofossils in the North Atlantic Ocean: implications for Heinrich events and ocean circulation. *Geology*, **23**, 487–490.

Rainio, H. (1991): The Younger Dryas ice-marginal formations of southern Finland. In *Eastern Fennoscandian Younger Dryas End Moraines* (edited by H. Rainio & M. Saarnisto), Geological Survey of Finland, Espoo, 25–72.

Raisbeck, G.M., Yiou, F., Jouzel, J., Petit, J.R., Barkov, N.I. & Bard, E. (1992): ^{10}Be deposition at Vostok, Antarctica during the last 50,000 years and its relationship to possible cosmogenic production variations during this period. In *The Last Deglaciation: Absolute and Radiocarbon Chronologies* (edited by E. Bard & W.S. Broecker), *NATO ASI Series* **1**, **2**, Springer-Verlag, Berlin, 127–139.

Ram, M. & Gayley, R.I. (1991): Long-range transport of volcanic ash to the Greenland ice sheet. *Nature*, **349**, 401–404.

Rampino, M.R. & Self, S. (1992): Volcanic winter and accelerated glaciation following the Toba super-eruption. *Nature*, **359**, 50–52.

Rampino, M.R. & Self, S. (1993): Climate-volcanism feedback and the Toba eruption of ~74,000 years ago. *Quaternary Research*, **40**, 269–280.

Ran, E.T.H. (1990): Dynamics of the vegetation and environment during the Middle Pleniglacial in the Dinkel Valley (The Netherlands). *Mededelingen rijks geologische dienst*, **44–3**, 139–205.

Ran, E.T.H., Bohncke, S.J.P., van Huissteden, K.J. & Vandenberghe, J. (1990): Evidence of episodic permafrost conditions during the Weichselian Middle Pleniglacial in the Hengelo Basin (The Netherlands). *Geologie en Mijnbouw*, **69**, 207–218.

Rasmussen, K.A., Macintyre, I.G. & Prufert, L. (1993): Modern stromatolite reefs fringing a brackish coastline, Chetumal Bay, Belize. *Geology*, **21**, 199–202.

Rasmussen, T.L. (1991): Benthonic and planktonic foraminifera in relation to the Early Holocene stagnation in the Ionian Basin, Central Mediterranean. *Boreas*, **20**, 357–376.

Raymo, M.E. & Ruddiman, W.F. (1992): Tectonic forcing of late Cenozoic climate. *Nature*, **359**, 117–122

Raymo, M.E., Ruddiman, W.F., Backman, J., Clement, B.M. & Martinson, D.G. (1989): Late Pleistocene evolution of the northern hemisphere ice sheets and North Atlantic deep water circulation. *Paleoceanography*, **4**, 413–446.

Raymo, M.E., Ruddiman, W.F., Shackleton, N.J. & Oppo, D. (1990): Evolution of Atlantic–Pacific δ^{13}C gradients over the last 2.5 myr. *Earth and Planetary Science Letters*, **97**, 353–368.

Raynaud, D., Barnola, J.M., Chappellaz, J., Zardini, D., Jouzel, J. & Lorius, C. (1992): Glacial–interglacial evolution of greenhouse gases as inferred from ice core analysis: a review of recent results. *Quaternary Science Reviews*, **11**, 381–386.

Raynaud, D., Jouzel, J., Barnola, J.M., Chappellaz, J., Delmas, R.J. & Lorius, C. (1993): The ice record of greenhouse gases. *Science*, **259**, 926–934.

Rea, D.K. (1994): The paleoclimatic record provided by eolian deposition in the deep sea: the geologic history of the wind. *Review of Geophysics*, **32**, 159–195.

Reading, H.G.(1978): Facies. In *Sedimentary Environments and Facies* (edited by H.G. Reading), Blackwell, Oxford, 4–14.

Reed, B., Galvin, C.J. & Miller, J.P. (1962): Some aspects of drumlin geometry. *American Journal of Science*, **260**, 200–210.

Reid, C. (1899): *The Origin of the British Flora*. Dulau, London.

Reid, I. & Frostick, L. (eds) (1987): *Desert Sediments: Ancient and Modern*. Blackwell, Oxford.

Reid, I. & Frostick, L.E. (1993): Late Pleistocene rhythmite sedimentation at the margin of the Dead Sea trough: a guide to palaeoflood frequency. In *Geomorphology and Sedimentology of Lakes and Reservoirs* (edited by J. McManus & R.W. Duck), John Wiley & Sons, Chichester, 259–273.

Reider, R.G. (1983): A soil catena in the Medicine Bow Mountains, Wyoming, U.S.A., with reference to palaeoenvironmental influences. *Arctic and Alpine Research*, **15**, 181–192.

Reille, M. (1992): *Pollen et Spores d'Europe et d'Afrique du Nord*. Laboratoire de Botanique Historique et Palynologie, Marseille.

Reille, M. & Lowe, J.J. (1993): A re-evaluation of the vegetation history of the eastern Pyrenees (France) from the end of the Last Glacial to the present. *Quaternary Science Reviews*, **12**, 47–77.

Reille, M., Guiot, J. & Beaulieu, J.-L. de (1992): The Montaigu Event: an abrupt climatic change during the Early Würm in Europe. In *Start of a Glacial* (edited by G.J. Kukla & E. Went), *NATO ASI Series* **1**, **3**, Springer-Verlag, Berlin & Heidelberg, 85–95.

Reineck, H.-E. & Singh, I.B. (1973): *Depositional Sedimentary Environments*. Springer-Verlag, Berlin & New York.

Renberg, I. (1981a): Formation, structure and visual appearance of iron-rich, varved lake sediments. *Verhandlungen Internationalen Vereinigung für Limnologie*, **21**, 94–101.

Renberg, I. (1981b): Improved methods for sampling, photographing and varve-counting of varved lake sediments. *Boreas*, **10**, 255–258.

Renberg, I. & Hellberg, T. (1982): The pH history of lakes in south-western Sweden, as calculated from the subfossil diatom-flora of the sediments. *Ambio*, **11**, 30–33.

Renberg, I., Korsman, T. & Birks, H.J.B. (1993): Prehistoric increases in the pH of acid-sensitive Swedish lakes caused by land-use changes. *Nature*, **362**, 824–826.

Rendall, H., Worsley, P., Green, F. & Parks, D. (1991): Thermoluminescence dating of the Chelford Interstadial. *Earth and Planetary Science Letters*, **103**, 182–189.

Rendell, H.M. & Townsend, P.D. (1988): Thermoluminescence dating of a 10m loess profile in Pakistan. *Quaternary Science Reviews*, **7**, 251–255.

Reneau, S.L. & Raymond, R.J. (1991): Cation-ratio dating of rock varnish: why does it work? *Geology*, **19**, 937–940.

Ribes, E. (1990): Astronomical determinations of solar variability. *Philosophical Transactions of the Royal Society*, **A330**, 487–497.

Rice, R.J. (1977): *Fundamentals of Geomorphology*. Longman, London.

Richard, P.J.H. (1994): Wisconsinan Late-glacial environmental change in Québec: a regional synthesis. *Journal of Quaternary Science*, **9**, 165–170.

Richards, D.A. & Smart, P.L. (1991): Potassium–argon and argon–argon dating. In *Quaternary Dating Methods – a User's Guide* (edited by P.L. Smart & P.D. Frances), Technical Guide 4, Quaternary Research Association, Cambridge, UK, 37–44.

Richards, D.A., Smart, P.L. & Edwards, R.L. (1994): Maximum sea levels for the last glacial period from U-series ages of submerged speleothems. *Nature*, **367**, 357–360.

Richards, K.S. (ed.) (1990): Part two: form (morphometry). In *Geomorphological Techniques* (edited by A. Goudie), Unwin Hyman, London & Boston, 31–108.

Richmond, G.M. (1986a): Tentative correlation of deposits of the Cordilleran Ice Sheet in the northern Rocky Mountians. *Quaternary Science Reviews*, **5**, 135–160.

Richmond, G.M. (1986b): Stratigraphy and chronology of glaciations in Yellowstone National Park. *Quaternary Science Reviews*, **5**, 83–98.

Richmond, G.M. & Fullerton, D.S. (1986a): Introduction to Quaternary glaciations in the United States of America. *Quaternary Science Reviews*, **5**, 3–10.

Richmond, G.M. & Fullerton, D.S. (1986b): Summation of Quaternary glaciations in the United States of America. *Quaternary Science Reviews*, **5**, 183–196.

Riehle, J.R., Bowers, P.M. & Ager, T.A. (1990): The Hayes Tephra deposits, an upper Holocene marker horizon in south-central Alaska. *Quaternary Research*, **33**, 276–290.

Riehle, J.R., Mann, D.H., Peteet, D.M., Engstrom, D.R., Brew, D.A. & Meyer, C.E. (1992): The Mount Edgecumbe tephra deposits, a marker horizon in southeastern Alaska near the Pleistocene–Holocene boundary. *Quaternary Research*, **37**, 183–202.

Riggs, A.C., Carr, W.J., Kolesar, P.T. & Hoffman, R.J. (1994): Tectonic speleogenesis of Devils Hole, Nevada, and implications for hydrogeology and the development of long, continuous palaeoenvironmental records. *Quaternary Research*, **42**, 241–254.

Rind, D. (1987): Components of ice age circulation. *Journal of Geophysical Research*, **92**, 4241–4281.

Rind, D. (1993): How will future climate changes differ from those of the past? In *Global Changes in the Perspective of the Past* (edited by J.A. Eddy & H. Oeschger), John Wiley, Chichester & New York, 39–49.

Rind, D. & Overpeck, J. (1993): Hypothesised causes of decade-to-century-scale climate variability: climate model results. *Quaternary Science Reviews*, **12**, 357–374.

Rind, D., Peteet, D., Broecker, W.S., McIntyre, A. & Ruddiman, W.F. (1986): The impact of cold North Atlantic sea surface temperatures on climate. Implications for the Younger Dryas cooling (11–10 k). *Climate Dynamics*, **1**, 3–33.

Ringberg, B. (1984): Cyclic lamination in proximal varves reflecting the length of summer during Late Weichsel in southernmost Sweden. In *Climatic Changes on a Yearly to Millennial Basis* (edited by N.-A. Mörner & W. Karlén), Reidel, Dordrecht, 57–62.

Ritchie, J.C. (1987): *Postglacial Vegetation of Canada*. Cambridge University Press, Cambridge.

Roberts, M.C., Pullan, S.E. & Hunter, J.A. (1992): Applications of land-based high-resolution seismic reflection analysis to Quaternary and geomorphic research. *Quaternary Science Reviews*, **11**, 557–568.

Roberts, N. (1989): *The Holocene: An Environmental History*. Basil Blackwell, Oxford.

Roberts, N., Taieb, M., Barker, P., Damnati, B., Icole, M. & Williamson, D. (1993): Timing of the Younger Dryas event in East Africa from lake-level changes. *Nature*, **366**, 146–148.

Robin, G. de Q. (1977): Ice cores and climatic change. *Philosophical Transactions of the Royal Society of London*, **B280**, 143–168.

Robin, G. de Q. (ed.) (1983): *The Climatic Record in Polar Ice Sheets*. Cambridge University Press, Cambridge.

Robinson, D.A. & Williams, R.B.G. (eds) (1994): *Rock Weathering and Landform Evolution*. John Wiley, Chichester & New York.

Robinson, J.E. (1980): The marine ostracod record from the Lateglacial period in Britain and NW Europe: a review. In *Studies in the Lateglacial of North-west Europe* (edited by J.J. Lowe, J.M. Gray & J.E. Robinson), Pergamon, Oxford, 115–122.

Robinson, M. & Ballantyne, C.K. (1979): Evidence for a glacial advance predating the Loch Lomond Advance in Wester Ross. *Scottish Journal of Geology*, **15**, 271–277.

Robinson, S.G. & McCave, I.N. (1994): Orbital forcing of bottom current enhanced sedimentation in Feni Drift, NE Atlantic, during the mid-Pleistocene. *Paleoceanography*, **9**, 943–972.

Robinson, S.G., Maslin, M.A. & McCave, I.N. (1995): Magnetic susceptibility variations in Upper Pleistocene deep-sea sediments of the N.E. Atlantic: implications for ice rafting and paleocirculation at the last glacial maximum. *Paleoceanography*, **10**, 221–250.

Rodbell, D.T. (1990): Soil–age relationships on Late Quaternary moraines, Arrowsmith Range, Southern Alps, New Zealand. *Arctic and Alpine Research*, **22**, 355–365.

Rodbell, D.T. (1992a): Late Pleistocene equilibrium-line reconstructions in the northern Peruvian Andes. *Boreas*, **21**, 43–52.

Rodbell, D.T. (1992b): Lichenometric and radiocarbon dating of Holocene glaciation, Cordillera Blanca, Peru. *The Holocene*, **2**, 1–10.

Rognon, P. (1980): Pluvial and arid phases in the Sahara: the role of non-climatic factors. In *Palaeoecology of Africa* (edited by

E.M. Van Zinderen Bakker & J.A. Coetzee), Balkema, Rotterdam, 45–62.

Rolph, T.C., Shaar, J., Derbyshire, E. & Wang, J.T. (1993): The magnetic mineralogy of a loess section near Lanzhou, China. In *The Dynamics and Environmental Context of Aeolian Sedimentary Systems* (edited by K. Pye), Geological Society of London, 311–323.

Rose, J. (1974): Small scale spatial variability of some sedimentary properties of lodgement and slumped till. *Proceedings of the Geologists' Association*, **85**, 223–237.

Rose, J. (1985): The Dimlington Stadial/Dimlington Chronozone: a proposal for naming the main glacial episode of the Late Devensian in Britain. *Boreas*, **14**, 225–230.

Rose, J. (1989): Stadial type sections in the British Quaternary. In *Quaternary Type Sections: Imagination or Reality* (edited by J. Rose & C. Schlüchter), Balkema, Rotterdam, 45–67.

Rose, J. (1990): Raised shorelines. In *Geomorphological Techniques*, 2nd edition (edited by A.Goudie), Allen & Unwin, London, 327–341.

Rose, J. & Allen, P. (1977): Middle Pleistocene stratigraphy in southeast Suffolk. *Journal of the Geological Society of London*, **133**, 83–102.

Rose, J. & Letzer, J.M. (1975): Drumlin measurements: a test of the reliability of data derived from 1:25,000 scale topographic maps. *Geological Magazine*, **112**, 361–371.

Rose, J. & Schlüchter, C. (eds) (1989): *Quaternary Type Sections: Imagination or Reality*. Balkema, Rotterdam.

Rose, J., Boardman, J., Kemp, R.A. & Whiteman, C.A. (1985): Palaeosols and the interpretation of the British Quaternary stratigraphy. In *Geomorphology and Soils* (edited by K.S. Richards, R.R. Arnett & S. Ellis), George Allen & Unwin, London, 348–375.

Rosen, M.R. (1991): Sedimentologic and geochemical constraints on the evolution of Bristol Dry Lake Basin, California, USA. *Palaeogeography, Palaeoclimatology, Palaeoecology*, **84**, 229–257.

Round, F.E. (1981) : *The Ecology of Algae*. Cambridge University Press, Cambridge.

Round, F.E., Crawford, R.M. & Mann, D.G. (1990): *The Diatoms*. Cambridge University Press, Cambridge.

Rousseau, D.-D. (1991): Climatic transfer function from Quaternary molluscs in European loess deposits. *Quaternary Research*, **36**, 195–209.

Rousseau, D.-D. (1992): Terrestrial molluscs as indicators of global aeolian dust fluxes during glacial stages. *Boreas*, **21**, 105–110.

Rousseau, D.-D. & Puissegur, J.-J. (1990): A 350,000-year climatic record from the loess sequence of Achenheim, Alsace, France. *Boreas*, **19**, 203–216.

Rousseau, D.-D., Puissegur, J.-J. & Lautridou, J.-P. (1990): Biogeography of the Pleistocene pleniglacial malacofaunas in Europe. Stratigraphic and climatic implications. *Palaeogeography, Palaeoclimatology, Palaeoecology*, **80**, 7–23.

Rousseau, D.-D., Limondin, N. & Puissegur, J.-J. (1993): Holocene environmental signals from mollusk assemblages in Burgundy (France). *Quaternary Research*, **40**, 237–253.

Rousseau, D.-D., Limondin, N., Magnin, F. & Puissegur, J.-J. (1994): Temperature oscillations over the last 10,000 years in western Europe estimated from terrestrial mollusc assemblages. *Boreas*, **23**, 66–73.

Rozanski, K., Goslar, T., Dulinski, M., Kuc, T., Pazdur, M.F. & Walanus, A. (1992): The Late Glacial–Holocene transition in central Europe derived from isotope studies of laminated sediments from Lake Gosciaz (Poland). In *The Last Deglaciation: Absolute and Radiocarbon Chronologies* (edited by E. Bard & W.S. Broecker), *NATO ASI Series* **1**, **2**, Springer-Verlag, Berlin, 69–79.

Ruddiman, W.F. (1987): Northern oceans. In *The Geology of North America*, Volume K-3, *North America and Adjacent Oceans During the Last Deglaciation* (edited by W.F. Ruddiman & H.E. Wright, Jr.), Geological Society of America, Boulder, Colorado, 137–153.

Ruddiman, W.F. & Kutzbach, J.E. (1990): Late Cenozoic plateau uplift and climate change. *Transactions of the Royal Society of Edinburgh: Earth Sciences*, **81**, 301–314.

Ruddiman, W.F. & McIntyre, A. (1973): Time-transgressive deglacial retreat of polar waters from the North Atlantic. *Quaternary Research*, **3**, 117–130.

Ruddiman, W.F. & McIntyre, A. (1976): Northeast Atlantic palaeoclimatic changes over the past 600,000 years. *Memoirs of the Geological Society of America*, **145**, 111–146.

Ruddiman, W.F. & McIntyre, A. (1977): Late-Quaternary surface ocean kinematics and climatic change in the high-latitude North Atlantic. *Journal of Geophysical Research*, **16**, 125–134.

Ruddiman, W.F. & McIntyre, A. (1981): The North Atlantic during the last deglaciation. *Palaeogeography, Palaeoclimatology, Palaeoecology*, **35**, 145–214.

Ruddiman, W.F. & McIntyre, A. (1982): Severity and speed of Northern Hemisphere glaciation pulses: the liimiting case? *Geological Society of America Bulletin*, **93**, 1273–1279.

Ruddiman, W.F. & Mix, A.C. (1993): The North and Equatorial Atlantic at 9000 and 6000 yr BP. In *Global Climates Since the Last Glacial Maximum* (edited by H.E. Wright, Jr., J.E. Kutzbach, T. Webb III, W.F. Ruddiman, F.A. Street-Perrott & P.J. Bartlein), University of Minnesota Press, Minneapolis.

Ruddiman, W.F. & Raymo, M. (1988): Northern Hemisphere climate regimes during the past 3 Ma: possible tectonic connections. *Philosophical Transactions of the Royal Society, London*, **B318**, 411–430.

Ruddiman, W.F., Sancetta, C. D. & McIntyre, A. (1977): Glacial/interglacial response rate of subpolar North Atlantic water to climatic change: the record in ocean sediments. *Philosophical Transactions of the Royal Society, London*, **B280**, 119–42.

Ruddiman, W.F., McIntyre, A., Niebler-Hunt, V. & Durazzi, J.T. (1980): Oceanic evidence for the mechanism of rapid northern hemisphere glaciation. *Quaternary Research*, **13**, 33–64.

Ruddiman, W.F., McIntyre, A.F. & Raymo, M.E. (1986): Matuyama 41,000-year cycle: North Atlantic Ocean and northern hemisphere ice sheets. *Earth and Planetary Science Letters*, **80**, 117–129.

Ruddiman, W.F., Cameron, D. & Clement, B.M. (1987): Sediment disturbance and correlation of offset holes drilled with the hydraulic piston corer. *Initial Reports of the Deep Sea Drilling Project*, **94**, 615–634.

Ruddiman, W.F., Raymo, M.E., Martinson, D.G., Clement, B.M. & Backman, J. (1989): Pleistocene evolution; Northern Hemisphere ice sheets and North Atlantic Ocean. *Paleoceanography*, **4**, 353–412.

Ruegg, G.H.J. (1994): Alluvial architecture of the Quaternary Rhine–Meuse river system in The Netherlands. *Geologie en Mijnbouw*, **72**, 321–330.

Ruggieri, C. (1971): Ostracodes as cold climate indicators in the Italian Quaternary. In *Paléoécologie Ostracodes* (edited by H.J. Oertli), PAU 1970, *Bull. Centre Recherches, Pau-SNPA*, **5**, 285–293.

Ruhe, R.V. (1965): Quaternary paleopedology. In *The Quaternary of the United States* (edited by H.E. Wright & D.G. Frey), Princeton University Press, Princeton, NJ, 755–764.

Rust, B.R. & Nanson, G.C. (1986): Contemporary and palaeochannel patterns and the Late Quaternary stratigraphy of Cooper Creek, south-west Queensland, Australia. *Earth Surface Processes and Landforms*, **11**, 581–590.

Rutter, N. (1992): Chinese loess and global change. *Quaternary Science Reviews*, **11**, 275–281.

Rutter, N. & Ding, Z. (1993): Palaeoclimates and monsoon variations interpreted from micromorphogenic features of the Baoji palaesols, China. *Quaternary Science Reviews*, **12**, 853–862.

Rutter, N., Ding, Z.L., Evans, E.M. & Wang, Y.C. (1990): Magnetostratigraphy of the Baoji-type pedostratigraphic section, Loess Plateau, north-central China. *Quaternary International*, **7–8**, 97–102.

Rutter, N., Ding, Z.L., Evans, M.E. & Liu, T.S. (1991): Baoji-type pedostratigraphic section, Loess Plateau, north-central China. *Quaternary Science Reviews*, **10**, 1–22.

Rutter, N.W. (1969): Comparison of moraines formed by normal and surging glaciers. *Canadian Journal of Earth Sciences*, **6**, 991–999.

Rymer, L. (1978): The use of uniformitarianism and analogy in palaeoecology. In *Biology and Quaternary Environments* (edited by D. Walker & J.C. Guppy), Australian Academy of Science, Canberra, 245–258.

Saarnisto, M. (1986): Annually laminated lake sediments. In *Handbook of Holocene Palaeoecology and Palaeohydrology* (edited by B.E. Berglund), John Wiley, Chichester & New York, 343–370.

Salinger, M.J. & McGlone, M.S. (1990): New Zealand climate – the past two million years. In *New Zealand Climate Report*, The Royal Society of New Zealand, Wellington, 13–17.

Salis, J. & Bonhommet, N. (1992): Variation of geomagnetic field intensity from 8–60 Ka BP, Massif Central, France. In *The Last Deglaciation: Absolute and Radiocarbon Chronologies* (edited by E. Bard & W.S. Broecker), *NATO ASI Series* **1**, **2**, Springer-Verlag, Berlin, 156–161.

Salmonsen, I. (1994): A seismic stratigraphic analysis of Lower Pleistocene deposits in the western Dutch sector of the North Sea. *Geologie en Mijnbouw*, **72**, 349–361.

Saltzman, B. (1990): Three basic problems of paleoclimatic modelling: a personal perspective and review. *Climate Dynamics*, **5**, 67–78.

Salvador, A. (ed.) (1994): *International Stratigraphic Guide: A Guide to Stratigraphic Classification, Terminology and Procedure*, 2nd edition. International Subcommission on Stratigraphy, Geological Society of America.

Sancetta, C. & Silvestri, S. (1986): Pliocene–Pleistocene evolution of the North Pacific ocean–atmosphere system interpreted from fossil diatoms. *Paleoceanography*, **1**, 163–180.

Sancetta, C., Villareal, T. & Falkowski, P. (1991): Massive fluxes of Rhizosolenid diatoms: a common occurrence? *Limnology and Oceanography*, **36**, 1452–1457.

Sandgren, P. & Fredskild, B. (1991): Magnetic measurements recording Late Holocene man-induced erosion in S. Greenland. *Boreas*, **20**, 315–331.

Sandgren, P., Björck, S., Brunnberg, L. & Kristiansson, J. (1988): Palaeomagnetic records from two varved clay sequences in the Middle Swedish marginal zone. *Boreas*, **17**, 215–228.

Sarnthein, M. (1978): Sand deserts during the glacial maximum and climatic optimum. *Nature*, **272**, 43–46.

Sarnthein, M., Tetzlaff, G., Koopmann, B., Wolter, K. & Pflaumann, U. (1981): Glacial and interglacial wind regimes over the eastern sub-Tropical Atlantic and North-west Africa. *Nature*, **293**, 193–196.

Sarnthein, M., Jansen, E., Duplessy, J.-C., Erlenkauser, H., Flatoy, A., Veum, T., Vogelsang, E. & Weinelt, M.S. (1992): $\delta^{18}O$ time-slice reconstructions of meltwater anomalies at Termination 1 in the North Atlantic between 50° and 80°N. In *The Last Deglaciation: Absolute and Radiocarbon Chronologies* (edited by E. Bard & W.S. Broecker), *NATO ASI Series* **1**, **2**, Springer-Verlag, Berlin, 183–200.

Sarnthein, M., Winn, K., Jung, S.J.A., Duplessy, J.-C., Labeyrie, L., Erlenkauser, H. & Ganssen, G. (1994): Changes in east Atlantic deepwater circulation over the last 30,000 years: eight time slice reconstructions. *Paleoceanography*, **9**, 209–267.

Sauramo, M. (1918): Geochronologische Studien über die spätglaciale Zeit im Südfinnland. *Bulletin de la Commission Géologique de Finlande*, **50**. 44 pp.

Sauramo, M. (1923): Studies on the Quaternary varve sediments in southern Finland. *Bulletin de la Commission Géologique de Finlande*, **60**. 164 pp.

Schirmer, W. (1988): Holocene valley development on the Upper Rhine and Maine. In *Lake, Mire and River Environments* (edited by G. Lang & C. Schlüchter), Balkema, Rotterdam, 153–160.

Schlüchter, C. (1986): The Quaternary glaciations of Switzerland, with special reference to the northern Alpine Foreland. *Quaternary Science Reviews*, **5**, 413–420.

Schmidt, M. (1986): Possible influences of solar radiation variations on the atmospheric circulation in the Northern hemisphere of the earth. *Climate Change*, **8**, 279–296.

Schoch, W.H., Pawlik, B. & Schweingruber, F.H. (1988): *Botanical Macro-remains*. Paul Haupt, Bern & Stuttgart.

Schönwiese, C.-D. (1988): Volcanism and air temperature variations in recent centuries. In *Recent Climatic Change* (edited by S. Gregory), Belhaven Press, London & New York, 20–29.

Schøtt, C., Waddington, E.D. & Raymond, C.F. (1992): Predicted time-scales for GISP2 and GRIP boreholes at Summit, Greenland. *Journal of Glaciology*, **38**, 162–167.

Schove, D.J. (1983): *Sunspot Cycles*. Benchmark Volume 38, Hutchinson & Ross, London.

Schramm, C.T. (1985): Implications of radiolarian assemblages for the late Quaternary paleoceanography of the eastern equatorial Pacific. *Quaternary Research*, **24**, 204–218.

Schubert, G. & Yuen, D.A. (1982): Initiation of ice ages by creep instability and surging of the East Antarctic ice sheet. *Nature*, **296**, 127–130.

Schumm, S.A. & Brackenridge, G.R. (1987): River responses. In *The Geology of North America*, Volume K-3, *North America and Adjacent Oceans during the Last Deglaciation* (edited by W.F. Ruddiman & H.E. Wright, Jr.), Geological Society of America, Boulder, Colorado, 221–240.

Schwarcz, H.P. (1986): Geochronology and isotope geochemistry of speleothems. In *Handbook of Environmental Isotope Geochemistry*, volume 2 (edited by J. Ch. Fontes & P. Fritz), Elsevier, Amsterdam, 271–303.

Schwarcz, H.P. (1989): Uranium series dating of Quaternary deposits. *Quaternary International*, **1**, 7–18.

Schwarcz, H.P. & Latham, A.G. (1989): Dirty calcites. I. Uranium series dating of contaminated calcites using leachates alone. *Chemical Geology (Isotopes Geosciences Section)*, **80**, 35–43.

Schwarcz, H.P. & Yonge, C. (1983): Isotopic composition of paleowaters as inferred from speleothem and its fluid inclusion. In *Palaeoclimates and Paleowaters – a Collection of Environmental Isotope Studies*, International Atomic Energy Agency, Vienna, 115–133.

Schwarcz, H.P., Gascoyne, M. & Harmon, R.S. (1982): Applications of U-series dating to problems of Quaternary climate. In *Uranium Series Disequilibrium: Applications to Environmental Problems* (edited by M. Ivanovich & R.S. Harmon), Clarendon Press, Oxford, 326–350.

Schweingruber, F.H. (1988): *Tree Rings. Basics and Applications of Dendrochronology*. Reidel, Dordrecht.

Schweingruber, F.H. (1990): *Anatomy of European Woods*. Paul Haupt, Bern & Stuttgart.

Schweingruber, F.H., Bartholin, T., Schär, E. & Briffa, K.R. (1988): Radiodensitometric–dendroclimatological conifer chronologies from Lapland (Scandinavia) and the Alps (Switzerland). *Boreas*, **17**, 559–566.

Schwert, D.P. (1992): Faunal transitions in response to an ice age: the late Wisconsinan record of Coleoptera in the north-central United States. *Coleopterists Bulletin*, **46**, 68–94.

Schwert, D.P. & Ashworth, A.C. (1988): Late Quaternary history of the northern beetle fauna of North America: a synthesis of fossil distributional evidence. *Memoirs of the Entomological Society of Canada*, **144**, 93–107.

Scoffin, T.P. (1987): *An Introduction to Carbonate Sediments and Rocks*. Blackie, London.

Scott, E.M., Long, A. & Kra, R. (1990): Proceedings of the Workshop on Intercomparison of Radiocarbon Laboratories. *Radiocarbon*, **32**, 253–397.

Scott, E.M., Harkness, D.D., Cook, G.T., Aitchison, T.C. & Baxter, M.S. (1991): Future quality assurance in ^{14}C dating. In *Radiocarbon Dating: Recent Applications and Future Potential* (edited by J.J. Lowe), *Quaternary Proceedings*, 1, Quaternary Research Association, Cambridge, **1** 1–4.

Scuderi, L.A. (1987): Glacier variations in the Sierra Nevada, California, as related to a 1200-year tree-ring chronology. *Quaternary Research*, **27**, 220–231.

Scuderi, L.A. (1990): Tree-ring evidence for climatically-effective volcanic eruptions. *Quaternary Research*, **34**, 67–85.

Seidenkrantz, M.-S. (1993): Benthic foraminiferal and stable isotope evidence for a 'Younger Dryas-style' cold spell at the Saalian–Eemian transition, Denmark. *Palaeogeography, Palaeoclimatology Palaeoecology*, **102**, 103–120.

Seidenkrantz, M.-S., Kristensen, P. & Knudsen, K.L. (1995): Marine evidence for climatic instability during the last interglacial in shelf records from northwest Europe. *Journal of Quaternary Science*, **10**, 77–82.

Sejrup, H.P. (1987): Molluscan and foraminiferal biostratigraphy of an Eemian–Early Weichselian section on Karmøy, southwestern Norway. *Boreas*, **16**, 27–42.

Sejrup, H.P. & Haugen, J.-E. (1994): Amino acid diagenesis in the marine bivalve *Arctica islandica* Linné from northwest European sites: only time and temperature? *Journal of Quaternary Science*, **9**, 301–309.

Sejrup, H.P. & Larsen, E. (1991): Eemian–Early Weichselian N–S temperature gradients: North Atlantic–NW Europe. *Quaternary International*, **10–12**, 161–166.

Sejrup, H.P., Aarseth, I., Ellingsen, K.L., Reither, E., Jansen, E., Løvlie, R., Bent, A., Brigham-Grette, J., Larsen, E. & Stoker, M. (1987): Quaternary stratigraphy of the Fladen area, central North Sea: a multidisciplinary study. *Journal of Quaternary Science*, **2**, 35–58.

Sejrup, H.-P., Sjøholm, J., Furnes, H., Beyer, J., Eide, L., Jansen, E. & Mangerud, J. (1989): Quaternary tephrachronology on the Iceland Plateau, north of Iceland. *Journal of Quaternary Science*, **4**, 109–114.

Sejrup, H.P., Aarseth, I. & Haflidason, H (1991): The Quaternary succession in the northern North Sea. *Marine Geology*, **101**, 103–111.

Sejrup, H.P., Haflidason, H., Aarseth, I., King, E., Forsberg, C.F., Long, D. & Rokoengen, K. (1994): Late Weichselian glaciation history of the northern North Sea. *Boreas*, **23**, 1–13.

Self, R. & Sparks, R.S.J. (eds) (1981): *Tephra Studies*. Reidel, Dordrecht.

Seppälä, M., Gray, J. & Ricard, J. (1991): Development of low-centred ice-wedge polygons in the northernmost Ungava Peninsula, Québec, Canada. *Boreas*, **20**, 259–282.

Seret, G., Guiot, J., Wansard, G., Beaulieu, J.-L. de & Reille, M. (1992): Tentative palaeoclimatic reconstruction linking pollen and sedimentology in La Grande Pile (Vosges, France). *Quaternary Science Reviews*, **11**, 425–430.

Servant-Vildary, S. & Roux, M. (1990): Multivariate analysis of diatoms and water chemistry in Bolivian saline lakes. *Hydrobiologia*, **197**, 267–290.

Shabtiae, S. & Bentley, C.R. (1988): Ice thickness map of the West Antarctic ice streams by radar sounding. *Annals of Glaciology*, **11**, 126–136.

Shackleton, N.J. (1967): Oxygen isotope analyses and Pleistocene temperatures re-assessed. *Nature*, **215**, 15–17.

Shackleton, N.J. (1969): The last interglacial in the marine and terrestrial records. *Proceedings of the Royal Society of London*, **B174**, 135–154.

Shackleton, N.J. (1977): The oxygen isotope record of the Late Pleistocene. *Philosophical Transactions of the Royal Society, London*, **B280**, 169–182.

Shackleton, N.J. (1987): Oxygen isotopes, ice volume and sea level. *Quaternary Science Reviews*, **6**, 183–190.

Shackleton, N.J. & Hall, M.A. (1990): Stable isotope history of the Pleistocene at ODP Site 677. In *Proceedings of the Ocean Drilling Program Scientific Results 111* (edited by K. Becker & H. Sakai), College Station, Texas (Ocean Drilling Program), 295–316.

Shackleton, N.J. & Opdyke, N.D. (1973): Oxygen isotope and palaeomagnetic stratigraphy of equatorial Pacific core V28–238: oxygen isotope temperatures and ice volume on a 10^5 and 10^6 year scale. *Quaternary Research*, **3**, 39–55.

Shackleton, N.J. & Opdyke, N.D. (1976): Oxygen isotope and palaeomagnetic stratigraphy of Pacific core V28–239, Late Pliocene to Holocene. *Geological Society of America Memoir*, **145**, 449–464.

Shackleton, N.J. & Opdyke, N.D. (1977): Oxygen isotope and palaeomagnetic evidence for early northern hemisphere glaciation. *Nature*, **261**, 547–550.

Shackleton, N.J. & Pisias, N.G. (1985): Atmospheric carbon dioxide, orbital forcing and climate. In *The Carbon Cycle and Atmospheric CO_2: Natural Variations Archean to Present* (edited by E. Sundqvist & W.S. Broecker), American Geophysical Union, *Geophysical Monograph* **32**, 303–317.

Shackleton, N.J., Hall, M.A., Line, J. & Cang, S. (1983): Carbon isotope data in core V19–30 confirm reduced carbon dioxide concentration in the ice age atmosphere. *Nature*, **306**, 319–322.

Shackleton, N.J., Backman, J., Zimmerman, H., Kent, D.V., Hall, A., Homrighausen, R., Huddlestun, P., Keene, J.B., Kaltenback, A.J., Krumsiek, K.A.O, Morton, A.C., Murray, J.W. & Westberg-Smith, J. (1984): Oxygen isotope calibration of the onset of ice-rafting in DSDP site 552A: history of glaciation in the North Atlantic region. *Nature*, **307**, 620–633.

Shackleton, N.J., Duplessy, J.-C., Arnold, M., Maurice, P., Hall, M.A. & Cartlidge, J. (1988): Radiocarbon age of last glacial Pacific deep water. *Nature*, **335**, 708–711.

Shackleton, N.J., Berger, A. & Peltier, W.R. (1990): An alternative astronomical calibration of the lower Pleistocene timescale based on ODP Site 677. *Transactions of the Royal Society of Edinburgh: Earth Sciences*, **81**, 251–261.

Shackleton, N.J., Le, J., Mix, A. & Hall, M.A. (1992): Carbon isotope records from Pacific surface waters and atmospheric carbon dioxide. *Quaternary Science Reviews*, **11**, 387–400.

Shakesby, R.A., Dawson, A.G. & Matthews, J.A. (1987): Rock glaciers, protalus ramparts and related phenomena, Rondane, Norway: a continuum of large-scale talus-derived landforms. *Boreas*, **16**, 305–317.

Shane, L.K.C. & Anderson, K.H. (1993): Intensity gradients and reversals in Late Glacial environmental change in east-central North America. *Quaternary Science Reviews*, **12**, 307–320.

Sharp, M., Dowdeswell, J.A. & Gemmell, J.C. (1989): Reconstructing past glacier dynamics and erosion from glacial geomorphic evidence: Snowdon, North Wales. *Journal of Quaternary Science*, **4**, 115–130.

Sharp, R.P. (1988): *Living Ice. Understanding Glaciers and Glaciation*. Cambridge University Press, Cambridge.

Shaw, G.E. (1989): Aerosol transport from sources to ice sheets. In *The Environmental Record in Glaciers and Ice Sheets* (edited by F. Oeschger & C.C.Langway), John Wiley, Chichester & New York, 13–28.

Shaw, J. (1994): A qualitative view of sub-ice-sheet landscape evolution. *Progress in Physical Geography*, **18**, 159–184.

Shaw, J. & Archer, J. (1978): Winter turbidity current deposits in late Pleistocene lacustrine varves, Okanagen Valley, British Columbia, Canada. *Boreas*, **7**, 123–130.

Shaw, J. & Sharpe, D.R. (1987): Drumlin formation by subglacial meltwater erosion. *Canadian Journal of Earth Sciences*, **24**, 2316–2322.

Shaw, P. (1988): After the flood: the fluvio-lacustrine landforms of Northern Botswana. *Earth-Science Reviews*, **25**, 449–456.

Shen, G.T. & Sandford, C.L. (1990): Trace element indicators of climate variability in reef-building corals. In *Global Ecological Consequences of the 1982–83 El Niño Southern Oscillation* (edited by P. Glynn), Elsevier, New York, 255–283.

Shennan, I. (1987): Global analysis and the correlation of sea-level data. In *Sea Surface Studies* (edited by R.J.N Devoy), Croom Helm, London, 198–230.

Shennan, I. (1989): Holocene crustal movements and sea-level changes in Great Britain. *Journal of Quaternary Science*, **4**, 77–89.

Shennan, I, Orford, J.D. & Plater, A.J. (1992): IGCP Project 274. Quaternary coastal evolution: case studies, models and regional patterns. *Proceedings of the Geologists' Association*, **103**, 163–272.

Shennan, I., Innes, J., Long, A.J. & Zong, Y. (1994): Late Devensian and Holocene sea-level at Loch nan Eala near Arisaig, northwest Scotland. *Journal of Quaternary Science*, **9**, 261–284.

Shimmield, G.B. (1992): Can sediment geochemistry record changes in coastal upwelling palaeoproductivity? Evidence from NW Africa and the Arabian Sea. In *Upwelling Systems Since the Early Miocene* (edited by C. Summerhayes, W.L. Prell & K.-C. Emeis), Geological Society Special Publication, The Geological Society, London.

Shiraiwa, T. & Watanabe, T. (1991): Late Quaternary glacial fluctuations in the Langtang Valley, Nepal Himalaya, reconstructed by relative dating methods. *Arctic and Alpine Research*, **23**, 404–416.

Shoemaker, E.M. (1995): On the meltwater genesis of drumlins. *Boreas*, **24**, 3–10.

Shopov, Y.Y., Ford, D.C. & Schwarcz, H.P. (1994): Luminescent microbanding in speleothems: high-resolution chronology and paleoclimate. *Geology*, **22**, 407–410.

Shore, J.S., Bartley, D.D. & Harkness, D.D. (1995): Problems encountered with the ^{14}C dating of peat. *Quaternary Geochronology (Quaternary Science Reviews)*, **14**, 373–384.

Shotton, F.W. *et al.* (1993): The Middle Pleistocene deposits of Waverley Wood Pit, Warwickshire, England. *Journal of Quaternary Science*, **8**, 293–325.

Shroba, R.R. & Birkeland, P.W. (1983): Trends in Late-Quaternary soil development in the Rocky Mountains and Sierra Nevada of the western United States. In *Late Quaternary Environments of the United States. 1. The Late Pleistocene* (edited by S.C. Porter), Longman, London, 145–156.

Shukla, J., Nobre, C. & Sellers, P.J. (1990): Amazon deforestation and climate change. *Science*, **247**, 1322–1325.

Šibrava, V. (1986): Correlation of European glaciations and their relation to the deep-sea record. *Quaternary Science Reviews*, **5**, 433–442.

Šibrava, V., Bowen, D.Q. & Richmond, G.M. (1986): Quaternary glaciations in the Northern Hemisphere. *Quaternary Science Reviews*, **5**, 1–514.

Siegenthaler, U. & Eicher, U. (1986): Stable oxygen and carbon isotope analyses. In *Handbook of Holocene Palaeoecology and Palaeohydrology* (edited by B.E. Berglund), John Wiley, Chichester& New York, 407–422.

Siesser, W.G. (1993): Calcareous nannoplankton. In *Fossil Prokaryotes and Protists* (edited J.H. Lipps), Basil Blackwell, London.

Sigvaldason, G., Annerts, K. & Nilsson, M. (1992): Effect of glacier loading/deloading on volcanism: postglacial volcanic production rate of the Dyngjufjöll area, central Iceland. *Bulletin Volcanologique*, **54**, 385–392.

Simola, H., Coard, M.A. & O'Sullivan, P.E. (1981): Annual laminations in the sediment of Loe Pool, Cornwall. *Nature*, **290**, 238–241.

Singhvi, A.K., Bronger, A., Pant, R.K. & Sauer, W. (1987): Thermoluminescence dating and its implications for chronostratigraphy of loess–palaeosol sequences in the Kashmir Valley (India). *Chemical Geology*, **65**, 45–56.

Sinka, K.J. (1993): Developing the Mutual Climate Range method of palaeoclimatic reconstruction (2 vols). Unpublished PhD thesis, University of East Anglia.

Sissons, J.B. (1967): *The Evolution of Scotland's Scenery*. Oliver & Boyd, Edinburgh.

Sissons, J.B. (1974): A Late-glacial ice cap in the central Grampians, Scotland. *Transactions of the Institute of British Geographers*, **62**, 95–114.

Sissons, J.B. (1976): The *Geomorphology of the British Isles: Scotland*. Methuen, London.

Sissons, J.B. (1979): The Loch Lomond Stadial in the British Isles. *Nature*, **280**, 199–202.

Sissons, J.B. (1980a): The Loch Lomond Advance in the Lake District. *Transactions of the Royal Society of Edinburgh; Earth Sciences*, **71**, 13–27.

Sissons, J.B. (1980b): Palaeoclimatic inferences from Loch Lomond Advance glaciers. In *Studies in the Lateglacial of North-west Europe* (edited by J.J. Lowe, J.M. Gray & J.E. Robinson), Pergamon, Oxford, 23–30.

Sissons, J.B. (1983): Quaternary. In *Geology of Scotland*, 2nd edition (edited by G.Y. Craig), Scottish Academic Press, Edinburgh.

Sissons, J.B. & Sutherland, D.G. (1976): Climatic inferences from former glaciers in the south-east Grampian Highlands. *Journal of Glaciology*, **17**, 325–346.

Sjøholm, J., Sejrup, H.P. & Furnes, H. (1991): Quaternary volcanic ash zones on the Icelandic Plateau, southern Norwegian sea. *Journal of Quaternary Science*, **6**, 159–174.

Sly, P.G. (1978): Sedimentary processes in lakes. In *Lakes: Chemistry, Geology, Physics* (edited by A. Lerman), Springer-Verlag, New York & Berlin, 65–90.

Smart, P.L. (1991a): Uranium series dating. In *Quaternary Dating Methods – a User's Guide* (edited by P.L. Smart & P.D. Frances), Technical Guide 4, Quaternary Research Association, Cambridge.

Smart, P.L. (1991b): Electron Spin Resonance (ESR) dating. In *Quaternary Dating Methods – a User's Guide* (edited by P.L. Smart & P.D. Frances), Technical Guide 4, Quaternary Research Association, Cambridge, 128–160.

Smart, P.L. & Richards, D.A. (1992): Age estimates for the Late Quaternary high sea-stands. *Quaternary Science Reviews*, **11**, 687–696.

Smirnova, T.Y. & Nikonov, A.A. (1990): A revised lichenometric method and its application to dating past earthquakes. *Arctic and Alpine Research*, **22**, 375–388.

Smith, A.G. & Cloutman, E.W. (1988): Reconstruction of Holocene vegetation history in three dimensions at Waun-Fignen-Felen, an upland site in South Wales. *Philosophical Transactions of the Royal Society, London*, **B322**, 159–219.

Smith, B.W., Rhodes, E.J., Stokes, S., Spooner, N.A. & Aitken, M.J. (1990): Optical dating of sediments: initial quartz results from Oxford. *Archaeometry*, **32**, 19–31.

Smith, D.E. & Dawson, A.G. (eds) (1983): *Shorelines and Isostasy*. Academic Press, London.

Smith, G.I. & Street-Perrott, A.F. (1983): Pluvial lakes of the western United States. In *Late-Quaternary Environments of the United States. 1. The Late Pleisocene* (edited by S.C. Porter), Longman, London, 190–212.

Smith, J. (1838): On the last changes in the relative levels of the land and sea in the British Islands. *Edinburgh New Philosophical Journal*, **25**, 378–394.

Smol, J.P. (1990): Freshwater algae. In *Methods in Quaternary Ecology* (edited by B.G. Warner), *Geoscience Canada, Reprint Series* **5**, St John's, Newfoundland, 3–14.

Smol, J.P. & Dixit, S.S. (1990): Patterns of pH change inferred from chrysophycean microfossils in Adirondack and Northern New England lakes. *Journal of Paleolimnology*, **4**, 31–41.

Smol, J.P., Brown, S.R. & McNeely, R.N. (1983): Cultural disturbances and trophic history of a small meromictic lake from central Canada. *Hydrobiologia*, **103**, 125–130.

Smol, J.P., Battarbee, R.W., Davis, R.B. & Merilinen, J. (eds) (1986): *Diatoms and Lake Acidity*. W. Junk, The Hague.

Snowball, I. & Thompson, R. (1990): A mineral magnetic study of Holocene sedimentation in Lough Catherine, Northern Ireland. *Boreas*, **19**, 127–146.

Snowball, I. & Thompson, R. (1992): A mineral magnetic study of Holocene sediment yields and deposition patterns in the Llyn Geirionydd catchment, north Wales. *The Holocene*, **2**, 238–248.

Sonett, C.P. & Finney, S.A. (1990): The spectrum of radiocarbon. *Philosophical Transactions of the Royal Society*, **A330**, 413–426.

Sorensen, C.J. (1977): Reconstructed Holocene bioclimates.

Annals of the Association of American Geographers, 67, 214–222.

Sorensen, C.J., Knox, J.C., Larsen, J.A. & Bryson, R.A. (1971): Paleosols and the forest border in Keewatin. *Quaternary Research*, **1**, 468–473.

Sørensen, R. (1979): Late Weichselian deglaciation in the Oslofjord area, south Norway. *Boreas*, **8**, 241–246.

Southon, J.E., Nelson, D.E. & Vogel, J.S. (1992): The determination of past ocean–atmosphere radiocarbon differences. In *The Last Deglaciation: Absolute and Radiocarbon Chronologies* (edited by E. Bard & W.S. Broecker), *NATO ASI Series* **1**, **2**, Springer-Verlag, Berlin, 219–227.

Sowers, T., Bender, N., Labeyrie, L., Martinson, D., Jouzel, J., Raynaud, R., Pichon, J.J. & Korotkevich, Y.S. (1993): A 135,000-year Vostok-Specmap common temporal framework. *Paleoceanography*, **8**, 737–766.

Sparks, B.W. (1961): The ecological interpretation of Quaternary non-marine Mollusca. *Proceedings of the Linnaean Society London*, **172**, 71–80.

Sparks, B.W. & West, R.G. (1970): Late Pleistocene deposits at Wretton, Norfolk. I. Ipswichian interglacial deposits. *Philosophical Transactions of the Royal Society, London*, **B258**, 1–30.

Sparks, B.W. & West, R.G. (1972): *The Ice Age in Britain*. Methuen, London & New York.

Spaulding, G.W. (1991): Pluvial climatic episodes in North America and Africa: correlation with global climate. *Palaeogeography, Palaeoclimatology, Palaeoecology*, **84**, 217–227.

Spaulding, W.G., Leopold, E.B. & Van Devender, T.R. (1983): Late Wisconsin palaeoecology of the American Southwest. In *Late Quaternary Environments of the United States. 1. The Late Pleistocene* (edited by S.C. Porter), Longman, London, 259–293.

Stabell, B. (1985): The development and succession of taxa within the diatom genus *Fragilaria* Lyngbye as a response to basin isolation from the sea. *Boreas*, **14**, 273–286.

Stace, C. (1991): *New Flora of the British Isles*. Cambridge University Press, Cambridge.

Starkel, L. (1987): The evolution of European rivers: a complex response. In *Palaeohydrology in Practice* (edited by K.J. Gregory, J. Lewin & J.B. Thornes), John Wiley, Chichester & New York, 333–339.

Starkel, L. (1988): Tectonic, anthropogenic and climatic factors in the history of the Vistula river valley downstream of Cracow. In *Lake, Mire and River Environments* (edited by G. Lang & C. Schlüchter), Balkema, Rotterdam, 161–170.

Starkel, L. (1991): Long-distance correlation of fluvial events in the temperate zone. In *Temperate Palaeohydrology* (edited by L. Starkel, K.J. Gregory & J.B. Thornes), John Wiley, Chichester & New York, 473–495.

Stauffer, B. (1989): Dating of ice by radioactive isotopes. In *The Environmental Record in Glaciers and Ice Sheets* (edited by H. Oeschger & C.C. Langway, Jr.), John Wiley, Chichester & New York, 123–139.

Stauffer, B., Hofer, H., Oeschger, H., Schwander, J. & Siegenthaler, U. (1984): Atmospheric CO_2 concentration during the last glaciation. *Annals of Glaciology*, **5**, 160–164.

Stauffer, B.E., Lochbronner, H., Oeschger, H. & Schwander, J. (1988): Methane concentration in the glacial atmosphere was only half that of the preindustrial Holocene. *Nature*, **332**, 812–814.

Stead, I.M., Bourke, J.B. & Brothwell, D. (1986): *Lindow Man: The Body in the Bog*. British Museum, London.

Steen-McIntyre, V., Fryxell, R. & Malde, H.E. (1981): Geologic evidence for age of deposits at Hueyatlaco archaeological site, Valsequillo, Mexico. *Quaternary Research*, **16**, 1–17.

Steffensen, J.P. (1988): Analysis of the seasonal variations in dust, Cl^-, NO_3^- andSO_4^{2-} in two central Greenland firn cores. *Annals of Glaciology*, **10**, 171–177.

Stevenson, A.C., Birks, H.J.B., Flower, R.J. & Battarbee, R.W. (1989a): Diatom-based pH reconstructions of lake acidification using canonical correspondence analysis. *Ambio*, **18**, 228–233.

Stevenson, C.M., Carpenter, J. & Scheetz, B.E. (1989b): Obsidian dating: recent advances in the experimental determination and application of hydration rates. *Archaeometry*, **31**, 193–206.

Stieglitz, R.D., Moran, J.M. & Quigley, D.P. (1978): Pre-Twocreekan age of the type Valders till, Wisconsin: comment. *Geology*, **6**, 136.

Stocker, T. & Wright, D. (1991): Rapid transitions of the ocean's deep circulation induced by changes in surface water fluxes. *Nature*, **351**, 729–732.

Stoermer, E.F., Taylor, S.M. & Callender, E. (1971): Palaeoecological interpretation of the Holocene diatom succession in Devil's Lake, North Dakota. *Transactions of the American Microscopical Society*, **90**, 195–206.

Storzer, D. & Wagner, G.A. (1969): Correction of thermally lowered fission-track ages of tektites. *Earth and Planetary Science Letters*, **5**, 463–468.

Stothers, R.B. (1984): The great Tambora eruption in 1815 and its aftermath. *Science*, **224**, 1191–1198.

Street, F.A. (1981): Tropical palaeoenvironments. *Progress in Physical Geography*, **5**, 157–185.

Street, F.A. & Grove, A.T. (1979): Global maps of lake-level fluctuations in Africa. *Quaternary Research*, **12**, 83–118.

Street-Perrott, F.A. (1991): General Circulation (GCM) modelling of palaeoclimates: a critique. *The Holocene*, **1**, 74–80.

Street-Perrott, F.A. & Perrott, R.A. (1990): Abrupt climatic fluctuations in the tropics: an oceanic feedback mechanism. *Nature*, **343**, 607–612.

Street-Perrott, F.A. & Perrott, R.A. (1993): Holocene vegetation, lake levels, and climate of Africa. In *Global Climates Since the Last Glacial Maximum* (edited by H.E. Wright, Jr., J.E. Kutzbach, T. Webb III, W.F. Ruddiman, F.A. Street-Perrott and P.J. Barthlein), University of Minnesota Press, Minneapolis & London.

Street-Perrott, F.A. & Roberts, N. (1983): Fluctuations in closed-basin lakes as an indicator of past atmospheric circulation patterns. In *Variations in the Global Water Budget* (edited by F.A. Street-Perrott, M.A. Beran & R.A.S. Ratcliffe), Reidel, Dordrecht, 365–383.

Street-Perrott, F.A., Beran, M. & Ratcliffe, R.A.S. (eds) (1983): *Variations in the Global Water Budget*. Reidel, Dordrecht.

Street-Perrott, F.A., Roberts, N. & Metcalfe, S.E. (1985): Geomorphic implications of late Quaternary hydrological and climatic changes in the Northern Hemisphere tropics. In *Environmental Change and Tropical Geomorphology*, (edited by I. Douglas & T. Spencer), George Allen & Unwin, London, 165–183.

Street-Perrott, F.A., Marchand, D.S., Roberts, N. & Harrison, S.P. (1989): Global lake-level variations from 18,000 to 0 years ago: a palaeoclimatic analysis. *US Department of Energy Technical Report (TR046)*, 1–213.

Stringer, C.B., Currant, A.P., Schwarcz, H.P. & Collcutt, S. (1986): Age of Pleistocene faunas from Bacon Hole, Wales. *Nature*, **320**, 59–62.

Strömberg, B. (1985a): Revision of the lateglacial Swedish varve chronology. *Boreas*, **14**, 101–106.

Strömberg, B. (1985b): New varve measurements in Västergötland, Sweden. *Boreas*, **14**, 111–116.

Stuart, A.J. (1974): Pleistocene history of the British vertebrate fauna. *Biological Reviews*, **49**, 225–266.

Stuart, A.J. (1979): Pleistocene occurrences of the European pond tortoise (*Emys orbicularis* L.) in Britain. *Boreas*, **8**, 359–371.

Stuart, A.J. (1980): The vertebrate fauna from the Interglacial deposits at Sugworth, near Oxford. *Philosophical Transactions of the Royal Society London*, **B289**, 87–97.

Stuart, A.J. (1982): *Pleistocene Vertebrates in the British Isles*. Longman, London.

Stuart, A.J. (1991): Mammalian extinctions in the Late Pleistocene of northern Eurasia and North America. *Biological Reviews*, **66**, 453–562.

Stuiver, M. (1970): Long-term C14 variations. In *Radiocarbon Variations and Absolute Chronology* (edited by I.U. Olsson), Wiley, New York & London, 197–213.

Stuiver, M. (1971): Evidence for the variation of atmospheric ^{14}C content in the Late Quaternary. In *The Late Cenozoic Glacial Ages* (edited by K.K. Turekian), Yale University Press, New Haven, 57–70.

Stuiver, M. (1982): A high-precision calibration of the AD radiocarbon timescale. *Radiocarbon*, **24**, 1–26.

Stuiver, M. & Becker B. (1993): High-precision bidecadal calibration of the radiocarbon timescale, AD 1950 – 6000 BP. *Radiocarbon*, **35**, 35–66.

Stuiver, M. & Brazunias, T.F. (1993): Sun, ocean, climate and atmospheric $^{14}CO_2$: an evaluation of causal and spectral relationships. *The Holocene*, **3**, 289–205.

Stuiver, M. & Pearson, G.W. (1986): High-precision calibration of the radiocarbon timescale, AD 1950–500 BC. *Radiocarbon*, **28**, 805–838.

Stuiver, M. & Pearson, G.W. (1993): High-precision bidecadal calibration of the radiocarbon timescale, AD 1950 – 500 BC. *Radiocarbon*, **35**, 1–24.

Stuiver, M. & Quay, P.D. (1980): Changes in atmospheric carbon-14 attributed to a variable sun. *Science*, **207**, 11–19.

Stuiver, M. & Reimer, P.J. (1993): Extended ^{14}C data base and revised CALIB 3.0 ^{14}C age calibration program. *Radiocarbon*, **35**, 215–230.

Stuiver, M., Kromer, B., Becker, B. & Ferguson, C.W. (1986): Radiocarbon age calibration back to 13,300 years BP and the ^{14}C age matching of the German oak and US bristlecone pine chronologies. *Radiocarbon*, **28**, 969–979.

Stuiver, M., Brazunias, T.F., Becker, B. & Kromer, B. (1991): Climatic, solar, oceanic and geomagnetic influences on Late-Glacial and Holocene atmospheric $^{14}C/^{12}C$ change. *Quaternary Research*, **35**, 1–24.

Sturchio, N.C., Pierce, K.L., Murrell, M.T. & Sorey, M.L. (1994): Uranium-series ages of travertines and timing of the last glaciation in the Northern Yellowstone area, Wyoming–Montana. *Quaternary Research*, **41**, 265–277.

Suess, H.E. (1970): Bristlecone-pine calibration of the radiocarbon time-scale 5,000 BC to the present. In *Radiocarbon Variations and Absolute Chronology* (edited by I.U. Olsson), John Wiley, Chichester & New York, 303–311.

Suess, H.E. & Linick, T.W. (1990): The ^{14}C record in bristlecone pine wood of the past 8000 years based on the dendrochronology of the late C.W. Ferguson. *Philosophical Transactions of the Royal Society*, **A330**, 403–412.

Sugden, D.E. (1977): Reconstruction of the morphology, dynamics and thermal characteristics of the Laurentide Ice Sheet at its maximum. *Arctic and Alpine Research*, **9**, 21–47.

Sugden, D.E. & Hulton, N. (1994): Ice volumes and climatic change. In *The Changing Global Environment* (edited by N. Roberts), Blackwell, Oxford, 150–172.

Sugden, D.E. & John, B.S. (1976): *Glaciers and Landscape*. Edward Arnold, London.

Sukumar, R., Ramesh, R., Pant, R.K. & Rajagopalan, G. (1993): A ^{13}C record of late Quaternary climatic change from tropical peats in southern India. *Nature*, **364**, 703–705.

Sullivan, T.J., Turner, R.S., Charles, D.F., Cumming, B.F., Smol, J.P., Schofield, C.L., Driscoll, C.T., Cosby, B.J., Birks, H.J.B., Uatala, A.J., Kingston, J.C., Dixit, S.S., Bernert, J.A., Ryan, P.F. & Marmorek, D.R. (1992): Use of historical assessment for evaluation of process-based model projections of future environmental change: lake acidification in the Adirondack Mountains. *Environmental Pollution*, **77**, 253–262.

Summerfield M.A. (1983): Silcrete as a palaeoclimatic indicator: evidence from Southern Africa. *Palaeogeography, Palaeoclimatology, Palaeoecology*, **41**, 65–79.

Sundquist, E.T. (1993): The global carbon dioxide budget. *Science*, **259**, 934–941.

Sutcliffe, A.J. (1960): Joint Mitnor Cave, Buckfastleigh. *Transactions of the Proceedings of Torquay Natural History Society*, **13**, 1–26.

Sutcliffe, A.J. (1985): *On the Track of Ice Age Mammals*. British Museum (Natural History), London.

Sutherland, D.G. (1980): Problems of radiocarbon dating of deposits from newly deglaciated terrain: examples from the Scottish Lateglacial. In *Studies in the Lateglacial of North-west Europe* (edited by J.J. Lowe, J.M. Gray & J.E. Robinson), Pergamon, Oxford, 139–149.

Sutherland, D.G. (1981): The high-level marine shell beds of Scotland and the build-up of the last Scottish ice sheet. *Boreas*, **10**, 247–254.

Sutherland, D.G. (1984a): Modern glacier characteristics as a basis for inferring former climates, with particular reference to the Loch Lomond Stadial. *Quaternary Science Reviews*, **3**, 291–309.

Sutherland, D.G. (1984b): The Quaternary deposits and landforms of Scotland and the neighbouring shelves: a review. *Quaternary Science Reviews*, **3**, 157–254.

Sutherland, D.G. (1986): A review of Scottish marine shell radiocarbon dates, and their standardisation and interpretation. *Scottish Journal of Geology*, **22**, 145–164.

Svendsen, J.E. & Mangerud, J. (1987): Late Weichselian and Holocene sea-level history for a cross section of western Norway. *Journal of Quaternary Science*, **2**, 113–132.

Svensson, G. (1988a): Bog development and environmental conditions as shown by the stratigraphy of Store Mosse mire in southern Sweden. *Boreas*, **17**, 89–111.

Svensson, H. (1988b): Ice wedge casts and relict polygonal patterns in Scandinavia. *Journal of Quaternary Science*, **3**, 57–67.

Swyitski, J.P.M. (1991): *Principles, Methods and Applications of Particle Size Measurements*. Cambridge University Press, New York.

Sykes, G. (1991): Amino acid dating. In *Quaternary Dating Methods – a User's Guide* (edited by P.L. Smart & P.D. Frances), Technical Guide 4, Quaternary Research Association, Cambridge, 161–176.

Szabo, B.J. (1990): Ages of travertine deposits in eastern Grand Canyon National Park, Arizona. *Quaternary Research*, **34**, 24–32.

Szabo, B.J., McHugh, W.P., Schaber, G.G., Haynes, C.V. & Breed, C.S. (1989): Uranium-series dates of authigenic carbonates and Acheulian sites in southern Egypt. *Science*, **243**, 1053–1056.

Szabo, B.J., Haynes, C.V. Jr. & Maxwell, T.A. (1995): Ages of Quaternary pluvial episodes determined by uranium-series and radiocarbon dating of lacustrine deposits of Eastern Sahara. *Palaeogeography, Palaeoclimatology, Palaeoecology*, **113**, 227–242.

Talbot, M.R. (1980): Environmental responses to climatic change in the West African Sahel over the past 20,000 years. In *The Sahara and the Nile* (edited by M.A.J. Williams & H. Faure), Balkema, Rotterdam, 37–62.

Talbot, M.R. (1990): A review of the palaeohydrological interpretation of carbon and oxygen isotopic ratios in primary lacustrine carbonates. *Chemical Geology*, **80**, 261–279.

Talma, A.S. & Vogel, J.C. (1992): Late Quaternary paleotemperatures derived from a speleothem from Cango Caves, Cape Province, South Africa. *Quaternary Research*, **37**, 203–213.

Tapley, I.J. (1988): The reconstruction of palaeodrainage and regional geologic structures in Australia's Canning and Officer Basins using NOAA-AVHRR satellite imagery. *Earth-Science Reviews*, **25**, 409–425.

Tarling. D.H. (1983): *Palaeomagnetism*. Chapman & Hall, London.

Tauber, H. (1965): Differential pollen dispersal and the interpretation of pollen diagrams. *Danmarks Geologiske Undersøgelse, ser. II*, **89**, 1–69.

Tauber, H. (1970): The Scandinavian varve chrononology and C14 dating. In *Radiocarbon Variations and Absolute Chronology* (edited by I.U. Olsson), John Wiley, Chichester & New York, 173–195.

Taylor, D.W. (1988): Aspects of freshwater mollusc ecological biogeography. *Palaeogeography, Palaeoclimatology, Palaeoecology*, **62**, 511–576.

Taylor, K.C., Lamorey, G.W., Doyle, G.A., Alley, R.B., Grootes, P.M., Mayewski, P.A., White, J.W.C.& Barlow, L.K. (1993a): The 'flickering switch' of late Pleistocene climate change. *Nature*, **361**, 432–436.

Taylor, K.C., Hammer, C.U., Alley, R.B., Clausen, H.B., Dahl-Jensen, D., Gow, A.J., Gundestrup, N.S., Kipfstuhl, J., Moore, J.C. & Waddington, E.D. (1993b): Electrical conductivity measurements from the GISP2 and GRIP Greenland ice-cores. *Nature*, **366**, 549–552.

Teller, J.T. (1995): History and drainage of large ice-dammed lakes along the Laurentide ice sheet. *Quaternary International*, **28**, 83–92.

Teller, J.T. & Last, W.M. (1990): Paleohydrological indicators in playas and salt lakes, with examples from Canada, Australia and Africa. *Palaeogeography, Palaeoclimatology, Palaeoecology*, **76**, 215–240.

Teller, J.T., Rutter, N. & Lancaster, N. (1990): Sedimentology and paleohydrology of Late Quaternary lake deposits in the northern Namib sand sea, Namibia. *Quaternary Science Reviews*, **9**, 343–364.

Thomas, D.S.G. & Shaw, P.A. (1991): 'Relict' desert dune systems: interpretations and problems. *Journal of Arid Environments*, **20**, 1–14.

Thomas, D.S.G. & Tsoar, H. (1990): The geomorphological role of vegetation in desert dune systems. In *Vegetation and Erosion* (edited by J.B. Thornes), John Wiley, Chichester & New York, 471–489.

Thomas, G.S.P. (1985): The Quaternary of the Northern Irish Sea Basin. In *The Geomorphology of North-west England* (edited by R.H. Johnson), Manchester University Press, Manchester, 143–158.

Thomas, G.S.P. & Summers, A.J. (1984): Glacio-dynamic structures from the Blackwater Formation, Co. Wexford, Ireland. *Boreas*, **13**, 5–12.

Thomas, M.F. & Thorp, M.B. (1995): Geomorphic response to rapid climatic and hydrologic change during the Late Pleistocene and early Holocene in the humid and sub-humid tropics. *Quaternary Science Reviews*, **14**, 193–207.

Thompson, A. & Jones, A. (1986): Rates and causes of proglacial river terrace formation in southeast Iceland: an application of lichenometric dating techniques. *Boreas*, **15**, 231–246.

Thompson, L.G., Mosley-Thompson, E. *et al.* (1995) Late glacial stage and Holocene tropical ice core records from Huascarán, Peru. *Science*, **269**, 46–50.

Thompson, M. & Walsh, J.N. (1983): *A Handbook of Inductively Coupled Plasma Spectrometry*. Blackie, Glasgow & London.

Thompson, R. (1978): European palaeomagnetic secular variations 13000 – 0 BP. *Polskie Archiwum Hydrobiologii*, **25**, 413–418.

Thompson, R. (1986): Palaeomagnetic dating. In *Handbook of Holocene Palaeoecology and Palaeohydrology* (edited by B.E. Berglund), John Wiley, Chichester & New York, 313–327.

Thompson, R. (1991): Palaeomagnetic dating. In *Quaternary Dating Methods – a User's Guide* (edited by P.L. Smart & P.D.

Frances), Technical Guide 4, Quaternary Research Association, Cambridge, 177–198.

Thompson, R. & Edwards, K.J. (1982): A Holocene palaeomagnetic record and geomagnetic master curve from Ireland. *Boreas*, **11**, 335–349.

Thompson, R. & Oldfield, F. (1986): *Environmental Magnetism*. Allen & Unwin, London.

Thompson, R.S., Benson, L.V. & Hattori, E.M. (1986): A revised chronology for the last Pleistocene lake cycle in the central Lahontan Basin. *Quaternary Research*, **25**, 1–9.

Thomsen, E. (1990): Application of brachiopods in palaeoceanographic reconstructions; *Macandrevia cranium* (Miller, 1976) from the Norwegian Shelf. *Boreas*, **19**, 25–38.

Thorarinsson, S. (1981): Greetings from Iceland. Ash-fall and volcanic aerosols in Scandinavia. *Geografiska Annaler*, **63A**, 109–118.

Thornes, J.B. & Gregory, K.J. (1991); Unfinished business: a continuing agenda. In *Temperate Palaeohydrology* (edited by L. Starkel, K.J. Gregory & J.B. Thornes), John Wiley, Chichester & New York, 521–536.

Thorp, P.W. (1986): A mountain icefield of Loch Lomond Stadial age, western Grampians, Scotland. *Boreas*, **15**, 83–97.

Thorson, R.M. & Hamilton, T.D. (1986): Glacial geology of the Aleutian Islands (based on contributions of Robert F. Black). In *Glaciation in Alaska* (edited by T.D. Hamilton, K.M. Reed & R.M. Thorson), Alaska Geological Society, 171–192.

Thouveny, N. & Williamson, D. (1988): Palaeomagnetic study of the Holocene and Upper Pleistocene sediments from Lake Barombi Mbo, Cameroon: first results. *Physics of the Earth and Planetary Interiors*, **52**, 193–206.

Thouveny, N., Beaulieu, J.-L. de, Bonifay, E., Creer, K.M., Guiot, J., Icole, M., Johnsen, S., Jouzel, J., Reille, M., Williams, T. & Williamson, D. (1994): Climatic variations in Europe over the past 140 kyr deduced from rock magnetism. *Nature*, **371**, 503–506.

Tiedemann, R., Sarnthein, M. & Stein, R. (1989): Climatic changes in the western Sahara: aeolo-marine sediment record of the last 8 million years (Sites 657–661). *Proceedings of the Ocean Drilling Program: Scientific Results*, **108**, 241–277.

Tilman, D., May, R.M., Lehman, C.L. & Nowak, M.A. (1994): Habitat destruction and the extinction debt. *Nature*, **371**, 65–66.

Tolònen, K. (1986): Rhizopod analysis. In *Handbook of Holocene Palaeoecology and Palaeohydrology* (edited by B.E. Berglund), John Wiley, Chichester & New York, 645–666.

Tooley, M.J. (1978): *Sea-Level Fluctuations in North-West England during the Flandrian Stage*. Clarendon Press, Oxford.

Tooley, M.J. & Shennan, I. (eds) (1987): *Sea Level Changes*. Basil Blackwell, Oxford.

Townsend, P.D. *et al.* (1988): Thermoluminescence and Electron-Spin-Resonance dating. *Quaternary Science Reviews*, **7**, 243–536.

Trembour, F. & Friedman, I. (1984): The present status of obsidian hydration dating. In *Quaternary Dating Methods* (edited by W.C. Mahaney), Elsevier, Amsterdam, 141–152.

Tricart, J. (1975): Influence des oscillations climatiques récentes sur le modelé en Amazonie orientale (Région de Sanarem) d'après les images de radar latéral. *Zeitschrift für Geomorphologie*, **19**, 140–163.

Trimble Navigation (1989): *GPS: A Guide to the Next Utility*. Trimble Navigation Ltd, Sunnyvale, California.

Trudgill, S. (1985): *Limestone Geomorphology*, Longman, London.

Tucker, M.E. (1988) *The Field Description of Sedimentary Rocks*. Open University Press, Milton Keynes.

Tudhope, A.W. (1994): Extracting high-resolution climate records from coral skeletons. *Geoscientist*, **4**, 17–20.

Tudhope, A.W., Chilcott, C., Fallick, A.E., Jebb, M. & Shimmield, G. (1994): Southern Oscillation-related variations in rainfall recorded in the stable oxygen isotopic composition of living and fossil massive corals in Papua New Guinea. *Mineralogical Magazine*, **58A**, 914–915.

Turner, C. (1975a): The correlation and duration of Middle Pleistocene interglacial periods in Northwest Europe. In *After the Australopithecines* (edited by K.W. Butzer & G.L. Isaac), Mouton, The Hague, 259–308.

Turner, J. (1975b): The evidence for land use by prehistoric farming communities: the use of three-dimensional pollen diagrams. In *The Effect of Man on the Landscape of the Highland Zone* (edited by J.G. Evans, S. Limbrey & H. Cleere), Council for British Archaeology Research Report, **11**, 86–95.

Turner, R.C. & Scaife, R.G. (eds) (1995): *Bog Bodies: New Discoveries and New Perspectives*. British Museum Press, London.

Tutin, T.G. *et al.* (1964–80): *Flora Europaea* (5 vols). Cambridge University Press, Cambridge.

Tyldesley, J.B. (1973): Long-range transmission of tree pollen to Shetland. I. Sampling and trajectories. *New Phytologist*, **72**, 175–181.

Tzedakis, P.C. (1993): Long-term tree populations in northwest Greece through multiple Quaternary climatic cycles. *Nature*, **364**, 437–440.

Urey, H.C. (1947): The thermodynamic properties of isotopic substances. *Journal of the Chemical Society*, **1974**, 562–581.

Vacher, H.L. & Hearty, P. (1989): History of Stage 5 sea level in Bermuda: review with new evidence of a brief rise to present sea level during Substage 5a. *Quaternary Science Reviews*, **8**, 159–168.

Valentine, K.W.G. & Dalrymple, J.B. (1975): The identification, lateral variation and chronology of two buried palaeocatenas at Woodhall Spa and West Runton, England. *Quaternary Research*, **4**, 551–590.

Valet, J.-P. & Meynadier, L. (1993): Geomagnetic field intensity and reversals during the past four million years. *Nature*, **366**, 234–238.

Valladas, H. (1992): Thermoluminescence dating of flint. *Quaternary Science Reviews*, **11**, 1–6.

Van der Hammen, T. & Absy, M.L. (1994): Amazonia during the last glacial. *Palaeogeography, Palaeoclimatology, Palaeoecology*, **109**, 247–261.

Van der Heijden, E. (1994): *A combined anatomical and pyrolysis mass spectrometric study of peatified plant tissues*. PhD Thesis, University of Amsterdam, Haarlem.

Van der Plicht, J. (1993): The Groningen radiocarbon calibration program. *Radiocarbon*, **35**, 231–237.

Van der Weil, A.M. & Wijmstra, T.A. (1987a): Palynology of the lower part (78–120 m) of the core Tenaghi Philippon II, Middle Pleistocene of Macedonia, Greece. *Review of Palaeobotany and Palynology*, **52**, 73–88.

Van der Weil, A.M. & Wijmstra, T.A. (1987b): Palynology of the 112.8–197.8 m interval of the core Tenaghi Philippon II, Middle Pleistocene of Macedonia, Greece. *Review of Palaeobotany and Palynology*, **52**, 89–117.

Van der Wijk, A., El-Daoushy, F., Arends, A. & Mook, W.G. (1986): Dating peat with U/Th disequilibrium: some geochemical considerations. *Chemical Geology (Isotope Geoscience Section)*, **59**, 283–292.

Van Donk J. (1976): An ^{18}O record of the North Atlantic Ocean for the entire Pleistocene. *Geological Society of America Memoir*, **145**, 147–164.

Van Geel, B. (1986): Application of fungal and algal remains and other microfossils in palynological analyses. In *Handbook of Holocene Palaeoecology and Palaeohydrology* (edited by B.E. Berglund), John Wiley, Chichester & New York, 497–505.

Van Geel, B., Klink, A.G., Pals, J.P. & Wigers, J. (1986): An Upper Eemian lake deposit from Twente, eastern Netherlands. *Review of Palaeobotany and Palynology*, **47**, 31–61.

Van Geel, B., Coope, G.R. & Van der Hammen, T. (1989): Palaeoecology and stratigraphy of the Lateglacial type section at Usselo (The Netherlands). *Review of Palaeobotany and Palynology*, **60**, 25–129.

Van Vliet-Lanöe, B. (1988): *Le Rôle de la Glace de Ségrégation dans les Formations Superficielles de l'Europe de l'Ouest*. Thèse de Doctorat d'Etat, Université de Paris, Sorbonne, 2 vols, 854 pp.

Vandenberghe, J. (1985): Paleoenvironment and stratigraphy during the last glacial in the Belgian–Dutch border region. *Quaternary Research*, **24**, 23–38.

Vandenberghe, J. (1988): Cryoturbations. In *Advances in Periglacial Geomorphology* (edited by M.J. Clark), John Wiley, Chichester & New York, 179–198.

Vandenberghe, J. (1992): Periglacial phenomena and Pleistocene environmental conditions in The Netherlands – an overview. *Permafrost and Periglacial Processes*, **3**, 363–374.

Vandenberghe, J. & Kasse, C. (1989): Periglacial environments during the Early Pleistocene in the southern Netherlands and northern Belgium. *Palaeogeography, Palaeoclimatology, Palaeoecology*, **72**, 133–139.

Vandenberghe, J. & Pissart, A. (1993): Permafrost changes in Europe during the last glacial. *Permafrost and Periglacial Processes*, **4**, 121–135.

Vandenberghe, J. & van den Broek, P. (1982): Weichselian convolution phenomena and processes in fine sediments. *Boreas*, **11**, 299–315.

Varvas, M. & Punning, J.-M. (1993): Use of the ^{210}Pb method in studies of the development and human-impact history of some Estonian lakes. *The Holocene*, **3**, 34–44.

Velichko, A.A. (1990): Loess–paleosol formation on the Russian Plain. *Quaternary International*, **7/8**, 103–114.

Veum, T., Jansen, E., Arnold, M., Beyer, I. & Duplessy, J.-C. (1992): Water mass exchange between the North Atlantic and the Norwegian Sea durinng the last 28,000 years. *Nature*, **356**, 783–785.

Villareal, T.A., Altabet, M.A. & Culver-Rymsza, K. (1993): Nitrogen transport by vertically migrating diatom mats in the North Pacific Ocean. *Nature*, **363**, 709–712.

Vincens, A. (1991): Late Quaternary vegetation history of the South-Tanganyika Basin. Climatic implications in South Central Africa. *Palaeogeography, Palaeoclimatology, Palaeoecology*, **86**, 207–226.

Vincent, J.-S. & Klassen, R.W. (1989): Quaternary geology of the Canadian Interior plains. In *Quaternary Geology of Canada and Greenland* (edited by R.J. Fulton), Geological Survey of Canada, Geology of Canada, No. 1 (also Geological Society of America, *The Geology of North America*, volume K-1), 98–174.

Vita-Finzi, C. (1973): *Recent Earth History*. MacMillan, London and New York.

Vita-Finzi, C. (1986): *Recent Earth Movements: an Introduction to Neotectonics*. Academic Press, London.

Vogel, J.C. & Kronfeld J. (1980): A new method of dating peat. *South African Journal of Science*, **76**, 557–558.

Völkel, J. & Grunert, J. (1990): To the problem of dune formation and dune weathering during the Late Pleistocene and Holocene in the southern Sahara and the Sahel. *Zeitschrift für Geomorphologie*, **34**, 1–17.

Von Post, L. (1916): Forest tree pollen in south Swedish peat bog deposits (translated into English by M.B. Davis & K. Faegri, 1967). *Pollen et Spores*, **9**, 375–401.

Vorren, T., Vorren, K.-D., Alm, T., Gulliksen, S. & Løvlie, R. (1988): The last deglaciation (20,000 to 11,000 B.P.) on Andøya, northern Norway. *Boreas*, **17**, 41–77.

Vorren, T.E., Edvardsen, M., Hald, M. & Thomsen, E. (1983): Deglaciation of the Continental shelf off Southern Troms, North Norway. *Norges Geologiske Undersøkelse*, **380**, 173–187.

Wadhams, P. (1988): The underside of Arctic sea ice imaged by side-scan sonar. *Nature*, **333**, 161–164.

Wagner, G.A. & van den Haute, P. (1992): *Fission Track Dating*. Kluwer, Dordrecht.

Wagner, G.A. (1978): Archaeological applications of fission-track dating. *Nuclear Track Detection*, **2**, 51–64.

Wagstaff, M. (1981): Buried assumptions: some problems in the interpretation of the 'Younger Fill' raised by recent data from Greece. *Journal of Archaeological Science*, **8**, 247–264.

Wahlen, M., Deck, B., Weyer, H., Kubik, P., Sharma, P. & Gove, H. (1991): ^{36}Cl in the stratosphere. *Radiocarbon*, **33**, 257–258.

Waitt, R.B. Jr. & Thorson, R.M. (1983): The Cordilleran Ice Sheet in Washington, Idaho and Montana. In *Late Quaternary Environments of the United States*, Volume 1, *The Late Pleistocene* (edited by S.C. Porter), Longman, London, 53–70.

Walden, J., Smith, J.P. & Dackombe, R.V. (1987): The use of mineral magnetic susceptibility analyses in the study of glacial diamicts: a pilot study. *Journal of Quaternary Science*, **2**, 73–80.

Walden, J., Smith, J.P. & Dackombe, R.V. (1992): Mineral magnetic analyses as a means of lithostratigraphic correlation and provenance indication of glacial diamicts: intra- and inter-unit variation. *Journal of Quaternary Science*, **7**, 257–270.

Walker, D. (1982a): Vegetation's fourth dimension. *New Phytologist*, **90**, 419–429.

Walker, D. & Guppy, J.C. (eds) (1978): *Biology and Quaternary Environments*, Australian Academy of Science, Canberra.

Walker, I.R. (1987): Chironomidae (Diptera) in palaeoecology. *Quaternary Science Reviews*, **6**, 29–40.

Walker, I.R. & Mathewes, R.W. (1987a): Chironomidae (Diptera) and postglacial climate at Marion Lake, British Columbia, Canada. *Quaternary Research*, **27**, 89–102.

Walker, I.R. & Mathewes, R.W. (1987b): Chironomids, lake trophic status and climate. *Quaternary Research*, **28**, 431–437.

Walker, I.R. & Paterson, C.G. (1985): Efficient separation of subfossil Chironomidae from lake sediments. *Hydrobiologia*, **122**, 189–192.

Walker, I.R., Smol, J.P., Engstrom, D.R. & Birks, H.J.B. (1991): An assessment of Chironomidae as quantitative indicators of past climatic change. *Canadian Journal of Fisheries and Aquatic Sciences*, **48**, 975–987.

Walker, M.J.C. (1982b): Early and mid-Flandrian environmental history of the Brecon Beacons, South Wales. *New Phytologist*, **91**, 147–165.

Walker, M.J.C. (1993): Holocene (Flandrian) vegetation change and human activity in the Carneddau area of upland mid-Wales. In *Climatic Change and Human Impact on the Landscape* (edited by F.M. Chambers), Chapman & Hall, London, 168–183.

Walker, M.J.C. (1995): Climatic changes in Europe during the last glacial–interglacial transition. *Quaternary International*, **28**, 63–76.

Walker, M.J.C. & Harkness D.D (1990): Radiocarbon dating the Devensian Lateglacial in Britain: new evidence from Llanilid, South Wales. *Journal of Quaternary Science*, **5**, 135–144.

Walker, M.J.C. & Lowe, J.J. (1990): Reconstructing the environmental history of the last glacial–interglacial transition: evidence from the Isle of Skye, Inner Hebrides, Scotland. *Quaternary Science Reviews*, **9**, 15–49.

Walker, M.J.C. & Lowe, J.J. (eds) (1993): Records of the last deglaciation around the North Atlantic. *Quaternary Science Reviews*, **12**, 597–738.

Walker, M.J.C., Merritt, J.W., Auton, C.A., Coope, G.R., Field, M.H., Heijnis, H. & Taylor, B.J. (1992): Allt Odhar and Dalcharn: two pre-Late Devensian/Late Weichselian sites in northern Scotland. *Journal of Quaternary Science*, **7**, 69–86.

Walker, M.J.C., Coope, G.R. & Lowe, J.J. (1993): The Devensian (Weichselian) Lateglacial palaeoenvironmental record from Gransmoor, East Yorkshire, England. *Quaternary Science Reviews*, **12**, 659–680.

Walker, M.J.C., Bohncke, S.J.P., Coope, G.R., O'Connell, M., Usinger, H. & Verbruggen, C. (1994): The Devensian/Weichselian Late-glacial in northwest Europe (Ireland, Britain, north Belgium, The Netherlands, northwest Germany). *Journal of Quaternary Science*, **9**, 109–118.

Walker, R.G. (ed.) (1984): *Facies Models*. Geological Association of Canada, Toronto.

Walter, R.C. (1989): Application and limitation of fission-track geochronology to Quaternary tephras. *Quaternary International*, **1**, 35–46.

Walter, R.C., Manega, P.C., Hay, R.L., Drake, L.E. & Curtis, G.H. (1991): Laser-fusion $^{40}Ar/^{39}Ar$ dating of Bed 1, Olduvai Gorge, Tanzania. *Nature*, **354**, 145–149.

Warner, B.G. (ed.) (1990a): *Methods in Quaternary Ecology*. Geoscience Canada, Reprint Series No. 5.

Warner, B.G. (1990b): Testate amoebae (Protozoa). In *Methods in Quaternary Ecology* (edited by B.G. Warner), Geoscience Canada, Reprint Series No. 5, 65–74.

Warren, C.R. (1991): Terminal environment, topographic control and fluctuations of West Greenland glaciers. *Boreas*, **20**, 1–16.

Warren, W.P. & Croot, D.G. (1994): *Formation and Deformation of Glacial Deposits*. Balkema, Rotterdam & Brookfield.

Warrick, R. & Oerlemans, J. (1990): Sea level rise. In *Climate Change. The IPCC Scientific Assessment* (edited by J.T. Houghton, G. J. Jenkins & J.J. Ephraums), Cambridge University Press, Cambridge, 257–281.

Washburn, A.L. (1979): *Geocryology: a Survey of Periglacial Processes and Environments*. Edward Arnold, London.

Wasson, R.J. (1984): Late Quaternary palaeoenvironments in the desert dunefields of Australia. In *Late Quaternary Palaeoclimates of the Southern Hemisphere* (edited by J.C. Vogel), Balkema, Rotterdam, 419–432.

Wasson, R.J. (1986): Geomorphology and Quaternary history of the Australian continental dunefields. *Geographical Review of Japan*, **59**, 55–67.

Wasson, R.J., Rajaguru, S.N., Misra, V.N., Agrawal, D.P., Dhir, R.P., Singhvi, A.K. & Kameswara, Rao, K. (1983): Geomorphology, Late Quaternary stratigraphy and palaeoclimatology of the Thar dunefield. *Zeitschrift für Geomorphologie* (Supplementband), **45**, 117–151

Wasson, R.J., Fitchett, K., Mackey, B. & Hyde, R. (1988): Large-scale patterns of dune type, spacing and orientation in the Australian continental dunefield. *Australian Geographer*, **19**, 80–104.

Watson, A. (1983): Gypsum crusts. In *Chemical Sediments and Geomorphology* (edited by A.S. Goudie & K. Pye), Academic Press, London & New York, 133–161.

Watson, A. (1988): Desert gypsum crusts as palaeoenvironmental indicators: a micropetrographic study from crusts from the central Namib Desert. *Journal of Arid Environments*, **15**, 19–42.

Watson, E. (1977): The periglacial environment of Britain during the Devensian. *Philosophical Transactions of the Royal Society, London*, **B280**, 183–198.

Watson, E. (1981): Characteristics of ice-wedge casts in west central Wales. *Biuletyn Peryglacjalny*, **28**, 164–177.

Wattez, J., Courty, M.A. & Macphail, R.I. (1989): Burnt organomineral deposits related to animal and human activities in prehistoric caves. In *Soil Micromorphology* (edited by L. Douglas), Elsevier Press, Amsterdam, 431–439.

Watts, W.A. (1967): Late-glacial plant macrofossil from Minnesota. In *Quaternary Palaeoecology* (edited by E.J. Cushing & H.E. Wright), Yale University Press, New Haven, CT, 89–97.

Watts, W.A. (1978): Plant macrofossils and Quaternary

palaeoecology. In *Biology and Quaternary Environments* (edited D. Walker & J.C. Guppy), Australian Academy of Science, Canberra, 53–68.

Watts, W.A. (1980): Regional variation in the response of vegetation of Lateglacial climatic events in Europe. In *Studies in the Lateglacial of North-west Europe* (edited by J.J. Lowe, J.M. Gray & J.E. Robinson), Pergamon, Oxford, 1–22.

Watts, W.A. (1985a): A long pollen record from Laghi di Monticchio, southern Italy: a preliminary account. *Journal of the Geological Society*, **142**, 491–499.

Watts, W.A. (1985b): Quaternary vegetation cycles. In *The Quaternary History of Ireland* (edited by K.J. Edwards & W.P. Warren), Academic Press, London, 155–185.

Watts, W.A. & Stuiver, M. (1980): Late Wisconsin climate of northern Florida and the origin of species-rich deciduous forest. *Science*, **210**, 325–327.

Watts, W.A. & Winter, T.C. (1966): Plant macrofossils from Kirchner Marsh, Minnesota – a palaeoecological study. *Bulletin of the Geological Society of America*, **77**, 1339–1360.

Watts, W.A., Hansen, B.C.S. & Grimm, E.C. (1992): Camel Lake: a 40,000–yr record of vegetational and forest history from northwestern Florida. *Ecology*, **73**, 156–1066.

Weaver, A.J. & Hughes, T.M.C. (1994): Rapid interglacial climate fluctuations driven by North Atlantic ocean circulation. *Nature*, **367**, 447–449.

Weaver, A.J., Sarachik, E.A. & Marotze, J. (1991): Freshwater flux forcing of decadal and interdecadal oceanic variability. *Nature*, **353**, 836–838.

Weaver, P.P.E. & Pujol, C. (1988): History of the last deglaciation in the Alboran Sea (western Mediterranean) and adjacent North Atlantic as revealed by coccolith floras. *Palaeogeography, Palaeoclimatology, Palaeoecology*, **64**, 35–42.

Webb, P.-N. & Harwood, D.N. (1991): Late Cenozoic glacial history of the Ross Embayment, Antarctica. *Quaternary Science Reviews*, **10**, 215–237.

Webb, T. III (1987): The appearance and disappearance of major vegetational assemblages: long-term vegetational dynamics in eastern North America. *Vegetation*, **69**, 17–187.

Webb, T. III, Bartlein, P.J., Harrison, S.P. & Andersen, K.H. (1993a): Vegetation, lake levels and climate in eastern North America for the past 18,000 years. In *Global Climates since the Last Glacial Maximum* (edited by H.E. Wright Jr., J.E. Kutzbach, T. Webb III, W.F. Ruddiman, F.A. Street-Perrott & P.J. Bartlein), University of Minnesota Press, Minneapolis, 415–467.

Webb, T. III, Ruddiman, W.F., Street-Perrott, F.A., Markgraf, V., Kutzbach, J.E., Bartlein, P.J., Wright, H.E. Jr. & Prell, W.L. (1993b): Climatic changes during the past 18,000 years: regional syntheses, mechanisms and causes. In *Global Climates since the Last Glacial Maximum* (edited by H.E. Wright, Jr., J.E. Kutzbach, T. Webb III, W.F. Ruddiman, F.A. Street-Perrott & P.J. Bartlein), University of Minnesota Press, Minneapolis, 514–535.

Wehmiller, J.F. (1984): Relative and absolute dating of Quaternary mollusks with amino acid racemisations: evaluations, applications, questions. In *Quaternary Dating Methods* (edited by W.C. Mahaney), Elsevier, Amsterdam, 171–194.

West, R.G. (1977): *Pleistocene Geology and Biology*, 2nd edition. Longman, London.

West, R.G. (1979): Further on the Flandrian. *Boreas*, **8**, 426.

West, R.G. (1980): *The Pre-glacial Pleistocene of the Norfolk and Suffolk Coasts*. Cambridge University Press, London.

West, R.G. (1989): The use of type localities and type sections in the Quaternary, with especial reference to East Anglia. In *Quaternary Type Sections: Imagination or Reality* (edited by J. Rose & C. Schlüchter), Balkema, Rotterdam, 3–10.

West, R.G., Andrew, R. & Pettit, M. (1992): Taphonomy of plant remains on floodplains of tundra rivers, present and Pleistocene. *New Phytologist*, **123**, 203–221.

Westgate, J. & Briggs, N. (1985): Tephrochronology and fission-track dating. In *Dating Methods of Pleistocene Deposits and their Problems* (edited by N. Rutter), Geosciences Canada, Toronto, 31–38.

Westgate, J.A., Walter, W.C., Pearce, G.W. & Gorton, M.P. (1985): Distribution, stratigraphy, petrochemistry and palaeomagnetism of the late-Pleistocene Old Crow tephra in Alaska and the Yukon. *Canadian Journal of Earth Sciences*, **22**, 893–906.

Whalley, W.B. (1990): Part Three: material properties. In *Geomorphological Techniques*, 2nd edition (edited by A.S. Goudie), Unwin Hyman, London & Boston, 109–192.

Whatley, R.C. (1993): Ostracoda as biostratigraphical indices in Cainozoic deep-sea sequences. In *High Resolution Stratigraphy* (edited by E.A. Hailwood & R.B. Kidd), Geological Society Special Publication 70, London, 155–167.

Whillans, I.M. & Grootes, P.M. (1985): Isotopic diffusion in cold snow and firn. *Journal of Geophysical Research*, **90**, 3910–3918.

Whitaker, W. (1889): *The geology of London and Parts of the Thames Valley*. Memoir of the Geological Survey of Great Britain.

White, J., Molfino, B., Labeyrie, L., Stauffer, B. & Farquhar, B. (1993): How reliable and consistent are paleodata from continents, oceans and ice? In *Global Changes in the Perspective of the Past* (edited by J.A. Eddy & H. Oeschger), John Wiley, Chichester & New York, 73–102.

White, J.W.C., Ciais, P., Figge, R.A., Kenny, R. & Markgraf, V. (1994): A high-resolution record of atmospheric CO_2 content from carbon isotopes in peat. *Nature*, **367**, 153–155.

Whiteman, C. & Kemp, R.A. (1990): Pleistocene sediments, soils and landscape evolution at Stebbing, Essex. *Journal of Quaternary Science*, **5**, 145–161.

Whiteman, C.A. (1992): The palaeography and correlation of pre-Anglian terraces of the River Thames in Essex and the London Basin. *Proceedings of the Geologists' Association*, **103**, 37–56.

Whiteman, C.A. & Rose, J. (1992): Thames river sediments of the Early and Middle Pleistocene. *Quaternary Science Reviews*, **11**, 363–375.

Whittaker, A., Cope, J.C.W., Cowie, J.W., Gibbons, W., Hailwood, E.A., House, M.R., Jenkins, D.G., Rawson, P.F., Rushton, A.W.A., Smith, D.G., Thomas, A.T., & Wimbledon, W.A. (1991): A guide to stratigraphical procedure. *Journal of the Geological Society, London*, 148, 813–824.

Wigley, T.M.L. & Kelly, P.M. (1990): Holocene climatic change, ^{14}C wiggles and variations in solar irradiance. *Philosophical Transactions of the Royal Society, London*, **A330**, 547–560.

Wijmstra, T.A., Young, R. & Witte, H.J.L. (1990): An evaluation of the climatic conditions during the Late Quaternary in northern Greece by means of multivariate analysis of palynological data and comparison with recent phytosociological and climatic data. *Geologie en Mijnbouw*, **69**, 243–251.

Wilkins, G.R., Delcourt, P.A., Delcourt, H.R., Harrison, F.W. & Turner, M.R. (1991): Paleoecology of central Kentucky since the last glacial maximum. *Quaternary Research*, **15**, 113–125.

Williams, D.F., Arthur, M.A., Jones, D.S. & Healy-Williams, N. (1982): Seasonality and mean annual sea surface temperatures from isotopic and sclerochronological records. *Nature*, **296**, 432–434.

Williams, D.F., Thunell, R.C., Tappa, E., Rio, D. & Raffi, I. (1988): Chronology of the oxygen isotope record, 0–1.88 million years before present. *Palaeogeography, Palaeoclimatology, Palaeoecology*, **64**, 221–240.

Williams, G.E. & Polach, H.A. (1969): The evaluation of ^{14}C ages for soil carbonates from the arid zone. *Earth and Planetary Science Letters*, **4**, 240–242.

Williams, M.A.J. & Clarke, M.F. (1984): Late Quaternary environments in north central India. *Nature*, **308**, 633–635.

Williams, M.A.J., Dunkerley, D.L., De Deckker, P., Kershaw, A.P. & Stokes, T. (1993a): *Quaternary Environments*. Edward Arnold, London.

Williams, P.H., Vane-Wright, R.I. & Humphries, C.J. (1993b): Measuring biodiversity for choosing conservation areas. In *Hymenoptera and Biodiversity* (edited by J. La Salle & I.D. Gauld), CAB International, Wallingford, 309–328.

Williams, P.J. & Smith, M.W. (1989): *The Frozen Earth*. Cambridge University Press, Cambridge.

Williams, P.W. (1991): Tectonic geomorphology, uplift rates and geomorphic response in New Zealand. *Catena*, **18**, 439–452.

Williams, R.B.G. (1969): Permafrost and temperature conditions in England during the last glacial period. In *The Periglacial Environment* (edited by T.L. Péwé), McGill–Queen's University Press, Montreal, 399–410.

Williams, R.B.G. (1975): The British climate during the last glaciation: an interpretation based on periglacial phenomena. In *Ice Ages Ancient and Modern* (edited by A.E. Wright & F. Moseley), Seel House Press, Liverpool, 95–120.

Williams, V.S. (1983): Present and former equilibrium-line altitudes near Mount Everest, Nepal and Tibet. *Arctic and Alpine Research*, **15**, 201–211.

Williamson, D., Thouveny, N., Hillaire-Marcel, C., Mondeguer, A., Taieb, M., Tiercelin, J-J. & Vincens, A. (1991): Chronological potential of palaeomagnetic oscillations recorded in Late Quaternary sediments from Lake Tanganyika. *Quaternary Science Reviews*, **10**, 351–361.

Willman, H.B. & Frye, J.C. (1970): Pleistocene stratigraphy of Illinois. *Illinois Geological Survey Bulletin*, **94**.

Wilson, A.T. (1964): Origin of ice ages: an ice shelf theory for Pleistocene glaciation. *Nature*, **201**, 147–149.

Wilson, E.O. (ed.) (1988): *Biodiversity*. National Academy Press, Washington, DC.

Wilson, S.E., Walker, I.R., Mott, R.J. & Smol, J.P. (1993): Climate and limnological changes associated with the Younger Dryas in Atlantic Canada. *Climate Dynamics*, **8**, 117–187.

Winkler, M.G. (1988): Effect of climate on development of two Sphagnum bogs in south-central Wisconsin. *Ecology*, **69**, 1032–1043.

Winograd, I.J., Szabo, B.J., Coplen, T.B. & Riggs, A.C. (1988): A 250,000-years climatic record from Great Basin vein calcite: implications for Milankovitch theory. *Science*, **242**, 1275–1280.

Winograd, I.J., Coplen, T.B., Landwehr, J.M., Riggs, A.C., Ludwig, K.R., Szabo, B.J., Kolesar, P.T. & Revesz, K.M. (1992): Continuous 500,000-year climate record from vein calcite in Devil's Hole, Nevada. *Science*, **258**, 255–260.

Winspear, N.R. & Pye, K. (1995): Textural, geochemical and mineralogical evidence for the origin of the Peoria Loess of central and southern Nebraska, USA. *Earth Surface Processes and Landforms*, **20**, 735–745.

Winter, A. & Siesser, W.G. (eds) (1994): *Coccolithophores*. Cambridge University Press, Cambridge.

Wintle, A.G. (1987): Thermoluminescence dating of loess. *Catena, Supplement* **9**, 103–115.

Wintle, A.G. (1990): A review of current research on TL dating of loess. *Quaternary Science Reviews*, **9**, 385–397.

Wintle, A.G. (1991): Luminescence dating. In *Quaternary Dating Methods – a User's Guide* (edited by P.L. Smart & P.D. Frances), Technical Guide 4, Quaternary Research Association, Cambridge, 108–127.

Wintle, A.G. & Huntley, D.J. (1982): Thermoluminescence dating of sediments. *Quaternary Science Reviews*, **1**, 31–53.

Wintle, A.G., Shackleton, N.J. & Lautridou, J.P. (1984): Thermoluminescence dating of periods of loess deposition and soil formation in Normandy. *Nature*, **310**, 491–493.

Wintle, A.G., Lancaster, N. & Edwards, S.R. (1994): Infrared stimulated luminescence (IRSL) dating of late-Holocene aeolian sands in the Mojave Desert, California, USA. *The Holocene*, **4**, 74–78.

Wise, S.M. (1980): Caesium-137 and lead-210: a review of the techniques and some applications in geomorphology. In *Timescales in Geomorphology* (edited by R.A. Cullingford, D.A. Davidson & J. Lewin), John Wiley, Chichester & New York, 109–127.

Wohlfarth, B., Björck, S., Possnert, G., Lemdahl, G., Brunnberg, L., Ising, J., Olsson, S. & Svensson, N.O. (1993): AMS dating Swedish varved clays of the last glacial/interglacial transition and the potential/ difficulties of calibrating Late Weichselian 'absolute' chronologies. *Boreas*, **22**, 113–128.

Woillard, G. (1979): Abrupt end to the last interglacial SS in north-east France. *Nature*, **281**, 558–562.

Woillard, G.M. (1978): Grande Pile peat bog: a continuous pollen record for the last 140,000 years. *Quaternary Research*, **9**, 1–21.

Woillard, G.M. & Mook, W.G. (1982): Carbon-14 dates at Grande Pile: correlation of land and sea chronologies. *Science*, **215**, 159–161.

Wollast, R. (1981): Interactions between major biogeochemical cycles in marine ecosystems. In *Some Perspectives of the*

Major Biogeochemical Cycles (edited by G.E. Likens), John Wiley, Chichester & New York, 125–142.

Wood, A.M. & Whatley, R.C. (1994): Northeastern Atlantic and Arctic faunal provinces based on the distribution of Recent ostracod genera. *The Holocene*, **4**, 174–192.

Wood, A.M., Whatley, R.C., Cronin, T.M. & Holtz, T. (1993): Pliocene palaeotemperature reconstruction for the southern North Sea based on Ostracoda. *Quaternary Science Reviews*, **12**, 747–768.

Woodroffe, C.D., Short, S.A., Stoddart, D.R., Spencer, T. & Harmon, R.S. (1991): Stratigraphy and chronology of late Pleistocene reefs in the southern Cook Islands, South Pacific. *Quaternary Research*, **35**, 246–263.

Woodward, F.I. (1987): Stomatal numbers are sensitive to increases in CO_2 from pre-industrial levels. *Nature*, **327**, 617–618.

Woodward, J.C., Macklin, M.G. & Lewin, J. (1994): Pedogenic weathering and relative-age dating of Quaternary alluvial sediments in the Pindus Mountains of Northwest Greece. In *Rock Weathering and Landform Evolution* (edited by D.A. Robinson & R.B.G. Williams), John Wiley, Chichester & New York, 259–283.

Wooldridge, S.W. (1938): The glaciation of the London basin and the evolution of the lower Thames drainage system. *Quarterly Journal of the Geological Society of London*, **94**, 627–667.

Wooldridge, S.W. & Linton, D.L. (1955): *Structure, Surface and Drainage in South-east England*. George Philip, London.

Worsley, P. (1984): Periglacial environment. *Progress in Physical Geography*, **8**, 270–276.

Worsley, P. (1991): Possible early Devensian glacial deposits in the British Isles. In *Glacial Deposits in Great Britain and Ireland* (edited by J. Ehlers, P.L. Gibbard & J. Rose), Balkema, Rotterdam, 47–52.

Worsley, P. (1992): Problems of permafrost engineering as exemplified by three communities in the North-western Canadian Arctic. *Quaternary Proceedings*, **2**, 67–78.

Wright, H.E. Jr. (1973): Tunnel valleys, glacial surges and subglacial hydrology of the Superior lobe, Minnesota. *Geological Society of America Memoir*, **136**, 251–276.

Wright, H.E. Jr. (1989): The amphi-Atlantic distribution of the Younger Dryas palaeoclimatic oscillation. *Quaternary Science Reviews*, **8**, 295–306.

Wright, H.E. Jr., Kutzbach, J.E., Webb, T. III, Ruddiman, W.F., Street-Perrott, A.F. & Bartlein, P.J. (eds) (1993): *Global Climates since the Last Glacial Maximum*. University of Minneapolis Press, Minnesota.

Wyckoff, R.W.G. (1980): Collagen in fossil bones. In *Biogeochemistry of Amino Acids* (edited by P.E. Hare, T.C. Hoering & K. King Jr.), John Wiley, Chichester & New York, 17–22.

Xiao, J., Porter, S.C., An, Z., Kumai, H. & Yoshikawa, S. (1995): Grain size of quartz as an indicator of winter monsoon strength on the Loess Plateau of central China during the last 130,000 yr. *Quaternary Research*, **43**, 22–29.

Yan, Z. & Petit-Maire, N. (1994): The last 140 ka in the Afro-Asian arid/semi-arid transitional zone. *Palaeogeography, Palaeoclimatology, Palaeoecology*, **110**, 217–233.

Yanchou, L., Mortlock, A.J., Price, D.M. & Redhead, M.L. (1987): Thermoluminescence dating of coarse-grain quartz from the Malan Loess at Zhaitang Section, China. *Quaternary Research*, **28**, 356–363.

Yatsu, E. & Shimoda, S. (1990): X-ray diffraction of clay minerals. In *Geomorphological Techniques*, 2nd edition (edited by A.S. Goudie), Unwin Hyman, London & Boston, 153–160.

Zagwijn, W.H. (1985): An outline of the Quaternary stratigraphy of the Netherlands. *Geologie en Mijnbouw*, **64**, 17–24.

Zagwijn, W.H. (1989): Vegetation and climate during warmer intervals in the Late Pleistocene of western and central Europe. *Quaternary International*, **3/4**, 57–67.

Zagwijn, W.H. (1994): Reconstruction of climate change during the Holocene in western and central Europe based on pollen records of indicator species. *Vegetation History and Archaeobotany*, **3**, 65–88.

Zahn, R. (1992): Deep ocean circulation puzzle. *Nature*, **356**, 744–746.

Zahn, R. (1994): Core correlations: linking ice-core records to ocean circulation. *Nature*, **371**, 289–291.

Zbinden, H., Andrée, M., Oeschger, H., Ammann, B., Lotter, A., Bonani, G. & Wölfli, W. (1989): Atmospheric radiocarbon at the end of the last glacial: an estimate based on AMS radiocarbon dates on terrestrial macrofossils from lake sediments. *Radiocarbon*, **31**, 795–804.

Zeuner, F.E. (1959): *The Pleistocene Period*, 2nd edition. Hutchinson, London.

Zhang, X., An, Z., Chen, T., Zhang, G., Arimoto, R. & Ray, B.J. (1994): Late Quaternary records of the atmospheric input of eolian dust to the center of the Chinese Loess Plateau. *Quaternary Research*, **41**, 35–43.

Zielinski, G.A. & McCoy, W.D. (1987): Palaeoclimatic implications of the relationship between modern snowpack and Late Pleistocene equilibrium-line altitudes in the mountains of the Great Basin, Western U.S.A. *Arctic and Alpine Research*, **19**, 127–134.

Zöller, L. & Wagner, G.A. (1990): Thermoluminescence dating of loess – recent developments. *Quaternary International*, **7/8**, 119–128.

Zreda, M.G., Phillips, F.M., Elmore, D., Kubik, P.W., Sharma, P. & Dorn, R.I. (1991): Cosmogenic chlorine-36 production rates in terrestrial rocks. *Earth and Planetary Science Letters*, **105**, 94–109.

Zumberge, J.E., Engel, M.H. & Nagy, B. (1980): Amino acids in bristlecone pine: an evaluation of factors affecting racemisation rates and palaeothermometry. In *Biogeochemistry of Amino Acids* (edited by P.E. Hare, T.C. Hoering & K. King, Jr.), John Wiley, Chichester & New York, 503–525.

Index

Geographical names are restricted to countries and certain selected localities (e.g. Loess Plateau), excluding the British Isles and United States which feature prominently in the book.